Food Nanotechnology

Principles and Applications

Contemporary Food Engineering

Series Editor

Professor Da-Wen Sun, Director

Food Refrigeration and Computerized Food Technology,

National University of Ireland,

Dublin, Ireland

http://www.ucd.ie/sun/

Food Nanotechnology: Principles and Applications, *edited by C. Anandharamakrishnan and S. Parthasarathi*

Computational Fluid Dynamics in Food Processing, Second Edition, *edited by Da-Wen Sun*

Food Biofortification Technologies, *edited by Agnieszka Saeid*

Trends in Fish Processing Technologies, *edited by Daniela Borda, Anca I. Nicolau, and Peter Raspor*

High Pressure Processing of Fruit and Vegetable Juices, *edited by Milan Houška and Filipa Vinagre Marques da Silva*

Advances in Meat Processing Technology, *edited by Alaa El-Din A. Bekhit*

Engineering Aspects of Membrane Separation and Application in Food Processing, *edited by Robert W. Field, Erika Bekassy-Molnar, Frank Lipnizki, and Gyula Vatai*

Edible Oils: Extraction, Processing, and Applications, *edited by Smain Chemat*

Advances in Heat Transfer Unit Operations: Baking and Freezing in Bread Making, *edited by Georgina Calderon-Dominguez, Gustavo F. Gutierrez-Lopez and Keshavan Niranjan*

Advances in Technologies for Producing Food-relevant Polyphenols, *edited by Jose Cuevas Valenzuela, Jose Rodrigo Vergara-Salinas and Jose Ricardo Perez-Correa*

Innovative Processing Technologies for Foods with Bioactive Compounds, *edited by Jorge J. Moreno*

Light Scattering Technology for Food Property, Quality and Safety Assessment, *edited by Renfu Lu*

Edible Food Packaging: Materials and Processing Technologies, *edited by Miquel Angelo Parente Ribeiro Cerqueira, Ricardo Nuno Correia Pereira, Oscar Leandro da Silva Ramos, Jose Antonio Couto Teixeira and Antonio Augusto Vicente*

Handbook of Food Processing: Food Preservation, *edited by Theodoros Varzakas and Constantina Tzia*

Handbook of Food Processing: Food Safety, Quality, and Manufacturing Processes, *edited by Theodoros Varzakas and Constantina Tzia*

Advances in Postharvest Fruit and Vegetable Technology, *edited by Ron B.H. Wills and John Golding*

Engineering Aspects of Food Emulsification and Homogenization, *edited by Marilyn Rayner and Petr Dejmek*

Food Engineering Handbook: Food Process Engineering, *edited by Theodoros Varzakas and Constantina Tzia*

Food Engineering Handbook: Food Engineering Fundamentals, *edited by Theodoros Varzakas and Constantina Tzia*

Food Nanotechnology

Principles and Applications

Edited by

C. Anandharamakrishnan and S. Parthasarathi

CRC Press
Taylor & Francis Group
Boca Raton London New York

CRC Press is an imprint of the
Taylor & Francis Group, an **informa** business

CRC Press
Taylor & Francis Group
6000 Broken Sound Parkway NW, Suite 300
Boca Raton, FL 33487-2742

CRC Press is an imprint of Taylor & Francis Group, an Informa business
No claim to original U.S. Government works

Printed on acid-free paper

International Standard Book Number-13: 978-1-4987-6717-0 (Hardback)

Visit the Taylor & Francis website at
http://www.taylorandfrancis.com

and the CRC Press website at
http://www.crcpress.com

Contents

Series Preface

CONTEMPORARY FOOD ENGINEERING

Food engineering is the multidisciplinary field of applied physical sciences combined with the knowledge of product properties. Food engineers provide the technological knowledge transfer essential to the cost-effective production and commercialization of food products and services. In particular, food engineers develop and design processes and equipment in order to convert raw agricultural materials and ingredients into safe, convenient, and nutritious consumer food products. However, food engineering topics are continuously undergoing changes to meet diverse consumer demands, and the subject is being rapidly developed to reflect market needs.

In the development of food engineering, one of the many challenges is to employ modern tools and knowledge, such as computational materials science and nanotechnology, to develop new products and processes. At the same time, improving quality, safety, and security remain critical issues in food engineering study. New packaging materials and techniques are being developed to provide more protection to foods, and novel preservation technologies are emerging to enhance food security and defense. Additionally, process control and automation regularly appear among the top priorities identified in food engineering. Advanced monitoring and control systems are developed to facilitate automation and flexible food manufacturing. Furthermore, energy saving and minimization of environmental problems continue to be important food engineering issues, and significant progress is being made in waste management, efficient utilization of energy, and reduction of effluents and emissions in food production.

Consisting of edited books, the *Contemporary Food Engineering* book series attempts to address some of the recent developments in food engineering. Advances in classical unit operations in engineering applied to food manufacturing are covered, as well as such topics as progress in the transport and storage of liquid and solid foods; heating, chilling, and freezing of foods; mass transfer in foods; chemical and biochemical aspects of food engineering and the use of kinetic analysis; dehydration, thermal processing, nonthermal processing, extrusion, liquid food concentration, membrane processes, and applications of membranes in food processing; shelf-life, electronic indicators in inventory management, and sustainable technologies in food processing; and packaging, cleaning, and sanitation. The books are aimed at professional food scientists, academics researching food engineering problems, and graduate level students.

The editors of the books are leading engineers and scientists from many parts of the world. All the editors were asked to present their books in a manner that will address the market need and pinpoint the cutting-edge technologies in food engineering. Furthermore, all contributions are written by internationally renowned experts who have both academic and professional credentials. All authors have attempted to provide critical, comprehensive, and readily accessible information on the art and science of a relevant topic in each chapter, with reference lists to be used by readers for further information. Therefore, each book can serve as an essential reference source to students and researchers in universities and research institutions.

Professor Da-Wen Sun, Director
Food Refrigeration & Computerized Food Technology
National University of Ireland, Dublin
(University College Dublin)
Dublin, Ireland
http://www.ucd.ie/sun/

Series Editor

Born in Southern China, Professor Da-Wen Sun is a global authority in food engineering research and education. He is a Member of the Royal Irish Academy, the highest academic honor in Ireland; a Member of Academia Europaea (the Academy of Europe), one of the most prestigious academies in the world; a Foreign Member of the Polish Academy of Sciences; a Fellow of International Academy of Food Science and Technology; a Fellow of International Academy of Agricultural and Biosystems Engineering; and a Full Member (Academician) of International Academy of Refrigeration. He is also the founder and Editor-in-Chief of *Food and Bioprocess Technology*, one of the most prestigious food science and technology journals; Series Editor of the *Contemporary Food Engineering* book series with already over 50 volumes published; and the Founding President of the International Academy of Agricultural and Biosystems Engineering (iAABE). In addition, he served as the President the International Commission of Agricultural and Biosystems Engineering (CIGR), the world's largest organization in the field, in 2013–2014, and is now Honorary President of CIGR. He has significantly contributed to the field of food engineering as a researcher, as an academic authority and as an educator.

His main research activities include cooling, drying, and refrigeration processes and systems, quality and safety of food products, bioprocess simulation and optimization, and computer vision/image processing and hyperspectral imaging technologies. His many scholarly works have become standard reference materials for researchers in the areas of hyperspectral imaging, computer vision, ultrasonic freezing, vacuum cooling, computational fluid dynamics modeling, etc. The results of his work have been published in over 900 papers including more than 500 peer-reviewed journal papers indexed by Web of Science, with an average citation of over 36 per paper (Web of Science h-index = 92, SCOPUS h-index = 95), among them, 54 papers have been selected by Thomson Reuters's Essential Science Indicators[SM] as highly-cited papers, ranking him No. 2 in the world in Agricultural Sciences. He has also edited 17 authoritative books. In addition, Professor Sun has been named Highly Cited Researcher in the last 4 consecutive years (2015–2018) by Clarivate Analytics (formerly Thomson Reuters).

He received a first class BSc Honours and MSc in Mechanical Engineering, and a PhD in Chemical Engineering in China before working in various universities in Europe. He became the first Chinese national to be permanently employed in an Irish University when he was appointed College Lecturer at National University of Ireland, Dublin (University College Dublin) in 1995, and was then continuously promoted in the shortest possible time to Associate Professor, Professor, and Full Professor. Dr Sun is now a Full Professor of Food and Biosystems Engineering and Director of the Food Refrigeration and Computerised Food Technology Research Group at University College Dublin (UCD).

As a leading educator in food engineering, Professor Sun has significantly contributed to the field of food engineering. He has trained many PhD students, who have made their own contributions to the industry and academia. He has also given lectures on advances in food engineering on a regular basis in academic institutions internationally and delivered keynote speeches at international conferences. In recognition of his significant contribution to food engineering worldwide, and for his outstanding leadership in the field, the International Commission of Agricultural and Biosystems Engineering (CIGR) awarded him the CIGR Merit Award in 2000, in 2006, and again in 2016, the Institution of Mechanical Engineers (IMechE) based in the UK named him Food Engineer of the Year 2004, in 2008 he was awarded CIGR Recognition Award in honor of his distinguished achievements as the top one percent of agricultural engineering scientists in the world. In 2007, he was presented with the only AFST(I) Fellow Award in that year by the Association of Food Scientists and Technologists (India). In 2010, he received the CIGR Fellow Award; the title of Fellow is the highest honor in CIGR, and is conferred on individuals who have made sustained, outstanding contributions worldwide. In March 2013, he was presented with the You Bring Charm to the World Award by Hong Kong-based Phoenix Satellite Television with other award recipients, including the 2012 Nobel Laureate in Literature, and the Chinese Astronaut Team for the Shenzhou IX Spaceship. In July 2013, he received the Frozen Food Foundation Freezing Research Award from the International Association for Food Protection (IAFP) for his significant contributions to enhancing the field of food freezing technologies; this is the first time that this prestigious award was presented to a scientist outside the USA. In June 2015, he was presented with the IAEF Lifetime Achievement Award. This IAEF (International Association of Engineering and Food) award highlights the lifetime contribution of a prominent engineer in the field of food, and in February 2018, he was conferred with the honorary doctorate degree by Universidad Privada del Norte in Peru.

Preface

Nanoparticles differ significantly from their corresponding bulk material in terms of their physical, chemical, and biological properties. Such extreme reductions in particle size with associated increase in surface area to volume ratio results in unique properties that can in turn provide scope to suit the numerous ranges of its applications into various fields of science and technology.

This book mainly focuses on the design, production, and utilization of nanoparticles, with specific applications for the food industry. Through several studies, it has been proven that nanotechnology can offer distinct advantages over conventional methods in terms of functionality, targeted delivery of food bioactive compounds, improved stability in the gastrointestinal tract, and controlled release profiles. This book is an exploratory text, presenting the latest of such data on all aspects of applications of nanotechnology in food systems.

To present these, this book is divided into three sections. The first section describes fundamentals of nanotechnology and the broad range of its application in the food industry. Often overlooked, possible harmful effects of nanoparticles, toxicity, regulatory aspects, and concerns spanning from the local environment to the global perspective are presented.

The second section describes products and processes of nanotechnology, including the current status of nanotechnology and how it is being applied in the food sector. Starting with an overview of approaches for nanomaterial production, dedicated chapters on natural polymeric nanoparticles, encapsulation (and multi-layered) systems, fabrication of nanomaterials using electrospraying/spinning technique, plasma processing and nanodelivery systems have been included.

For each of these approaches, stability and release mechanisms, and interactions with the human digestive system have been elaborated upon. With the enormous scope of applying nano-based concepts in food packaging, a critical summary of such applications and the potential of biopolymers to revolutionize food packaging applications are presented.

The third section describes diagnostic methods and characterization techniques. An exclusive report on the application of nanosensors and biosensors for detection of food contaminants (including pesticides, organic pollutants, and toxic metals), and adulterants is included.

In a nutshell, this book will be an excellent foundation for academicians and researchers working in the field of food nanotechnology. With a perfect blend of science and engineering aspects, each technology is presented with up-to-date information on the subject, limitations, and directions for future research. Food security and sustainability are concerns of today and the future, and we earnestly hope that this book will serve as a well-documented technical resource that would provide in-depth scientific information, and ideas to come up with novel approaches to use nanotechnology to alleviate growing global concerns.

C. Anandharamakrishnan
S. Parthasarathi

Editors

C. Anandharamakrishnan is the Director of the Indian Institute of Food Processing Technology (IIFPT), Ministry of Food Processing Industries, Government of India. With two decades of experience in research and development (R&D) and administration at the Council of Scientific and Industrial Research—Central Food Technological Research Institute (CSIR-CFTRI), a prestigious Food Research Institution in India, he is a recognized expert in the field of nanoencapsulation techniques for the delivery of food bioactives, probiotics, and flavors. He also works on computational modeling of food processes and biological systems and is recognized for his work on the fabrication of an artificial gastrointestinal system to evaluate bioaccessibility and bioavailability of food, nanoencapsulation of micronutrients, and polyphenols. He received his PhD from Loughborough University, United Kingdom, for his work on spray-freeze drying systems. Dr. C. Anandharamakrishnan has been an active member of the Association of Food Scientists and Technologists, India. He is an elected Fellow of the Royal Society of Chemistry (FRSC) and the Institute of Engineers (FIE) and is a recipient of several awards, including the Professor Jiwan Singh Sidhu Award 2010 from the Association of Food Scientists and Technologies (India) (AFST[I]). He has published more than 67 peer-reviewed papers in food science and engineering journals, 4 books, and 17 book chapters. He also has 9 patents to his credit and has supervised e8 PhD theses and more than 40 bachelor's and master's theses.

S. Parthasarathi is presently a postdoctoral fellow at Riddet Institute, Massey University, New Zealand. He received his PhD from CSIR—Central Food Technological Research Institute, Mysore, India and his Master of Technology in Chemical Engineering from Anna University, Chennai, India. His research interests include nano- and microencapsulation of food bioactive compounds for the protection and controlled delivery of bioactives and the investigation of the influence of food nanoparticles on *in vivo* bioavailability.

Contributors

Uchangi Satyaprasad Akshath
Academy of Scientific and Innovative Research (AcSIR)
Microbiology and Fermentation Technology Department
CSIR—Central Food Technological Research Institute
Mysore, India

C. Anandharamakrishnan
Indian Institute of Food Processing Technology
Thanjavur, India

Praveena Bhatt
Academy of Scientific and Innovative Research (AcSIR)
Microbiology and Fermentation Technology Department
CSIR—Central Food Technological Research Institute
Mysore, India

L. Bhavani Devi
Department of Food Engineering
CSIR—Central Food Technological Research Institute
Mysore, India

Sayantani Dutta
Computational Modeling and Nanoscale Processing Unit
Indian Institute of Food Processing Technology
Thanjavur, India

R. Gopirajah
Food Engineering Lab
Food Science Department
Cornell University
Ithaca, New York

Heera Jayan
Department of Food Engineering
CSIR—Central Food Technological Research Institute
Mysore, India

P. Karthik
School of Food Science and Biotechnology
Zhejiang Gongshang University
Hangzhou, People's Republic of China

S. Kritika
Department of Food Engineering
CSIR—Central Food Technological Research Institute
Mysore, India

Maria Leena
Computational Modeling and Nanoscale Processing Unit
Indian Institute of Food Processing Technology
Thanjavur, India

J.A. Moses
Computational Modeling and Nanoscale Processing Unit
Indian Institute of Food Processing Technology
Thanjavur, India

Monali Mukherjee
Academy of Scientific and Innovative Research (AcSIR)
Microbiology and Fermentation Technology Department
CSIR—Central Food Technological Research Institute
Mysore, India

S. Parthasarathi
Riddet Institute
Massey University
Palmerston North, New Zealand

S. Priyanka
Computational Modeling and Nanoscale Processing Unit
Indian Institute of Food Processing Technology
Thanjavur, India

Preethi R.
Computational Modeling and Nanoscale Processing Unit
Indian Institute of Food Processing Technology
Thanjavur, India

B. Rohini
Department of Food Engineering,
CSIR—Central Food Technological Research Institute
Mysore, India

Sivakama Sundari S.K.
Computational Modeling and Nanoscale Processing Unit
Indian Institute of Food Processing Technology
Thanjavur, India

Vimala Bharathi S.K.
Computational Modeling and Nanoscale Processing Unit
Indian Institute of Food Processing Technology
Thanjavur, India

Das Trishitman
Department of Food Engineering
CSIR—Central Food Technological Research Institute
Mysore, India

K.S. Yoha
Computational Modeling and Nanoscale Processing Unit
Indian Institute of Food Processing Technology
Thanjavur, India

Section 1

Perspectives

1 Introduction

S. Parthasarathi and C. Anandharamakrishnan

CONTENTS

Nanotechnology is a multidisciplinary science combining biotechnology, chemical, mechanical and electronics engineering to understand, manipulate and fabricate devices/systems at the atomic/molecular/supramolecular levels with extraordinary functions and properties (de Francisco and García-Estepa, 2018 Jia, 2005). The discovery of nanotechnology has been attributed to Dr. Richard Phillips Feynman, an American physicist and Nobel laureate who presented a paper entitled "There's Plenty of Room at the Bottom" (Feynman, 1960) at the American Physical Society's annual meeting in the California Institute of Technology. However, Dr. Feynman did not specify the term "nanotechnology" in his talk. Later, in 1974, Dr. Taniguchi used the term "nanotechnology" in a paper entitled "On the Basic Concept of Nanotechnology" (Taniguchi, 1974).

US-based National Nanotechnology Initiative (NNI) defined nanotechnology as research and technology development at the atomic, molecular, or macromolecular scale leading to the controlled creation and use of structures, devices, and systems with a length scale of approximately 100 nm. The British Standards Institution defined nanotechnology as the design, characterization, production, and application of structures, devices, and systems by controlling the shape and size at the nanoscale (Bawa et al., 2005; Ezhilarasi et al., 2013b). Generally, the nanoparticle is a particle of 100 nm or less in size, and there is an enormous difference between its properties and those of the same material in bulk. Because, introducing nano-sized particles allows an interaction on the molecular level, and makes the material based on nanoparticles more efficient than those based on bulk materials (Li et al., 2008). Further, high surface to core ratio is a unique characteristic of nanoparticles representing more atoms at the surface of the nanoparticle than deep within its core. These surface atoms have the potential of creating new and strong bonds which result in more reactive properties than in micro and macro particles (Binns, 2010; AlKahtani, 2018). To illuminate the effect of the size difference, relative sizes of familiar materials and the sizes of several nanostructures are given in Figure 1.1.

Nanotechnologies have the huge potential to create a revolution in many industries, including the aeronautical, drug development, information and communications, infrastructure, agriculture, and food industries. The application of nanotechnology in the food sector leads to a considerable number of new products with improved food quality characteristics such as texture, taste, sensory attributes, stability, etc. Researchers are developing a wide range of nanomaterials for application in the food sector, from delivery of nutrients to packaging. Figure 1.2 depicts the potential application of various nanomaterials in the food industry.

1.1 APPLICATION OF NANOTECHNOLOGY IN THE FOOD SECTOR

Over the past decade, research and subsequent development of nanomaterial-based products have created great excitement and enormous interest. In 2010, the Science and Technology Committee of the House of Lords in the UK stressed the need for serious research into the implication of

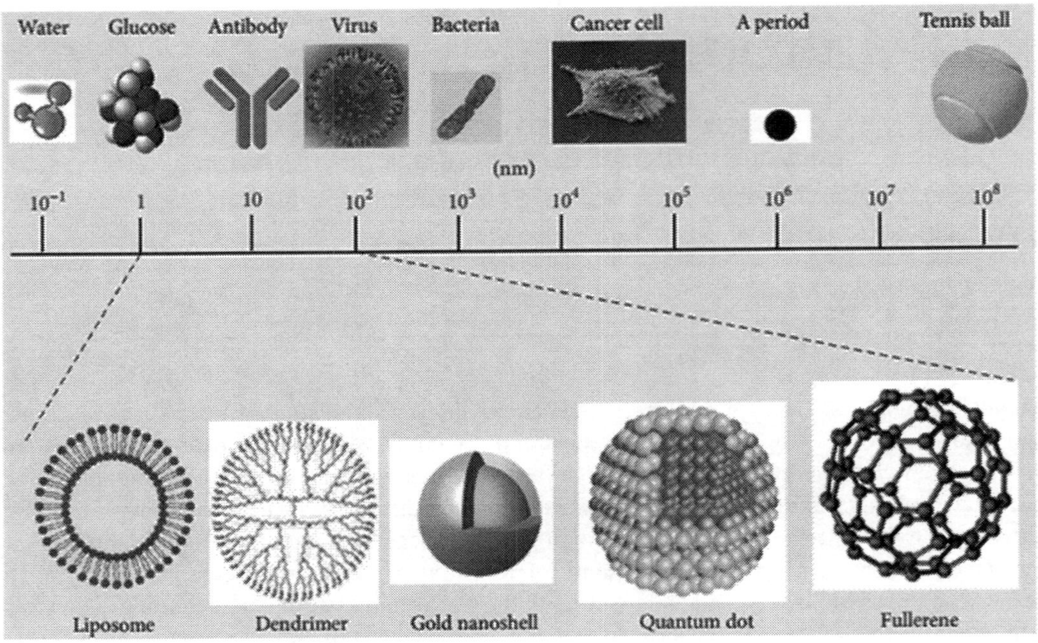

FIGURE 1.1 Relative size of nanoparticles and other familiar materials. (Pradhan et al., 2015.)

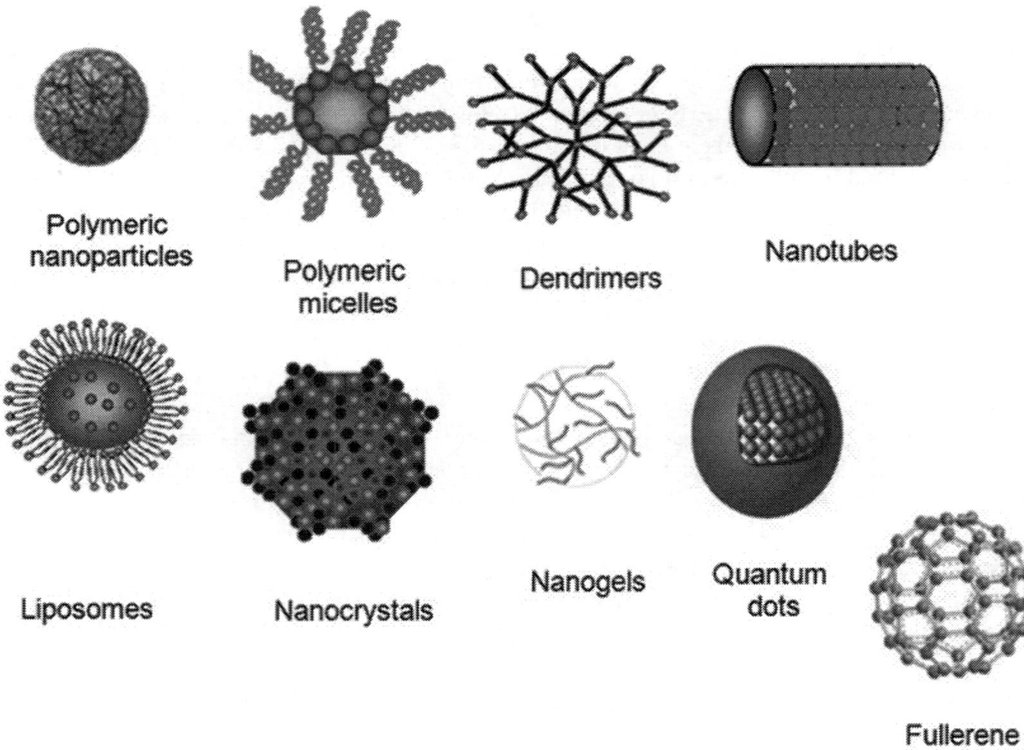

FIGURE 1.2 Nanomaterials for various applications in food industries. (Ojha and Kumar, 2017.)

nanotechnology in the food sector. Moreover, the House of Lords report highlights the potential for public backlash given certain similarities between food nanotechnologies and GMOs (Dudo et al., 2011, p. 8):

> Consumers are particularly sensitive about new technologies involving the scientific manipulation of food and understandably cautious about their introduction. The public response to the development of genetically modified food illustrates how quickly the views of some sectors of the public can change if action is not taken to meet concerns they may have about a new food technology.

The applications of nano-based technology in the food industry may include nanoparticulate delivery systems (e.g. micelles, liposomes, nanoemulsion, biopolymeric nanoparticles, and cubosomes), food safety and biosecurity (e.g. nanosensors), and toxicological studies (Pradhan et al., 2015). Nano-based food products can have improved nutritional properties, physical stability, and bioavailability. There is great interest among researchers in developing novel nano-based food products, and there has been an increasing number of publications over the last decade (see Figure 1.3).

Nanoencapsulation of bioactive compounds is one of the most promising technologies in pharmaceutical, nutraceutical and food industries (Anandharamakrishnan, 2014 and 2015). Encapsulation is a process by which small quantities of core materials (nutrient/therapeutic compounds and bioactive compounds) are encapsulated within the wall material to form capsules. Many researchers improved the stability of nutrient/bioactive compounds using the encapsulation technique for *Lactobacillus plantarum* (Rajam and Anandharamakrishnan, 2015; Rajam et al., 2012), hydroxycitric acid (Ezhilarasi et al., 2013a), DHA (Karthik and Anandharamakrishnan, 2013), vanillin (Hundre et al., 2015), and vitamin E (Parthasarathi and Anandharamakrishnan, 2016; Parthasarathi et al., 2016).

Nanotechnology offers great potential to revolutionize conventional food science and the food industry. However, the current scenario of nanotechnology in the food sector is relatively small when compared to nano-based products in the automobile, mechanical, and electronic industries. In the next 5–10 years, research on implications of nanotechnology in food will lead to revolutionary advances. Nano-based applications in the food industry may include (Chen et al., 2014; Handford et al., 2014):

- Smart Agriculture – nanofertilizer (containing nano zinc, silica, iron, etc. to endorse control release and improve quality), and nanopesticide (nanoencapsulated pesticides to reduce dosage and increase pest control efficacy);

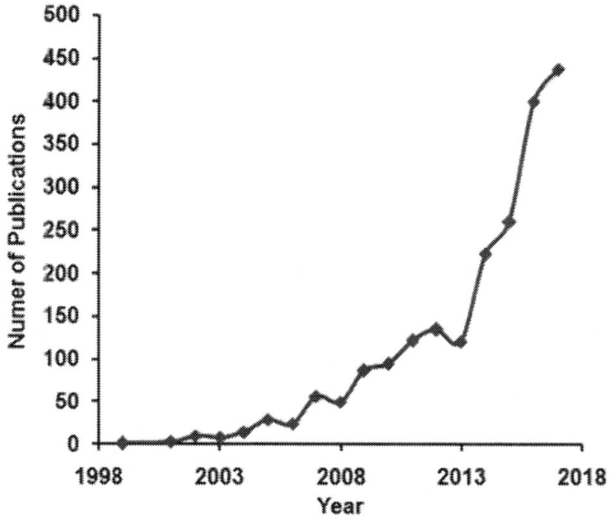

FIGURE 1.3 Number of published manuscripts till 2017. The keywords used were the combination of "food" and "nanotechnology" in PubMed.

TABLE 1.1
Potential Application of Nanotechnology in Food Science and Related Industries

Area	Application	References
Agriculture (nano modification of seed and fertilizers/ pesticides)	Pesticides	Zhao et al. (2016) and Tan et al. (2018)
	Fertilizers	Tarafdar et al. (2014) and Wang and Nguyen (2018)
	Nanosensor to monitor soil quality	Sangeetha et al. (2017)
Food processing (interactive smart food)	Nanoencapsulation of flavor and aroma	Anandharamakrishnan and Ishwarya (2015), Ezhilarasi et al. (2013b) and Hundre et al. (2015)
	Nanoemulsions	Karthik and Anandharamakrishnan (2016a)
	Electrospraying	Bhushani and Anandharamakrishnan (2014) and Bhushani et al. (2017)
	Solid lipid nanoparticles	Ezhilarasi et al., (2016)
Nutrition (food fortification and modification)	Nutraceutical and nutrient delivery	Karthik and Anandharamakrishnan (2016b) and Bhushani et al. (2016)
	Mineral and vitamin fortification	Fathima et al., (2017) and Fan et al. (2017)
	Drinking water purification	Bolisetty and Mezzenga (2016) and Momba et al. (2017)
Packaging material (smart packaging and food tracking)	UV protection	Goudarzi et al., (2017) and Lizundia et al. (2016)
	Antimicrobials	Nakazato et al. (2017)
	Contaminant sensors	Kulshreshtha et al., (2017) and Patra et al. (2017)

- Nanobiosensors – detection of pathogens, toxins, and contamination in foods;
- Smart field sensing systems – monitoring crop growth and other field conditions including nutritional status, temperature, soil fertility, and moisture level;
- Nano-based food products with less fat, salt and sugar, and lower calories, but maintaining the textural and flavor qualities;
- Nano-based nutrient delivery system – effective delivery of micronutrients and sensitive bioactive compounds;
- Nanotechnology-based equipment-insulation coating material to reduce heat loss in the food processing industries;
- Biodegradable nanocomposite-based food packaging material with improved mechanical and barrier properties (Table 1.1).

The consumer should be cautious with but not afraid of the applications of nanotechnology in food science and its allied industries. Many internationally recognized regulatory bodies such as the United States Environmental Protection Agency, the National Institute for Occupational Safety and Health, the Food and Drug Administration (FDA), the Health and Consumer Protection Directorate of the European Commission, and international organizations such as the International Organization for Standardization and the Organization for Economic Cooperation and Development, have issued guidance documents with respect to the potential risks posed by nanomaterials. Although the fate and potential toxicity of nanomaterials are not fully understood, it is clear that there has been a noteworthy improvement in the application of nanotechnology in the food industry.

This book is mainly concerned with the design, production, and utilization of nanoparticles within the food industry. The numerous reports on nanotechnology reveal that nanoparticles offer distinct advantages over conventional methods for the delivery of bioactive food compounds, including stability in the gastrointestinal tract, controlled release of bioactive compounds, targeted delivery and functionality. This book also attempts to gather and present the latest data on all aspects of the application of nanotechnology in the food industry.

REFERENCES

AlKahtani, Rawan N. 2018. "The implications and applications of nanotechnology in dentistry: A review." *The Saudi Dental Journal*, 30:107–116.

Anandharamakrishnan, Chinnaswamy. 2014. *Techniques for Nanoencapsulation of Food Ingredients.* New York, NY: Springer.

Anandharamakrishnan, Chinnaswamy. 2015. *Spray Drying Techniques for Food Ingredient Encapsulation.* New York, NY: John Wiley & Sons.

Anandharamakrishnan, C. and S. Padma Ishwarya. 2015. "Spray drying for nanoencapsulation of food components." In *Spray Drying Techniques for Food Ingredient Encapsulation*, 180–197. New York, NY: John Wiley & Sons Ltd.

Bawa, Raj, S.R. Bawa, Stephen B. Maebius, Ted Flynn, and Chiming Wei. 2005. "Protecting new ideas and inventions in nanomedicine with patents." *Nanomedicine: Nanotechnology, Biology and Medicine*, 1 (2):150–158

Bhushani, J. Anu and C. Anandharamakrishnan. 2014. "Electrospinning and electrospraying techniques: Potential food based applications." *Trends in Food Science and Technology*, 38 (1):21–33.

Bhushani, J. Anu, P. Karthik, and C. Anandharamakrishnan. 2016. "Nanoemulsion based delivery system for improved bioaccessibility and Caco-2 cell monolayer permeability of green tea catechins." *Food Hydrocolloids*, 56:372–382.

Bhushani, J. Anu, Nawneet Kumar Kurrey, and C. Anandharamakrishnan. 2017. "Nanoencapsulation of green tea catechins by electrospraying technique and its effect on controlled release and in-vitro permeability." *Journal of Food Engineering*, 199:82–92.

Binns, Chris. 2010. *Introduction to Nanoscience and Nanotechnology.* Vol. 14: John Wiley & Sons, New Jersey.

Bolisetty, Sreenath and Raffaele Mezzenga. 2016. "Amyloid–carbon hybrid membranes for universal water purification." *Nature Nanotechnology*, 11 (4):365.

Chen, Hongda, James N. Seiber, and Matt Hotze. 2014. *ACS Select on Nanotechnology in Food and Agriculture: A Perspective on Implications and Applications.* ACS Publications.

de Francisco, Elena Villena and Rosa M. García-Estepa. 2018. "Nanotechnology in the agrofood industry." *Journal of Food Engineering*, 238:1–11.

Dudo, Anthony, Doo-Hun Choi, and Dietram A. Scheufele. 2011. "Food nanotechnology in the news. Coverage patterns and thematic emphases during the last decade." *Appetite*, 56 (1):78–89.

Ezhilarasi, P.N., D. Indrani, B.S. Jena, and C. Anandharamakrishnan. 2013a. "Freeze drying technique for microencapsulation of Garcinia fruit extract and its effect on bread quality." *Journal of Food Engineering*, 117 (4):513–520.

Ezhilarasi, P.N., P. Karthik, N. Chhanwal, and C. Anandharamakrishnan. 2013b. "Nanoencapsulation techniques for food bioactive components: A review." *Food and Bioprocess Technology*, 6 (3):628–647.

Ezhilarasi, P.N., S.P. Muthukumar, and C. Anandharamakrishnan. 2016. "Solid lipid nanoparticle enhances bioavailability of hydroxycitric acid compared to a microparticle delivery system." *RSC Advances*, 6 (59):53784–53793.

Fan, Taotao, Xiaoyan Yu, Bing Shen, and Leming Sun. 2017. "Peptide self-assembled nanostructures for drug delivery applications." *Journal of Nanomaterials*, 2017.

Fathima, Syeda Juveriya, Ilaiyaraja Nallamuthu, and Farhath Khanum. 2017. "Vitamins and minerals fortification using nanotechnology: Bioavailability and recommended daily allowances." In *Nutrient Delivery* 457–496. London: Elsevier.

Feynman, R.P. 1960. "There's plenty of room at the bottom." *Engineering and Science*, 23 (5):22–36.

Goudarzi, Vahid, Iman Shahabi-Ghahfarrokhi, and Amin Babaei-Ghazvini. 2017. "Preparation of eco-friendly UV-protective food packaging material by starch/TiO2 bio-nanocomposite: Characterization." *International Journal of Biological Macromolecules*, 95:306–313.

Handford, Caroline E., Moira Dean, Maeve Henchion, Michelle Spence, Christopher T. Elliott, and Katrina Campbell. 2014. "Implications of nanotechnology for the agri-food industry: Opportunities, benefits and risks." *Trends in Food Science and Technology*, 40 (2):226–241.

Hundre, Swetank Y., P. Karthik, and C. Anandharamakrishnan. 2015. "Effect of whey protein isolate and β-cyclodextrin wall systems on stability of microencapsulated vanillin by spray-freeze drying method." *Food Chemistry*, 174:16–24.

Jia, Lee. 2005. "Global governmental investment in nanotechnologies." *Current Nanoscience*, 1 (3):263–266.

Karthik, P. and C. Anandharamakrishnan. 2013. "Microencapsulation of docosahexaenoic acid by spray-freeze-drying method and comparison of its stability with spray-drying and freeze-drying methods." *Food and Bioprocess Technology*, 6 (10):2780–2790.

Karthik, P. and C. Anandharamakrishnan. 2016a. "Enhancing omega-3 fatty acids nanoemulsion stability and in-vitro digestibility through emulsifiers." *Journal of Food Engineering*, 187:92–105.

Karthik, P. and C. Anandharamakrishnan. 2016b. "Fabrication of a nutrient delivery system of docosahexaenoic acid nanoemulsions via high energy techniques." *RSC Advances*, 6 (5):3501–3513.

Kulshreshtha, Niha Mohan, Divya Shrivastava, and Prakash Singh Bisen. 2017. "Contaminant sensors: Nanotechnology-based contaminant sensors." In *Nanobiosensors*, 573–628. London: Elsevier.

Li, Li, Haihua Pan, Jinhui Tao, Xurong Xu, Caiyun Mao, Xinhua Gu, and Ruikang Tang. 2008. "Repair of enamel by using hydroxyapatite nanoparticles as the building blocks." *Journal of Materials Chemistry*, 18 (34):4079–4084.

Lizundia, Erlantz, Leire Ruiz-Rubio, José Luis Vilas, and Luis Manuel León. 2016. "Poly (l-lactide)/zno nanocomposites as efficient UV-shielding coatings for packaging applications." *Journal of Applied Polymer Science*, 133, 42426(1-7).

Momba, Maggy N.B., Lerato Baloyi, Lizzy Mpenyana-Monyatsi, and Ilunga Kamika. 2017. "Nanotechnology-based filters for cost-effective drinking water purification in developing countries." In *Water Purification*, ed. Grumezescu AM, 169–208. London: Academic Press.

Nakazato, Gerson, Renata K.T. Kobayashi, Amedea B. Seabra, and Nelson Duran. 2017. "Use of nanoparticles as a potential antimicrobial for food packaging." In *Food Preservation*, 413–447. London: Elsevier.

Ojha, Smriti and Babita Kumar. 2017. "A review on nanotechnology based innovations in diagnosis and treatment of multiple sclerosis." *Journal of Cellular Immunotherapy*.

Parthasarathi, S. and C. Anandharamakrishnan. 2016. "Enhancement of oral bioavailability of vitamin E by spray-freeze drying of whey protein microcapsules." *Food and Bioproducts Processing*, 100:469–476.

Parthasarathi, S., S.P. Muthukumar, and C. Anandharamakrishnan. 2016. "The influence of droplet size on the stability, in vivo digestion, and oral bioavailability of vitamin E emulsions." Food and Function, 7 (5):2294–2302.

Patra, Santanu, Ekta Roy, Rashmi Madhuri, and Prashant K Sharma. 2017. "A technique comes to life for security of life: The food contaminant sensors." In *Nanobiosensors*, 713–772. London: Elsevier.

Pradhan, Neha, Surjit Singh, Nupur Ojha, Anamika Shrivastava, Anil Barla, Vivek Rai, and Sutapa Bose. 2015. "Facets of nanotechnology as seen in food processing, packaging, and preservation industry." *BioMed Research International*, 2015, 365672.

Rajam, R. and C. Anandharamakrishnan. 2015. "Microencapsulation of Lactobacillus plantarum (MTCC 5422) with fructooligosaccharide as wall material by spray drying." *LWT-Food Science and Technology*, 60 (2):773–780.

Rajam, R., P. Karthik, S. Parthasarathi, G.S. Joseph, and C. Anandharamakrishnan. 2012. "Effect of whey protein–alginate wall systems on survival of microencapsulated Lactobacillus plantarum in simulated gastrointestinal conditions." *Journal of Functional Foods*, 4 (4):891–898.

Sangeetha, Jeyabalan, Devarajan Thangadurai, Ravichandra Hospet, Etigemane Ramappa Harish, Prathima Purushotham, Mohammed Abdul Mujeeb, Jadhav Shrinivas, Muniswamy David, Abhishek Channayya Mundaragi, and Shivasharana Chandrabanda Thimmappa. 2017. "Nanoagrotechnology for soil quality, crop performance and environmental management." In *Nanotechnology*, 73–97. Singapore: Springer.

Tan, Wenjuan Qin Gao, Chaoyi Deng, Yi Wang, Wen-Yee Lee, Jose A. Hernandez-Viezcas, Jose R. Peralta-Videa, and Jorge L. Gardea-Torresdey. 2018. "Foliar exposure of Cu $(OH)_2$ nanopesticide to basil (ocimum basilicum): variety-dependent copper translocation and biochemical responses." *Journal of Agricultural and Food Chemistry*, 66 (13):3358–3366.

Taniguchi, Norio. 1974. "On the basic concept of nano-technology." Paper read at Proceedings of International Conference on Production, London.

Tarafdar, J.C., Ramesh Raliya, Himanshu Mahawar, and Indira Rathore. 2014. "Development of zinc nanofertilizer to enhance crop production in pearl millet (Pennisetum americanum)." *Agricultural Research*, 3 (3):257–262.

Wang, San-Lang, and Anh Dzung Nguyen. 2018. "Effects of Zn/B nanofertilizer on biophysical characteristics and growth of coffee seedlings in a greenhouse." *Research on Chemical Intermediates*: 44(8),4889-4901.

Zhao, Lijuan, Cruz Ortiz, Adeyemi S. Adeleye, Qirui Hu, Hongjun Zhou, Yuxiong Huang, and Arturo A. Keller. 2016. "Metabolomics to detect response of lettuce (Lactuca sativa) to Cu $(OH)_2$ nanopesticides: Oxidative stress response and detoxification mechanisms." *Environmental Science and Technology*, 50 (17):9697–9707.

2 Fundamentals of Nanotechnology

S. Parthasarathi and C. Anandharamakrishnan

CONTENTS

> The impact of nanotechnology on health, wealth, and lives of people will be at least the equivalent of the combined influences of microelectronics, medical imaging, computer-aided engineering, and man-made polymers developed in this century.
>
> **Dr. Richard Smalley, Nobel Laureate in Chemistry (1996)**

2.1 WHAT IS A NANOMETER?

The prefix "nano" comes from a Greek word referring to one billionth. It indicates a factor of 10^{-9}; therefore, a nanometer is one billionth of a meter. This branch of science and technology deals with materials having at least one spatial dimension in the size range of 1 to 100 nm. Our eye could visualize the smallest objects of 1 mm i.e. the 1000th part of a meter. A nanometer is one thousand millionth of a meter. For comparison, the approximate size of an apple is 76 mm; A dog flea (species of flea that lives on a wide variety of mammals) is approximately 1 mm; the maximum size of pollen grains is 90–100 micrometer (μm); a single strand of a spider web is approximately 4–5 μm; virus is

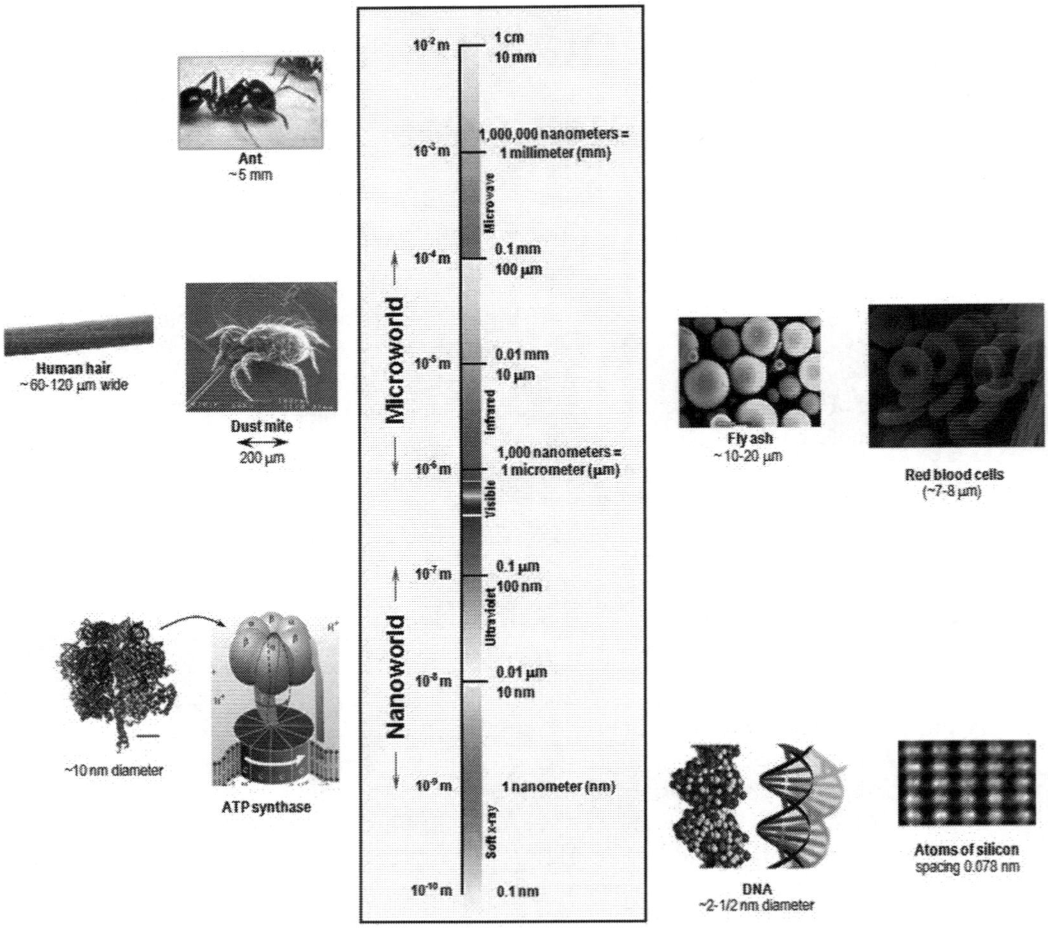

FIGURE 2.1 Overview of micro- and nanoworld. (Department of Energy Office of Basic Energy Sciences, www.sc.doe.gov/bes/scale_of_things.html.)

approximately 20 nanometer (nm), and the diameter of a helium molecule is 0.1 nm. To get a more precise idea about the nanoscale, a list of some naturally occurring objects is given in Figure 2.1.

2.1.1 Size of a Protein Molecule

Assuming partial specific volume of protein 0.73 cm^3/g (v_2), the volume of a protein of mass M (Dalton, Da) can be calculated using the following equations (Erickson, 2009):

$$V\left(\text{nm}^3\right) = \frac{\left(0.73\,\text{cm}^3/\text{g}\right)\times\left(10^{21}\,\text{nm}^3/\text{cm}^3\right)}{\left(6.023\times10^{23}\,\text{Da}/\text{g}\right)}\times \text{M}\left(\text{Da}\right) \tag{2.1}$$

$$V\left(\text{nm}^3\right) = 1.212\times10^{-3}\left(\text{nm}^3/\text{Da}\right)\times \text{M}\left(\text{Da}\right) \tag{2.2}$$

Assuming protein is a spherical-shaped molecule, the radius (R_{min}) can be calculated by,

$$V = \frac{4}{3}\pi R_{min}^3 \tag{2.3}$$

TABLE 2.1
R_{min} for Proteins of Different Mass

Protein M (kDa)	5	10	20	50	100	200	500
R_{min} (nm)	1.1	1.42	1.78	2.4	3.05	3.84	5.21

Source: Erickson (2009).

$$R_{\min} = \left({}^{3V}\!/\!{}_{4\pi} \right)^{3} \tag{2.4}$$

$$R_{\min} = 0.066\ \mathrm{M}^{1/3} \tag{2.5}$$

where R_{min} is the minimal radius (nm) of protein containing the given mass (M in Dalton). The radius of protein goes from 5,000 to 500,000 Da are given in Table 2.1.

2.2 WHAT IS NANOTECHNOLOGY?

Nanoscience and nanotechnology are the study and application of extremely small things (1–100 nm) widely in the field of chemistry, physics, engineering, material science, and biology. The concepts of nanoscience and nanotechnology were introduced by Nobel Laureate Richard Feynman's 1960 article "There's Plenty of Room at the Bottom." In his visionary talk, Feynman stated, "The principles of physics, as far as I can see, do not speak against the possibility of manoeuvring things atom by atom," and "the entire 24 volumes of the Encyclopedia Britannica on the head of a pin." (Feynman, 1960). However, the term nanotechnology was not used until 1974, when Professor Norio Taniguchi of the Tokyo Science University first mentioned it in his paper "On the Basic Concept of 'Nanotechnology'" presented at the Japan Society of Precision Engineering. Prof. Norio Taniguchi coined "nanotechnology" to describe the semiconductor process: "Nano-technology mainly consists of the processing of, separation, consolidation and deformation of materials by one atom or one molecule" (Taniguchi, 1974).

2.2.1 Definitions of Nanotechnology

Nanotechnology has been evolving for the past six decades and there are several definitions of nanotechnology generated for specific purposes. The British Standards Institution (BSI, 2005) proposed the following definitions for the general terms of nanotechnology.

- *Nanoscience*: The study of phenomena and manipulation of materials at atomic, molecular, and macromolecular scales, where properties differ significantly from those at a larger scale.
- *Nanotechnology*: The design, characterization, production, and application of structures, devices, and systems by controlling shape and size at the nanoscale.
- *Nanomaterial*: Material with one or more external dimensions, or an internal structure, which could exhibit novel characteristics compared to the same material without nanoscale features.
- *Nanoparticle*: Particle with one or more dimensions at the nanoscale.

The Foresight Institute includes various fields of science that come into play with nanotechnology: "Structures, devices, and systems having novel properties and functions due to the arrangement of their atoms on the 1 to 100 nanometer scale. Many fields of endeavor contribute to nanotechnology, including molecular physics, materials science, chemistry, biology, computer science, electrical engineering, and mechanical engineering" (The Foresight Institute, n.d.).

According to The European Commission, "nanotechnology is the study of phenomena and fine-tuning of materials at atomic, molecular and macromolecular scales, where properties differ significantly from those at a larger scale. Products based on nanotechnology are already in use and analysts expect markets to grow by hundreds of billions of euros during this decade" (The European Commission, n.d.).

The National Nanotechnology Initiative (NNI) includes certain activities such as measuring and manipulating nanoscale matter in the definition. NNI defines nanotechnology as "the understanding and control of matter at dimensions between approximately 1 and 100 nanometers, where unique phenomena enable novel applications. Encompassing nanoscale science, engineering, and technology, nanotechnology involves imaging, measuring, modeling, and manipulating matter at this length scale" (NNI, n.d.).

2.3 HISTORY OF NANOTECHNOLOGY

Nanotechnology seems to be a new research field, but surprisingly Romans used nanometals for decorating glasses and cups in the fourth century AD. The focus on the optical properties of metallic colloids started only in the early twentieth century with Gustav Mie's work (Mie, 1908). However, the application of the outstanding properties of metal colloids dates back to several millennia ago (Sciau, 2012; García, 2011).

2.3.1 Lycurgus Cup

The best-known example of the Romans' work on nanotechnology is the Lycurgus Cup – a mythological frieze depicting King Lycurgus, exposed in the British Museum in London.

The most remarkable aspect of the Lycurgus Cup is not only the cut-work design which stands out proudly from the body of the vessel but also the dichroic characteristics of the glass. The cup resembles jade with an opaque greenish-yellow tone in direct light. When the light is transmitted through the glass, it turns into a translucent ruby color (see Figure 2.2). The distinct visual appearance of the glass is due to the presence of metal nanoparticles. A detailed report on the Lycurgus Cup by Brill, 1965, and TEM was employed to investigate the origin of the color (Barber and Freestone, 1990). TEM revealed the presence of metal nanoparticles typically in the range of 50–100 nm (Figure 2.3a). X-ray analysis showed the presence of silver-gold alloy with a ratio of 7:3% and 10% copper. Furthermore, numerous nanoparticles (15–100 nm) of sodium chloride were identified in the study (Figure 2.3b); chlorine may be derived from the mineral salts, which were used to supply the alkali during the glass manufacture. The detailed chemical composition is given in Table 2.2.

2.3.2 Luster

Gold and other metallic nanoparticles were used not only as colorants for glasses and ceramics but also to create particular optical effects in glass and potteries (Schaming and Remita, 2015). Luster is a thin coating of unoxidized metal applied on glass, potteries, and ceramics to give an iridescent graze. It was first developed in Mesopotamia during the ninth century and diffused throughout the Mediterranean basin, and became popular in the Islamic culture. Luster decoration is one of the most complex techniques involving a high-temperature process (up to 600°C) and the use of chemical materials (a mixture of copper and silver salts and oxides, vinegar, ochre, and clay) (Padovani et al., 2003; Caiger-Smith, 1985). Cipriano Piccolpasso described the preparation of luster in his work *I tre libri dell'arte del vasaio*, 1557. Archaeologists recovered ancient lusterwares made of glass and ceramics from Europe and Middle East dating back to the period between the ninth and seventeenth centuries. The color of the luster varies according to the relative angle between light

FIGURE 2.2 Lycurgus Cup. (Wikipedia, Creative Commons.)

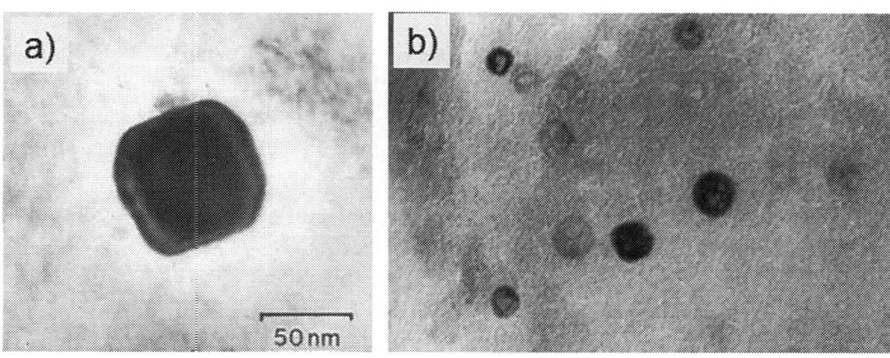

FIGURE 2.3 TEM images of (a) silver-gold alloy nanoparticles, (b) sodium chloride particles in the Lycurgus Cup. (Barber and Freestone, 1990.)

sources, pottery surface, and the observer (Colomban, 2009). Padovani et al. (2003) revealed the relation between color and glaze composition. The golden color is mainly due to the presence of Ag nanoparticles and red color to Cu nanoparticles, although some Ag nanoparticles and Ag^+ and Cu^+ ions were present.

2.3.3 Damascus Blades

Till the nineteenth century, swords were used as weapons in the battlefield, and that is where the Damascus sword was first encountered by Crusaders. The main characteristics of the Damascus sword are the wavy banding pattern (known as damask) on the sword surface and the extremely sharp cutting edge (able to slice even a silk handkerchief floating in the air). Figure 2.4 depicts

TABLE 2.2
Composition of the Lycurgus Cup

SiO_2	73.5	Fe_2O_3	1.5	SnO_2	<0.01
Na_2O	13–15	P_2O_5	0.2	B_2O_3	0.1
CaO	6.5	MnO	0.45	TiO_2	0.07
K_2O	0.9	Sb_2O_3	0.3	Ag	0.03
MgO	0.5–0.6	CuO	0.04	Au	0.004
Al_2O_3	2.5	PbO	0.2		

Source: Barber and Freestone (1990).

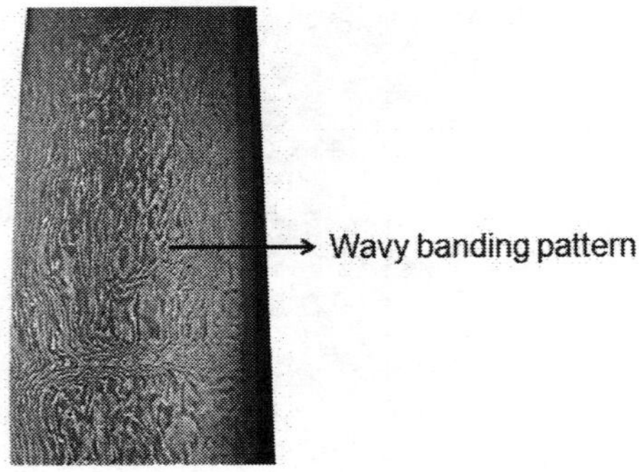

FIGURE 2.4 Eighteenth-century Persian-forged Damascus steel sword. (Wikipedia, Creative Commons.)

an eighteenth-century Persian-forged Damascus steel sword. It has been reported that Damascus swords were manufactured in southern India through the thermomechanical treatment of forging and annealing process. However, European bladesmiths were unable to replicate the process, so the manufacturing procedures were lost at the end of the eighteenth century (Reibold et al., 2009). The carbon composition in iron plays a vital role in the sword's characteristics. If 2% carbon is added to iron, it will produce hard and brittle steel; on the other hand, soft and malleable steel is obtained by the composition of 0.5% carbon in iron (Srinivasan, 2007).

Researchers identified the presence of rare-earth elements including cementite nanowires in the microstructure of Damascus steel (Levin et al., 2005; Kochmann et al., 2004). With the aid of high-resolution transmission electron microscopy (HR-TEM), Reibold et al. (2006) revealed the presence of carbon nanotubes in Damascus sabre. Furthermore, carbon nanotubes encapsulate and protect the cementite (Fe_3C) nanowires in steel. In Damascus steel, cementite is embedded and cementite veins run close to the edge to provide good cutting properties (Levin et al., 2005).

2.4 SELF-ASSEMBLED NANOSTRUCTURES

Molecular self-assembly is a powerful tool in the synthesis of functional nanostructures through bottom-up fabrication technique. Self-assembly can be defined as the spontaneous organization of disordered molecular units into ordered structures as a consequence of specific, local interactions among the components themselves (Lehn, 2002). In other terms, self-assembly is a powerful approach for the construction of novel supramolecular architectures with the aid of local noncovalent interactions such as hydrogen bonds, water-mediated hydrogen bonds, hydrophobic interactions, and electrostatic interactions (Li et al., 2008). Although the individual contribution of these forces may be weak, their collective interactions could produce stable structures. Researchers developed many self-assembling nanomaterials from carbohydrates, proteins, peptides, nucleic acids, organic polymers, etc., to understand the phenomenon and apply those materials from molecular devices to delivery systems (Zhang, 2003; Panda and Chauhan, 2014). Self-assembled biomolecules have been evaluated in disease diagnostics, drug delivery, tissue engineering, and regenerative medicine. Various self-assembled structures such as lipids, proteins, and polysaccharides have been explored as nutraceutical delivery systems. The first approval of nanomedicine for cancer therapy (Taxol, anticancer drug) by the US Food and Drug Administration (USFDA) in 2005 recognized the efforts of nanotechnology in drug delivery. The selection of nanoparticle materials should be based on the following parameters (Lohcharoenkal et al., 2014):

- required size of nanoparticles
- inherent properties of bioactive compounds such as stability and solubility
- release characteristics
- surface charge and hydrophobicity of the nanoparticles
- biocompatibility and biodegradability of nanomaterials

These biopolymer-based nanoparticle delivery systems recently gained considerable attention due to their desirable properties, such as low toxicity and biodegradability.

2.4.1 Proteins and Peptides

Nature is an inspiration for researchers in developing complex functional materials, and we can find many simple and complex self-assembled nanostructures in a biological system. Lipid membrane structure is one of the simple self-assembled structures in the biological system. The primary force for the formation of the lipid membrane structure is the simple and iterative hydrophobic interactions among amphiphilic lipid molecules (Lim et al., 2009). The best example for complex self-assembly in the biological system is protein. Proteins are linear biopolymers (polymer chain) composed of hundreds of building blocks, α-L amino acids joined together in a linear sequence. In nature, there are 20 different amino acids, which produce different proteins based on the number and order of amino acids in the polymer chain (Vendruscolo et al., 2003). Researchers commonly differentiate peptides and proteins based on their size and structure (see Figure 2.5). When a polymer chain consists of 2–50 amino acids it is called a peptide, whereas a longer chain with more than 50 amino acids is called protein. Understanding protein self-assembly is fundamental for many industrial and physiological processes, including fibrillization of Aβ, which is a feature of Alzheimer's disease (Cohen et al., 2013), protein condensation diseases such as cataract formation (Benedek 1997), and Parkinson's disease (Uversky and Eliezer, 2009).

Peptides consist of natural or synthetic amino acids, which they use as building blocks (repeating units) for the construction of molecular assemblies. Peptides can be classified based on their structures, such as dipeptides, cyclic peptides, amphiphilic peptides, α-helical peptides, and β-sheet peptides. Based on their structures and self-assembly characteristics, these peptides can be assembled into nanotubes, nanofibers, nanospheres, nanobelts, and hydrogels and nanovesicles (Fan et al., 2017).

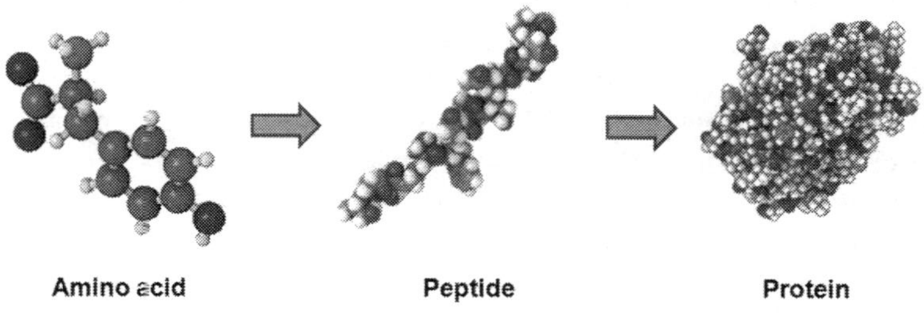

FIGURE 2.5 Structure of amino acid, peptide, and protein.

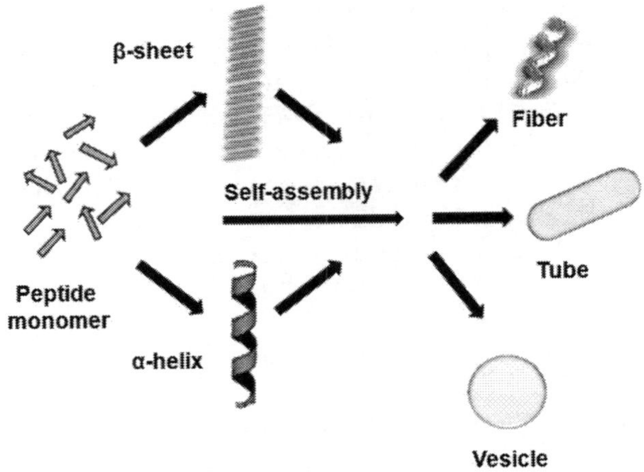

FIGURE 2.6 Self-assembly of peptides. (Panda and Chauhan, 2014.)

Figure 2.6 depicts the self-assembly of peptides. Understanding the self-assembly mechanism of proteins and peptides helps in developing food-grade nanoparticles for the efficient delivery of poor water-soluble bioactive compounds. Readers can refer to McClements (2018) for more details on bioactive proteins and peptides.

2.4.2 Polysaccharides

Polysaccharides are composed of a long chain of monosaccharides and available from different resources, including plants (arabinoxylans, pectin, guar gum), algal (alginate), and animal origin (chitosan, chondroitin). Based on the charge, these polysaccharides can be classified as polyelectrolytes (positively charged polysaccharides-chitosan) and non-polyelectrolytes (negatively charged polysaccharides – alginate, hyaluronic acid, pectin). Structurally, polysaccharides possess various derivable groups, and can be easily synthesized for different polysaccharide derivatives (Liu et al., 2008). Polysaccharides are natural biopolymers which are highly stable, safe, non-toxic, hydrophilic and biodegradable.

In future, polysaccharides offer credible answers to the demand for global sustainability. Polysaccharides such as starch and inulin provide stored solar energy in the form of sugar for fueling cells. To focus on the research on the application of polysaccharides in the food and health sector, the European Commission (EC) established the European Polysaccharide Network of Excellence (EPNOE). EPNOE is focused on polysaccharide and polysaccharide-related research and business, comprising more than 30 research and academic institutions and 10 industries (Persin et al., 2011).

Self-assembly of polysaccharides is quite easy (aqueous procedure), and does not require any specific reaction conditions or solvent. First, water-soluble polymers are hydrophobically modified by hydrophobic groups such as alkyl, aralkyl, and deoxycholic acid. When the hydrophobically modified polymers dissolve in the aqueous phase, polymeric amphiphiles spontaneously form self-aggregates by undergoing intra- and/or inter-molecular association between hydrophobic moieties (Yang et al., 2015; Kataoka et al., 2001). Nanoparticles exhibit unique characteristics including unusual rheological features, dynamic stability, and encapsulation of hydrophobic bioactive compounds; moreover, they offer monodispersed nanoparticles without energy input (Yang et al., 2015).

***Sodium alginate*:** It is a hydrophilic biopolymer obtained from brown seaweeds that has shown promise in pharmaceutical and food delivery systems. Sodium alginate is chemically composed of D-mannuronic acid (M) and L-guluronic acid (G) structures in varied proportions in MM or GG blocks interspersed with MG blocks (Aslani and Kennedy, 1996). The chemical structure of sodium alginate is depicted in Figure 2.7. Sodium alginate is widely applied for nanofibers (Mokhena and

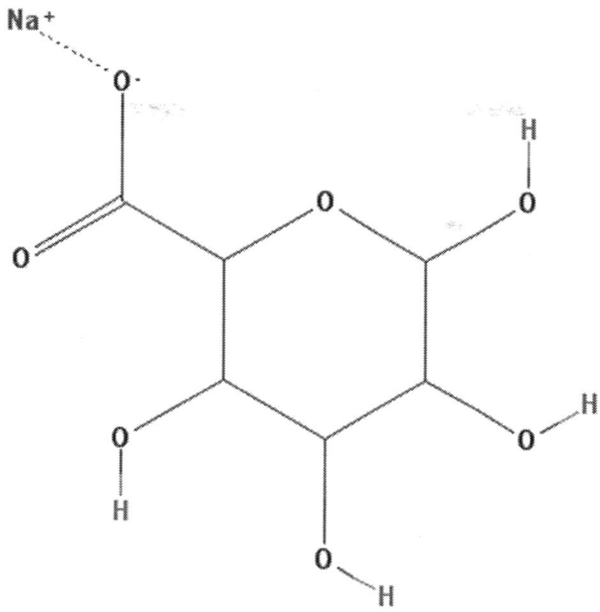

FIGURE 2.7 Chemical structure of sodium alginate.

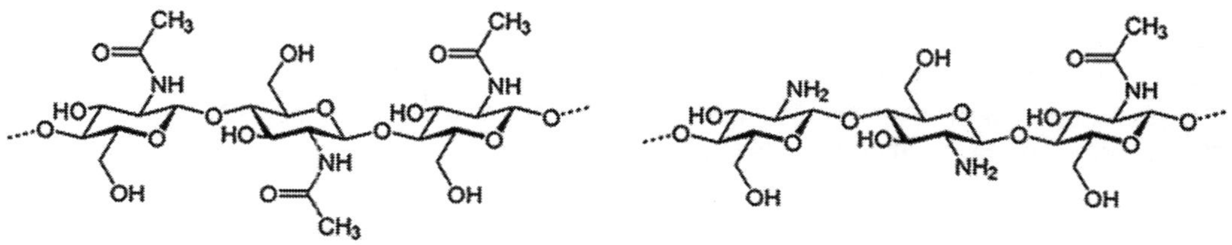

FIGURE 2.8 Chemical structure of chitin and chitosan. (Van den Broek et al., 2015.)

Luyt, 2017), edible packaging films (Tavassoli-Kafrani et al., 2016; Sirviö et al., 2014), nanogels (Sun et al., 2017), and nanoencapsulation (Purwanti et al., 2018) of food bioactive compounds.

Chitosan: Chitin is a precursor of chitosan, first discovered in 1811 by French chemist Henry Braconnot in mushrooms. During the synthesis, Henry obtained another group of compounds by partial or full deacetylation through treatment with a concentrated alkaline solution, which is called chitosan (Ma et al., 2017). Structurally, chitosan is composed of glucosamine and N-acetylglucosamine units linked together by β (1,4) glycosidic links (Quiñones et al., 2018). The chemical structures of chitin and chitosan are depicted in Figure 2.8. Chitosan could be a promising encapsulation and delivery system for many food bioactive compounds—epigallocatechin gallate (Liang et al., 2017), and anthocyanin (Ge et al., 2018). In a recent study, Kumar et al. (2018) reported that nanocomposite film of silver nanoparticles, chitosan, and gelatin improved the shelf life of red grapes for 14 days.

2.5 APPLICATION OF NANOTECHNOLOGY IN FOOD INDUSTRY

> Nanotechnology has the potential of application in the food industry and processing as new tools for pathogen detection, disease treatment delivery systems, food packaging, and delivery of bioactive compounds to target sites.
>
> **Rashidi and Khosravi-Darani (2011)**

TABLE 2.3
Nanomaterials in the Food Sector

Nanostructures	Size (in nanometer)
DNA	12
Glucose	21–75
Liposome	30–10000
LDH	40–300
Amylopectin	44–200
Casein micelle	60–100
PLA nanosphere	100–300
Zein	200
Cubosome	500
Nanosensors	<1000

Source: Sozer and Kokini (2009).

Nanotechnology emerged as one of the most promising technologies to revolutionize conventional food science and the food industry (He and Hwang, 2016). By, 2020, nanotechnology will generate at least $3 trillion through worldwide economy and require 6 million technicians/employers in different sectors (Duncan, 2011; Youssef and El-Sayed, 2018). The application of nanotechnology in the agriculture and food industries was first addressed by the United States Department of Agriculture (USDA) roadmap published in 2003. A five-year project (2003–2008) on "Application of Nanotechnology, Antimicrobial, and Polymer Films in Food Safety and Quality" was carried out at Clemson University with the following three objectives:

1. developing nanotechnology applications for food safety and quality
2. optimizing antimicrobial and antioxidant packaging films for foods
3. developing biopolymer film applications for foods

Nanotechnology can be applied by two different approaches in food sectors, viz., bottom up or top down. The sizes of different nanoparticles applied in current food sectors are given in Table 2.3. The top-down approach is the preferred technique, as it is achieved by means of physical processing of food materials including grinding, milling, and high-pressure homogenization. By contrast, the bottom-up approach utilizes self-assembly and self-organization concepts. Examples include the organization of casein micelles or starch and the folding of globular proteins and protein aggregates (Ravichandran, 2010). Food scientists are exploring the potential application of nanotechnology to encapsulate and deliver nutrients/bioactive compounds (Bhushani et al., 2016; Anandharamakrishnan, 2014; Ezhilarasi et al., 2013; Hundre et al., 2015; Karthik and Anandharamakrishnan, 2016a; 2013; Parthasarathi and Anandharamakrishnan, 2016; Parthasarathi et al., 2016; Rajam and Anandharamakrishnan, 2015), for pathogen detection through nanosensors, to create high-barrier packaging materials through nanocomposites (Anandharamakrishnan and Usha Kiran, 2014), and to obtain antimicrobial agents. The potential applications of nanotechnology in the food industry are depicted in Figure 2.9.

The revolution of nanotechnology in food processing has its roots in the era of the pasteurization process, employed to kill spoilage bacteria (1000 nm). Later, the discovery of DNA structure (2.5 nm) opened the gateway of applications in the agricultural, biomedical, and biotechnology industries (Chellaram et al., 2014).

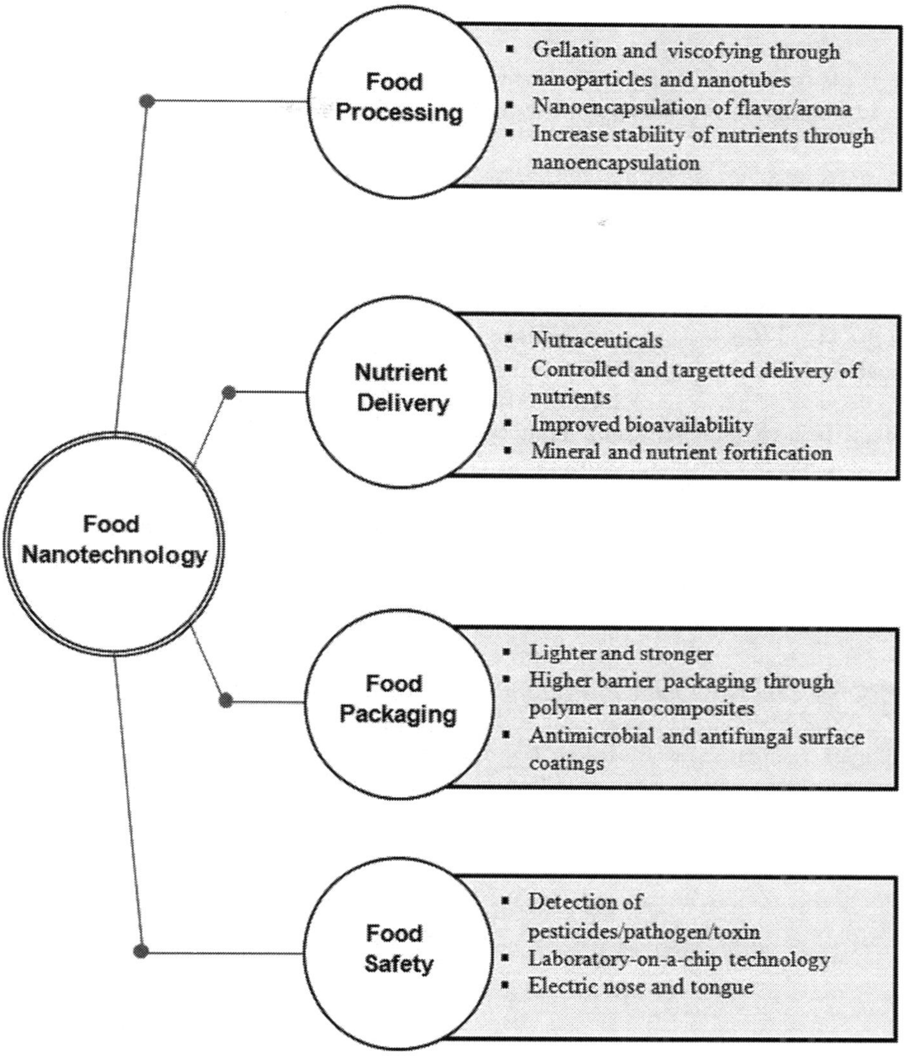

FIGURE 2.9 Application of nanotechnology in food.

2.5.1 Nutrient Delivery

Nanotechnology is helping to considerably improve and even revolutionize the food industry by offering new horizons in nutrient delivery all over the world. It offers a suitable means of delivering a vast spectrum of nutrients from micro to macromolecules such as vitamins, proteins, carbohydrates, etc. The application of nanomaterials for the delivery of nutrients could improve the bioavailability of bioactive compounds. The bioavailability of a compound is generally defined as the fraction that can be absorbed and used for physiological functions and/or storage. Mathematically, the fraction of a bioactive compound absorbed for systemic circulation (bioavailability) can be expressed as follows (McClements and Li, 2010):

$$F = F_a * \left(1 - E_g\right) * \left(1 - E_{\text{liver}}\right) \tag{2.6}$$

$$F = F_a * F_g * F_h \tag{2.7}$$

where:

F_a is the fraction absorbed
F_g is the fraction escaping intestinal metabolism ($1\text{-}E_g$)
F_h is the fraction escaping hepatic metabolism ($1\text{-}E_{liver}$)

Intestinal absorption is the prerequisite for oral bioavailability. However, based on food structure, the absorption of the nutrient may result in one of four possible outcomes: delayed absorption, decreased absorption, increased absorption, and sometimes unaffected absorption. The process of intestinal absorption of bioactive compounds and its metabolism are depicted in Figure 2.10.

To overcome the drawbacks associated with the conventional dosage forms, nutrient delivery built on the nanotechnology platform reported effective delivery. Formulations including nanoparticles, nanocapsules, nanocrystalline structures, and micellar systems were successfully used for the effective delivery of nutrients. These formulations are known to enhance the transport of nutrients across many biological membranes and improve the stability of nutrients. Nanoformulations of bioactive compounds in food increase overall bioavailability due to their specialized intestinal uptake mechanisms, such as circumventing the P-gp efflux, preventing the degradation of nutrients during gastrointestinal transport, and minimizing first-pass metabolism (Zakeri-Milani et al., 2014).

2.5.1.1 Intestinal Absorption of Nanoparticles

The intestinal tract is highly absorptive and is composed of villi with a total absorptive surface area of 300–400 m^2 (Schenk and Mueller, 2008). Gastrointestinal (GI) transit time is an important factor that determines oral absorption, as most of the bioactive compounds are absorbed from the upper part of the small intestine, i.e. duodenum and jejunum (Abuhelwa et al., 2017). Nanoparticle absorption in the GI tract can be well explained through various microscopic routes. Two transepithelial routes allow the passage of substances from the lumen to the basolateral side: transcellular transport (i.e. through an epithelial cell) or paracellular transport (i.e. between adjacent epithelial cells) (Bellmann et al., 2015). Most of the orally administered bioactive compounds are not retained in the

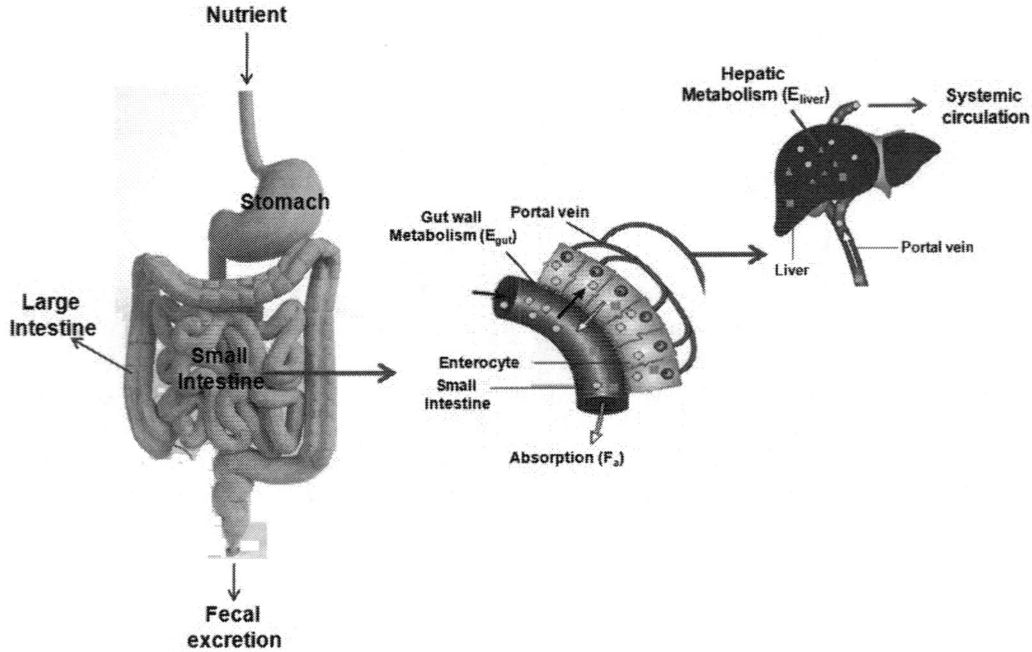

FIGURE 2.10 Schematic representation of bioavailability.

GI tract, but undergo direct transit. Nanoparticles have an improved residence time in the GI tract due to its mucoadhesive ability. To protect the intestinal epithelium, mucus is continuously secreted to remove pathogens as well as lubricate the epithelial surface. Mucus is composed of proteins, carbohydrates, lipids, salts, antibodies, bacteria, and cellular debris. The main protein present in mucus is mucin, and there are at least 20 proteins encoded in the MUC gene family (Ensign et al., 2012). Mucus production in an adult human is approximately 1 kg/day, 10–200 μm thick (Lundquist and Artursson, 2016).

2.5.1.2 Strategies to Improve Oral Bioavailability

In recent years, different strategies including solubility, stability in the GI tract and intestinal permeability have been used in enhancing the oral bioavailability of bioactive compounds. The nanodelivery system is an emerging processing technique in the food and nutraceutical industries, used to overcome poor aqueous solubility, dissolution, and/or bioavailability (Yen et al., 2010; Karthik and Anandharamakrishnan, 2016b; Anandharamakrishnan and Ishwarya, 2015).

There was a prominent difference in the bioavailability for the nanoencapsulated bioactive compound. The nanoformulation of vitamin E showed a threefold increase in oral bioavailability (area under the curve – AUC) when compared to the conventional emulsion (Parthasarathi et al., 2016). Curcumin-loaded rice bran albumin nanoparticles showed a tenfold increase in bioavailability compared to free curcumin (Liu et al., 2018), which indicates that nanoformulation could improve the bioavailability of nutrients. The shape of the nanoparticles has an impact in the transport and absorption in the GI tract. Li et al. (2017) disclosed the effect of GI absorption when nanorods and nanospheres were orally administered. Nanorods had significantly longer retention time in the gastrointestinal tract than nanospheres. Both nanospheres and nanorods were intubated to animals orally, and transport of nanoparticles was effected using ex vivo imaging – IVIS Spectrum Live Imaging System. At the end of the experiment (8 h), approximately 30% of the nanorods were present in the small intestine, against only 10% of the nanospheres. Nanorods exhibited much longer retention time in the jejunum and ileum due to their higher penetration ability into the villi compared to nanospheres. Different methods were followed to study the in vivo gastric digestion and intestinal absorption of nutrients, and many researchers were developing an effective alternative to in vivo studies as they are time-consuming, not economical and more prone to errors. However, selecting a suitable method depends on operating conditions and on how closely the model predicts the actual phenomenon (Gopirajah and Anandharamakrishnan, 2014 and 2016). Recently, Karthik and Anandharamakrishnan (2018) proposed a methodology to investigate the rheological transformations of nanoemulsions in response to the simulated conditions of small intestinal mechanical forces.

2.5.2 Food Packaging

Nanotechnology is an important tool for improving material characteristics and provides a significant solution in the packaging sector for satisfying consumers' requirements. Normally, plastic is the preferred packing material in the food sector due to its low weight, ease of production, and processing. However, the low thermal and mechanical stability, minor gas or vapor barrier, and insufficient light protection of plastic materials restrict their application (Bott et al., 2014). Nano-food packaging has the potential to overcome those disadvantages and represents an effective alternative to conventional food packaging. Reinforcement of engineered nanoparticles such as titanium dioxide (TiO_2), silver nanoparticles, silica, and carbon nanotubes could considerably increase functional properties such as tensile strength and heat resistance. In the current scenario, food packaging needs not only to facilitate supply but also to simplify its end-use suitability and communication at consumer level (Youssef and El-Sayed, 2018).

Polymer nanocomposite is a polymer matrix bonded with nanoparticles to provide multifunctional properties such as strength and toughness, and environmental and thermal stability. Polymer nanocomposite packaging can be categorized into four types based on their application (Chaudhry et al., 2008; Huang et al., 2015):

1. Improved packaging: this category includes polymers with nanofillers (polymer nanocomposites) which improve packaging properties such as moisture stability, temperature control, barrier properties, and ultraviolet-blocking properties. Improved packaging materials can improve the shelf life of fresh and processed food, poultry, meat and seafood products by delaying moisture loss, reducing lipid oxidation and discoloration, and enhancing product appearance.
2. Active packaging: this includes incorporation of nanofillers with antimicrobial or antioxidant activities, which results in inhibiting microbial growth and food spoilage. Carbon nanotubes, chitosan nanoparticles, and nanoclays are known for their antibacterial properties (Ghanbarzadeh et al., 2015; Souza and Fernando, 2016).
3. Intelligent packaging: for biosecurity purposes, nanosensors are included in the food packaging materials, which improve detection and tracking of the food condition during storage and transportation. This helps to monitor the external/internal changes in the product's environment.
4. Degradable or compostable biopolymers: Incorporation of nanofillers in biopolymers results in improving the physical and chemical characteristics of the biopolymer.

2.5.2.1 Migration of Nanoparticles

Though incorporation of nanoparticles in the packaging material improves tensile strength, heat resistance, and shelf life, there is anxiety amongst consumers about the migration of nanoparticles from the packaging material to the food content. Therefore, it is crucial to understand the potential migration of nanoparticles to food matrices and their toxicity inside the human body (de Azeredo et al., 2011). Migration tests should be performed for the nano-packaging material and approval should be based on the migration assessments. Bott et al., (2014) developed a mathematical migration model based on the mobility of the molecule which is assessed by the diffusion coefficient of the molecule. Authors also reported that no TiN (Titanium nitride) nanoparticles migrated from low-density polyethylene (LDPE) films and remained undetected even with inductively coupled plasma – mass spectrometry (ICP-MS). In a recent study, Salarbashi et al. (2018) developed a soluble soybean polysaccharide/TiO2 bionanocomposite film with zero migration of TiN nanoparticles from the film surface into the food model (bread).

2.5.3 Food Safety

Food safety is a global concern as foodborne illness or food poisoning is a major health concern in most countries. Food poisoning is an illness (diarrhea, vomiting) caused by consumption of food/drink contaminated with bacteria, virus, parasites, molds, allergens, or toxins. It is essential to assure that the consumption of food by more than 7,500 million of habitants is safe (Socas-Rodríguez et al., 2017).Conventional analytical techniques such as immunological assays and chemical/microbial analysis are time-consuming, expensive, and require high sample quantity; thus, the in-sensitivity of analytical equipment leads to the application of nano in food safety. Nanomaterials including nanotubes, nanoparticles, fullerenes, nanofibers, nanocylinders, and nanosheets are employed to determine the quality of food (Cushen et al., 2012). The unique properties (optical and electronic properties) of nanomaterials are used for the developement of nano processors, nanobattery, and nanosensors (Fuertes et al., 2016). Foster (2005) defined nanosensors as any biological, chemical or surgical sensory points used to convey information about nanoparticles to the macroscopic world. Nanobiosensors are considered as an effective tool for monitoring quality aspects of food content, including pH, humidity, and microbial content. A biosensor is defined as "an analytical device which incorporates a biologically active element with an appropriate physical transducer to generate

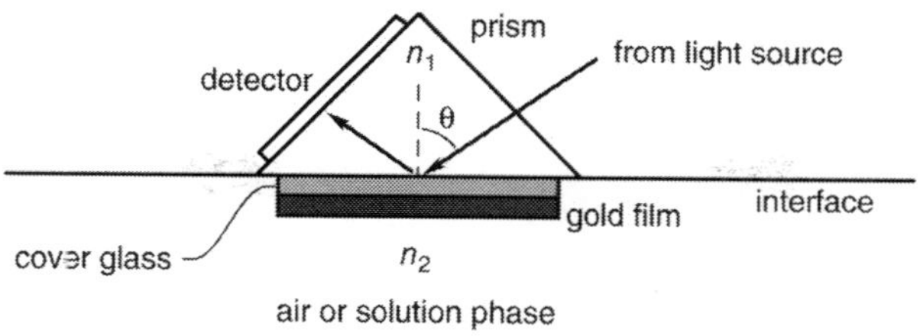

FIGURE 2.11 Schematic representation of an SPR setup. (Tang et al., 2010.)

a measurable signal proportional to the concentration of chemical species in any type of sample" (Touhami 2014; Arnold and Meyerhoff, 1988). Nanobiosensors can be classified into two categories based on their working mechanism: (i) optical and (ii) electrochemical sensors (Wang and Duncan, 2017).

2.5.3.1 Optical Sensors

Optical biosensors are developed based on optoelectronic and microelectronic technologies. These biosensors work by exploiting the interaction of the optical field with a biorecognition element. They can be broadly classified as label-free and label-based biosensors. In the label-free mode, the optical signal is generated by the interaction of the analyzing material with the transducer. In label-based sensors, the optical signal is generated by a colorimetric, fluorescent or luminescent method with the use of labels (Damborský et al., 2016). Optical nanosensors are widely applied in food quality analysis due to their rapid detection of analytes with a minimal sample. Typical biosensors are based on surface plasmon resonance (SPR) technology commonly used for label-free biological sensing (see Figure 2.11). Readers can refer to Krishna et al. (2018) and Tang et al. (2010) for a better understanding of the working mechanism of SPR technology.

These SPR-based optical sensors are employed as an effective alternative for the polymerase chain reaction (PCR)-based techniques for detecting microbial pathogens in food. PCR-based techniques are time-consuming and require expensive equipment. The limit of detection (LOD) for bacteria in SPR should be in the order of 10^4 CFU/mL or higher, where no preconcentration or amplification steps are employed (Vaisocherová-Lísalová et al., 2016). In order to improve the LODs, different strategies were used by researchers, such as modifying carbohydrates by introducing phenyl residues and aliphatic chains to mannose in order to detect *E. coli*, *Citrobacter freundii*, and *Staphylococcus epidermidis* even at 3.0 CFU/mL (Yazgan et al., 2014).

2.5.3.2 Electrochemical Sensors

The US Centers for Disease Control and Prevention reported that nearly 76 million US citizens are affected by food poisoning every year. Hospitalized victims are estimated to be 300,000 and 5,000 people die of this every year. Not all food poisoning cases are same. If the food is contaminated with harmful substances such as bacteria, viruses, parasites, toxins, and chemicals, this will cause a critical situation (Health, 2015). Scientists successfully developed nanosensors to detect food adulterants, pathogens, and toxins, and even to monitor the freshness of foods.

Electrochemical biosensors are self-contained biosensors which measure the changes in electrical current, potential, conductance, or impedance at the interface of the electrode and sample matrix based on electrochemical transducers (Wang and Duncan, 2017; Neethirajan et al., 2018). These unique properties (inherent sensitivity, simplicity, speed, and compatibility) offer new approaches in the detection of pathogens in food. Electrochemical biosensors can be categorized into four types: amperometric, potentiometric, impedance, and voltammetric biosensors. In order to improve the transducer response characteristics, nanomaterials such as carbon-based nanomaterials, metal and metal oxide nanoparticles, magnetic nanoparticles and molecularly imprinted polymers were used in the electrochemical biosensors (Zeng et al., 2016). These biosensors are highlighting the wide application of nanomaterials in the detection of pathogens (*Salmonella*, *E-coli*), insecticides, and other contaminants.

2.6 SUMMARY

Nanotechnology has achieved tremendous progress in most engineering sectors and is considered as an emerging field in food processing. It is anticipated that nanotechnology will change food processing industries more deeply than many developments of the past. This chapter is focused on how nanotechnology might impact traditional research in food processing and its potential application. In the last 10 years, there has been an increase in the awareness and understanding of the properties and functionalities of nanostructures. Nanostructures include nanotubes, fullerenes, nanofibers, nanocylinders, nanosheets, and self-assembled nanostructures, polymer nanocomposites that have demonstrated potential application in the food industry from processing to packaging. Though the food processing industry could benefit from the applications of nanotechnology, this might have unintended effects on the environment. There is an urgent need for cooperation between industries, public, and governments to find economically viable solutions. It is imperative that governments increase the funding concerning research into new technologies.

REFERENCES

Abuhelwa, Ahmad Y., Desmond B. Williams, Richard N. Upton, and David J.R. Foster. 2017. "Food, gastrointestinal pH, and models of oral drug absorption." *European Journal of Pharmaceutics and Biopharmaceutics*, 112:234–248.

Anandharamakrishnan, Chinnaswamy. 2014. *Techniques for Nanoencapsulation of Food Ingredients*. Springer: New York.

Anandharamakrishnan, C. and S. Padma Ishwarya. 2015. "Spray drying for nanoencapsulation of food components." *Spray Drying Techniques for Food Ingredient Encapsulation*: 180–197: Wiley-Blackwell: Chichester.

Anandharamakrishnan, C. and K. Usha Kiran. 2014. "Bionanocomposites and their potential applications in food packaging." *Polymers for Packaging Applications*, 229–262 : CRC Press: Boca Raton.

Arnold, Mark A. and Mark E. Meyerhoff. 1988. "Recent advances in the development and analytical applications of biosensing probes." *Critical Reviews in Analytical Chemistry*, 20 (3):149–196.

Aslani, Parisa and Ross A. Kennedy. 1996. "Studies on diffusion in alginate gels. I. Effect of cross-linking with calcium or zinc ions on diffusion of acetaminophen." *Journal of Controlled Release*, 42 (1):75–82.

Barber, D.J. and Ian C. Freestone. 1990. "An investigation of the origin of the colour of the Lycurgus Cup by analytical transmission electron microscopy." *Archaeometry*, 32 (1):33–45.

Bellmann, Susann, David Carlander, Alessio Fasano, Dragan Momcilovic, Joseph A. Scimeca, W. James Waldman, Lourdes Gombau, Lyubov Tsytsikova, Richard Canady, and Dora I.A. Pereira. 2015. "Mammalian gastrointestinal tract parameters modulating the integrity, surface properties, and absorption of food-relevant nanomaterials." *Wiley Interdisciplinary Reviews: Nanomedicine and Nanobiotechnology*, 7 (5):609–622.

Benedek, George B. 1997. "Cataract as a protein condensation disease: The proctor lecture." *Investigative Ophthalmology and Visual Science*, 38 (10):1911–1921.

Bhushani, J. Anu, P. Karthik, and C. Anandharamakrishnan. 2016. "Nanoemulsion based delivery system for improved bioaccessibility and Caco-2 cell monolayer permeability of green tea catechins." *Food Hydrocolloids*, 56:372–382.

Bott, J., A. Störmer, and R. Franz. 2014. "A model study into the migration potential of nanoparticles from plastics nanocomposites for food contact." *Food Packaging and Shelf Life*, 2 (2):73–80.

Brill, Robert. H. 1965. "The chemistry of the Lycurgus cup." *Proceedings of the 7th International Congress on Glass, Comptes Rendus 2*, Paper 223, 1–13.

BSI. 2005. Large scale gains for small scale work (Press release). https://www.bsigroup.com/en-GB/about-bsi/media-centre/press-releases/2005/6/Large-scale-gains-for-small-scale-work/

Caiger-Smith, Alan. 1985. *Lustre Pottery: Technique, Tradition and Innovation in Islam and the Western World* Faber and Faber: London.

Chaudhry, Qasim, Michael Scotter, James Blackburn, Bryony Ross, Alistair Boxall, Laurence Castle, Robert Aitken, and Richard Watkins. 2008. "Applications and implications of nanotechnologies for the food sector." *Food Additives and Contaminants*, 25 (3):241–258.

Chellaram, C., G. Murugaboopathi, A.A. John, R. Sivakumar, S. Ganesan, S. Krithika, and G. Priya. 2014. "Significance of nanotechnology in food industry." *APCBEE Procedia*, 8:109–113.

Cohen, Samuel I.A., Sara Linse, Leila M. Luheshi, Erik Hellstrand, Duncan A. White, Luke Rajah, Daniel E. Otzen, Michele Vendruscolo, Christopher M. Dobson, and Tuomas P.J. Knowles. 2013. "Proliferation of amyloid-β42 aggregates occurs through a secondary nucleation mechanism." *Proceedings of the National Academy of Sciences*, 110 (24):9758–9763.

Colomban, Philippe. 2009. "The use of metal nanoparticles to produce yellow, red and iridescent colour, from bronze age to present times in lustre pottery and glass: Solid state chemistry, spectroscopy and nanostructure." *Paper Read at Journal of Nano Research*, 8: 109–132.

Cushen, M., J. Kerry, M. Morris, M. Cruz-Romero, and E. Cummins. 2012. "Nanotechnologies in the food industry–recent developments, risks and regulation." *Trends in Food Science and Technology*, 24 (1):30–46.

Damborský, Pavel, Juraj Švitel, and Jaroslav Katrlík. 2016. "Optical biosensors." *Essays in Biochemistry*, 60 (1):91–100.

de Azeredo, Henriette Monteiro Cordeiro, Luiz Henrique Capparelli Mattoso, and Tara Habig McHugh. 2011. "Nanocomposites in food packaging–a review." In *Advances in Diverse Industrial Applications of Nanocomposites, Ed. Boreddy Reddy.* InTech: London.

Duncan, Timothy V. 2011. "Applications of nanotechnology in food packaging and food safety: Barrier materials, antimicrobials and sensors." *Journal of Colloid and Interface Science*, 363 (1):1–24.

Ensign, Laura M., Richard Cone, and Justin Hanes. 2012. "Oral drug delivery with polymeric nanoparticles: The gastrointestinal mucus barriers." *Advanced Drug Delivery Reviews*, 64 (6):557–570.

Erickson, Harold P. 2009. "Size and shape of protein molecules at the nanometer level determined by sedimentation, gel filtration, and electron microscopy." *Biological Procedures Online*, 11 (1):32.

The European Commission. n.d. "Nanotechnologies." http://ec.europa.eu/health/scientific_committees/opinions_layman/en/nanotechnologies/l-3/1-introduction.htm.

Ezhilarasi, P.N., P. Karthik, N. Chhanwal, and C. Anandharamakrishnan. 2013. "Nanoencapsulation techniques for food bioactive components: A review." *Food and Bioprocess Technology*, 6 (3):628–647.

Fan, Taotao, Xiaoyan Yu, Bing Shen, and Leming Sun. 2017. "Peptide self-assembled nanostructures for drug delivery applications." *Journal of Nanomaterials*, 2017:4562474.

Feynman, Richard P. 1960. "There's plenty of room at the bottom." Lecture at the annual meeting of the American Physical Society, California Institute of Technology, December 29, 1959. Transcript published in *Caltech Magazine, Engineering and Science* in 1960. http://new2.foresight.org/nano/whatisnano.html.

Foster, Lynn E. 2005. *Nanotechnology: Science, Innovation, and Opportunity*: Prentice Hall PTR: New Jersey.

Fuertes, Guillermo, Ismael Soto, Raúl Carrasco, Manuel Vargas, Jorge Sabattin, and Carolina Lagos. 2016. "Intelligent packaging systems: Sensors and nanosensors to monitor food quality and safety." *Journal of Sensors*, 2016:4046061.

García, Miguel A. 2011. "Surface plasmons in metallic nanoparticles: Fundamentals and applications." *Journal of Physics D: Applied Physics*, 44 (28):283001.

Ge, Jiao, Pengxiang Yue, Jinpeng Chi, Jin Liang, and Xueling Gao. 2018. "Formation and stability of anthocyanins-loaded nanocomplexes prepared with chitosan hydrochloride and carboxymethyl chitosan." *Food Hydrocolloids*, 74:23–31.

Ghanbarzadeh, Babak, Seyed Amir Oleyaei, and Hadi Almasi. 2015. "Nanostructured materials utilized in biopolymer-based plastics for food packaging applications." *Critical Reviews in Food Science and Nutrition*, 55 (12):1699–1723.

Gopirajah, R. and C. Anandharamakrishnan. 2014. "Methods integrating physical mechanisms underlying the food digestion and release of nutrients in human stomach." *Journal of Nutritional Science*, 1: 1–13.

Gopirajah, R. and C. Anandharamakrishnan. 2016. "Advancement of imaging and modeling techniques for understanding gastric physical forces on food." *Food Engineering Reviews*, 8 (3):323–335.

He, Xiaojia and Huey-Min Hwang. 2016. "Nanotechnology in food science: Functionality, applicability, and safety assessment." *Journal of Food and Drug Analysis*, 24 (4):671–681.

Health, Bureau of Environmental. 2015. "Food poisoning: What it is and how to report it in Massachusetts." Massachusetts Department of Public Health: Boston.

Huang, Jen-Yi, Xu Li, and Weibiao Zhou. 2015. "Safety assessment of nanocomposite for food packaging application." *Trends in Food Science and Technology*, 45 (2):187–199.

Hundre, Swetank Y., P. Karthik, and C. Anandharamakrishnan. 2015. "Effect of whey protein isolate and β-cyclodextrin wall systems on stability of microencapsulated vanillin by spray-freeze drying method." *Food Chemistry*, 174:16–24.

Karthik, P. and C. Anandharamakrishnan. 2013. "Microencapsulation of docosahexaenoic acid by spray-freeze-drying method and comparison of its stability with spray-drying and freeze-drying methods." *Food and Bioprocess Technology*, 6 (10):2780–2790.

Karthik, P. and C. Anandharamakrishnan. 2016a. "Enhancing omega-3 fatty acids nanoemulsion stability and in-vitro digestibility through emulsifiers." *Journal of Food Engineering*, 187:92–105.

Karthik, P. and C. Anandharamakrishnan. 2016b. "Fabrication of a nutrient delivery system of docosahexaenoic acid nanoemulsions via high energy techniques." *RSC Advances*, 6 (5):3501–3513.

Karthik, P. and C. Anandharamakrishnan. 2018. "Droplet coalescence as a potential marker for physicochemical fate of nanoemulsions during in-vitro small intestine digestion." *Colloids and Surfaces A: Physicochemical and Engineering Aspects*, *553*: 278–287.

Kataoka, Kazunori, Atsushi Harada, and Yukio Nagasaki. 2001. "Block copolymer micelles for drug delivery: Design, characterization and biological significance." *Advanced Drug Delivery Reviews*, 47 (1):113–131.

Kochmann, Werner, Marianne Reibold, Rolf Goldberg, Wolfgang Hauffe, Alexander A Levin, Dirk C. Meyer, Thurid Stephan, Heide Müller, André Belger, and Peter Paufler. 2004. "Nanowires in ancient Damascus steel." *Journal of Alloys and Compounds*, 372 (1–2):L15–L19.

Krishna, Venkatramana D., Kai Wu, Diqing Su, Maxim C.J. Cheeran, Jian-Ping Wang, and Andres Perez. 2018. "Nanotechnology: Review of concepts and potential application of sensing platforms in food safety." *Food Microbiology*, *75*: 47–57.

Kumar, Santosh, Ankita Shukla, Partha Pratim Baul, Atanu Mitra, and Dipankar Halder. 2018. "Biodegradable hybrid nanocomposites of chitosan/gelatin and silver nanoparticles for active food packaging applications." *Food Packaging and Shelf Life*, 16:178–184.

Lehn, Jean-Marie. 2002. "Toward self-organization and complex matter." *Science*, 295 (5564):2400–2403.

Levin, A.A., D.C. Meyer, M. Reibold, W. Kochmann, N. Pätzke, and P. Paufler. 2005. "Microstructure of a genuine Damascus sabre." *Crystal Research and Technology*, 40 (9):905–916.

Li, X.H., A.F. Jalbout, and W.D. Xiang. 2008. "Self-assembly of supermolecular species directed by hydrogen bonding and aromatic π-π stacking interactions." *Eclética Química*, 33 (2):13–19.

Li, Dong, Jie Zhuang, Haisheng He, Sifan Jiang, Amrita Banerjee, Yi Lu, Wei Wu, Samir Mitragotri, Li Gan, and Jianping Qi. 2017. "Influence of particle geometry on gastrointestinal transit and absorption following oral administration." *ACS Applied Materials and Interfaces*, 9 (49):42492–42502.

Liang, Jin, Hua Yan, Xiulan Wang, Yibin Zhou, Xueling Gao, Pradeep Puligundla, and Xiaochun Wan. 2017. "Encapsulation of epigallocatechin gallate in zein/chitosan nanoparticles for controlled applications in food systems." *Food Chemistry*, 231:19–24.

Lim, Yong-beom, Kyung-Soo Moon, and Myongsoo Lee. 2009. "Recent advances in functional supramolecular nanostructures assembled from bioactive building blocks." *Chemical Society Reviews*, 38 (4):925–934.

Liu, Zonghua, Yanpeng Jiao, Yifei Wang, Changren Zhou, and Ziyong Zhang. 2008. "Polysaccharides-based nanoparticles as drug delivery systems." *Advanced Drug Delivery Reviews*, 60 (15):1650–1662.

Liu, Chun, Xiaoquan Yang, Wei Wu, Zhao Long, Huaxi Xiao, Feijun Luo, Yingbin Shen, and Qinlu Lin. 2018. "Elaboration of curcumin-loaded rice bran albumin nanoparticles formulation with increased in vitro bioactivity and in vivo bioavailability." *Food Hydrocolloids, 77*:834–842.

Lohcharoenkal, Warangkana, Liying Wang, Yi Charlie Chen, and Yon Rojanasakul. 2014. "Protein nanoparticles as drug delivery carriers for cancer therapy." *BioMed Research International*, 2014:180549.

Lundquist, Patrik and Per Artursson. 2016. "Oral absorption of peptides and nanoparticles across the human intestine: Opportunities, limitations and studies in human tissues." *Advanced Drug Delivery Reviews*, 106:256–276.

Ma, Zhengxin, Alejandro Garrido-Maestu, and Kwangcheol Casey Jeong. 2017. "Application, mode of action, and in vivo activity of chitosan and its micro-and nanoparticles as antimicrobial agents: A review." *Carbohydrate Polymers*, 176:257–265.

McClements, David Julian. 2018. "Encapsulation, protection, and delivery of bioactive proteins and peptides using nanoparticle and microparticle systems: A review." *Advances in Colloid and Interface Science*, *253*: 1–22.

McClements, David Julian, and Yan Li. 2010. "Structured emulsion-based delivery systems: Controlling the digestion and release of lipophilic food components." *Advances in Colloid and Interface Science*, 159 (2):213–228.

Mie, G. 1908. Beiträge zur Optik trüber Medien, speziell kolloidaler Metallösungen. *Annalen der physik*, 330 (3): 377–445.

Mokhena, T.C. and A.S. Luyt. 2017. "Electrospun alginate nanofibres impregnated with silver nanoparticles: Preparation, morphology and antibacterial properties." *Carbohydrate Polymers*, 165:304–312.

Neethirajan, S., X. Weng, A. Tah, J.O. Cordero, and K.V. Ragavan. 2018. "Nano-biosensor platforms for detecting food allergens–New trends." *Sensing and Bio-Sensing Research*, *18*: 13–30.

NNI. n.d. "What it is and how it works." https://www.nano.gov/nanotech-101/what.

Padovani, S., C. Sada, P. Mazzoldi, B. Brunetti, I. Borgia, A. Sgamellotti, A. Giulivi, F. d'Acapito, and G. Battaglin. 2003. "Copper in glazes of Renaissance luster pottery: Nanoparticles, ions, and local environment." *Journal of Applied Physics*, 93 (12):10058–10063.

Panda, Jiban Jyoti, and Virander Singh Chauhan. 2014. "Short peptide based self-assembled nanostructures: Implications in drug delivery and tissue engineering." *Polymer Chemistry*, 5 (15):4418–4436.

Parthasarathi, S. and C. Anandharamakrishnan. 2016. "Enhancement of oral bioavailability of vitamin E by spray-freeze drying of whey protein microcapsules." *Food and Bioproducts Processing*, 100:469–476.

Parthasarathi, S., S.P. Muthukumar, and C. Anandharamakrishnan. 2016. "The influence of droplet size on the stability, in vivo digestion, and oral bioavailability of vitamin E emulsions." *Food and Function*, 7 (5):2294–2302.

Persin, Zdenka, Karin Stana-Kleinschek, Timothy J. Foster, Jan EG Van Dam, Carmen G. Boeriu, and Patrick Navard. 2011. "Challenges and opportunities in polysaccharides research and technology: The EPNOE views for the next decade in the areas of materials, food and health care." *Carbohydrate Polymers*, 84 (1):22–32.

Purwanti, Nanik, Azmi Syahrian Zehn, Eka Dian Pusfitasari, Nauman Khalid, Erfin Yundra Febrianto, Sutrisno Suro Mardjan, Andreas, and Isao Kobayashi. 2018. "Emulsion stability of clove oil in chitosan and sodium alginate matrix." *International Journal of Food Properties*, 21 (1):566–581.

Quiñones, Javier Pérez, Hazel Peniche, and Carlos Peniche. 2018. "Chitosan based self-assembled nanoparticles in drug delivery." *Polymers*, 10 (3):235.

Rajam, R. and C. Anandharamakrishnan. 2015. "Microencapsulation of Lactobacillus plantarum (MTCC 5422) with fructooligosaccharide as wall material by spray drying." *LWT-Food Science and Technology*, 60 (2):773–780.

Rashidi, Ladan and Kianoush Khosravi-Darani. 2011. "The applications of nanotechnology in food industry." *Critical Reviews in Food Science and Nutrition*, 51 (8):723–730.

Ravichandran, R. 2010. "Nanotechnology applications in food and food processing: Innovative green approaches, opportunities and uncertainties for global market." *International Journal of Green Nanotechnology: Physics and Chemistry*, 1 (2):P72–P96.

Reibold, M., P. Paufler, A.A. Levin, W. Kochmann, N. Pätzke, and D.C. Meyer. 2006. "Materials: Carbon nanotubes in an ancient Damascus sabre." *Nature*, 444 (7117):286.

Reibold, Marianne, Peter Paufler, Aleksandr A. Levin, Werner Kochmann, Nora Pätzke, and Dirk C. Meyer. 2009. "Discovery of nanotubes in ancient Damascus steel." In *Physics and Engineering of New Materials, Eds:* Cat. D.T., Pucci, A., Wandelt, K, 305–310. Springer: Berlin.

Salarbashi, D., M. Tafaghodi, and B. S. F. Bazzaz. 2018. "Soluble soybean polysaccharide/TiO2 bionanocomposite film for food application." *Carbohydr Polym*, 186:384–393. doi:10.1016/j.carbpol.2017.12.081.

Schaming, Delphine and Hynd Remita. 2015. "Nanotechnology: From the ancient time to nowadays." *Foundations of Chemistry*, 17 (3):187–205.

Schenk, Mirjam and Christoph Mueller. 2008. "The mucosal immune system at the gastrointestinal barrier." *Best Practice and Research Clinical Gastroenterology*, 22 (3):391–409.

Sciau, Philippe. 2012. "Nanoparticles in ancient materials: The metallic lustre decorations of medieval ceramics." In *The Delivery of Nanoparticles, Ed. Abbass A. Hashim, 525-540.* InTech: London.

Sirviö, Juho Antti, Aleksi Kolehmainen, Henrikki Liimatainen, Jouko Niinimäki, and Osmo E.O. Hormi. 2014. "Biocomposite cellulose-alginate films: Promising packaging materials." *Food Chemistry*, 151:343–351.

Socas-Rodríguez, Bárbara, Javier González-Sálamo, Javier Hernández-Borges, and Miguel Ángel Rodríguez-Delgado. 2017. "Recent applications of nanomaterials in food safety." *TrAC Trends in Analytical Chemistry*, *96*: 172–200.

Souza, Victor Gomes Lauriano, and Ana Luisa Fernando. 2016. "Nanoparticles in food packaging: Biodegradability and potential migration to food—A review." *Food Packaging and Shelf Life*, 8:63–70. doi:10.1016/j.fpsl.2016.04.001.

Sozer, Nesli and Jozef L. Kokini. 2009. "Nanotechnology and its applications in the food sector." *Trends in Biotechnology*, 27 (2):82–89.

Srinivasan, C. 2007. "Do damascus swords reveal India's mastery of nanotechnology?" *Current Science*, 92 (3):279–280.

Sun, Zhe, Zeng Yi, Huaiying Zhang, Xiaomin Ma, Wen Su, Xiaoyu Sun, and Xudong Li. 2017. "Bio-responsive alginate-keratin composite nanogels with enhanced drug loading efficiency for cancer therapy." *Carbohydrate Polymers*, 175:159–169.

Taniguchi, Norio. 1974. On the basic concept of "nano-technology." *Proceedings of the International Conference on Production Engineering, Tokyo*, Part II. Japan Society of Precision Engineering, pp. 18–23.

Tang, Yijun, Xiangqun Zeng, and Jennifer Liang. 2010. "Surface plasmon resonance: An introduction to a surface spectroscopy technique." *Journal of Chemical Education*, 87 (7):742–746.

Tavassoli-Kafrani, Elham, Hajar Shekarchizadeh, and Mahdieh Masoudpour-Behabadi. 2016. "Development of edible films and coatings from alginates and carrageenans." *Carbohydrate Polymers*, 137:360–374.

Touhami, Ahmed. 2014. "Biosensors and nanobiosensors: Design and applications." *Nanomedicine*, 15374–400.

Uversky, Vladimir N. and David Eliezer. 2009. "Biophysics of Parkinson"s disease: Structure and aggregation of α-synuclein." *Current Protein and Peptide Science*, 10 (5):483–499.

Vaisocherová-Lísalová, Hana, Ivana Víšová, Maria Laura Ermini, Tomáš Špringer, Xue Chadtová Song, Jan Mrázek, Josefína Lamačová, N. Scott Lynn Jr, Petr Šedivák, and Jiří Homola. 2016. "Low-fouling surface plasmon resonance biosensor for multi-step detection of foodborne bacterial pathogens in complex food samples." *Biosensors and Bioelectronics*, 80:84–90.

van den Broek, Lambertus A.M., Rutger J.I. Knoop, Frans H.J. Kappen, and Carmen G. Boeriu. 2015. "Chitosan films and blends for packaging material." *Carbohydrate Polymers*, 116:237–242.

Vendruscolo, Michele, Jesús Zurdo, Cait E. MacPhee, and Christopher M. Dobson. 2003. "Protein folding and misfolding: A paradigm of self–assembly and regulation in complex biological systems." *Philosophical Transactions of the Royal Society of London A: Mathematical, Physical and Engineering Sciences*, 361 (1807):1205–1222.

Wang, Yun and Timothy V. Duncan. 2017. "Nanoscale sensors for assuring the safety of food products." *Current Opinion in Biotechnology*, 44:74–86.

Yang, Jisheng, Suya Han, Haicheng Zheng, Hongbiao Dong, and Jiubing Liu. 2015. "Preparation and application of micro/nanoparticles based on natural polysaccharides." *Carbohydrate Polymers*, 123:53–66.

Yazgan, Idris, Naumih M. Noah, Ousmane Toure, Siyi Zhang, and Omowunmi A. Sadik. 2014. "Biosensor for selective detection of E. coli in spinach using the strong affinity of derivatized mannose with fimbrial lectin." *Biosensors and Bioelectronics*, 61:266–273.

Yen, Feng-Lin, Tzu-Hui Wu, Cheng-Wei Tzeng, Liang-Tzung Lin, and Chun-Ching Lin. 2010. "Curcumin nanoparticles improve the physicochemical properties of curcumin and effectively enhance its antioxidant and antihepatoma activities." *Journal of Agricultural and Food Chemistry*, 58 (12):7376–7382.

Youssef, Ahmed M., and Samah M. El-Sayed. 2018. "Bionanocomposites materials for food packaging applications: Concepts and future outlook." *Carbohydrate Polymers*, 193: 19–27.

Zakeri-Milani, Parvin, Ziba Islambulchilar, Fatemeh Majidpour, Ensieh Jannatabadi, Farzaneh Lotfipour, and Hadi Valizadeh. 2014. "A study on enhanced intestinal permeability of clarithromycin nanoparticles." *Brazilian Journal of Pharmaceutical Sciences*, 50 (1):121–129.

Zeng, Yan, Zhihong Zhu, Dan Du, and Yuehe Lin. 2016. "Nanomaterial-based electrochemical biosensors for food safety." *Journal of Electroanalytical Chemistry*, 781:147–154.

Zhang, Shuguang. 2003. "Fabrication of novel biomaterials through molecular self-assembly." *Nature Biotechnology*, 21 (10):1171.

3 Characteristics and Behavior of Nanofluids

Vimala Bharathi S.K., Sayantani Dutta, J.A. Moses, and C. Anandharamakrishnan

CONTENTS

3.1 INTRODUCTION

Conventional fluids are routinely used in various applications and instruments; their prime role has been to reduce surface to surface friction, facilitate motion of particles, and promote the better transfer of heat and power, even as shock absorbers and balancers (within a living cell). They have been advantageous to almost all domains of known science, especially where there is a certain mechanization involved, be it mechanical, electrical, chemical, or biological. Fluids have been an aid in reducing the complexity of process engineering for quite some time. There have been developments in fluids at different times, and their characteristics have been specifically improved each time. The initiatives include altering chemical nature, chemical combination, and preparation method; employing alternative collaterals, and modulation in an operational environment, to name a few. Many times these modifications have indeed succored the difficulties encountered; however, the scope of improvements still persisted. Lately, there has been a steady demand for the development of technologies to combat the difficulties that conventional fluids involved in process engineering put forth, and that has led to designing a new class of fluids altogether. This new class has been specifically designed to bridge the existing fallacies and limitations of conventional fluids, and with a meticulous approach, so that these fluids offer the improved scope of applying engineering ingenuity and process optimization. The process scientists and engineers thus have developed a neoteric class of fluids, the "nanofluids."

Nanofluids are two-phase fluids with nano-sized (1 to 100 nm) solid materials (nanofibers, nanoparticles, nanorods, nanowires, nanotubes, nanosheets, and nanodroplets) dispersed in liquid phase (water, oil, toluene, ethylene glycol, and so on) (Reddy and Chamkha, 2018; Chitra and Sendhilnathan, 2014; Sridhara and Satapathy, 2011; Vajravelu et al., 2013). Nano-sized materials can be metals, non-metals or carbons. These nanofluids exhibit properties (physical, chemical, electrical, and optical) that are different from those of the base fluid (Chamsa-ard et al., 2017).

Nanofluids are especially preferred for their thermal conductivity. In general, the addition of solids to the base liquid enhances thermal property (Maxwell 1873). According to Choi and Eastman (1995), the surface of the particle plays a major role in the heat transfer characteristics of the fluid. Explicitly, the higher the surface to volume ratio, the higher the heat transfer rates. Moreover, millimeter-sized solid particles lead to problems such as clogging and sedimentation of particles and low-pressure drop, and also damage equipment through corrosion (Sridhara and Satapathy, 2011). Thus, the use of nanoparticles as the solid phase in nanofluids can enhance thermal conductivity without clogging the particles. Besides providing benefits in terms thermal properties, nanofluids also possess other properties such as an antimicrobial action, depending on the nature of the solid nanomaterial used (Zhang et al., 2007). Moreover, magnetic nanofluids are explored more in biomedical and engineering applications (Vekas et al., 2007).

Nanofluids can be prepared mainly through two methods: single-step direct evaporation and two-step method. The former involves the formation of nanomaterials directly in the liquid phase through evaporation and condensation; whereas the latter involves the production of nanomaterials externally and then dispersion into the liquid phase (Sridhara and Satapathy, 2011). Nanofluids are found to have potential application in various fields, including, but not limited to, micro-electronics, medical, manufacturing, and transportation (Paul et al., 2010). The objective of this chapter is to give an overview of the physicochemical and thermo-physical characteristics of nanofluids. Further, the applications of nanofluids are also discussed to provide in-depth understanding in this area.

3.2 AGGREGATION AND DISPERSION BEHAVIOR OF NANOMATERIALS

The basic nature of fluids per se is that they have neither attraction nor repulsion intra-particular. They only respond to external stimuli, be they mechanical, chemical, electrical, or any other equivalent. On the other hand, when the nanomaterials in the nanofluids are brought into physical contact, the materials either tend to aggregate or repel each other depending on the nature of the materials and the medium in which they are dispersed. The attractive force that occurs as a result of dipole interaction of the molecules is known as Van der Waals force. The Van der Waals force of attraction and an electrostatic double layer of interaction are the two major forces that play significant roles in the aggregation of nanomaterials (Krishna et al., 2015). The effect of these forces changes with the size of the nanomaterials. For nanomaterials of size 100 nm or lower, the major cause for the physical contact is their Brownian movement (Hotze et al., 2010). Moreover, an increase in concentration can also increase the tendency of nanomaterials to aggregate at a faster rate (Das et al., 2003a). Also, nanofluids tend to agglomerate over time (Krishna et al., 2015).

The size of the nanomaterials also influences their aggregation behavior (Timofeeva et al., 2007). In general, aggregation can be homoaggregation (aggregation of similar materials) and heteroaggregation (aggregation of different materials). Homoaggregates are often seen in nanofluids prepared in laboratory conditions; while heteroaggregates often occur in nanofluids obtained in practical conditions (Hotze et al., 2010).

Uniform dispersion of nanoparticles without aggregation can lead to enhanced performance and stability. Moreover, aggregation of nanomaterials leads to denser particles and, hence, results in the sedimentation of the particle Agglomeration can be prevented by reducing the density difference between solids and liquids, or by enhancing the viscosity of the fluids in a nanofluid system (Sridhara and Satapathy, 2011). Compared to fluids with micro-sized solids, nanofluids show better dispersion and stability (Paul et al., 2010). This property of the nanofluid can be measured by analyzing the zeta potential value.

In general, to prevent the agglomeration of nanoparticles in nanofluids, sonication and/or mechanical stirring followed by the addition of a stabilizing agent are performed. Here, sonication and mechanical stirring help with uniform dispersion; whereas the stabilizing agent maintains the dispersion of the nanoparticles by preventing the aggregation of nanomaterials in the nanofluids. A better understanding of the properties of the base liquid, nanoparticle, and the stabilizing agent is necessary, as a stabilizing agent can alter the properties of nanofluids. Moreover, the stabilizing agent, liquid and nanomaterials react to form an interfacial layer. This interfacial layer also affects the properties of the nanofluid (Chamsa-ard et al., 2017). Other ways to prevent agglomeration are the introduction of a surface charger, which changes the surface properties of the nanomaterials, and the generation of repulsive forces between the nanomaterials through the formation of polymer layers around them (Krishna et al., 2015). For instance, Lee et al. (2006) modified the surface charge density and surface density by altering the pH of water-based CuO nanofluids and quantified the stability through Derjaquin–Landau–Verwey–Overbeek (DLVO) linearized mean field theory. The study concluded that with a decrease in pH, the stability, as well as the thermal conductivity of nanofluids, was found to be improved. This is related to the increase in the surface charge of the nanomaterials in the nanofluid, as the pH moves away from the point of zero charge. This increase may be due to the increased presence of potential-determining ions (H^+ and OH^-) (Lee et al., 2006). In an experiment conducted on the influence of pH on the stability of water-based copper nanofluids at various pH ranges from 2 to 12, the optimum pH with higher stability was observed at 9.5. The lower the pH, the lower the electrostatic force observed. When the pH is increased beyond the optimum, the reagent used to adjust pH (NaOH) causes a disturbance in the electrostatic double layer, hence reducing stability (Li et al., 2007).

The state of dispersion and stability of nanofluids depends on the properties of the solid material dispersed and of the base fluid, such as morphology, chemical structure, interaction, and the pre-treatment. It also depends on the presence/absence of stabilizers and surfactants, their characteristics, interaction with nanofluids, and so on (Hwang et al., 2007). Other factors such as pH, electrolytic ionic concentration, and zeta potential also affect the stability of nanofluids (Krishna et al., 2015; Lee et al., 2006). The aforesaid unique behavior of nanofluids could aid in developing equipment/instruments with a specific operational module in order to achieve the desired results; especially in process planning and machine design aspects.

3.3 PROPERTIES OF NANOFLUIDS

The efficacy of nanofluids depends on various characteristics such as thermal conductivity, density, viscosity, specific heat, heat capacity, and the interaction with the surrounding environment in which the heat transfer takes place. Hence, to understand the behavior of the fluid during heat transfer application, it is necessary to have knowledge of these properties.

3.3.1 Thermal Conductivity

The conventional or base fluids have a well-known thermal conductive behavior and, more often than not, they function in that spectrum alone. At times, for process development, this becomes painstaking since the scope of alteration is narrow, and there is a risk of losing out on the existing properties, in case the fluid is overstretched in its thermal nature. This flexibility is offered by nanofluids, whose behavior is dependent on many factors, antithetical to those of conventional fluids.

Thermal conductivity is a chief factor that differentiates the properties of nanofluids from those of the base fluid. The thermal conductivity of nanofluids depends on volume fraction, particle size, shape of the solid phase, thermal conductivity of the nanomaterial and the base liquid, pH and temperature of the nanofluids, Brownian movement, interfacial layer, and additives (Sridhara and Satapathy, 2011; Hwang et al., 2007). Several studies have been developed to evaluate the influence of the above-mentioned factors on the thermal conductivity of nanofluids (Xuan et al., 2003; Jang and Choi, 2004; Das et al., 2003b; Duangthongsuk and Wongwises, 2009; Hong et al., 2006). Amani et al. (2016) assessed the effect of the magnetic field (100, 200, 300 and 400 G intensity) on the thermal conductivity of water-based $MnFe_2O_4$ nanofluids at various concentrations (0.25% to 3%).

Thermal conductivity was found to increase when increasing the temperature of the nanofluid in the absence of a magnetic field. However, conductivity decreased when augmenting the temperature in the presence of a magnetic field. On the other hand, thermal conductivity was found to increase when increasing the intensity of the magnetic field. The researches claim that the reason for these effects may be the inverse relation between the temperature and the magnetic field (Amani et al., 2016). Hong et al. (2006) proved that with an increase in aggregation/cluster size of nanomaterials in the nanofluids, the thermal conductivity of the nanofluids decreases. The thermal conductivity of nanofluids is very high with the addition of a very low concentration of nanoparticles (Lee et al., 2008; Bowers et al., 2018). For instance, Patel et al. (2003) showed that the thermal conductivity enhancement can be up to 14% even with 0.011% toluene-based Au nanoparticles loading. Yu and Choi (2003) claimed that the layer of liquid molecules surrounding the solid surface would have different thermal properties than that of pure liquid. Hence, this layer must be considered while predicting the thermal behavior of nanofluids. The impact of a formed layer on thermal conductivity is higher when the size of the nanomaterial is less than 10 nm. Researchers modified the Maxwell equation and included the thermal property of the layer that has intermediate property to that of solid nanomaterial and the base liquid. Equation (3.1) shows the effective conductivity explained by Maxwell (Maxwell, 1873), and Equation (3.2) explains the equation modified by Yu and Choi (2003).

$$k_{\text{Maxwell}} = \frac{\left(k_p + 2k_l + 2\left(k_p - k_l\right)\phi\right)}{\left(k_p + 2k_l - \left(k_p - k_l\right)\phi\right)} k_l \tag{3.1}$$

$$k_{\text{Modified}} = \frac{\left(k_{pe} + 2k_l + 2\left(k_{pe} - k_l\right)\left(1+\beta\right)^3 \phi\right)}{\left(k_{pe} + 2k_l - \left(k_{pe} - k_l\right)\left(1+\beta\right)^3 \phi\right)} k_l \tag{3.2}$$

where:

- k_p is the thermal conductivity of dispersed particles
- k_l is the thermal conductivity of the liquid
- ϕ is the volume concentration of the particle
- k_{pe} is the equivalent thermal conductivity of the equivalent particle
- β is the ratio of layer thickness (h) to nanomaterial radius (r)

For large particles, the nanolayer formed is very small compared to that of the nanomaterial ($r >> h$) and β tends to zero, and the modified equation transforms to the original Maxwell equation.

In 2009, the International Nanofluid Property Benchmark Exercise was executed by several researchers from 34 organizations worldwide. The experiment was conducted for different nanofluids with a wide variety in properties of base fluids (aqueous and non-aqueous), nanomaterials (metals and metal oxides), and shape of nanomaterials (near-spherical and elongated) at various concentrations using various experimental approaches (such as transient hot-wire method, steady-state method, and optical method). The study observed that the thermal conductivity of the tested nanofluids increased with particle concentration and aspect ratio. The deviation for the samples tested at various places by various researchers was observed to be 10% or even lower. Though a systematic difference in thermal conductivity was visible when measurement of thermal conductivity was carried out with a similar type of measurement technique at the same temperature, the values were similar and, furthermore, the data of various experimental techniques on normalization and that of base fluid yielded similar results. Moreover, the study revealed the results were in accordance with those predicted by the Maxwell theory of effective medium. This study helped to resolve the conflicting results to some extent (Jacopo Buongiorno et al., 2009). Though some of the studies have proved that the theoretical models work well for nanofluids, a few other researchers have also proved the inability of the model to explain the experimental data. For instance, oil-based nanotubes at 1.0% volume concentration showed thermal conductivity enhancement by 160%; while the models predicted the enhancement to be less than 10% (Choi et al., 2001). Thus, the reliability of the models developed for predicting thermal conductivity varies.

The thermal conductivity of nanofluids can be measured using techniques such as the cylindrical cell method, temperature oscillation method, steady-state parallel-plate method, 3 w method, thermal constants analyzer method, thermal comparator method, and transient hot-wire method. Among these techniques, the transient hot-wire method is adopted by many researchers because of its accuracy, repeatability, feasibility, and possibility of measuring instantaneously (Sridhara and Satapathy, 2011; Paul et al., 2010). Having said that, thermal conductivity is a turnkey aspect of nanofluids, and it is therefore quintessential to better exploit the merits of these special fluids in designing thermally durable equipment, and/or process kinetics. This non-classical behavior could be especially made to use in miniature thermal approaches that have been in demand for quite some time, and therefore the augmented ability of nanofluids could support such developmental projects.

3.3.2 Viscosity

The viscosity of nanofluids defines their suitability for use in a commercial application. The viscosity of a fluid affects heat transfer behavior as well. Also, viscosity governs some of the properties of the nanofluids such as pumping power and pressure drop in convective and laminar flow heat transfer (Mishra et al., 2014). In a study, pressure loss was found to be proportional to the viscosity at a constant flow rate and geometry (Rea et al., 2009). The viscosity of nanofluids varies significantly upon addition of nanomaterials. For instance, the viscosity of SiC nanofluids was increased by 68% to 102% compared to that of the base fluid (deionized water) (Lee et al., 2011).

Several researchers have studied the rheological behavior of nanofluids as a function of various factors such as their temperature, concentration, size, shape, and so on (Chen et al., 2007; He et al., 2007). Lee et al. (2008) showed that viscosity of water-based Al_2O_3 (alumina) nanofluids decreased with an increase in temperature. Though the thermal conductivity of alumina nanofluids increased linearly with concentration, the viscosity of nanofluids did not show a linear relationship at a concentration below 0.3%. Yet, the viscosity trend was found to be increasing with an increase in concentration (Lee et al., 2008). Similarly, titanium dioxide (TiO_2) nanofluids also showed an increase in viscosity with an increase in nanomaterial size and concentration (He et al., 2007). Contradictorily, when comparing the viscosity of CuO nanoparticles obtained from two different sources, the lower-sized particle showed higher viscosity at the same concentration and temperature (Pastoriza-Gallego et al., 2011).

TiO_2 nanofluids initially showed shear thinning behavior (decrease in viscosity with an increase in shear rate), followed by constant viscosity above the shear rate of 100 s^{-1} (He et al., 2007). A similar study has been carried out by Chen et al., (2007) with ethylene glycol-based TiO_2 nanofluids consisting of 0.5–0.8 weight % spherical nanomaterials. The temperature range chosen was 20–60°C. The shear viscosity was found to be temperature dependent, while relative viscosity was found to be temperature independent. The reason for the increase in shear viscosity was found to be due to the formation of aggregates. Moreover, the shear viscosity of the nanofluids was predicted by the Krieger-Dougherty equation. The study found that the theoretical and experimental data were similar for spherical particles when the aggregate size was almost three times that of the initial nanomaterial (Chen et al. 2007).

The theory to estimate viscosity was initially developed by Einstein (1906). A number of formulae and theories have been developed by researchers over time (Mooney, 1951; Krieger and Dougherty, 1959; Nielsen, 1970; Frankel and Acrivos, 1967). Later, these models were modified by various researchers including Avsec and Oblak (2007); Batchelor (1977); Bicerano et al. (1999); Brinkman (1952); Chen et al. (2007); Graham (1981); Kitano et al. (1981); and Lundgren (1972). Certain theories developed can be used for one particular nanofluid only. For instance, models developed by Abu-Nada (2009) and Masoumi et al. (2009) can be used only for alumina-water nanofluids. However, none of the developed models can be used to exactly predict the viscosity of nanofluids (Mishra et al., 2014; Murshed et al., 2008).

3.3.3 Specific Heat

The specific heat capacity of nanofluids is crucial to analyze energy performances. Specific heat aids in analyzing energy and exergy (Shahrul et al., 2014). No pertinent studies have been conducted on the specific heat of nanofluids. Researches opine the reason could be the need of a wide range of studies to be conducted in order to analyze the specific heat of nanofluids (Sekhar and Sharma, 2013). Since the basic nature of particles and basic fluid could also be dissimilar, this may hinder the correct estimation of specific heat. Also, being a composite per se, the components may have antagonistic effects on specific heat and its exhibition in many processes and conditions; that might be difficult to account for and/or express quantitatively. Hence, it is hard to completely understand the specific heat of nanofluids.

Specific heat is measured using various types of differential scanning calorimeters (Zhou and Ni, 2008). Figure 3.1 represents a schematic diagram of such a calorimeter. A few literature works have reported the setups developed by researchers for measuring specific heat, and some have reported the simulation studies used (Vajjha and Das, 2009; Namburu et al., 2007; Sonawane et al., 2011; Kulkarni et al., 2008). Pak and Cho (1998) were pioneers in developing the theoretical equation for estimating the specific heat of nanofluids. Later, several researchers developed empirical and numerical relationships for identifying the specific heat of the given nanofluids (Xuan and Roetzel, 2000; Buongiorno, 2006; Zhou et al., 2010).

Reported studies were mostly on the effect of volume concentration and temperature on the specific heat of nanofluids (Murshed, 2012; Teng and Hung, 2014; Shahrul et al., 2014). For instance, Zhou and Ni (2008) analyzed the specific heat capacity of Al_2O_3/water nanofluid. The temperature range was considered from 20°C to 45°C. The study found that with an increase in volume

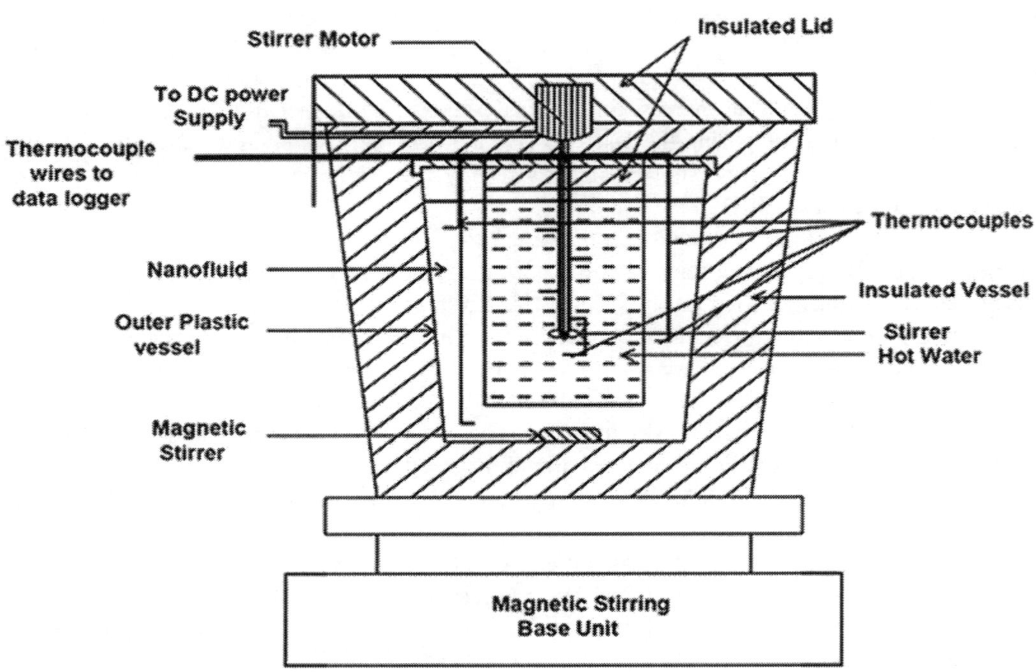

FIGURE 3.1 Schematic diagram of calorimeter for specific heat measurement. (Sonawane et al., 2011.)

fraction from 1.43% to 21.7%, the specific heat of the nanofluid was found to decrease by 6% to 45%. Moreover, the study found that the experimental results were in agreement with the result predicted using a thermal equilibrium model, rather than that of a simple mixing model (Zhou and Ni, 2008). Similar results were obtained by Zhou et al. (2010) in their review of the specific heat capacity of CuO nanofluid. Vajjha and Das (2009) measured the specific heat of three nanofluids: ethylene glycol and water (60:40 by mass)-based aluminum oxide, ethylene glycol and water (60:40 by mass)-based zinc oxide, and deionized water-based silicon dioxide. The temperature range considered was the range used during the operation of automobile coolants and the fluids used in the heating of buildings in cold regions (315–363 K). A maximum of 10% (v/v) nanoparticles were taken for analysis. The study found that specific heat decreased with an increase in concentration; whereas the specific heat of nanofluids increased with an increase in temperature. Also, the authors developed a correlation equation (with an average error of 2.7%) based on concentration (vol %), temperature and specific heat of nanomaterial, and the base liquid for analyzing the specific heat of the nanofluids (Vajjha and Das, 2009).

3.4 HEAT AND MASS TRANSFER IN NANOFLUIDS

Heat and mass transfer capacities of fluids are yet another aspect that determines the suitability of fluids for process engineering. In basic fluids, aka conventional fluids, the heat transfer rate of liquids is enhanced with an increase in pumping rate. However, nanofluids containing nano-sized solids with conductivity three times (approximately) that of liquids, and this would increase the heat transfer rate by two times (double) when used in the same heat transfer equipment. To obtain doubled heat transfer rate, the flow rate of the base fluid should be increased ten times (Stephen U. S. Choi and Eastman, 1995). This shows a remarkable difference between the existing conventional fluids and the new nanofluids. Studies have been carried out to examine the effect of temperature on the heat transfer rate up to 400 K; whereas the effect of sub-zero temperatures on heat transfer has not been explored yet (Sridhara and Satapathy, 2011). Patel et al. (2003) found that the thermal conductivity enhancement of Au-citrate nanofluid amounted to more than half that of Au-thiolate nanofluid at room temperature; whereas at 60°C, the enhancement effect shown is found to be similar. Heat transfer characteristics also depend on various other factors such as Brownian movement. Theoretical models developed by Jang and Choi (2004) proved that the Brownian motion of nanomaterials in nanofluids is one of the essential factors to be considered in order to understand the heat transfer behavior of nanofluids. The model developed also predicts the effect of particle size and temperature on the heat transfer property of the nanofluids.

Alumina nanofluids can be extensively used to enhance the heat transfer of nanofluids (Sridhara and Satapathy, 2011; Sandeep et al., 2017; Rea et al., 2009; Ho et al., 2010). Wen and Ding (2005) found that the addition of alumina nanofluids (1.25% by weight) can enhance thermal conductivity by 40% (approximately). However, some studies have shown contradictory results. Bang and Chang (2005) showed that the addition of alumina nanoparticles decreased the thermal conductivity of nanofluids and, with an increase in concentration, thermal conductivity was found to decrease linearly. This contradiction could be attributed to the heat transfer characteristics and behavior of nanofluids, which depend not only on the nanoparticle added but also on the base liquid, boiling surface, and interaction between the surface and the fluid (Wen and Ding, 2005). Heat transfer characteristics also depend on the surrounding material to which the heat transfer takes place, and its geometry (Bergman, 2009). When measuring the heat transfer coefficients of the alumina-water nanofluids, the increase in heat transfer coefficients was 17% and 27% in the entrance region and 6 vol % of nanomaterials in the fully developed region. Similarly, for zirconia-water nanofluids, the increase in the heat transfer coefficient was 2% and 3% in the entrance region and 1.32 vol % in the fully developed region (Rea et al., 2009).

Analogous to heat transfer, nanofluids enhance mass transfer as well. Though a limited number of studies have been carried out to examine the effect of nanofluids on mass diffusivity, the results obtained were contradicting (Ashrafmansouri and Esfahany, 2014). Fang et al. (2009) studied the effect of nanomaterial volume fraction and temperature on the mass diffusion coefficient of fluorescent Rhodamine B in Cu-water nanofluid. The study revealed that 0.5% of nanomaterial-incorporated nanofluid showed 10.71 and 26 times higher diffusion coefficient of Rhodamine B than that of deionizied water for the temperatures (fluid) of 15°C and 25°C, respectively. Thus, the mass transfer of nanofluids depends on temperature as well as on the concentration of the nanomaterial. However, Turanov and Tolmachev (2009) showed that though quasi-monodisperse spherical silica nanoparticles-based aqueous suspensions showed enhanced thermal conductivity with an increase in volume fraction, the self-diffusion coefficient was found to decrease with a volume concentration of nanomaterials.

A few researchers observed nil effect on diffusion with the addition of nanomaterials (Ozturk et al., 2010; Subba-Rao et al., 2011; Feng and Johnson, 2012). These studies reveal that mass transfer in nanofluids depends on various factors, such as Brownian movement, type of nanomaterial used, and so on. The increase in the mass diffusion coefficient may be attributed to the disturbance caused by moving nanomaterials (Krishnamurthy et al., 2006); while a decrease in mass transfer in a few cases may be caused by the creation of tortuous pathways by the nanomaterials, which hinders the diffusion/movement of mass (Gerardi et al., 2009). Ashrafmansouri and Esfahany (2014) reviewed and explained clearly the mass transfer enhancement and degradation mechanisms of nanofluids. According to the authors, the mechanisms responsible for the improved mass transfer are the Brownian motion of nanomaterials and consequent micro-convection, the grazing effect of nanomaterials (nanomaterials adsorb the mass/solute from the gas and transfer it to the base liquid), the creation of a larger interfacial area by breaking the bubble size (due to collision with nanomaterials), and its maintenance (by covering the bubble and preventing coalescence). The shearing action induced by nanomaterials results in a reduction in the thickness of the film and surface tension. The mechanisms for reduced mass transfer are agglomeration, increased elasticity, reduced diffusion coefficient, and high viscosity. By considering the above-mentioned factors, the design of nanomaterials can be done in such a way as to facilitate high mass transfer (Ashrafmansouri and Esfahany, 2014).

The unique behavior of nanofluids in heat and mass transfer, as exhibited by numerous studies, assert their viability as preferred alternatives to existing fluids. The pliability of these fluids asserts their versatile nature and concretes their performance expectations for safe process deliverables. Such a nature is particular beneficial to research and development because many industrial needs are influenced by the behavior of fluids heat and mass transfer applications. This behavior of nanofluids also offers scope for improvement of techniques for process optimization.

3.5 APPLICATION OF NANOFLUIDS

Nanofluids are explored in more depth in the areas of engineering and biomedical applications due to their high heat transfer capacity. Nanofluids can reduce the need for high flow rate and pump power and protect the equipment from wear and tear; moreover, they reduce damage to the equipment. The other advantages of nanofluids include improved energy and cost savings, and also the feasibility of their application in microchannels (Choi and Eastman, 1995). Nanofluids are mostly used as coolants in the field of biomedical applications.

3.5.1 NANOFLUID COOLANT

Owing to their high heat transfer rate and the ability to eliminate fins and turbulent flow in cooling systems, nanofluids have been identified to have potential application as coolants in cooling electronic systems, micro-devices, welding cooling, domestic refrigerators and coolers, engine cooling, diesel-electric generators, boiler flue gas cooling, to detect knock occurrence in gas engines, and so on (Saidur et al., 2011). Rafati et al. (2012) evaluated the effects of different nanofluids (silica, alumina and titanium as nanomaterials) with a mixture of water and ethylene glycol as base fluid with three different volumetric concentrations (0.1%, 0.25%, and 0.5%) at three different flow rates (0.5, 0.75, and 1.0 L/min) for the cooling of computer microchips. The study found that a maximum decrease in the operating temperature was obtained for alumina-added nanofluid with 1.0% volumetric concentration at the rate of 1 L/min. The decrease in temperature went from 49.4 to 43.9°C. Moreover, the study concluded that a balance between the concentration of nanomaterials and the flow rate should be considered to reduce the power consumption of the cooling system (Rafati et al., 2012). A similar study has found that the addition of 6.8 vol % of water-based Al_2O_3 nanofluid can enhance the heat transfer coefficient by 40% more than that of base fluid when tested in a closed system used for cooling microprocessors and various other electronic components (Nguyen et al., 2007).

The incorporation of nanomaterials into the coolants can reduce not only energy consumption but also the size of the equipment required for cooling (Lee and Choi, 1996). A radiator is one of the important components that need a compact size (Sidik et al., 2015). Ali et al. (2014) investigated the effect of water-based Al_2O_3 nanofluid in the actual vehicle radiator of a 2007 Toyota Yaris at five different concentrations (0.1%, 0.5%, 1%, 1.5%, and 2% by volume). The study found that the optimum concentration was 1%, beyond which deterioration of the heat transfer occurs (Ali et al., 2014). Besides, nanofluid coolants are also used in military systems such as submarines, military vehicles, high-power laser diodes, etc. – those involving a high rate of cooling to the level of MW/m^2.

3.5.2 Biomedical Applications

The foremost biomedical applications of magnetic nanofluids include drug delivery, MRI (Magnetic Resonance Imaging) contrast enhancement, hyperthermia, magnetic cell separation, and so on (Vekas et al., 2007). Magnetic nanofluids, especially ferrites (Figure 3.2) possess certain advantages such as low inherent toxicity suitable magnetic properties,stability, ease of synthesis, and so on, which aid in the potential application for hyperthermia treatment (Sharifi et al., 2012). The instability of organic antibacterial materials lead to the exploration of inorganic antibacterial nanofluids. For instance, zinc oxide (ZnO) nanomaterial-based nanofluids are found to possess antibacterial

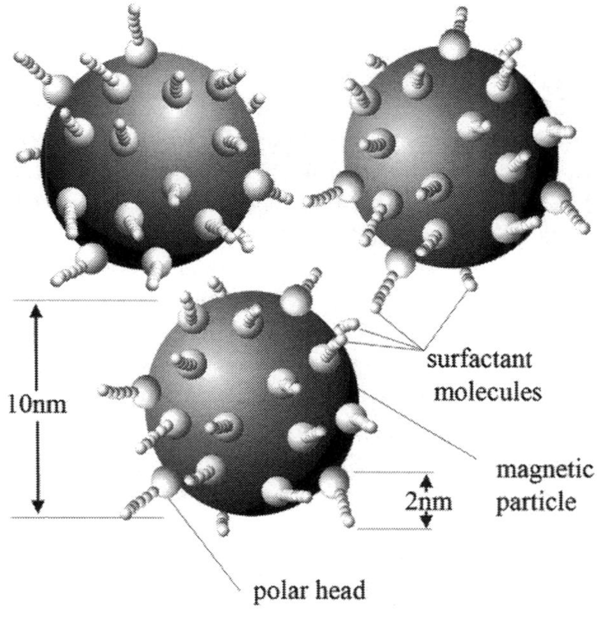

FIGURE 3.2 Schematic representation of coated magnetic particles in a ferrofluid. (Sharifi et al., 2012.)

properties against *E.coli* (Zhang et al., 2007). Similarly, copper oxide-based nanoparticles were found to have antibacterial properties against *Klebsiella pneumonia, Pseudomonas aeruginosa, Salmonella paratyphi* and *Shigella* strains (Mahapatra et al., 2008).

Nanofluids are also used in drug delivery applications. For instance, gold nanoparticles are preferred as carrier material owing to their non-toxic nature and controlled intra-cellular release upon interaction with thiols (Ghosh et al., 2008). Apart from the applications discussed in this section, nanofluids are also used in nano-cryosurgery and cancer therapies (Chitra and Sendhilnathan, 2013). However, nanofluids in applications involving human lives involve a high amounts of risk. Hence, lots of research has to be carried out prior to actual use.

3.6 LIMITATIONS OF NANOFLUIDS

Though nanofluids have gained popularity lately due to their diverse applications, a few limitations do exist that need to be overcome during practical applications. The major limitation of nanofluids is their stability. Strong Van der Waals force forms aggregates, thus affecting the properties of nanofluids. With better stability, the heat transfer rate increases (Wen et al., 2009). Moreover, the stability of nanofluids changes over time. For instance, Jaeseon Lee and Mudawar (2007) assessed the stability of nanofluid Al_2O_3 at 1% and 2% visually by sealing the sample in glass beakers. The study showed that the stability of the nanofluids drastically decreased after 30 days. Moreover, the authors explained that the settling of nanoparticles in the cooling system's reservoir, besides degrading the thermal performance, can also lead to clogging.

In general, the viscosity of nanofluids is higher than that of pure liquid. Also, their viscosity increases with the concentration of nanomaterials (Saidur et al., 2011). Moreover, high-pressure drop and pumping power requirements have been observed for the nanofluids with high viscosity (Jaeseon Lee and Mudawar, 2007; Yu et al., 2007). A pressure drop occurs due to friction and it also increases with an increase in concentration (Peng et al., 2009). For applications as a coolant, the specific heat of the coolant should be higher. The higher the specific heat, the higher the heat-holding capacity. As explained in previous sections, the specific heat of the nanofluid decreases with concentration. Furthermore, the production of nanofluids requires sophisticated equipment and, hence, the cost of nanofluids is high (Saidur et al., 2011). These are the major limitations of nanofluids to be considered while selecting solutions for commercial application.

3.7 FUTURE TRENDS

The major hindrance for the commercial application of nanofluids is the unreliability in the outcome reported by various researchers for an identical fluid (Paul et al., 2010; Wen and Ding, 2005). Moreover, the effect of added nanoparticles on the thermal properties of the fluids is vast (Zhu et al., 2009; Amrollahi et al., 2009; Guangbin et al., 2016). Several approaches developed by researchers such as Maxwell (1873) and Hamilton and Crosser (1962) and numerical approaches (Kalteh 2013; Reddy and Chamkha, 2018) have been used in predicting the thermal properties of nanofluids. Still, studies are being performed for the complete evaluation of the effect of the solid phase on nanofluids. It is necessary for future researchers to focus on and identify common techniques and methods to evaluate the properties of nanofluids. Also, the effect of nanofluids in phase-change applications such as refrigerants needs to be better understood.

It would be rather germane to emphasize that despite the difficulties that may have been observed, the foreground is that nanofluids offer some of the best characteristics that science has developed in recent times. The merits set off the difficulties in many situations, and the difficulties are more of a transient nature than permanent since this is the future of process fluids. The acclaim nanofluids have received clearly demonstrates that even though engineers can at times be circumspect about nanofluids in practice, the fluids' productivity would definitely attract research and industrial proclivity, and aid in the development of further *avant-garde* endeavors in this field, in the very near future.

REFERENCES

Abu-Nada, Eiyad. 2009. "Effects of variable viscosity and thermal conductivity of Al_2O_3-water nanofluid on heat transfer enhancement in natural convection." *International Journal of Heat and Fluid Flow*, 30 (4): 679–690. https://doi.org/10.1016/j.ijheatfluidflow.2009.02.003.

Ali, M., A. M. El-Leathy, and Z. Al-Sofyany. 2014. "The effect of nanofluid concentration on the cooling system of vehicles radiator." *Advances in Mechanical Engineering*, 962510: 1–13. https://doi.org/10.1155/2014/962510.

Amani, Mohammad, Pouria Amani, Alibakhsh Kasaeian, Omid Mahian, and Somchai Wongwises. 2016. "Thermal conductivity measurement of spinel-type ferrite $MnFe_2O_4$ nanofluids in the presence of a uniform magnetic field." *Journal of Molecular Liquids*. 230:121–128. https://doi.org/10.1016/j.molliq.2016.12.013.

Amrollahi, A., A. M. Rashidi, M. Emami Meibodi, and K. Kashefi. 2009. "Conduction heat transfer characteristics and dispersion behaviour of carbon nanofluids as a function of different parameters." *Journal of Experimental Nanoscience*, 4 (4): 347–363. https://doi.org/10.1080/17458080902929929.

Ashrafmansouri, Seyedeh-saba and Mohsen Nasr Esfahany. 2014. "Mass transfer in nano fluids: A review." *International Journal of Thermal Sciences*, 82: 84–99. https://doi.org/10.1016/j.ijthermalsci.2014.03.017.

Avsec, Jurij and Maks Oblak. 2007. "The calculation of thermal conductivity, viscosity and thermodynamic properties for nanofluids on the basis of statistical nanomechanics." *International Journal of Heat and Mass Transfer*, 50 (21–22): 4331–4341. https://doi.org/10.1016/j.ijheatmasstransfer.2007.01.064.

Bang, In Cheol and Soon Heung Chang. 2005. "Boiling heat transfer performance and phenomena of Al_2O_3 – water nano-fluids from a plain surface in a pool" 48: 2407–2419. https://doi.org/10.1016/j.ijheatmasstransfer.2004.12.047.

Batchelor, G. K. 1977. "The effect of Brownian-motion on bulk stress in a suspension of spherical-particles." *Journal of Fluid Mechanics*, 83 (1): 97–117.

Bergman, T. L. 2009. "Effect of reduced specific heats of nanofluids on single phase, laminar internal forced convection." *International Journal of Heat and Mass Transfer*, 52 (5–6): 1240–44. https://doi.org/10.1016/j.ijheatmasstransfer.2008.08.019.

Bicerano, Jozef, Jack F. Douglas, and Douglas A. Brune. 1999. "Model for the viscosity of particle dispersions." *Journal of Macromolecular Science*, 39 (4): 561–642.

Bowers, James, Hui Cao, Geng Qiao, Qi Li, Gan Zhang, Ernesto Mura, and Yulong Ding. 2018. "Progress in natural science : Materials international flow and heat transfer behaviour of nano Fl uids in microchannels." *Progress in Natural Science: Materials International*, 28 (2): 225–234. https://doi.org/10.1016/j.pnsc.2018.03.005.

Brinkman, H. C. 1952. "The viscosity of concentrated suspensions and solutions." *The Journal of Chemical Physics*, 20 (4): 571. https://doi.org/10.1063/1.1700493.

Buongiorno, J. 2006. "Convective transport in nanofluids." *Journal of Heat Transfer*, 128 (3): 240. https://doi.org/10.1115/1.2150834.

Buongiorno, Jacopo, David C. Venerus, Naveen Prabhat, Thomas Mckrell, Jessica Townsend, Rebecca Christianson, Yuriy V. Tolmachev, et al. 2009. "A benchmark study on the thermal conductivity of nanofluids." *Journal of Applied Physics*, 106 (094312): 1–14.

Chamsa-ard, Wisut, Sridevi Brundavanam, Chun Fung, Derek Fawcett, and Gerrard Poinern. 2017. "Nanofluid types, their synthesis, properties and incorporation in direct solar thermal collectors: A review." *Nanomaterials*, 7 (6): 131. https://doi.org/10.3390/nano7060131.

Chen, Haisheng, Yulong Ding, and Chunqing Tan. 2007. "Rheological behaviour of nanofluids." *New Journal of Physics*, 9 (367): 1–24. https://doi.org/10.1088/1367-2630/9/10/367.

Chitra, S.R. and S. Sendhilnathan. 2013. "A theoretical research on the biological applications of nanofluids/ferrofluids due to the amazing properties of nanofluids/ferrofluids." *International Journal of Scientific and Engineering Research*, 4 (5): 278–283. https://doi.org/ISSN 2229–5518.

Chitra, S.R. and S. Sendhilnathan. 2014. "Investigation on thermal studies of nanofluids related to their applications." *Heat Transfer-Asian Research*, 44 (5): 420–49. https://doi.org/10.1002/htj.21129.

Choi, Stephen U. S. and J. A. Eastman. 1995. "Enhancing thermal conductivity of fluids with nanoparticles." *ASME International Mechanical Engineering Congress and Exposition*, 66 (March): 99–105. https://doi.org/10.1115/1.1532008.

Choi, S. U. S., Z. G. Zhang, W. Yu, F. E. Lockwood, and E. A. Grulke. 2001. "Anomalous thermal conductivity enhancement in nanotube suspensions." *Applied Physics Letters*, 79 (14): 2252–2254. https://doi.org/10.1063/1.1408272.

Das, Sarit K., Nandy Putra, and Wilfried Roetzel. 2003a. "Pool boiling characteristics of nano-fluids." *International Journal of Heat and Mass Transfer*, 46: 851–862.

Das, Sarit Kumar, Nandy Putra, Peter Thiesen, and Wilfried Roetzel. 2003b. "Temperature dependence of thermal conductivity enhancement for nanofluids." *Journal of Heat Transfer*, 125 (August): 567–574. https://doi.org/10.1115/1.1571080.

Duangthongsuk, Weerapun and Somchai Wongwises. 2009. "Measurement of temperature-dependent thermal conductivity and viscosity of TiO2–water nanofluids." *Experimental Thermal and Fluid Science*, 33 (4): 706–714. https://doi.org/10.1016/j.expthermflusci.2009.01.005.

Einstein, Albert. 1906. "Eine Neue Bestimmung Der Moleküldimensionen." *Annalen Der Physik*, 324 (2): 289–306. https://doi.org/10.3929/ETHZ-B-000225616.

Fang, Xiaopeng, Yimin Xuan, and Qiang Li. 2009. "Experimental investigation on enhanced mass transfer in nanofluids." *Applied Physics Letters*, 95 (2013108): 1–3. https://doi.org/10.1063/1.3263731.

Feng, Xuemei and Drew W. Johnson. 2012. "Mass transfer in SiO_2 nanofluids: A case against purported nanoparticle convection effects." *International Journal of Heat and Mass Transfer*, 55 (13–14): 3447–3453. https://doi.org/10.1016/j.ijheatmasstransfer.2012.03.009.

Frankel, N. A. and Andreas Acrivos. 1967. "On the viscosity of a concentrated suspension of solid spheres." *Chemical Engineering Science*, 22 (6): 847–853. https://doi.org/10.1016/0009–2509(67)80149-0.

Gerardi, Craig, David Cory, Jacopo Buongiorno, Lin Wen Hu, and Thomas McKrell. 2009. "Nuclear magnetic resonance-based study of ordered layering on the surface of alumina nanoparticles in water." *Applied Physics Letters*, 95 (25): 8–11. https://doi.org/10.1063/1.3276551.

Ghosh, Partha, Gang Han, Mrinmoy De, Chae Kyu Kim, and Vincent M. Rotello. 2008. "Gold nanoparticles in delivery applications." *Advanced Drug Delivery Reviews*, 60 (11): 1307–1315. https://doi.org/10.1016/j.addr.2008.03.016.

Graham, A. L. 1981. "On the viscosity of suspensions of solide spheres." *Applied Science Research*, 37 (June): 275–286.

Guangbin, Yu, Gao Dejun, Chen Juhui, Dai Bing, Liu Di, Song Ye, and Chen Xi. 2016. "Experimental research on heat transfer characteristics of CuO nanofluid in adiabatic condition." *Journal of Nanomaterials*, Vol. 2016. 1–7. https://doi.org/10.1155/2016/3693249.

Hamilton, R. L. and O. K. Crosser. 1962. "Thermal conductivity of heterogeneous two-component systems." *Industrial and Engineering Chemistry Fundamentals*, 1 (3): 187–191. https://doi.org/10.1021/i160003a005.

He, Yurong, Yi Jin, Haisheng Chen, Yulong Ding, Daqiang Cang, and Huilin Lu. 2007. "Heat transfer and flow behaviour of aqueous suspensions of TiO_2 nanoparticles (nanofluids) flowing upward through a vertical pipe." *International Journal of Heat and Mass Transfer*, 50: 2272–2281. https://doi.org/10.1016/j.ijheatmasstransfer.2006.10.024.

Ho, C. J., W. K. Liu, Y. S. Chang, and Lin C. C. 2010. "Natural convection heat transfer of alumina-water nanofluid in vertical square enclosures : An experimental study." *International Journal of Thermal Sciences*, 49 (August): 1345–1353. https://doi.org/10.1016/j.ijthermalsci.2010.02.013.

Hong, K. S., Tae-keun Hong, Ho-soon Yang, and K. S. Hong. 2006. "Thermal conductivity of Fe nanofluids depending on the cluster size of nanoparticles," 031901: 1–4. https://doi.org/10.1063/1.2166199.

Hotze, Ernest M., Tanapon Phenrat, and Gregory V. Lowry. 2010. "Nanoparticle aggregation: Challenges to understanding transport and reactivity in the environment." *Journal of Environmental Quality*, 39 (6): 1909–1924. https://doi.org/10.2134/jeq2009.0462.

Hwang, Y., J. K. Lee, C. H. Lee, Y. M. Jung, S. I. Cheong, C. G. Lee, B. C. Ku, and S. P. Jang. 2007. "Stability and thermal conductivity characteristics of nanofluids." *Thermochimica Acta*, 455: 70–74. https://doi.org/10.1016/j.tca.2006.11.036.

Jang, Seok Pil and Stephen U. S. Choi. 2004. "Role of Brownian motion in the enhanced thermal conductivity of nanofluids." *Applied Physics Letters*, 84 (21): 4316–4318. https://doi.org/10.1063/1.1756684.

Kalteh, Mohammad. 2013. "Investigating the effect of various nanoparticle and base liquid types on the nanofluids heat and fluid flow in a microchannel." *Applied Mathematical Modelling*, 37 (18–19): 1–10. https://doi.org/10.1016/j.apm.2013.03.067.

Kitano, T., T. Kataoka, and T. Shirota. 1981. "An empirical equation of the relative viscosity of polymer melts filled with various inorganic fillers." *Rheologica Acta*, 20 (2): 207–209. https://doi.org/10.1007/BF01513064.

Krieger, Irvin M. and Thomas J. Dougherty. 1959. "A mechanism for non-newtonian flow in suspensions of rigid spheres." *Transactions of the Society of Rheology*, 3 (1): 137–152. https://doi.org/10.1122/1.548848.

Krishna, K, Hari, S, Neti, A. Oztekin, S. Mohapatra, K. Hari Krishna, S. Neti, A. Oztekin, and S. Mohapatra. 2015. "Modeling of particle agglomeration in nanofluids." *Journal of Applied Physics*, 117 (094304): 1–8. https://doi.org/10.1063/1.4913874.

Krishnamurthy, S., P. Bhattacharya, P. E. Phelan, and R. S. Prasher. 2006. "Enhanced mass transport in nanofluids." *Nano Letters*, 6 (3): 419–423. https://doi.org/10.1021/nl0522532.

Kulkarni, Devdatta P., Ravikanth S. Vajjha, Debendra K. Das, and Daniel Oliva. 2008. "Application of aluminum oxide nanofluids in diesel electric generator as jacket water coolant." *Applied Thermal Engineering*, 28 (14–15): 1774–1781. https://doi.org/10.1016/j.applthermaleng.2007.11.017.

Lee, Shinpyo and Stephen U. S. Choi. 1996. "Application of Metallic Nanoparticle Suspensions in Advanced Cooling Systems."

Lee, Donggeun, Jae-won Kim, and Bog G. Kim. 2006. "A new parameter to control heat transport in nanofluids: Surface charge state of the particle in suspension." *The Journal of Physical Chemistry B*, 110 (9): 4323–4328.

Lee, Jaeseon and Issam Mudawar. 2007. "Assessment of the effectiveness of nanofluids for single-phase and two-phase heat transfer in micro-channels." *International Journal of Heat and Mass Transfer*, 50: 452–463. https://doi.org/10.1016/j.ijheatmasstransfer.2006.08.001.

Lee, Ji-hwan, Kyo Sik Hwang, Seok Pil Jang, Byeong Ho Lee, Jun Ho Kim, Stephen U. S. Choi, and Chul Jin Choi. 2008. "Effective viscosities and thermal conductivities of aqueous nanofluids containing low volume concentrations of Al_2O_3 nanoparticles." *Journal of Heat and Mass Transfer*, 51: 2651–2656. https://doi.org/10.1016/j.ijheatmasstransfer.2007.10.026.

Lee, Seung Won, Sung Dae Park, Sarah Kang, In Cheol Bang, and Ji Hyun Kim. 2011. "Investigation of viscosity and thermal conductivity of SiC nanofluids for heat transfer applications." *International Journal of Heat and Mass Transfer*, 54 (1–3): 433–438. https://doi.org/10.1016/j.ijheatmasstransfer.2010.09.026.

Li, Xinfang, Dongsheng Zhu, and Xianju Wang. 2007. "Evaluation on dispersion behavior of the aqueous copper nano-suspensions." *Journal of Colloid and Interface Science*, 310: 456–463. https://doi.org/10.1016/j.jcis.2007.02.067.

Lundgren, T. S. 1972. "Slow flow through stationary random beds and suspensions of spheres." *Journal of Fluid Mechanics*, 51 (2): 273–299. https://doi.org/10.1017/S002211207200120X.

Mahapatra, Ojas, Megha Bhagat, C. Gopalakrishnan, and Kantha D. Arunachalam. 2008. "Ultrafine dispersed CuO nanoparticles and their antibacterial activity." *Journal of Experimental Nanoscience*, 3 (3): 185–193. https://doi.org/10.1080/17458080802395460.

Masoumi, N., N. Sohrabi, and A. Behzadmehr. 2009. "A new model for calculating the effective viscosity of nanofluids." *Journal of Physics D: Applied Physics*, 42 (5): 05501. https://doi.org/10.1088/0022-3727/42/5/055501.

Maxwell, James Clark. 1873. *A Treatise On Electricity and Magnetism*. Vol. 1. London:Clarendon Press.

Mishra, Purna Chandra, Mukherjee Sayantan, Santosh Kumar Nayak, and Arabind Panda. 2014. "A brief review on viscosity of nanofluids." *International Nano Letters*, 4: 109–120. https://doi.org/10.1007/s40089-014-0126-3.

Mooney, M. 1951. "The viscosity of a concentrated suspension of spherical particles." *Journal of Colloid Science*, 6 (2): 162–170. https://doi.org/10.1016/0095-8522(51)90036-0.

Murshed, S. M. Sohel. 2012. "Simultaneous measurement of thermal conductivity, thermal diffusivity, and specific heat of nanofluids." *Heat Transfer Engineering*, 33 (8): 722–731. https://doi.org/10.1080/01457632.2011.635986.

Murshed, S. M. S., K. C. Leong, and C. Yang. 2008. "Thermophysical and electrokinetic properties of nanofluids – a critical review." *Applied Thermal Engineering*, 28 (17–18): 2109–2125. https://doi.org/10.1016/j.applthermaleng.2008.01.005.

Namburu, P. K., D. P. Kulkarni, A. Dandekar, D. K. Das. 2007. "Experimental investigation of viscosity and specific heat of silicon dioxide nanofluids." *Micro and Nanoletters*, 2 (3): 67–71. https://doi.org/10.1049/mnl.

Nguyen, Cong Tam, Gilles Roy, Christian Gauthier, and Nicolas Galanis. 2007. "Heat transfer enhancement using Al_2O_3-water nanofluid for an electronic liquid cooling system." *Applied Thermal Engineering*, 27 (8–9): 1501–1506. https://doi.org/10.1016/j.applthermaleng.2006.09.028.

Nielsen, Lawrence E. 1970. "A generalized equation for the elastic moduli of composite materials." *Journal of Applied Physics*, 41 (11): 4626–4627. https://doi.org/10.1002/nav.3800080206.

Ozturk, Serdar, Yassin A. Hassan, and Victor M. Ugaz. 2010. "Interfacial complexation explains anomalous diffusion in nanofluids." *Nano Letters*, 10 (2): 665–671. https://doi.org/10.1021/nl903814r.

Pak, Bock Choon, and Young I. Cho. 1998. "Hydrodynamic and heat transfer study of dispersed fluids with submicron metallic oxide particles." *Experimental Heat Transfer*, 11 (2): 151–170. https://doi.org/10.1080/08916159808946559.

Pastoriza-Gallego, M.J., C. Casanova, J. L. Legido, and M. M. Pineiro. 2011. "Fluid phase equilibria CuO in water nanofluid : Influence of particle size and polydispersity on volumetric behaviour and viscosity." *Fluid Phase Equilibrium*, 300: 188–196. https://doi.org/10.1016/j.fluid.2010.10.015.

Patel, Hrishikesh E., Sarit K. Das, T. Sundararajan, A. Sreekumaran Nair, Beena George, and T. Pradeep. 2003. "Thermal conductivities of naked and monolayer protected metal nanoparticle based nanofluids: Manifestation of anomalous enhancement and chemical effects." *Applied Physics Letters*, 83 (14): 2931–2933. https://doi.org/10.1063/1.1602578.

Paul, G., M. Chopkar, I. Manna, and P. K. Das. 2010. "Techniques for measuring the thermal conductivity of nanofluids: A review." *Renewable and Sustainable Energy Reviews*, 14 (7): 1913–1924. https://doi.org/10.1016/j.rser.2010.03.017.

Peng, Hao, Guoliang Ding, Weiting Jiang, Haitao Hu, and Yifeng Gao. 2009. "Measurement and correlation of frictional pressure drop of refrigerant-based nanofluid flow boiling inside a horizontal smooth tube." *International Journal of Refrigeration*, 32 (7): 1756–1764. https://doi.org/10.1016/j.ijrefrig.2009.06.005.

Rafati, M., A. A. Hamidi, and M. Shariati Niaser. 2012. "Application of nanofluids in computer cooling systems (heat transfer performance of nanofluids)." *Applied Thermal Engineering*, 45–46:9–14. https://doi.org/10.1016/j.applthermaleng.2012.03.028.

Rea, Ulzie, Tom Mckrell, Lin-wen Hu, and Jacopo Buongiorno. 2009. "Laminar convective heat transfer and viscous pressure loss of alumina – water and zirconia – water nanofluids." *International Journal of Heat and Mass Transfer*, 52 (7–8): 2042–2048. https://doi.org/10.1016/j.ijheatmasstransfer.2008.10.025.

Reddy, P. Sudarsana, and Ali J. Chamkha. 2018. "Heat and mass transfer characteristics of nanofluid over horizontal circular cylinder." *Ain Shams Engineering Journal*, 9 (4): 707–716. https://doi.org/10.1016/j.asej.2016.03.015.

Saidur, R., K. Y. Leong, and H. A. Mohammad. 2011. "A review on applications and challenges of nanofluids." *Renewable and Sustainable Energy Reviews*, 15 (3): 1646–1668. https://doi.org/10.1016/j.rser.2010.11.035.

Sandeep, N., Ram Prakash Sharma, and M. Ferdows. 2017. "Enhanced heat transfer in unsteady magnetohydrodynamic nanofluid flow embedded with aluminum alloy nanoparticles." *Journal of Molecular Liquids*, 234: 437–443. https://doi.org/10.1016/j.molliq.2017.03.051.

Sekhar, Y. Raja and K. V. Sharma. 2013. "Study of viscosity and specific heat capacity characteristics of water-based Al_2O_3 nanofluids at low particle concentrations." *Journal of Experimental Nanoscience*, 10 (2): 86–102. https://doi.org/10.1080/17458080.2013.796595.

Shahrul, I. M., I. M. Mahbubul, S. S. Khaleduzzaman, R. Saidur, and M. F. M. Sabri. 2014. "A comparative review on the specific heat of nanofluids for energy perspective." *Renewable and Sustainable Energy Reviews*, 38: 88–98. https://doi.org/10.1016/j.rser.2014.05.081.

Sharifi, Ibrahim, H. Shokrollahi, and S. Amiri. 2012. "Ferrite-based magnetic nanofluids used in hyperthermia applications." *Journal of Magnetism and Magnetic Materials*, 324 (6): 903–915. https://doi.org/10.1016/j.jmmm.2011.10.017.

Sidik, Nor Azwadi Che, Muhammad Noor Afiq Witri Mohd Yazid, and Rizalman Mamat. 2015. "A review on the application of nanofluids in vehicle engine cooling system." *International Communications in Heat and Mass Transfer*, 68: 85–90. https://doi.org/10.1016/j.icheatmasstransfer.2015.08.017.

Sonawane, Sandipkumar, Kaustubh Patankar, Ankit Fogla, Bhalchandra Puranik, Upendra Bhandarkar, and S. Sunil Kumar. 2011. "An experimental investigation of thermo-physical properties and heat transfer performance of Al_2O_3-aviation turbine fuel nanofluids." *Applied Thermal Engineering*, 31 (14–15): 2841–2849. https://doi.org/10.1016/j.applthermaleng.2011.05.009.

Sridhara, Veeranna and Lakshmi Narayan Satapathy. 2011. "Al_2O_3-based nanofluids: a review." *Nanoscale Research Letters*, 6 (456): 1–16.

Subba-Rao, Venkatesh, Peter M. Hoffmann, and Ashis Mukhopadhyay. 2011. "Tracer diffusion in nanofluids measured by fluorescence correlation spectroscopy." *Journal of Nanoparticle Research*, 13 (12): 6313–6319. https://doi.org/10.1007/s11051-011-0607-5.

Sudarsana Reddy, P. and Ali J. Chamkha. 2016. "Influence of size, shape, type of nanoparticles, type and temperature of the base fluid on natural convection MHD of nanofluids." *Alexandria Engineering Journal*, 55 (1): 331–341. https://doi.org/10.1016/j.aej.2016.01.027.

Teng, Tun Ping and Yi Hsuan Hung. 2014. "Estimation and experimental study of the density and specific heat for alumina nanofluid." *Journal of Experimental Nanoscience*, 9 (7): 707–718. https://doi.org/10.1080/17458080.2012.696219.

Timofeeva, Elena V., Alexei N. Gavrilov, James M. Mccloskey, Yuriy V. Tolmachev, Samuel Sprunt, Lena M. Lopatina, and Jonathan V. Selinger. 2007. "Thermal conductivity and particle agglomeration in alumina nanofluids: Experiment and theory." *Physical Review E*, 76 (061203): 1–16. https://doi.org/10.1103/PhysRevE.76.061203.

Turanov, A. N. and Yuriy V. Tolmachev. 2009. "Heat- and mass-transport in aqueous silica nanofluids." *Heat and Mass Transfer*, 45 (12): 1583–1588. https://doi.org/10.1007/s00231-009-0533-6.

Vajjha, Ravikanth S. and Debendra K. Das. 2009. "Specific heat measurement of three nanofluids and development of new correlations." *Journal of Heat Transfer*, 131 (7): 071601. https://doi.org/10.1115/1.3090813.

Vajravelu, Kuppalapalle, Kerehalli Vinayaka Prasad, and Chiu-on Ng. 2013. "The effect of variable viscosity on the flow and heat transfer of a viscous Ag-water and Cu-water nanofluids." *Journal of Hydrodynamics*, 25 (1): 1–9. https://doi.org/10.1016/S1001-6058(13)60332-7.

Vekas, Ladislau, Doina Bica, and Mikhail V. Avdeev. 2007. "Magnetic nanoparticles and concentrated magnetic nanofluids: Synthesis, properties and some applications." *China Particuology*, 5: 43–49. https://doi.org/10.1016/j.cpart.2007.01.015.

Wen, Dongsheng and Yulong Ding. 2005. "Experimental investigation into the pool boiling heat transfer of aqueous based c -alumina nanofluids." *Journal of Nanoparticle Research*, 7: 265–274. https://doi.org/10.1007/s11051-005-3478-9.

Wen, Dongsheng, Guiping Lin, Saeid Vafaei, and Kai Zhang. 2009. "Review of nanofluids for heat transfer applications." *Particuology*, 7: 141–50. https://doi.org/10.1016/j.partic.2009.01.007.

Xuan, Yimin and Wilfried Roetzel. 2000. "Conceptions for heat transfer correlation of nanofluids." *International Journal of Heat and Mass Transfer*, 43 (19): 3701–3707. https://doi.org/10.1016/S0017-9310(99)00369-5.

Xuan, Yimin, Qiang Li, and Weifeng Hu. 2003. "Aggregation structure and thermal conductivity of nanofluids" 49 (4):1038-1043.

Yu, W. and S. U. S. Choi. 2003. "The role of interfacial layers in the enhanced thermal conductivity of nanofluids: A renovated Maxwell model." *Journal of Nanoparticle Research*, 5: 167–171.

Yu, W., D. M. France, S. U. Choi, and J. L. Routbort. 2007. "Review and assessment of nanofluid technology for transportation and other applications." *Argonne National Lab Argonne*, I (May 31).

Zhang, Lingling, Yunhong Jiang, Yulong Ding, Malcolm Povey, and David York. 2007. "Investigation into the antibacterial behaviour of suspensions of ZnO nanoparticles (ZnO nanofluids)." *Journal of Nanoparticle Research*, 9: 479–489. https://doi.org/10.1007/s11051-006-9150-1.

Zhou, Sheng-Qi, and Rui Ni. 2008. "Measurement of the specific heat capacity of water-based Al_2O_3 nanofluid." *Applied Physics Letters*, 92 (9): 093123. https://doi.org/10.1063/1.2890431.

Zhou, Le PingWang, Bu Xuan, Xiao Feng Peng, Xiao Ze Du, and Yong Ping Yang. 2010. "On the specific heat capacity of CuO nanofluid." *Advances in Mechanical Engineering*. Vol. 2010. Article ID 172085. 1-4. https://doi.org/10.1155/2010/172085.

Zhu, Dongsheng, Xinfang Li, Nan Wang, Xianju Wang, Jinwei Gao, and Hua Li. 2009. "Dispersion behavior and thermal conductivity characteristics of Al_2O_3-H_2O nanofluids." *Current Applied Physics*, 9 (1): 131–139. https://doi.org/10.1016/j.cap.2007.12.008.

4 Understanding the Risk

Sivakama Sundari S.K., J.A. Moses, and C. Anandharamakrishnan

CONTENTS

4.1 INTRODUCTION

Nanotechnology is a rapidly growing science that is known to have numerous applications in almost all existing industrial sectors. Nanotechnology also plays an important role in delivering various types of bioactive components with improved bioavailability and stability. Due to its various unique properties, nanomaterial has received a lot of attention in the research sector. Before it gains acceptance in the consumer's shelf, however, safety evaluation of nanoparticles has to be carried out. The World Health Organization (WHO) emphasizes the safety of consumers with respect to nanoscience applications. A clear view of the applications of nanotechnology in food systems and its safety before incorporation into the food system is essential (Anandharamakrishnan, 2014 and 2015; Bhushani and Anandharamakrishnan, 2017; Karthik et al., 2017). Despite numerous benefits, the risks associated with nano-based applications cannot be ignored. In order to overcome the associated risks, a proper understanding of nanotoxicity is essential. Various international, governmental, and regulating organizations such as the British Standards Institution (BSI), the Organization for Economic Co-operation and Development (OECD), and the International Organization for Standardization (ISO) are involved in understanding the risks associated with nanotechnology. They also strive to establish safety measures in developing as well as using concepts of nanotechnology (Donaldson and Poland, 2013; Chaudhry and Castle, 2011; Anandharamakrishnan, 2014), explaining three different concerns for nanofood after its consumption. These include: "least concern," "some concern" and "major concern." Foods that have been digested or solubilized in the gastrointestinal tract (GI) with non-bio persistent nanomaterials will cause "least concern." "Some concern" is caused by non-bio persistent nanomaterials in food that do not get digested but get carried through the GI tract. "Major concern" includes nanofoods containing nano-additives that are indigestible, insoluble and potentially bio persistent. The risk assessment in nanotechnology considers routes of exposure of the nanoparticles, their translocation, and their toxicity (Lakshmanan et al., 2012). It is a concern that nanoparticles possess a certain unknown mode of toxicity often termed as "nano-specific effect" which has led to the urge of its identification (Donaldson and Poland, 2013). The available literature lacks in scientific proof of toxicity of nanoparticles, emphasizing the need for risk assessment of nanoparticles before their application in different sectors.

Various forms of nanoparticles such as carbon nanotubes, gold nanoparticles, silver nanoparticles, nanospheres, nanosomes, quartz and mineral dust nanoparticles, metal oxide nanoparticles, polymer-based nanoparticles, and iron nanoparticles exhibit various levels of toxicity (Mocan et al., 2010). Consumption of nanoparticles by the average person is around 10^{12} particles through daily diet, majorly consisting of TiO_2 and aluminosilicates from food additives (Lomer et al., 2002). A lot of experimental data proved that nanoformulation of bioactive compounds could improve physiochemical and gastrointestinal stability (Karthik and Anandharamakrishnan, 2016a,b; Bhushani et al., 2016; Hundre et al., 2015), and oral bioavailability (Bhushani et al., 2017; Parthasarathi et al., 2016; Parthasarathi and Anandharamakrishnan, 2016). Certain studies involving nanoparticles have indicated that they can produce toxic effects to human health and can possibly result in various concerns including cardiovascular diseases and pneumoconiosis (Dockery et al., 1993; Schwartz, 1994; Neas, 2000; Peters et al., 2006). Common sources for the entry of nanoparticles include wood smoke, and automobiles and furnace exhaust (Barregard et al., 2006). In other terms, the routes of exposure to nanoparticles are generally inhalation (Jaques and Kim, 2000), ingestion (Behrens et al., 2002), dermal exposure (Tinkle et al., 2003), and parenteral modes (Lockman et al., 2003), see Figure 4.1. Entry of nanoparticles in the environment happens through three modes: anthropogenic sources, unintentional sources, and engineered nanoparticles production and usage (Farré et al., 2009).

Metallic nanoparticles such as gold, silver, copper, titanium oxide, and palladium possess unique chemical, physical, and optical properties. They are used in a wide variety of applications including catalysis, display devices, microelectronics, light emitting diodes, biological purposes, and solar cells (Das et al., 2011; Slouf et al., 2012). Silver nanoparticles have a proven antibacterial activity due to which they increase the shelf life of various food materials. Several studies proved both positive and negative aspects of silver nanoparticles. Silver nanoparticles, when used as dermatological ointments, have proved to be cytotoxic to human fibroblasts and skin, thereby resulting in oxidative stress, apoptosis, necrosis of the skin, and lipid oxidation (Duncan, 2011). Gold nanoparticles are used as a delivery channel for various drugs (Everts et al., 2006). They are also used in imaging systems (Goodman et al., 2004) and DNA binding applications. Due to their smaller size, they have the capacity to penetrate through the skin (Jia et al., 2008). Apart from this capability, nanoparticles possesses the potential to induce oxidative stress and toxicity (Li et al., 2008). In certain studies, gold nanoparticles have been reported to permeate through human skin and to be present in diseased cells. Their lethality was understood through studies conducted by (Chen et al., 2009).

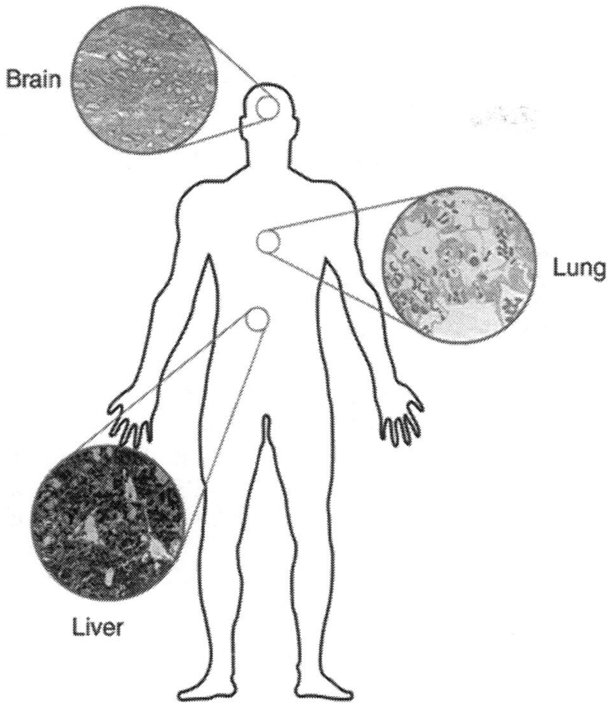

FIGURE 4.1 Routes for the entry of nanoparticles where brain is the vulnerable site. (Elsaesser and Howard, 2012.)

4.1.1 Factors Influencing Nanotoxicity

The levels of nanotoxicity are influenced by various factors such as size of nanoparticles, their shape and surface coating, crystalline structure and its dosage, surface area, and opsonization. These are explained in detail below.

4.1.2 Particle Size

Reducing bulk materials to nanosize brings out the unique and characteristic properties of nanomaterials. However, particle size influences the level of toxicity of nanoparticles. Based on size and type of nanoparticles used, toxicity varies. For example, gold nanoparticles of size smaller than 2 nm (Schmid, 2008) exhibit higher levels of toxicity than particles bigger than 13 nm (Jahnen-Dechent and Simon, 2008). Due to their smaller size, nanoparticles can also cross the barrier between blood and brain, through either passive or active diffusion (Geiser et al., 2005). Also, gold nanoparticles of size 1.4 nm are toxic to cells, as they lead to interactions at DNA levels. However, such trends are not observed in smaller and larger nanoparticles (Pan et al., 2007). It has also been found that quantum dots and silica nanoparticles exhibit different functionalities based on their size. Therefore, nanoparticles toxicity will be exhibited differently upon different particle size. Apart from size, the shape of nanoparticles also contributes to toxicity. A study conducted by Forest et al. (2016) explained the affect of shape upon *in vitro* toxicity of cerium oxide nanoparticles. It was observed from the study that nanotoxicity of cerium oxide particles varied significantly based on their shape, which included rod, cubic, and octahedral nanoparticles. Similar results were observed by Ispas et al. (2009) and Carnovale et al. (2016) where the authors explained the shape dependent toxicity of various metal nanoparticles.

4.1.3 Surface Coating

Surface coating over nanoparticles also influences their toxic effects. For instance, when carbon nanotubes are used for specific applications contaminated with ferrous iron, this leads to the generation of reactive oxygen species (ROS) in the biological system (Nel et al., 2001). It is also found that there are certain possibilities for nanoparticles by which bacterial endotoxin may get adsorbed, thereby resulting in certain anomalies (Shevoda et al., 2003).

4.1.4 Crystalline Structure

TiO_2 exists in three crystalline forms, namely anatase, rutile, and brookite (Fadeel and Bennett, 2010). The reactivity of TiO_2 depends mainly upon the crystalline surface structure and surface chemistry. It was found that TiO_2 of size ranging from 3 to 10 nm in its anatase crystallization form possesses 100 times more toxicity than its rutile crystalline form. This is because the anatase crystalline form absorbs water dissociatively (like H^+ and OH^-), whereas the rutile form absorbs water non-dissociatively (as H_2O). After absorption, when the material is exposed to UV illumination, the anatase forms OH^- ions upon reaction. However, the rutile form remains inactive as explained by (Sayes et al., 2007).

4.1.5 Dosage of Nanoparticles

The dosage of nanoparticles also contributes to the nanotoxicity level. When the concentrations of nanoparticles are higher, they aggregate with each other and hence their toxicity can be reduced. This may be due to this aggregation effect: their properties get altered and they are not capable of entering inside certain body parts. However, at lower concentrations, nanoparticles exhibit greater toxicity (Buzea et al., 2007).

4.1.6 Opsonization

Nanoparticles encapsulated inside a specific wall material enter a biological system to interact only at the targeted site. For this interaction, they can be modified using chemical techniques such as tethering, or with coupling agents (Fadeel and Bennett, 2010). In some instances, nanoparticles may bind to proteins present inside the biological system, which in turn affects their efficiency. Such effects can also modulate the toxicity of the nanoparticles involved (Dutta et al., 2007; Cedervall et al., 2007).

4.1.7 Surface Area-Dependent Toxicity

The surface of nanoparticles contains more atoms owing to a higher surface to volume ratio than in corresponding microparticles. Microparticles occupy less than 1% of atoms in the surface position, whereas particles of size around 10 nm occupy more than 10% of the surface position. This contributes to the changes accompanied by various physical and chemical properties that will lead to associated toxic effects (Jones and Grainger, 2009).

A proper understanding and assessment of the risks of nanomaterials should be formulated prior to product application and commercialization. This requires an understanding of the underlying mechanisms, their properties, and their associated risks. This chapter briefly deals with the risks involved with nanoparticles based on conclusions from scientific findings and published works to date.

4.2 THE NANOPARTICLE AND THE ENVIRONMENT

It is important to assess the interaction between nanoparticles and the environment in order to understand toxic effects, and it is only after prolonged usage and interaction with the ecosystem that these can be seen. For this reason, it is essential to determine whether engineered nanoparticles retain their size, shape and reactivity at nanoscale levels. Fractions of products such as paints, cosmetics, and sun screens deposit on land and water surfaces, where they have the potential to contaminate the soil and groundwater (Ray, 2009).

In some instances, engineered nanoparticles may agglomerate from nanoscale to a few micrometers, which leads to certain changes in their properties. They are also capable of entering into the human body when such changes happen in their properties (Stahlhofen et al., 1989; Lam et al., 2004; Lee et al., 2007). It is also reported in many cases that patients with different diseases were found to carry the deposition of different nanoparticles (Ray, 2009). With its rapid growth, nanotechnology increases the risk of engineered nanoparticles entering the environment. Possible routes through which nanoparticles enter the environment include burning of fossil fuels, water treatment processes, burning of biomass, and waste incineration. Importantly, all these possible routes are due to human activities (Joo et al., 2009; Fang et al., 2009).

4.2.1 Nanotoxicity and Aquatic Species

Aquatic invertebrates were first studied for nanotoxicity since they are the ultimate recipients of nanotoxins that come in contact with the environment (Baun et al., 2008). Many researchers have reported various toxic effects of engineered nanoparticles by carrying out acute and chronic toxicity trials in various aquatic invertebrates. They have also conducted studies on the bioaccumulation of nanotoxins in invertebrates. Aquatic invertebrates on which the experiments were conducted include *Daphnia magna, Chironomusriparius, Vibrio fischeri* and *Oryziaslatipes* (Peralta-Videa et al., 2011). Lee et al. (2009) conducted genotoxicity and ecotoxicity studies of CeO_2, SiO_2, and TiO_2 nanoparticles in *D. magna* and *C. riparius* larvae. Their study concluded that CeO_2 nanoparticles of size 15 and 30 nm are capable of retarding reproduction characteristics of *D. magna,* whereas SiO_2 nanoparticles are capable of increasing mortality rates in both species. Conversely, TiO_2 nanoparticles did not show any toxic effects on these species. Carbon nanoparticles are also known to be responsible for toxicity effects among various aquatic species. They exhibit certain indirect toxic effects that are nonspecific, including occlusion of surface tissues and physical irritation on some aquatic species (Farré et al., 2009). Carbon nanotubes were found to accumulate on the surfaces of aquatic organisms' gills, which later resulted in lesions and skin irritation (Smith et al., 2007). It is also known that carbon nanoparticles can result in various toxic effects for the respiratory tract of rainbow trout (Smith et al., 2007). TiO_2 nanoparticles exhibit various respiratory toxic effects and certain sub-lethal injuries in rainbow trout species (Federici et al., 2007).

4.2.2 Nanotoxicity and Terrestrial Organisms

Terrestrial organisms are not immune from nanotoxicity. The nanotoxicity of ZnO, Al_2O_3, and TiO_2 was studied. Tests conducted on *Caenorhabditis elegans* using bulk and nanoparticles of ZnO exhibited toxic effects resulting from bulk material at higher levels as compared to nanoparticles. It was concluded that nanoparticles exhibit toxicity, but the toxic level varies according to the properties of the nanomaterials used (Pipan-Tkalec et al., 2010). Researches also reported that bioaccumulation of zinc was observed in *C. elegans* due to the dissolution of zinc molecules from zinc oxide nanoparticles.

4.2.3 Nanotoxicity and Plants

Plants play a very important role in the food chain. Lei et al. (2008) explained oxidative stress in spinach due to the presence of TiO_2 nanoparticles. They noted that antioxidant stress was found to reduce gradually in its chloroplast due to a reduction in hydrogen peroxide, malonyldialdehyde, and superoxide radicals, thereby enhancing superoxide dismutase, ascorbate peroxidase, guaiacol peroxidase, and catalase activities. It was also observed that bioaccumulation and root elongation were decreased in ryegrass species due to the presence of ZnO nanoparticles in the size range 20 nm (Lin and Xing, 2008).

4.2.4 Nanotoxicity and Mammalian Cells

Nanoparticles also exhibit toxic effects against mammalian cells, and these effects were assessed on rat liver cells for mitochondrial function, cell line morphology, permeability of plasma membrane, and apoptosis in the presence of numerous metal oxide nanoparticles. Mitochondrial toxicity was observed representing the cytotoxic effects. The cellular membrane of rat liver cells also showed damages in the presence of metal oxide nanoparticles. This study concluded that some cells attained apoptosis due to dysfunction of mitochondrial activity. Accordingly, oxidative stress will occur leading to cell apoptosis (Jeng and Swanson, 2006).

Therefore, it is essential to have a detailed understanding of the fate of nanoparticles in the environment. Assessment has to be done on the distribution of nanoparticles that are responsible for toxic effects. Target organelles should also be assessed for nanotoxicity effects, and possible mechanisms have to be evaluated. In recent years, only a very small amount of research has been conducted on the toxic effects of various engineered nanoparticles. Many studies have concluded that engineered nanoparticles do not lead to acute toxicity. However, they are one among the factors that are capable of causing sub-lethal injuries in all possible methods and causing long-term effects on the environment. The reason is mainly that nanoparticles act as synergists. They combine with environmental toxins by getting attached to them and aggregating, after which they are delivered to the environment and also to the human physiology.

4.3 *IN-VITRO* AND *IN-VIVO* TOXICOLOGICAL STUDIES

4.3.1 Skin Penetration of Nanoparticles

The largest organ in the human body is the skin, contributing to 10% of the total body mass (Crosera et al., 2009). It is the protective layer which prevents the body from external damage caused by the environment (Lam et al., 2004). The human skin is composed of three layers with an approximate surface area of 18,000 cm^2 (Zhao et al., 2012). It consists of epidermis, dermis and subcutaneous layer (Hoet et al., 2004), Figure 4.2. The skin of most species grows from hair follicles. Possible routes for the entry of nanoparticles into the body include the digestive tract through inhalation or absorption (Chen and Schluesener, 2008). Nanoparticles also enter biological systems by injection, absorption, and implantation for the delivery systems (Sartorelli et al., 2007). Apart from this, skin acts as the major route for the entry of foreign particles (Wu et al., 2009) with its larger exposure area. Based on physicochemical properties of materials, four pathways through which the particles enter through the skin have been identified. They are the intercellular route, transcellular route, and two transappendageal routes that happen *via* hair follicles and sweat glands (Scheuplein, 1967). Several *in-vitro* and *in-vivo* toxicological studies have been conducted by researchers, and possible effects on skin and its exposure to nanoparticles are known (Crosera et al., 2009). Kreilgaard (2002) indicated that TiO_2 particles of size ranging from 5 to 20 nm have the possibility to penetrate the skin and are able to interact with the immune system. Nanoparticles enter the skin by penetrating it. But the extent of penetration depends on interactions that occur between the nanoparticles and the skin (Baroli et al., 2007).

Larese Filon et al. (2011) studied the skin penetration effects of gold nanoparticles through the intact and damaged skin of humans in an *in-vitro* diffusion system. He carried out experiments to evaluate permeation rates of gold nanoparticles over an exposure period of 24 h. *In-vitro* percutaneous absorption studies using static diffusion "Franz cells" considered a skin exposure area of 3.29 cm^2 and an average skin thickness of 1 mm. Results concluded that penetration of gold nanoparticles is a dose-dependent mechanism. The concentration of gold nanoparticles was found

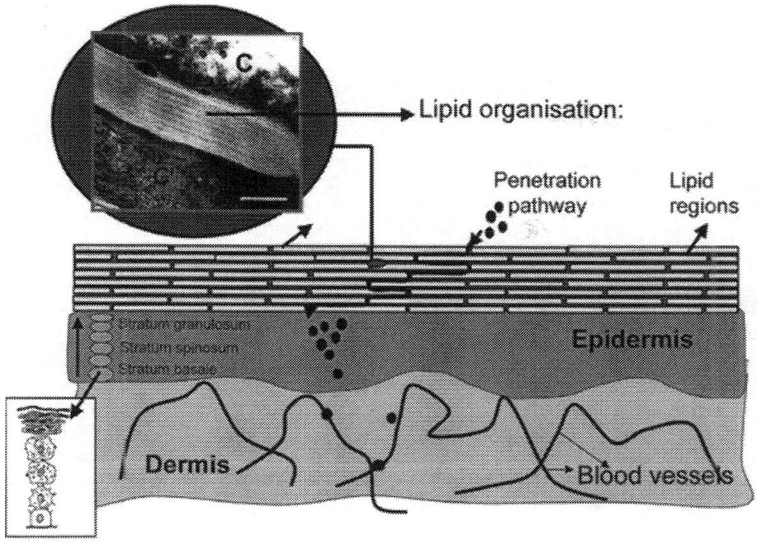

FIGURE 4.2 Structure of human skin. (Bouwstra, 2002.)

to be higher in damaged cells than in intact cells. However, gold nanoparticles have the ability to penetrate through both cell types. The study also stated that skin penetration of gold nanoparticles is a greater concern when compared with other nanoparticle types.

4.3.1.1 *In-vitro* Toxicological Studies of Single-Walled Carbon Nanotubes in Human Keratinocytes

Carbon nanotubes exist as single-walled or multi-walled nanotubes. Fe and Ni transition metal catalysts are generally used in their preparation. These nanotubes possess unique properties that might also be responsible for their toxicity (Borm, 2002). Shvedova et al. (2003) evaluated the *in-vitro* toxic effects of single-walled carbon nanotubes in human epidermal keratinocytes (HacaT). Keratinocytes are the predominant cell type in the epidermis, the outermost layer of the skin, also referred as "basal cells." On exposure to single-walled carbon tubes, studies observed oxidative stress, depletion of vitamin E and anti-oxidant reserve, decrease in intracellular glutathione, and oxidation of SH group in the protein. For this, 96 well plates were used for growing cell cultures and incubated with single-walled carbon nanotubes of varying concentrations for 2, 4, 6, 8, and 18 h at 37°C. Cells were washed twice in phosphate buffered saline (PBS) medium and trypsinized. Test results indicated that exposure of HacaT cells to single-walled carbon nanotubes resulted in cellular toxicity and oxidative stress indicated by the formation of free radicals, loss of cell viability, and depletion of antioxidants and accumulation of peroxidase products within 18 h of exposure. Human cells also showed morphological changes on exposure. Single-walled carbon nanotubes containing 30% of iron in the unrefined form sometimes overload toxicity, resulting in cancer, diabetes, immune system anomalies and skin, liver, and lung diseases. They can also result in the acceleration of oxidative stress because of catalytic toxic effects associated with iron present in unrefined single-walled carbon nanotubes.

4.3.1.2 *In-vitro* Cytotoxic Study on the Nanocrystalline Silver on Keratinocytes

About 90% of the outermost skin layer comprises of keratinocytes. Lam et al. (2004) reported the cytotoxic effects of Acticoat silver dressing, which is generally used for dressing wounds. The experiments were carried out by applying Acticoats on cultured keratinocyte cells for 30 min at 37°C. In the experiment, laser skin was used as the proliferation medium for the keratinocyte cells. The results of the experimental study confirmed that Acticoat with nanocrystalline silver is cytotoxic to keratinocyte cells. This is because Acticoat kills normal cells in addition to bacterial

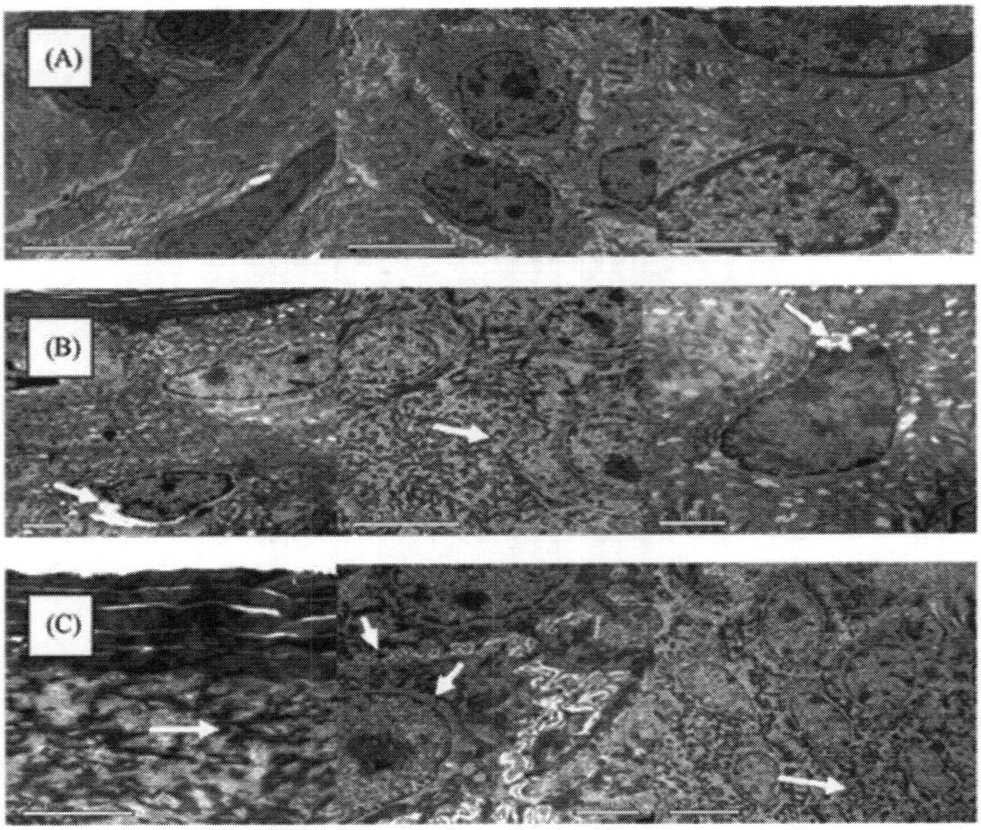

FIGURE 4.3 TEM images confirming the penetration of nanoparticles in the layer of porcine skin (A) control; (B) TEM image of 4 nm of TiO_2; (C) TEM image of 60 nm TiO_2 (arrows represent the accumulation of nanoparticles). (Wu et al., 2009.)

cells. Thus it results in damage to healthy normal cells, and its usage should be avoided. Therefore, *in-vitro* studies confirmed the toxic effects of nanoparticles in dermal applications.

4.3.1.3 *In-vivo* Toxicological Study of Titanium Dioxide Nanoparticles on the Skin of Hairless Mice and Porcine

Wu et al. (2009) conducted experimental studies on the *in-vivo* toxicity of TiO_2 nanoparticles of sizes 4 nm and 60 nm. A porcine ear was treated with nanoparticles by topical application, and the effects were studied for 30 days. Results indicated that nanoparticles get located in the deepest layer of the epidermis. The rate of penetration of nanoparticles depends upon their size, for instance, a nanoparticle of size 4 nm is capable of penetrating and reaching the epidermal layer of the skin (see Figure 4.3). Penetration of nanoparticles in porcine skin results in various abnormalities including cellular structure changes, as confirmed by TEM images.

An in-vivo study on the dorsal skin of hairless mice over a period of 60 days was conducted with nanoparticles of varying size. Results indicated that nanoparticles penetrated through the skin and also got distributed over various organs through the penetrated skin, resulting in the formation of lesions. Nanoparticles of size 21 nm showed greater distribution effects and were also observed to reach the brain. Among the distributed organs, changes were observed in skin and liver. Collagen levels in mice skin were found to reduce at higher exposure periods and thereby result in aging. This also induces oxidative stress in cells. Changes that occurred in the mice organs are represented in Figure 4.4 (Wu et al., 2009).

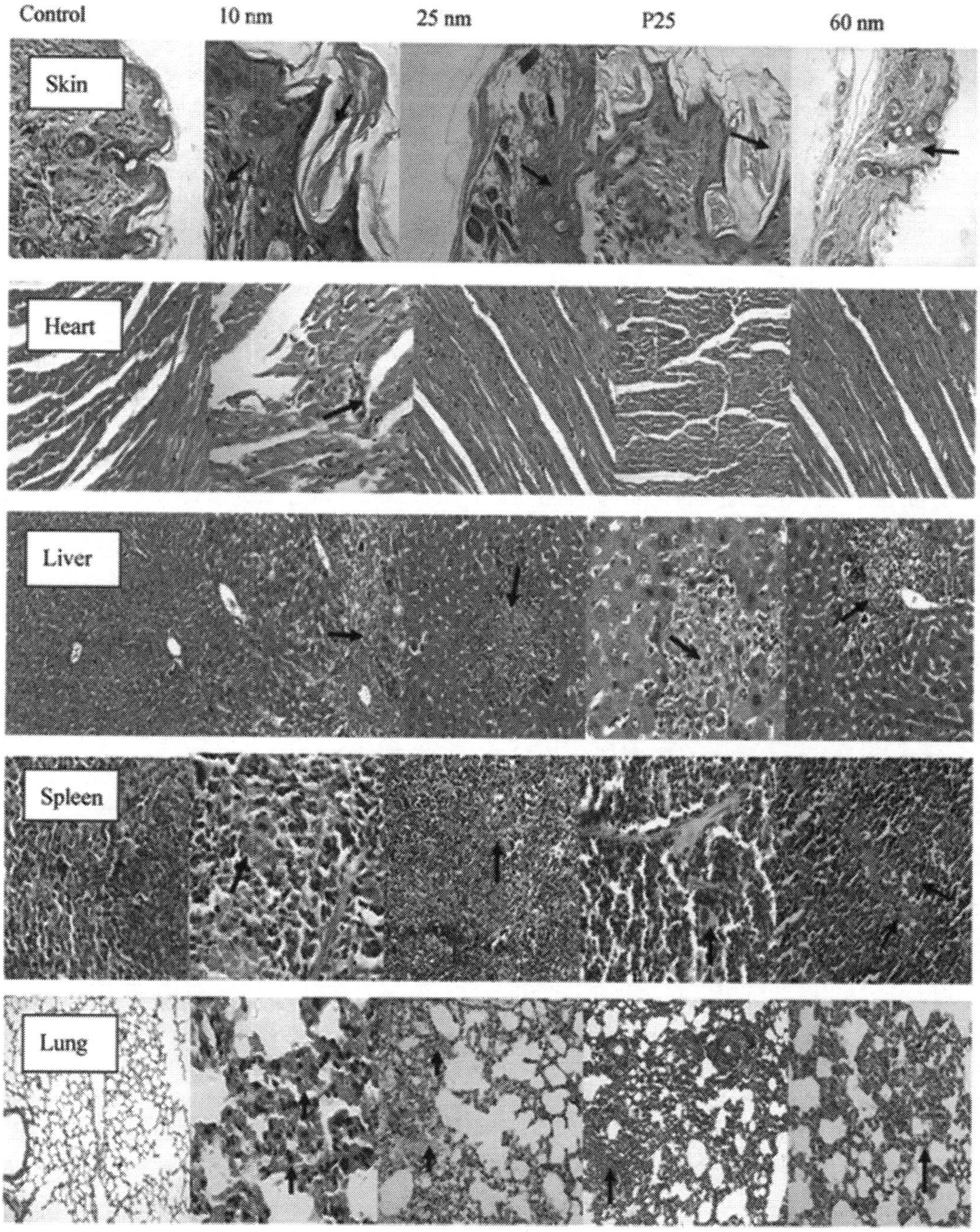

FIGURE 4.4 Various organelle changes on exposure to TiO_2 nanoparticles in mice skin of varying sizes (arrows illustrate changes that occurred on exposure to the nanoparticles). (Wu et al., 2009.)

4.3.2 Inhalation of Nanoparticles

The respiratory system is a key targeted region for the entry of nanoparticles that may lead to nanotoxicity effects. The pathway through which nanoparticles enter lungs is breathing, but other exposure routes such as oral administration also lead to their distribution in the lungs. Nanoparticles enter the lungs either intentionally or non-intentionally (Card et al., 2008). They are designed in such way that agglomeration can be prevented by increasing their entry rate towards lung cells through inhalation and deposition of contents (Stern and McNeil, 2008). Oberdorster et al. (2004) noted a translocation of inhaled nanoparticles of around 50% towards the liver, which occurred based on studies conducted in a rat model over an exposure period of 24 h. Similarly, Kreyling et al. (2002) conducted experiments on mice with iridium nanoparticles of size 15–20 nm. It was reported that nanoparticles were found to have deposited in the lungs, after which they were translocated to other organs such as spleen, kidneys, brain, and heart. In the same way, Sung et al. (2008) conducted experiments and concluded that prolonged exposure to silver nanoparticles could result in accumulation in the lung and liver tissues, which makes them the most vulnerable organs. Therefore, it has been inferred from the literature that lungs are the easiest target for the entry of nanoparticles. Upon entry, nanoparticles can translocate towards the brain through the nasopharyngeal track (Oberdorster et al., 2004).

Lanthanide cerium oxide nanoparticles are used in catalysts, solar cells, solar fuel cells, and ultraviolet absorbents (Zheng et al., 2005). Srinivas et al. (2011) determined the acute toxicity of cerium oxide nanoparticles in rats by exposing head and nose regions to nanoparticles. The toxicity experiment was carried out for 24 h, 48 h, and 14 days. It was observed that, after 24 h of exposure, malondialdehyde (MDA) levels, which are responsible for oxidative stress, were elevated, and intracellular glutathione (GSH) levels, which provide antioxidant effects, were decreased (see Figure 4.5). Alveolar macrophages and neutrophils showed the presence of phagocytized and non-phagocytized cerium oxide nanoparticles. The latter were observed to be deposited in the alveolar, bronchiole, and bronchi regions of the lungs. This experimental study confirmed that exposure to cerium oxide nanoparticles leads to cytotoxic effects through oxidative stress. Sung et al. (2008) also studied the effect of silver nanoparticles on changes in the pulmonary functions of rats over a period of 90 days. Silver nanoparticles, known for their excellent antimicrobial activity, induced changes in lung functioning and inflammatory effects, even when consumed at lower doses as compared to sub-micrometer particles. Authors concluded that nanoparticles enter the human system either through intentional or non-intentional pathways and exhibit various changes associated with lung functions, often resulting in various kind of toxicity by getting accumulated in various cells.

4.4 UPTAKE AND EFFECT OF NANOPARTICLES IN THE BRAIN

The brain of a human is an organ that makes part of the central nervous system (CNS). It is located in the head and protected by the skull. The neurovascular system presents two barriers: the blood-nerve barrier and the blood-brain barrier. They play an important role by limiting the entry of bioactive molecules, acting as a physiological barrier for components entering the peripheral and central nervous systems (Oberdorster et al., 2009). It is observed that large pharmaceutical molecules such as peptides, proteins, and nucleic acids can enter the brain, while 98% of small pharmaceutical molecules cannot (Pardridge, 2007). Various nanoparticles were designed for uptake in the brain (van Rooy et al., 2011). Endothelial cells that line the microvessels of the brain are responsible for the formation of the blood-brain barrier, and this is buoyed with astrocytes and pericytes (De Boer et al., 2003). These cells are responsible for various barrier functions that join the endothelial cells very closely and thus act as a physical barrier (van Rooy et al., 2011). Apart from being a physical barrier, this also acts as a transport and metabolic barrier. These barriers are mainly responsible for the inward and outward movement of molecules that occur between the blood and brain interface. Transfers of hydrophilic molecules by paracellular transport mechanisms are not possible because of the tight junctions that are present in the brain endothelial cells. However, different routes are available for the entry of molecules through the blood-brain barrier, as shown in Figure 4.6.

It is known that the central nervous system is the passage through which bioactive drugs can go through the blood-brain barrier and reach the brain, whereas various components with pharmacological activities will be restricted. This is because of the presence of the barrier system. A possible solution is the packing of pharmacological components inside a nanoparticle large enough to be able to enter the brain by crossing the blood-brain barrier. Nanoparticles enter the brain through a receptor or adsorptive-mediated endocytosis and then undergo transcytosis in the brain endothelial

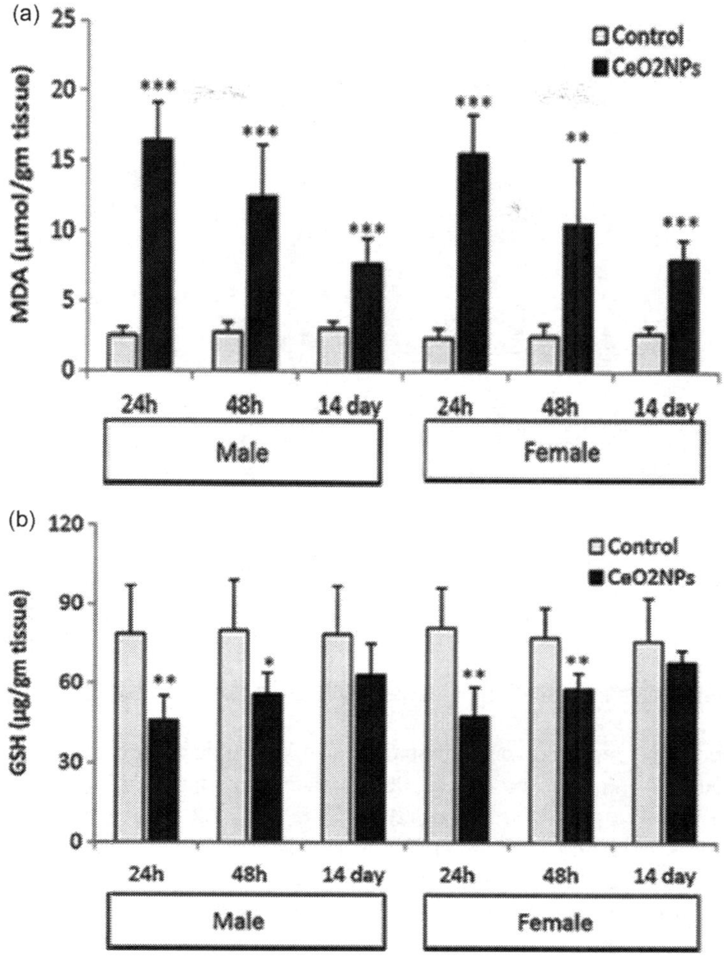

FIGURE 4.5 (a) Changes in the level of malondialdehyde after exposing rat to cerium oxide nanoparticles; (b) changes in the level of intracellular glutathione after exposing rat to cerium oxide nanoparticles. (Srinivas et al., 2011.)

cells to enter the brain parenchyma (De Boer et al., 2003). Particles that have been used to pack bioactive components that enter the brain include solid- lipid nanoparticles, liposomes, poly (lactic-co-glycolic acid), poly (butyl cyanoacrylate) nanoparticles, and nanogels.

4.4.1 Nanoparticles and their interaction with the brain

The entry of nanoparticles inside the brain leads to various negative effects, hence detailed understanding is required to enable precautionary steps. Engineered nanoparticles less than 100 nm in size are generally used for the delivery of various bioactive molecules and gene delivery applications across the blood-brain barrier (Kreuter, 2001). Metal nanoparticles such as copper, silver, or aluminum of size ranging from 50 to 60 nm have the capacity to disrupt the blood-brain barrier (Sharma and Sharma, 2007). Shim et al. (2014) proposed that zinc oxide nanoparticles affect proteins present in the brain and plasma, thereby resulting in toxic effects. Various studies suggested that nanoparticles have the ability to reach the brain and can be detected using techniques such as electron microscopy and capillary detection (Chang et al., 2009). Nanoparticles can cross the blood-barrier present between blood and brain due to its high surface area and reactivity (Hartung, 2010). This may result in neurotoxic effects at higher potential rates (Shubayev et al., 2009; Oberdorster et al., 2009).

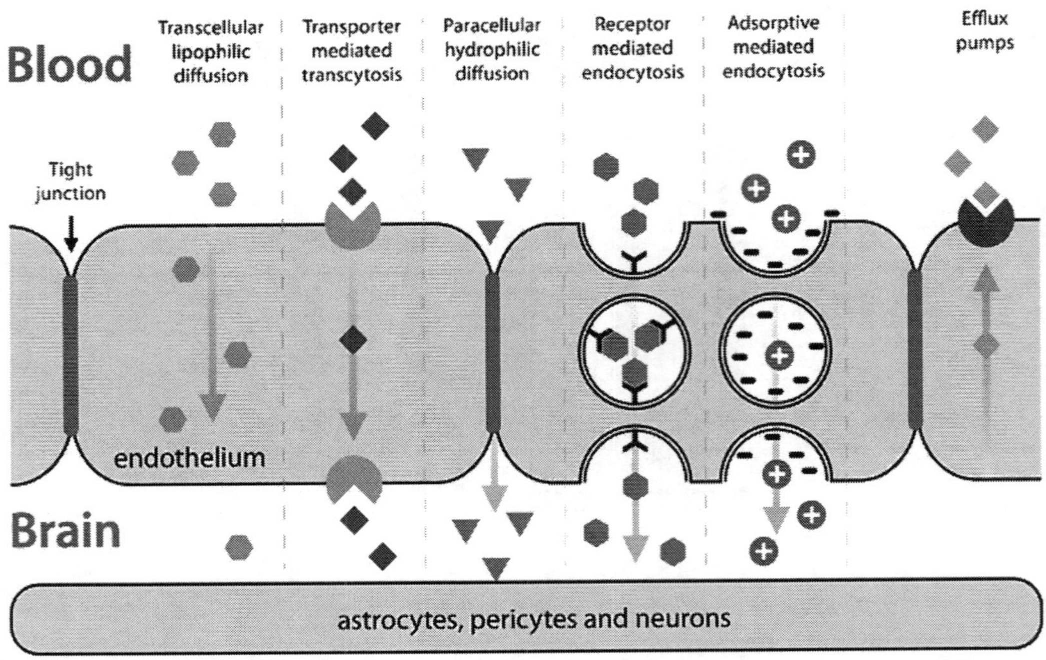

FIGURE 4.6 Transport of molecules across the blood-brain barrier. (Van Rooy et al., 2011.)

Other researchers also reported that nanoparticles, like single-walled carbon nanotubes (Yang et al., 2007a), quantum dots (Yang et al., 2007b), and other materials can cross the blood-brain barrier and enter the brain (Koziara et al., 2004). Xia et al. (2009) reported that alumino oxide nanoparticles exhibit inflammatory effects in rat brains. This is because nanoparticles cross the blood-brain barrier and can reside in the parenchyma, or they can interact with the blood-brain barrier resulting in brain dysfunction. Tang et al. (2008) studied the effect of silver nanoparticles and microparticles on neurons and the blood-brain barrier by injecting them subcutaneously in rats. Ravishankar Rai and Jamuna Bai (2011) explained the antimicrobial activity of silver nanoparticles and their potential application in food products. Silver nanoparticles can move into the barrier in the lungs and thus can be deposited throughout the body (Takenaka et al., 2001). Experimental results of Tang et al. (2008) evidenced that loading rat brains with microparticles results in no abnormal substances in the brain. However, silver nanoparticles caused pyknotic and necrotic neurons within 2–24 weeks. Silver nanoparticles can also move to other neurons. Their toxic effects may result in the destruction of neurons, or else, cell membrane thrombolysis might occur (Figure 4.7). Apart from this, silver nanoparticles also have the ability to accumulate inside the brain over a longer period and can result in destructive changes inside the brain.

4.5 INTERFACE BETWEEN THE CELL MEMBRANE AND NANOPARTICLES

The cell membrane is composed of phospholipid bilayers that are responsible for partitioning various intracellular compartments, each of which executes a particular function. There are two segments of the cell membrane in which the outer cell membrane acts as an interface for the cell with the external environment. The inner part of the cell membrane isolates vesicles, mitochondria, nucleus and other components from the cytosol (Vasir and Labhasetwar, 2008). The interaction of nanoparticles with the cell membrane depends largely upon its surface properties. Importantly, the adhesion force and surface pressure on the cell membrane depend on the size of nanoparticles (Peetla and Labhasetwar, 2008). Nanoparticles affect the cell membrane either directly by physical damage, or indirectly through oxidation (Elsaesser and Howard, 2012), thereby resulting in cell death. Certain nanoparticles

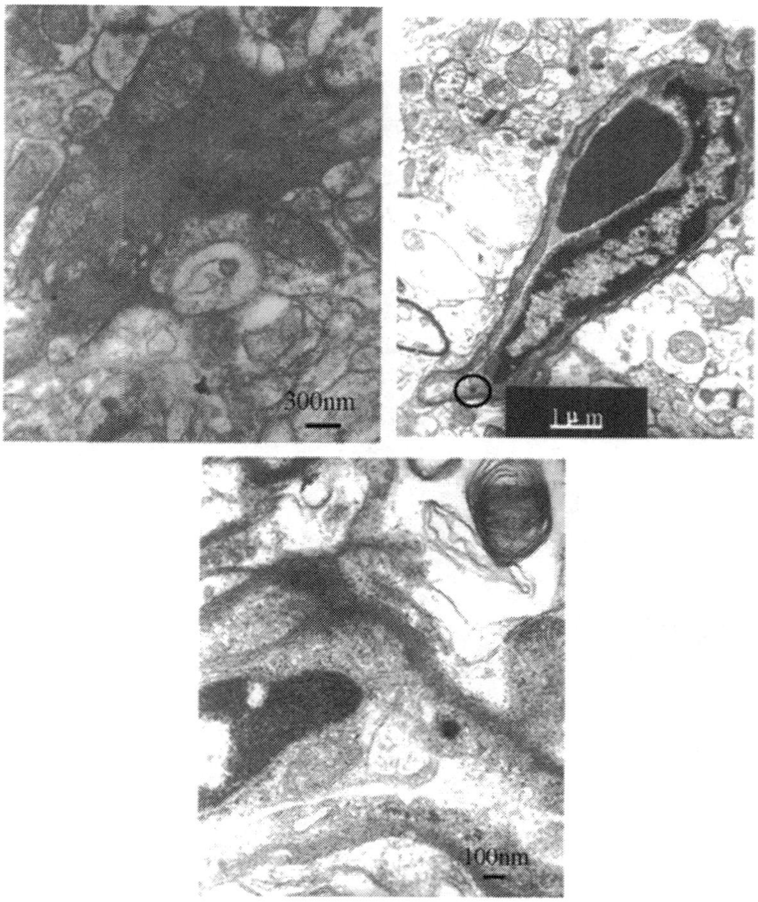

FIGURE 4.7 Accumulation of silver nanoparticles in the brain showing abnormalities across the blood-brain barrier. (Tang et al., 2008.)

such as TiO_2, carbon nanotubes, silver and polystyrene are responsible for mitochondrial damage and associated with apoptosis (Hussain et al., 2005; Jia et al., 2005; Xia et al., 2006). The nuclear membrane transports constituents inside and outside through pores (Parfenov et al., 2006). However, nanoparticles get transported by a "receptor-mediated mechanism" or through diffusion (Williams et al., 2009). Various effects of nanoparticles on cells upon interaction are represented in Figure 4.8.

Ruenraroengsak et al. (2012) determined membrane damage by amine-modified nanoparticles using human alveolar type-I-like cells. Polystyrene nanoparticles, unmodified nanoparticles, carboxyl, and amine-modified nanoparticles of 50 to 100 nm were used in the experiment. The results confirmed that amine-modified nanoparticles were capable of causing cellular damage by creating holes on type-I-like cells. This confirmed that epithelial cells are highly vulnerable to engineered nanoparticles. The same researches also concluded that membrane toxicity caused by nanoparticles depends on size and surface chemistry. In order to avoid such toxicity, a detailed evaluation of nanoparticles has to be carried out, possibly before exposure to living cells.

4.6 CASE STUDY

4.6.1 Toxicity of Nanosilica in Food Materials

Silica exists in its amorphous or crystalline forms. Among the crystalline forms of silica, quartz is the best known form in natural and synthetic formulae. Like crystalline silica, amorphous silica

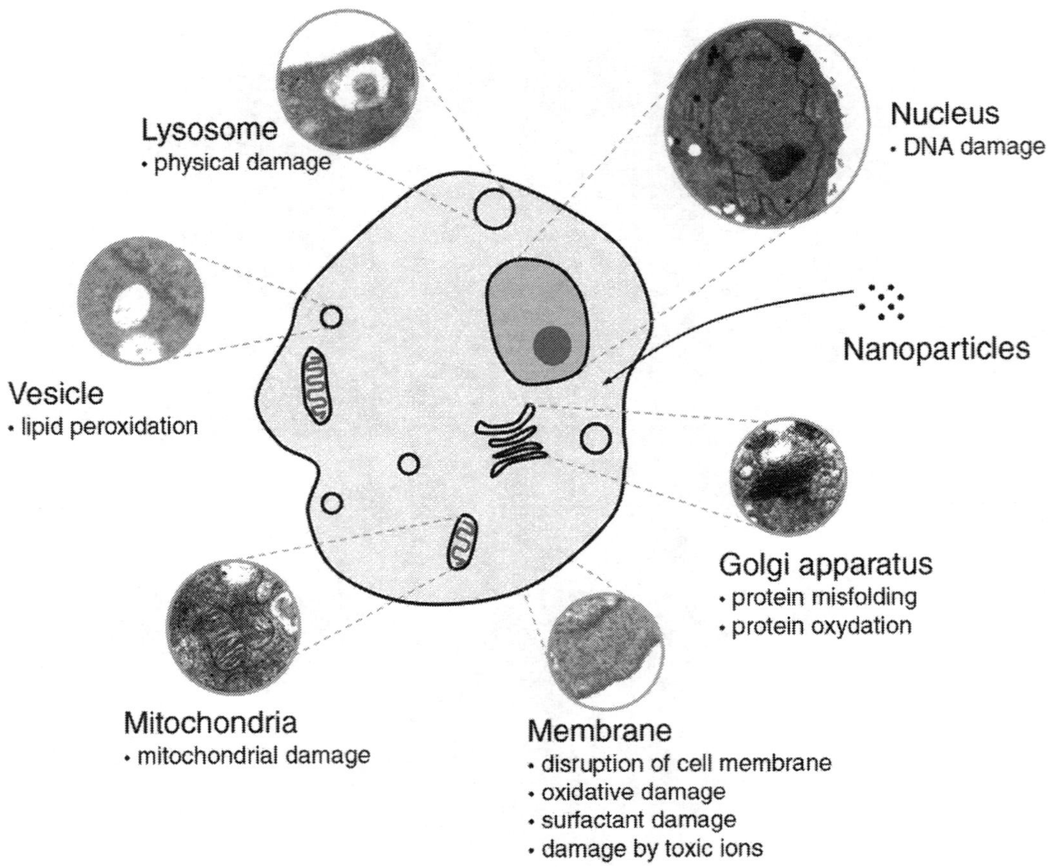

FIGURE 4.8 Effects of nanoparticle interaction on cells. (Elsaesser and Howard, 2012.)

also exists in two different forms, namely, natural and human-made formulae (Napierska et al., 2010). Silica nanoparticles are generally used in various types of food products, including wine and beer for clarification, and as powders for preventing caking in various food products (Dekkers et al., 2013). Apart from food applications, silica nanoparticles are also used in the field of biotechnology for cancer therapy, drug delivery, and enzyme immobilization (Kumar et al., 2004). With their innumerable applications, entry of these nanoparticles can easily occur through skin absorption, ingestion or injection (Ye et al., 2010b). It is very common that once the size of the particle is brought down to the nanoscale, its property changes.

Ye et al. (2010a) investigated the *in-vitro* toxicity of silica nanoparticles. In this study, the toxicity of silica nanoparticles was analyzed based on their dosage, size, and time of exposure in rat myocardial cells. The control used in this study was myocardial cells treated with ultra-pure sterile water; these cells were exposed to silica nanoparticles of sizes 21 and 48 nm. Nanoparticles were prepared using the sol-gel method. The concentrations of silica nanoparticles used ranges from 0.1 to 0.6 mg/ml with exposure time of 12, 24, 36, and 48 h. Studies were conducted to determine toxicity, cell injury, and generation of reactive oxygen species responsible for oxidative stress. Apart from this analysis, cell cycle distribution was also assessed after exposure to silica nanoparticles. Results concluded that silica nanoparticles were capable of producing cytotoxic effects in *in-vitro* cell cultures by inducing oxidative stress that can eventually result in cell injuries. It was also observed that silica nanoparticles lead to the G1 phase arrest in the cell cycle mechanism with increased levels of p53 (tumor protein 53) and p21 (cyclin-dependent kinase inhibitor), which can cause cell death. Researchers also observed that toxicity is purely a size- and dose-dependent mechanism.

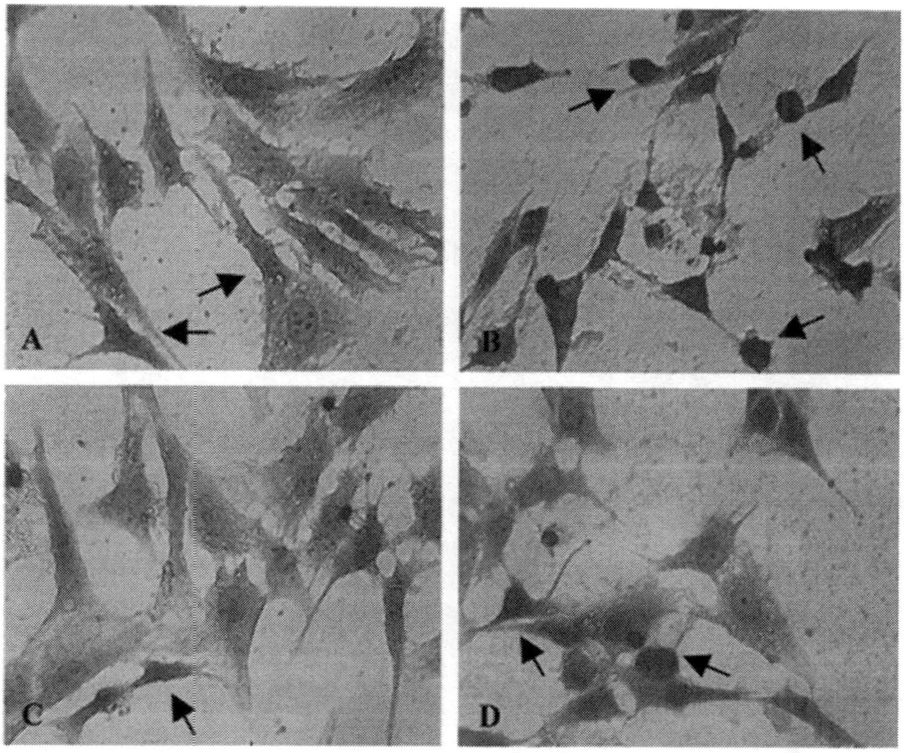

FIGURE 4.9 Light microscopic images showing the toxic effects of silica nanoparticles on the myocardial cell line (arrows represent the condensation and irregular shapes of the cells). (A) Represents cells exposed to a concentration of 0.1 mg/ml; (B) represents cells exposed to a concentration of 0.3 mg/ml; (C) represents cells exposed to a concentration of 0.1 mg/ml; and (D) represents the cells exposed to a concentration of 0.3 mg/ml. (Ye et al., 2010a.)

Figure 4.9 represents the changes that occur in cell lines after exposure to silica nanoparticles of two different sizes over a period of 24 h.

Similarly, Ye et al. (2010b) observed apoptosis of nanosilica through *in-vitro* studies on the human hepatic cell line. Nanoparticles of sizes 21, 48, and 86 nm were used in the study. Concentrations of nanoparticles used were 0.2, 0.4, and 0.6 mg/ml over exposure periods of 12, 24, 36, and 48 h. Results showed that nanoparticles of size 21 nm with a concentration of 0.6 mg/ml were capable of inducing oxidative stress and apoptosis, thereby resulting in DNA damage (Figure 4.10). Changes such as endoplasmic reticulum ectasis, cell swelling, microvilli disappearance, and mitochondrial vacuolization were observed in TEM images once the cells were exposed to silica nanoparticles. It is definite that cytotoxicity is exhibited by nanoparticles in size-, dose- and time-dependent functions. Choi et al. (2011) studied the genotoxicity of nanosilica particles in the cell lines of a mammal. Silica nanoparticles of size 10 nm were taken and assessed for genotoxicity and cytotoxicity. For this, they used bronchial epithelial cells and lymphoma cells of mice that had been cultivated. The results concluded that nanosilica of size 10 nm can induce toxicity, causing issues such as injury to liver cells and induction of inflammation. Researchers also stated that nanosilica is capable of causing damage to DNA at the primary level along with cytotoxic effects. However, there was no sign of mutation in the genes of the mammalian cells. Athinarayanan et al. (2014) studied the toxicity of silica nanoparticles responsible for oxidative stress and altered cell cycle mechanism in the lung fibroblast cells of a human, *in-vitro*. For this study, isolation of silica nanoparticles from food and characterization were performed. The results highlighted the presence of silica nanoparticles of spherical shape in food; their sizes were found to be 10–50 nm. Toxicological studies such as intracellular reactive oxygen

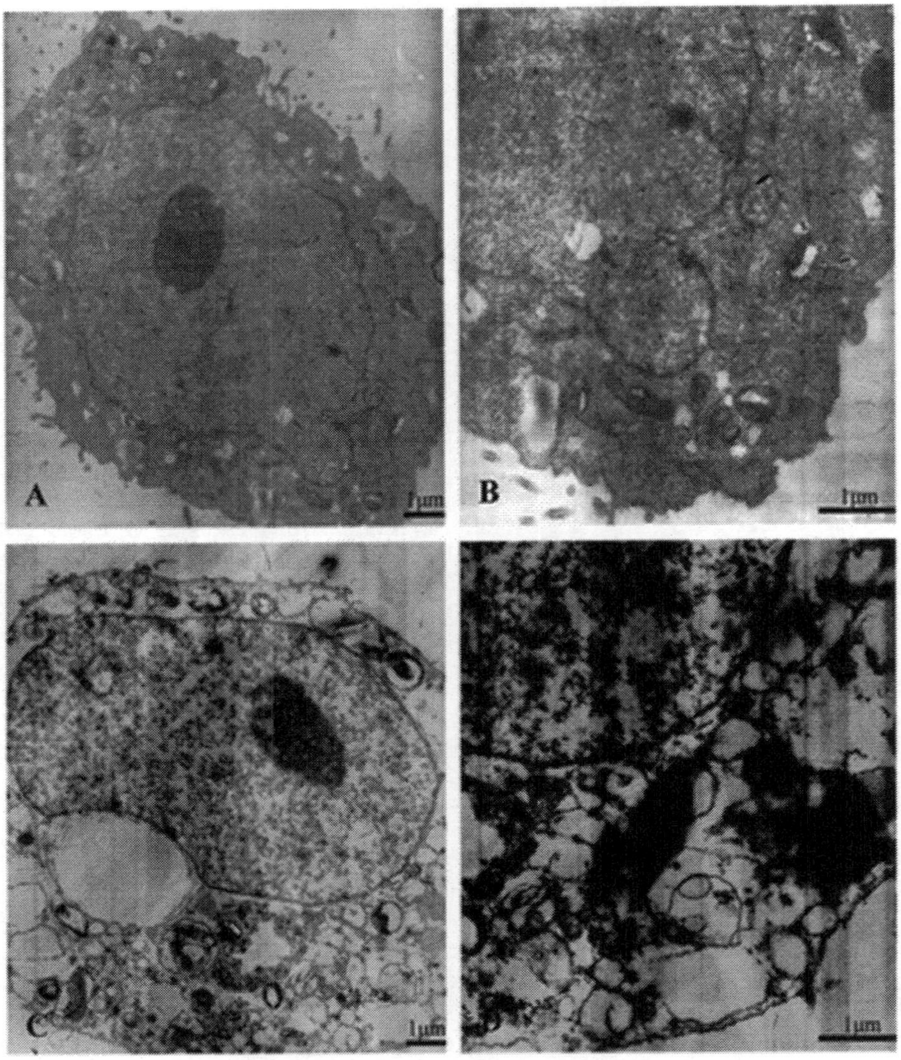

FIGURE 4.10 Transmission electron micrographs of L-02 cells after 24 h exposure to silica nanoparticles (21 nm) at the concentration of 0.6 g/ml: (A) control group at 6000x; (B) control group at 11,500x; (C) exposed group at 6000x; (D) exposed group at 11,500x. (Ye et al., 2010b.)

species generation, fibroblast lung cells viability, cell cycle phases, and metabolic stress levels were carried out to evaluate the toxicity. These studies obtained shocking results, showing that nanoparticles obtained from food material resulted in various changes occurring in the cell line of the human lung fibroblasts (Figure 4.11). They also resulted in exhaustion of mitochondrial membrane, caused damage to cells, and generated reactive oxygen species. Moreover, they created stress where the antioxidant levels were reduced abruptly by altering the cell cycle mechanisms. Authors suggested that higher concentrations of nanoparticles will lead to greater toxic effects among various species.

The above case studies confirmed that silica nanoparticles are highly responsible for toxic effects. However, toxic effects are dependent on size, dosage, concentration, and time of exposure. Studies also confirmed that when silica nanoparticles were used in lower concentration, this did not lead to toxic effects, but when the concentration increased, various toxic effects were produced (Yang et al., 2011). Therefore, when nanoparticles are used in food applications, safety evaluation should be carried out in a detailed manner.

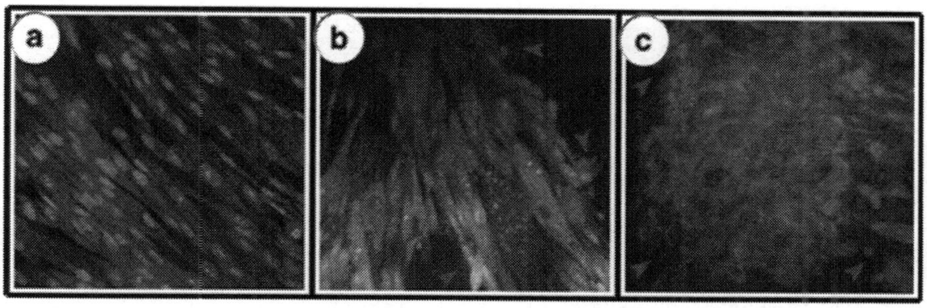

FIGURE 4.11 Fluorescence microscopic images showing morphological changes in lung fibroblast cells (WI-38) treated with different concentrations of silica nanoparticles for 24 h: (a) control (without exposure); (b) WI-38 cells exposed to 50 μg/ml of silica nanoparticles; (c) WI-38 cells exposed to 200 μg/ml of silica nanoparticles (arrows represent abnormal changes in the cell morphology). (Athinarayanan et al., 2014.)

4.7 CONCLUSION

Nanotechnology supports various applications and commercialization, but it also increases risk. Toxicity evaluation of nanoparticles is a positive way to advance the application of engineered nanoparticles, apart from providing safety guidelines. An actual scientific basis is required for clear acceptance of toxicity in nanoparticles. *In-vivo* studies for engineered nanoparticles are lacking in this research era. New organizations created between agencies, stakeholders, researchers, and governments will be required to address various risks

REFERENCES

Anandharamakrishnan, C. 2014. *Techniques for Nanoencapsulation of Food Ingredients*. New York. Springer.

Anandharamakrishnan, C. 2015. *Spray Drying Techniques for Food Ingredient Encapsulation*. United Kingdom, UK: John Wiley & Sons.

Athinarayanan, J., Periasamy, V. S., Alsaif, M. A., Al-Warthan, A. A., and Alshatwi, A. A. 2014. "Presence of nanosilica (E551) in commercial food products: TNF-mediated oxidative stress and altered cell cycle progression in human lung fibroblast cells." *Cell Biology and Toxicology*, 30(2): 89–100.

Baroli, B., Ennas, M. G., Loffredo, F., Isola, M., Pinna, R., and López-Quintela, M. A. 2007. "Penetration of metallic nanoparticles in human full-thickness skin." *Journal of Investigative Dermatology*, 127(7): 1701–1712.

Barregard, L., Sällsten, G., Gustafson, P., Andersson, L., Johansson, L., Basu, S., and Stigendal, L. 2006. "Experimental exposure to wood-smoke particles in healthy humans: effects on markers of inflammation, coagulation, and lipid peroxidation." *Inhalation Toxicology*, 18(11): 845–853.

Baun, A., Hartmann, N. B., Grieger, K., and Kusk, K. O. 2008. "Ecotoxicity of engineered nanoparticles to aquatic invertebrates: A brief review and recommendations for future toxicity testing." *Ecotoxicology*, 17(5): 387–395.

Behrens, I., Pena, A. I. V., Alonso, M. J., and Kissel, T. 2002. "Comparative uptake studies of bioadhesive and non-bioadhesive nanoparticles in human intestinal cell lines and rats: the effect of mucus on particle adsorption and transport." *Pharmaceutical Research*, 19(8): 1185–1193.

Bhushani, J. Anu and Anandharamakrishnan, C. 2017. "Food-grade nanoemulsions for protection and delivery of nutrients." In Shivendu Ranjan, Nandita Dasgupta and Eric Lichtfouse (Eds), *Nanoscience in Food and Agriculture*, 4, 99–139. Springer cham, Switzerland.

Bhushani, J. Anu, Karthik, P. and Anandharamakrishnan, C. 2016. "Nanoemulsion based delivery system for improved bioaccessibility and Caco-2 cell monolayer permeability of green tea catechins." *Food Hydrocolloids*, 56:372–382.

Bhushani, J. Anu, Kurrey, Nawneet Kumar, and Anandharamakrishnan, C. 2017. "Nanoencapsulation of green tea catechins by electrospraying technique and its effect on controlled release and in-vitro permeability." *Journal of Food Engineering*, 199:82–92.

Borm, P. J. 2002. "Particle toxicology: From coal mining to nanotechnology." *Inhalation Toxicology*, 14(3): 311–324.

Bouwstra, J. A. and Honeywell-Nguyen, P. L. 2002. "Skin structure and mode of action of vesicles." *Advanced Drug Delivery Reviews*, 54: S41–S55.

Buzea, C., Pacheco, I. I., and Robbie, K. 2007. "Nanomaterials and nanoparticles: Sources and toxicity." *Biointerphases*, 2(4): MR17–MR71.

Card, J. W., Zeldin, D. C., Bonner, J. C., and Nestmann, E. R. 2008. "Pulmonary applications and toxicity of engineered nanoparticles." *American Journal of Physiology-Lung Cellular and Molecular Physiology*, 295(3): L400–L411.

Carnovale, C., Bryant, G., Shukla, R., and Bansal, V. 2016. "Size, shape and surface chemistry of nano-gold dictate its cellular interactions, uptake and toxicity." *Progress in Materials Science*, 83: 152–190.

Cedervall, T., Lynch, I., Lindman, S., Berggård, T., Thulin, E., Nilsson, H., Dawson, K.A., and Linse, S. 2007. "Understanding the nanoparticle–protein corona using methods to quantify exchange rates and affinities of proteins for nanoparticles." *Proceedings of the National Academy of Sciences*, 104(7): 2050–2055.

Chang, J., Jallouli, Y., Kroubi, M., Yuan, X.B., Feng, W., Kang, C.S., Pu, P.Y., and Betbeder, D. 2009. "Characterization of endocytosis of transferrin-coated PLGA nanoparticles by the blood–brain barrier." *International Journal of Pharmaceutics*, 379(2): 285–292.

Chaudhry, Q. and Castle, L. 2011. "Food applications of nanotechnologies: An overview of opportunities and challenges for developing countries." *Trends in Food Science & Technology*, 22(11): 595–603.

Chen, X., and Schluesener, H. J. (2008). "Nanosilver: A nanoproduct in medical application." *Toxicology Letters*, 176(1): 1–12.

Chen, Y. S., Hung, Y. C., Liau, I., and Huang, G. S. 2009. "Assessment of the in vivo toxicity of gold nanoparticles." *Nanoscale Research Letters*, 4(8): 858.

Choi, H. S., Kim, Y. J., Song, M., Song, M. K., and Ryu, J. C. 2011. "Genotoxicity of nano-silica in mammalian cell lines." *Toxicology and Environmental Health Sciences*, 3(1): 7–13.

Crosera, M., Bovenzi, M., Maina, G., Adami, G., Zanette, C., Florio, C., and Larese, F. F. 2009. "Nanoparticle dermal absorption and toxicity: A review of the literature." *International Archives of Occupational and Environmental Health*, 82(9): 1043–1055.

Das, M. R., Sarma, R. K., Saikia, R., Kale, V. S., Shelke, M. V., and Sengupta, P. 2011. "Synthesis of silver nanoparticles in an aqueous suspension of graphene oxide sheets and its antimicrobial activity." *Colloids and Surfaces B: Biointerfaces*, 83(1): 16–22.

De Boer, A. G., Van Der Sandt, I. C. J., and Gaillard, P. J. 2003. "The role of drug transporters at the blood-brain barrier." *Annual Review of Pharmacology and Toxicology*, 43(1): 629–656.

Dekkers, S., Bouwmeester, H., Bos, P. M., Peters, R. J., Rietveld, A. G., and Oomen, A. G. 2013. "Knowledge gaps in risk assessment of nanosilica in food: Evaluation of the dissolution and toxicity of different forms of silica." *Nanotoxicology*, 7(4): 367–377.

Dockery, D.W., Pope, C.A., Xu, X., Spengler, J.D., Ware, J.H., Fay, M.E., Ferris, B.G., Jr, and Speizer, F.E. 1993. "An association between air pollution and mortality in six US cities." *New England Journal of Medicine*, 329(24): 1753–1759.

Donaldson, K. and Poland, C. A. 2013. "Nanotoxicity: Challenging the myth of nano-specific toxicity." *Current Opinion in Biotechnology*, 24(4): 724–734.

Duncan, T. V. 2011. "Applications of nanotechnology in food packaging and food safety: Barrier materials, antimicrobials and sensors." *Journal of Colloid and Interface Science*, 363(1): 1–24.

Dutta, D., Sundaram, S.K., Teeguarden, J.G., Riley, B.J., Fifield, L.S., Jacobs, J.M., Addleman, S.R., Kaysen, G.A., Moudgil, B.M. and Weber, T.J., 2007. "Adsorbed proteins influence the biological activity and molecular targeting of nanomaterials." *Toxicological Sciences*, 100(1): 303–315.

Elsaesser, A. and Howard, C. V. 2012. "Toxicology of nanoparticles." *Advanced Drug Delivery Reviews*, 64(2): 129–137.

Everts, M., Saini, V., Leddon, J.L., Kok, R.J., Stoff-Khalili, M., Preuss, M.A., Millican, C.L., Perkins, G., Brown, J.M., Bagaria, H., and Nikles, D.E. 2006. "Covalently linked Au nanoparticles to a viral vector: Potential for combined photothermal and gene cancer therapy." *Nano Letters*, 6(4): 587–591.

Fadeel, B. and Garcia-Bennett, A. E. 2010. "Better safe than sorry: Understanding the toxicological properties of inorganic nanoparticles manufactured for biomedical applications." *Advanced Drug Delivery Reviews*, 62(3): 362–374.

Fang, J., Shan, X. Q., Wen, B., Lin, J. M., and Owens, G. 2009. "Stability of titania nanoparticles in soil suspensions and transport in saturated homogeneous soil columns." *Environmental Pollution*, 157(4): 1101–1109.

Farré, M., Gajda-Schrantz, K., Kantiani, L., and Barceló, D. 2009. "Ecotoxicity and analysis of nanomaterials in the aquatic environment." *Analytical and Bioanalytical Chemistry*, 393(1): 81–95.

Federici, G., Shaw, B. J., and Handy, R. D. 2007. "Toxicity of titanium dioxide nanoparticles to rainbow trout (Oncorhynchusmykiss): Gill injury, oxidative stress, and other physiological effects." *Aquatic Toxicology*, 84(4): 415–430.

Forest, V., Leclerc, L, Hochepied, J. F., Trouvé, A., Sarry, G., and Pourchez, J. 2017. "Impact of cerium oxide nanoparticles shape on their in vitro cellular toxicity." *Toxicology in Vitro*, 38: 136–141.

Geiser, M., Rothen-Rutishauser, B., Kapp, N., Schürch, S., Kreyling, W., Schulz, H., Semmler, M., Hof, V.I., Heyder, J. and Gehr, P., 2005. "Ultrafine particles cross cellular membranes by nonphagocytic mechanisms in lungs and in cultured cells." *Environmental Health Perspectives*, 113(11):1555–1560.

Goodman, C. M., McCusker, C. D., Yilmaz, T., and Rotello, V. M. 2004. "Toxicity of gold nanoparticles functionalized with cationic and anionic side chains." *Bioconjugate Chemistry*, 15(4): 897–900.

Hartung, T. 2010. "Food for thought on alternative methods for nanoparticle safety testing." *Altex*, 27(2): 87–95.

Hoet, P. H., Brüske-Hohlfeld, I., and Salata, O. V. 2004. "Nanoparticles–known and unknown health risks." *Journal of Nanobiotechnology*, 2(1), 12.

Hundre, Swetank Y., Karthik P., and Anandharamakrishnan, C. 2015. "Effect of whey protein isolate and β-cyclodextrin wall systems on stability of microencapsulated vanillin by spray-freeze drying method." *Food Chemistry*, 174:16–24.

Hussain, S. M., Hess, K. L., Gearhart, J. M., Geiss, K. T., and Schlager, J. J. 2005. "In vitro toxicity of nanoparticles in BRL 3A rat liver cells." *Toxicology in Vitro*, 19(7): 975–983.

Ispas, C., Andreescu, D., Patel, A., Goia, D. V., Andreescu, S., and Wallace, K. N. 2009. "Toxicity and developmental defects of different sizes and shape nickel nanoparticles in zebrafish." *Environmental science & technology*, 43(16): 6349–6356.

Jahnen-Dechent, W. and Simon, U. 2008. "Function follows form: Shape complementarity and nanoparticle toxicity." Nanomedicine, 3(5) 601–603.

Jaques, P. A. and Kim, C. S. 2000. "Measurement of total lung deposition of inhaled ultrafine particles in healthy men and women." *Inhalation Toxicology*, 12(8): 715–731.

Jeng, H. A. and Swanson, J. 2006. "Toxicity of metal oxide nanoparticles in mammalian cells." *Journal of Environmental Science and Health Part A*, 41(12): 2699–2711.

Jia, G., Wang, H., Yan, L., Wang, X., Pei, R., Yan, T., Zhao, Y., and Guo, X. 2005. "Cytotoxicity of carbon nanomaterials: Single-wall nanotube, multi-wall nanotube, and fullerene." *Environmental Science and Technology*, 39(5): 1378–1383.

Jia, H. Y., Liu, Y., Zhang, X. J., Han, L., Du, L. B., Tian, Q., and Xu, Y. C. 2008. "Potential oxidative stress of gold nanoparticles by induced-NO releasing in serum." *Journal of the American Chemical Society*, 131(1): 40–41.

Jones, C. F. and Grainger, D. W. 2009. "In vitro assessments of nanomaterial toxicity." *Advanced Drug Delivery Reviews*, 61(6): 438–456.

Joo, S. H., Al-Abed, S. R., and Luxton, T. 2009. "Influence of carboxymethyl cellulose for the transport of titanium dioxide nanoparticles in clean silica and mineral-coated sands." *Environmental Science & Technology*, 43(13): 4954–4959.

Karthik, P., and Anandharamakrishnan, C. 2016a. "Enhancing omega-3 fatty acids nanoemulsion stability and in-vitro digestibility through emulsifiers." *Journal of Food Engineering*, 187:92–105.

Karthik, P. and Anandharamakrishnan, C. 2016b. "Fabrication of a nutrient delivery system of docosahexaenoic acid nanoemulsions via high energy techniques." *RSC Advances*, 6 (5):3501–3513.

Karthik, P., Ezhilarasi, P.N., and Anandharamakrishnan, C. 2017. "Challenges associated in stability of food grade nanoemulsions." *Critical Reviews in Food Science and Nutrition*, 57 (7):1435–1450.

Koziara, J. M., Lockman, P. R., Allen, D. D., and Mumper, R. J. 2004. "Paclitaxel nanoparticles for the potential treatment of brain tumors." *Journal of controlled release*, 99(2): 259–269.

Kreilgaard, M. 2002. "Influence of microemulsions on cutaneous drug delivery." *Advanced Drug Delivery Reviews*, 54: S77–S98.

Kreuter, J. 2001. "Nanoparticulate systems for brain delivery of drugs." *Advanced Drug Delivery Reviews*, 47(1): 65–81.

Kreyling, W.G., Semmler, M., Erbe, F., Mayer, P., Takenaka, S., Schulz, H., Oberdörster, G., and Ziesenis, A. 2002. "Translocation of ultrafine insoluble iridium particles from lung epithelium to extrapulmonary organs is size dependent but very low." *Journal of Toxicology and Environmental Health Part A*, 65(20): 1513–1530.

Kumar, M.N.V., Sameti, M., Mohapatra, S.S., Kong, X., Lockey, R.F., Bakowsky, U., Lindenblatt, G., Schmidt, C.H., and Lehr, C.M. 2004. "Cationic silica nanoparticles as gene carriers: Synthesis, characterization and transfection efficiency *in vitro* and *in vivo*." *Journal of Nanoscience and Nanotechnology*, 4(7): 876–881.

Lakshmanan, A., Latha, P., and Subramanian, K. S. 2012. "Nanotoxicity—an overview." *International Journal of Advanced Life Sciences*, 5(1): 1–11.

Lam, C. W., James, J. T., McCluskey, R., and Hunter, R. L. 2004. "Pulmonary toxicity of single-wall carbon nanotubes in mice 7 and 90 days after intratracheal instillation." *Toxicological Sciences*, 77(1): 126–134.

Larese Filon, F., Crosera, M., Adami, G., Bovenzi, M., Rossi, F., and Maina, G. 2011. "Human skin penetration of gold nanoparticles through intact and damaged skin." *Nanotoxicology*, 5(4): 493–501.

Lee, K. J., Nallathamby, P. D., Browning, L. M., Osgood, C. J., and Xu, X. H. N. 2007. "In vivo imaging of transport and biocompatibility of single silver nanoparticles in early development of zebrafish embryos." *ACS Nano*, 1(2): 133–143.

Lee, S. W., Kim, S. M., & Choi, J. 2009. "Genotoxicity and ecotoxicity assays using the freshwater crustacean Daphnia magna and the larva of the aquatic midge Chironomusriparius to screen the ecological risks of nanoparticle exposure." *Environmental Toxicology and Pharmacology*, 28(1): 86–91.

Lei, Z., Mingyu, S., Xiao, W., Chao, L., Chunxiang, Q., Liang, C., Hao, H., Xiaoqing, L. and Fashui, H., 2008. "Antioxidant stress is promoted by nano-anatase in spinach chloroplasts under UV-B radiation." *Biological Trace Element Research*, 121(1): 69–79.

Li, J. J., Zou, L. I., Hartono, D., Ong, C. N., Bay, B. H., and Lanry Yung, L. Y. 2008. "Gold nanoparticles induce oxidative damage in lung fibroblasts in vitro." *Advanced Materials*, 20(1): 138–142.

Lin, D. and Xing, B. 2008. "Root uptake and phytotoxicity of ZnO nanoparticles." *Environmental Science and Technology*, 42(15): 5580–5585.

Lockman, P. R., Oyewumi, M. O., Koziara, J. M., Roder, K. E., Mumper, R. J., and Allen, D. D. 2003. "Brain uptake of thiamine-coated nanoparticles." *Journal of Controlled Release*, 93(3): 271–282.

Lomer, M. C., Thompson, R. P., and Powell, J. J. 2002. "Fine and ultrafine particles of the diet: Influence on the mucosal immune response and association with Crohn's disease." *Proceedings of the Nutrition Society*, 61(01): 123–130.

Mocan, T., Clichici, S., Agoşton-Coldea, L., Mocan, L., Şimon, Ş., Ilie, I., Biriş, A., and Mureşan, A. 2010. "Implications of oxidative stress mechanisms in toxicity of nanoparticles (review)." *Acta Physiologica Hungarica*, 97(3): 247–255.

Napierska, D., Thomassen, L. C., Lison, D., Martens, J. A., and Hoet, P. H. 2010. "The nanosilica hazard: Another variable entity." *Particle and Fibre Toxicology:* 7(1), 1.

Neas, L. M. 2000. "Fine particulate matter and cardiovascular disease." *Fuel Processing Technology:* 65, 55–67.

Nel, A. E., Diaz-Sanchez, D., and Li, N. 2001. "The role of particulate pollutants in pulmonary inflammation and asthma: Evidence for the involvement of organic chemicals and oxidative stress." *Current Opinion in Pulmonary Medicine*, 7(1): 20–26.

Oberdörster, G., Sharp, Z., Atudorei, V., Elder, A., Gelein, R., Kreyling, W., and Cox, C. 2004. "Translocation of inhaled ultrafine particles to the brain." *Inhalation Toxicology*, 16(6–7): 437–445.

Oberdörster, G., Elder, A., and Rinderknecht, A. 2009. "Nanoparticles and the brain: Cause for concern." *Journal of Nanoscience and Nanotechnology*, 9(8): 4996–5007.

Pan, Y., Neuss, S., Leifert, A., Fischler, M., Wen, F., Simon, U., Schmid, G., Brandau, W. and Jahnen-Dechent, W., 2007. "Size-dependent cytotoxicity of gold nanoparticles." *Small*, 3(11): 1941–1949.

Pardridge, W. M. 2007. "Blood–brain barrier delivery." *Drug Discovery Today*, 12(1): 54–61.

Parfenov, A. S., Salnikov, V., Lederer, W. J., and Lukyanenko, V. 2006. "Aqueous diffusion pathways as a part of the ventricular cell ultrastructure." *Biophysical Journal*, 90(3): 1107–1119.

Parthasarathi, S. and Anandharamakrishnan, C. 2016. "Enhancement of oral bioavailability of vitamin E by spray-freeze drying of whey protein microcapsules." *Food and Bioproducts Processing*, 100: 469–476.

Parthasarathi, S., Muthukumar, S.P., and Anandharamakrishnan, C. 2016. "The influence of droplet size on the stability, in vivo digestion, and oral bioavailability of vitamin E emulsions." *Food and Function*, 7 (5):2294–2302.

Peetla, C. and Labhasetwar, V. 2008. "Biophysical characterization of nanoparticle–endothelial model cell membrane interactions." *Molecular Pharmaceutics*, 5(3): 418–429.

Peralta-Videa, J. R., Zhao, L., Lopez-Moreno, M. L., de la Rosa, G., Hong, J., and Gardea-Torresdey, J. L. 2011. "Nanomaterials and the environment: A review for the biennium 2008–2010." *Journal of Hazardous Materials*, 186(1): 1–15.

Peters, A., Veronesi, B., Calderón-Garcidueñas, L., Gehr, P., Chen, L.C., Geiser, M., Reed, W., Rothen-Rutishauser, B., Schürch, S., and Schulz, H. 2006. "Translocation and potential neurological effects of fine and ultrafine particles a critical update." *Particle and Fibre Toxicology*, 3(1): 13.

Pipan-Tkalec, Z., Drobne, D., Jemec, A., Romih, T., Zidar, P. and Bele, M., 2010. "Zinc bioaccumulation in a terrestrial invertebrate fed a diet treated with particulate ZnO or $ZnCl_2$ solution." *Toxicology*, 269(2–3), 198–203.

Ravishankar Rai, V. and Jamuna Bai, A. 2011. "Nanoparticles and their potential application as antimicrobials." In A. Méndez-Vilas (Ed.) *Science Against Microbial Pathogens, Communicating Current Research and Technological Advances*. Formatex, Badajoz: 197–209, Mysore.

Ray, P. C., Yu, H., and Fu, P. P. 2009. "Toxicity and environmental risks of nanomaterials: Challenges and future needs." *Journal of Environmental Science and Health Part C*, 27(1): 1–35.

Ruenraroengsak, P., Novak, P., Berhanu, D., Thorley, A.J., Valsami-Jones, E., Gorelik, J., Korchev, Y.E., and Tetley, T.D. 2012. "Respiratory epithelial cytotoxicity and membrane damage (holes) caused by amine-modified nanoparticles." *Nanotoxicology*, 6(1): 94–108.

Sartorelli, P., Ahlers, H.W., Alanko, K., Chen-Peng, C., Cherrie, J.W., Drexler, H., Kezic, S., Johanson, G., Filon, F.L., Maina, G., and Montomoli, L. 2007. "How to improve skin notation. Position paper from a workshop." *Regulatory Toxicology and Pharmacology*, 49(3): 301–307.

Sayes, C. M., Reed, K. L., and Warheit, D. B. 2007. "Assessing toxicity of fine and nanoparticles: Comparing in vitro measurements to in vivo pulmonary toxicity profiles." *Toxicological Sciences*, 97(1): 163–180.

Scheuplein, R.J., 1967. "Mechanism of percutaneous absorption: II. Transient diffusion and the relative importance of various routes of skin penetration." *Journal of Investigative Dermatology*, 48(1), 79–88.

Schmid, G. 2008. "The relevance of shape and size of Au 55 clusters." *Chemical Society Reviews*, 37(9): 1909–1930.

Schwartz, J. 1994. "Air pollution and daily mortality: A review and meta-analysis." *Environmental Research*, 64(1): 36–52.

Sharma, H. S. and Sharma, A. 2007. "Nanoparticles aggravate heat stress induced cognitive deficits, blood–brain barrier disruption, edema formation and brain pathology." *Progress in Brain Research*, 162: 245–273.

Shim, K. H., Hulme, J., Maeng, E. H., Kim, M. K., and An, S. S. A. 2014. "Analysis of zinc oxide nanoparticles binding proteins in rat blood and brain homogenate." *International Journal of Nanomedicine*, 9(Suppl 2): 217.

Shubayev, V. I., Pisanic, T. R., and Jin, S. 2009. "Magnetic nanoparticles for theragnostics." *Advanced Drug Delivery Reviews*, 61(6): 467–477.

Shvedova, A., Castranova, V., Kisin, E., Schwegler-Berry, D., Murray, A., Gandelsman, V., Maynard, A. and Baron, P. 2003. "Exposure to carbon nanotube material: Assessment of nanotube cytotoxicity using human keratinocyte cells." *Journal of Toxicology and Environmental Health Part A*, 66(20): 1909–1926.

Slouf, M., Hruby, M., Bakaeva, Z., Vlkova, H., Nebesarova, J., Philimonenko, A. A., and Hozak, P. 2012. "Preparation of stable Pd nanocubes and their use in biological labeling." *Colloids and Surfaces B: Biointerfaces*, 100, 205–208.

Smith, C. J., Shaw, B. J., and Handy, R. D. 2007. "Toxicity of single walled carbon nanotubes to rainbow trout (*Oncorhynchusmykiss*): Respiratory toxicity, organ pathologies, and other physiological effects." *Aquatic Toxicology*, 82(2): 94–109.

Srinivas, A., Rao, P. J., Selvam, G., Murthy, P. B., and Reddy, P. N. 2011. "Acute inhalation toxicity of cerium oxide nanoparticles in rats." *Toxicology Letters*, 205(2): 105–115.

Stahlhofen, W. G. A. C., Rudolf, G., and James, A. C. 1989. "Intercomparison of experimental regional aerosol deposition data." *Journal of Aerosol Medicine*, 2(3): 285–308.

Stern, S. T. and McNeil, S. E. 2008. "Nanotechnology safety concerns revisited." *Toxicological Sciences*, 101(1): 4–21.

Sung, J.H., Ji, J.H., Yoon, J.U., Kim, D.S., Song, M.Y., Jeong, J., Han, B.S., Han, J.H., Chung, Y.H., Kim, J., and Kim, T.S. 2008. "Lung function changes in Sprague-Dawley rats after prolonged inhalation exposure to silver nanoparticles." *Inhalation Toxicology*, 20(6): 567–574.

Takenaka, S., Karg, E., Roth, C., Schulz, H., Ziesenis, A., Heinzmann, U., Schramel, P., and Heyder, J. 2001. "Pulmonary and systemic distribution of inhaled ultrafine silver particles in rats." *Environmental Health Perspectives*, 109(Suppl 4): 547.

Tang, J., Xiong, L., Wang, S., Wang, J., Liu, L., Li, J., Wan, Z., and Xi, T. 2008. "Influence of silver nanoparticles on neurons and blood-brain barrier via subcutaneous injection in rats." *Applied Surface Science*, 255(2): 502–504.

Tinkle, S. S., Antonini, J. M., Rich, B. A., Roberts, J. R., Salmen, R., DePree, K., and Adkins, E. J. 2003. "Skin as a route of exposure and sensitization in chronic beryllium disease." *Environmental Health Perspectives*, 111(9): 1202–1208.

van Rooy, I., Cakir-Tascioglu, S., Hennink, W. E., Storm, G., Schiffelers, R. M., and Mastrobattista, E. 2011. "In vivo methods to study uptake of nanoparticles into the brain." *Pharmaceutical Research*, 28(3): 456–471.

Vasir, J. K. and Labhasetwar, V. 2008. "Quantification of the force of nanoparticle-cell membrane interactions and its influence on intracellular trafficking of nanoparticles." *Biomaterials*, 29(31): 4244–4252.

Williams, Y., Sukhanova, A., Nowostawska, M., Davies, A.M., Mitchell, S., Oleinikov, V., Gun'ko, Y., Nabiev, I., Kelleher, D., and Volkov, Y. 2009. "Probing cell-type-specific intracellular nanoscale barriers using size-tuned quantum dots." *Small*, 5(22): 2581–2588.

Wu, J., Liu, W., Xue, C., Zhou, S., Lan, F., Bi, L., Xu, H., Yang, X., and Zeng, F.D. 2009. "Toxicity and penetration of TiO_2 nanoparticles in hairless mice and porcine skin after subchronic dermal exposure." *Toxicology Letters*, 191(1): 1–8.

Xia, T., Kovochich, M., Brant, J., Hotze, M., Sempf, J., Oberley, T., Sioutas, C., Yeh, J.I., Wiesner, M.R., and Nel, A.E., 2006. "Comparison of the abilities of ambient and manufactured nanoparticles to induce cellular toxicity according to an oxidative stress paradigm." *Nano Letters*, 6(8):1794–1807.

Xia, T., Li, N., and Nel, A. E. 2009. "Potential health impact of nanoparticles." *Annual Review of Public Health*, 30: 130–150.

Yang, R.S., Chang, L.W., Wu, J.P., Tsai, M.H., Wang, H.J., Kuo, Y.C., Yeh, T.K., Yang, C.S., and Lin, P. 2007a. "Persistent tissue kinetics and redistribution of nanoparticles, quantum dot 705, in mice: ICP-MS quantitative assessment." *Environmental Health Perspectives*: 115(9) 1339–1343.

Yang, S.T., Guo, W., Lin, Y., Deng, X.Y., Wang, H.F., Sun, H.F., Liu, Y.F., Wang, X., Wang, W., Chen, M., and Huang, Y.P. 2007b. "Biodistribution of pristine single-walled carbon nanotubes in vivo." *The Journal of Physical Chemistry C*, 111(48): 17761–17764.

Yang K, Wan J, Zhang S, Zhang Y, Lee ST, Liu Z. 2011. In vivo pharmacokinetics, long term biodistribution, and toxicity of PEGylated graphene in mice. ACS Nano 5(1):516–522.

Ye, Y., Liu, J., Chen, M., Sun, L., and Lan, M. 2010a. "In vitro toxicity of silica nanoparticles in myocardial cells." *Environmental Toxicology and Pharmacology*, 29(2): 131–137.

Ye, Y., Liu, J., Xu, J., Sun, L., Chen, M., and Lan, M. 2010b. "Nano-SiO_2 induces apoptosis via activation of p53 and Bax mediated by oxidative stress in human hepatic cell line." *Toxicology in Vitro*, 24(3): 751–758.

Zhao, Y., Wang, B., Feng, W., and Bai, C. 2012. "Nanotoxicology: Toxicological and biological activities of nanomaterials." *Nanoscience and Nanotechnologies*. Encyclopedia of Life Support Systems (EOLSS) Publishers, Oxford, UK: 1–53.

Zheng, X., Zhang, X., Wang, X., Wang, S., and Wu, S. 2005. "Preparation and characterization of CuO/CeO_2 catalysts and their applications in low-temperature CO oxidation." *Applied Catalysis A: General*, 295(2): 142–149.

5 Ethical and Regulatory Issues in Applications of Nanotechnology in Food

Maria Leena, J.A. Moses, and C. Anandharamakrishnan

CONTENTS

5.1 INTRODUCTION

Nanotechnology has been defined as the control or restructuring of matter at the atomic and molecular levels in the size range of about 1–100 nanometers (nm) ("ISO/TS 27687:2008 – Nanotechnologies – Terminology and Definitions for Nano-Objects – Nanoparticle, Nanofibre and Nanoplate" 2016; Lövestam et al., 2010). Nanotechnology is remaking the world at an alarmingly rapid pace. Nanomaterials are an integral part of the natural food system, as the building blocks of many food components (i.e. protein, milk, lipid structures, etc.) are nanometer-sized components which determine their properties. In addition, recent technological developments lead the way for manufactured nanoparticles to be added to foods and thus influence the food processing sector. These could be finely divided forms of existing ingredients, or completely novel chemical structures (examples include homogenized milk, nanoemulsion formulations of food ingredients such as co-enzyme Q_{10} in food supplements, etc.) (FSA-Food standards agency, 2017; FDA, 2017d).

Nanotechnology is having an impact on all stages of food science from farm to fork, such as the way food is grown, enhancing food taste, improving health benefits of foods, food processing, food packaging, food safety, etc. For example, one of the nanotechnology applications in the farm field is a network of nanosensors and dispensers which sense when plants need nutrients or water and release nutrients, fertilizer, and water as needed, thus optimizing the growth of each plant (Fraceto et al., 2016). Nanotechnology applications in food processing include, (1) enhanced packaging material (clay nanocomposites) which presents gas barrier properties (Tas et al., 2017; Anandharamakrishnan and Usha Kiran, 2014), moisture barrier (silicate nanoparticles), and UV protection, in order to prevent food spoilage and drying out; (2) antibacterial food storage containers to minimize health risks from harmful bacteria (silver nanoparticles in milk containers (Griffiths, 2015), zinc oxide nanoparticles (Espitia et al., 2012; Sirelkhatim et al., 2015)); (3) sensors in food packaging to detect food spoilage (Fuertes et al., 2016); (4) interactive food packaging materials (color-changing packaging); (5) interactive foods – nanocapsules that contain flavor and color enhancers when embedded in food provide an option to select the desired flavor and color, and improve the absorption of nutrients (Anandharamakrishnan, 2014a,b; Anandharamakrishnan and Ishwarya, 2015a–c; Bhushani and Anandharamakrishnan, 2017; Bhushani et al., 162017; Ezhilarasi et al., 2013).

The potential applications of nanomaterials in food can be categorized as (Handford et al., 2014 and 2015; Dasgupta et al., 2015):

1. Processed or formulated food ingredients in nanoscale range for improved processing efficiency and health benefits;
2. Delivery system: nano-encapsulated food additives for the delivery of bioactive molecules;
3. Smart packaging materials: nanomaterials added to food packaging materials to create innovative food contact materials;
4. Food quality and safety: nanomaterials for the removal of undesirable components from food using a nano filtration system;
5. Nanomaterials as pesticides, agrochemicals, micronutrients, and veterinary medicines for improved food production.

As a result of the changes made to food additives and food processing via nanotechnology, we have novel foods called nanofoods. Though nanofoods are different from genetically modified foods, their production raises questions about health and environmental risks. Will these "nano"-labeled food products really be safe for humans and the environment?

The "unnaturalness" of nanofoods raises the main ethical objection about them. To address ethical issues, we need to analyze the benefits and unintended risks to human health and environment which can occur in current and future scenarios. These issues can be analyzed by taking the points below into consideration:

1. Do we actually need an innovative food if our food is all right as it is?
   - People would not need nutrients and fortifications if we had good diets.
   - Benefit of exceptional efficacy of nanotechnology – nanoencapsulation of bioactives like omega 3 fatty acids, vitamins, and polyphenols for improving their bioavailability, provided safety and metabolic change associated with nanoencapsulation is properly addressed.
   - In case one is not clear about the risk of a nanoparticle which could also have beneficial health effects, it would be better to analyze it as a medicine and not as a food.
2. Who is driving the development of nanofoods – do we agree with their motives, goals, and reasons?
   - Fear of opposition and competitive market leads companies to maintain secrecy which results in lack of reliable public information about nanotechnology and food.

Environmental, health, and safety (EHS) issues that can arise with nanomaterials necessitate significant research to understand the risks associated with them. The responsible development of nanotechnology should be the main objective for every nanotechnology stakeholder.

The essential social objectives for every nanotechnology stakeholder towards responsible nanotechnology development are as follows (Khan, 2012):

1. Understanding local and global forces and issues that affect people and societies
2. Alerting societies of technological risks and failures

3. Guiding societies to utilize the technology appropriately
4. Taking ethical decisions and leadership to solve technological problems to save society

Since nanotechnology is an emerging field, it poses challenges and uncertainty about what the risks might be and about how the government might regulate it in the future. As there is an absence of straightforward regulations for the use of nanotechnology, most of the countries across the globe accepted the risk assessment method to deal with uncertainty risks in nanotechnology. Risk assessment is defined as "a process intended to calculate or estimate the risk to a given target organism, system, or [sub]population, including the identification of attendant uncertainties, following exposure to a particular agent, taking into account the inherent characteristics of the agent of concern as well as the characteristics of the specific target system" (World Health Organization [WHO] 2004 and 2009). Most of the countries have given guidance on risk assessment as a preliminary regulatory approach to the use of nanotechnology. Since nanotechnology is a multidisciplinary field, there is a major question on who should create and enforce policies regarding nanotechnology. Who can enforce the law? Who decides what is safe and what is not? This requires a collaborative effort from nanotechnology stakeholders to develop standards for measurement and nomenclature, in order to help assess and address these risks.

In this context, this chapter presents the social and ethical issues with nanotechnology and international regulatory approaches towards the responsible development of nanotechnology in food.

5.2 SOCIAL AND ETHICAL ISSUES ASSOCIATED WITH NANOTECHNOLOGY

Advances in nanotechnology also present numerous health and environmental challenges. Due to their minuscule size, nanoparticles are airborne and in the range of respirable-sized particles. So they have the potential to enter the body through the respiratory system or penetrate the skin when they come into contact with it. As a result, nanoparticles may get deposited in the respiratory tract or lungs or enter the bloodstream, and translocate to other organs.

Since most nanoparticles first migrate to the lungs, the chief human toxicity concern surrounding nanotechnology is lung damage. For example, one of the most widely used nanoparticle in clinical applications is polyamidoamine dendrimers (PAMAMs); which induce acute lung injury *in vivo*. Notably, PAMAMs trigger lung damage by a type of programmed cell death known as autophagic through the Akt-TSC2-mTOR signaling pathway (Li et al., 2009). PAMAMs induce overactivity of autophagy (process intended to degrade damage materials in a cell during cell growth and renewal process) and cause cell death (Li et al., 2009). Various experimental studies revealed that equivalent mass doses of insoluble incidental nanoparticles are more potent than large particles of similar composition in causing pulmonary inflammation and lung tumors (Ken Donaldson et al., 2005; Renwick et al., 2004; Arick et al., 2015). Nanoparticles that come into contact with skin penetrate intravenously and escape the human body's opsonification processes of foreign bodies, which may cause unwanted health issues (Dobrovolskaia et al., 2008; Li and Huang, 2008). Though nanomaterials are derived from conventional chemicals, changes in the size of particles, chemical composition, and crystal structure influence their oxidant generation properties and thus their cytotoxicity.

Apart from exposure issues, the nonspecific activity of engineered nanomaterials is threatening to human health. For example, silver nanoparticles (Ag Nps, one of the most widely studied antibacterial nanoparticles) cause apoptosis of cells by the nonspecific effect on cell membranes, mitochondria, and genetic material. Hence they do not discriminate between bad bacteria, good bacteria, and non-bacteria cells. The toxic effect of Ag Nps on mammalian cells derived from skin, liver, lung, brain, vascular system, reproductive organs, etc., is confirmed through various *in vitro* studies. Even non-cytotoxic doses of Ag Nps induce genes associated with cell cycle progression, DNA damage, and apoptosis in human cells (Ahamed et al., 2010).

The potential impact of nanomaterials exposure to the environment also persists. Studies from the University of California show that exposure to TiO_2 nanomaterials impacts freshwater foodwebs (like algae, herbivores, and fish) (Kulacki and Cardinale, 2012; Kulacki et al., 2012).

While the above points emphasize the concerns about human health and environmental issues, safety issues also persist with nanotechnology. Compared to the coarser combustible material, nanoscale combustible materials present a higher risk of explosion at similar mass concentration due to their high surface area. For instance, nanoscale Al/MoO_3 thermites ignite 300 times faster than corresponding micrometer-scale material (Granier and Pantoya, 2004). Some nanomaterials may initiate catalytic reactions depending on their composition and structure that would not otherwise be anticipated based on their chemical composition (Bouillard et al., 2010).

To handle these safety and health issues of nanotechnology, there is a strong need for the development of risk assessment methods and protocols by the regulatory bodies across the world. However, there are some societal challenges that exist in dealing with nanotechnology, which include (Khan, 2012):

1. Challenge of control over technological development
2. Challenge of prediction of boundaries of technological development
3. Challenge of prediction of future technological possibilities and reliability
4. Challenge of formulating common regulatory policies

Role of ethics in responsible nanotechnology development (Ronald Sandler, 2009):

1. Case-by-case ethical assessment: Nanotechnology is a multidisciplinary field, which has diversified applications in multiple fields such as energy, agriculture, computing, medicine, weapons, textiles, building materials, and environmental remediation. Therefore, the ethical profiles of nanotechnology applications are various. For example, if we compare the development of carbon nanotube-enabled memory chips (sited in a suburb) project with a synthetic biology research project as part of a biological defense program (sited in an urban area), both will have different ethical profiles in terms of their objectives, risks, benefits and beneficiaries, control, oversight, regulation, and degree to which they involve a controversial moral practice. While the former does not have major ethical issues, the latter raises ethical problems with sanctity-of-life issues, biological weapon issues, public health and safety issues, public funding issues and transparency/oversight issues. Thus a *case-by-case ethical assessment is important.* Application of specific social and ethical evaluation will contribute more to specific decision making in policy and regulatory designs and help avoid public or regulatory reactions.
2. Ethical capacity: Ethical capacity such as (a) ethical frameworks, (b) professional codes of conduct, (c) well-developed case studies and historical precedents, and (d) individuals and organizations with expertise and experience in identifying, analyzing and addressing relevant ethical issues are essential to make ethical decisions. Social and governmental capacity in terms of access to information, resources, legal authority, expertise, educational institutions, media and communications, forums for public discourse, professional organizations, public interest/advocacy organizations, etc. are necessary for the responsible development of emerging nanotechnologies.
3. Ethics identify the limits: Ethics would also help in identifying goals and limits and the way we pursue them.

Considering the extent to which nanotechnology will contribute to human flourishing in just and sustainable ways, it may also be susceptible to many of the following issues. The possible social and ethical issues that arise from nanotechnology are characterized as, (a) contested moral issues which arise from morally controversial practices or activities of nanotechnology (e.g. any activities that a substantial number of citizens believes should be prohibited, such as genetic modification of living organisms, use of embryonic stem cells, biological and chemical weapons, genetic patterning and construction of artificial organisms, etc.); (b) technoculture issues which arise from problematic aspects of the role of technology within social systems and structures (such as overconfidence and reliance on technology, interference of technology with human–nature relationship, overestimation of our capacity to predict and control technologies, etc.); (c) social context issues which arise from the interaction of nanotechnologies with problematic features of social or institutional contexts (e.g. inadequate information, privacy and consumer safety protection, unequal access to education, healthcare and technology, under-representation of women and minority groups in engineering and academia, etc.); (d) form of life issues which arise as a result of impact of nanotechnology on social standards, practices, and institutions (e.g. nanotechnology impact on human health, family structures, social networks and life trajectories, etc.); (e) transformational issues arise from transformation of human life style in terms of artificial intelligence, alteration of our relationship with the natural environment, molecular manufacturing, etc., which arise from nanotechnology in combination with other technological advances such as information technology, computer science, cognitive science, biotechnology, and robotics (Ronald Sandler, 2009). These types are not mutually exclusive: a particular issue might fall within more than one type and some aspects cut across all the types.

5.3 INTERNATIONAL REGULATORY APPROACHES

The growing need for stronger legislation on nanotechnology has convinced every country to make regulatory approaches for the use of nanotechnology. These regulatory approaches can be classified into two categories, namely, horizontal legislation and vertical legislation. The existing legislation on new product development which is not specifically intended to regulate nanotechnology yet happens to include the attributes of nanotechnology is classified as horizontal legislation. Vertical legislation is specifically made to regulate the use of nanotechnology and the areas that are likely to use nanotechnology.

The U.S., the leading R&D player in nanotechnology, has different regulatory bodies for the use of nanotechnology. In the U.S., the Food and Drug Administration (FDA) has the major responsibility to review any new food and pharmaceutical products. FDA has given guidelines and stands on when to consider the regulatory implications on nanotechnology applications in FDA-regulated products (FDA, 2014a). It encourages industries to consult with FDA early in the product development process to address questions related to safety, effectiveness, or public health impact and regulatory status of products that use nanotechnology. Significant New Use Rule (SNUR) has been implemented by the Environmental Protection Agency (EPA) to ensure nanoscale materials undergo appropriate regulatory reviews. According to this, Significant New Use Notice (SNUN) has to be submitted to EPA based on chemical substances listed on the Toxic Substances Control Act inventory. Any persons/industries intending to manufacture/process or import new nanoscale materials must submit SNUN to EPA at least 90 days prior to commencing activity. EPA will evaluate the intended uses of nanoscale materials based on information provided by SNUR and will take necessary action to prohibit or limit activities that may present an unreasonable risk to human health or the environment.

In Canada, every new substance being manufactured or imported into the country must undergo a risk assessment for its potential effects on the environment and human health. The Canadian Environment Protection Agency (CEPA) has issued guidelines to help determine if a nanomaterial is considered a new substance or not (Canadian Environmental Protection Act, 2007). According to these guidelines, if the substance is already listed in the Domestic Substances List (DSL) and its nanoscale does not have any unique structure or molecular arrangements then it is considered as existing material and it is not subjected to regulations. For example, titanium dioxide is listed on the DSL and its nanoscale form does not have unique structures or molecular arrangements, so it is not subjected to the nano regulations. If a substance listed on the DSL has unique structures or molecular arrangements in its nanoscale then it is treated as "new" and subjected to notification under regulations. Any nanomaterials manufactured or imported into Canada that are not listed on the DSL are considered as new materials. For example, the nanomaterial fullerene (CAS No. 99685-96-8) is not listed on the DSL and is considered a "new" substance under the regulations.

In Europe, Classification, Labeling and Packaging (CLP) regulation state that if the form or physical state of a substance is changed, an evaluation must be done to determine if there is a need for change in hazard classification. This could result in a different classification and labeling requirements for bulk forms and nano forms of the same chemical substances. CLP regulation indicates that the classification and labeling of nanomaterials will be done on a case-by-case basis.

In Hong Kong, the Centre for Food Safety refers to the World Health Organization's (WHO) requirements for risk assessment on nanoscale materials for assessing nanoparticles before they can be used in food (Shuk man CHOW Centre for food safety Hong Kong, 2011). While the Public Health and Municipal Services Ordinance of Hong Kong dictates that all foods sold must be fit for human consumption, there is no specific legislation for monitoring nanotechnology applications. Additionally, there is a lack of comprehensive and compulsory danger assessment schemes to manage the potential risks of nanotechnology to public and environmental health.

The major international groups involved in the regulation of nanotechnology R&D towards safety development are listed in Table 5.1.

TABLE 5.1
International Working Groups Toward Safety Developments of Nanotechnology

International Organizations/Working Groups	About
National Nanotechnology Initiative (NNI)	The National Nanotechnology Initiative (NNI) is a U.S. government research and development (R&D) initiative involving 20 departments and independent agencies. The NNI was officially created in 2003 based on Nanotechnology Research and Development Act, (P.L. 108–153). NNI works towards the shared vision of "a future in which the ability to understand and control matter at the nanoscale leads to a revolution in technology and industry that benefits society" (National Nanotechnology Initiative).
OECD Working Party on Nanotechnology (WPN)	A forum for advising on emerging policy issues in science, technology, and innovation related to the responsible development and use of nanotechnology. It is the premier multilateral forum that brings governments together to discuss and create policy perspectives relating to nanotechnology (OECD).
International Organization for Standardization (ISO) as coordinated by the American National Standards Institute's U.S. Technical Advisory Group to ISO TC229 (Nanotechnologies)	The world's largest developer of voluntary international standards. International standards give state-of-the-art specifications for products, services, and good practice, helping to make the industry more efficient and effective. Developed through global consensus, they help to break down barriers to international trade (International Organization for Standardization).
International Alliance for NanoEHS Harmonization (IANH)	Founded in September 2008, IANH is a voluntary initiative of labs and other stakeholders to create toxicology-test protocols on representative nanomaterials in a "round robin" fashion.
International Council on Nanotechnology (ICON)	An international, multi-stakeholder organization whose mission is to develop and communicate information regarding potential environmental and health risks of nanotechnology, thereby fostering risk reduction while maximizing societal benefit. ICON is one of many sponsors of the Good Nano Guide, which is intended to be a community platform to share workplace practices for handling nanomaterials (International Council on Nanotechnology (ICON): CBEN).
Asia Pacific Nanotechnology Forum (APNF)	This group fosters regional cooperation on nanotechnology R&D, with nation members including Australia, China, Korea, Japan, Malaysia, New Zealand, Singapore, Taiwan, Thailand, and Vietnam (Asia Pacifc Nanotechnology Forum).
Asian Nano Forum	Founded in 2004, this is a collaborative network supported by many countries in the Asia Pacific Region. The forum creates mechanisms for sharing information and helps to support regional development through joint projects (Asia Nano Forum).

TABLE 5.2
Reports on Nanotechnology Regulations: Need, Approach, Challenges and Status across the World

S.No	From	Report	Description
1	Royal Institute of International Affairs	Securing the Promise of Nanotechnologies: Towards Transatlantic Regulatory Cooperation	The approaches toward regulation of nanomaterials in the EU and the U.S. in an attempt to foster trans-Atlantic cooperation and consistency in the regulation of nanomaterials (Breggin et al., 2009).
2	United States Congress	Engineered Nanoscale Materials and Derivative Products: Regulatory Challenges	Regulatory challenges for nanomaterials (Engineered Nanoscale Materials and Derivative Products: Regulatory Challenges, 2017).
3	United States Environmental Protection Agency (EPA)		Research needed to address environmental risks of nanomaterials.
4	National Science and Technology Council, U.S.	Environmental, Health, and Safety Research Needs for Engineered Nanoscale Materials	Identified the research and information needs in understanding and managing risks of engineered nanomaterials in various applications.
5	National Institute for Occupational Safety and Health (NIOSH), U.S.	Managing the Health and Safety Concerns Associated with Engineered Nanomaterials	Overview of the potential hazards of engineered nanoparticles and measures that can be taken to minimize workplace exposures (NIOS—National Institute for Occupational Safety and Health, 2009).
6	The Canadian Environmental Protection Act, 1999	Regulatory Framework for Nanomaterials	NanoPortal—Proposed Regulatory Framework for Nanomaterials Under the Canadian Environmental Protection Act (1999 and 2017)
7	Institute for Science, Society and Policy (ISSP)	Timeline Nanotechnology – Policy and Regulation in Canada, Australia, the European Union, the United Kingdom, and the United States	Alin Charriere; Beth Dunning (2014)

Some of the most important reports on nanotechnology regulations: need, approach, challenges, and status across the globe, are listed in Table 5.2.

Apart from the legislative approach, there are several guidelines provided by different authorities across the world for the sustainable and responsible development of nanotechnology. For example, the National Institute for Occupational Safety and Health (NIOSH) of the U.S. has created a "Nanotechnology Field Research Effort" (2017) to assess the workplace practices, potential occupational exposure to nanomaterials, and the available safety measures and controlling technologies to prevent the exposure.

NIOSH has published interim guidelines for managing the health and safety concerns associated with engineered nanomaterials and working with nanomaterials (NIOS – National Institute for Occupational Safety and Health, 2009). Other countries also provided guidelines for safety handling to prevent adverse effects to health and environment during preparation/handling/use of nanomaterials. Some of them are listed in Table 5.3.

5.4 FOOD AND DRUG ADMINISTRATION'S APPROACH TO REGULATION OF NANOPRODUCTS

The U.S. Food and Drug Administration (FDA) regulates nanotechnology products under its existing statutory and regulatory authorities, in accordance with the specific legal standards applicable to each type of product under its jurisdiction. FDA is working with the White House, U.S. government

TABLE 5.3
Various Guidelines on Nanotechnology Usage

S.No	Country	Who	Guidelines
1	Australia	Commonwealth Scientific and Industrial Research Organisation (CSIRO)	Safe handling and use of carbon nanotubes (Safe Handling and use of Carbon Nanotubes, 2012).
2	Australia	Safe Work Australia	Work health and safety assessment tool for handling engineered nanomaterials (Work Health and Safety Assessment Tool for Handling Engineered Nanomaterials, 2017).
3	Germany	Federal Institute for Occupational Safety (BAuA)	Safe handling of nanomaterials and other advanced materials at workplaces (Safe Handling of Nano Materials and Other Advanced Materials in Workplace, 2015).
4	Germany	Federal Institute for Occupational Safety (BAuA)	Safety Management and Nanomaterial (Aart Rouw and Federal Institute for Occupational Safety and Health, 2015).
5	Germany, France, Switzerland, UK, Italy	NanoIndex project (Project partners – Germany, France, Switzerland, UK, Italy)	Assessment of Personal Exposure to Airborne Nanomaterials (Assessment of Personal Exposure to Airborne Nanomaterials, 2017).
6	U.S.	FDA	Guidance for Industry: Considering whether an FDA-Regulated Product Involves the Application of Nanotechnology (FDA, 2014a).
7	U.S.	FDA	Guidance for Industry: Safety of Nanomaterials in Cosmetic Products (FDA, 2017b).
8	U.S.	FDA	Guidance for Industry: Assessing the Effects of Significant Manufacturing Process Changes, Including Emerging Technologies, on the Safety and Regulatory Status of Food Ingredients and Food Contact Substances, Including Food Ingredients that are Color Additives (FDA, 2017a).
9	U.S.	FDA	Guidance for Industry: Use of Nanomaterials in Food for Animals (FDA, 2015).
10	Danish	Danish Environmental Protection Agency	Guideline for the Danish Inventory of Nanoproducts (Danish Environmental and Protection Agency, 2014).
11	Netherlands	Dutch Ministry of Social Affairs and Employment	Guidance for working safely with nanomaterials and nanoproducts (Ralf Cornelissen et al., 2011).
12	Japan	Ministry of the Environment	Guidelines for preventing the environmental impact of manufactured nanomaterials (Guidelines for Preventing the Environmental Impact of Manufactured Nanomaterials, 2009)

agencies, National Nanotechnology Initiative, and international regulators to generate data and coordinate policy approaches to ensure the safety and effectiveness of nanotechnology products. FDA's nanotechnology regulatory science research portfolio focuses on understanding interactions of nanomaterials with biological systems, and on the adequacy of testing approaches for assessing safety, effectiveness, and quality of products containing nanomaterials (FDA, 2012 and 2013). FDA's regulatory policy approach is consistent with relevant overarching U.S. government policy principles and supports innovation under appropriate oversight. FDA ensures transparent and predictable regulatory pathways grounded in the best available science.

FDA does not make a categorical judgment that nanotechnology is inherently safe or harmful. FDA's regulatory approaches are adaptive and flexible to the specific characteristics and the effects of nanomaterials in the particular biological context of each product and its intended use.

The major attributes of FDA's regulatory approach towards nanotechnology products are as follows (FDA, 2016c and 452017):

1. "Product-focused, the science-based regulatory policy."

   FDA regulatory approaches are product-specific. Based on product-specific guidance, FDA follows particular approaches for each product area. Furthermore, technical assessments are product-specific, taking into account the effects of nanomaterials in the particular biological and mechanical context of each product and its intended use.
2. "Variations in legal standards for different product-classes."

   FDA follows differing legal standards for different product classes, even if two products present the same level of risk. For example, food additives are considered safe when there is a reasonable certainty of no harm from their intended use (*Federal Food Drug and Cosmetic Act (FFDCA)* 2017). Whereas, drugs are evaluated not only on the basis of their risk profile but also of their predicted benefit (FD&C Act, 2017). Thus, FDA follows divergent regulatory approaches for different product classes and different applications of nanomaterials, even where objective measures of risk are similar.
3. "Where premarket review authority exists, attention to nanomaterials is being incorporated into standing procedures."

   The FDA imposes premarket review requirements for some products such as new drugs, new animal drugs, biologics, food additives, color additives, certain human devices, and certain new dietary ingredients in dietary supplements. Applicants need to submit data to answer questions related to the safety, effectiveness (where applicable), or regulatory status of the new products. These premarket review procedures also incorporate attention to nanomaterials and suggest whether additional safety or effectiveness data are required.
4. "Where a statutory authority does not provide for premarket review, consultation is encouraged to reduce the risk of unintended harm to human or animal health."

   For the products which are not subjected to premarket reviews such as dietary supplements (except certain new dietary ingredients) and food (except food or color additives), FDA depends on available information collected from public or voluntary submission, adverse event reporting (where applicable), and on post-market surveillance activities, to provide oversight. In such cases, FDA provides consultancy services to nanotechnology manufacturers on safety reviews, designing post-market safety oversight, as applicable.
5. "FDA will continue post-market monitoring."

   FDA continues to monitor the marketplace for products containing nanomaterials and will take actions, as needed, to protect consumers.
6. "Industry remains responsible for ensuring that its products meet all applicable legal requirements, including safety standards."

   According to FDA guidelines, it is industries' responsibility to ensure that their products meet all standards and safety requirements, regardless of the emerging nature of the technology involved in the manufacturing. For any questions related to the safety, effectiveness, or other attributes of products that contain nanomaterials, or about the regulatory status of such products, the industry may contact FDA at an early stage of the product development, which facilitates a mutual understanding of the specific scientific and regulatory issues for nanotechnology products.
7. "FDA collaborates with domestic and international counterparts on regulatory policy issues."

   Through policy dialogues with other agencies, the FDA contributes to policies in the field of nanotechnology and suggests modalities to implement policy activities. FDA also works with numerous other foreign counterparts who deal with regulatory aspects to share information on applications and regulatory issues of nano-based products.

8. "FDA gives technical guidance to help industries meet its regulatory and statutory obligations."

   FDA gives technical guidance to meet the standards issued by regulatory authorities. These guidance documents are upgraded over time and (depending on product class) will address the interpretation of relevant statutory and regulatory standards. These guidelines basically describe the FDA's interpretation of or policy on a regulatory issue.

5.5 FDA'S SCIENTIFIC AND TECHNICAL GUIDANCE

The U.S. Food and Drug Administration (FDA) formed the Nanotechnology Task Force, in 2006, with the intention to identify and address the potential effects of FDA-regulated nanotechnology products on health. In 2007, the Nanotechnology Task Force recommended FDA to issue guidelines to industry and take steps to address the potential risks and benefits of drugs, medical devices, cosmetics, and other products that incorporate nanotechnology (U.S. Food and Drug Administration, 2007). In June 2011, FDA issued its first draft guidelines on evaluation and use of nanomaterials in FDA-regulated products (FDA, 2016c). In April 2012, FDA issued two product-specific (food and cosmetic) draft guidances, addressing the use of nanotechnology by the foods and cosmetics industries for public comment. Considering the public comments received, in June 2014, FDA finalized all three guidance documents. Additionally, in June 2014, FDA issued draft guidance on the use of nanomaterials in food for animals. In August 2015, after considering public comments, FDA finalized this guidance as well.

Finally, the latest guidance documents that are available for the use of nanotechnology in food are as follows:

- "Final Guidance for Industry: Considering Whether an FDA-Regulated Product Involves the Application of Nanotechnology" (FDA, 2014a)
- "Final Guidance for Industry: Assessing the Effects of Significant Manufacturing Process Changes, Including Emerging Technologies, on the Safety and Regulatory Status of Food Ingredients and Food Contact Substances, Including Food Ingredients that are Color Additives" (FDA, 2014b)
- "Final Guidance for Industry: Use of Nanomaterials in Food for Animals" (FDA, 2015)

5.5.1 FDA's Guidance for Industry

FDA's guidance for industry (FDA, 2014a) clearly explains when to consider whether an FDA-regulated product involves the application of nanotechnology:

1. "When a material or end product is engineered to have at least one external dimension, or an internal or surface structure, in the nanoscale range (approximately 1 nm to 100 nm)";
2. "Whether a material or end product is engineered to exhibit properties or phenomena, including physical or chemical properties or biological effects, that are attributable to its dimension(s), even if these dimensions fall outside the nanoscale range, up to one micrometer (1,000 nm). This is because materials or end products can also exhibit related properties or phenomena attributable to a dimension(s) outside the nanoscale range of approximately 1 nm to 100 nm that are relevant to evaluations of safety, effectiveness, performance, quality, public health impact, or regulatory status of products."

According to this guidance, the term engineered materials represents the products that are deliberately manipulated by the application of nanotechnology to produce specific properties. These products are different from products that contain materials that naturally occur in the nanoscale range.

In a size range of approximately 1 nm to 100 nm, materials can exhibit new or altered physicochemical properties that may enable novel applications. Because of deliberate manipulation, engineered nanomaterials have new properties or phenomena that raise questions about their safety, performance, effectiveness, quality or public health impact and thus demand further evaluation.

This guidance includes both finished products (e.g. a drug tablet for administration to a patient) as well as materials that are intended for use in a finished product (e.g. a food additive added to a food during processing). FDA will consider whether a material or end product contains or involves in its manufacture the use of materials that meet either Point 1 or Point 2.

Point 1 (dimension-based approach) focuses on size-based definition; accordingly, if a material or end product is engineered to have at least one external dimension in the range of 1 nm to 100 nm, or is engineered to have an internal or surface structure in the range of 1 nm to 100 nm, they are considered as nanomaterials. For example, (1) primary particles engineered with at least one external dimension within the nanoscale range, (2) any aggregates or agglomerates formed by such nanoscale primary particles, (3) coated, functionalized, or hierarchically assembled engineered structures that include functional nanoscale entities, embedded or attached to the surface, are covered under Point 1. In these cases, industry and FDA should consider any unique characteristics or biological effects exhibited by the product that may influence its safety, effectiveness, public health impact, or regulatory status.

Point 2 (property-based approach) states that the identification and assessment of specific dimension-dependent properties and phenomena are ultimately more relevant for purposes of FDA regulatory review and oversight. Point 2, therefore, focuses on the properties of the material and its behavior in biological systems.

For example, as noted above, dimension-dependent properties or phenomena may be used for various functional effects such as increased bioavailability or decreased toxicity of drug products, better detection of pathogens, improved food packaging materials, or improved delivery of nutrients. These effects may derive from altered or unique characteristics of materials in the nanoscale range that are not normally observed or expected in larger-scale materials with the same chemical composition. Hence the evaluation of safety, effectiveness, public health impact, or regulatory status of products under FDA's jurisdiction should include a consideration of the specific tests (whether traditional, modified, or new) to determine the physicochemical properties and biological effects of a product that involves the application of nanotechnology.

Even outside the nanoscale range (1 nm to 100 nm), materials or end products can exhibit properties or phenomena attributable to a dimension(s). Available scientific information does not establish a uniform upper boundary above 100 nm where novel properties and phenomena similar to those seen in materials with dimensions in the nanoscale range cease for all potential materials or end products. Hence, Point 2 focuses on the importance of considering properties or phenomena attributable to dimensions, even outside the nanoscale range (approximately 1 nm to 100 nm) to dimensions up to 1,000 nm.

Therefore, this includes materials with dimension(s) outside the nanoscale range that may exhibit dimension-dependent properties or phenomena associated with the application of nanotechnology and distinct from those of macro-scaled materials; and excludes macro-scaled materials that may have properties attributable to their dimension(s) but are not likely associated with the application of nanotechnology.

All these considerations apply not only to new products, but also when changes to manufacturing processes alter the dimensions, properties, or effects of an FDA-regulated product or any of its constituent parts.

5.5.2 FDA's Guidance to Manufacturers

The manufacturing process of a food substance is considered for the purposes of safety assessment as it may affect the properties and safety of the finished product. The manufacturing process may affect the identity of the food substance, the purity of a food substance (such as the amounts of impurities and contaminants in the food substance), or its conditions of use. For example:

1. Particle size is a critical decider of a material's bio-availability and migration characteristics (i.e. from packaging materials to food). For example, nano-engineered food substances can considerably alter bioavailability compared to their traditionally manufactured counterparts. When a food substance is manufactured to include a particle size distribution shifted more fully into the nanometer range, safety assessments should be based on data relevant to the nanometer version of the food substance.
2. In the process of manufacturing polymers that will be later used for food-based applications, it is essential to consider the substance's chemical identity and the presence of any prospective contaminant responsible for migratory effects.
3. Preparation of an enzyme from two different sources (namely animal and microbial sources) contains distinctly different constituents derived from the production organism and constituents derived from the manufacturing process (e.g. components of the fermentation media or the residues of processing aids).
4. Levels of polycyclic aromatic hydrocarbons (PAHs) that can be present in fish oil obtained from fish from a natural marine environment can be reduced by activated carbon treatment of fish oil.

FDA's guidance to industry on manufacturing processes (FDA, 2016c), describes the factors manufacturers should consider when determining whether there has been a significant change in manufacturing process for a food substance already in the market. Such change would:

- affect the identity of the food substance
- affect the safety of the use of the food substance
- affect the regulatory status of the use of the food substance and
- warrant a regulatory submission to FDA

FDA's guidance recommends manufacturers consult with FDA regarding a significant change in the manufacturing process for a food substance already in the market.

5.6 FOOD STANDARDS AGENCY'S REGULATIONS ON NANOFOODS

In the United Kingdom, the assessment of novel foods is taken care of by the Food Standards Agency (FSA). FSA is responsible for food safety, nutrition, and protecting the interests of consumers in relation to food. FSA takes a lead in regulations for food safety, consumer protection in relation to food and animal feed, and advises the UK government and public on risks that arise from food.

FSA's regulatory implications and risk assessment in relation to nanotechnologies and food are as follows (FSA, 2008):

1. Novel foods and processes

   FSA's Novel Foods Regulation (EC) 258/97 mandates the premarket approval for all foods or ingredients (other than food additives) and processes which are not consumed (or in practice) within the EU before *15th May 1997*. This regulation is applicable to all EU member states.

   The assessment of novel food or food ingredients does not consider the particle size; however, it includes the details of composition, nutritional value, metabolism, the level of microbiological and chemical contaminants, toxicological studies, allergenicity, and intended use. Since food production processes can alter the final composition of the food and make the food novel, details of the manufacturing process are also considered for assessment.

Decisions regarding approval of novel foods are based on a qualified majority vote by the member states at the EC Standing Committee on the Food Chain and Animal Health.

2. Food additives

   In the UK, Food Additives are controlled by the Colours in Food Regulations 1995 (as amended), the Sweeteners in Food Regulations 1995 (as amended), and the Miscellaneous Food Additives Regulations 1995 (as amended), with smoke flavorings being specifically controlled by the Smoke Flavourings (England) Regulations 2005. The use of food additives in terms of lists of permitted additives, the foods in which they can be used, and maximum levels of use are controlled by the European Parliament and Council legislation. All permitted additives are assessed for safety and comply with specific purity criteria laid down in corresponding European Commission directives.

   The EU food additives legislation states that when a food additive which is already included in a community list is changed by a novel production method or undergoes a change in particle size (through nanotechnology), or in starting materials, then it is considered as a novel additive material. Thus, it requires a new entry in the community lists or change in the specifications. All new nanomaterials would need to undergo safety assessments by EFSA before they get included in the relevant list and so be permitted in foods and on the market.

   In food additive legislation, the minimum particle size is specified only in the case of microcrystalline cellulose (E460 (i)), where the presence of particles <5 microns (5000 nm) is limited because of uncertainties over their safety. There is a limit on the molecular weight distribution of carrageenan (which could be regarded as a size limitation) based on concerns over potential toxicity in the gut associated with the smaller "degraded" components.

3. Food contact materials

   All food contact materials and articles that are brought into contact with food and those that might reasonably be expected to transfer their constituents to food are controlled by Regulation (EC) 1935/2004.

   The Food Contact Materials (FCM) legislation (European Commission) 1935/2004 deals with the migration of nanocomponents into food from food contact materials and articles, and requires that these materials and articles not transfer their constituents to foods under normal and foreseeable conditions of use in quantities that could endanger human health, or bring about an unacceptable change in the composition of the food or deterioration in the organoleptic properties of the food.

   This legislation imposes FCM to indicate whether a packaging is "active" or "intelligent." In the case of active packaging material, the nanocomponents are intended to migrate into the food, but must only do so to improve shelf life or maintain or improve the condition of the food. In such cases, change to the food must comply with EC regulations applicable to food. Similarly, in intelligent packaging, nanocomponents are intended to monitor the condition of the food in the packaging or the environment around the food, and should not give any misleading information about the condition of the food.

Nanocomponents incorporated into some specific materials such as plastic, ceramic, and regenerated cellulose films are subject to different kinds of regulations. Plastic materials and articles are subjected to policy 2002/72/EC implemented in Great Britain by The Plastic Materials and Articles in Contact with Food Regulations 1998 (as amended). This legislation does not differentiate between nanoscale components and others. Ceramic food contact materials and articles are regulated by the ceramic ware legislation which controls the migration of lead and cadmium into foodstuffs. Other issues affecting the safety, quality, and nature of food with which these materials come into contact are covered by the provisions of the European Regulation No. (EC) 1935/2004. Regenerated cellulose film can only be manufactured using substances listed in Directive 2002/72/EC on food contact plastics.

Based on the above-listed regulations, FSA assesses the food safety implications of food produced using nanotechnology and authorizes the company to market assessed food products. During safety assessment of the use of nanotechnology in the food chain, the FSA will consult an independent advisory committee, the Advisory Committee on Novel Foods and Processes (ACNFP). ACNFP is a non-statutory, independent body of scientific experts that advises the FSA on any matters relating to novel foods (including genetically modified foods), novel processes (including food irradiation), and food technologies.

ACNFP enforces preapproval under the novel foods regulation (Regulation (EC) No 258/97) if nanotechnology is used to develop novel foods and processes (Food Standards Agency, 2016; Regulation (EC) No 258/97 of the European Parliament and of the Council of 27 January 1997 Concerning Novel Foods and Novel Food Ingredients 2017). For the assessment of nanomaterials in food, ACNFP follows the guidance issued by the European Food Safety Authority (EFSA). ACNFP carries out a thorough safety evaluation based on rigorous scrutiny of scientific data (i.e. toxicological, allergenicity, and nutritional information), as well as a scientific safety assessment, and deals with issues of consumer concern and ethical issues.

EFSA published its first opinion on the potential risks arising from nanoscience and nanotechnologies in food and feed in 2009 (Barlow et al., 2009). Building on its first opinion, in May 2011 EFSA published a guidance document (EFSA Scientific Committee, 2011). EFSA's guidance will be used whenever products of nanotechnology are evaluated for food or feed applications in the EU. EFSA's guidance document provides a practical approach for assessing potential risks arising from applications of nanoscience and nanotechnologies in the food and feed chain (EFSA Scientific Committee, 2011).

For the risk assessment of engineered nanomaterials in food and food contact materials, EFSA provides guidance on the below-mentioned aspects (EFSA Scientific Committee, 2011):

1. The physicochemical characterization of engineered nanomaterials (ENM)

   Nanomaterial may be in the form of food additives, enzymes, flavorings, food contact materials, novel foods, feed additives, and pesticides. Physico-chemical characterization provides guidance on the identification, detection, and characterization of ENM. Parameters and currently available methods for characterization and identification of ENM are listed in the guidance document.

2. Exposure assessments

   Based on the proposed uses, exposure scenarios should be outlined prior to commencing risk assessment. These exposure scenarios will be the deciding factor for the extent of hazard characterization required.

3. Hazards characterization

   Testing approaches to identify and characterize hazards arising from the nano properties.

   This should be based on information from toxicity testing, *in vitro* digestion and genotoxicity testing, and *in vivo* studies which include ADME (absorption, distribution, metabolism, and excretion) studies, genotoxicity testing, and 90-day repeated oral dose toxicity studies in rodents.

4. Risk characterization

   Risk characterization is an iterative process. The final result of risk characterization should be the overall assessment of the safety of the ENM in its intended use together with the parameters under which the assessment is valid and the uncertainties associated with the assessment. It should provide qualitative and if possible quantitative guidance to risk managers. Assumptions made during the risk assessment and nature and magnitude of any uncertainties should be clarified.

At every stage of information assessment, the weight-of-evidence process decides whether a risk assessment is required or not. The weight-of-evidence approach takes into account all available sources of information and types of data. At each evaluation step, decisions depend on the amount and quality of the information available at that particular stage and the validity of the tests used to generate the data. If the totality of the available information is considered suitable at a particular stage, then a risk assessment can be performed and no further testing would be required. However, if this is not considered possible, then the default presumption is that a sequence of further testing should be undertaken.

5. Uncertainty analysis

   This involves analyzing the possibility of uncertainty in carrying out the abovementioned assessments.

Some of the possible reasons for uncertainty in assessing ENM are:

- nonavailability of standard methods for physicochemical characterization of all the various ENM structures/properties
- reproducibility and accuracy of the available characterization methods depend on the sample preparation procedures and calibration of the analytical equipment
- differences in the physical principles applied by various measurement techniques
- aggregation/agglomeration behavior of ENM, sample handling/preparation procedures, and other factors such as dilutions or dispersions required for different methods

The guidance allows for reduced information to be provided when no exposure to the ENM is verified by data indicating no migration from food contact materials or when complete degradation/dissolution is demonstrated with no absorption of engineered nanomaterials as such. The guidance indicates uncertainties that should be considered to perform a risk assessment (EFSA Scientific Committee, 2011).

The EFSA guidance released in 2011 has some drawbacks as it focuses only on the oral route of exposure and does not include environmental hazard evaluation and worker exposure to nanomaterials. Since this field is a fast-developing sector, the EFSA guidance document will be updated during 2016–2018; this will involve a multidisciplinary working group of experts; a supporting data collection of existing studies to provide information about best practice; and public consultations. The EFSA Scientific Committee with the new working group will develop a *de novo* guidance for environmental risk assessment for nanomaterials in agri/food/feed products in 2018 (OECD, 2016b).

5.7 THE ORGANIZATION FOR ECONOMIC CO-OPERATION AND DEVELOPMENT (OECD) APPROACH TO NANOFOOD PRODUCTS

The Organisation for Economic Co-operation and Development (OECD) is an inter-governmental economic organization with 35 member countries, founded in 1960 to stimulate economic progress and world trade. The mission of the OECD is to promote policies that will improve the economic and social well-being of people around the world. The OECD provides a forum in which governments can work together to share experiences and seek solutions to common problems. The OECD works with governments to understand what drives economic, social and environmental change. The OECD analyzes and compares data to predict future trends. It measures productivity and global flows of trade and investment. Drawing on facts and real-life experience, the OECD sets international standards on a wide range of things, from agriculture and tax to the safety of chemicals (OECD, 2017a).

Considering the vital need for responsible development of nanotechnology, the OECD has established various working groups, namely, 1) the Working Party on Manufactured Nanomaterials (WPMN, subsidiary to the Chemicals Committee); 2) the Working Party on Biotechnology (WPB); 3) the group of National Experts for Scientific and Technological Indicators (NESTI) and their parent committees, and 4) the Working party on Nanotechnology (WPN, subsidiary to the Committee for Scientific and Technological Policy (CSTP)), to advise upon emerging policy issues of nanoscience, technology, and innovation.

The OECD's approach towards the safe and responsible development of nanomaterials is through (OECD, 2011):

1. The establishment of an OECD database of manufactured nanomaterials (www.oecd.org/env/nanosafety/database) to promote collaboration in research activities and strategies on environmental, human health, and safety (EHS) issues of nanomaterials
2. Ensuring appropriate testing methods and testing of specific nanomaterials for their effect on human health and safety evaluation (*in vivo* and *in vitro*)
3. Facilitating international cooperation on risk assessment strategies and cooperating voluntary schemes and regulatory programmes
4. Guidance on exposure measurement and exposure mitigation (workplace, consumers, and the environment)
5. Enhancing the knowledge base about life cycle aspects of manufactured nanomaterials (at various stages of development and application) and their impacts on human health and environmental safety

Since the safety of nanomaterials is an important concern for the regulatory bodies across the world, OECD's WPMN launched a sponsorship testing programme, to verify the testing methods used on manufactured nanomaterials. Selected lists of manufactured nanomaterials (which are in, or close to, commercial use, such as fullerenes (C_{60}), SWCNTs, MWCNTs, silver nanoparticles, TiO_2, CeO, ZnO, SiO_2, dendrimers, nanoclays and gold nanoparticles) are tested for approximately 111 endpoints (identified from the International Uniform Chemical Information Database (IUCLID)) relevant to environmental safety and human health (WPMN, 2017).

These keys nanomaterials have been tested for their:

- nanomaterial information/identification
- physical-chemical properties and material characterization
- environmental fate
- residues in food and feeding stuffs
- toxicological and eco-toxicological effects
- environmental toxicology
- mammalian toxicology and
- material safety

Dossier and endpoints of these tested nanomaterials can be found in the OECD website (WPMN, 2017). The outcomes of this testing program provide information on the "intrinsic properties" of nanomaterials.

Foods that may contain nanomaterials or otherwise involve the application of nanotechnology are generally governed by existing national and/or regional legislative and regulatory frameworks. OECD's WPN conducted a survey to find out the regulatory frameworks for nanotechnology in food and medical products. The outcome of this survey provides a summary of current legislation and regulations among twelve OECD-WPN delegations (not all OECD countries responded to this survey) as listed in Table 5.4.

TABLE 5.4
Summary of Legislation for Foods across OECD Countries

Delegation (Responsible Organization)	Existing Legislation	Summary
Australia (Foods Standards Australia New Zealand)	Food Standards Australia New Zealand Act	Established the Food Standards Australia New Zealand and authorized it to regulate food
Canada (Health Canada)	Food and Drugs Act	Authorizes Health Canada to regulate, among other products, food and drugs
European Union (European Commission [EC], European Food Safety Authority [EFSA])	General Food Legislation (EC 178/2002)	Established the EFSA and food safety requirements.
	Novel Food (EC 258/97),	Established safety requirements for foods which had no significant consumption prior to 1997
	Food Additives Regulation (EC 1333/2008)	Regulates the use and labeling of food additives
	Plastics Regulation (EC 10/2011)	Regulates the use of plastic materials and articles intended to come into contact with food
	Active and Intelligent Materials Regulation (EC 450/2009)	Regulates the use of active and intelligent materials and articles intended to come into contact with food
	Labelling Regulation (EC 1169/2011)	Regulates the labeling of foods for the purposes of consumer protection
France (Ministry of Agriculture/ General Directorate on Food, European Commission, European Food Safety Authority)	Grenelle II Law Decree n° 2012–232 Order of 6 August 2012	Established a system of compulsory declaration for all nanomaterials
	Applicable European Commission Regulations (see EC legislation above)	See EC legislation above
Korea (Korea Food and Drug Administration)	Food Sanitation Act	Authorizes the Ministry of Health and Welfare to regulate foods for the purposes of national health
	Health and Functional Food Act	Authorizes the Ministry of Health and Welfare to regulate and license the production of health and functional foods
	Framework Act on Food Safety	Authorizes the regulation of food for the purpose of safety
Germany (Federal Ministry of Food, Agriculture and Consumer Protection, European Commission, European Food Safety Authority)	German Food and Feed Code (LFGB)	Implements the provisions of EU Regulation 178/2002.
	Applicable EC Regulations (see EC legislation above)	See EC legislation above
	Cellulose Film Directive (Directive 2007/42/EC)	Regulates the use of cellulose film-based materials and articles that come into contact with food
Japan (Ministry of Health, Labor, and Welfare)	Food Sanitation Law (Law no. 233)	Authorizes the regulation of foods by the Ministry of Health, Labor, and Welfare for the purposes of health and safety
Netherlands (European Commission, European Food Safety Authority)	Applicable EC Regulations (see EC legislation above)	See EC legislation above

(*Continued*)

TABLE 5.4 (CONTINUED)
Summary of Legislation for Foods across OECD Countries

Delegation (Responsible Organization)	Existing Legislation	Summary
Norway (European Commission, Norwegian Food Safety Authority, European Food Safety Authority)	Applicable EC Regulations (see EC legislation above)	See EC legislation above
Poland (European Commission, European Food Safety Authority	Applicable EC Regulations (see EC legislation above)	See EC legislation above
Russian Federation (Federal Service for Surveillance on Consumer Rights Protection and Human Wellbeing, Ministry of Health and Social Development of the Russian Federation)	"Federal Law of 21 of November 2011, on the Fundamentals of Protection of the Public Health" and others	Implements the following provisions, among others: methods of classification of products and nano industry technologies in terms of their potential danger; guidelines on selection order for the nanomaterials control sampling; control for nanomaterials applied in food products, used in the chemical industry, in packaging materials, and applied in agriculture; and various testing for nanomaterials in the tissues of animals and plants
United States (U.S.) (Food and Drug Administration [FDA], Department of Health and Human Services [HHS]; and Food Safety and Inspection Service [FSIS], Department of Agriculture [USDA])	Federal Food, Drug, and Cosmetic Act, as amended	Authorizes the Secretary of Health and Human Services (who acts through the FDA) to regulate, among other products (most relevant to this report), foods, drugs, and medical devices.
	Public Health Service Act	Authorizes HHS to license biological products, and also authorizes the conduct and support of public health research by HHS
	Federal Meat Inspection Act	Authorizes the FSIS to regulate livestock and related meat products
	Poultry Products Inspection Act	Authorizes the FSIS to regulate poultry products
	Egg Products Inspection Act	Authorizes the FSIS to regulate egg products

Note: The U.S. Consumer Product Safety Commission (CPSC) has regulatory authority over consumer products, including products with food contact surfaces (such as food containers, and food cooking, eating, and preparation articles). However, FDA regulates such products with respect to any migration of a substance from the contact surface to the food.

Source: Regulatory Frameworks for Nanotechnology in Food and Medical Products, (2013).

5.8 CASE STUDY: SAFETY FEATURES IN THE PRODUCTION OF NANOPARTICLES: FROM LABORATORY TO COMMERCIAL PRODUCTION

5.8.1 Exposure Risks

Gold nanoparticle occupational exposure assessment in a pilot-scale facility (OECD, 2016a):

To assess the gold nanoparticle exposure risk to human health, the Organisation for Economic Co-operation and Development (OECD) conducted a study (OECD, 2016a) in the most widely used reduction method of gold (III) derivatives to gold nanoparticles (AuNPs) for the assessment of nanoparticle emission during synthesis. Authors compared the nanoparticle synthesis at lab-scale (200 ml) and pilot-scale level (80 liter) and identified the tasks that result in the emission of AuNPs into the environment during their synthesis.

The three-tiered exposure assessment approaches followed in this study are as follows:

Tier 1: information gathering on the synthesis of AuNPs.
Tier 2: assessment of nanoparticle emission during AuNP synthesis using handheld particle counters.
Tier 3: assessment of nanoparticle emission during AuNP synthesis using a combination of mobility particle sizers focusing on particle number concentration and size, and personal and area air sampling for particle characterization.

Measurement methods:
Combination of real-time particle counters: measurement of nanoparticles emission before (background levels) and during the preparation, synthesis, and post-synthesis phase using particle counters capable of detecting temporal changes in particle number concentration (PNC) and particle size distribution (PND) in the range of 5 nm–20 μm.
Instruments and sampling methods used:

1. Hand-Held Particle Counter (HHPC-6): Measures PNC expressed as the total number of particles per liter of air (particles/L) (upper limit of detection of 70 particles/cm^3).
2. Ultrafine Particle Counter (UPC): Measures PNC in the size range 10–1,000 nm, expressed as the total number of particles per cubic centimeter (particles/cm^3) of sampled air, and has an upper limit of detection of 100,000 particles/cm^3.
3. Desk top particle counters: Scanning Mobility Particle Counter (SMPS) and Aerodynamic Particle Sizer (APS) are placed as close as possible to the reactor (about 30 to 40 cm away from the main reactor opening).
4. Personal sampling: 25 mm mixed cellulose ester (MCE) filters (pore size 0.8 μm) and standard volume pump operated at 4 L/min were attached at breathing zones of three laboratory workers directly involved in the synthesis process and one worker indirectly involved in the synthesis process.
5. Area sampling: Area samples were collected by placing open-faced MCE filters (25 mm diameter and 0.8 μm pore size) before sampling as close as possible to the source and in other areas within the facility, drawing air using pumps at a flow rate of 7 L/min.

Samples collected from personal and area sampling filters were analyzed using inductively coupled plasma mass spectrometer (ICP-MS) and transmission electron microscopy (TEM) for elemental and morphology respectively.
Result:
This is the first report that illustrated the emission of gold nanoparticles during synthesis using the citrate reduction method (OECD, 2016a). The AuNP synthesis via the citrate reduction method occurs at temperatures above 100°C, and may cause the emission of nanoparticles as a result of evaporation. It was therefore assumed that exposure may occur if no control measures are in place to contain evaporation. A walk-through survey was carried out to investigate the possibility of nanoparticle emissions as measured by considering local background particle concentration and the concentration of particles generated by the metal nanoparticle synthesis. Real-time particle number concentrations were measured using real-time particle number counters (see Table 5.5). As the real-time particle counters could not distinguish between background (e.g. incidental) and engineered nanoparticles, TEM and ICP-MS were employed for offline analysis methods for the particles collected on filters from personal and area sampling. ICP-MS showed the presence of gold in the sampled filters. It was found that gold concentration ranged from 0.015 to 0.15 μg/cm^3 with the highest concentration near the reactor opening (30 cm) for the area samples.

ICP-MS analyses of personal samples are listed in Table 5.6. They show that the person who had the higher time-weighted average (TWA), namely person responsible for the addition of citrate and

TABLE 5.5
Nanoparticle Emission during AuNPs Synthesis

Activities Profile	7–290 nm (SMPS)		10 - <300 nm (P-Trak)	
	Avg+SD (#/cm^3)	Range (#/cm^3)	Avg+SD (#/cm^3)	Range (#/cm^3)
Background	7901+297	7450–8330	7171+3	7167–7177
Preparation phase	7361+210	6860–7610	6593+144	6369–6857
Synthesis phase	10295+216	7700–13600	8449+2419	6524–13690
Post-synthesis phase	9535+1870	7070–13900	7671+1326	5904–10358

TABLE 5.6
ICP-MS Analysis of Personal Samples–Gold Concentration (Au^{3+})

	Job Description	Air Concentration (μg/m^3)	Time Weighted Average TWA (μg/cm^3)
Employee A	Addition of $HAuCl_4.3H_2O$	0.00225	0.000281
Employee B	Addition of citrate and monitoring of color changes	0.102468	0.0128
Employee C	Supervising synthesis process	0.127133	0.0158
Employee D	Decanting of AuNP	0.0706	0.00235

supervising the synthesis process, had higher exposed concentration. TWA levels were very low and particle emission was less than the proposed nano reference values (NRV) and below the OECD excursion guidance criteria.

From this study, it was observed that the highest emission occurs during the addition of citrate to the heated chloroauric acid solution (above 100°C) during synthesis of AuNPs by citrate reduction method. During the upscaled production of 80 L AuNPs there was a significant increase ($p<0.0000$) in nanoparticle number concentration from background (7901 particles/cm^3) level to (10295 particles/cm^3) during synthesis stage. Exposure during synthesis was not just limited to the pristine AuNPs (i.e. primary particles) but it extended to their aggregates/agglomerates as well. While this study has laid the basis for future exposure assessment of other nanoparticles/nanomaterials, further research needs to be undertaken to assess the adverse health effects of exposure to AuNPs.

5.8.2 Explosion Risks

Nanomaterials have very different properties such as greater reactivity, strength, fluorescence, and conduction compared to micrometer-sized materials. This significant property is due to their large specific surface area which results in the presence of a higher number of atoms on the surface of nanoparticles. Thus, nanomaterials have different properties to their respective micrometer-sized counterparts. It has been reported that particles in the size range of nm (nanomaterials) may remain suspended in the air for days or even weeks (Hallock et al., 2009). Hence nanomaterials present a dust explosion hazard.

Possible reasons for explosion risk with nanomaterials:

1. Minimum ignition temperature

   The decrease in the particle size results in minimum ignition temperature which increases the probability of nanomaterials explosions over larger particles (Vignes et al., 2012). Huang et al. reported that as a result of the oxidation of aluminum, aluminum

TABLE 5.7
Comparison of Nano and Micrometer-Sized Aluminum Powders Explosibility Results

Particle Size	P_{max} (bar)	K_{st} (bar m/s)	MEC (g/m^3)	MIE (mJ)
40 μm Al	5.9	77	35	59.7
100 nm Al	12.5	296	50	<1
35 nm Al	7.3	349	40	<1

Source: Wu et al. (2010).

TABLE 5.8
Dust Explosion Class

Class	K_{St} (bar m/s)	Explosion Characteristics
0	$K_{St}=0$	No explosion
1	$0<K_{St}\leq 200$	Weak or medium-scale explosion
2	$200<K_{St}\leq 300$	Large-scale explosion
3	$K_{St}>300$	Extremely large-scale explosion

Source: Wu et al. (2010).

nanoparticles ignite at a lower temperature (approximately 900 K), which is below the melting point for aluminum (933.47 K). Similarly, Dreizin et al. (2009) reported that the initiation temperature for micron size thermites was 610°C, whereas the nanosize thermites ignited even at 100°C.

2. Minimum ignition energy

   Wu et al. presented the explosion characteristics of nano- and micro-sized aluminum (Al) powder (Wu et al., 2010). The comparisons of explosibility in nano- and micrometer-sized aluminum powders are presented in Table 5.7.

   From this data, we can observe that as the particle size decreases from micro to nano range, the explosion pressure appears to increase. However, once within the nano range, the explosion pressure decreases. The maximum explosion pressure (K_{St}) appears to increase as it transitions from the micrometer to nano range. Based on (K_{St}), explosion severity will be determined (Table 5.8) (Wu et al., 2010).

   The minimum ignition energy (MIE) of micro-sized Al particles is ~60 mJ. When the particle size decreases to nano range, MIE is less than 1 mJ. These low ignition energies (<1 mJ) may therefore allow nanomaterials to ignite due to electrostatic sparks, collision, or mechanical friction. Due to these very low minimum ignition energies, metallic nanomaterials have a high possibility of explosion.

3. Agglomeration reduces the explosion severity

   In general, as particle size decreases, collision rate increases. Hence the rate of collision for nanoparticles will be higher. Collision between nanoparticles results in agglomeration and formation of larger size particles (10–200 μm). The high rate of agglomeration of nanoparticles limits the dispersion of fine, cohesive powders into a cloud of individual particles. Hence these two physical processes, namely, high coagulation rate and limited dispersibility, reduce the explosion severity of nanoparticles. For example, 100 nm aluminum particles were found to explode less violently than aluminum particles with a diameter of

200 nm. This may be due to a greater impact through agglomeration for the 100 nm sample (Bouillard et al., 2010).

4. Changes in oxide shell

   Particle combustion rate and ignition are affected by the oxide shell. Ignition reaction rate is indirectly proportional to particle fuel and oxide shell at a particle size lower than 10 nm (Yetter et al., 2009). Combustion reaction of nanosized particles is kinetically controlled, unlike micrometer-sized particles whose reaction rate is controlled by diffusion. The severity of nanomaterial explosions will not be controlled by the particle size but rather by the combustion of the pyrolysis gas/air mixture (Eckhoff, 2011).

Due to these reasons, nanomaterials present explosion hazards in many processes such as mixing, grinding, drilling, cleaning, shifting, etc. Much more research is needed to gain a better understanding of explosion behavior of nanomaterials, with specific focus on the effect of agglomeration on ignition.

ACKNOWLEDGMENT

The author Maria Leena thanks the Department of Science and Technology (DST), Government of India for the award of the Women Scientist Fellowship (WOS-A).

REFERENCES

Aart, Rouw, and Federal Institute for Occupational Safety and Health (BAuA). 2015. "Safety Management and Nanomaterial." Accessed March 21. http://www.nanosafetycluster.eu/nanoToGo/Nano to go!/General/1 Safety management and nanomaterial_v1-0.pdf.

Ahamed, Maqusood, Mohamad S. AlSalhi, and M.K.J. Siddiqui. 2010. "Silver nanoparticle applications and human health." Clinica Chimica Acta, 411 (23): 1841–1848. doi:10.1016/j.cca.2010.08.016.

Alin Charrierre and Beth Dunning. 2014. "Timeline Nanotechnology April 2014 ; Policy and Regulation in Canada, Australia, the European Union, the United Kingdom, and the United States." http://issp.uottawa.ca/sites/issp.uottawa.ca/files/issp2014-nanotechnologytimeline.pdf.

Anandharamakrishnan, C. 2014a. "Electrospraying and electrospinning techniques for nanoencapsulation." In Techniques for Nanoencapsulation of Food Ingredients, Chapter 5, pp 43–49. Springer Briefs in Food, Health, and Nutrition, New York, Springer.

Anandharamakrishnan, C. 2014b. "Liquid-based nanoencapsulation techniques." In Techniques for Nanoencapsulation of Food Ingredients,Chapter 4, 29–41. Springer Briefs in Food, Health, and Nutrition, New York, Springer.

Anandharamakrishnan, C. and S. Padma Ishwarya. 2015a. "Introduction to encapsulation of food ingredients." In Spray Drying Techniques for Food Ingredient Encapsulation, Chapter 2, pp 37–64, Chichester, John Wiley & Sons.

Anandharamakrishnan, C. and S. Padma Ishwarya. 2015b. "Selection of wall material for encapsulation by spray drying." In Spray Drying Techniques for Food Ingredient Encapsulation, Chapter 4, 77–100, Cambridge, John Wiley & Sons.

Anandharamakrishnan, C. and S. Padma Ishwarya. 2015c. "Spray drying for nanoencapsulation of food components." In Spray Drying Techniques for Food Ingredient Encapsulation, Chapter 8, pp 180–197. 9, Cambridge, John Wiley & Sons.

Anandharamakrishnan, C. and K. Usha Kiran. 2014. "Bionanocomposites and their potential applications in food packaging." In J. Varghese, S. Alavi, S. Yaragalla, S. Thomas (eds.), Polymers for Packaging Applications, Chapter 10, pp 229. Boca Raton, CRC Press.

Arick, Davis Q., Yun Hwa Choi, Hyun Chang Kim, and You-Yeon Won. 2015. "Effects of nanoparticles on the mechanical functioning of the lung." Advances in Colloid and Interface Science, 225: 218–228. doi:10.1016/j.cis.2015.10.002.

"Asia Nano Forum." 2017. Accessed March 22. http://www.asia-anf.org/.

"Asia Pacific Nanotechnology Forum." 2017. Accessed March 22. http://www.apnf.org/.

"Assessment of Personal Exposure to Airborne Nanomaterials." 2017. Accessed March 21. http://www.nanoindex.eu/wp-content/uploads/2016/06/NanoBrosch%C3%BCre.pdf.

Barlow, Sue, Andrew Chesson, John D. Collins, Albert Flynn, Anthony Hardy, Ada Knaap, Harry Kuiper, et al. 2009. The potential risks arising from nanoscience and nanotechnologies on food and feed safety. The EFSA Journal, 958. doi:10.2903/j.efsa.2009.958.

Bhushani, Anu and C. Anandharamakrishnan. 2017. "Food-grade nanoemulsions for protection and delivery of nutrients." In s. Ranjan, N. Dasgupta, and E. Lichtfouse (eds.), Nanoscience in Food and Agriculture, 4, 99–139. New York, Springer.

Bhushani, Anu, Udayakumar Harish, and Chinnaswamy Anandharamakrishnan. 2017. "Nanodelivery of nutrients for improved bioavailability." In Nutrient Delivery, 369–411. Elsevier.

Bouillard, J., A. Vignes, O. Dufaud, L. Perrin, and D. Thomas. 2010. "Ignition and explosion risks of nanopowders." Journal of Hazardous Materials, 181 (1): 873–880. doi:10.1016/j.jhazmat.2010.05.094.

Breggin, Linda, Robert Falkner, Nico Jaspers, John Pendergrass, and Read Porter. 2009. "Securing the Promise of Nanotechnologies Towards Transatlantic Regulatory Cooperation." https://www.chathamhouse.org/sites/files/chathamhouse/public/Research/Energy%2C Environment and Development/r0909_nanotechnologies.pdf.

Canadian Environmental Protection Act. 2007. "New Substances Program Advisory Note 2007–06." https://www.ec.gc.ca/subsnouvelles-newsubs/default.asp?lang=En&n=3C32F773-1.

"CDC - Nanotechnology - Field Research Effort - NIOSH Workplace Safety and Health Topic." 2017. Accessed February 13. https://www.cdc.gov/niosh/topics/nanotech/field.html.

Danish Environmental and Protection Agency. 2014. "Guideline for the Danish Inventory of Nanoproducts." http://www2.mst.dk/Udgiv/publications/2014/08/978-87-93178-89-2.pdf.

Dasgupta, Nandita, Shivendu Ranjan, Deepa Mundekkad, Chidambaram Ramalingam, Rishi Shanker, and Ashutosh Kumar. 2015. "Nanotechnology in agro-food: From field to plate." Food Research International, 69: 381–400. doi:10.1016/j.foodres.2015.01.005.

Dobrovolskaia, Marina A., Parag Aggarwal, Jennifer B. Hall, and Scott E. McNeil. 2008. "Preclinical studies to understand nanoparticle interaction with the immune system and its potential effects on nanoparticle biodistribution." Molecular Pharmaceutics, 5 (4): 487–495. doi:10.1021/mp800032f.

Dreizin, Edward L. 2009. "Metal-based reactive nanomaterials." Progress in Energy and Combustion Science, 35 (2): 141–167. doi:10.1016/j.pecs.2008.09.001.

Eckhoff, R. K. 2011. "Are enhanced dust explosion hazards to be foreseen in production, processing and handling of powders consisting of nano-size particles?" In Journal of Physics: Conference Series, 304 (1): 012075. doi:10.1088/1742-6596/304/1/012075.

EFSA Scientific Committee. 2011. "Scientific opinion on guidance on the risk assessment of the application of nanoscience and nanotechnologies in the food and feed chain." EFSA Journal, 9 (5): 2140. doi:10.2903/j.efsa.2011.2140.

"Engineered Nanoscale Materials and Derivative Products: Regulatory Challenges - EveryCRSReport.com." 2017. Accessed March 21. https://www.everycrsreport.com/reports/RL34332.html.

Espitia, Paula Judith Perez, Nilda de Fátima Ferreira Soares, Jane Sélia dos Reis Coimbra, Nélio José de Andrade, Renato Souza Cruz, and Eber Antonio Alves Medeiros. 2012. "Zinc oxide nanoparticles: Synthesis, antimicrobial activity and food packaging applications." Food and Bioprocess Technology, 5 (5): 1447–1464. doi:10.1007/s11947-012-0797-6.

Ezhilarasi, P.N., P. Karthik, N. Chhanwal, and C. Anandharamakrishnan. 2013. "Nanoencapsulation techniques for food bioactive components: A review." Food and Bioprocess Technology, 6 (3):628–647.FDA. 2012. "FDA's Regulatory Science Program in Nanotechnology." https://www.fda.gov/downloads/advisorycommittees/committeesmeetingmaterials/scienceboardtothefoodanddrugadministration/ucm322212.pdf.

FDA. 2013. "Nanotechnology Regulatory Science Research Plan." Office of the Commissioner. https://www.fda.gov/ScienceResearch/SpecialTopics/Nanotechnology/ucm273325.htm.

FDA. 2014a. "Guidance for Industry Considering Whether an FDA-Regulated Product Involves the Application of Nanotechnology." https://www.fda.gov/downloads/RegulatoryInformation/Guidances/UCM401695.pdf.

FDA. 2014b. "Guidance on the Safety and Regulatory Status of Food Ingredients and Food Contact Substances Including Food Ingredients That Are Color Additives." https://www.fda.gov/downloads/Cosmetics/GuidanceRegulation/GuidanceDocuments/UCM300927.pdf.

FDA. 2015. "Guidance for Industry Use of Nanomaterials in Food for Animals." https://www.fda.gov/downloads/AnimalVeterinary/GuidanceComplianceEnforcement/GuidanceforIndustry/UCM401508.pdf.

FDA. 2017a. "Guidance for Industry: Assessing the Effects of Significant Manufacturing Process Changes, Including Emerging Technologies, on the Safety and Regulatory Status of Food Ingredients and Food Contact Substances, Including Food Ingredients That Are Color Additives." Center for Food Safety and Applied Nutrition. Accessed March 21. https://www.fda.gov/Food/GuidanceRegulation/GuidanceDocumentsRegulatoryInformation/ucm300661.htm.

FDA. 2017b. "Guidance for Industry: Safety of Nanomaterials in Cosmetic Products." Accessed March 21. https://www.fda.gov/Cosmeics/GuidanceRegulation/GuidanceDocuments/ucm300886.htm.

FDA. 2017c. "Nanotechnology – Nanotechnology Fact Sheet." Office of the Commissioner. Accessed November 23. http://www.fda.gov/ScienceResearch/SpecialTopics/Nanotechnology/ucm402230.htm#guidancefood.

FDA. 2017d. "Nanotechnology | Food Standards Agency." Accessed February 13. https://www.food.gov.uk/science/novel/nano.

FD&C Act. 2017. *FD&C Act 505 [21 USC 355] New Drugs : Importation of Active Pharmaceutical Ingredients*. Accessed February 24. https://www.fda.gov/downloads/AboutFDA/CentersOffices/CDER/ucm118004.pdf.

FDA U.S. "FDA's Approach to Regulation of Nanotechnology Products." Accessed April 24, 2018. https://www.fda.gov/ScienceResearch/SpecialTopics/Nanotechnology/ucm301114.htm#_ftn1.

Federal Food Drug and Cosmetic Act (FFDCA). 2017. Accessed February 24. http://www.law.cornell.edu/uscode/uscprint.html.

Food Standards Agency. 2016. "Nanotechnology | Food Standards Agency." Accessed December 15. https://www.food.gov.uk/science/novel/nano.

Fraceto, Leonardo F., Renato Grillo, Gerson A. de Medeiros, Viviana Scognamiglio, Giuseppina Rea, and Cecilia Bartolucci. 2016. "Nanotechnology in agriculture: Which innovation potential does it have?" Frontiers in Environmental Science, 4 (March). Frontiers: 20. doi:10.3389/fenvs.2016.00020.

Frameworks Regulatory for Nanotechnology and in Foods. 2013. "Regulatory Frameworks for Nanotechnology in Foods and Medical Products," no. 2012. doi:10.1787/5k47w4vsb4s4-en.

FSA. 2008. "REPORT OF FSA REGULATORY REVIEW A Review of Potential Implications of Nanotechnologies for Regulations and Risk Assessment in Relation to Food." https://www.food.gov.uk/sites/default/files/multimedia/pdfs/nanoregreviewreport.pdf.

FSA-Food standards agency. 2017. "Nanotechnology-Enabled Foods and Food Contact Materials on the UK Market." Accessed February 25. https://www.food.gov.uk/science/novel/nano/monitoring.

Fuertes, Guillermo, Ismael Soto, Raúl Carrasco, Manuel Vargas, Jorge Sabattin, and Carolina Lagos. 2016. "Intelligent packaging systems: Sensors and nanosensors to monitor food quality and safety." Journal of Sensors 2016. Hindawi Publishing Corporation: 1–8. doi:10.1155/2016/4046061.

Granier, John J. and Michelle L. Pantoya. 2004. "Laser Ignition of nanocomposite thermites." Combustion and Flame, 138 (4): 373–383. doi:10.1016/j.combustflame.2004.05.006.

Griffiths, S. 2015. "Silver nanoparticles extend shelf life of milk." The Journal of the Institute of Food Science and Technology, 2. Retrieved from https://www.fstjournal.org/news/silver-nanoparticles-extend-shelf-life-milk/730

"Guidelines for Preventing the Environmental Impact of Manufactured Nanomaterials (abstract)." 2009. https://www.env.go.jp/chemi/nanomaterial/eibs-conf/guideline_0903_enab.pdf.

Hallock, Marilyn F., Pam Greenley, Lou DiBerardinis, and Dan Kallin. 2009. "Potential risks of nanomaterials and how to safely handle materials of uncertain toxicity." Journal of Chemical Health and Safety, 16 (1): 16–23. doi:10.1016/j.jchas.2008.04.001.

Handford, Caroline E., Moira Dean, Maeve Henchion, Michelle Spence, Christopher T. Elliott, and Katrina Campbell. 2014. "Implications of nanotechnology for the agri-food industry: opportunities, benefits and risks." Trends in Food Science and Technology. doi:10.1016/j.tifs.2014.09.007.

Handford, Caroline E., Moira Dean, Michelle Spence, Maeve Henchion, Christopher T. Elliott, and Katrina Campbell. 2015. "Awareness and attitudes towards the emerging use of nanotechnology in the agri-food sector." Food Control, 57: 24–34. doi:10.1016/j.foodcont.2015.03.033.

"International Council on Nanotechnology (ICON): CBEN." 2017. Accessed March 22. http://cben.rice.edu/industry/icon.aspx.

"International Organization for Standardization." 2017. Accessed March 22. https://www.iso.org/home.html.

"ISO/TS 27687:2008 – Nanotechnologies – Terminology and Definitions for Nano-Objects – Nanoparticle, Nanofibre and Nanoplate." 2016. Accessed December 15. http://www.iso.org/iso/catalogue_detail?csnumber=44278.

Ken Donaldson, Lang Tran, Luis Albert Jimenez, Rodger Duffin, David E. Newby, Nicholas Mills, William MacNee, and Vicki Stone. 2005. "Carbon black nanoparticles induce type II epithelial cells to release chemotaxins for alveolar macrophages." Particle and Fibre Toxicology, 2 (1). BioMed Central: 10. doi:10.1186/1743–8977-2-10.

Khan, Ahmed S. 2012. Nanotechnology : Ethical and Social Implications. Boca Raton, CRC Press.

Kulacki, Konrad J. and Bradley J. Cardinale. 2012. "Effects of nano-titanium dioxide on freshwater algal population dynamics." PLoS ONE, 7 (10). doi:10.1371/.

Kulacki, Konrad J., Bradley J. Cardinale, Arturo A. Keller, Raven Bier, Helen Dicksony, and Keywords—nanoparticle Tio. 2012. "How do stream organisms respond to, and influence, the concentration of titanium dioxide nanoparticles? A mesocosm study with algae and herbivores." Environmental Toxicology and Chemistry, 31 (10): 2414–2422. doi:10.1002/etc.1962.

Li, C., H. Liu, Y. Sun, H. Wang, F. Guo, S. Rao, J. Deng, et al. 2009. "PAMAM nanoparticles promote acute lung injury by inducing autophagic cell death through the akt-TSC2-mTOR signaling pathway." Journal of Molecular Cell Biology, 1 (1). Oxford University Press: 37–45. doi:10.1093/jmcb/mjp002.

Li, Shyh-Dar and Leaf Huang. 2008. "Pharmacokinetics and biodistribution of nanoparticles." Molecular Pharmaceutics, 5 (4): 496–504. doi:10.1021/mp800049w.

Lövestam, Göran, Hubert Rauscher, Gert Roebben, Birgit Sokull Klüttgen, Neil Gibson, Jean-Philippe Putaud, and Hermann Stamm. 2010. "Considerations on a Definition of Nanomaterial for Regulatory Purposes." https://ec.europa.eu/jrc/sites/jrcsh/files/jrc_reference_report_201007_nanomaterials.pdf.

"NanoPortal – Proposed Regulatory Framework for Nanomaterials under the Canadian Environmental Protection Act, 1999. 2017. Accessed March 21. http://nanoportal.gc.ca/default.asp?lang=En&xml=5F29EE34-5014-4EA3-9101-A937237E7A50.

"National Nanotechnology Initiative." Accessed March 22, 2017. http://www.nano.gov/about-nni.

NIOS – National Institute for Occupational Safety and Health. 2009. "Approaches to Safe Nanotechnology." https://www.cdc.gov/niosh/docs/2009-125/pdfs/2009-125.pdf.

OECD. 2011. "Nanosafety at the OECD: The First Five Years 2006 –2010 OECD." https://www.oecd.org/science/nanosafety/47104296.pdf.

OECD. 2017a. "About the OECD – OECD." (Accessed February 26, 2017) http://www.oecd.org/about/.

OECD. 2017b. Emerging Technologies. (Accessed March 22.2017) http://www.oecd.org/sti/nano/.

OECD. 2016a. "Nanomaterials Exposure Case Study Series on the Safety of Manufactured Nanomaterials." No 77, ENV/JM/MONO(2016)60 http://www.oecd.org/officialdocuments/publicdisplaydocumentpdf/?cote=env/jm/mono(2016)60&doclanguage=en

OECD. 2016b. "OECD Environment, Health and Safety Publications, Series on the Safety of Manufactured Nanomaterials, No. 53. DOSSIER ON SILVER NANOPARTICLES." 1 (168): 1–59. doi:ENV/JM/MONO(2007)10.

Ralf Cornelissen, IVAM UvA bv, IndusTox Consult Frans Jongeneelen, IVAM UvA bv Pieter van Broekhuizen, and IVAM UvA bv Fleur van Broekhuizen. 2011. "Guidance Working Safely with Nanomaterials and –products, the Guide for Employers and Employees." http://www.industox.nl/Guidance on safe handling nanomats&products.pdf.

Regulation (EC) No. 258/97 of the European Parliament and of the Council of 27 January 1997 Concerning Novel Foods and Novel Food Ingredients. 2017. Official Journal L 043, 14/02/1997 P. 0001 – 0006; OPOCE. Accessed March 22. http://eur-lex.europa.eu/LexUriServ/LexUriServ.do?uri=CELEX%3A31997R0258%3AEN%3AHTML.

Renwick, L. C., D. Brown, A. Clouter, and K. Donaldson. 2004. "Increased inflammation and altered macrophage chemotactic responses caused by two ultrafine particle types." Occupational and Environmental Medicine, 61 (5). BMJ Publishing Group Ltd: 442–47. doi:10.1136/OEM.2003.008227.

Ronals Sandler. 2009. "Nanotechnology: The Social and Ethical Issues." http://www.pewtrusts.org/~/media/legacy/uploadedfiles/wwwpewtrustsorg/reports/nanotechnologies/nanopen16finalpdf.pdf.

"Safe Handling and Use of Carbon Nanotubes." 2012. http://www.safeworkaustralia.gov.au/sites/SWA/about/Publications/Documents/664/Safe Handling and Use of Carbon Nanotubes.pdf.

"Safe Handling of Nano Materials and Other Advanced Materials at Workplaces." 2017. Accessed March 21. http://www.nanosafetycluster.eu/nanoToGo/Nano to go!/Brochure/Safe handling of nanomaterials and other advanced materials at workplaces_v1-0.pdf.

Shuk Man CHOW Centre for Food Safety Hong Kong. 2011. "Centre for Food Safety – Food Safety Focus – Nanotechnology and Food Safety." Hong Kong. http://www.cfs.gov.hk/english/multimedia/multimedia_pub/multimedia_pub_fsf_54_02.html.

Sirelkhatim, Amna, Shahrom Mahmud, Azman Seeni, Noor Haida Mohamad Kaus, Ling Chuo Ann, Siti Khadijah Mohd Bakhori, Habsah Hasan, and Dasmawati Mohamad. 2015. "Review on zinc oxide nanoparticles: Antibacterial activity and toxicity mechanism." Nano-Micro Letters, 7 (3): 219–242. doi:10.1007/s40820-015-0040-x.

Tas, C. Erdinc, Saman Hendessi, Mustafa Baysal, Serkan Unal, Fevzi C. Cebeci, Yusuf Z. Menceloglu, and Hayriye Unal. 2017. "Halloysite Nanotubes/Polyethylene Nanocomposites for Active Food Packaging Materials with Ethylene Scavenging and Gas Barrier Properties." *Food and Bioprocess Technology*, January. Springer US, 1–10. doi:10.1007/s11947-017-1860-0.

US Food and Drug Administration. 2007. "A Report of the U.S. Food and Drug Administration Nanotechnology Task Force July 25, 2007." http://www.fda.gov/downloads/ScienceResearch/SpecialTopics/Nanotechnology/ucm110856.pdf.

Vignes, A., F. Muñoz, J. Bouillard, O. Dufaud, L. Perrin, A. Laurent, D. Thomas. 2012. "Risk assessment of the ignitability and explosivity of aluminum nanopowders." Process Safety and Environmental Protection, 90 (4): 304–310. doi:10.1016/j.psep.2011.09.008.

"Work Health and Safety Assessment Tool for Handling Engineered Nanomaterials." 2017. Accessed March 21. http://www.safeworkaustralia.gov.au/sites/SWA/about/Publications/Documents/547/Work_health_safety_tool_handling_engineered_nanomaterials.pdf.

World Health Organization (WHO). 2004. "IPCS Risk Assessment Terminology IPCS; International Programme on Chemical Safety- Inter Organization Programme for the Sound Management of Chemicals – a Cooperative Agreement among UNEP, ILO, FAO, WHO, UNIDO, UNITAR and OECD."

World Health Organization (WHO). 2009. "Chapter 2- Risk Assessment and Its Role in Risk Analysis; IPCS Environmental Health Criteria 240 Principles and Methods for the Risk Assessment of Chemicals in Food." http://www.inchem.org/documents/ehc/ehc/ehc240_chapter2.pdf.

WPMN. 2017. "Testing Programme of Manufactured Nanomaterials – Dossiers and Endpoints – OECD." Accessed February 26. http://www.oecd.org/chemicalsafety/nanosafety/dossiers-and-endpoints-testing-programme-manufactured-nanomaterials.htm.

Wu, Hong-Chun, Hsin-Jung Ou, Hsiao-Chi Hsiao, and Tung-Sheng Shih. 2010. "Explosion characteristics of aluminum nanopowders." Aerosol and Air Quality Research, 10: 38–42. doi:10.4209/aaqr.2009.06.0043.

Yetter, Richard A., Grant A. Risha, and Steven F. Son. 2009. "Metal particle combustion and nanotechnology." Proceedings of the Combustion Institute, 32 (2): 1819–1838. doi:10.1016/j.proci.2008.08.013.

Section 2

Product and Processes

6 Fabrication of Nanomaterials

Preethi R., Maria Leena, J.A. Moses, and C. Anandharamakrishnan

CONTENTS

6.1 INTRODUCTION

Nanomaterials provide a unique outlook to the design and construction of materials due to their size range within the 100 nm scale, which has an impact on various fields such as energy, medical, electronics, and recently the food industry. The world of unseen tiny particles is dominated by the forces of physics and chemistry. These particles possess unusual properties not present in traditional or ordinary materials. Recent advances of nanotechnology bestow a platform to offer a momentous revolution in the food processing and preservations sector with novel approaches that combine active molecules to create nanostructures. Nanomaterials in the food industry deliver on the promise of improving the quality of food through an excellent performance in controlling microbial growth and retaining flavor and nutrients. Nanotechnology is becoming an astonishing process in the food revolution with its use of nano supplements, nanofood processing, nanobar codes, organic nanoadditives, self-sanitizing surfaces, food freshness indicators, nanocoating, and nanoflim in kitchenware, which require the fabrication of nanoparticles with various methods. This chapter provides a comprehensive overview of the physical, chemical and biological methods used to fabricate nanoscale particles to meet the demand for potent nanofood ingredients.

6.2 FABRICATION OF NANOPARTICLES

Synthesis of nanoparticles falls into two categories, a top-down approach, and a bottom-up approach. Converting materials from a larger size to nanoscale is known as a top-down method; a nanomaterial starting from the atomic level is the bottom-up method. The properties of nanomaterials, such as their electrical, mechanical, chemical, and luminescent properties, are highly dependent on size, shape, morphology, composition, crystallinity, etc. The selection of a well-defined method is required for the synthesis of nanomaterials with the desired properties.

6.2.1 Chemical Synthesis of Nanomaterials

The chemical synthesis method is one of the simplest and most established routes for nanomaterial synthesis. Traditionally in the chemical synthesis, nanoparticles fabrication involves many steps. The reaction medium for the synthesis of nanomaterials contains three major components: precursors, reducing agents, and stabilizing agents or capping agents. Initially, in the reduction method, the formation of metal atoms can succeed by reduction of metal precursors using chemicals as reducing agents. The developed metal atoms then undergo a nucleation process through slow growth, and finally, are built as nanoparticles. Sodium citrate, ascorbic acid, alcohol, and hydrazine compounds are widely used in the synthesis of nanoparticles through chemical methods (Yu et al., 2008; Thatai et al., 2014). Generally, chemical synthesis routes can be grouped into two categories: liquid-phase and vapor-phase synthesis.

6.2.1.1 Vapor-Phase Fabrication

In this chemical synthesis route, a precursor in vapor state undergoes condensation, nucleation, and growth stages and thus forms nanostructures. The vapor form of the desired precursor material may be generated by different forms of energy source such as thermal, laser, evaporation, plasma, electron beam, etc. The thermodynamically unstable supersaturated vapor-phase molecules react to form a condensed state followed by nucleation. After the nucleation, vapor-phase molecules condense or react with nucleated particles and this results in particle growth. This also prevents further supersaturation of the vapor phase (Figure 6.1) (Swihart, 2003). Precursor composition, catalytic activity, vapor pressure, temperature, and particle retention time control the size and morphology of the nanoparticle synthesized (Sakthivel and Venkatesh, 2012).

Azam et al. (2018) synthesized graphene nanoplatelets using palm-based waste cooking oil as the source of carbon and nickel oxide powder as a catalyst in catalytic a chemical vapor deposition process under argon gas at atmospheric pressure. A thin graphitic structure composed of less than 50 stacks and irregular surface with an approximate size range of 20 nm to 1.5 μm were produced (Azam et al., 2018). As the palm-based cooling oil is generally used in food industries, this study provides the best recycling and waste management process to produce high value-added products. A similar process can be used for synthesized monolayer graphene from waste chicken fat Rosmi et al., 2016). Fatty acids present in chicken fat decompose into short hydrocarbon chains and serve

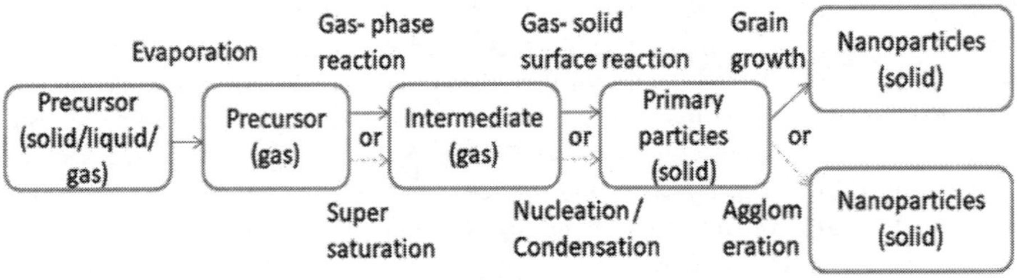

FIGURE 6.1 Concepts of bottom-up vapor-phase nanoparticle synthesis process. (Sakthivel and Venkatesh, 2012.)

as carbon source for graphene formation through a low-pressure chemical vapor deposition method in a multistep process, where the chicken fat is first converted to oil by heating it at 200°C, and then the oil is heated up to 1080°C at vacuum pressure of 2 Pa and 100 SCCM of pure H_2 gas. Then to the pressure of 150 Pa for 60 min in LPCVD system for 60 min followed by cooling down to room temperature. The formation of a uniform, smooth graphene structure was noticed in the cooper substrate and exhibited high conductivity which could be used on sensing applications (Rosmi et al., 2016).

6.2.1.2 Liquid-Phase Fabrication

Liquid-phase synthesis of nanomaterials is based on the wet chemical method, where the precursors in a liquid medium nucleate and undergo crystal growth. A schematic picture of nanoparticle formation from liquid-phase synthesis is shown in Figure 6.2. Compared to the gas-phase method, this is a low-temperature process and provides good control over the particle size and crystal shape compared to gas-phase synthesis. Also, it produces highly porous and monodisperse particles. Other methods in wet chemical synthesis are precipitation, sol-gel process, thermal decomposition, hydro or solvent thermal synthesis, microemulsion, and microwave-assisted and ionic liquid-based syntheses (Gai et al., 2014). Both organic and inorganic nanoparticles are produced through the wet chemical synthesis route.

1. *Co-precipitation method*

   Among the various chemical routes, co-precipitation from aqueous solutions is one of the easiest, most frequently used, and less energy consuming methods for nanoparticles production. Various inorganic nanoparticles and organic nanocarrier systems are produced by precipitation method. Synthetic hydroxyapatite (HAp) inorganic nanoparticles, which have a wide range of medical applications in bone and dental implant coatings, are prepared via wet chemical precipitation method. For example, Sibte et al., (2014) prepared HAp nanoparticles with 8.47–24.47 nm crystallite size range through precipitation of $Ca(OH)_2$ and H_3PO_4 precursors in an aqueous medium. Leena et al. (2016) produced biomimetic, bone-like β-type carbonated nano HAp with rod shape in 27×7 nm dimensions through wet chemical precipitation method. The higher concentration (3.5×) of precursors $CaCl_2$ and Na_2HPO_4 in stimulated body fluid precipitated and produced nano HAp in a shorter time (3 h).

   Fessi et al. (1989) referred to interfacial deposition followed by de-solvation as nanoprecipitation method for polymeric nanoparticle production. Initially, it was attempted to produce drug-loaded polymeric nanoparticles from poly-(D,L-lactide) polymer at the oil in water interface following acetone displacement from the oily nanodroplets (interfacial polymer deposition followed by solvent displacement). Through this approach, various drugs, such as Dexamethasone, Taxol, Indo-methacin, and loaded PLA nanoparticles with a particle size range of 40 to 100 nm, were produced. Sanna et al. (2013) developed trans resveratrol-loaded poly(epsilon-caprolactone) and poly(D,L-lactic-co-glycolic

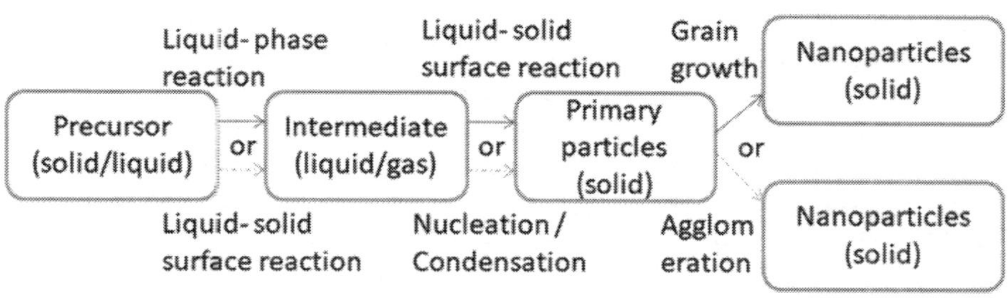

FIGURE 6.2 Concept of liquid-phase fabrication. (Sakthivel and Venkatesh, 2012.)

acid)–poly(ethylene glycol) blend polymer nanoparticles through nanoprecipitation method. Similarly, resveratrol-loaded core-shell nanoparticles were prepared by antisolvent precipitation of resveratrol and zein polymer in ethanol solution in water. After the removal of ethanol by evaporation using a rotary evaporator, the core-shell nanoparticles were produced by dispersing zein nanoparticles in the pectin solution (Huang et al., 2017). Similarly, many food active molecules such as curcumin, quercetin, tocopherol, etc., -loaded nanoparticles were produced by nanoprecipitation method (Patel et al., 2010).

2. *Sol-Gel process*

Sol-gel is the process in which the liquid precursors, usually metal alkoxides, undergo hydrolysis and condensation and form a colloidal suspension referred to as "sol." The formed "sol" undergo network formation, referred to as "gel" of inorganic polymers and ceramics. There are several types of gel networks that form in sol-gel process: (a) colloidal gel were the particles are connected by van der Waals or hydrogen bonding, (b) inorganic polymers interconnected via covalent or intermolecular bonding referred to as metal–oxane polymers, (c) weakly interconnected metal complex, (d) *in situ* polymerization through covalent and co-ordinate bonding of organic polymers (Pechini method), and (e) polymer networks forming through intermolecular bonding.

The key steps of the sol-gel process are as follows (Danks et al., 2016):

a. Synthesis of the colloidal solution or "sol" of precursor particles, usually metal alkoxides/hydrated metal species in very small particle size (0.1–1 μm) which are distributed in liquid phase through Brownian motion.
b. Gelation: metal–oxo–metal or metal–hydroxy–metal bonds formation though polycondensation.
c. Syneresis or "aging" where condensation continues within the gel network, often shrinking it and resulting in the expulsion of solvent.
d. Drying the gel either to form a dense "xerogel" via the collapse of the porous network or an aerogel for example through supercritical drying.
e. Removal of surface M–OH groups through calcination at high temperature up to 800°C (if required).

Sol-gel is a conventional process for metal oxide micro and nanoparticles preparation from solution state precursors (Danks et al., 2016). Various inorganic nanostructures with superior properties such as photocatalytic, magnetic, conductivity, UV protection, etc., have been conventionally prepared through a sol-gel process. It offers advantages such as the molding of inorganic materials into complex geometries and thin coating formation at low temperature due to gelation process. (Petrova et al., 2015) produced zinc-iron oxide nanoparticles with superparamagnetic behavior at room temperature. Through the sol-gel process, nanoparticles with the highest magnetization of 59 emu/g at 300 K for $Zn_{0.18}Fe_{2.82}O_4$ and 86 emu/g at 5 K for $Zn_{0.45}Fe_{2.55}O_4$ particles, both exceeding the maximum magnetization of the pure magnetite, were produced. Azimi-Foulad et al. (2018) reported TiO_2-CdO-Ag nanocomposite with superior photocatalytic property produced through sol-gel method. The prepared photocatalytic nanocomposite effectively degraded the methylene blue, an organic dye in water, through the photocatalytic process. Synthetic and biopolymers have also been effectively used for *in situ* polymerization around metal ions to ensure the metals are mixed and stabilized homogeneously within a covalent matrix or "gel" (Pechini synthesis). Mirzaei et al. (2016) reported highly stable α-Fe_2O_3 nanoparticles formation through Pechini sol-gel process where metallic citrate is polymerized by ethylene glycol in a two-step reaction process: 1) the formation of a complex between an organic acid, such as citric acid, with the precursor metal (Fe $(NO_3)_3.9H_2O$) and, 2) an esterification reaction with ethylene glycol. Here the synthetic polymer poly(vinylpyrrolidone) was used for *in situ* polymerization. The developed nanoparticle was employed for sensing ethanol vapor at ppm level, one of the most important parameters to be sensed in the food industry. This ethanol sensor exhibited 8-fold higher response towards ethanol vapor at

50 ppm and 2.5 times higher selectivity compared to other vapors and gases. The sol-gel process offers extremely thin layers of nanoparticles such as gold, aluminum, SiO_2, TiO_2 on a metal or ceramic surface, which have been prepared for various applications such as superior conducting materials for sensing application, antireflection, refraction, etc.

3. *Thermal decomposition*

The chemical reaction of precursors in the aqueous medium is performed in an autoclave or reactor at high temperature (>200°C) and high pressure (>2000 psi). The hydrothermal process is suitable for a substance which is insoluble in atmospheric condition (temp < 100°C and < 1 atm). Usually, hydrothermal treatment produces monodisperse nanoparticles in a short time. Precursor concentration, temperature and residence time influence the final particle size and morphology (Hao and Teja, 2003). The advantage of hydrothermal synthesis under supercritical water is also explored for the synthesis of metal oxides nanomaterials. Above the critical point (374°C and 22.1 Mpa for water) the dielectric constant and density of solvent changes drastically, which affects the solubility of metal oxides or any other solutes. Supercritical water exhibits low dielectric constant (<10) which allows high solubility of organics and low solubility of inorganics, allowing the synthesis of hybrid nanoparticles. The solubility of metal oxide and hydrolysis rate affect the particle size. Under supercritical conditions, the hydrothermal reaction rate is enhanced more than 10^3 times than that of conventional hydrothermal reaction. Additionally, supercritical water can reduce the alkaline concentration, thus making the hydrothermal process free toxic organic solvents in the nanoparticle synthesis suitable for green and sustainable chemistry (Hayashi and Hakuta, 2010; Xu et al., 2008). Nano hydroxyapatite particles have been prepared through hydrothermal degradation of various natural sources, including egg shells, cuttlebone, etc. For example, cuttlebone in $(NH_4)_2HPO_4$ solution was sealed in Teflon-lined stainless-steel reactors and run at 180°C for 96 h (Hongmin et al., 2014). This hydrothermal treatment formed porous hydroxyapatite bone substitute with a nanocrystalline surface. Copper nanoparticles possess various applications with catalytic and optical properties. Therefore, Togashi et al. (2018) fabricated copper nanoscale particles through oleylamine as precursor under the thermal decomposition process. This research work clearly indicates that decomposition temperature strongly affected size distribution (Figure 6.3).

Recently Sharma et al. (2017) synthesized spongy carbon nanoglobules with a diameter in the range of 150 ± 25 nm through hydrothermal treatment of pineapple juice and ethanol

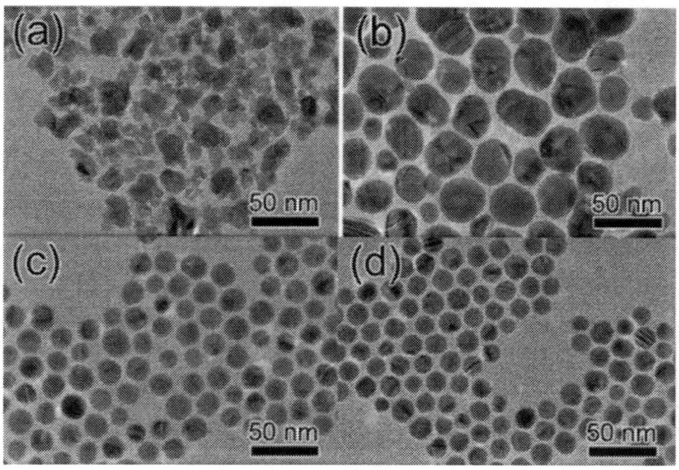

FIGURE 6.3 TEM images of copper nanoparticles synthesized at (a) 200, (b) 220, (c) 240, (d) 260°C. (Togashi et al., 2018.)

in a Teflon-lined stainless-steel autoclave for 3 h at 150°C. Highly stable fluorescent hydrophilic carbon nanoglobules formed in this study exhibited photo luminance peak at 453 nm upon excitation at 370 nm. This fluorescent CNG was investigated for sensing of toxic metal ions. When exposed to chromium (VI) ions, a drastic decrease in the fluorescence intensity of CNG was observed over the linear range of concentrations 0–18 μM.

4. *Microwave-assisted fabrication*

Microwaves (MWs) are electromagnetic waves in the wavelength range of 1 mm and 1 m and frequencies ranging from 0.3 to 300 GHz. At the conventionally used frequency of 2.45 GHz with a wavelength of about 12.24 cm the energy of microwave photons is just about 1 J mol^{-1} (i.e. 1.0×10^5 eV) which is too low to cleave chemical bonds, even lower than the Brownian motion, but matches the rotational energies of molecules. Thus, microwaves only affect the rotation of the molecule and heat the materials without inducing any chemical reaction (Zhu and Chen, 2013). MWs generally heat any materials containing mobile electric charges, such as polar molecules in a solvent or conducting ions in a solid. Heating of materials occurs when rotating polar molecules lose energy in a collision with neighboring molecules. Whereas, in the case of conducting and semi-conducting materials, ions or electrons form an electric current and energy is lost due to the electrical resistance of the material (Tsuji, 2017). Unlike high-energy vibration and electronic excitation of molecules caused by IR, UV- Vis waves, MWs cause only low-energy dielectric heating.

MW heating allows rapid and direct heating of reactants which permits higher reaction rate, shorter reaction time, and high yield compared to conventional heating through furnace or oil bath in which heating occurs only through convection or conduction. Thus, MWs demonstrate significant advantages in terms of reduced processing time and less environmental impact (Gawande et al., 2014). Dunn et al. (2017) used microwave heating to produce optically transparent highly stable lipid nanoparticles from Brij 78 and vitamin E TPGS, and trilaurin, trimyristin, or miglyol 812 by assessing phase inversion temperature. Lipid nanoparticles with a size range of 20–30 nm and low polydispersity were attained in one minute. Kahrilas et al. (2014) reported MW-assisted synthesis of silver nanospheres with 7–8 nm diameter from silver nitrate precursor using orange peel extract as reducing and capping agent. MW heating assisted the formation of silver nanoparticles within 15 min. Similarly, El-Nagga et al. (2016) reported gold nanoparticle and silver-gold core-shell nanoparticle formation through MW irradiation. Here, curdlan biopolymer is used as both reducing and capping agent for tetrachloroauric acid and silver nitrate under MW irradiation for 1 min for nanoparticle formation. For core-shell nanoparticle production, a portion of $AgNO_3$ was initially reduced by MW irradiation and then an equivalent portion of $HAuCl_4$ was added and allowed to reduce through MW heating and formed as core-shell nanoparticles of 45 nm size.

5. *Sonochemical fabrication*

In sonochemical synthesis, application of powerful ultrasound radiation (20 kHz–10 MHz) creates bubbles (acoustic cavitation) which grow due to the diffusion of solute vapor into the vacuum of the bubble and collapse when the bubble size reaches its maximum. The range of events that a bubble could undergo in ultrasonic fields, such as bubble nuclei, dissolution or coalescence, diffusion, pressure node, resonance, and collapse, is elsewhere reported by Ashokkumar et al. (2007) (Figure 6.4). Upon collapse of the bubble (which usually occurs in less than nanoseconds), a very high temperature (5000–25000 K) is reached, which creates a hot spot and high pressure ~ 1000 bar followed by rapid cooling rates > 10^{11} K/s. This fast kinetics (rapid cooling and heating rate) does not allow the growth of nuclei and hinders the formation of crystallization, producing only amorphous nanoparticles. In the case of nonvolatile compounds, a chemical reaction occurs in 200 nm radius near the collapsing bubble which may form nano crystalline or nanoamorphous materials (Gedanken, 2004). The sonochemical effect can be categorized as (a) primary

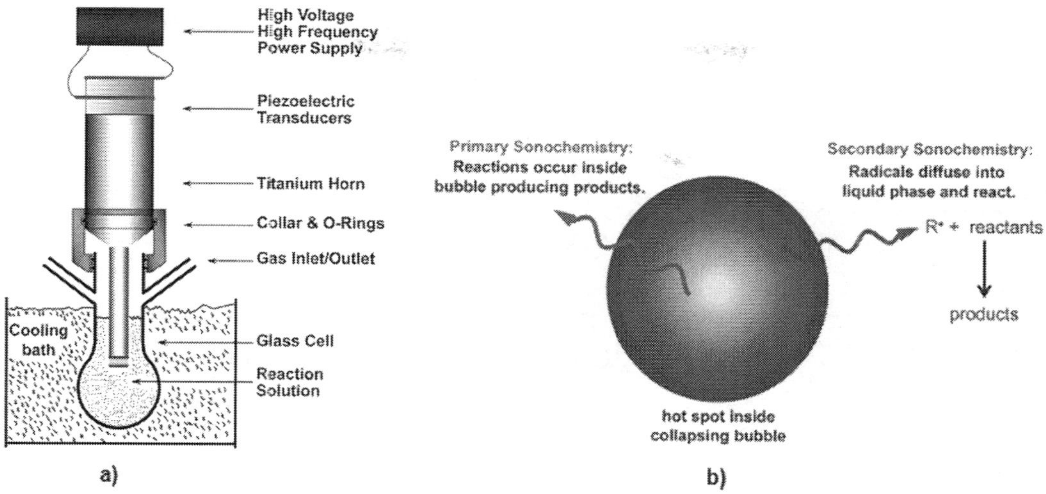

FIGURE 6.4 (a) A typical laboratory rig for the sonochemical synthesis of nanomaterials (b) Primary and secondary sonochemistry for preparation of nanomaterials. (Bang and Suslick, 2010; Xu et al., 2013.)

effect where gas-phase chemistry occurs inside collapsing bubbles, (b) secondary effect where solution-phase chemistry occurs outside the bubbles, and (c) physical effect caused by high-speed jets or shock waves derived from bubble collapse.

The physical effects of ultrasound may cause surface damage, high-speed inter-particle collisions, and fragmentation of friable solids. For example, Tanya et al. (2004) reported two Zn particles melting at the point of contact due to high-intensity ultrasound. Jiang et al. (2004) reported the conversion of chemical precursors, namely silver nitrate and chloroauric acid, in the presence of PVP as a capping agent to uniform triangular silver nanoplates and ring-like gold nanostructure using ultrasonic intensity at 60 W/cm^2 for 30 min, induced through ultrasonic probe at ambient conditions. Similarly, many metal nanoparticles including iron, silver, gold, silica, Se, ZnO, etc. have been prepared by sonochemical route (Xu et al., 2013; Bang and Suslick, 2010).

6.2.2 Physical Method

6.2.2.1 Ball Milling

Nanoparticles fabricated by milling have a broad application in the food industry. Macro or micro food particles in the form of liquid or solid are reduced into nanoscale by applying shear forces generated by moving beads as well as particle collision. Bulk materials with dispersion media are placed in the mixing chamber with an agitator or by rotating the chamber. Beads such as steel, glass, polystyrene resin or ceramics are placed in the milling chamber (Figure 6.5). Ball or bead milling is used for the reduction of particles size. Cornstarch powder with 100 nm was achieved from 0.26–9.61 μm. Similarly, chitosan, wheat bran dietary fiber, and yam were milled to nano size from submicrometer particles. Chitosan is a promising functional food ingredient: different molecular weights of chitosan were pulverized to nanoscale using media milling to improve their physicochemical properties (Khandel and Kumar Shahi, 2016). To alter the chemical nature of metals such as aluminum, magnesium, and carbon, it is possible to blend them in the ball milling. Both organic and inorganic nanoparticles size reduction can be obtained from this system. The structural and chemical changes are produced by mechanical energy, with 2 to 20 nm in size. The size depends on the speed of the rotation of the balls. The disadvantages of this method are the formation of irregular shapes and crystal defects (Saleh and Gupta, 2016).

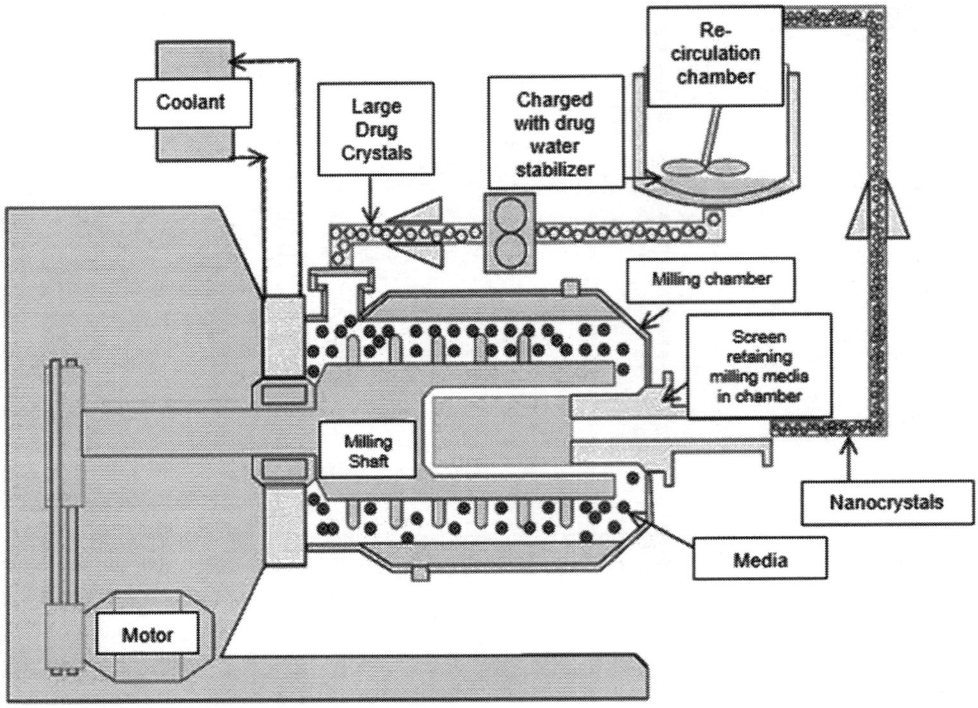

FIGURE 6.5 Schematic diagram of wet media milling. (Wais et al., 2016.)

6.2.2.2 High-Pressure Homogenization

The high-pressure homogenizer is the conventional method of producing small droplets in the food industry. Coarse emulsions are fed into the inlet of a high-pressure value homogenizer using a high shear mixer (Figure 6.6), then the coarse emulsion is transferred into a chamber through a pump, and forced through narrow value to the end of the chamber. The disruption of the droplet depends on the nozzle designs, pressure, and viscosity ratio. During the homogenization, it is important to monitor the presence of a sufficient amount of emulsifier, which helps to cover the surface of all the newly formed droplets to prevent re-coalescence. Table 6.1 summarized the nanoemulsion used in the food industry fabricated using a high-pressure homogenizer.

The microfluidizer (Figure 6.7) is applied in the food industry to fabricate flavor emulsions and nutraceutical emulsions, which have potential application in beverages and dairy emulsions. Similarly to the high-pressure homogenizer, the microfluidizer uses high pressure (up to 1700 bar) to pass a premix medium (emulsion) through an orifice to facilitate the nanoparticles through droplet disruption. However, the emulsion is flowing within the homogenizer. Various nanoemulsions were formed using microfluidizer peanut oil-Tween 80-SDS, Hexadecane-Tween 80-SDS, Corn oil-whey protein and MCT-Tocopherol-Glycerol, which has broad application in the food processing field (Patel et al., 2008; McClements, 2011).

In the ultrasonic homogenizer, based on high-intensity ultrasonic waves, compelling forces are created to disintegrate oil and water phases into nano-sized particles. The sonicator probe with piezoelectric quartz crystals uses a liquid medium, which provides the necessary energy to break down the particles through intense mechanical vibrations. The efficiency of synthesis is based on an amount of emulsifier, oil-phase viscosity, and an aqueous phase. The significant disadvantages of this process are oxidation of lipids, protein denaturation, and polysaccharide depolymerization at high-intensity sonication. Sunflower oil, flaxseed oil, olive oil, sesame oil, soybean oil, and canola oil are converted into nanoemulsion using the ultra homogenizer (McClements, 2011; Wais et al., 2016).

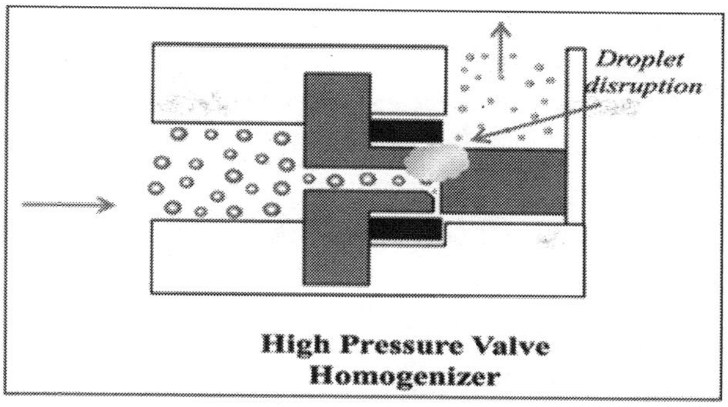

FIGURE 6.6 Schematic representation of high-pressure valve homogenizer. (McClements and Rao, 2011.)

TABLE 6.1
List of Nanoemulsion Fabrication Using High-Pressure Valve Homogenizers.

Oil Phase	Bioactive Component	Surfactant/Co-Surfactant	Particle Diameter (nm)
MCT	β–carotene	Tween 20,40,60,80	132–148
MCT	β–carotene	Tween 20, decaglycerol monolaurate, whey protein isolate, modified starch	115–178
Oleoresin capsicum/ ethyl acetate	Poly-caprolactone	Poly-caprolactone Pluronic F68	320–460
MCT	Curcumin	Tween 20	160

Source: McClements and Rao (2011).

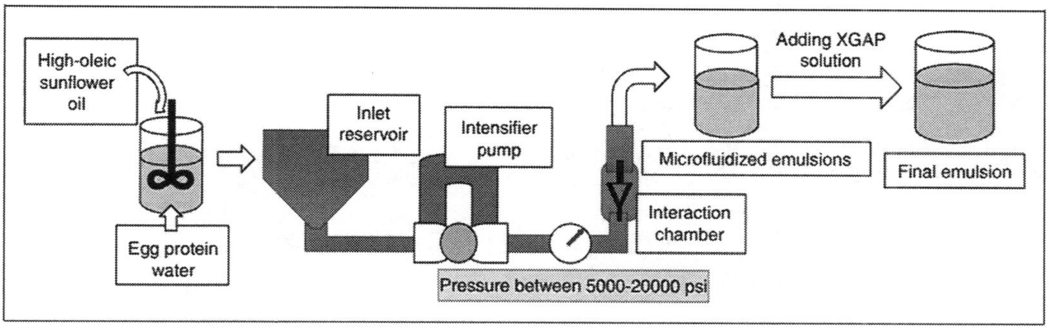

FIGURE 6.7 Schematic representation of microfluidizer. (Santos et al., 2018.)

6.2.2.3 Laser Ablation Method

This is a significant method when compared to liquid or gas-phase evaporation methods, due to the fabrication of nanoparticles with a low level of impurities and narrow size distribution. The laser is utilized as a source of energy for ablating solid or bulk materials to the nanoscale material. Bulk materials are exposed to high-energy laser, at a specific point on a solid surface, where light-absorbing materials evaporate (Figure 6.8). Ablation refers to the amputation of atoms by a breakdown of the chemical bonds as well as thermal evaporation (multiphoton excitation). The plasma structure depends upon the target materials and ambient media (liquid or gas). Recently, an advanced laser ablation technique has developed to regulate the size, shape, and aggregation

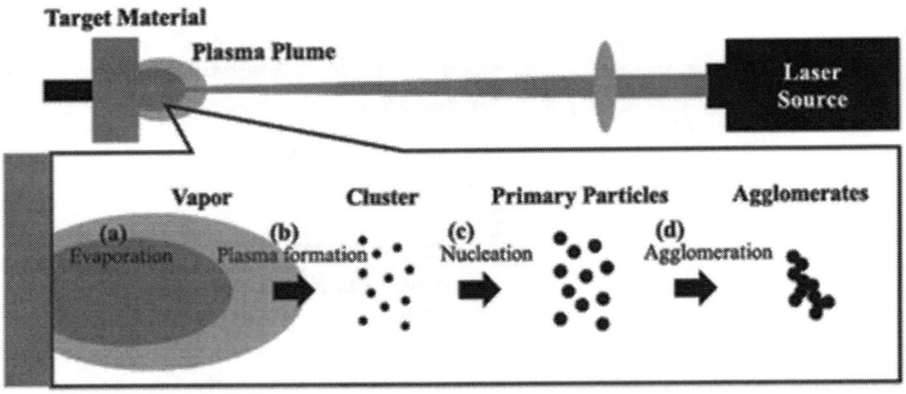

FIGURE 6.8 Schematic diagram of laser ablation method. (Kim et al., 2016.)

of the nanoscale particles. Figure 6.8 illustrates the mechanism behind this technique (Ghorbani, 2014; Kim et al., 2016). The various metal nanoparticles syntheses obtained through this method have fewer impurities, since metal and metal oxide nanoparticles have a unique feature for optical, catalytic, electrical, and magnetic properties, which is adapted for the fabrication of nanosensors.

6.2.2.4 Lithography

This is a well-established process for making nanostructured materials and patterns. The term lithography derives from the Greek words lithos – rock and grapho – to write, giving the meaning "small writing on a rock." Depending on the desired output, lithography can be performed using four different sources, such as light (optical or photolithography), electrons (e-beam lithography), ions (i-beam lithography), or X-ray (X-ray lithography, LIGA), based on depositing, etching or writing on a surface (Figure 6.9). The patterning of both organic and inorganic materials possesses applications in the field of electronics, integrated circuits, microfluidics devices, ion-selective electrodes, membranes, and hydrogel. All lithographic methods are commonly used to fabricate nanostructures with controlled shape and size, and spacing. The most common technique is photolithography. Due

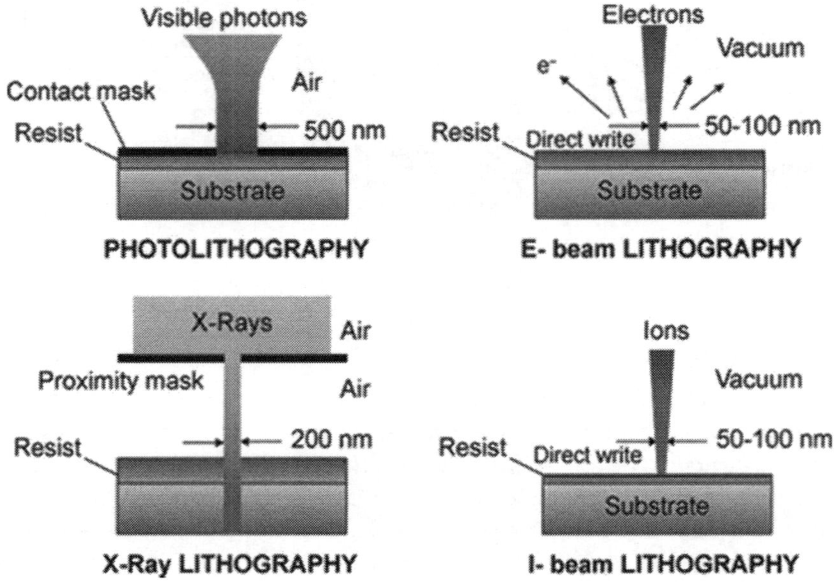

FIGURE 6.9 Lithography technique. (Daraio and Jin, 2012.)

to the limited diffraction resolution of photolithography, this is not applicable for nanoparticles fabrication.

Nanosphere lithography is the most suitable tool for nanofabrication. It is a low-cost tool to produce well-ordered, 2D periodic arrays of nanoparticles on the substrate. Size-tunable silver nanoparticles arrays developed through the nanosphere method, which has potent optical properties and is used as a filter, biosensors, chemical sensors, and bioprobes (Hulteen et al., 1999). Similarly, gold nanoparticles arrays are also fabricated by using this technique (Tan et al., 2005).

Based on stereolithography, 3D printing technology has developed and has been introduced by 3D Systems company (US-based). It is a laser-based technology; an initial layer formed through a pre-polymeric resin is treated with UV light in the shape of a 2D cross-section; similarly, with the same procedure a final layer is built to construct the 3D structure. Stereolithography provides programmed and systematized fabrication technique of 3D geometries. Therefore, it is a more convenient, cost-efficient and faster method. This method helps to convert sinter sugars and sugar-rich powders into complex shapes at a range of scales (Wegrzyn, Golding, and Archer, 2012; Yazdi et al., 2016). Multiple colors, flavors, and structures can be achieved through 3D printing. Due to their strong viscosity, resins are not used, but by implementing 3D printing techniques, silver nanoparticles along with resin developed as a potent nanocomposite with antimicrobial effects, which is used in the field of food packaging (Taormina et al., 2018).

6.2.2.5 Electrospinning

The electrospinning process is divided into two stages. In the first stage, polarization takes place in the polysaccharide solution under the impact of a high electric field (typically 1 kV/cm), the polymers form Taylor cones and jets. A Taylor cone is an aspect of the deformation of an immobile or pending droplet in an electric field, which brings down the droplet into a pointed shape. This instability of droplets may alter into a series of microscopic beads or a microscopically thin fluid jet at an appropriately high electric field strength. There is a formation of fiber in the second stage, which is collected in the collector phase. In the initial stage, the structure of the solution at the tip of the needle will likely determine the possibility to form a jet. The jet is elongated while traveling to the collector plate, due to the evaporation of the water, and then turns into a fiber (Figure 6.10) (Stijnman et al., 2011).

Electrospinning is a developing technology which opens up new outlets in food science. It effects the encapsulation of active molecules and flavor using food-grade biopolymers which are biocompatible, derived from plants, seaweeds, and microbes. Galactomannans, glucomannans, starch, methylcellulose, carrageenans, alginate, pullulan, and dextran are widely applied in the food industry. The electrospun nanofibers are determined based on the bend ratio on the solution such as viscosity, surface tension, and electrical conductivity. Addition of pullan provides a uniform nanofiber, to increase the glass transition temperature of the whey protein isolates. Other than that, this method is widely used in the immobilization of enzymes, vitamins, antimicrobials, nutraceutical delivery systems, controlled release of biomaterials (for beta-carotene and curcumin), filtration, and

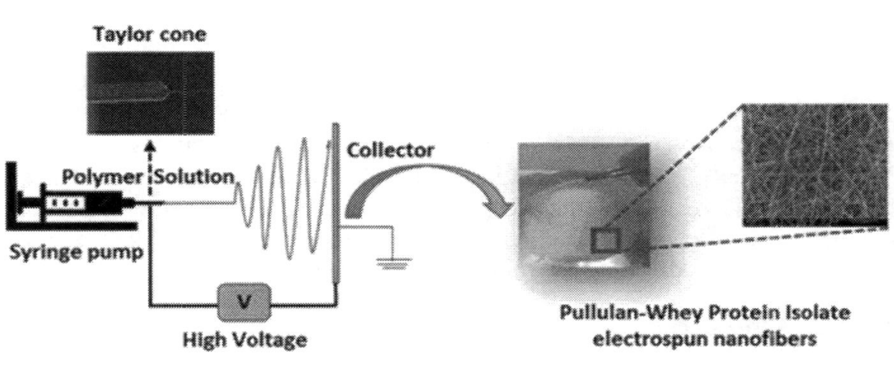

FIGURE 6.10 Schematic diagram of the electrospinning method. (Drosou et al., 2018.)

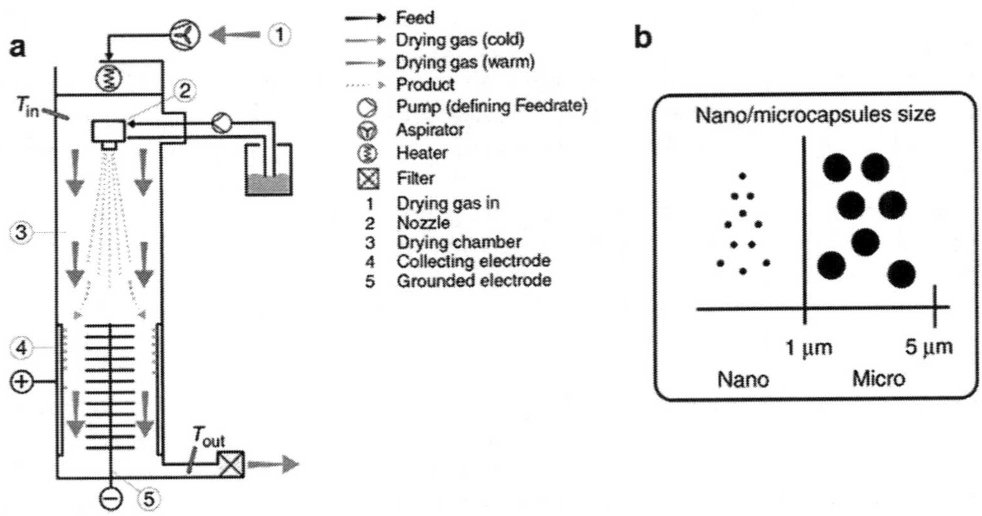

FIGURE 6.11 (a) Flow diagram of nano spray dryer, (b) difference of particles size of nano spray drying. (Arpagaus et al., 2017.)

biosensors. Silver nanoparticles incorporated into cellulose acetate nanofiber exert potent antibacterial activity and prevent bacterial growth on food surfaces (Razaei et al., 2015; Drosou et al., 2018; Stijnman et al., 2011).

6.2.2.6 Nanospray Drying

Spray drying has been used to fabricate both micro-sized and nano-sized particles or dry the nanoparticle liquid suspension into a dry powder, which has been used in pharmaceutical applications for controlled drug release. In the last few decades, the implementation of nanotechnology in the food sector has increased due to the enhanced physical, chemical, and biological properties of food products in nanoscale. The conventional spray dryer was modified for the efficient production of nanoparticles, and is termed as nanospray dryer.

The nanospray dryer differed from the conventional dryer due to the presence of vibration mesh, which produces tiny droplets. Droplet generation depends on the vibration induced by the piezoelectric driven actuator at an ultrasonic frequency (≈60 kHz) controlling vibration of a slim perforated membrane with 4 to 7 μm diameter size (Figure 6.11). The functional property of the food ingredients also showed high stability, solubility, controlled release of bioactive components, bioavailability, and organoleptic properties due to the greater surface area. Based on the mesh size, temperature, feed rate, flow rate and concentration of wall material, the final product size varies (Figure 6.11) (Anandharamakrishnan and Ishwarya, 2015; Arpagaus et al., 2017).

Nanofood is also related to the improvement of food color, prolongation of shelf life and preservation. Besides, many strategies have been applied to generate nanocapsules, different applications, complexity, and scalability, with the physicochemical nature of the bioactive compound being one of the essential features of the manufacturing method. Essential oils, vitamins, flavors, coffee, tea, milk, eggs, omega-3 fatty acids, coenzyme Q10, plant polyphenols, etc., are converted into nanoscale ingredient through this technique (Table 6.2) (Arpagaus et al., 2017). The application of nanospray in the food sector is emerging and expected to proliferate because customer attitudes and choice of food products vary nowadays: they are looking for nutritious food enriched with natural biocompounds.

6.2.2.7 Physical Vapor Deposition (PVD)

This method is also called thin film process, and is an atomistic deposition process, where the material is vaporized from liquid or solid source to form atoms and transferred in the form of vapor

TABLE 6.2
List of Bioactive Compounds Fabricated in Nanoscale

Bioactive Substance	Size	Application
Vitamin 12	0.2–5.5	Vitamin supplement
Folic acid B9, vitamin E	0.4–1.1, 0.4–1.1	Vitamin supplement
Curcumin	0.2–07	Food additive, antioxidant, nutritional supplemtn
Leucine	2.1–5.4	Amino acid, nutritional supplement
Beta-galactosidase	1–5	Enzyme catalyzing hydrolysis saccharides
Lycopene, β-carotene	–	Antioxidant
Citrus peel oil, cardamom essential oil	–	Flavoring substance

through low-pressure plasma or vacuum environment to the substrate, and then it condenses. This method is commonly used to deposit films with various thicknesses in the range of nanometers. The shape can vary from flat to geometries with a variety of 1–10 nm. Vacuum deposition, sputter deposition, arc vapor deposition, and ion plating are various PVD methods. Vacuum deposition is used to form permeation barrier films on flexible packaging materials and water- and corrosion-resistant coatings in the food industry. In this vacuum deposition process, the substrate after thermal vaporization crosses with little or no collision with a gas molecule, which reduces the gaseous contamination at a low level.

Sputter deposition involves the deposition of particles vaporized from the surface by the physical sputtering process. The surface atoms physically ejected from the solid surface through momentum transfer from an atomic-sized energetic bombarding particle, which is a gaseous ion, accelerated from plasma. This deposition can be done in a vacuum condition using low- (<5 m Torr) or high- (5–30 m Torr) pressure plasma. This method is widely adopted in the production of thin film semiconductor materials, coatings on architectural glass, and reflective surfaces on compact discs, magnetic film, and dry film lubricants.

Arc vapor deposition is mainly used in decorative coatings, by using a high-current, low-voltage arc turn to vapor a cathodic or anodic electrode and deposit the materials on a substrate. In ion plating, there is a concurrent or periodic bombardment of the depositing film used by atomic-size energetic particles to alter and control the property of the film. The depositing material is sometimes vaporized by other PVD methods as mentioned in the previous method. An inert or reactive gas is used during the process, which uses energetic particles for bombardment. The ion plating can be done in a plasma or vacuum environment. The use of reactive gas under plasma produces a dense coating at relatively high gas pressures, which enhances the surface coverage. VPD have various applications, such as the production of single and multilayer coatings, nanocomposite materials, nanophase coatings, thin-walled freestanding structures and foils, and gold-silver-copper electrical conductors. They also provide permeation barriers for moisture and gases (Mattox, 2010).

6.2.3 Biological Synthesis

Conventionally, nanoparticles have been fabricated and stabilized via expensive chemical and physical processes, where the use of toxic materials with potentially hazardous properties lead to environmental toxicity, cytotoxicity, and carcinogenicity. The addition of chemicals to avoid the agglomeration in the colloids prevents their biomedical application (Shah et al., 2015). There has been an increased demand for various customized nanoparticles over the past decades, to produce nanoparticles in different sizes, shapes, and compositions, where biological synthesis methods assist in fabricating the desired particles. Nanoparticles synthesized from biological entities are non-toxic, biocompatible and eco-friendly, which has opened a new streamline to use in the food industry, in

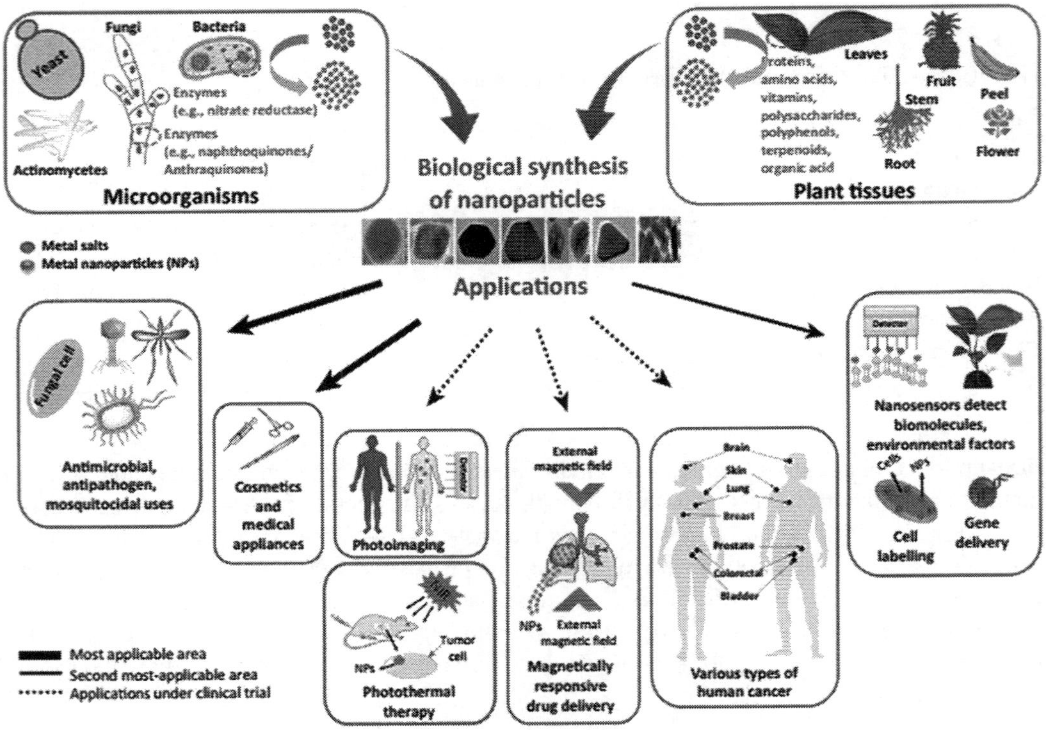

FIGURE 6.12 Biological synthesis of nanoparticles and its application. (Singh et al., 2016.)

the area of food packaging, wastewater treatment, and in various other industries (Figure 6.12). Food waste is an excellent source for supplying reducing/stabling agents in the synthesis of nanoparticles.

6.2.3.1 Polysaccharide Method

Nanoparticles are synthesized from polysaccharides, which act as a capping agent. In the production of starch-based nanoparticles, starch acts as a capping agent and β-D glucose as a reducing agent. The nanoparticles synthesis based on polysaccharides is easily integrated with the application in food and pharmaceutical industries. Polysaccharides are natural biopolymers, and due to their properties of biocompatibility and their biodegradable nature they act as a promising host in the synthesis of metallic nanoparticles. Natural polysaccharides have excellent chemical and structural features, and differences ranging in charge, chain lengths, and stereochemistry, which synthesize advanced functionalized materials with antimicrobial activity. Hydrophilic groups of polysaccharides create non-covalent bonds with the tissue cells and from cell-cell recognition and adhesion. Hence the nutraceutical products loaded in polysaccharides will reach the target of 100% bioavailability. Over the last few decades, remarkable progress has been achieved in the food industry in relation to biodegradable food packaging. Polysaccharide sources such as chitosan, pullulan, carboxymethyl polysaccharides, gum, and arabinogalactan are extracted from biological sources such as microorganisms (Table 6.3). Irradiation techniques have also been applied for the synthesis of metal nanoparticles from polysaccharide sources to achieve the desired nanoparticles.

6.2.3.2 Tollens Process

It is a one-step method, a synthesis of nanoparticles with a controlled size, which is environmentally green due to the use of non-toxic chemicals. The chemical reaction of Tollens' reagent consists of ammonia and silver nitrate, by an aldehyde group producing silver particles. However, the formation of silver nanoparticles depends on several factors including temperature, pH, and the concentration

TABLE 6.3
List of Polysaccharides Synthesis Metal Nanoparticles

Polysaccharides	Metals	Diameter (nm)	Shape
Chitosan	Ag/ZnO	10.0–65.0	Spherical/uneven distribution
Pectin	Ag	5.4–10.6	Spherical
Agarose/dextran/gelatin	Fe_2O_3	10.0	Dumbbell shape
Guar gum	Ag	16.0	Spherical
Carboxymethyl polysaccharides	Ag	20.0–40.0	Spherical/polygonal
Pullulan	Ag	2.0–30.0	Spherical/oval-shaped
Hydroxypropylcellulose	Ag	25.0–55.0	Spherical
6-O-chitosan sulfate	Au	15.0	Spherical
Fucoidan	Ag	17.0	Spherical
β-glucan	Ag	15.0	–
Xanthan gum/chitosan	Ag	5.0–20.0	Spherical
Chitosan-carboxymethyl cellulose	Ag	5.0–20.0	Irregular shape

Source: Wang et al. (2017).

of substrate (aldehydes or ketones), water activity (aw), moisture, and time. Saccharides used as a reducing agent, in the presence of ammonia, produce a nanoparticle film or hydrogel with a size of 50–200 nm particles. In the modified Tollens method, oleic acid is used a stabilizer and glucose as a reducing agent under the UV radiation attribute-controlled size of particles. A trace amount of aldehyde flavor compounds can be detected and screened based on Tollens-process-synthesized silver nanoparticles (Basiuk and Basiuk, 2015; Ayob et al., 2001). AgNps composites were fabricated using biopolymer xylan as the stabilizing and reducing agent with the Tollens agent under microwave irradiation. The fabricated silver nanoparticles act as a selective and sensitive detector of heavy metals in water treatment (Ta et al., 2018).

6.2.3.3 Irradiation

Microwave heating of particles has several advantages for their application in the food industry. Microwave irradiation is based on the principle of dielectric heating, based on two mechanisms: dipolar polarization and ionic conduction. Based on electromagnetic radiations between the frequencies of 0.3 and 300 GH, which produce an oscillating filed, these dipoles or irons try to realign themselves within this field and generate heat. In green synthesis, heating the sample fabricates desirable features with shorter reaction periods and better product yield.

Microwave heating replaces chemical synthesis and has more advantages in the synthesis of inorganic nanoparticles using biological molecules as a reducing and stabilizing agent. In general, conductive heating is processed by an external source, which is slow and inefficient; however, microwave irradiation creates internal heating and uniform mixing of heat in the medium, bringing out efficient nanoparticles with controlled size. Metal nanoparticles, such as silver nanoparticles, have been distinguishable in the field of food industries due to their application. Hence, the demand for silver nanoparticles increased similarly to that of other particles such as palladium and gold, which have been synthesized by microwave-assisted process within a few minutes. Nanoparticles of various shapes, like spherical, polygonal, rod, and wire, can be synthesized by this method (Razavi et al., 2015). AgNps are synthesized from carboxymethylcellulose sodium, and silver nitrates form monodisperse particles, which are highly stable for two months at ambient temperature. Pulse radiolysis technique, ionizing radiation, and radiolysis have been also applied to study the reactions of inorganic and organic molecules in nanoparticle fabrication, which fabricate particles with controlled shapes and sizes (Sharma et al., 2009).

6.2.3.4 Polyoxometalates (POMs) Method

Polyoxometalate (POM) is a polyatomic ion, a combination between oxygen and early transition metals at their highest oxidation states. POMs are the potential candidate in synthesizing nanoparticles, notably silver nanoparticles, because they are soluble in water and can go through a stepwise process and multielectron redox reactions without interrupting their structure. This is a one-step synthesis method, where POMs act as a reducing and stabilizing agent. The particle size and shape can be varied based on the selection of different POMs. POM-based nanoparticles are used to construct various electrochemical cells and photoelectronic sensor devices (Bhosale et al., 2014).

6.2.3.5 Fabrication of Nanoparticles from Natural Sources

Biological molecules of extracts from plants and microorganisms act as reductant and stabilizer or capping agent. Phenolic compounds, enzymes, amino acids, proteins, polysaccharides, vitamins, and secondary metabolites of plants and microorganisms are responsible for flavor, color, nutritional value, and defense mechanisms. They have a vital, essential application in food, nutraceutical, and pharmaceutical sciences. But they have a low absorption rate, low aqueous solubility, poor gastrointestinal stability, and low permeation due to high molecular weight and low bioavailability. Utilizing the bioactive molecules in a formed nanoparticle creates new possibilities to enhance their properties and overcome their limitations.

a. Nanoparticle Fabrication using Plants

The use of plants as renewable resources in the fabrication of nanoparticles has more advantages than other methods. Various polyphenols in plants act as a potent antioxidant, which helps to reduce oxidative stress in cells. The phytochemicals and functional groups involved in the synthesis of various nanoparticles elucidate the desired effect in the reduction of metal, aggregation, and stabilization. Three different periods take place in plant-mediated synthesis: incubation period, growth phase, and termination period. The growth ratio of the particle is slower than the rapid reduction, which increases the number of small-size nanoparticles (Figure 6.13). The presence of carbonyl groups, amino acids, dehydrogenase, aldehydes, hydroxyl moieties, carboxylic acid ketones, intracellular CO_2, membrane proteins, NADP reductase, peptide, sulfhydryl, heterocyclic compounds, hydrogenase, tannic acid, tartaric acid, quercetin, amine functional group, phenolics, reducing sugars, saponin, secondary metabolites, and sesquiterpenes in the medium reacts with the precursor to fabricate bioinspired nanoparticles and bind on the surface of the particles (Bhaumik et al., 2015; Akhtar et al., 2013; Rajan et al., 2015).

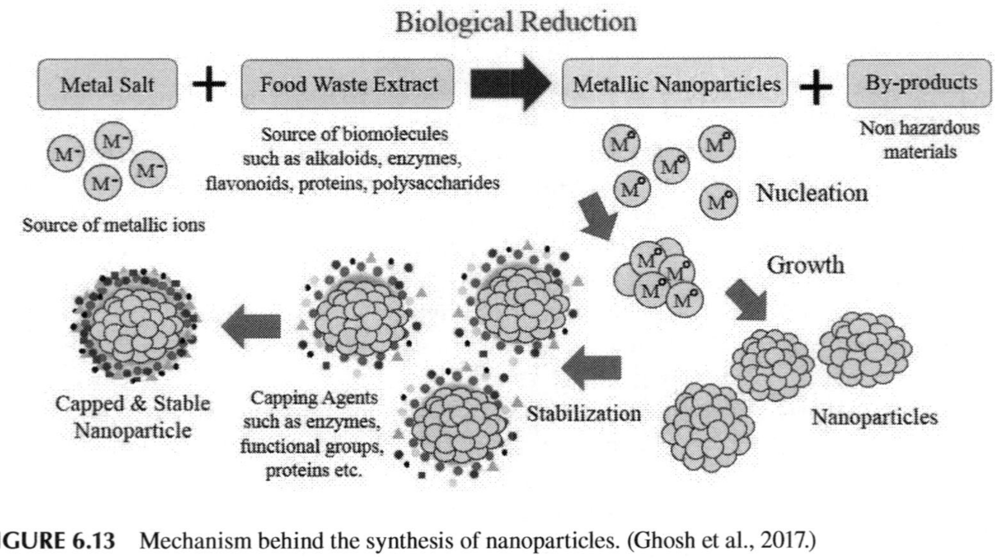

FIGURE 6.13 Mechanism behind the synthesis of nanoparticles. (Ghosh et al., 2017.)

Whole plants extracts were also utilized for the fabrication of nanoparticles. The shape of the particles varies according to the various metabolites and concentration present in the different parts of the plants. The shape might be irregular, tetrahedral, triangular, hexagonal, spherical, decahedral platelet, rod, nanocrystal, or nanofibers (Park et al., 2011). Pure compounds identified and isolated from plants have also been applied for the synthesis of metallic nanoparticles. Compounds such as sorbic acid, ellagic tannins, purple acid phosphatase (wood apple), cyclic peptides, citric acid, ascorbic acid, retinoic acid, euphol, and polyhydroxy limonoids derived from plants sources have a potential pattern in nanoscale synthesis. The stability of the particle is high with high antioxidant and antimicrobial properties. Food sources and their by-product waste, such as peel of *Citrus sinesis* (Konwarh et al., 2011), kiwifruit (Gao et al., 2014), *Psidium guajava* fruit (Lokina et al., 2015), and seeds of *Trigonella foenumgraecum* have been established in nanofabrication (Aromal and Philip, 2012) (Table 6.4).

b. Nanoparticle Fabrication Using Microorganisms

Next to plants, microbes are the best potential source for the synthesis of nanoparticles. The production of nanoparticles is either intracellular or extracellular. The enzyme present within the cell, based on electrostatic interaction, assists the synthesis of nanoparticles. Bacteria are easy to manipulate, which gives a better performance in the synthesis pattern. Many theories have been proposed to explain the mechanism of synthesis of metal nanoparticles from microbes because the nanoparticles are stable for more than three months. The nature of the reducing agent varies with the species, hence tracking and identification of particular components from the complex cellular environment of microbes are extremely challenging.

The major biomolecules of microbes are proteins, polyols, heterocyclic compounds, and peptide, which play an important role in synthesis through an electron shuttle or charge capping mechanism.

TABLE 6.4
Nanoparticles Fabricated from Various Food Sources

Food Source	Size and Morphology	Nanoparticle
Ananas comosus **(Pineapple)**	5 – 35 nm, spherical	Ag
Tanacetum vulgare **(tansy fruit)**	Ag: 16 nm, spherical Au: 11 nm, triangular	Ag and Au
Citrus sinensis **(orange) peel**	3 – 12 nm, spherical	Ag
Citrus unshiu **(mandarin) peel**	5 – 20 nm, spherical	Ag
Sorghum spp. (bran) (aw)	10 nm, quasi-spherical	Ag
Emblica officinalis **(Indian gooseberry)**	Ag: 10 – 20 nm, spherical Au: 15 – 25 nm, spherical	Ag and Au
Carica papaya **(pawpaw)**	60 to 80 nm, spherical	Ag
Daucus carota L. **(black carrot)**	4.32 – 17.65 nm, spherical	Ag
Allium sativum **(garlic clove)**	4 – 22 nm, spherical	Ag
Citrullus lanatus **rind**	17.96, spherical	Ag
Musa paradisiac **(banana) peel**	50 nm, crystalline, irregular shape	Pd
Various commercially available tea/ coffee extracts	20 – 60 nm, spherical	Pd and Ag
Lignin	16 – 20 nm, spherical	Pd and Pt
Citrullus lanatus **(watermelon) rind**	96 nm, spherical	Pd
Tea waste	5 – 25 nm, cubes and pyramids	Fe_3O_4
Citrus sinensis **(orange) peel**	29 nm, spherical	MgO
Musa paradisiac **(banana) peel**	20 – 50 nm, spherical	Mn_3O_4

Source: Ghosh et al. (2017).

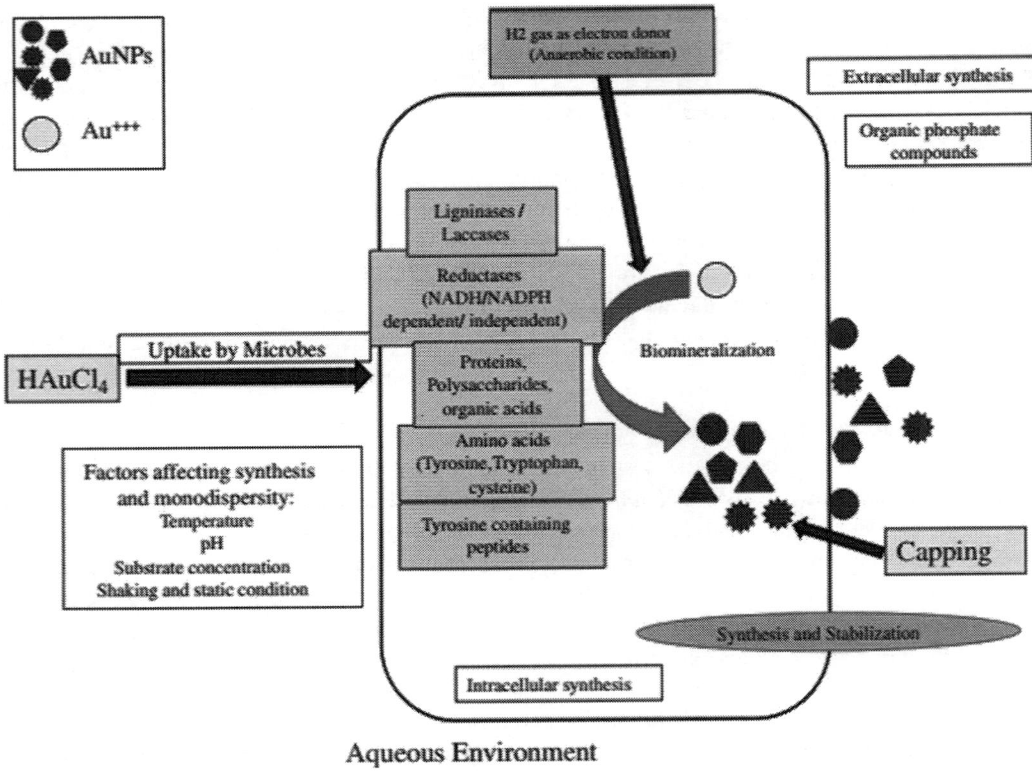

FIGURE 6.14 Mechanism of microbial-mediated synthesis. (Shedbalkar et al., 2014.)

For example, peptide molecules bind with the surface of metal precursors under a localized reducing environment and start their nucleation process in order to grow. Amino acid residues such as arginine, cysteine, lysine, methionine, tyrosine, and tryptophan were found be initiators for the synthesis of nanoparticles (Figure 6.14) (Shedbalkar et al., 2014).

A difference in particle size is observed when there are differences in cell growth and metal concentration in the different incubation periods. Buttermilk, which contains *Lactobacillus* strains, exposed to metal ions produces silver-gold alloy crystals, silver, and gold particles (Thakkar et al., 2010). Similarly, various shapes and sizes of nanoparticle synthesis are produced by various microorganisms (Figure 6.15).

Fungi acted as promising candidates in the fabrication of nanoscale materials. The biosynthesis pattern is different depending on whether it is based on the intra or extracellular mechanism. During intracellular fabrication, the metal ions bind to the surface and generate an opposite charge to that present on the metal ion surface due to the electrostatic interaction force and absorb the metal ion inside the cell. The enzyme present inside the cell reduces the metal ions and generates nanoparticles. Whereas, in extracellular synthesis the presence of nitrate reductase enzymes reduces metal ions into highly stable silver particles (Figure 6.16).

Various studies in nanoscale development have been carried out with yeast. The nano-sized particles are generated by reduction of precursors through chelating with extracellular peptides and polysaccharides. Spherical antimony trioxide (Sb_2O_3) particles are synthesized from baker's yeast (*Saccharomyces cerevisiae*) used in the manufacturing of pet jars with antimicrobial properties. (Lei et al., 2016; Khandel et al., 2016). Fabrication has been extended using a virus, a biological substance such as protein, biolipid cylinders, viral capsules, and viral scaffolds to induce the nucleation and assembly of inorganic materials. cowpea mosaic virus, tobacco mosaic virus, papaya mosaic virus, potato virus X, tomato mosaic virus, turnip mosaic virus, and flock house virus

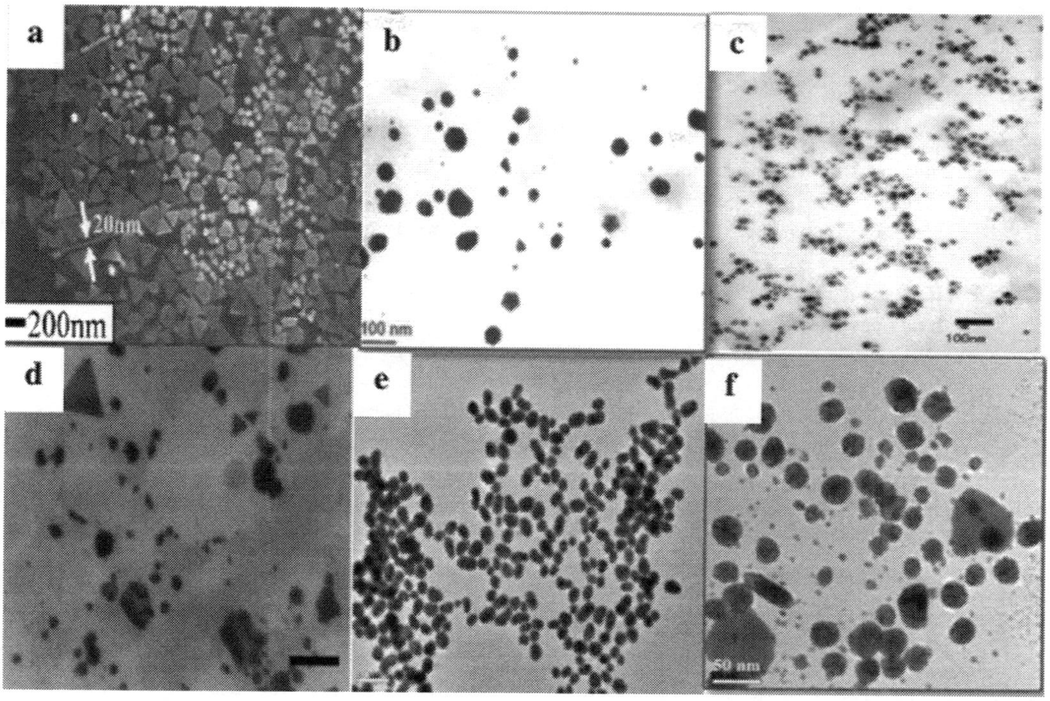

FIGURE 6.15 TEM images of microbes-mediated nanoparticles: (a) triangle shape of Au nanoplates by *Aspergillus niger*, (b) spherical, triangular and hexagonal-shaped AuNPs by *Penicillium brevicompactum*, (c) spherical AuNPs by *Rhodopseudomonas capsulate* (d) triangle, spherical AuNPs of *Verticillium luteoalbum*, (e) oval shaped AuNPs by *Stenotrophomonas maltophilia*, and (f): AuNPs by *Sclerotium rolfsii*. (Shedbalkar et al., 2014.)

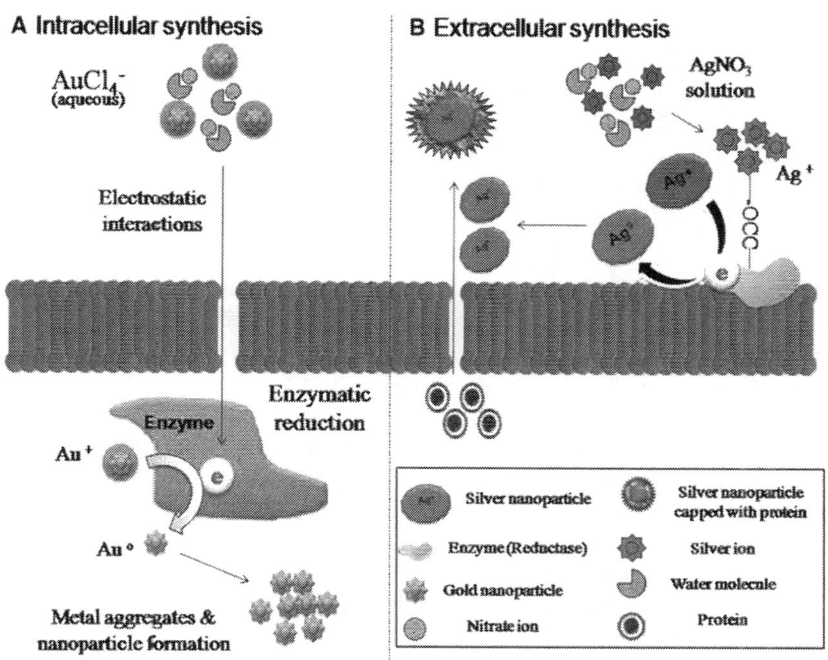

FIGURE 6.16 Mechanisms behind the extracellular and intracellular synthesis metal nanoparticles of fungi. (Kashyap et al., 2013.)

TABLE 6.5
Microbes Produce More Nanoparticles

Microbial Source	Size	Mode of Synthesis	Nanoparticle
Bacillus licheniformis	50	Extracellular	Ag
Enterobacteria	52.5	Extracellular	Ag
Fusarium oxysporum	5–15	Extracellular	Ag
Cladosporium cladosporioides	10–100	Extracellular	Ag
Penicillium fellutanum	5–25	Extracellular	Ag
Trichoderma viride	2–4	Extracellular	Ag
Streptomyces sp.	10–100	Extracellular	Ag
Geobacillus sp.	5–50	Intracellular	Au
Morganella sp.	15–20	Intracellular	Cu
Stereum hirsutum	5–20	Extracellular	Cu
Lactobacillus sporogens	5–15	Intracellular	Zn
Lactobacillus crispatus	70–114	Extracellular	Ti
Schizosaccharomyces pombe	200	Intracellular	CdS
Sargassum wightii	8–12	Extracellular	Au
Chlorella vulgaris	9–20	–	Au

Source: Thakkar et al. (2010) and Ghosh et al. (2017).

produced promising and efficient nanoparticles which have significant effectors in targeted therapeutics. No research has yet been done in the field of food processing or perseveration (Narayanan et al., 2017; Destito et al., 2018). Table 6.5 lists the microorganisms reported in the fabrication of metal nanoparticles.

6.2.3.6 Factors Influencing Biological Fabrication of Nanoparticles

The biological synthesis of nanoparticles from either plants or microorganism sources depends on various factors such as pH, temperature, time, and biomolecules concentration to achieve a high yield of nanoparticles with defined shape and structure. Still, the mechanism underlying biological synthesis is not fully understood; research is still in progress. The schematic diagram in Figure 6.17 represents the overall synthesis process and the various factors for the successful synthesis of nanoparticles described as follows.

1. *pH*

   pH plays a very important role in the reaction medium, which determines the size and shape of nanoparticle fabrication. The nanoparticles' size can be altered or controlled by adjusting the pH of the reaction medium. Singh et al. (2009) have described the size and shape changes of silver nanoparticles due to the effect of pH. AgNPs have been generated using glucose as a reducing agent and starch as stabilizer though green route synthesis method. Under acidic conditions, the size of particles decreases and, due to the insufficient protection of the surface, larger particles with finer crystallite size are achieved. The behavior of surface plasmon resonance (SPR) differs according to the alteration of pH by addition of an alkaline base solution (Figure 6.18 and Table 6.6). Gold nanoparticles synthesized using *Avenasativa* form larger size nanoparticles from 25 to 85 nm at pH 2, though the smaller size was observed at pH 3 and 4 (Kuchibhatla et al., 2012).
2. *Temperature*

   Temperature is an essential factor in all three synthesis methods (chemical, physical, and biological). Both physical and chemical methods require more or less 350°C, whereas

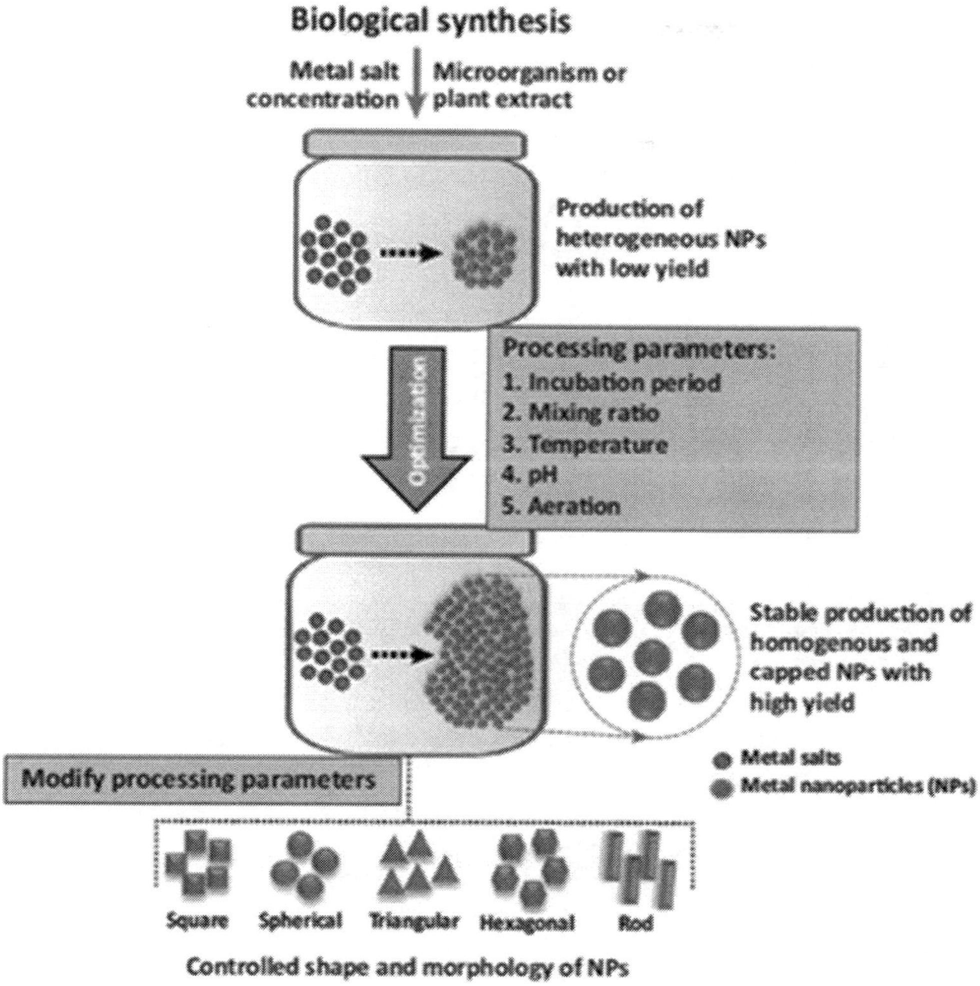

FIGURE 6.17 Factors influencing stable and high yield nanoparticles. (Singh et al., 2016.)

in the green synthesis method, the nanoparticle synthesis takes place at ambient temperature or less than 100°C (Singh et al., 2009). Temperature acts as a critical factor in influencing the shape of the nanoparticles. Gold nanoparticles synthesis using lemongrass extract showed a spherical shape at a higher temperature, where nanotriangles were relatively reduced to spherical shapes and triangle shapes at a lower temperature (Figure 6.19) (Akhilesh Rai et al., 2005). Higher temperature promotes nucleation rate and a high number of spherical particles, but at a lower temperature, the low rate of reduction of metal ions accelerates the oriented growth of nuclei, which promotes triangular-shaped nanoparticles. Silver nanoparticles synthesis using sweet orange peel extract produced about 35 nm in size at 25°C, whereas, when the temperature increased, the size of the particle was reduced to 10 nm (Kaviya et al., 2011). When the temperature increases the reaction rate, and particle formation rate becomes faster, however, the particle size decreases as the temperature increases steadily.

3. *Pressure*

The rate of synthesis of nanoparticles is high at ambient pressure, which is the most advantageous factor in the green or biological synthesis method. But both chemical and physical methods apply pressure to generate nanoparticles. Chemical vapor synthesis and

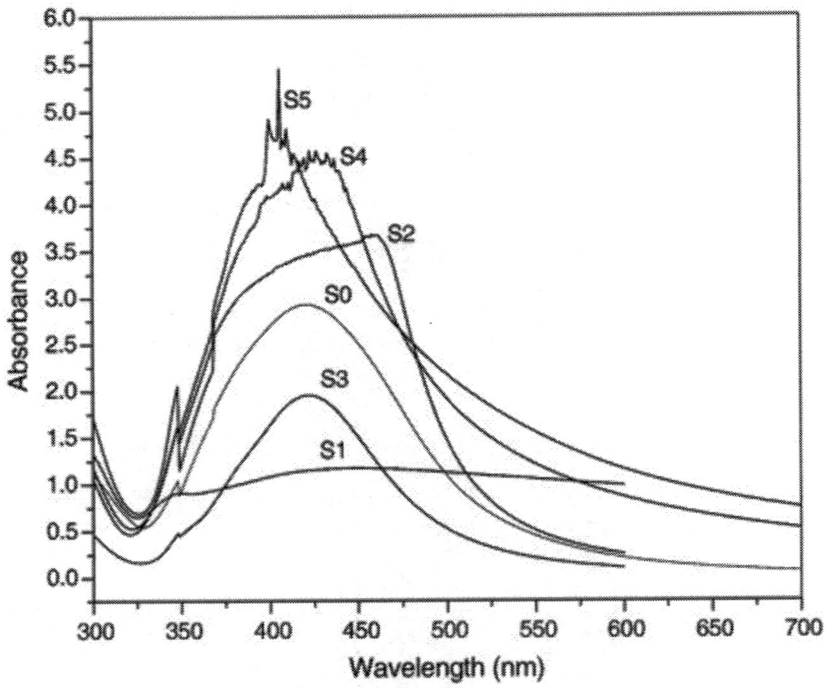

FIGURE 6.18 UV-visible spectra of sol of AgNPs prepared at different pH. (Singh et al., 2009.)

supercritical hydrothermal synthesis methods require about 30 Mpa for the synthesis of zinc oxide (Tran et al., 2013; Abhilash and Pandey, 2012).

4. *Time*

The rate of reaction and the stability of the nanoparticles depends upon the reaction time. The stability of the nanoparticles was found to be strongly dependent on particle size. During the nucleation and growth, the particle size and chemical state change according to the time duration. The time-dependent oxidation state of ceria nanoparticles in the reaction medium showed an impact on both physical and chemical properties of the particles (Figure 6.20). The nanoparticles suspension showed a significant shift in the absorption which indirectly proved the stability of the nanoparticles (Kuchibhatla et al., 2012). Ahmad and Sharma (2012) revealed the importance of time during the silver synthesis using *Ananas comosus*. Rapid color changes occurred within 2 min and continued up to 5 min with slight variations in color with a 12 nm particle size. Increasing the reaction time produces massive-size particles. The AgNPs produced by leaf extract of *Azadirachta indica*, showed particle sizes ranging from 10 to 35 nm, through the reaction time varied from 30 min to 4 h (Kuchibhatla et al., 2012).

5. *Reactant concentration*

The concentration of biomolecules in the reaction medium extensively influences the formation/construction of nanoparticles. Increasing the concentration of *Aloe vera* extract changes the shape of the gold nanoparticles from triangular to spherical (Chandran et al., 2006). The presence of carbonyl compounds in the extract assisted in defining the shape of the particle and its growth. The particle size of 50 to 350 nm was produced by increasing the concentration of the biomolecules, and the shape could be hexagonal, triangular, spherical, or decahedral, formed based on the strength of the extract in the reaction medium (Kuchibhatla et al., 2012). In some cases, for example, silver nanoparticles synthesis from different concentrations (5%–50%) of *Magnolia* leaf showed high reaction rate at

TABLE 6.6
Characteristics of AgNPs at Different pH

Sample	Volume of NaOH (ml)	pH (after Formation of Ag Nanoparticles)	λ_{max} (nm)	Half Width ($\Delta\ \lambda_{hw}$) (nm)	Average Crystallite Size (nm)
S0	0.0	9.4	420	105	–
S1	0.5	7.5	449	157	16
S2	1.0	4.2	459	110	11
S3	1.5	2.9	422	78	19
S4	5	1.5	426	91	25
S5	10	6.0	405	87	18

Source: Singh et al. (2009).

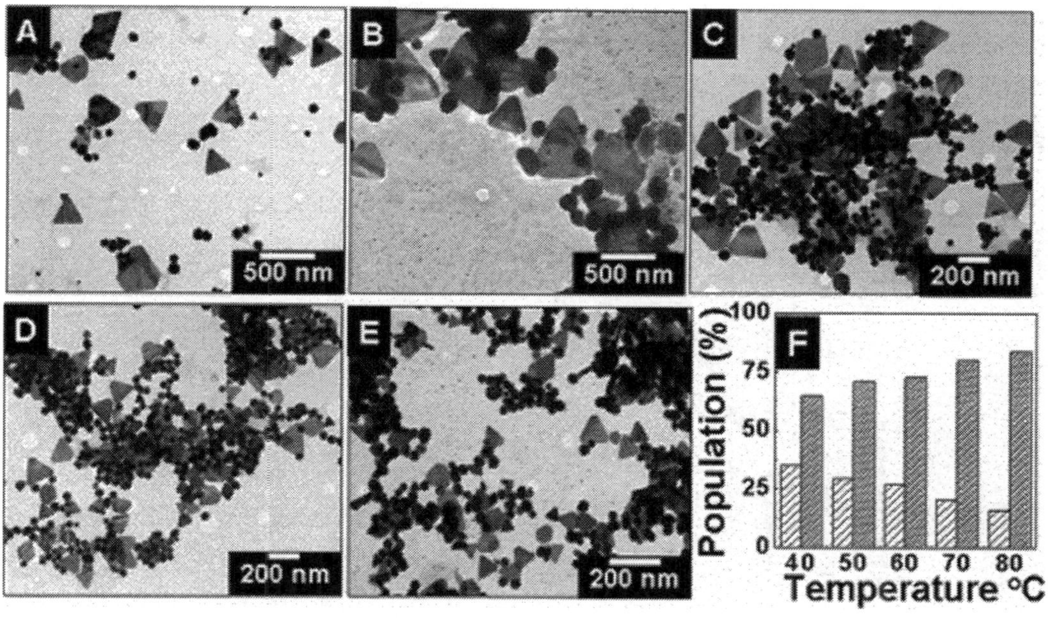

FIGURE 6.19 TEM images of gold nanoparticles at (A) 40, (B) 50, (C) 60, (D) 70, (E) 80°C. (F) Histogram of gold triangle-shaped nanoparticles (light lines) and spherical-shaped gold nanoparticles (dark lines) at different temperatures. (Akhilesh Rai et al., 2005.)

20% concentration, meanwhile the size of particles increases as the extract concentration increases (Figure 6.21). High concentration level of reducing agent in the reaction medium leads to aggregation of particles with a size range from 100 to 800 nm (Song and Kim, 2009).

In the biological process, factors such as temperature, pH, concentration, time, etc., play an essential role in achieving the desired size, shape and monodispersed nanoparticles distribution. Nanoparticles fabricated through the green chemistry method have broad application in the food industry. Metal nanoparticles such as silver are highly heat-conducting materials used in mechanical devices. Due to their antimicrobial activity, silver nanoparticles can be applied in wastewater treatment. To detect and monitor spoiled food and the presence of microorganisms, bio-gold and silver nanoparticles are used to develop biosensors (Kuppusamy et al., 2016).

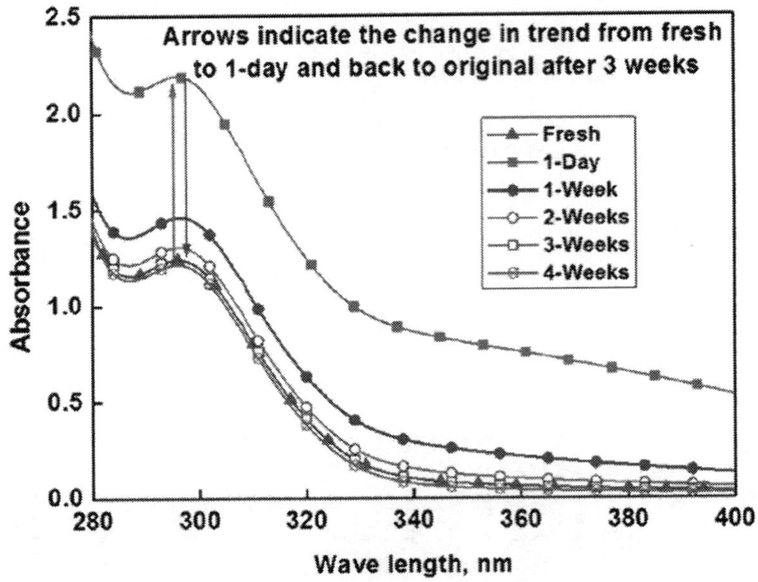

FIGURE 6.20 Spectrum of ceria nanoparticles at different time periods. (Kuchibhatla et al., 2012.)

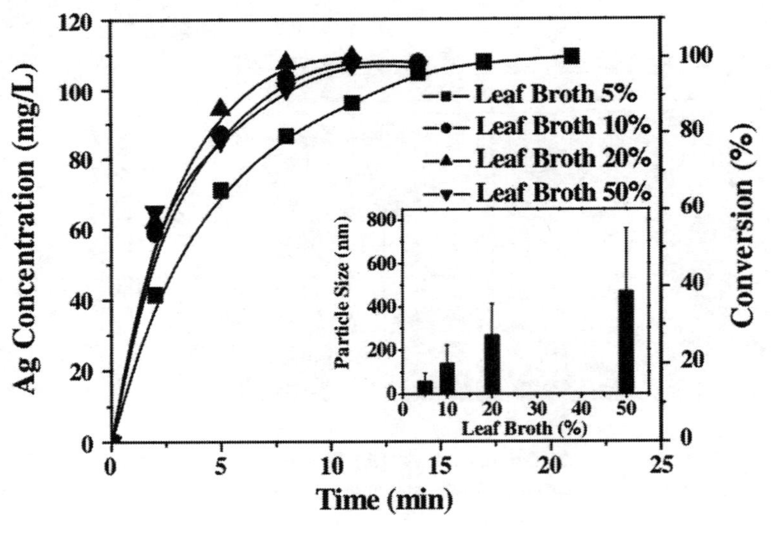

FIGURE 6.21 Influence of various concentrations of extract on particle size formation. (Song and Kim, 2009.)

6.3 CONCLUSION

The applications of nanotechnology in the food industry are increasing day by day, which creates more demand for nanoparticles. Hence, scientists are developing various physical, chemical, and biological techniques to satisfy the need for nanoscale food products with multiple functional properties. A diverse range of methods is available to fabricate the desired nanoparticles with specific reactions medium, size, shape, and application. This chapter provides the overall methods employed in the fabrication of nanoparticles for the food processing and preservation fields. On account of their extensive use in the food industry, specific disadvantages occurred during the scalable process, which should be optimized. Research should focus on exploring cost-effective and eco-friendly technologies.

REFERENCES

Abhilash and B.D. Pandey. 2012. "Synthesis of zinc-based nanomaterials: A biological perspective." *IET Nanobiotechnology*, 6 (4): 144–148. doi:10.1049/iet-nbt.2011.0051.

Ahmad, Naheed and Seema Sharma. 2012. "Green synthesis of silver nanoparticles using extracts of *Ananas comosus*." *Green and Sustainable Chemistry*, 2 (November): 141–147. doi:10.4236/gsc.2012.24020.

Akhtar, Mohd Sayeed, Jitendra Panwar, and Yeoung Sang Yun. 2013. "Biogenic synthesis of metallic nanoparticles by plant extracts." *ACS Sustainable Chemistry and Engineering*, 1 (6): 591–602. doi:10.1021/sc300118u.

Anandharamakrishnan, C. and Padma Ishwarya. 2015. *Spray Drying Technique for Food Ingredient Encapsulation*. Chichester, John Wiley & Sons.

Aromal, S. Aswathy and Daizy Philip. 2012. "Green synthesis of gold nanoparticles using *Trigonella foenum-graecum* and its size-dependent catalytic activity." Spectrochimica Acta Part A : Molecular and Biomolecular Spectroscopy, 97: 1–5.

Arpagaus, Cordin, Philipp John, Andreas Collenberg, and David Rütti. 2017. "Nanocapsules formation by nano spray drying." *Nanoencapsulation Technologies for the Food and Nutraceutical Industries*, 346–401. doi:10.1016/B978-0-12-809436-5.00010-0.

Ashokkumar, M., J. Lee, S. Kentish, and F. Grieser. 2007. Bubbles in an acoustic field: An overview. *Ultrasonics Sonochemistry*, 14(4), 470–475. doi: 10.1016/j.ultsonch.2006.09.016.

Ayob, Mohd Khan, Osman Hassan, and Mazlah Othman. 2001. "A rapid method for detection of aldehyde-based flavour compounds." *Polygonumminus Cultured Tissue*, 7 (1): 29–33.

Azam, M. A., N. E. S. A. A. Mudtalib, and R. N. A. R. Seman. 2018. "Synthesis of graphene nanoplatelets from palm-based waste chicken frying oil carbon feedstock by using catalytic chemical vapour deposition." *Materials Today Communications*, 15, 81–87. doi: 10.1016/j.mtcomm.2018.02.019.

Azimi-Fouladi, A., S. A. Hassanzadeh-Tabrizi, and Saffar-Teluri, A. 2018. "Sol-gel synthesis and characterization of TiO_2-CdO-Ag nanocomposite with superior photocatalytic efficiency." *Ceramics International*, 44(4), 4292–4297. doi: 10.1016/j.ceramint.2017.12.016.

Bang, J. H. and K. S. Suslick. 2010. Applications of ultrasound to the synthesis of nanostructured materials. *Advanced Materials*, 22(10), 1039–1059. doi: 10.1002/adma.200904093.

Basiuk, Vladimir A. and Elena V. Basiuk. 2015. "Green processes for nanotechnology: From inorganic to bioinspired nanomaterials." *Green Processes for Nanotechnology: From Inorganic to Bioinspired Nanomaterials*, February: 1–446. doi:10.1007/978-3-319-15461-9.

Bhaumik, Jayeeta, Neeraj S. Thakur, Pavan K. Aili, Amit Ghanghoriya, Amit K. Mittal, and Uttam C. Banerjee. 2015. "Bioinspired nanotheranostic agents: Synthesis, surface functionalization, and antioxidant potential." *ACS Biomaterials Science and Engineering*, 1 (6): 382–392. doi:10.1021/ab500171a.

Bhosale, R.R., Kulkarni, A.S., Gilda, S.S., Aloorkar, N.H., Osmani, R.A. and Harkare, B.R., 2014. Innovative eco-friendly approaches for green synthesis of silver nanoparticles. *International Journal of Pharmaceutical Sciences and Nanotechnology*, 7(1): 2328-2337.

Chandran, S. Prathap, Minakshi Chaudhary, Renu Pasricha, Absar Ahmad, and Murali Sastry. 2006. "Synthesis of gold nanotriangles and silver nanoparticles using aloe vera plant extract." *Biotechnology Progress,* 22 (2): 577–583. doi:10.1021/bp0501423.

Danks, A. E., S. R. Hall, and Z. Schnepp. 2016. The evolution of "sol-gel" chemistry as a technique for materials synthesis. *Materials Horizons*, 3(2), 91–112. doi: 10.1039/c5mh00260e.

Daraio, Chiara and Sungho Jin. 2012. "Synthesis and patterning methods for nanostructures useful for biological applications." In *Nanotechnology for Biology and Medicine*, ed. Gabriel A. S and Parpura V, pp. 27–44. Springer, New York, .

Destito, G., A. Schneemann, and M. Manchester. "Biomedical nanotechnology using virus-based nanoparticles." *SpringerLink*. January 01, 1970. Accessed July 25, 2018. https://link.springer.com/content/pdf/10.1007/978-3-540-69379-6_5.

Drosou, Christina, Magdalini Krokida, and Costas G. Biliaderis. 2018. "Composite pullulan-whey protein nanofibers made by electrospinning: Impact of process parameters on fiber morphology and physical properties." *Food Hydrocolloids,* 77: 726–735.

Dunn, S. S., Beckford Vera, D. R., Benhabbour, S. R., and Parrott, M. C. 2017. "Rapid microwave-assisted synthesis of sub-30 nm lipid nanoparticles." *Journal of Colloid and Interface Science*, 488, 240–245. https://doi.org/10.1016/J.JCIS.2016.10.093.

El-Naggar, M. E., Shaheen, T. I., Fouda, M. M. G., and Hebeish, A. A. 2016. "Eco-friendly microwave-assisted green and rapid synthesis of well-stabilized gold and core-shell silver-gold nanoparticles." *Carbohydrate Polymers*, 136, 1128–1136. https://doi.org/10.1016/j.carbpol.2015.10.003.

Fessi, H., Puisieux, F., Devissaguet, J. P., Ammoury, N., and Benita, S. 1989. "Nanocapsule formation by interfacial polymer deposition following solvent displacement." *International Journal of Pharmaceutics*, 55(1), 1–4. https://doi.org/10.1016/0378-5173(89)90281-0.

Gai, S., Li, C., Yang, P., and Lin, J. 2014. "Recent Progress in Rare Earth Micro /Nanocrystals: Soft Chemical Synthesis, Luminescent Properties, and Biomedical Applications" *Chemical reviews*, 114(4), 2343-2389.

Gao, Ying, Qiang Huang, Qingping Su, and Rui Liu. 2014. "Green synthesis of silver nanoparticles at room temperature using kiwifruit juice." Spectroscopy Letters, 47 (10): 790–95. doi:10.1080/00387010.2013.848898.

Gawande, M. B., Shelke, S. N., Zboril, R., and Varma, R. S. 2014. "Microwave-Assisted Chemistry: Synthetic Applications for Rapid Assembly of Nanomaterials and Organics." *Accounts of Chemical Research*, 47(4), 1338-1348.

Gedanken, A. 2004. "Using sonochemistry for the fabrication of nanomaterials." *Ultrasonics Sonochemistry*, 11(2), 47–55. https://doi.org/10.1016/j.ultsonch.2004.01.037.

Ghorbani, Hamid Reza. 2014. "A Review of methods for synthesis of Al nanoparticles." *Oriental Journal of Chemistry*, 30 (4), 1941–1949.

Ghosh, Purabi R., Derek Fawcett, Shashi B. Sharma, and Gerrard E. J. Poinern. 2017. "Production of high-value nanoparticles via biogenic processes using aquacultural and horticultural food waste." *Materials*, 10 (8). 852. doi:10.3390/ma10080852.

Hao, Y. and A. S. Teja. 2003. "Continuous hydrothermal crystallization of α–Fe_2O_3 and Co_3O_4 nanoparticles." *Journal of Materials Research*, 18(02), 415–422. doi:org/10.1557/JMR.2003.0053.

Hayashi, H. and Y. Hakuta. 2010. "Hydrothermal synthesis of metal oxide nanoparticles in supercritical water." *Materials*, 3(7), 3794–3817. doi:org/10.3390/ma3073794.

Hongmin, L., Z. Wei, Y. Xingrong, W. Jing, G. Wenxin, C. Jihong, C. Fulin. 2014. "Osteoinductive nanohydroxyapatite bone substitute prepared via in situ hydrothermal transformation of cuttlefish bone." *Journal of Biomedical Materials Research Part B: Applied Biomaterials*, 103(4), 816-824. doi:10.1002/jbm.b.33261.

Huang, X., Y. Dai, J. Cai, N. Zhong, H. Xiao, D. J. McClements, and K. Hu. 2017. "Resveratrol encapsulation in core-shell biopolymer nanoparticles: Impact on antioxidant and anticancer activities." *Food Hydrocolloids*, 64, 157–165. doi:10.1016/j.foodhyd.2016.10.029.

Hulteen, John C., David A. Treichel, Matthew T. Smith, Michelle L. Duval, Traci R. Jensen, and Richard P. Van Duyne. 1999. "Nanosphere lithography: Size-tunable silver nanoparticle and surface cluster arrays." *Journal of Physical Chemistry B*, 103 (19): 3854–3863. doi:10.1021/jp9904771.

Jiang, L., S. Xu, J. Zhu, J. Zhang, J. Zhu, and H. Chen. 2004. "Ultrasonic-assisted synthesis of monodisperse single-crystalline silver nanoplates and gold nanorings." *Inorganic Chemistry*, 43(19), 5877–5883.

Kahrilas, G. A., L. M. Wally, S. J. Fredrick, M. Hiskey, A. L. Prieto, and J. E. Owens. 2014. "Microwave-assisted green synthesis of silver nanoparticles using orange peel extract." *ACS Sustainable Chemistry and Engineering*, 2(3), 367–376. doi: 10.1021/sc4003664.

Kashyap, Prem Lal, Sudheer Kumar, Alok Kumar Srivastava, and Arun Kumar Sharma. 2013. "Myconanotechnology in agriculture: A perspective." *World Journal of Microbiology and Biotechnology*, 29 (2): 191–207. doi:10.1007/s11274-012-1171-6.

Kaviya, S., J. Santhanalakshmi, B. Viswanathan, J. Muthumary, and K. Srinivasan. 2011. "Biosynthesis of Silver nanoparticles using *Citrussinensis* peel extract and its antibacterial activity." *Spectrochimica Acta - Part A: Molecular and Biomolecular Spectroscopy*, 79 (3). Elsevier B.V.: 594–598. doi:10.1016/j.saa.2011.03.040.

Khandel, Pramila and Sushil Kumar Shahi. 2016. "Microbes mediated synthesis of metal nanoparticles: Current status and future prospects." *International Journal of Nanomaterials and Biostructures*, 6 (C): 1–24.

Kim, Myungjoon, Saho Osone, Taesung Kim, Hidenori Higashi, and Takafumi Seto. 2016. "Synthesis of nanoparticles by laser ablation : A review," *KONA Powder and Particle Journal*, 34, 80-90. doi:10.14356/kona.2017009.

Konwarh, Rocktotpal, Biswajit Gogoi, Ruby Philip, M. A. Laskar, and Niranjan Karak. 2011. "Biomimetic preparation of polymer-supported free radical scavenging, cytocompatible and antimicrobial 'green' silver nanoparticles using aqueous extract of *Citrussinensis* peel." *Colloids and Surfaces B: Biointerfaces*, 84 (2): 338–345. doi:10.1016/j.colsurfb.2011.01.024.

Kuchibhatla, Satyanarayana V. N. T., A. S. Karakoti, D. R. Baer, S. Samudrala, M. H. Engelhard, J. E. Amonette, S. Thevuthasan, and S. Seal. 2012. "Influence of aging and environment on nanoparticle chemistry: Implication to Confinement effects in nanoceria." *The Journal of Physical Chemistry*, 116(26), 14108-14114.

Kuppusamy, Palaniselvam, Mashitah M. Yusoff, Gaanty Pragas Maniam, and Natanamurugaraj Govindan. 2016. "Biosynthesis of metallic nanoparticles using plant derivatives and their new avenues in pharmacological applications – an updated report." *Saudi Pharmaceutical Journal*, 24 (4). King Saud University: 473–484. doi:10.1016/j.jsps.2014.11.013.

Leena, M. M., D. Rana, T. J. Webster, and M. Ramalingam. 2016. "Accelerated synthesis of biomimetic nano hydroxyapatite using simulated body fluid." *Materials Chemistry and Physics,* (180), 166–172. doi:org/10.1016/j.matchemphys.2016.05.060.

Lei, Z. H. U., Zhu Tian Wang, Hai Bin Xu, Ru Bao Sun, Hong Zhang, and Jian Bo Zhang. 2016. "Exposure assessment of Sb_2O_3 in PET food contact materials." *Biomedical and Environmental Sciences*, 29, 4: 305–313.

Lokina, S., A. Stephen, V. Kaviyarasan, C. Arulvasu, and V. Narayanan. 2015. "Cytotoxicity and antimicrobial studies of silver nanoparticles synthesized using *Psidium guajava L.* extract." *Synthesis and Reactivity in Inorganic, Metal-Organic, and Nano-Metal Chemistry*, 45 (3): 426–432. doi:10.1080/15533174.2013.831881.

Mattox, D.M., 2010. *Handbook of Physical Vapor Deposition (PVD) Processing*. Burlington, William Andrew Applied Science Publishers.

McClements, D.J. 2011. "Edible nanoemulsions: Fabrication, properties, and functional performance." *Soft Matter*, 7 (6): 2297–2316. doi:10.1039/c0sm00549e.

McClements, David Julian and Jiajia Rao. 2011. "Food-grade nanoemulsions: Formulation, fabrication, properties, performance, biological fate, and potential toxicity." *Critical Reviews in Food Science and Nutrition*, 51 (4): 285–330. doi:10.1080/10408398.2011.559558.

Mirzaei, A., K. Janghorban, B. Hashemi, M. Bonyani, S. G. Leonardi, and G. Neri. 2016"Highly stable and selective ethanol sensor based on α-Fe_2O_3 nanoparticles prepared by Pechini sol-gel method." *Ceramics International*, 42(5), 6136–6144. doi:10.1016/j.ceramint.2015.12.176.

Narayanan, Kannan Badri and Sung Soo Han 2017. "Helical plant viral nanoparticles — bioinspired synthesis of nanomaterials and nanostructures." *Bioinspiration and Biomimetics*, 12. IOP Publishing: 031001. doi:10.1088/1748-3190/aa6bfd.

Park, Y., Y.N. Hong, A. Weyers, Y. S. Kim, and R. J. Linhardt. 2011. "Polysaccharides and phytochemicals: A natural reservoir for the green synthesis of gold and silver nanoparticles." *IET Nanobiotechnology*, 5 (3): 69. doi:10.1049/iet-nbt.2010.0033.

Patel, R. P., Ashok H. Baria, and Nikunjana A. Patel. 2008. "An overview of size reduction technologies in the Field of pharmaceutical manufacturing." *Asian Journal of Pharmaceutics*, 2 (4): 216–220. http://asiapharmaceutics.info/index.php/ajp/article/view/202.

Patel, A., Y. Hu J. K. Tiwari, and K. P. Velikov. 2010. "Synthesis and characterization of zein-curcumin colloidal particles." *Soft Matter*, 6(24), 6192–6199. doi:10.1039/c0sm00800a.

Petrova, E., D. Kotsikau, and V. Pankov. 2015. "Structural characterization and magnetic properties of sol-gel derived $Zn_xFe_{3-x}O_4$ nanoparticles." *Journal of Magnetism and Magnetic Materials*, 378, 429–435. doi:10.1016/j.jmmm.2014.11.076.

Rai, A., Amit Singh, Absar Ahmad, and Murali Sastry. 2005. "Role of halide ions and temperature on the morphology of biologically synthesized gold nanotriangles," *Langmuir*, 22(2), 736–741. doi:10.1021/LA052055Q.

Rajan, R., Chandran, K., Harper, S.L., Yun, S.I. and Kalaichelvan, P.T., 2015. Plant extract synthesized silver nanoparticles: an ongoing source of novel biocompatible materials. *Industrial Crops and Products*, 70, 356-373.

Razavi M., E. Salahinejad, M. Fahmy, M. Yazdimamaghani, D. Vashaee, and L. Tayebi. 2015. "Green chemical and biological synthesis of nanoparticles and their biomedical applications." In: Basiuk V., Basiuk E. (eds) *Green Processes for Nanotechnology*. Springer, Cham.

Rosmi, M. S., S. M. Shinde, N. D. A. Rahman, A. Thangaraja, S. Sharma, K. P. Sharma, and M. Tanemura. 2016. "Synthesis of uniform monolayer graphene on re-solidified copper from waste chicken fat by low pressure chemical vapor deposition." *Materials Research Bulletin*, 83, 573–580. doi:10.1016/j.materresbull.2016.07.010.

Sakthivel, S. and R. P. Venkatesh. 2012. "Solid state synthesis of nano-mineral particles." *International Journal of Mining Science and Technology*, 22(5), 651–655. doi:10.1016/j.ijmst.2012.08.010.

Saleh, T. A. and Gupta, V. K. 2016. *Nanomaterial and Polymer Membranes: Synthesis, Characterization, and Applications*. Elsevier, Amsterdam.

Santos, J., N. Calero, J. Muñoz, and M. T. Cidade. 2018. "Development of food emulsions containing an advanced performance xanthan gum by microfluidization technique." *Food Science and Technology International*, 24(5), 373-381. doi:10.1177/1082013218756140.

Sanna, V., Siddiqui, I. A., Sechi, M., & Mukhtar, H. 2013. "Resveratrol-loaded nanoparticles based on poly(epsiloncaprolactone) and poly(D,L-lactic-co-glycolic acid)-poly(ethylene glycol) blend for prostate cancer treatment." Molecular Pharmaceutics, (10), 3871–3881. https://doi.org/10.1021/mp400342f.

Shah, Monaliben, Derek Fawcett, Shashi Sharma, Suraj Kumar Tripathy, and Gérrard Eddy Jai Poinern. 2015. "Green synthesis of metallic nanoparticles via biological entities." *Materials*. 8. doi:10.3390/ma8115377.

Sharma, V. K., Ria A. Yngard, and Yekaterina Lin. 2009. "Silver nanoparticles: Green synthesis and their antimicrobial activities." *Advances in Colloid and Interface Science*, 145 (1–2). Elsevier B.V.: 83–96. doi:10.1016/j.cis.2008.09.002.

Sharma, S., A. Umar, S. K. Mehta, and S. K. Kansal. 2017. "Fluorescent spongy carbon nanoglobules derived from pineapple juice: A potential sensing probe for specific and selective detection of chromium (VI) ions." *Ceramics International*, 43. Elsevier. doi:org/10.1016/j.ceramint.2017.02.127.

Shedbalkar, Utkarsha, Richa Singh, Sweety Wadhwani, Sharvari Gaidhani, and B. A. Chopade. 2014. "Microbial synthesis of gold nanoparticles: Current status and future prospects." *Advances in Colloid and Interface Science*, 209. Elsevier B.V.: 40–48. doi:10.1016/j.cis.2013.12.011.

Sibte, S., A. Abidi, and Q. Murtaza. 2014. "Synthesis and characterization of nano-hydroxyapatite powder using wet chemical precipitation reaction." *Journal of Materials Science and Technology*, 30(4), 307–310. doi: org/10.1016/j.jmst.2013.10.011.

Singh, M., Sinha, I. and Mandal, R.K., 2009. Role of pH in the green synthesis of silver nanoparticles. *Materials Letters*, 63(3-4): 425-427

Singh, Priyanka, Yu Jin Kim, Dabing Zhang, and Deok Chun Yang. 2016. "Biological synthesis of nanoparticles from plants and microorganisms." *Trends in Biotechnology*, 34 (7). Elsevier Ltd: 588–599. doi:10.1016/j.tibtech.2016.02.006.

Song, Jae Yong and Beom Soo Kim. 2009. "Rapid biological synthesis of silver nanoparticles using plant leaf extracts." *Bioprocess and Biosystems Engineering*, 32 (1): 79–84. doi:10.1007/s00449-008-0224-6.

Stijnman, Ann C., Igor Bodnar, and Hans Tromp, R. 2011. "Electrospinning of food-grade polysaccharides." *Food Hydrocolloids* 25(5): 1393–1398.

Swihart, M. T. 2003. "Vapor-phase synthesis of nanoparticles." *Current Opinion in Colloid and Interface Science, 8, 8*, 127–133. https://doi.org/10.1016/S1359-0294.

Ta, Hang, Nina Arndt, Yuao Wu, Hui Jean Lim, Shea Landeene, Run Zhang, Danielle Kamato, Peter Little, Andrew Whittaker, and Zhi Ping Gordon Xu. 2018. "Activatable magnetic resonance nanosensor as a potential imaging agent for detecting and discriminating thrombosis." *Nanoscale*, 10, 15013-15115. doi:10.1039/C8NR05095C.

Tan, B. J. Y., C. H. Sow, T. S. Koh, K. C. Chin, A. T.S. Wee, and C. K. Ong. 2005. "Fabrication of size-tunable gold nanoparticles array with nanosphere lithography, reactive ion etching, and thermal annealing." *Journal of Physical Chemistry B*, 109 (22): 11100–11109. doi:10.1021/jp045172n.

Tanya Prozorov, Ruslan Prozorov, and Kenneth S. Suslick, 2004. "High velocity interparticle collisions driven by ultrasound." *Journal of the American Chemical Society*, 126(43): 13890-13891. doi:org/10.1021/JA049493O.

Taormina, Gabriele, Corrado Sciancalepore, Federica Bondioli, and Massimo Messori. 2018. "Special resins for stereolithography: In situ generation of silver nanoparticles." *Polymers*, 10 (2). doi:10.3390/polym10020212.

Thakkar, Kaushik N., Snehit S. Mhatre, and Rasesh Y. Parikh. 2010. "Biological synthesis of metallic nanoparticles." *Nanomedicine: Nanotechnology, Biology, and Medicine*, 6 (2). Elsevier Inc.: 257–262. doi:10.1016/j.nano.2009.07.002.

Thatai, Sheenam, Parul Khurana, Jyoti Boken, Surendra Prasad, and Dinesh Kumar. 2014. "Nanoparticles and core–shell nanocomposite based new generation water remediation materials and analytical techniques: A review." *Microchemical Journal*, 116: 62–76. doi:10.1016/j.microc.2014.04.001.

Togashi, Takanari, Masato Nakayama, Atsuki Hashimoto, Manabu Ishizaki, Katsuhiko Kanaizuka, and Masato Kurihara. 2018. "Solvent-free synthesis of monodisperse Cu nanoparticles by thermal decomposition of an oleylamine-coordinated Cu oxalate complex." *Dalton Transactions,* 47 (15). Royal Society of Chemistry: 5342–5347. doi:10.1039/c8dt00345a.

Tran, Quang Huy, Van Quy Nguyen, and Anh Tuan Le. 2013. "Silver nanoparticles: Synthesis, properties, toxicology, applications and perspectives." *Advances in Natural Sciences: Nanoscience and Nanotechnology*, 4 (3). doi:10.1088/2043-6262/4/3/033001.

Tsuji, M. 2017. "Microwave-assisted synthesis of metallic nanomaterials in liquid phase." *Chemistry Select*, 2(2), 805–819. doi:org/10.1002/slct.201700011.

Wais, Ulrike, Alexander W. Jackson, Tao He, and Haifei Zhang. 2016. "Nanoformulation and encapsulation approaches for poorly water-soluble drug nanoparticles." *Nanoscale*, 8 (4): 1746–1769. doi:10.1039/c5nr07161e.

Wang, Cong, Xudong Gao, Zhongqin Chen, Yue Chen, and Haixia Chen. 2017. "Preparation, characterization and application of polysaccharide-based metallic nanoparticles: A review." *Polymers*, 9 (12). doi:10.3390/polym9120689.

Wegrzyn, Teresa F., Matt Golding, and Richard H. Archer. 2012. "Food layered manufacture: A new process for constructing solid foods." *Trends in Food Science and Technology*, 27 (2). Elsevier Ltd: 66–72. doi:10.1016/j.tifs.2012 04.006.

Xu, C., J. Lee, and A. S. Teja. 2008. "Continuous hydrothermal synthesis of lithium iron phosphate particles in subcritical and supercritical water." *The Journal of Supercritical Fluids*, 44(1), 92–97. doi:10.1016/J.SUPFLU.2007.09.001.

Xu, H., B. W. Zeiger, and K. S. Suslick. 2013. "Sonochemical synthesis of nanomaterials." *Chemical Society Reviews*, 42(7), 2555–2567. doi:10.1039/c2cs35282f.

Yazdi, Alireza Ahmadian, Adam Popma, William Wong, Tammy Nguyen, Yayue Pan, and Jie Xu. 2016. "3D printing: An emerging tool for novel microfluidics and lab-on-a-chip applications." *Microfluidics and Nanofluidics*, 20 (3). Springer Berlin Heidelberg: 1–18. doi:10.1007/s10404-016-1715-4.

Yu, C. H., Kin Tam, and Edman S.C. Tsang. 2008. "Chapter 5 chemical methods for preparation of nanoparticles in solution." *Handbook of Metal Physics*, 5: 113–141. doi:10.1016/S1570-002X(08)00205-X.

Zhu, Y. and F. Chen. 2013. "Microwave-assisted preparation of inorganic nanostructures in liquid phase." *Chemical Reviews*, 114(12): 6462–6555. doi:10.1021/cr400366s.

7 Protein- and Polysaccharide-Based Nanoparticles

S. Priyanka, S. Kritika, J.A. Moses, and C. Anandharamakrishnan

CONTENTS

7.1 INTRODUCTION

Nature designed the elementary blocks of life, such as amino acids, DNA, hormones, and sugars, in nanolevel (Weiss et al., 2006). In 1959, the concept of nanostructures was proposed by Richard Feynman and in 1974, Norio Taniguchi coined the term "nanotechnology" for the manipulation of particles at the submicron level. The term "nano" refers to a magnitude of 10^{-9} m (Quintanilla-Carvajal et al., 2010). This field involves study at the atomic, molecular, and macromolecular levels, specifically in the size range of 1 to 100 nm. At such minuscule level, the properties and functionalities of nanomaterials differ greatly from the bulk matter. Research in the recent past has reported the benefits of nanoparticles over microparticles (those with diameter > 1 μm (McClean et al., 1998).

Nanotechnology is one of the most promising and emerging technologies in scientific fields of research for the past few decades. The reduction of particle size to the nanoscale range results in an increase of surface-to-volume ratio, which in turn increases reactivity many folds with alterations in various properties such as optical, mechanical, and electrical properties. These new properties exhibited by nanoparticles offer many unique and novel applications in various fields (Neethirajan and Jayas, 2011). Bioactive nanoparticles are in the nanometer size range 10–100 nm, and are derived from natural sources, namely from plants, animals, microorganisms, etc. An endless list of polysaccharides, proteins and synthetic polymers is available for the production of nanoparticles. Those of natural origin are termed "biomaterials"; defined as materials intended to communicate with biological systems.

Furthermore, compared to microencapsulated particles, nanoparticles have the ability to provide improved bioavailability and controlled release profile, and enable precision targeting of bioactive compounds to a greater extent (Mozafari et al., 2006). In addition to the improved bioavailability on nanoparticles, the food and pharmaceutical industries employ nanotechnology to safeguard bioactive compounds such as polyphenols, micronutrients, enzymes, antioxidants, and nutraceuticals from adverse environmental factors, and to support the controlled release of core materials at the targeted site (Ezhilarasi et al., 2013; Gouin, 2004).

A wide range of functional properties of these biopolymers such as polysaccharides, proteins, synthetic polymers, and their derivatives or modified forms, overlays numerous ways in which the nanoparticle can be fabricated. Instead of opting for a single polysaccharide or protein to fabricate nanoparticles, scientists have experimented with using blends of various polysaccharides or proteins (Figure 7.1).

A wide range of interesting and health-promoting food ingredients known as nutraceutical are being studied extensively as they possess the significant potential to combat diseases and promote better health. The efficacy of all nutraceutical compounds in diseases prevention is associated with the bioavailability of bioactive ingredients (Chen et al., 2006). After oral intake, only a fraction of this compound is available in the biological tissue for quite a number of reasons. Instability in the gastrointestinal tract (low residence time, low permeability, pH, and enzymes), instability under food-processing conditions (heat, light, oxygen, and the presence of co-factors) reduce the bioactivity of nutraceutical compounds and in turn their health benefits (Bell, 2001).

The next step after the production of nanoparticles involves characterization in terms of size, morphology, solubility, surface charge, chemical interaction, encapsulation efficiency, loading efficiency, kinetic release in *in vitro* or *in vivo* conditions, etc. Studies are also directed to ensure their storage and processing stability, ability to withstand external environmental conditions, cytotoxicity, and biocompatibility. Yet another aspect which needs to be mentioned is that the materials used for nanoparticle fabrication must be GRAS (generally recognized as safe). Since the drug/nutrient loaded into the nanoparticle passes through the human body, the products should not in any way cause harm; i.e. they must be biocompatible and non-toxic. For example, unlike synthetic polymers, proteins are metabolizable into peptides by digestive enzymes. Above a certain molecular weight, synthetic polymers accumulate in the body and give rise to toxic degradation products (Elzoghby et al., 2012). Our discussion in this chapter will hence be limited to the polysaccharide- and protein-based nanoparticles which are natural and non-toxic.

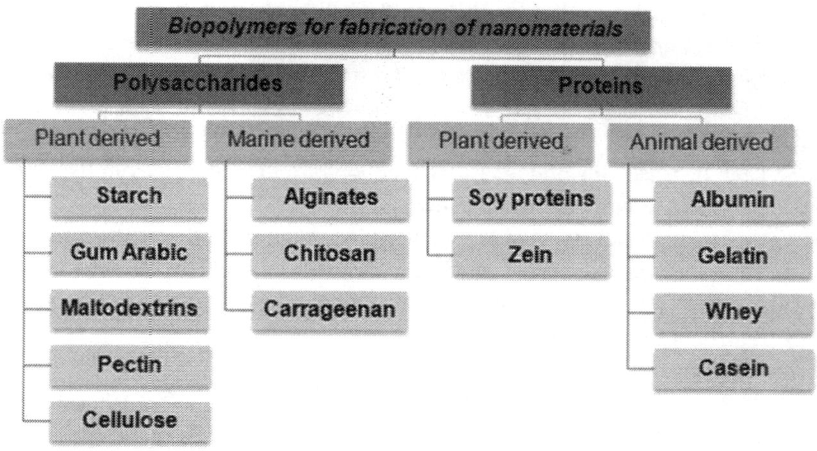

FIGURE 7.1 Commonly used biopolymers for fabrication of nanomaterials.

TABLE 7.1
Characteristic Differences of Protein and Polysaccharide Polymers

Proteins	Polysaccharides
Wide-ranging structures	Similar structures
Monodisperse	Polydisperse
Linear chain	Linear/branched chain
Flexible structure	Rigid structure
Amphiphilic	Hydrophilic
Polyelectrolyte	Non-ionic or charged
Thermo-sensitive	Thermo-insensitive
Strong surfactant binding	Weak surfactant binding
Emulsifying/foaming	Thickening/water-holding

Source: Dickinson (2003).

Instead of opting for a single polysaccharide or protein to fabricate nanoparticles, scientists have experimented with various blends of polysaccharides or proteins. The main aim of such work has been to explore the most competent combination of polysaccharide and protein in achieving the desirable nanoparticle characteristics. Further, it is imperative that the polysaccharide or protein do not alter the drug/nutrient it carries within; i.e. the functional and therapeutic properties are retained until it reaches the targeted site or its end use is achieved. Both proteins and polysaccharides are natural polymers and, hence, environment-friendly. A close look at their structure indicates that they are complex with proteins having functional groups. This is the reason proteins are more reactive than polysaccharides. However, both exhibit complex aggregation behavior and are used as gelling/stabilizing agents in food and drug formulations (Dickinson, 2003) (Table 7.1).

7.1.1 Advantages of Biopolymeric Nanoparticles

(1) Nanoparticles improve the efficacy of drug/nutrient as they fit into colloidal drug delivery systems and enhance cellular uptake.

(2) To overcome the problems related to the stability of liposomes, during storage and in biological fluids, bio-nanoparticles are developed that can serve as an alternative in delivery systems.
(3) The small size of the particle presents a greater surface area for cross-linking or interaction with the functional groups on the surface (i.e. covalent binding or layer-by-layer assembly) (Yu et al., 2007).
(4) Sustained cell or tissue-specific release and action of nutrient/drug from the nanoparticles can be achieved by use of bio-nanoparticles.
(5) Biopolymeric nanoparticles are biodegradable, biocompatible and pose low immunogenicity.
(6) Other merits of nanoparticles over microparticles are their ability to enhance drug/nutrient encapsulation, pharmacokinetics, pharmacodynamics, bioavailability, and therapeutic efficacy (Sharma et al., 2004).

The following sections of this chapter discuss (1) the physicochemical properties of various proteins and polysaccharides that have been used for the synthesis of nanoparticles, (2) the potential applications and fabrication of nanoparticles, (3) the most commonly used techniques for the preparation of nanoparticles, and (4) the existing drawbacks, if any, in the use of these proteins and polysaccharides, so that in future alternative materials may be sought. An exhaustive list of proteins and polysaccharides used for incorporation in nutraceuticals, the particle size of the nanoparticle, the method used for their preparation, and other salient features have been provided. This will provide an overview as to which material and what technique would lead to a particular nanoparticle.

7.2 POLYSACCHARIDE-BASED NANOPARTICLES

Carbohydrates are a class of macromolecules widely present in nature which can be further classified into monosaccharides and polysaccharides. Monosaccharides are the the building blocks of all other polysaccharides. Polysaccharides are formed from monosaccharides by series of complex condensation reactions. Due to a wide range of sugars, their isomers, and corresponding complex linkages, polysaccharides are significantly diversified (Swierczewska et al., 2016). Polysaccharides are a class of natural biopolymers along with proteins, nucleic acid, and lipids. They are highly harnessed because of their stability and sustainable nature; they are also extremely safe, non-toxic, biodegradable, and eco-friendly (Yang et al., 2015).

Polysaccharides are an assorted group of polymeric materials of either natural or synthetic origin. The glycosidic linkages between the monosaccharide units form the backbone of polysaccharides. Natural polysaccharides are of plant, animal, algal, or microbial origin, whereas synthetic polysaccharides are products obtained via one or a multi-step chemical synthesis. Structurally, polysaccharides may be linear or branched based on the degree of ramification and they further bear a number of functional groups. These lead to varied physical properties such as water solubility, viscosity, and gelling behavior. Due to their amazing properties, polysaccharide-based nanoparticles can play a part in various applications in promising sectors, namely pharmaceuticals, cosmetics, agricultural engineering, food engineering, and technology (Yang et al., 2015).

Polysaccharide nanoparticles act as proficient vehicles for sustained and targeted drug release with the twin role of improving the therapeutic effects and minimizing the side effects of formulated drugs (Lemarchand et al., 2006). In the further sections, the characteristics of certain polysaccharides of plant and animal origin and the techniques used for fabrication of polysaccharide nanoparticles have been discussed.

7.2.1 Plant-Derived Polysaccharides

7.2.1.1 Starch

Starch is the most copious storage material in plants, which makes it a natural, biodegradable, and sustainable biopolymer. Moreover, it is of great interest owing to its physicochemical properties such as pasting profile, gelatinization, retrogradation, film-forming ability, and rheological features. It is a naturally occurring polymer possessing bio-adhesion properties. It has been extensively used in food, agriculture, and pharmaceutical sectors because of its low cost, availability, safety, edibility, and ease of modification (Perez and Agama-Acevedo, 2018; Razavi and Amini, 2016). Chemically, starch is a semi-crystalline polysaccharide (granule size ranging from 1 to 100 μm) comprising straight chain glucose polymer (amylose) and branched chain polymer (amylopectin). The latter is a double helix forming stacks of alternating crystalline and amorphous lamellae. Starch granules are of different size and shape based on their botanical origin (Razavi and Amini, 2016). Raw starch has poor solubility in cold water, high viscosity post-gelatinization and a tendency to retrograde (Shi et al., 2011).

Native starches can be modified effortlessly through various physical, chemical or enzymatic techniques, or their combination. Some kind of alteration or modification is required to improve their functionality, such as physical or chemical treatment of starch with a focus to preserve its crystalline native structure. Haaj et al. (2013) have described the hierarchical structure of starch from which it is possible to obtain platelet-like nanoparticles referred to as starch nanocrystals. Starch granules themselves are made up of various nano-sized semi-crystalline blocklets. Starch nanoparticles can be categorized into various types based on their morphology, physical properties, and so on. Moreover, they are usually classified based on their shapes or crystalline nature, namely nanocrystals (SNCs), nanoparticles (SNPs), and nanofibers (SNFs), to distinguish amongst different types of particles (Razavi and Amini, 2016).

7.2.1.2 Maltodextrins

The derivatives of starches after partial hydrolysis by acids or enzymes are known as hydrolyzed starches, and maltodextrin (MDX) is one such product. It has low viscosity at high solids ratio and its DE (dextrose equivalent) values range from 4 to 42. The DE represents the percentage of reducing sugars calculated as glucose on a dry substance basis. Several functional and physical characteristics such as freezing temperature, viscosity, and solubility are influenced by the DE (Garnero et al., 2013). The molecular weight decreases with increasing DE values. Such a feature affords varied wall formulations for encapsulation purposes (Desobry et al., 1997). But as molecular weight decreases, the glass transition temperature of maltodextrin lowers; making it more hygroscopic and leading to instability of the product (Chuy and Labuza, 1994). Hence, the DE value of maltodextrins has to be chosen based on the core material to be entrapped. The possibilities of using maltodextrin as an excellent film-forming agent was demonstrated (Selmin et al., 2015). Some interesting properties listed by Kenyon and Anderson (1988) include low bulk density, anti-caking, bland taste, oxygen barrier, and good binding.

7.2.1.3 Gum Arabic

Gum arabic or gum acacia (GA) is the oldest and most widely used tree gum exudate of all the available gums. It is a natural plant polysaccharide of the *Acacia seyal* or *Acacia senegal* or *Acacia polyacantha* tree well known for its film-forming, emulsifying, low viscosity, and volatile-retention properties. The emulsifying property of GA is attributed to the presence of protein fraction (1.5%–2.6%) (Idris et al., 1998). The hydrophobic polypeptide moiety gets intensely adsorbed at the oil-water interface. The emulsion is stabilized by steric and electrostatic repulsion due to the presence of carbohydrates. It is neutral or slightly acidic in nature with a maximum viscosity at pH 6 (Verbeken et al., 2003). It easily dissolves both in cold and hot water up to 50% solids content, which permits its wide use. GA exhibits Newtonian behavior even at concentrations close to 50% solids and hence finds extensive applications in food. It is very attractive as a biopolymer for coating nanoparticles because of its unique properties, such as stabilizing nature, hydrocolloid emulsification, and biocompatibility. GA consists mainly of three fractions, each with unique chemical structures, but the most significant one is the highly branched polysaccharide whose molecular weight is about 3×10^5 g/mol (Dror et al., 2006). However, high cost and a limited supply of GA limit its wide applications in food and pharma industries. This is one reason for a shift towards the production of modified and derived starches, celluloses, and such other polymers having similar functionalities as those of GA.

7.2.1.4 Cellulose

Cellulose is the most abundant structural polysaccharide present in plants, some microbes, and invertebrate animals (Lavoine et al., 2012; Eichhorn et al., 2010). The elementary structure of cellulose is made up of a linear chain of β-(1,4)-linked D-glucose units which are joined to form cellobiose repeating units. Due to the parallel arrangement and side-by-side packing of cellulose glucan chains, there is a formation of microfibrils, which account for the stability and limited flexibility of its structure. The physical and chemical properties of the chains are affected by the extent of molecular organization and degree of polymerization. Furthermore, cellulose-based biopolymers are known for their limited inflammatory reactions, which makes them biocompatible (dos Santos and Grenha, 2015).

Cellulose can go through a great extent of chemical modification due to the presence of hydroxyl groups. Some common examples of cellulose are ether derivatives such as methylcellulose (MC), hydroxypropyl cellulose (HPC), hydroxyethyl cellulose (HEC), hydroxypropylmethylcellulose (HPMC), and carboxymethylcellulose (CMC). Moreover, these cellulose derivatives are famous for their controlled and sustained delivery of bioactives. Thus, the ability to modify viscosity, gelation properties, and molecular weight of cellulose-based materials allows them to be used for various applications in the pharmaceutical and biomedical industries (Moura et al., 2013). MC, HPMC, and CMC have been awarded GRAS (generally recognized as safe) status by the American Food and Drug Administration (FDA) since 1973 (FDA, 2014).

Cellulose acetate phthalate is one of the non-toxic cellulose derivatives used in the pharmaceutical industry for encapsulation of bioactives (Hanafi et al., 2013). Furthermore, its insolubility in acidic pH ($pH < 5.0$) and solubility at higher pH makes it an appropriate wall material for protecting bioactive substances in the stomach's acidic conditions, but it must be released in a controlled manner under the neutral condition of the small intestine. Carboxymethylcellulose (CMC) is a cellulose derivative that is hydrophilic polyanionic in nature. CMC is commonly used as a building block for assembling food-grade biopolymer particles (Tripathy and Raichur, 2013).

Microbial cellulose has grabbed the attention of everyone as a novel biomedical material due to its excellent strength, biocompatibility, and swelling capacity. Moreover, bacterial cellulose is very pure so it is considered an exceptional competitor for the fabrication of cellulose-based hydrogels (Mangalam et al., 2009).

7.2.1.5 Pectin

Pectin is a plant-based complex polysaccharide which is abundantly present in the cell wall of plants, especially in fruit peels. Pectin is available at very cheap cost because it is the major by-product from the fruit- and vegetable-processing industries (Beneke et al., 2009). The fundamental structure of pectin consists of a linear chain of α-(1, 4)-linked D-galacturonic acid units, presenting highly branched regions. The pectin is abundant in neutral sugars such as L-rhamnose, L-arabinose, D-galactose, D-xylose, and D-glucose, like other galacturonic acid-based polysaccharides (Beneke et al., 2009; Ridley et al., 2001). The negative charge is attributed to the presence of carboxyl groups, some of which naturally present as methyl esters.

Moreover, pectin is classified based on the degree of esterification – the percentage of esterified carboxyl groups (Sungthongjeen et al., 2004). Pectin has an average molecular weight of about 30–100 kDa and was awarded GRAS status in 1977 (FDA, 2018). Pectin is mostly used in the food industry as a gelling and thickening agent (Opanasopit et al., 2008). Furthermore, in pharmaceutical industries, it is used as an adsorbent, bulking agent, and as an emulsion stabilizer. At lower degrees of esterification (<50%), pectin reveals gel-forming abilities. The formation of rigid gels is due to the action of calcium or multivalent cations on pectin, which efficiently cross-links the galacturonic acid chains (Sungthongjeen et al., 2004). Pectin was found to be appropriate for the delivery of bioactive compounds due to its resistance to the proteases and amylases in the upper gastrointestinal tract, and can be specifically conveyed into the colonic region for the delivery of drugs due to its selective digestion by colonic microflora (Alvarez-Lorenzo et al., 2013).

7.2.2 Marine-Derived Polysaccharides

Marine polysaccharides have extensive applications in modern medicine, food processing, nanobiotechnology, and biomedical engineering for encapsulating bioactives, gene delivery, and drug delivery (Raveendran et al., 2013). Marine algae produce a wide variety of biopolymers with different structures and fascinating properties. Marine algae such as alginate, carrageenan, and agarose have been documented as a rich source of sulfated polysaccharides (Senni et al., 2011). Marine polysaccharides such as chitin, chitosan, and oligosaccharides are derived from the exoskeleton of marine crustaceans. Marine seaweeds (macroalgae) are multicellular organisms, which are classified based on their pigmentation into red, green, and blue algae. These marine-based polysaccharides are extensively utilized in the fields of food, medicine, fertilizer, and bioethanol production (Horn, 2009).

7.2.2.1 Alginate

Alginates are natural polymeric substances obtained from brown seaweed, which is also known as marine kelp. Commercially, alginate production involves an alkaline extraction process. It is an anionic polymer and forms a gel in the presence of divalent cations such as Ca^{2+}. The industrial application of alginates is extensive, mainly as stabilizers and thickeners. Structurally, alginates consist of α-L-guluronic acid and β-D-mannuronic acid linearly linked by 1–4 linkages. Epimerase interconversion of mannuronic acid to guluronic acid leads to the formation of urinates (Panikkar and Brasch, 1996). The physical properties of alginates are governed by the concentration of cross-linking cation, molecular weight of the polymer, composition, and arrangement of uronate residues (Draget et al., 1990).

Recently, alginates have been playing a major role in the biomedical fields, being employed as tableting agents (Rahim et al., 2015) and as the matrix for immobilizing systems (Jain and Bar-Shalom, 2014), as well as in tissue engineering (Venkatesan, Bhatnagar, and Kim, 2014) and drug delivery (Agarwal et al., 2015). They are bioadhesive in nature and can offer a site-specific sustained release of the entrapped compound. Release kinetics can be monitored on the basis of the type of alginate selected, the method used, and the pore size of alginate beads.

7.2.2.2 Chitosan

Chitosan has its source in the shells of crabs, shrimps, and krill (Kas, 1997), and it is produced by the deacetylation of chitin. It is the second most abundant cationic polysaccharide available in nature, after cellulose. It has a linear structure consisting of acetylated units (N-acetyl-D-glucosamine) and deacetylated units (β-(1–4)-linked D-glucosamine). It is hydrophilic and positively charged. The high charge density of chitosan is due to the amino acid group and hence it chemically reacts to form a salt with acids. It is biocompatible, biodegradable and non-toxic, thus it also has an antimicrobial and antifungal action (Darmadji and Izumimoto, 1994). Chitosan, being mucoadhesive, opens the tight junctions between epithelial cells leading to enhanced intestinal permeation of drugs, nutrients, and proteins (Cano-Cebrian et al., 2005). Molecular weight, degree of acetylation, and viscosity contribute to the overall properties of chitosan (van der Lubben et al., 2001). The degree of acetylation, in turn, affects its solubility, hydrophobicity, and electrostatic interaction with polyanions. A general trend observed regarding the properties of chitosan is that lower molecular weights and a lower degree of acetylation display greater solubility and faster degradation than high molecular weights (Bowman and Leong, 2006). Besides, numerous studies have reported site-specific, controlled release of chitosan nanoparticles in the stomach and intestines and also enhanced permeability across the mucosal epithelia.

7.3 PROTEIN-BASED NANOPARTICLES

Proteins are versatile macromolecules and have many different active functional groups. They are characterized based on amino acid composition, sequence, size, shape, net charge, isoelectric point, and solubility (Nehete et al., 2013) They are biodegradable, biocompatible, non-toxic, non-immunogenic, etc. They also possess emulsifying, water binding, and gel-forming capacities. The amphiphilic nature of proteins interacts with both solvent the drug/nutrient resulting in improved solubility, better tissue penetration, and enhanced cellular uptake (Jahanshahi and Babaei, 2008). They can be broadly divided into two categories: plant-derived and animal-derived proteins.

7.3.1 Plant-Derived Proteins

7.3.1.1 Soy Protein

The enriched form of soybean (*Glycine max* L.) protein known as soy protein isolate (SPI) has a good nutrient profile as well as remarkable functional properties that can be used in food formulations. SPI chiefly consists of glycinin and β-conglycinin (Tang et al., 2009). A balanced composition of polar, non-polar and charged amino acids in SPI is ideal for loading drugs/nutrients with various functional groups. Hydrophobic amino acids indulge in a strong interaction with encapsulated compounds, while the polar and charged groups facilitate protein-drug association via electrostatic attraction and hydrogen bonding (Shen and Tang, 2012).

7.3.1.2 Zein

Zein is a major class of protein comprising water-insoluble but alcohol-soluble proteins existing in the endosperm of corn kernels (35%–60% of total proteins) (Shukla and Cheryan, 2001). It is a mixture of four types of proteins: α-zein (75%–85%), β-zein (10%–15%), γ-zein (5%–10%), and δ-zein. All of these differ in molecular weight, solubility, and amino acid sequence (Zhu et al., 2007). α-zein is the major commercially available protein in the market. The solubility of zein is enhanced in the presence of alcohol, a high concentration of urea, alkaline solutions and anionic detergents (Shukla and Cheryan, 2001). The typical solubility of zein is due to the amino acid sequence, which leads to the easy production of zein micro/nanoparticles that serve as a vehicle for drug/bioactive delivery. Loss of zein solubility due to increased concentration of water and simultaneous precipitation and solidification to form particles is the major mechanism behind the formation of zein micro/nanoparticles (Luo and Wang, 2014).

7.3.2 Animal-Derived Proteins

7.3.2.1 Albumin

Albumin is a protein found in blood plasma, and is biodegradable, biocompatible, and less immunogenic (Rubino et al., 1993). Albumin of both human and bovine origins has been reported to have desirable properties for nanoparticle formation. Albumin carries reactive groups such as thiol, and amino and carboxylic groups, which makes it versatile, allowing it to be used for surface modification and/or ligand binding by covalent linkage (Jahanshahi and Babaei, 2008).

7.3.2.2 Gelatin

Gelatin is a water-soluble macromolecule obtained as a result of heat dissolution and controlled partial hydrolysis of collagen. Acid hydrolysis with pI between 7 and 9 gives type A gelatin whereas alkaline hydrolysis with pI between 4.8 and 5 gives type B gelatin (Sundar et al., 2010). Gelatin is rich in glycine, hydroxyproline, and alanine amino acids. The properties that make gelatin a unique carrier vehicle over other synthetic polymers are as follows: non-irritability, biocompatibility, biodegradability (Schwick and Heide, 1969), non-toxicity, non-carcinogenicity, low immunogenicity, and antigenicity (Bajpai and Choubey, 2006).

7.3.2.3 Whey Protein

Whey protein is chiefly obtained as a by-product from the cheese industry and exhibits a very good nutrient profile. These are globular proteins and are vulnerable to denaturation and gelation due to heat, but stable over a wide pH range (Morr and Ha, 1993). They have good gelling and emulsification properties. Many studies report the use of WPI or WPC for micro/nanoparticle production for drug/nutrient delivery (Hundre, Karthik, and Anandharamakrishnan, 2015; Rajam and Anandharamakrishnan, 2015; Beaulieu et al., 2002; Picot and Lacroix, 2004; Zhu et al., 2007).

7.4 NANOPARTICLE FABRICATION TECHNIQUES

Nowadays, tailor-made nanoparticles are ruling various industrial sectors, including pharmaceuticals, cosmetics, agricultural, and food. This is because it has become easy to alter their bioavailability, release profile, targeted delivery, various nutritional properties, flavor profile, appearance, rheological characteristics, stability, and processability. Nanoparticles are frequently fabricated from food-grade biopolymers, namely proteins and polysaccharides. Food-grade biopolymers exhibit a varied range of molecular, functional and physicochemical properties (e.g. molecular weight, charge, branching, flexibility, polarity, and solubility) which facilitate their assembly of colloidal particles that show evidence of a broad range of functional attributes. By cautious selection of appropriate biopolymers and fabrication methods, biopolymer particles can be precisely fabricated with tailored behaviors or features (Joye and McClements, 2014).

In this section, we review the current developments in the fabrication and design of functional biopolymer nanoparticles. Their importance and some of the challenges that hinder their usage and technical solutions to overcome those challenges will be the focus of future research. Nanoparticle fabrication techniques can be broadly classified into two main categories: the "top-down" approach and the "bottom-up" approach. In the "top-down" approach, the size of the particles is reduced from bulk to nanoscale during the fabrication process, for example by using various mechanical forces as in ball milling. On the contrary, in the "bottom-up" approach, the particle size is increased from atomic scale to nanoscale by various approaches such self-assembly, coacervation, etc. Moreover, in some cases, a combination of both techniques has been used (Jafari, 2017) (Figure 7.2).

7.4.1 Fabrication Techniques Used for Proteins and Polysaccharide-Based Nanomaterials

The most promising areas of current research are in identifying the most suitable biopolymers and feasible techniques to fabricate them to produce low-cost, eco-friendly, biodegradable, food-grade and high-throughput biopolymer particles. Biopolymer nanoparticles can be produced from various proteins and polysaccharides that have GRAS status, which aids in creating delivery systems with different functional attributes. This section will discuss the current research trends in fabrication techniques commonly used to prepare nanomaterials based on protein and polysaccharide in a nutshell (Joye and McClements, 2014).

Among nanoparticles, both protein and polysaccharide-based nanoparticles have received special attention because they are naturally available, biodegradable in nature, easily metabolizable, and can be manipulated effortlessly for the specific applications (targeted delivery and sustained release). Moreover, there are numerous possibilities for altering and modifying their surface properties for effective covalent drug attachment (Weber et al., 2000).

7.4.1.1 Fabrication of Protein-Based Nanoparticles

There are numerous techniques for the fabrication of protein-based nanoparticles, namely desolvation, precipitation, coacervation, and self-assembly (Elzoghby et al., 2012). The selection of techniques for fabrication depends on numerous phenomena such as the functional properties, physicochemical properties, and amino acid composition of the protein to be used as wall material, and

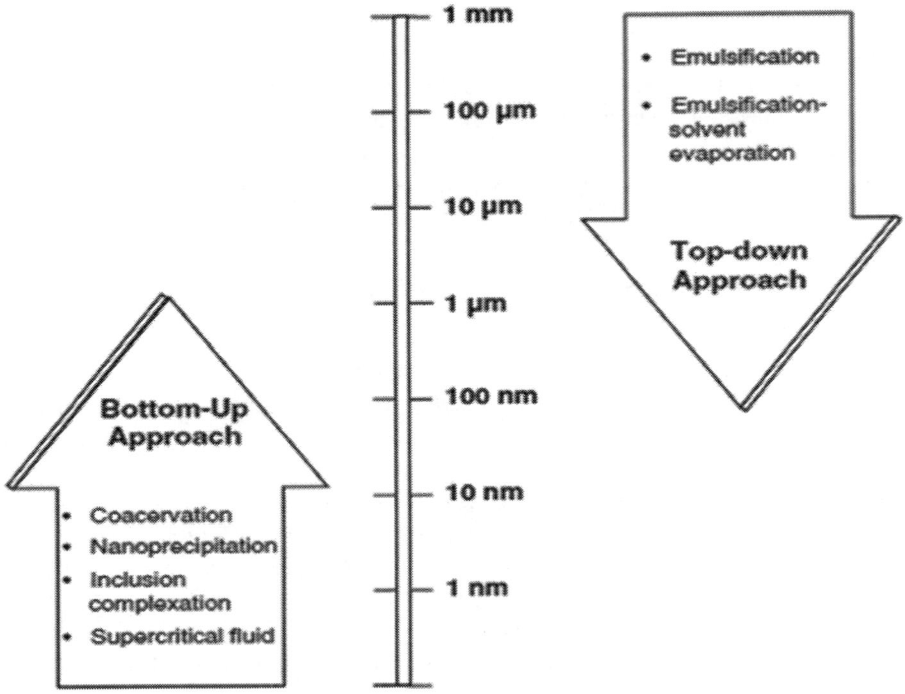

FIGURE 7.2 Classification of nanoencapsulation: "top-down" and "bottom-up" approaches. (Ezhilarasi et al., 2013.)

the properties of the core material (bioactive compound) to be encapsulated (Tarhini et al., 2017). Moreover, the development of nanoparticles depends on the ability of the protein to unfold and the reduction in the hydrophobic intramolecular interactions in the selected protein biopolymer. During the formation of nanoparticles, the protein biopolymer goes through a huge conformational change based on its amino acid composition, concentration, preparation conditions (pH, ionic strength, and solvent), and cross-linking methods (Pathak and Thassu, 2016).

In spite of the dissimilarities in the various techniques adopted to fabricate protein nanoparticles, all the encapsulation methods are based on the precipitation process of the used macromolecules. The precipitation of polymers and biomolecules happens as a result of a decrease in the solubility of macromolecules in the given solvent. The solubility of macromolecules can be reduced either by the addition of a non-solvent or by altering their physiochemical properties such as salinity, pH, and temperature of the protein biopolymer solution (Miladi et al., 2014). The unfolding of the proteins during the fabrication process uncovers the disulfides and thiol groups present in them. After that, thermal or chemical cross-linking occurs which results in the entrapment of bioactive compounds in the cross-linked nanoparticles. The most frequently used techniques for the fabrication of protein-based nanoparticles are self-assembly, electrohydrodynamic processes (electrospinning and electrospraying), coacervation, desolvation, and precipitation (Pathak and Thassu, 2016).

7.4.1.2 Fabrication of Polysaccharide-Based Nanoparticles

Polysaccharides have received an enormous amount of attention and have recently been recognized as the most promising carrier material for encapsulation because of their outstanding qualities such as abundance, non-toxic nature, biocompatibility, and biodegradability. Considerable investigations have been directed towards the development of polysaccharide-based nanoparticles. The most common fabrication methods used for the production of polysaccharide-based nanoparticles include self-assembly, complex coacervation, emulsification, desolvation, etc. (Yang et al., 2015).

7.5 FABRICATION TECHNIQUES

7.5.1 Desolvation

Desolvation is a commonly used fabrication technique for the production of biopolymer-based nanoparticles. Desolvation of polysaccharides is usually done either by changing charges or by adding desolvating agents. The most frequently used desolvating agents are salts or alcohols which are chosen based the characteristics of the bioactive molecule to be encapsulated in the biopolymer (Yang et al., 2015). In a nutshell, the desolvation process for protein is done by the addition of desolvating agents such as alcohol and acetone – whichever is suitable to the protein solution under constant stirring. The inclusion of a desolvating agent results in protein dehydration and hence leads to a conformational change from stretched to coil conformation (Elzoghby et al., 2015).

Denser protein nanoparticles are made by cross-linking the amino acid functional group of the proteins, which in turn protects the coacervates (Tarhini et al., 2017). Rao et al. (2012) produced a nanocomposite in which silver is encapsulated in chitosan. The chitosan was desolvated with ethanol, which was followed by the addition poly (ethylene glycol-di-aldehyde) for cross-linking in the presence of silver nitrate solution. The fabricated nanocomposites were characterized to learn about their physicochemical properties.

Gelatin nanoparticles are preferably used for selective drug delivery of didanosine, an anti-HIV drug, to various target organs. The gelatin nanoparticles, coupled with mannose, are prepared through a two-step desolvation technique. The nanoparticles were analyzed for their particle size, zeta potential, and drug entrapment percentage, which were found to be in the range of 248–325 nm, 6.2 ± 0.12 mV, and 40.2%–48.5% respectively. The coupling of mannose with nanoparticles profoundly increased the uptake of the drug in various vital organs such as lung, liver, and lymph nodes. However, the coupling with mannose increased the particle size during drug loading and the value of zeta potential was decreased (Jain et al., 2008) (Figure 7.3).

The other biopolymers that can be fabricated by the desolvation technique are albumin, silk, whey, chitosan, and gliadin. In some special cases, to prepare a nanoparticle of small and uniform size, an additional desolvation step is necessary. In spite of its widespread use in the fabrication of biopolymer-based nanoparticles, the desolvation technique has two drawbacks in the use of toxic cross-linkers and organic solvents (Elzoghby et al., 2015).

7.5.2 Coacervation

The coacervation technique involves the encapsulation of bioactive molecules in a polymer-rich solution. Coacervation is a novel technique which involves the phase separation of either a single

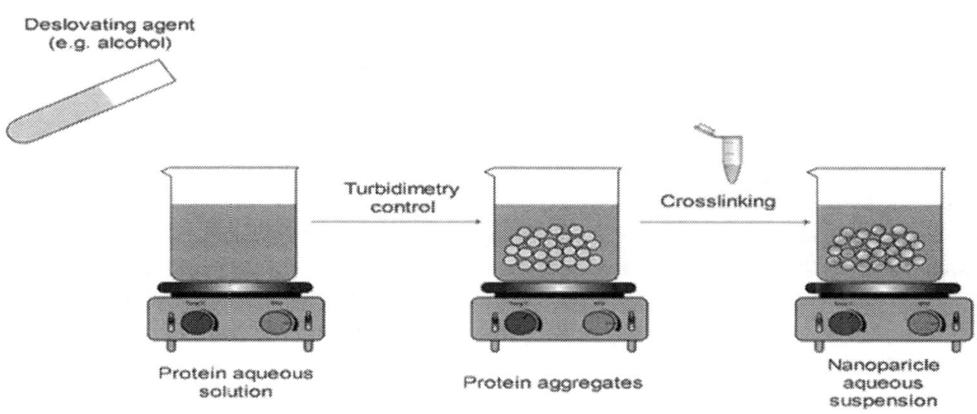

FIGURE 7.3 Nanoparticle fabrication by desolvation technique. (Tarhini et al., 2017.)

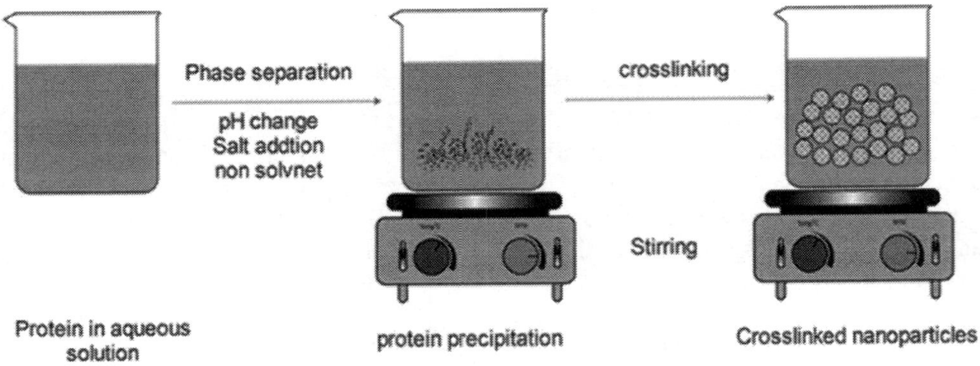

FIGURE 7.4 Nanoparticle fabrication by coacervation technique. (Tarhini, Greige-Gerges, and Elaissari, 2017.)

polyelectrolyte or a mixture of polyelectrolytes from a solution and the subsequent deposition of the newly formed coacervate phase around the active ingredient (Anandharamakrishnan, 2014). Furthermore, this technique is based on the interaction between two oppositely charged polyelectrolytes in water to form the coacervate (Anandharamakrishnan, 2015). Moreover, to increase the robustness and stability of the coacervates, the hydrocolloid shell can also be cross-linked either by a suitable chemical cross-linker such as glutaraldehyde, or by an enzymatic cross-linker such as transglutaminase (Zuidam and Shimoni, 2010).

In the coacervation technique, when salt is added to the protein solution, there is a liquid-liquid phase separation, which in turn produces a transparent solution at the top and a polymer-rich dense phase at the bottom. Hence the desired nanoparticles are formed (Elzoghby et al., 2015). Various biopolymer nanoparticles such as gelatin, zein, chitosan and maltodextrin can also be fabricated through the coacervation technique (Figure 7.4).

The coacervation process can be classified into two types based on the number of polymer types used. If only one type of polymer is employed, it is known as simple coacervation, and if more than one or two polymers are employed, it is known as complex coacervation. The degree of interaction between the biopolymer and the nature of the coacervate complex depends on various factors which include the nature of the biopolymer and its properties such as charge, molar mass, and flexibility; the properties of solutions such as pH and ionic strength, and the concentration and ratio of the biopolymers (Tolstoguzov, 2003; De Kruif et al., 2004; Turgeon et al., 2007).

Wang et al. (2008) encapsulated capsaicin in gelatin using the simple coacervation process which was cross-linked by using glutaraldehyde and dried using a vacuum oven. The nanocapsules obtained had a mean particle size of 100 nm. Furthermore, it was established that the nanoencapsulates of capsaicin produced using cross-linked gelatin as wall material had a superior melting point and thermal pyrolysis temperatures in comparison with uncovered capsaicin (Figure 7.5).

The core ingredient (a) is subsequently deposited, suspended or emulsified in the coacervate solution (b), such that the polymer entraps the active component (c). Coacervate is induced to enclose the core by changing the temperature or the pH, or by addition of a concentrated ionic salt solution (d). The resultant solution can be cooled to form a gel network of the core and the polymer present in it, or the same can be induced through appropriate chemical or enzyme cross-linkers (Anandharamakrishnan, 2015). Coacervation is based on the differential solubility of proteins in solvents depending on the solvent polarity, pH, ionic strength, or presence of electrolytes. The coacervation process reduces the solubility of the protein leading to phase separation."

Xing et al. (2005) used complex coacervation techniques to encapsulate capsaicin using two oppositely charged biopolymer solutions such as gelatin and acacia. The wet nanocapsules were treated with hydrolyzable tannins because of its unique reaction with gelatin aids in precipitation. Furthermore, the nanocapsules were cross-linked with glutaraldehyde subjected to freeze-drying.

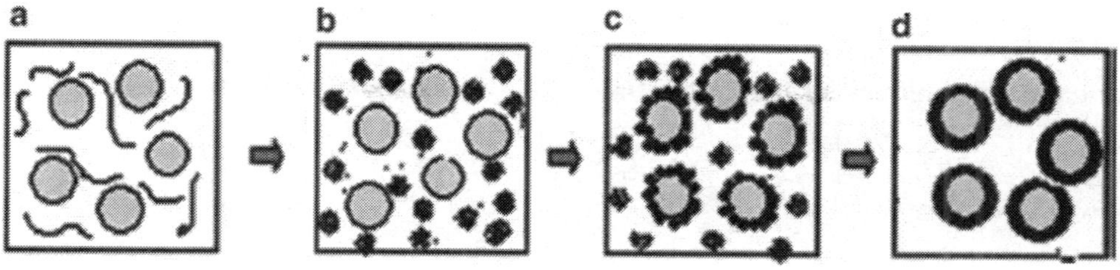

FIGURE 7.5 Simple coacervation of capsaicin in cross-linked gelatin (a) dispersion of capsaicin in gelatin solution; (b) coacervation of gelatin with acacia in the solution; (c) coacervation of insoluble complex on the surface of the capsaicin; and (d) shell formation by the addition of glutaraldehyde solution. (Jincheng et al., 2010.)

The mean diameter of nanocapsules was found to be of 300–600 nm and encapsulation efficiency (81.17%), spherical morphology, and good dispersion properties were observed in this study. Moreover, the incorporation of hydrolyzable tannins into the wet encapsulate had an imperative effect on the morphology and particle size distribution of the nanocapsules due to the synergistic actions of hydrogen bonding and hydrophobic effects.

Gan and Wang (2007) encapsulated bovine serum albumin (BSA) using polyionic coacervation fabrication process in chitosan by either incorporation or incubation method using polyanion tripolyphosphate (TPP) as the cross-linker that aids in coacervation. The BSA-loaded chitosan–TPP nanoencapsulates were characterized by their particle size which was in the range of 200–580 nm. The BSA encapsulates were analyzed for their morphology, zeta potential, and encapsulation efficiency. They showed a positive zeta potential.

By using the coacervation technique, we can usually fabricate biopolymer nanocapsules of size range generally varying from 100 to 600 nm. This method depends on an appropriate drying technique such as vacuum or freeze-drying. The most commonly used wall materials for this fabrication technique are maltodextrin, gelatin, chitosan, and gum acacia. Moreover, it was found that treating the coacervates with tannins favorably modifies morphological characteristics, such as shape, and provides good dispersion and particle distribution. Additionally, treatment with chemical cross-linker glutaraldehyde improves the thermal stability and melting point of the encapsulates (Ezhilarasi et al., 2013) (Table 7.2).

7.5.3 Nanoprecipitation

Nanoprecipitation is popularly known as the solvent displacement technique. The nanoprecipitation technique usually consists of two solvents which are miscible, wherein the wall (biopolymer) and core material (bioactive compound) should be soluble in either one of the solvent and insoluble in the other. Nanoprecipitation takes place due to rapid desolvation of the biopolymer instantaneously when it is added to the non-solvent. Certainly, the biopolymer precipitates by entrapping the bioactive compound straightaway, the moment the biopolymer containing solvent diffuses into the dispersing medium (Fessi et al., 1989).

The nanoprecipitation process depends on the reduction of solvent quality in which most of the nanoparticles are dissolved. Moreover, various factors such as solubility conditions, pH, modification of salt concentration, and incorporation of the non-solvent phase will result in the reduction of solvent quality (Miladi et al., 2014). The non-solvent based precipitation technique comprises the following stages (Figure 7.6):

(1) Production of the supersaturated solution by the addition of non-solvent
(2) Nucleation process is stimulated as the supersaturation reaches the critical value
(3) Nuclei growth occurs by condensation or coagulation (Joye and McClements, 2014)

TABLE 7.2
Fabrication Techniques for the Production of Bio-Based Nanoparticles

Fabrication Method	Protein/ Polysaccharide Wall Material	Bioactive Compound	Particle Size	Function	References
Coacervation	Gelatin, maltodextrin, and tannins	Capsaicin	100 nm	Conceals pungent odor	Wang et al. (2008)
	Gelatin, acacia, and hydrolyzable tannins	Capsaicin	300–600 nm	Delayed release and enhancement in efficiency	Xing et al. (2005)
	Gelatin, acacia, and tannins	Capsaicin	100 nm	Conceals pungent odor and increases stability	Jincheng et al., (2010)
	Chitosan	BSA –bovine serum albumin	200–580 nm	Controlled release of encapsulated protein	Gan and Wang (2007)
Nanoprecipitation	Ethyl cellulose and methyl cellulose	Curcumin	117 and 218 nm	Enhances oral sustainability and bioavailability	Suwannateep et al. (2011)
	Chitosan/poly (ε-caprolactone)	Curcumin	220 and 360 nm	Sustained release and increase in cellular uptake	Liu et al. (2012)

Source: Ezhilarasi et al. (2013).

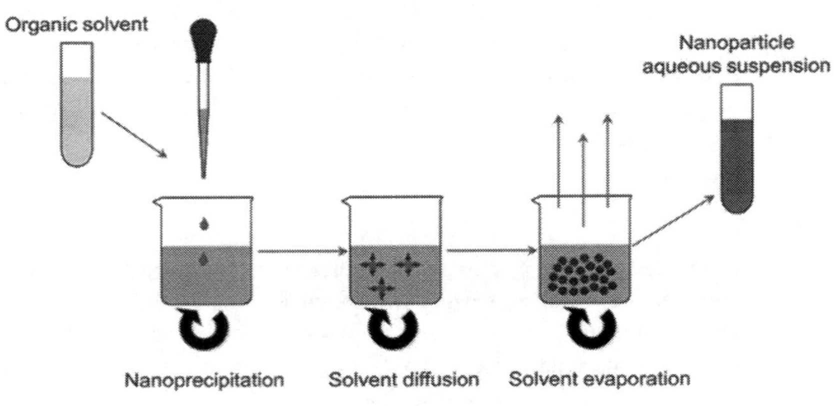

FIGURE 7.6 Preparation of nanoparticles by nanoprecipitation. (Tarhini et al., 2017.)

The spontaneous emulsification of the organic internal phase comprising the dissolved biopolymer, bioactive compound, and organic solvent into the aqueous external phase is the base for nanoprecipitation (Ezhilarasi et al., 2013).

Lee et al. (2012) used the nanoprecipitation technique to fabricate gelatin-based nanoencapsulates. The nanoparticles were produced using water as solvent and ethanol as non-solvent. It was revealed that those nanoparticles that are produced by excluding the cross-linker had irregular morphology (shape) owing to particle aggregation. On the contrary, the nanoparticles produced with cross-linker yielded particles with uniform round shape, low polydispersity index (0.096), and the unimodal size of 251 nm. The above study clearly reveals that nanoprecipitation is the appropriate method for fabrication of gelatin nanoparticles.

In another study, curcumin-zein nanoparticles were synthesized by an antisolvent precipitation technique. The study reveals that the solvent system and the zein/curcumin ratio plays a major role in controlling the mean particle size in the range of 100–150 nm. This formulation enhanced the stability of curcumin at all physiological pH and following UV irradiation. Furthermore, it was found that all the particles produced were spherical in shape. The synthesized nanoparticles showed excellent photostability of curcumin which was confirmed by UV irradiation technique. Additionally, the particles had good colloidal stability in simulated gastrointestinal conditions and a wide range of pH. Moreover, the curcumin-zein nanoparticle formulation was recommended for oral delivery due to the edible nature of zein (Patel et al., 2010).

Whey protein isolate was used to produce nanoparticles by a nanoprecipitation technique in which the non-solvent used was ethanol, which was added at a speed rate of 1 ml/min in order to attain 1:5 solvent/non-solvent volume ratio. The nanoparticles obtained were of spherical shape with size range down to 10–100 nm. Particle properties such as size showed no change at pH 3, a favorable pH for food applications. Additionally, it was revealed that a combination of homogenization and heating can be used for controlling the mean particle size (Gülseren et al., 2012) (Table 7.3).

Qin et al. (2016) used the nanoprecipitation technique to fabricate starch nanoparticles with seven different types of native starches whose amylose content varies from 0.8% to 69.0%. The native starches taken for the study were waxy corn, sweet potato, high amylose corn, normal corn, tapioca, pea, and potato starches. The study compared the mean particle size of native starch at micro and nanoscale levels. They were found to be 15–49 μm and 30–75 nm respectively. Fascinatingly, it was found that the native starch granules were small and the obtained starch nanoparticles were also small.

Chitosan was used to produce nanoparticles by nanoprecipitation technique. Chitosan was directly dissolved in aqueous acetic acid or in water as N-(methyl sulfonic acid) chitosan. In this study methanol was used as the non-solvent diffusing phase. Nanoparticles prepared using N-(methyl sulfonic acid) chitosan are slightly smaller than those obtained using acidic media (200 vs. 400 nm approximately). Furthermore, the size of the particles increased drastically on replacing methanol with water in the suspension. On the other hand, the addition of Tween-80 in little amounts to the non-solvent phase produced nanoparticles of 300 nm size, regardless of the solvent to non-solvent volume ratio (Luque-Alcaraz et al., 2016). The above-mentioned studies give a strong indication that nanoprecipitation is a promising and useful approach for the fabrication of nanoparticles using natural, biodegradable, non-toxic polysaccharide and protein-based biopolymers (Figure 7.7).

TABLE 7.3
Commonly Employed Biopolymers in the Fabrication of Nanoparticles by Nanoprecipitation Technique

Class	Name of Biopolymer	References
Polysaccharide	Starch	Qin et al. (2016)
	Chitosan	Luque-Alcaraz et al. (2016)
Protein	Gelatin	Han et al. (2013)
	Zein	Patel et al. (2010)

Source: Rivas et al. (2017).

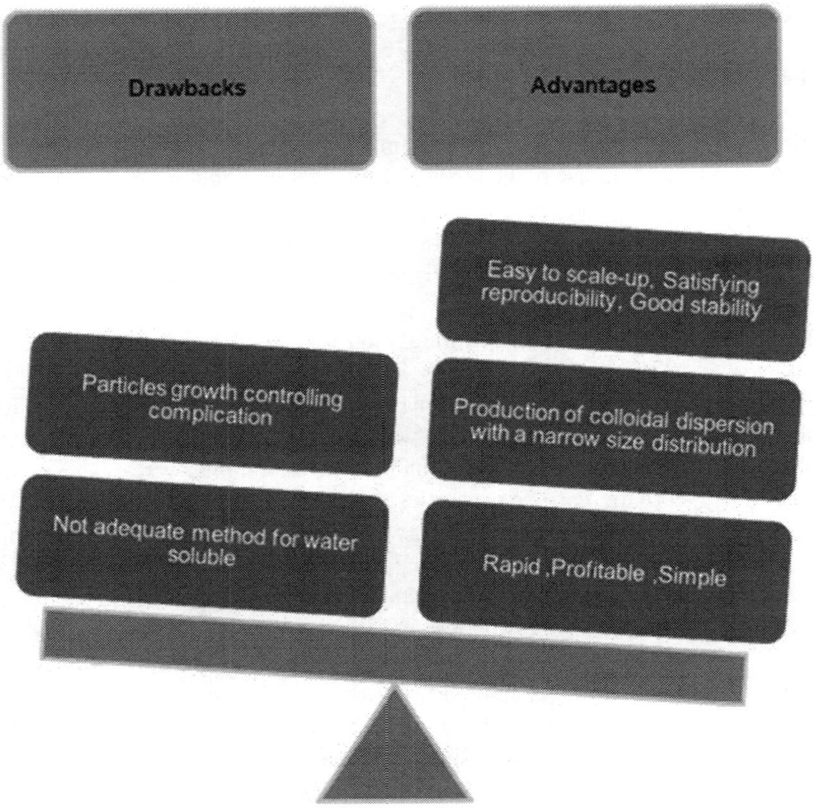

FIGURE 7.7 Nanoprecipitation method: advantages and drawbacks. (Miladi et al., 2014; Paliwal et al., 2014.)

7.5.4 Nanospray Drying

> Spray drying is the transformation of feed from a fluid state into a dried particulate form by spraying the feed into a hot drying medium.
>
> **Masters (1991)**

The principle of spray drying for the purpose of moisture elimination is based on subjecting the feed to be dried to heat and simultaneously adjusting the humidity level of the drying medium. The spray dryer usually utilizes the convection mode of heat transfer (Anandharamakrishnan and Ishwarya, 2015a). The distinctiveness of the spray drying process is in the atomization of feed using an atomizer into the drying chamber which enhances the evaporation rate of moisture and in turn results in a superior drying rate. Before becoming powder, the liquid feed entering the spray dryer experiences a series of transformations. Spray drying involves four stages, namely: (1) atomization of the feed solution, (2) interaction of feed spray with hot gas, (3) evaporation of moisture, and (4) separation of the particles (Anandharamakrishnan, 2015).

Spray drying is a unique drying process that atomizes the inlet feed solution into a drying chamber containing the hot gas at a temperature ranging from 100°C to 200°C to transform the feed liquid into a powder. The feed solution supplied to the spray dryer can be an emulsion, biopolymer solution, or a suspension. For the production of a diverse range of biopolymer particles of hydrophilic and hydrophobic nature, spray drying can be employed. However, only those that are not soluble in water can be exploited in aqueous-based foods without undergoing desolvation, e.g. zein (Davidov-Pardo et al., 2015).

For the fabrication of heat-labile molecules such as proteins, phenolic compounds, and lipids, the spray drying technique is most suitable because it prevents them from losing their biological activity and denaturation. Spray drying is extensively used in commercial production of bio-based nanoparticles owing to the flexibility of its system parameters, which in turn control flow properties, size of the particles, bulk density, and throughput (Tarhini et al., 2017).

Spray drying can encapsulate active components within a protective outer layer while it transforms the liquid feed into a dry, stable form. The difference in drying characteristics between the wall and core components governs the encapsulation during spray drying. Due to its innate film-forming properties, the wall material dries out at a much faster rate than the medium (usually water) in which the core is suspended for the preparation of feed solution. This enables the wall material to form a crust or coating around the droplet containing the core. The dried particles of bioactive nanoencapsulates are obtained on completion of the spray drying process (Anandharamakrishnan, 2015). Nanospray drying was employed effectively to fabricate bovine serum albumin (BSA) nanoparticles. This study employed the novelty Nano Spray Dryer B-90 (furnished with vibrating mesh spray and electrostatic spray collector) to produce smooth spherical nanoparticles with a mean size of 460 nm. Furthermore, this fabrication technique can be successfully employed for the production of protein nanoparticles for controlled pulmonary, nasal, and oral delivery (Lee et al., 2011).

Jafari et al. (2007, 2008) encapsulated oil through the production of nanoemulsions in conjunction with spray drying, using whey protein concentrate and modified starch (Hi-Cap) as wall material. In both studies, it was found that only the emulsion droplets were of nano-sized (200–800 nm), but while being subjected to the spray drying process they were converted into micron size (above 20 μm) (Table 7.4).

Ferreira et al. (2007) used carbohydrate matrix such as starch and maltodextrin to encapsulate catechin, which is well known for its antioxidant activity, using homogenization followed by spray drying at an inlet temperature of 150–190°C. The obtained nanoparticles had a smooth surface with a spherical shape and a diameter of 80 nm. It was also found that the stability of the nanoencapsulates was due to the extremely negative zeta potential. The release of catechins from the nanocapsules was observed at pH 5 after heating the solution to a temperature above 80°C. The stability of the nanoparticles was not altered by reducing the pH as the catechin release was equivalent at pH 3 and 5. Moreover, encapsulating catechins in carbohydrate matrix contributed to their enhanced bioavailability and prohibited oxidation (Ferreira et al., 2007).

Nanoencapsulation using spray drying depends on other nanoencapsulation techniques such as nano emulsification preceding spray drying. Therefore, the conventional spray drying technique itself may not be regarded as a nanoencapsulation technique (Anandharamakrishnan (2015); Anandharamakrishnan and Ishwarya (2015b); Bhushani and Anandharamakrishnan (2017). However, in spray drying, it is possible to control morphology and particle size by changing process parameters and formulations (Anandharamakrishnan et al., 2008). Hence, spray drying can be

TABLE 7.4
Wall Materials and Their Properties for Encapsulation by Spray Drying Technique

Property	Wall Material(s)
Solubility	Maltodextrin, corn syrup solids, modified starches
Film formation	Maltodextrin, corn syrup solids, methyl cellulose, gelatin
Emulsification	Modified starches, gum arabic, gelatin, cyclodextrin, lecithin, whey protein, skimmed milk powder
Low viscosity at a high solids concentration	Maltodextrin
Higher glass transition temperature	Maltodextrin (DE: 5–15), gum arabic, trehalose
Bland taste	Maltodextrin, gum arabic
The barrier to oxygen and water	Hydrogenated fats, hydrolyzed starch with higher DE

Source: Anandharamakrishnan (2015).

TABLE 7.5
Fabrication Technique for the Production of Bio-Based Nanoparticles

Fabrication Method	Protein/ Polysaccharide Wall Material	Bioactive Compound	Particle Size	Function	References
Spray drying	Carbohydrate matrix and maltodextrin	Catechin	80 nm	Improves stability and protects from oxidation	Ferreira et al. Coelho (2007)
	Modified *n* – octenylsuccinate-starch	β-Carotene	300–600 nm (droplet size); 12 μm (particle size)	Improves dispersibility, bioavailability, and pigmentation strength	De Paz et al. (2012)
	Maltodextrin	d-Limonene	0.2–1.2 μm (emulsion droplet size); 21–53 μm (dried particle size)	Improves the retention and stability during the processing	Jafari et al. (2007)
	Maltodextrin	Fish oil	0.21–5.9 μm (droplet size); 25–41 μm (particle size)	Minimizes the unencapsulated oil at the surface and maximizes the encapsulation efficiency	Jafari et al. (2008)

Source: Ezhilarasi et al. (2013).

successfully used for producing particles of nanometer size by making appropriate modifications in the drying of nanoemulsion and suspensions (Ezhilarasi et al., 2013) (Table 7.5).

7.5.5 Self-Assembly

Self-assembly is a fabrication technique in which biomolecules, especially proteins, under certain conditions, instantaneously and convergently assemble to form nanomaterials and even nanodevices. The molecules generally have non-polar moieties in their tails and polar moieties in their heads. When these molecules are added to water, they spontaneously self-assemble to form micelles due to the van der Waals forces between the heads of the molecules.

Protein micelles can be produced spontaneously when the individual protein chains are dissolved in an aqueous solution beyond the critical micelle concentration (threshold concentration) and critical solution temperature (CMT) to form nano-sized aggregates (Jafari et al., 2007). Furthermore, it was found that by cross-linking the polymer chains, the stability and toughness of the micellar structures can be improved. Moreover, the obtained nano-sized cross-linked polymer micelle structures are stable upon dilution and resistant to variations in experimental conditions such as pH, ionic strength, solvents, etc. (Batrakova et al., 2006) (Figure 7.8).

Casein micelles are considered as a biomimetic approach for encapsulating the bioactive compounds in them. In nature, calcium phosphate, which is vital for the regular growth and functioning of infants, is encapsulated in the submicellar structures contained in casein micelles. The outer layer of casein micelles is made of κ-casein, which aids in stabilizing the molecules by producing a strong steric repulsion, while the core is primarily formed by αs1-, αs2-, and β-caseins (Kaya-Celiker and Mallikarjunan, 2012).

Casein micelles can be fabricated by means of self-assembly technique *in vitro* by self-assembly of casein units surrounded by hydrophobic compounds. Moreover, for the efficient encapsulation of the hydrophobic bioactive compounds, the maintenance of proper conditions such as ionic strength and pH is essential. This is usually done by using different salts (such as potassium phosphate, calcium chloride, and/or potassium citrate) (Semo et al., 2007).

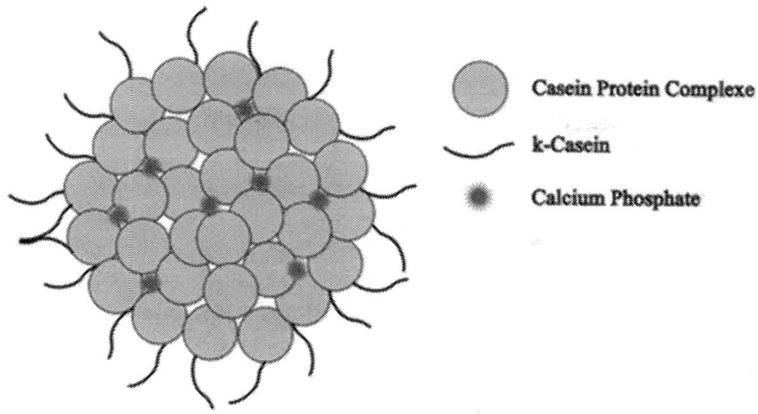

FIGURE 7.8 Schematic representation of casein micelle. (Tarhini et al., 2017.)

Moreover, it is necessary to concentrate on the fact that some portions of the bioactive compounds that are attached to casein units do not take part in micelle formation. Hence, it is very important to maximize the self-assembly of the casein units into micelle structures by strictly optimizing the reaction parameters (Davidov-Pardo et al., 2015).

Another study focused on the opportunity for using sodium caseinate self-assembly followed by dissociation and reassembly to encapsulate curcumin by altering the pH of the medium (pH 12 @ 21°C for 30 min), followed by neutralization. Thus, the reassociation of sodium caseinate happens after the pH is readjusted to around 7.0 (neutralization). The sodium caseinate units encapsulate the bioactive compounds by reassembling into micelles. Furthermore, the stability and bioavailability of encapsulated lipophilic bioactive compounds such as curcumin and DHA was significantly enhanced (Pan et al., 2014).

Mandal and Kundu (2009) fabricated a self-assembled nanoparticle using silk sericin protein from *Antheraea mylitta*, tropical tasar silk cocoons blended with poloxamer. It is interesting to note that the blend of silk sericin protein and poloxamer has the ability to encapsulate both hydrophobic compounds (paclitaxel) and hydrophilic compounds (inulin). Moreover, an increase in particle size by 7 nm was observed when inulin was encapsulated inside the hydrophilic core of the particle. Apart from this, the polydispersity index and average diameter size of the inulin encapsulates were found to be 0.27 and 123 ± 5 nm respectively. In addition, a storage study of about 10 days showed that the particles were stable, and an increase of 11 nm in size was observed. Moreover, a size increase of 15 nm was shown by encapsulated hydrophobic paclitaxel. This study discovered the significance of silk protein as a potent drug carrier.

Kim et al. (2005) encapsulated deoxycholic acid in modified glycol chitosan self-aggregates by covalent attachment. They positively exploited the amphiphilicity of deoxycholic acid to embed deoxycholic acid into the glycol chitosan wall material to initiate self-association to form self-aggregates. The mean particle size of the self-assembled deoxycholic acid-glycol chitosan encapsulates was in the range of 245–450 nm.

7.5.6 Electrospinning and Electrospraying

Electrospinning and electrospraying are electrohydrodynamic processes, where a polymer solution can be spun or sprayed by the application of a high potential electric field to obtain fibers or particles at micron, submicron and nanoscale respectively. Electrospinning and electrospraying are considered as "sister" technologies, which are simple, cost-effective and flexible. Electrospun fibers and electrosprayed particles possess many structural and functional advantages.

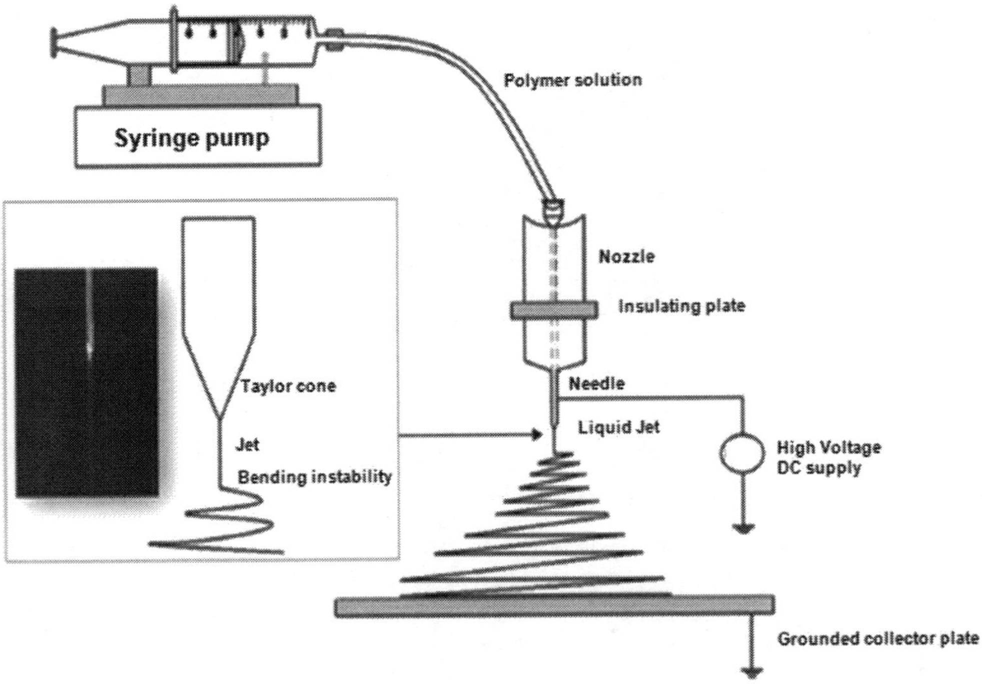

FIGURE 7.9 A typical electrospinning setup. (Bhushani and Anandharamakrishnan, 2014.)

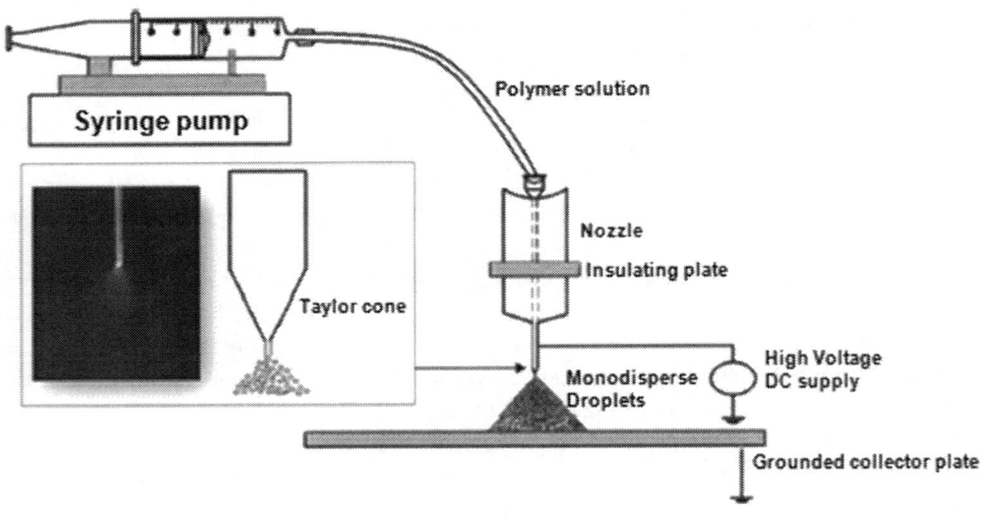

FIGURE 7.10 A typical electrospraying setup. (Bhushani and Anandharamakrishnan, 2014.)

The typical setup for electrospinning and electrospraying consists of four main components (Bhushani and Anandharamakrishnan, 2014) (Figures 7.9 and 7.10):

(1) A high voltage source (1–30 kV) usually operated in a direct current mode, though alternating current mode is also possible (Kessick et al., 2004)
(2) A blunt-ended stainless steel needle or capillary
(3) A syringe pump
(4) A grounded collector, either flat plate or rotating drum

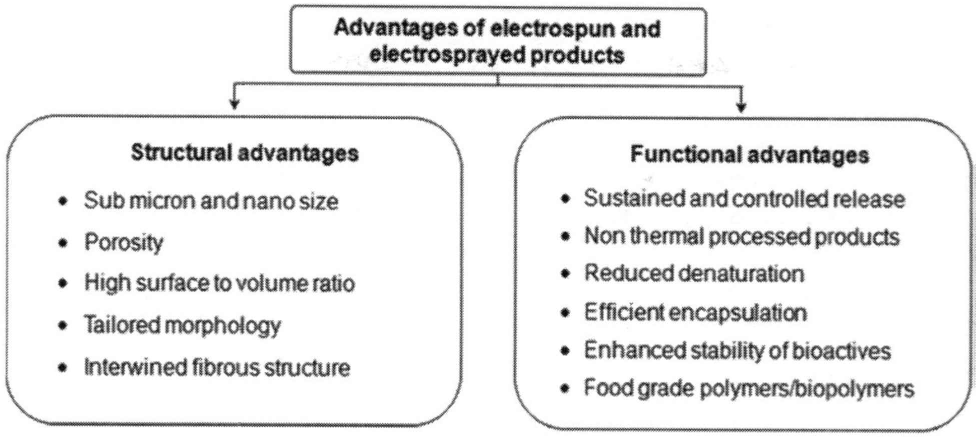

FIGURE 7.11 Advantages of electrosprayed and electrospun products. (Bhushani and Anandharamakrishnan, 2014.)

The particle size of the final product can be manipulated by controlling the principal factors, such as: (1) system parameters (molecular weight, microstructural features of polymers), (2) solution (viscosity, pH, surface tension, and conductivity), (3) instrumental (flow rate of the solution, applied electrical potential, distance between the collector plate and the tip of the needle, and occasionally nature of collector material), and (4) ambient parameters (air velocity, temperature, humidity in the process chamber, evaporation rate of solvent) (Raghavan et al., 2012). In a study, Bhushani, Kurrey, and Anandharamakrishnan (2017) encapsulated catechins by electrospraying method using zein as wall material with 160–190 nm and reported that the permeability and stability of catechin was greatly influenced by core-to-wall ratio (Figure 7.11 and Table 7.6).

Torres-Giner et al. (2010) used the electrospraying technique in order to prepare docosahexaenoic acid (DHA)-loaded zein nanoparticles. An optical microscope was used to show bright field and fluorescence images of the ultrathin matrices of blank zein and zein-DHA nanoparticles. From the obtained results, it was found that the size of encapsulated nanoparticles was either the same as the blank, or in the shortest dimension approximately 530 nm for pure zein and 490 nm for DHA-zein (Torres-Giner et al., 2010). Furthermore, López-Rubio and Lagaron (2012) prepared whey protein (WP) nanoparticles for encapsulating the antioxidant β-carotene using the electrospinning method. The bioactive molecule was encapsulated successfully with a high encapsulation efficiency of about 90%. The obtained dispersions were reasonably stable and the electrospraying technique established its feasibility in the field of drug delivery (López-Rubio and Lagaron, 2012).

For instance, the production of nanofibers (average diameter 140 nm) from hydrolyzed chitosan was optimized with respect to the electrospinning conditions and concentration of the acetic acid solution (Homayoni et al., 2009). Polysaccharides were electrosprayed for the production of polyelectrolyte complex (polycation: sodium alginate solution; polyanion: chitosan or calcium chloride aqueous solution) micro (80–230 mm) and nanocapsules (80 nm) suitable for bioactive encapsulation (Fukui et al., 2010). However, a major drawback of the electrospraying and electrospinning techniques is that many proteins cannot be used separately to create electrospun fibers due to their complex macromolecular and 3D structures in combination with their strong inter- and intramolecular forces. To prevail over this problem, surfactants, plasticizers, or reducing agents can be incorporated in the polymer solutions (Elzoghby et al. 2015). The advantages and possible applications of electrospun and electrosprayed products outshine their main limitation, namely low throughput. However, this limitation restricts their commercial exploitation at a large scale (Bhushani and Anandharamakrishnan, 2014) (Figure 7.12).

TABLE 7.6
Production of Bio-Based Nanoparticles Using Electrospinning and Electrospraying

Fabrication Method	Protein/ Polysaccharide Wall Material	Bioactive Compound	Particle Size	Function	References
Electrospraying	Zein ultrathin fibers	DHA	500–700 nm	Improved stability	Torres-Giner et al. (2010)
	Whey protein concentrate	Carotene	<100 nm	Whey protein explored as the coat material to deliver bioactives	López-Rubio and Lagaron (2012)
	Zein	Curcumin	175–250 nm	Enhanced stability and dispersion in aqueous food matrix	Gomez-Estaca et al. (2012)
Electrospinning	Zein prolamine	Beta-carotene	1,140 nm	Enhanced oxidative stability and light stability	Fernandez et al. (2009)
	Zein	Curcumin	310 nm	Improved sustained release and effective free radical scavenging ability	Brahatheeswaran et al. (2012)
	Cellulose acetate	Curcumin	314–340 nm	Stable chemical integrity and the product was confirmed to be non-toxic to human dermal fibroblasts	Suwantong et al. (2010)

Source: Anandharamakrishnan (2014).

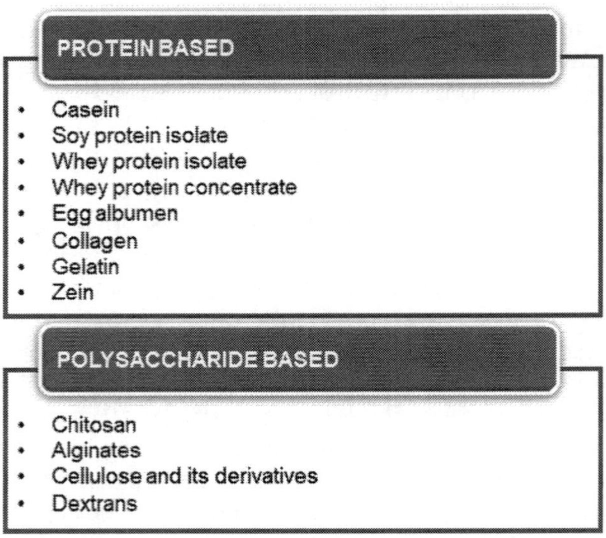

FIGURE 7.12 Common protein and polysaccharide-based encapsulating material used in electrospinning and electrospraying. (Anandharamakrishnan, 2014.)

7.6 CONCLUSION AND FUTURE PERSPECTIVES

Bio-based nanoparticles are regarded as the quantum leap in various fields, namely food technology, pharmaceuticals, biotechnology, cosmetics, agriculture, etc. Bio-based nanoparticles such as proteins and polysaccharides are well recognized for their ability to protect the encapsulated core material, targeted delivery, and sustained release in the system. Furthermore, they are harnessed for their abundance, non-toxic nature, biocompatibility, and biodegradability. But the utilization of these bio-based nanoparticles is in the initial experimental stage. Bio-based nanofabricated materials are perfect for designing and fabricating desired applications, namely food structuring, i.e. to customize food in various ways by manipulating its taste and texture, and improving the bioavailability of nutrients. Furthermore, it is easy to tune the physiochemical and functional properties of biopolymers as required.

Each bio-based polymer, namely protein or polysaccharide-based, has its unique physiochemical or functional properties and particular fabrication method. Further, each ingredient and fabrication technique have their own pros and cons and uniqueness. Therefore, the selection criteria for fabrication techniques are crucial and depend on the type of biomaterial and the end product properties. It is interesting to note that by altering the experimental parameters such as polymer concentration, pH, solvent, and cross-linker, we can manipulate the end product properties and functionality.

A flourishing area for extensive future research is the identification of a novel and unexplored natural source for biopolymers and an innovative fabrication technique with diversified applications. On the other hand, it is mandatory to do a thorough study to ensure the economic feasibility for industrial production. Paramount attention should be given to toxicological studies to ensure that products are safe for widespread application in the various aforementioned fields.

REFERENCES

Agarwal, Tarun, SN Gautham Hari Narayana, Kunal Pal, Krishna Pramanik, Supratim Giri, and Indranil Banerjee. 2015. "Calcium alginate-carboxymethyl cellulose beads for colon-targeted drug delivery." *International Journal Of Biological Macromolecules*, 75:409–417.

Alvarez-Lorenzo, Carmen, Barbara Blanco-Fernandez, Ana M. Puga, and Angel Concheiro. 2013. "Crosslinked ionic polysaccharides for stimuli-sensitive drug delivery." *Advanced Drug Delivery Reviews,* 65 (9):1148–1171.

Anandharamakrishnan, Chinnaswamy. 2014. *Techniques for Nanoencapsulation of Food Ingredients.* New York: Springer.

Anandharamakrishnan, Chinnaswamy. 2015. *Spray Drying Techniques for Food Ingredient Encapsulation.* Chichester, UK: John Wiley & Sons.

Anandharamakrishnan, C. and S. Padma Ishwarya. 2015a. "Spray drying for nanoencapsulation of food components." *Spray Drying Techniques for Food Ingredient Encapsulation.* Chichester, UK: John Wiley & Sons:180–197.

Anandharamakrishnan, C. and S. Padma Ishwarya. 2015b. "Introduction to encapsulation of food ingredients." *Spray Drying Techniques for Food Ingredient Encapsulation.* Chichester, UK: John Wiley & Sons, Ltd.:37–64.

Anandharamakrishnan, C., Chris D. Rielly, and A.G.F. Stapley. 2008. "Loss of solubility of α-lactalbumin and β-lactoglobulin during the spray drying of whey proteins." *LWT-Food Science and Technology,* 41 (2):270–277.

Bajpai, A.K. and Jyoti Choubey. 2006. "In vitro release dynamics of an anticancer drug from swellable gelatin nanoparticles." *Journal of Applied Polymer Science*, 101 (4):2320–2332.

Batrakova, Elena V., Tatiana K. Bronich, Joseph A. Vetro, and Alexander V. Kabanov. 2006. "Polymer micelles as drug carriers." In V.P. Torchilin (ed.), *Nanoparticulates as Drug Carriers.* London: Imperial College Press, 57–93.

Beaulieu, Lucie, Laurent Savoie, Paul Paquin, and Muriel Subirade. 2002. "Elaboration and characterization of whey protein beads by an emulsification/cold gelation process: Application for the protection of retinol." *Biomacromolecules,* 3 (2):239–248.

Bell, L.N. 2001. "Stability testing of nutraceuticals and functional foods." *Handbook of Nutraceuticals and Functional Foods.* Boca Raton, FL: CRC Press: 501–516.

Beneke, Carien E., Alvaro M. Viljoen, and Josias H. Hamman. 2009. "Polymeric plant-derived excipients in drug delivery." *Molecules* 14 (7):2602–2620.

Bhushani, J. Anu and C. Anandharamakrishnan. 2014. "Electrospinning and electrospraying techniques: Potential food based applications." *Trends in Food Science & Technology,* 38 (1):21–33.

Bhushani, Anu, and C. Anandharamakrishnan. 2017. "Food-Grade Nanoemulsions for Protection and Delivery of Nutrients." In Shivendu Ranjan, Nandita Dasgupta, Eric Lichtfouse (eds.), *Nanoscience in Food and Agriculture 4*. New York: Springer: 99–139.

Bhushani, J. Anu, Nawneet Kumar Kurrey, and C. Anandharamakrishnan. 2017. "Nanoencapsulation of green tea catechins by electrospraying technique and its effect on controlled release and in-vitro permeability." *Journal of Food Engineering,* 199:82–92.

Bowman, Katherine and Kam W. Leong. 2006. "Chitosan nanoparticles for oral drug and gene delivery." *International Journal of Nanomedicine,* 1 (2):117.

Brahatheeswaran, Dhandayuthapani, Anila Mathew, Ravindran Girija Aswathy, Yutaka Nagaoka, K. Venugopal, Yasuhiko Yoshida, Toru Maekawa, and D. Sakthikumar. 2012. "Hybrid fluorescent curcumin loaded zein electrospun nanofibrous scaffold for biomedical applications." *Biomedical Materials,* 7 (4):045001.

Cano-Cebrian, Maria J., Teodoro Zornoza, Luis Granero, and Ana Polache. 2005. "Intestinal absorption enhancement via the paracellular route by fatty acids, chitosans and others: A target for drug delivery." *Current Drug Delivery,* 2 (1):9–22.

Chen, Lingyun, Gabriel E. Remondetto, and Muriel Subirade. 2006. "Food protein-based materials as nutraceutical delivery systems." *Trends in Food Science and Technology,* 17 (5):272–283.

Chuy, Leticia E. and Theodore P. Labuza. 1994. "Caking and stickiness of dairy-based food powders as related to glass transition." *Journal of Food Science*, 59 (1):43–46.

Darmadji, Purnama and Masatoshi Izumimoto. 1994. "Effect of chitosan in meat preservation." *Meat Science,* 38 (2):243–254.

Davidov-Pardo, Gabriel, Iris J. Joye, and David Julian McClements. 2015. "Food-grade protein-based nanoparticles and microparticles for bioactive delivery: Fabrication, characterization, and utilization." *Advances in Protein Chemistry and Structural Biology*, 2015 (98): 293–325.

De Kruif, Cornelus G., Fanny Weinbreck, and Renko de Vries. 2004. "Complex coacervation of proteins and anionic polysaccharides." *Current Opinion in Colloid and Interface Science*, 9 (5):340–349.

De Paz, Esther, Ángel Martín, Antonio Estrella, Soraya Rodríguez-Rojo, Ana A. Matias, Catarina M.M. Duarte, and María José Cocero. 2012. "Formulation of β-carotene by precipitation from pressurized ethyl acetate-on-water emulsions for application as natural colorant." *Food Hydrocolloids*, 26 (1):17–27.

Desobry, Stephane A., Flavia M. Netto, and Theodore P. Labuza. 1997. "Comparison of spray-drying, drum-drying and freeze-drying for β-carotene encapsulation and preservation." *Journal of Food Science*, 62 (6):1158–1162.

Dickinson, Eric. 2003. "Hydrocolloids at interfaces and the influence on the properties of dispersed systems." *Food Hydrocolloids,* 17 (1):25–39.

dos Santos, Marlise A. and Ana Grenha. 2015. "Polysaccharide nanoparticles for protein and peptide delivery: Exploring less-known materials." In *Advances in Protein Chemistry and Structural Biology*, 2015 (98): 223–261.

Draget, Kurt Ingar, Kjetill Østgaard, and Olav Smidsrød. 1990. "Homogeneous alginate gels: A technical approach." *Carbohydrate Polymers,* 14 (2):159–178.

Dror, Yael, Yachin Cohen, and Rachel Yerushalmi-Rozen. 2006. "Structure of gum arabic in aqueous solution." *Journal of Polymer Science Part B: Polymer Physics,* 44 (22):3265–3271.

Eichhorn, Stephen J., A. Dufresne, M. Aranguren, N.E. Marcovich, J.R. Capadona, S.J. Rowan, Christoph Weder, Wim Thielemans, M. Roman, and S. Renneckar. 2010. "Current international research into cellulose nanofibres and nanocomposites." *Journal of Materials Science,* 45 (1):1.

Elzoghby, Ahmed O., Wael M. Samy, and Nazik A. Elgindy. 2012. "Albumin-based nanoparticles as potential controlled release drug delivery systems." *Journal of Controlled Release,* 157 (2):168–182.

Elzoghby, Ahmed O., Mayada M. Elgohary, and Nayra M. Kamel. 2015. "Implications of protein-and peptide-based nanoparticles as potential vehicles for anticancer drugs." In Rossen Donev (ed.), *Advances in Protein Chemistry and Structural Biology*, Oxford, Elsevier, 169–221.

Ezhilarasi, P.N., P. Karthik, N. Chhanwal, and C. Anandharamakrishnan. 2013. "Nanoencapsulation techniques for food bioactive components: A review." *Food and Bioprocess Technology,* 6 (3):628–647.

FDA. 2018. "Part 184: Direct food substances affirmed as generally recognized as safe." *Subpart B: Listing of specific substances affirmed as GRAS*, 184.1588-pectins. Title 21, Volume 3: 21CFR184.1588

Fernandez, Avelina, Sergio Torres-Giner, and Jose Maria Lagaron. 2009. "Novel route to stabilization of bioactive antioxidants by encapsulation in electrospun fibers of zein prolamine." *Food Hydrocolloids,* 23 (5):1427–1432.

Ferreira, Ivone, S. Rocha, and M. Coelho. 2007. "Encapsulation of antioxidants by spray-drying." *Chemical Engineering Transactions,* 11 (9):713–717.

Fessi, H.P.F.D., François Puisieux, J. Ph. Devissaguet, Nicolas Ammoury, and Simon Benita. 1989. "Nanocapsule formation by interfacial polymer deposition following solvent displacement." *International Journal of Pharmaceutics,* 55 (1):R1–R4.

Fukui, Yu, Tatsuo Maruyama, Yuko Iwamatsu, Akihiro Fujii, Tsutomu Tanaka, Yoshikage Ohmukai, and Hideto Matsuyama. 2010. "Preparation of monodispersed polyelectrolyte microcapsules with high encapsulation efficiency by an electrospray technique." *Colloids and Surfaces A: Physicochemical and Engineering Aspects*, 370 (1–3):28–34.

Gan, Quan, and Tao Wang. 2007. "Chitosan nanoparticle as protein delivery carrier—systematic examination of fabrication conditions for efficient loading and release." *Colloids and Surfaces B: Biointerfaces*, 59 (1):24–34.

Garnero, Claudia, Ana Karina Chattah, and Marcela Longhi. 2013. "Supramolecular complexes of maltodextrin and furosemide polymorphs: A new approach for delivery systems." *Carbohydrate Polymers*, 94 (1):292–300.

Gomez-Estaca, J., M.P. Balaguer, R. Gavara, and P. Hernandez-Munoz. 2012. "Formation of zein nanoparticles by electrohydrodynamic atomization: Effect of the main processing variables and suitability for encapsulating the food coloring and active ingredient curcumin." *Food Hydrocolloids,* 28 (1):82–91.

Gouin, Sebastien. 2004. "Microencapsulation: Industrial appraisal of existing technologies and trends." *Trends in Food Science and Technology,* 15 (7–8):330–347.

Gülseren, İbrahim, Yuan Fang, and Milena Corredig. 2012. "Whey protein nanoparticles prepared with desolvation with ethanol: Characterization, thermal stability and interfacial behavior." *Food Hydrocolloids,* 29 (2):258–264.

Haaj, Sihem Bel, Albert Magnin, Christian Pétrier, and Sami Boufi. 2013. "Starch nanoparticles formation via high power ultrasonication." *Carbohydrate Polymers*, 92 (2):1625–1632.

Han, Siyuan, Mingjun Li, Xiangang Liu, Hongxia Gao, and Yan Wu. 2013. "Construction of amphiphilic copolymer nanoparticles based on gelatin as drug carriers for doxorubicin delivery." *Colloids and Surfaces B: Biointerfaces,* 102:833–841.

Hanafi, Aimi, Nadine Nograles, Syahril Abdullah, Mariana Nor Shamsudin, and Rozita Rosli. 2013. "Cellulose acetate phthalate microencapsulation and delivery of plasmid DNA to the intestines." *Journal of Pharmaceutical Sciences*, 102 (2):617–626.

Homayoni, Homa, Seyed Abdolkarim Hosseini Ravandi, and Masoumeh Valizadeh. 2009. "Electrospinning of chitosan nanofibers: Processing optimization." *Carbohydrate Polymers,* 77 (3):656–661.

Horn, Svein Jarle. 2009. *Seaweed Biofuels: Production of Biogas and Bioethanol from Brown Macroalgae.* Riga: VDM Verlag Dr. Müller.

Idris, O.H.M., P.A. Williams, and G.O. Phillips. 1998. "Characterisation of gum from Acacia senegal trees of different age and location using multidetection gel permeation chromatography." *Food Hydrocolloids,* 12 (4):379–388.

Jafari, Seid Mahdi. 2017. "An overview of nanoencapsulation techniques and their classification." In Seid Mahdi (ed.), *Nanoencapsulation Technologies for the Food and Nutraceutical Industries.* London: Academic Press:1–34.

Jafari, Seid Mahdi, Yinghe He, and Bhesh Bhandari. 2007. "Production of sub-micron emulsions by ultrasound and microfluidization techniques." *Journal of Food Engineering*, 82 (4):478–488

Jafari, Seid Mahdi, Elham Assadpoor, Bhesh Bhandari, and Yinghe He. 2008. "Nano-particle encapsulation of fish oil by spray drying." *Food Research International,* 41 (2):172–183.

Jahanshahi, Mohsen and Zahra Babaei. 2008. "Protein nanoparticle: A unique system as drug delivery vehicles." *African Journal of Biotechnology,* 7 (25): 4926-4934.

Jain, Dharmendra and Daniel Bar-Shalom. 2014. "Alginate drug delivery systems: Application in context of pharmaceutical and biomedical research." *Drug Development and Industrial Pharmacy*, 40 (12):1576–1584.

Jain, Sanjay K., Yashwant Gupta, Anekant Jain, Asheesh R. Saxena, Piush Khare, and Aviral Jain. 2008. "Mannosylated gelatin nanoparticles bearing an anti-HIV drug didanosine for site-specific delivery." *Nanomedicine: Nanotechnology, Biology and Medicine*, 4 (1):41–48.

Jincheng, Wang, Zheng Xiaoyu, and Chen Sihao. 2010. "Preparation and properties of nanocapsulated capsaicin by complex coacervation method." *Chemical Engineering Communications,* 197 (7):919–933.

Joye, Iris Julie and David Julian McClements. 2014. "Biopolymer-based nanoparticles and microparticles: Fabrication, characterization, and application." *Current Opinion in Colloid and Interface Science,* 19 (5):417–427.

Kas, H. Süheyla. 1997. "Chitosan: Properties, preparations and application to microparticulate systems." *Journal of Microencapsulation*, 14 (6):689–711.

Kaya-Celiker, Hande and Kumar Mallikarjunan. 2012. "Better nutrients and therapeutics delivery in food through nanotechnology." *Food Engineering Reviews*, 4 (2):114–123.

Kenyon, Melanie M and Robert J. Anderson. 1988. "Maltodextrins and low-dextrose-equivalence corn syrup solids: Production and technology for the flavor industry." *ACS Symposium Series*, 370:7–11.

Kessick, Royal, John Fenn, and Gary Tepper. 2004. "The use of AC potentials in electrospraying and electrospinning processes." *Polymer*, 45 (9):2981–2984.

Kim, Kwangmeyung, Seunglee Kwon, Jae Hyung Park, Hesson Chung, Seo Young Jeong, Ick Chan Kwon, and In-San Kim. 2005. "Physicochemical characterizations of self-assembled nanoparticles of glycol chitosan– deoxycholic acid conjugates." *Biomacromolecules*, 6 (2):1154–1158.

Lavoine, Nathalie, Isabelle Desloges, Alain Dufresne, and Julien Bras. 2012. "Microfibrillated cellulose–Its barrier properties and applications in cellulosic materials: A review." *Carbohydrate Polymers*, 90 (2):735–764.

Lee, Sie Huey, Desmond Heng, Wai Kiong Ng, Hak-Kim Chan, and Reginald B.H. Tan. 2011. "Nano spray drying: A novel method for preparing protein nanoparticles for protein therapy." *International Journal of Pharmaceutics*, 403 (1–2):192–200.

Lee, Eun Ju, Saeed Ahmad Khan, Joong Kon Park, and Kwang-Hee Lim. 2012. "Studies on the characteristics of drug-loaded gelatin nanoparticles prepared by nanoprecipitation." *Bioprocess and Biosystems Engineering*, 35 (1–2):297–307.

Lemarchand, Caroline, Ruxandra Gref, Catherine Passirani, Emmanuel Garcion, Boris Petri, Rainer Müller, Dominique Costantini, and Patrick Couvreur. 2006. "Influence of polysaccharide coating on the interactions of nanoparticles with biological systems." *Biomaterials*, 27 (1):108–118.

Liu, Jinsong, Lihua Xu, Chuantong Liu, Dafeng Zhang, Siqian Wang, Zhennan Deng, Weiwei Lou, Haihong Xu, Qing Bai, and Jianfeng Ma. 2012. "Preparation and characterization of cationic curcumin nanoparticles for improvement of cellular uptake." *Carbohydrate Polymers*, 90 (1):16–22.

López-Rubio, Amparo and Jose M. Lagaron. 2012. "Whey protein capsules obtained through electrospraying for the encapsulation of bioactives." *Innovative Food Science and Emerging Technologies*, 13:200–206.

Luo, Yangchao and Qin Wang. 2014. "Zein-based micro-and nano-particles for drug and nutrient delivery: A review." *Journal of Applied Polymer Science*, 131 (16): 40696.

Luque-Alcaraz, A.G., J. Lizardi-Mendoza, F.M. Goycoolea, I. Higuera-Ciapara, and W. Argüelles-Monal. 2016. "Preparation of chitosan nanoparticles by nanoprecipitation and their ability as a drug nanocarrier." *RSC Advances*, 6 (64):59250–59256.

Mandal, Biman B. and S.C. Kundu. 2009. "Self-assembled silk sericin/poloxamer nanoparticles as nanocarriers of hydrophobic and hydrophilic drugs for targeted delivery." *Nanotechnology*, 20 (35):355101.

Mangalam, Anand P., John Simonsen, and Albert S. Benight. 2009. "Cellulose/DNA hybrid nanomaterials." *Biomacromolecules*, 10 (3):497–504.

Masters, K. 1991. *Spray Drying*, Longman Scientific & Technical and John Wiley & Sons Inc., Essex, UK.

McClean, Siobhan, Ena Prosser, Eucharia Meehan, Denise O"Malley, Nuala Clarke, Zeibun Ramtoola, and David Brayden. 1998. "Binding and uptake of biodegradable poly-DL-lactide micro- and nanoparticles in intestinal epithelia." *European Journal of Pharmaceutical Sciences*, 6 (2):153–163.

Miladi, K., D. Ibraheem, M. Iqbal, S. Sfar, H. Fessi, and A. Elaissari. 2014. "Particles from preformed polymers as carriers for drug delivery." *EXCLI Journal*, 13:28.

Morr, C.V. and E.Y.W. Ha. 1993. "Whey protein concentrates and isolates: Processing and functional properties." *Critical Reviews in Food Science and Nutrition*, 33 (6):431–476.

Moura, Liane I.F., Ana M.A. Dias, Eugénia Carvalho, and Hermínio C. de Sousa. 2013. "Recent advances on the development of wound dressings for diabetic foot ulcer treatment—a review." *Acta Biomaterialia*, 9 (7):7093–7114.

Mozafari, M. Reza, John Flanagan, Lara Matia-Merino, Ajay Awati, Abdelwahab Omri, Zacharias E. Suntres, and Harjinder Singh. 2006. "Recent trends in the lipid-based nanoencapsulation of antioxidants and their role in foods." *Journal of the Science of Food and Agriculture*, 86 (13):2038–2045.

Neethirajan, Suresh and Digvir S Jayas. 2011. "Nanotechnology for the food and bioprocessing industries." *Food and Bioprocess Technology*, 4 (1):39–47.

Nehete, J.Y., Bhambar, R.S., Narkhede, M.R., and Gawali, S.R., 2013. "Natural proteins: Sources, isolation, characterization and applications." *Pharmacognosy Reviews*, 7 (14): 107.

Opanasopit, Praneet, Auayporn Apirakaramwong, Tanasait Ngawhirunpat, Theerasak Rojanarata, and Uracha Ruktanonchai. 2008. "Development and characterization of pectinate micro/nanoparticles for gene delivery." *AAPS PharmSciTech*, 9 (1):67–74.

Paliwal, Rishi, R. Jayachandra Babu, and Srinath Palakurthi. 2014. "Nanomedicine scale-up technologies: Feasibilities and challenges." *AAPS PharmSciTech,* 15 (6):1527–1534.

Pan, Kang, Yangchao Luo, Yundi Gan, Seung J. Baek, and Qixin Zhong. 2014. "pH-driven encapsulation of curcumin in self-assembled casein nanoparticles for enhanced dispersibility and bioactivity." *Soft Matter,* 10 (35):6820–6830.

Panikkar, Rajeendernath and Donald J. Brasch. 1996. "Composition and block structure of alginates from New Zealand brown seaweeds." *Carbohydrate Research,* 293 (1):119–132.

Patel, Ashok, Yingchun Hu, Jyoti Kumar Tiwari, and Krassimir P. Velikov. 2010. "Synthesis and characterisation of zein–curcumin colloidal particles." *Soft Matter,* 6 (24):6192–6199.

Pathak, Yashwant and Deepak Thassu. 2016. *Drug Delivery Nanoparticles Formulation and Characterization.* Vol. 191. Boca Raton, FL: CRC Press.

Perez, Luis A Bello and Edith Agama-Acevedo. 2018. "Starch." In Marcelo A. Villar, Silvia E. Barbosa, M. Alejandra García, Luciana A. Castillo, and Olivia V. López (eds.), *Starch-Based Materials in Food Packaging.* London: Academic Press: 1–18.

Picot, Arnaud and Christophe Lacroix. 2004. "Encapsulation of bifidobacteria in whey protein-based microcapsules and survival in simulated gastrointestinal conditions and in yoghurt." *International Dairy Journal,* 14 (6):505–515.

Qin, Yang, Chengzhen Liu, Suisui Jiang, Liu Xiong, and Qingjie Sun. 2016. "Characterization of starch nanoparticles prepared by nanoprecipitation: Influence of amylose content and starch type." *Industrial Crops and Products,* 87:182–190.

Quintanilla-Carvajal, María Ximena, Brenda Hildeliza Camacho-Díaz, Lesvia Sofia Meraz-Torres, José Jorge Chanona-Pérez, Liliana Alamilla-Beltrán, Antonio Jimenéz-Aparicio, and Gustavo F Gutiérrez-López. 2010. "Nanoencapsulation: A new trend in food engineering processing." *Food Engineering Reviews,* 2 (1):39–50.

Raghavan, Prasanth, Du-Hyun Lim, Jou-Hyeon Ahn, Changwoon Nah, David C. Sherrington, Ho-Suk Ryu, and Hyo-Jun Ahn. 2012. "Electrospun polymer nanofibers: The booming cutting edge technology." *Reactive and Functional Polymers,* 72 (12):915–930.

Rajam, R., and C. Anandharamakrishnan. 2015. "Spray freeze drying method for microencapsulation of Lactobacillus plantarum." *Journal of Food Engineering,* 166:95–103.

Rahim, Safwan Abdel, Paul A. Carter, and Amal Ali Elkordy. 2015. "Design and evaluation of effervescent floating tablets based on hydroxyethyl cellulose and sodium alginate using pentoxifylline as a model drug." *Drug Design, Development and Therapy,* 9:1843.

Rao, K.S.V. Krishna, P. Ramasubba Reddy, Yong-Ill Lee, and Changdae Kim. 2012. "Synthesis and characterization of chitosan–PEG–Ag nanocomposites for antimicrobial application." *Carbohydrate Polymers,* 87 (1):920–925.

Raveendran, Sreejith, Brahatheeswaran Dhandayuthapani, Yutaka Nagaoka, Yasuhiko Yoshida, Toru Maekawa, and D. Sakthi Kumar. 2013. "Biocompatible nanofibers based on extremophilic bacterial polysaccharide, Mauran from Halomonas maura." *Carbohydrate Polymers,* 92 (2):1225–1233.

Razavi, Seyed Mohammad Ali, and Asad Mohammad Amini. 2016. "Starch nanomaterials: A state-of-the-art review and future trends." In Alexandru Grumezescu (ed.), *Novel Approaches of Nanotechnology in Food.* Amsterdam: Academic Press: 237–269.

Ridley, Brent L., Malcolm A. O"Neill, and Debra Mohnen. 2001. "Pectins: Structure, biosynthesis, and oligogalacturonide-related signaling." *Phytochemistry,* 57 (6):929–967.

Rivas, Claudia Janeth Martínez, Mohamad Tarhini, Waisudin Badri, Karim Miladi, Hélène Greige-Gerges, Qand Agha Nazari, Sergio Arturo Galindo Rodríguez, Rocío Álvarez Román, Hatem Fessi, and Abdelhamid Elaissari. 2017. "Nanoprecipitation process: From encapsulation to drug delivery." *International Journal of Pharmaceutics,* 532 (1):66–81.

Rubino, Orapin P., Richard Kowalsky, and James Swarbrick. 1993. "Albumin microspheres as a drug delivery system: Relation among turbidity ratio, degree of cross-linking, and drug release." *Pharmaceutical Research,* 10 (7):1059–1065.

Schwick, H.G. and K. Heide. 1969. "Immunochemistry and immunology of collagen and gelatin." In A. Hassig, P. Lundsgaard-Hansen, and H Nitschmann (eds.), *Modified Gelatins as Plasma Substitutes.* New York: Karger Publishers:111–125.

Selmin, Francesca, Ilaria Franceschini, Irma E. Cupone, Paola Minghetti, and Francesco Cilurzo. 2015. "Aminoacids as non-traditional plasticizers of maltodextrins fast-dissolving films." *Carbohydrate Polymers,* 115:613–616.

Semo, Efrat, Ellina Kesselman, Dganit Danino, and Yoav D. Livney. 2007. "Casein micelle as a natural nanocapsular vehicle for nutraceuticals." *Food Hydrocolloids,* 21 (5–6):936–942.

Senni, Karim, Jessica Pereira, Farida Gueniche, Christine Delbarre-Ladrat, Corinne Sinquin, Jacqueline Ratiskol, Gaston Godeau, Anne-Marie Fischer, Dominique Helley, and Sylvia Colliec-Jouault. 2011. "Marine polysaccharides: A source of bioactive molecules for cell therapy and tissue engineering." *Marine Drugs,* 9 (9):1664–1681.

Sharma, Anjali, Rajesh Pandey, Sadhna Sharma, and G.K. Khuller. 2004. "Chemotherapeutic efficacy of poly (DL-lactide-co-glycolide) nanoparticle encapsulated antitubercular drugs at sub-therapeutic dose against experimental tuberculosis." *International Journal of Antimicrobial Agents,* 24 (6):599–604.

Shen, Lan and Chuan-He Tang. 2012. "Microfluidization as a potential technique to modify surface properties of soy protein isolate." *Food Research International,* 48 (1):108–118.

Shi, Ai-min, Dong Li, Li-jun Wang, Bing-zheng Li, and Benu Adhikari. 2011. "Preparation of starch-based nanoparticles through high-pressure homogenization and miniemulsion cross-linking: Influence of various process parameters on particle size and stability." *Carbohydrate Polymers,* 83 (4):1604–1610.

Shukla, Rishi and Munir Cheryan. 2001. "Zein: The industrial protein from corn." *Industrial Crops and Products,* 13 (3):171–192.

Sundar, Sushmitha, Joydip Kundu, and Subhas C. Kundu. 2010. "Biopolymeric nanoparticles." *Science and Technology of Advanced Materials,* 11 (1):014104.

Sungthongjeen, Srisagul, Pornsak Sriamornsak, Tasana Pitaksuteepong, Atawit Somsiri, and Satit Puttipipatkhachorn. 2004. "Effect of degree of esterification of pectin and calcium amount on drug release from pectin-based matrix tablets." *Aaps Pharmscitech,* 5 (1):50.

Suwannateep, Natthakitta, Wijit Banlunara, Supason P. Wanichwecharungruang, Khajeelak Chiablaem, Kriengsak Lirdprapamongkol, and Jisnuson Svasti. 2011. "Mucoadhesive curcumin nanospheres: biological activity, adhesion to stomach mucosa and release of curcumin into the circulation." *Journal of Controlled Release,* 151 (2):176–182.

Suwantong, Orawan, Uracha Ruktanonchai, and Pitt Supaphol. 2010. "In vitro biological evaluation of electrospun cellulose acetate fiber mats containing asiaticoside or curcumin." *Journal of Biomedical Materials Research Part A,* 94 (4):1216–1225.

Swetank, Y. Hundre, P. Karthik, and C. Anandharamakrishnan. 2015. "Effect of whey protein isolate and b-cyclodextrin wall systems on stability of microencapsulated vanillin by spray–freeze drying method." *Food Chemistry,* 174:16–24.

Swierczewska, M., H.S. Han, K. Kim, J.H. Park, and S. Lee. 2016. "Polysaccharide-based nanoparticles for theranostic nanomedicine." *Advanced Drug Delivery Reviews,* 99:70–84.

Tang, Chuan-He, Xiao-Yan Wang, Xiao-Quan Yang, and Lin Li. 2009. "Formation of soluble aggregates from insoluble commercial soy protein isolate by means of ultrasonic treatment and their gelling properties." *Journal of Food Engineering,* 92 (4):432–437.

Tarhini, Mohamad, Hélène Greige-Gerges, and Abdelhamid Elaissari. 2017. "Protein-based nanoparticles: From preparation to encapsulation of active molecules." *International Journal of Pharmaceutics,* 522 (1–2):172–197.

Tolstoguzov, Vladimir. 2003. "Some thermodynamic considerations in food formulation." *Food Hydrocolloids,* 17 (1):1–23.

Torres-Giner, Sergio, Antonio Martinez-Abad, Maria J. Ocio, and Jose M. Lagaron. 2010. "Stabilization of a nutraceutical omega-3 fatty acid by encapsulation in ultrathin electrosprayed zein prolamine." *Journal of Food Science,* 75 (6):N69–N79.

Tripathy, Jasaswini and Ashok M. Raichur. 2013. "Designing carboxymethyl cellulose based layer-by-layer capsules as a carrier for protein delivery." *Colloids and Surfaces B: Biointerfaces,* 101:487–492.

Turgeon, S.L., C. Schmitt, and C. Sanchez. 2007. "Protein–polysaccharide complexes and coacervates." *Current Opinion in Colloid and Interface Science,* 12 (4–5):166–178.

van der Lubben, Inez M., J. Coos Verhoef, Gerrit Borchard, and Hans E Junginger. 2001. "Chitosan and its derivatives in mucosal drug and vaccine delivery." *European Journal of Pharmaceutical Sciences,* 14 (3):201–207.

Venkatesan, Jayachandran, Ira Bhatnagar, and Se-Kwon Kim. 2014. "Chitosan-alginate biocomposite containing fucoidan for bone tissue engineering." *Marine Drugs,* 12 (1):300–316.

Verbeken, Dierckx, S. Dierckx, and Koen Dewettinck. 2003. "Exudate gums: Occurrence, production, and applications." *Applied Microbiology and Biotechnology,* 63 (1):10–21.

Wang, Jin Cheng, Si Hao Chen, and Zi Cheng Xu. 2008. "Synthesis and properties research on the nanocapsulated capsaicin by simple coacervation method." *Journal of Dispersion Science and Technology,* 29 (5):687–695.

Weber, C., C. Coester, J. Kreuter, and K. Langer. 2000. "Desolvation process and surface characterisation of protein nanoparticles." *International Journal of Pharmaceutics,* 194 (1):91–102.

Weiss, Jochen, Paul Takhistov, and D Julian McClements. 2006. "Functional materials in food nanotechnology." *Journal of Food Science,* 71 (9):R107-R116.

Xing, Fubao, Guoxiang Cheng, Kejing Yi, and Linrong Ma. 2005. "Nanoencapsulation of capsaicin by complex coacervation of gelatin, acacia, and tannins." *Journal of Applied Polymer Science,* 96 (6):2225–2229.

Yang, Jisheng, Suya Han, Haicheng Zheng, Hongbiao Dong, and Jiubing Liu. 2015. "Preparation and application of micro/nanoparticles based on natural polysaccharides." *Carbohydrate Polymers,* 123:53–66.

Yu, Da-Guang, Wen-Ching Lin, and Ming-Chien Yang. 2007. "Surface modification of poly (L-lactic acid) membrane via layer-by-layer assembly of silver nanoparticle-embedded polyelectrolyte multilayer." *Bioconjugate Chemistry,* 18 (5):1521–1529.

Zhu, Fangyi, Aniket V. Kale, and Munir Cheryan. 2007. "Fractionation of zein by size exclusion chromatography." *Journal of Agricultural and Food Chemistry,* 55 (10):3843–3849.

Zuidam, Nicolaas Jan and Eyal Shimoni. 2010. "Overview of microencapsulates for use in food products or processes and methods to make them." In Nicolaas Jan Zuidam and Viktor A. Nedovic (eds.), *Encapsulation Technologies for Active Food Ingredients and Food Processing.* New York: Springer: 3–29.

8 Nanoemulsions
Preparation, Stability, and Application in Food

P. Karthik, Sayantani Dutta, and C. Anandharamakrishnan

CONTENTS

8.1 INTRODUCTION

Nanotechnology has the potential to revolutionize scientific research in the multidisciplinary field of pharmaceutical, food, biotechnology, medical, mechanical, and electrical engineering. In 1959, the concept of nanotechnology was introduced by Richard Feynman at the American Physical Society meeting (Khademosseini and Langer, 2006). The term "nano" refers to the size of an object or substance specified at 10^{-9} m (Quintanilla-Carvajal et al., 2010). Formulation of particle size within nanoscale can increase surface-to-volume ratio that results in better reactivity (Neethirajan and Jayas, 2011). The applications of nanotechnology in food systems include nanoencapsulation and nanoemulsion of nutraceuticals (Ezhilarasi et al., 2013; Bhushani et al., 2017; Anandharamakrishnan, 2014a; Anandharamakrishnan and Ishwarya, 2015), nanocomposites for

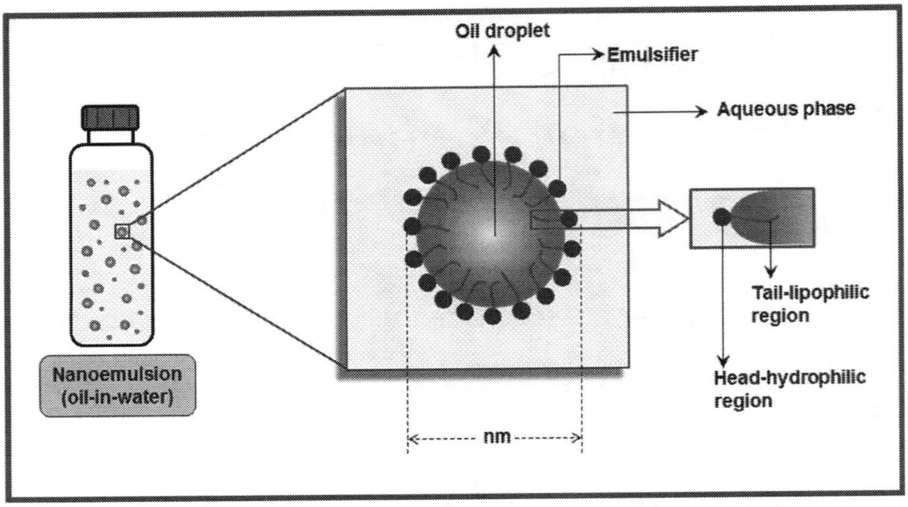

FIGURE 8.1 Schematic representation of oil-in-water nanoemulsion. (Karthik et al., 2017.)

food packaging (Anandharamakrishnan and Usha Kiran, 2014), nano-biosensors for detection of contamination and quality deterioration (da Costa Silva et al., 2013).

Nanoemulsion is a promising technique in the food sector that encapsulates food bioactive compounds and protects them from the surrounding environmental conditions. Food-grade nanoemulsions are extensively used as a delivery vehicle for host nano-sized sensitive bioactives (i.e. DHA algae oil, fish oil, green tea catechins, α-tocopherol, enzymes and flavors, etc.) to provide stability and improve bioavailability (Anandharamakrishnan, 2014b; Karthik et al., 2017). Figure 8.1 shows the schematic representation of a nanoemulsion. Nanoemulsions are colloidal systems comprising of two immiscible liquids (oil or water), with one phase being dispersed into another phase as fine droplets in the size range of 50 to 1000 nm (Sanguansri and Augustin, 2006). The use of the principal materials (i.e. oil and water) may decide the type of emulsion. Hence, oil that is being dispersed in the aqueous system refers to oil-in-water emulsion; whereas water that is being dispersed into the oil phase refers to water-in-oil emulsion. The appearance of nanoemulsions can be transparent, translucent, or milky white. The fabricated bioactive nanoemulsion offers a kinetically stable system for longer periods of storage. The preparation of nanoemulsion is mainly categorized by high-energy and low-energy approaches (McClements and Rao, 2011). Emulsification approaches differ based on the methods and instruments used for the preparation of nanoemulsion.

Emulsifiers or co-emulsifiers are one of the most important parts of nanoemulsion for the design of appropriate formulations. The emulsifier reduces the interfacial tension between the oil and water phases. Furthermore, it aids in decreasing the amount of energy required to disrupt the oil droplets during emulsion formation (Silvestre et al., 1999). If two immiscible phases are homogenized together, the emulsion system may undergo several instability mechanisms such as flocculation, creaming, sedimentation, coalescence, Ostwald ripening, and phase separation (Karthik et al., 2017). In addition, storage of nanoemulsion for a long period or at elevated temperature may influence its physicochemical stability and structural characteristics. This instability of nanoemulsions is addressed by the optimization of suitable emulsifiers and emulsification techniques, which is used for the formation of stable emulsions. Hence, it is essential to study the behavioral characteristics of emulsions in order to achieve a better understanding of their preparation, processing, and storage. In this chapter, preparation, stability, and application of food-grade nanoemulsion are discussed. This chapter mainly focuses on the concept of nanoemulsion formulation by high energy and low energy approaches, types of instability mechanisms and application in the delivery system and antimicrobial activity.

8.2 FORMULATION OF NANOEMULSION

Fabrication of food-grade nanoemulsions is generally categorized by two approaches depending on the underlying principle: high-energy or low-energy approaches. A schematic representation of a nanoemulsion is shown in Figure 8.1. Different types of equipment and methods were developed for the preparation of nanoemulsion in the past decade (McClements and Rao, 2011; Tadros, 2004). In order to achieve minimum droplet size, each approach depends on different factors, which are described in the following sections.

8.2.1 High-Energy Approaches (HEA)

High-energy approaches are administered by mechanical devices capable of generating intense disruptive forces that breakdown the macroscopic or oil phases into tiny droplets to form nanoemulsion. The high-energy mechanical devices include high-pressure homogenization, microfluidization, and ultrasonication techniques (Leong et al., 2009; Wooster et al., 2008). Numerous studies have suggested that the formation of minimum droplet size using HEA relies on homogenizer type, operating conditions (i.e. energy intensity, time, and temperature), composition (i.e. oil type, emulsifier type, and their concentration), and physicochemical properties (i.e. interfacial tension and viscosity) of components (Qian and McClements, 2011).

8.2.1.1 High-Pressure Homogenizer

Nanoemulsification finds a variety of applications in the food processing and pharmaceutical industries. Nanoemulsion can be produced by different systems; however, emulsification through high-pressure homogenization is one of the most important techniques for attaining higher mean particle size distribution and physical stability. A schematic representation of a high-pressure homogenizer is shown in Figure 8.2. A combination of intense shear, cavitation, and turbulent flow is used to break the larger droplets into nano-sized droplets. Very high pressure and multiple passages are needed to obtain a nanoemulsion with the desired droplet size (McClements and Rao, 2011; McClements and Li, 2010). Therefore, the oil droplet break-up occurs once the applied shear is greater than the Laplace pressure of the emulsion. In case of low-oil volume fraction and negligible continuous phase viscosity, Taylor proposed a mathematical model to predict the oil droplets' radius: $r \sim \gamma/(\eta_c \dot{y})$, where γ denotes interfacial tension, η_c denotes continuous phase viscosity and $\dot{y}$ denotes shear rate (Taylor, 1934; Meleson et al., 2004; Walstra, 1983). According to Taylor's equation, surfactant plays an important role that lowers the resistance to droplet break-up by decreasing interfacial tension (Wooster et al., 2008). However, this is not always the case because different surfactants have their own characteristics and ability to form an emulsion. The preparation of nanoemulsion is carried out in a two-step operation. The first step is the preparation of coarse emulsion with the help of a high-speed blender or mixer; the second step involves feeding it into the high-pressure homogenizer. The high-pressure homogenizer is highly efficient in reducing the droplet size of a coarse emulsion rather than formulating a nanoemulsion directly from oil phase and aqueous phase (McClements and Rao, 2011).

High-speed homogenizers are also used to prepare food-grade nanoemulsions. These devices generate heat while operating at high rpm and reduce efficiency. Moreover, they do not deliver noble dispersion in terms of lower mean droplet size, polydispersity index (PDI), and uniform droplet size distribution when compared to other high-energy approaches (Silva et al., 2012). Recently, Karthik and Anandharamakrishnan (2016b) prepared DHA nanoemulsion and compared the efficiency of a high-pressure homogenizer and a high-speed homogenizer in terms of emulsion stability and physicochemical properties. The authors suggested that high-pressure homogenization techniques had better physicochemical properties and stability over a longer period of storage. Shear force and turbulence are the main high-pressure emulsification factors which contribute to the lower particle size in the preparation of nanoemulsions. During the high-pressure homogenization process, the

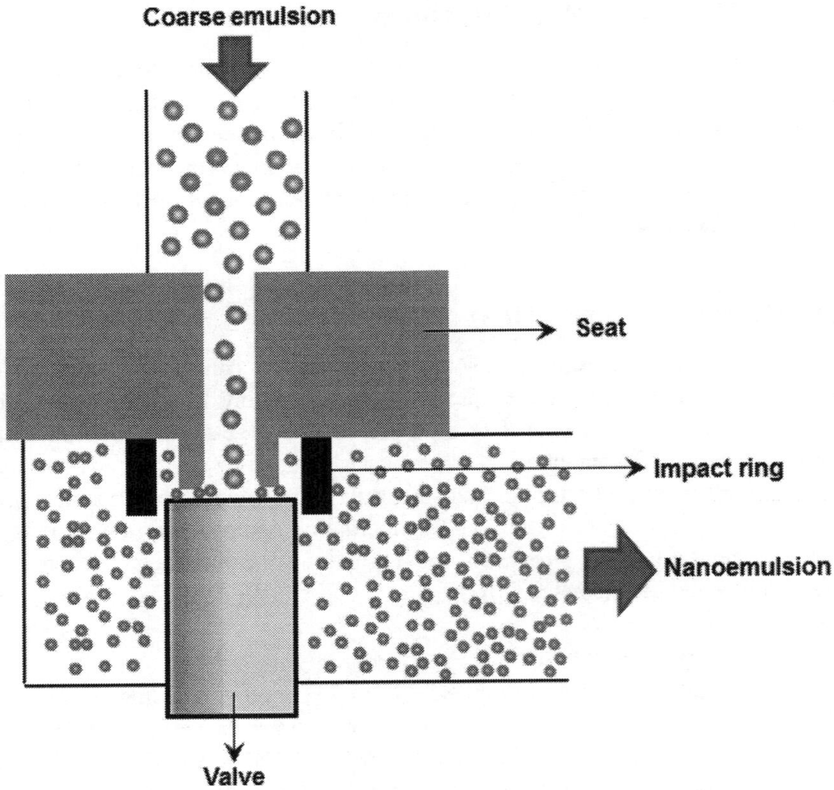

FIGURE 8.2 Schematic representation of high-pressure homogenizer.

size of the droplets is reduced and their distribution is significantly changed. The smaller mean particle size provides more surface area and this helps to achieve more physical stability due to its unique morphological and functional characteristics. Many researchers successfully used the high-pressure homogenizer to produce nanoemulsion using different bioactive compounds with improved stability, bioavailability, and bioaccessibility (Bhushani and Anandharamakrishnan, 2017; Rao and McClements, 2012; Liang et al., 2013; Ahmed et al., 2012).

8.2.1.2 Microfluidizer

The microfluidizer is suitable for industrial operations due to its flexible control of emulsion droplet size distributions and ability to yield fine emulsions from a large variety of materials (McClements, 2015). The invention of microfluidic technology was patented by Cook and Lagace (1985) and the first scientific publication was reported in 1987 (Washington, 1987). The microfluidizer contains a mainly air-driven intensifier pump and a fixed-geometry interaction chamber (Figure 8.3). Within the microfluidizer, the intensifier pump helps to deliver high pressure to force the sample through the interaction chamber. The microfluidic technique involves higher turbulent shear force generated by high pressure along with cavitation so as to rupture the coarse emulsion into extremely fine emulsion droplets (Dalgleish et al., 1996; Maa and Hsu, 1999; Olson et al., 2004). The process stream is delivered through a pneumatically powered pump which pressurizes the in-house compressed air, i.e. 150–650 kPa to about 150 MPa. The micro fluidizer works on the principle of dividing an emulsion flow through a channel into two process streams, feeding each stream through separate fine channels and leading the two streams into each other in an interaction chamber. When emulsion flows through the interaction chamber, intense disruptive forces are generated and produce efficient droplet disruption (McClements and Rao, 2011).

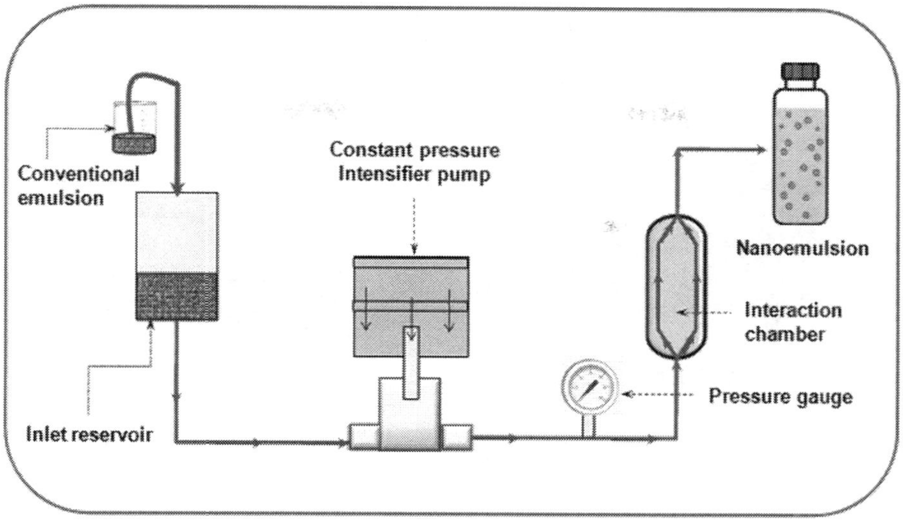

FIGURE 8.3 Schematic representation of microfluidizer.

Many researchers studied the major factors that influence the droplet size of nanoemulsions. These studies suggested that droplet size decreases with increasing homogenization pressure, number of passes, and emulsifier concentration, and with decreasing dispersed-to-continuous phase viscosity ratio and type of emulsifier (Karthik and Anandharamakrishnan, 2016a; Mahdi Jafari et al., 2006; Wooster et al., 2008; Karthik et al., 2018; Parthasarathi and Anandharamakrishnan, 2016). Moreover, scientific reports have suggested that microfluidization is a superior emulsifying device that produces narrower droplet size distribution compared to those obtained using conventional emulsifying devices to produce emulsion (Dalgleish et al., 1996; Pinnamaneni et al., 2003).

8.2.1.3 Ultrasonicator

The ultrasonication technique is an important processing tool in the area food product development with enhanced functional properties. The combination of the oil phase, aqueous phase, and surfactant is exposed to high-energy sound waves which make fine droplets by cavitation phenomenon. This technique is extensively utilized for the preparation of nanoemulsion. In the ultrasonication technique, electrical signals are transformed into mechanical vibrations. These mechanical vibrations generate sinusoidal pressure variations in the medium. In one of the low-pressure cycles, bubbles are formed as a consequence of cavitation. These bubbles contract and enlarge till they implode, creating disturbance in the flow field that eventually breaks down the oil droplets (Abbas et al., 2013; Leong et al., 2009)(see Figure 8.4). High-intensity ultrasonic waves (frequency > 20 kHz) are required to produce nanoemulsion. Conversely, increasing sonication time, power level and emulsifier concentration can also reduce oil droplet size (Kentish et al., 2008). Moreover, the design of the ultrasonic reaction chamber, the intensity of ultrasonic waves, and the product composition (viscosity of oil and aqueous phase, and type of emulsifier) also decide the emulsion droplet size (Leong et al., 2009).

Compared to other energy techniques, i.e. high-pressure homogenizer and microfluidizer, the method of producing nanoemulsions through ultrasonication is highly appreciated by numerous researchers (Delmas et al., 2011; Li et al., 2016). The main advantages of ultrasonication are smaller droplet size, surfactant consumption, lower polydispersity, and higher stability of nanoemulsions (Kentish et al., 2008; Tadros, 2004). Kentish et al. (2008) achieved 135 ± 5 nm sized droplets through ultrasonication for flax seed oil using Tween 40 emulsifier. Soybean oil nanoemulsion was prepared by Delmas et al. (2011), who reported a 150 nm droplet size varying with higher stability. Likewise, Kaltsa et al. (2014) prepared olive oil nanoemulsion and achieved higher stability. Recently, Li et al. (2016)

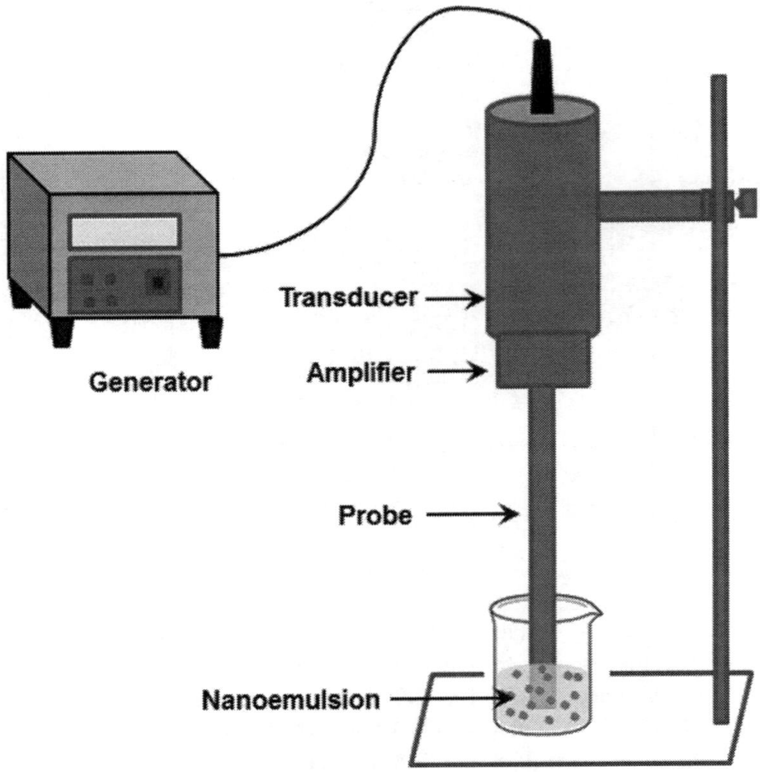

FIGURE 8.4 Schematic representation of ultrasonicator.

prepared curcumin nanoemulsion through the ultrasonication method and obtained a droplet size of 113.93 nm with higher curcumin loading ability. From the above studies, it is confirmed that ultrasonication can be used efficiently to produce nanoemulsions with lower mean droplet size distribution and higher stability (Table 8.1).

8.2.2 Low-Energy Approaches (LEA)

Nanoemulsions can also be produced without using high energy (high pressure, high speed), and the emulsion process then is called low-energy approach. Low-energy nanoemulsification methods involve the spontaneous formation of very fine oil droplets under the influence of a specific system, such as spontaneous emulsification, phase inversion, and solvent displacement methods (McClements, 2011; Anton and Vandamme, 2009). Low-energy emulsification utilizes either low external energy or the intrinsic physicochemical properties of the surfactants, co-surfactants, oil, and water to form nanoemulsions.

The advantages of low-energy methods are their formulation and low cost; they are also energy efficient, non-aggressive to encapsulated compounds, and capable of producing nanoemulsions with small particle sizes compared to high-energy techniques. Low-energy techniques divert the intrinsic physicochemical properties of surfactants, including co-surfactants and excipients, and generate emulsions with nano-sized droplets (Anton and Vandamme, 2009; Bhushani and Anandharamakrishnan, 2017).

Low-energy approaches can be broadly classified into isothermal (depending on a change in composition – spontaneous emulsification and emulsion phase inversion) and thermal (depending on a change in temperature – phase inversion temperature) methods. It is clear that the application of isothermal techniques may be cost effective as there is no requirement of temperature source.

TABLE 8.1
Studies on Different High-Energy and Low-Energy Emulsification Approaches of Food-Grade Nanoemulsion

High-Energy and Low-Energy Emulsification	Compounds	Emulsifiers	Operational Condition	DropLet Diameter (nm)	Advantage	References
High-pressure homogenization	DHA algae oil	Tween 40	800 bar for 8 cycles	less than 88	Increase in physiochemical stability	Karthik and Anandharamakrishnan (2016b)
	Green tea catechins	Soy protein isolate	80 MPa for 5 cycles	245–269	Improvement in stability and bioaccessibility	Bhushani et al. (2016)
	Lemon oil	Tween 80	9000 psi	112–296	Achieved physical stability	Rao and McClements (2012)
	Curcumin	β-lactoglobulin	9000 psi	Less than 200	Higher physical stability and improved bioaccessibility	Ahmed et al. (2012)
	β-carotene	Modified starch	150 MPa for 10 cycles	142–157	Achieved stable emulsion and significant improvement of bioaccessibility	Liang et al. (2013)
Microfluidization	DHA algae oil	Tween 40, sodium caseinate and soya lecithin	1000 bar for 5 cycles 900 bar for 7 cycles 1000 bar for 7 cycles	148, 206 and 760	Increase in physicochemical stability and FFA release	Karthik and Anandharamakrishnan (2016a)
	Vitamin E	Saponin	12,000 psi for four cycles	277	Improved the oral bioavailability of vitamin E	Parthasarathi and Anandharamakrishnan (2016)
	d-limonene	Modified starch Sodium caseinate Whey protein hydrolysate Whey protein concentrate	60 MPa for one cycle	150–600	Improved emulsion stability	Jafari et al. (2007)
	Peanut oil	Sodium dodecylsulfate (SDS) Tween 80 Polyethylene glycol	1000 bar for five cycles	40	Attained physically stable nanoemulsion with excellent optical clarity	Wooster et al.(2008)
	DHA algae oil	Tween 40, sodium caseinate and soya lecithin	1000 bar for 5 cycles 900 bar for 7 cycles 1000 bar for 7 cycles	148, 206 and 760	Achieved sustainable free fatty acids release	Karthik et al.(2018)

(*Continued*)

TABLE 8.1 (CONTINUED)
Studies on Different High-Energy and Low-Energy Emulsification Approaches of Food-Grade Nanoemulsion

High-Energy and Low-Energy Emulsification	Compounds	Emulsifiers	Operational Condition	DropLet Diameter (nm)	Advantage	References
Ultrasonication	Lemongrass oil	Sodium alginate	150 MPa for 10 cycles	1.3 to 1410	Achieved 1.37, 5.29 and 7.07log-reductions on E-coli growth	Salvia-Trujillo et al. (2014)
	Soybean oil	PEG-surfactants (Myrj s40, Myrj s50, or Myrj s100)	–	150	Achieved higher nanoemulsion stability	Delmas et al. (2011)
	Olive oil	whey protein isolate and xanthan gum	20 kHz	1140 to 890	Improved emulsion stability by significant reduction of droplet size and viscosity.	Kaltsa et al. (2014)
	Curcumin	Tween 80 and lecithin		113.93	Higher curcumin loading capacity was achieved and water dispersibility of curcumin in nanoemulsionincreased by 1400 fold than that of raw curcumin	Li et al. (2016)
PIT	Lemon oil	Sodium caseinate and Tween 20		100	Combination of sodium caseinate and Tween 20 reduced emulsion turbidity and droplet size were stable for 15 days	Su and Zhong (2016)
	Cinnamon oil	Non-ionic surfactant		101	Increasing surfactant concentration significantly decreased the mean droplet diameter of the nanoemulsions	Chuesiang et al. (2018)
PIC	Paraffin oil	Span 80-Brij 35		82	Nanoemulsions were stable without obvious change in droplet size in two months	Pan et al. (2014)
	Monomer	Brij 78		30–110	Polydispersity of the polymer nanoparticles was found to depend on the O/S ratio	Elgammal et al.(2015)
Solvent displacement	Lutein	Tween 20, 40, 60, and 80		< 100	Minimum particle size is achieved at low surfactant (0.1%) concentration	Tan et al. (2016)
	Ergocalciferol	Lecithin and sodium caseinate		56–112	Both lecithin and sodium caseinate formulated nanodispersions with longer storage stability	Shu et al. (2017)
	β-carotene	Tween 80		500	Nanostructured lipid carriers improved loading capacity of bioactive compound.	Oliveira et al. (2016)

Although low-energy approaches are economical and effective in producing small droplet sizes, they have some limitations such as being specific to particular oils and emulsifiers, and requiring a higher quantity of emulsifiers.

There are three broad categories investigated under low-energy approaches, viz. spontaneous emulsification, phase inversion, and solvent displacement techniques, which are elaborated below.

8.2.2.1 Spontaneous Emulsification Method

> If the two bulk liquids are not initially at equilibrium, it is conceivable that dynamic processes such as diffusion could produce emulsification when the two liquids are brought into contact without stirring.
>
> **Prof. Clarence A. Miller, Louis Calder Professor Emeritus, Rice University**

The spontaneous emulsification (SE) phenomenon was first discovered by Johannes Gad in 1879 and thereafter continuous research efforts have been devoted to its development and applications. Herein, the emulsion is produced without the requirement of any external energy input, i.e. it is produced spontaneously when both liquids (organic phase and aqueous phase) are put in contact in the presence of surfactants. This phenomenon is referred to as spontaneous emulsification (Santana-Solano et al., 2012). In this method, nanoemulsions are produced without external energy by putting in contact two immiscible liquids that are not in equilibrium (Solans et al., 2016).

The process can occur at room temperature, without the need of any special device. It utilizes variations in interfacial tension, phase transition region, the viscosity of interface and bulk, surfactant structure, and surfactant concentration for nanoparticle formation (Setya et al., 2014).

True SE occurs when two immiscible liquids in contact mix without any external influence (thermal or mechanical); it is influenced by solvents, with or without the presence of surfactants (Komaiko, 2016). When this mixing occurs with oil, water, and water-miscible solvents, the process is called "Ouzo effect" (Ganachaud and Katz, 2005; Komaiko, 2016). A schematic flow sheet of preparation of nanoparticles by the said technique has been summarized in Figure 8.5.

Komaiko and McClements (2015) reported the following protocols for achieving emulsions through the spontaneous emulsification method:

- by simply mixing oil, water, and water-miscible solvent together (Ganachaud and Katz, 2005)
- by putting in contact an oil, hydrophobic surfactant, and water miscible solvent mixture with an aqueous phase (Bouchemal et al., 2004)
- by adding oil and hydrophilic surfactant mixture into an aqueous phase (Anton and Vandamme, 2009)

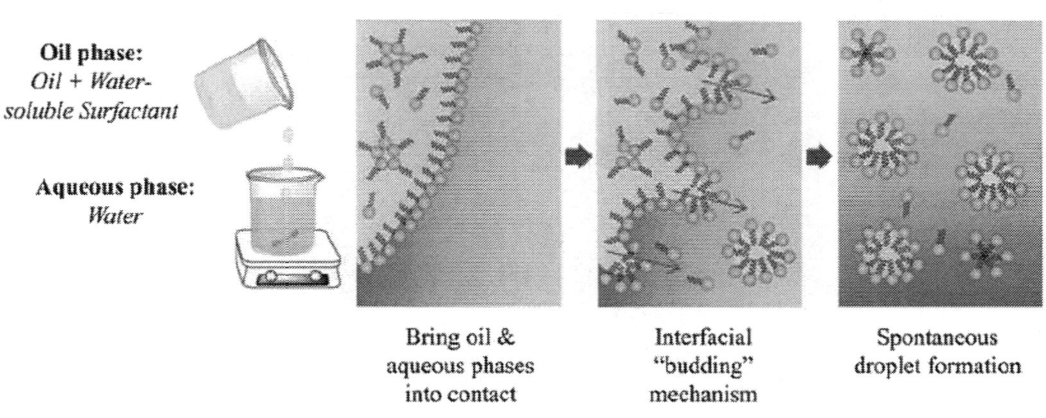

FIGURE 8.5 Schematic representation of spontaneous emulsification. (McClements and Rao, 2011.)

The plethora of possibilities of spontaneous emulsion could be better realized by understanding the findings of some researchers in diverse fields of food science. The key to stable particle formation has been attributed to surfactant-oil-water (SOR) ratio, which needs to be optimized as per the system investigated.

Spontaneous emulsification has been found to be useful in the encapsulation of eucalyptus oil, which has the antioxidant and antimicrobial rich 1,8-cineole as chief active constituent (Sugumar et al., 2015). The authors describe nano delivery of the active constituent that has been in high demand lately, owing to healthful benefits and antimicrobial effects on food and dairy spoilers, such as *Listeria monocytogenes*. The oil-in-water (O/W) emulsion was optimized at the ratio of oil: surfactant (Tween 20) of 1:2, a 50–100 nm particle size, and polydispersability index of <0.02. This stable food-grade nanoemulsion could be well employed as a preservative in food and dairy products, to circumvent risks of microbial attack. An atomic force microscope image of microbial cells has been depicted in Figure 8.6.

In another study, researchers demonstrated nano-based delivery of curcumin-loaded PLGA nanoparticles. These particles have shown enhanced *in-vitro* anti-tumor activity on HL60 Hep G2 cell lines. The nanoparticles had mean diameter 189.7 nm,11% drug loading, and 33.5% encapsulation efficiency (Chen et al., 2014). A SEM image has been illustrated in Figure 8.7.

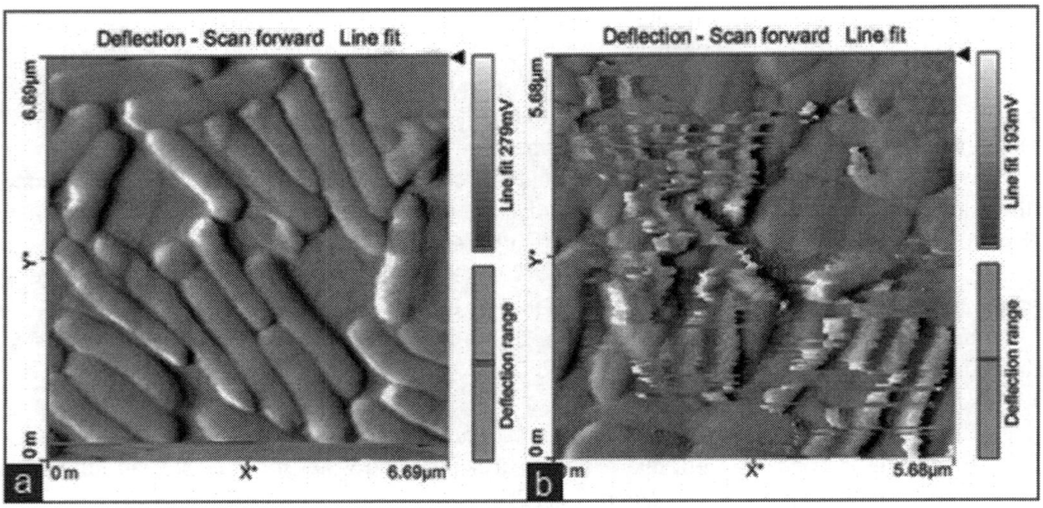

FIGURE 8.6 Atomic force microscopy images of *Listeria monocytogenes* (a) untreated bacteria showed intact cell morphology (b) bacteria treated with 10-fold nanoemulsion for 30 min showed membrane damage. (Sugumar et al., 2015.)

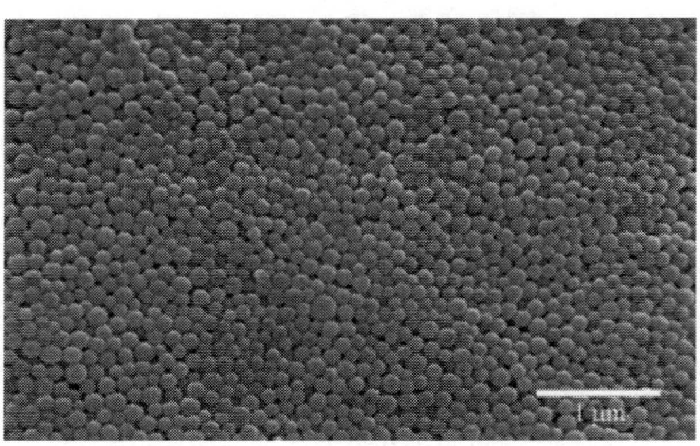

FIGURE 8.7 SEM image of curcumin-loaded PLGA nanoparticles. (Chen et al., 2014.)

In a separate find, investigators encapsulated β-carotene as nanoparticles, using a mix of long-chain triglycerides with medium-chain triglycerides, or flavor oil. Optimum results were obtained with long-chain triglycerides: orange oil, 1:1, particle diameter 109 mm. β-carotene being lipophilic, its bioavailability increased in simulated gastrointestinal tract conditions, indicating, first, that the mode of encapsulation was suited to the conditions of the digestive system, and, second, that the oil phase is suited for lipophilic nutraceuticals (Zhang et al., 2016).

Another fat-soluble biomolecule, lutein, was encapsulated in medium-chain triglycerides. The optimized composition was 0.12% lutein, 9.88% lipid, 10% surfactant (Tween 80), and 80% aqueous phase. The particles had mean diameter of 200 nm and polydispersity index of 0.23. The particles were immune to droplet aggregation and gravitational separation, and were stable at room temperature for a month (Surh et al., 2017).

Nutraceutical resveratrol has been successfully nanoencapsulated by this technique in grape seed oil and orange oil lipid composite (1:1), with 10% oil phase, 10% surfactant, and 80% aqueous phase and resveratrol at 120 μg/ml. The particles were found to be resilient against UV exposure (Davidov-Pardo and McClements, 2015). In a similar attempt, D-alpha-tocopherol was encapsulated and the resulting nanoparticles were of size 150–250 nm, and this nanoemulsion could further allow a very good drug loading molecule: hydrophobic molecule estradiolvalerate (Esmaeili et al., 2007).

Polyunsaturated fatty acids (ω-3 rich) of fish oil have also been nanoencapsulated by this technique. The optimized composition of 10% oil (50% fish oil/50% lemon oil), 10% surfactant (Tween 80) and 80% aqueous phase (10% glycerol in citrate buffer) The mean particle diameter was 51 nm, polydispersity index 0.15, with a stability of 30 days at room temperature (Gulotta et al., 2014).

8.2.2.2 Phase Inversion Method

The phase inversion method includes phase inversion temperature (PIT) and phase inversion composition (PIC). The phase inversion technique involves inversion of emulsion type from O/W to W/O and vice versa, as required, by continuous addition of the phase that is ultimately required to be the bulk phase. The process requires mild heat exposure (at PIT) to the chemical aggregate with constant stirring/agitation, followed by a cooling period. To avert phase separation, the process routinely employs a surfactant such as Tween 80/20 (with high HLB value), and the process has been in use for encapsulation of volatile bioactives and food supplements. Typically, the addition of the desired bulk phase is recommended to be made slowly, with constant agitation (to avoid phasing out) (Li et al., 2013). A schematic diagram of the phase inversion method has been illustrated in Figure 8.8 (Komaiko and McClements, 2016).

This technique has been explored for many bioactives encapsulations. Nanoencapsulation by PIT has been successful for polyphenols such as quercetin and epigallocatechin-3-gallate, a flavonoid-loaded nanocapsule, using polyethylene glycol as a surfactant. For epigallocatechin-3-gallate, Barras et al., (2009) observed significantly enhanced entrapment efficiency (95%), and higher stability of the capsules (4 weeks vs. 4 h when non-encapsulated).

In another attempt to elucidate the worthiness of nanoencapsulation by PIT, researchers have comprehensively demonstrated the heightened stability of epigallocatechin-3-gallate and its anti-atherogenic bioactives in macrophages (Zhang et al., 2013). The find highlights the potential of the said encapsulate in countering atherosclerotic lesion, through decrement in macrophage cholesterol content.

A variant of classical PTI, catastrophic PIT (CPIT) has been found to enhance the stability of natural D-limonene nanoemulsion. CPTI is typically a low-energy process; O/W system retracts to O/W, and vice versa. An optimized value of surfactant-to-oil (olive oil) ratio of 1.5 was worked out, when nanoemulsion (mean diameter 40.13 nm) could be produced. Phase inversion has been reported at 32% water, forming D-limonene O/W emulsion (Li et al., 2013).

Despite high efficiency, caution is warranted in PIT against droplet coalescence, post processing. Considering the need of an endeavour in this field, Rao and McClements have worked out a method involving surfactant displacement to minimize the said risk (Rao and McClements, 2010).

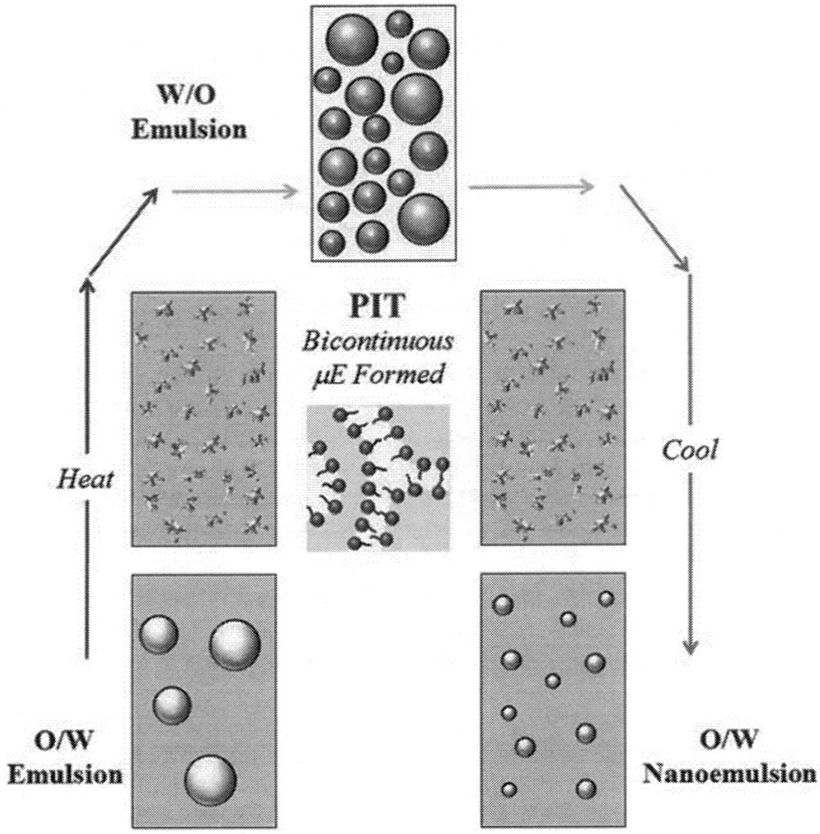

FIGURE 8.8 Illustration of phase inversion method of nanoencapsulation. (Komaiko and McClements, 2016.)

The authors used water/surfactant (Brij 30, C12E4)/oil (tetradecane) and revealed that for optimum stability, storage temperature should be about 27°C, lower than PIT (that was at 40°C). Higher temperature resulted in the coalescence of particles while lower temperature leads to unwanted gelation. Tween 80 (0.2%) or SDS (1%) could be used to displace the emulsifier, Brij 30, from the droplets' surfaces.

8.2.2.3 Solvent Displacement Method

The solvent displacement or nanoprecipitation method is reportedly preferred in pharmaceutical sciences for preparation of nanoemulsion, mainly for the major advantage of its simple preparation procedure at room temperature (Lovelyn and Attama, 2011). As described by McClements (2011), in the solvent displacement method, the organic phase consists of a lipid and an organic solvent. It is necessary that the lipid gets dissolved in the organic solvent without homogenization. This can be a liquid oil (e.g. ω-3 or CLA rich oil), a solid fat (e.g. hydrogenated palm oil), a lipophilic nutraceutical (such as β-carotene), or a mixed system. The organic solvent is selected carefully depending on its water miscibility, boiling point, and safety issues. This organic phase is then homogenized with an aqueous phase containing a hydrophilic emulsifier. In the solvent displacement method, an amphiphilic (partially miscible in water) organic solvent is used and the aqueous phase is initially saturated with the solvent. In the last step, the organic solvent is removed from the nanodispersion under reduced pressure.

The one step of rapid diffusion of the organic phase in the aqueous phase produces nanoemulsion at low-energy input with high yield of encapsulation (Silva et al., 2012). During the procedure, reduction of size of the droplets occurs when the emulsion is diluted with water; the addition of water causes displacement of the organic solvent from the oil droplets into the aqueous phase,

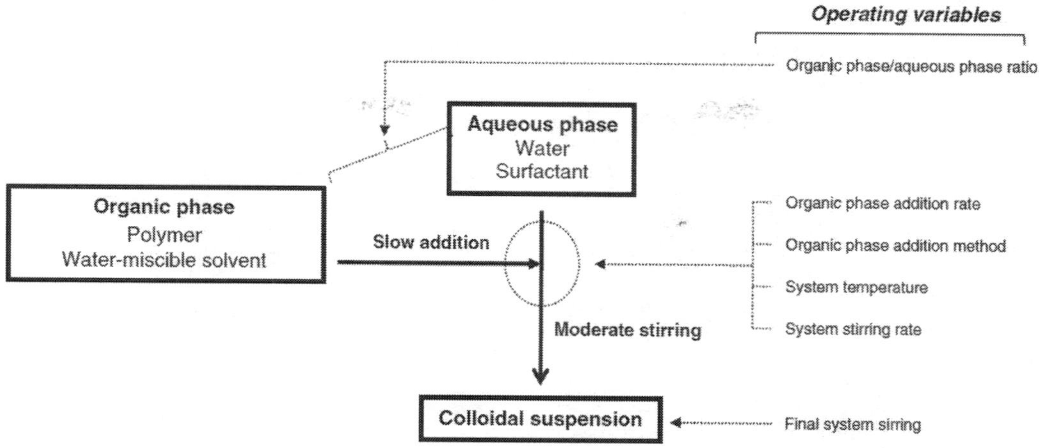

FIGURE 8.9 Schematic procedure and operating variables of solvent displacement method. (Mora-Huertas et al., 2011.)

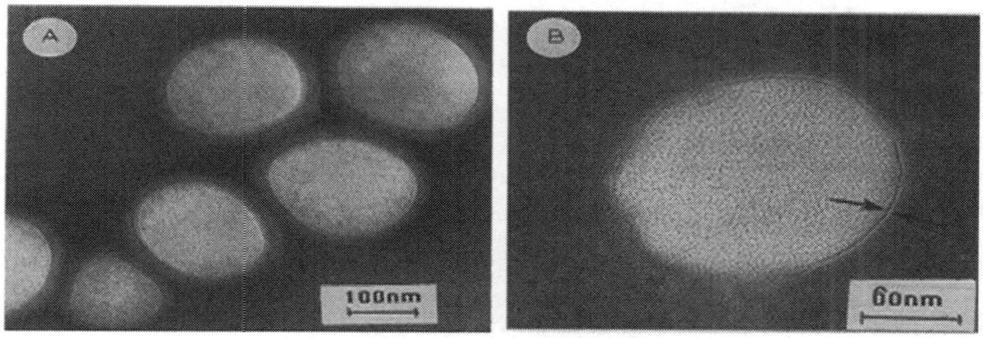

FIGURE 8.10 TEM photographs of PLA nanocapsules prepared by solvent displacement method. (Fessi et al., 1989.)

resulting in shrinkage in the size of the droplets. Researchers have reported a reduction in the size of raspberry-like nanocapsules from μm range to nm range by mere stirring employing sonication and "Ostwald ripening" (Schrade et al., 2011) (Figure 8.9).

Figure 8.10 displays the brief procedure of preparation of submicron particles by solvent displacement method. The procedure of preparing submicron particles by this method was first reported by Fessi et al. in 1989 (Mora-Huertas et al., 2011). They formulated indomethacin-loaded nanocapsules by the deposition of poly-(D,L.-lactide) polymer using acetone as an organic solvent, which was displaced from the oil nanodroplets (Fessi et al., 1989) (Figure 8.11). The formation of submicron particles by solvent displacement method can be described by two approaches: mechanical mechanisms that include dispersion mechanisms or spinodal decomposition and system chemical instability, i.e. condensation mechanism or nucleation. The ternary phase diagram illustrates the formation of particle dispersion in the metastable region, situated between the bimodal (miscibility-limit curve) and the spinodal (stability-limit curve) compositions (Figure 8.12). The mechanical mechanism proposed to understand the phenomenon concentrates on the spinodal region towards the metastable region. On the other hand, nucleation theory adopts a bimodal curve towards the metastable region to understand the process (Mora-Huertas et al., 2011).

The dispersion mechanism approach states that the interfacial turbulence and thermal inequalities caused by the differences in surface tension between the organic solvent and aqueous solvent

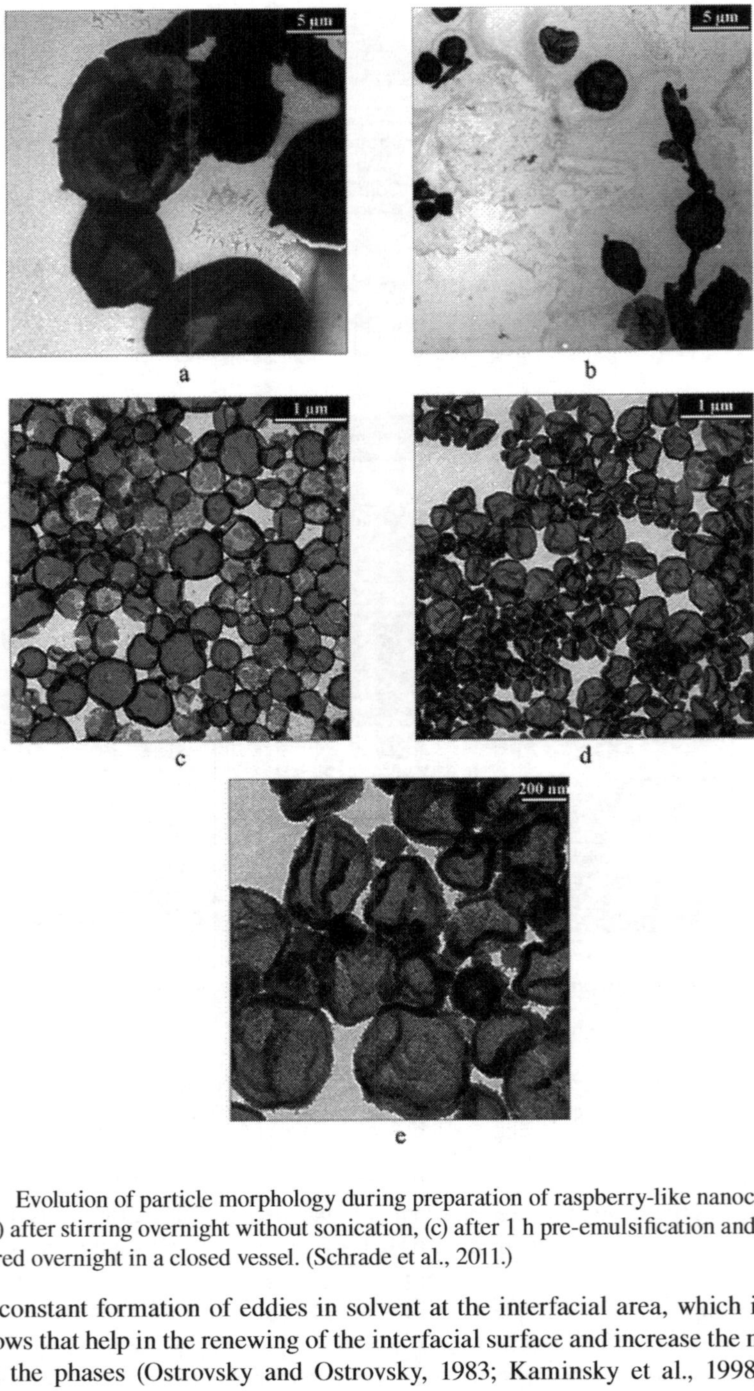

FIGURE 8.11 Evolution of particle morphology during preparation of raspberry-like nanocapsules (a) after stirring 1 h, (b) after stirring overnight without sonication, (c) after 1 h pre-emulsification and sonication, and (d) and (e) stirred overnight in a closed vessel. (Schrade et al., 2011.)

generate the constant formation of eddies in solvent at the interfacial area, which in turn creates convective flows that help in the renewing of the interfacial surface and increase the mass-exchange rate between the phases (Ostrovsky and Ostrovsky, 1983; Kaminsky et al., 1998). The violent spreading occurred due to the miscibility of both the solvents turn by turn breaks down the organic phase into "submicron droplets"; after that, the solvent is displaced from the regions of low surface tension precipitating the polymer particles (Mora-Huertas et al., 2011).

Since the use of a surfactant is optional for this method, several researchers opine that the formation of particles may not critically depend on the interfacial tension variations. Therefore, another approach states that when the organic phase containing polymer is homogenized with the aqueous phase, solvent diffuses into water resulting in supersaturation of the polymer which eventually forms nanoparticles by spontaneous nucleation. The assumption of this approach is that the rates

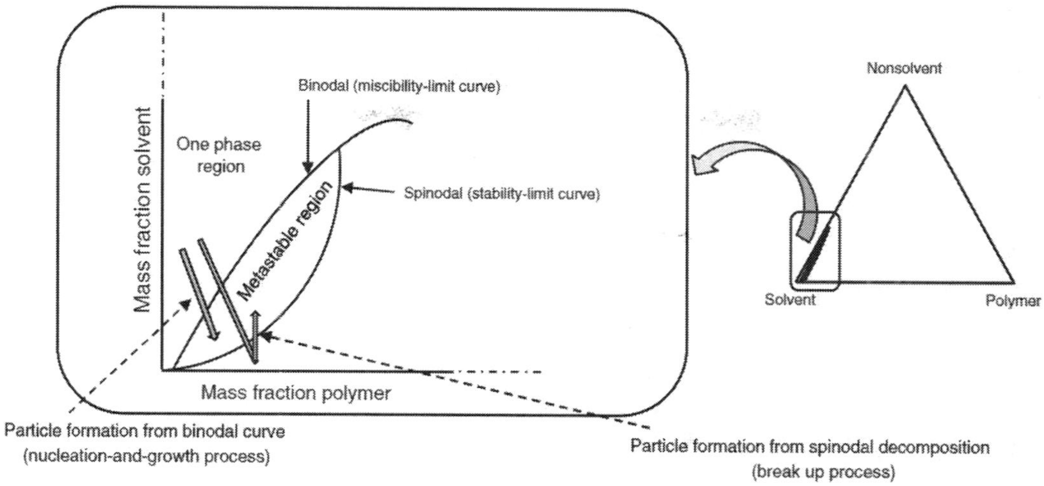

FIGURE 8.12 An illustration from a ternary phase diagram for the polymer/solvent/nonsolvent system related to particle formation by solvent displacement method. (Mora-Huertas et al., 2011.)

of homogenization and molecular diffusion are extremely rapid compared to the nucleation rate (Mora-Huertas et al., 2011).

Several types of research have been reported on the application of the solvent displacement method. Polymeric nanoparticles were formulated by this technique for PLGA, PLA, PCL, and poly (methyl vinyl ether-co-maleic anhydride) (PVM/MA); and also for delivery of cyclosporin A, bifonazole and clotrimazole (Nagavarma et al., 2012). In spite of several advantages and wide applications, the solvent displacement method has a few major limitations. The most significant among them is the use of organic solvent and removal of the same. The solvent-to-oil ratio is required to be high to achieve a desirable nanodroplet size (Lovelyn and Attama, 2011). Therefore, pilot scale application of this method is difficult.

Despite widespread prevalence of high energy processes for miniaturization, low energy techniques also provide similar size reduction by employing chemicals and their intrinsic nature for achieving the desired feat. This greatly aids as an effective alternative to physical techniques and has consequently gained much acclaim being a more benign approach.

8.3 STABILITY OF NANOEMULSION

Nanoemulsion instability is undesirable and creates various problems while in storage for a long period and during application. Maintaining nanoemulsion stability with desired properties and characteristics is important. The main instability mechanisms of nanoemulsion are flocculation, coalescence, creaming, sedimentation, Ostwald ripening, and phase separation (Figure 8.13, Step 2). All the main nanoemulsion instability mechanism studies have been presented in Table 8.1.

8.3.1 Flocculation

Flocculation occurs when two or more droplets associate with each other but retain their individual integrity. It takes place when attractive interactions between droplets dominate long-range repulsive interactions, but not short-range repulsive interactions. It is an important destabilization process that occurs in the emulsions system during storage for longer periods (Figure 8.14). The attractive interactions in food emulsions are due to van der Waals, depletion, and hydrophobic forces; however, repulsive interactions are electrostatic and steric forces (McClements, 2005).

Recently, Karthik and Anandharamkrishnan (2016a) studied the effect of flocculation on DHA nanoemulsion and reported that flocculation was not observed in the emulsions prepared through

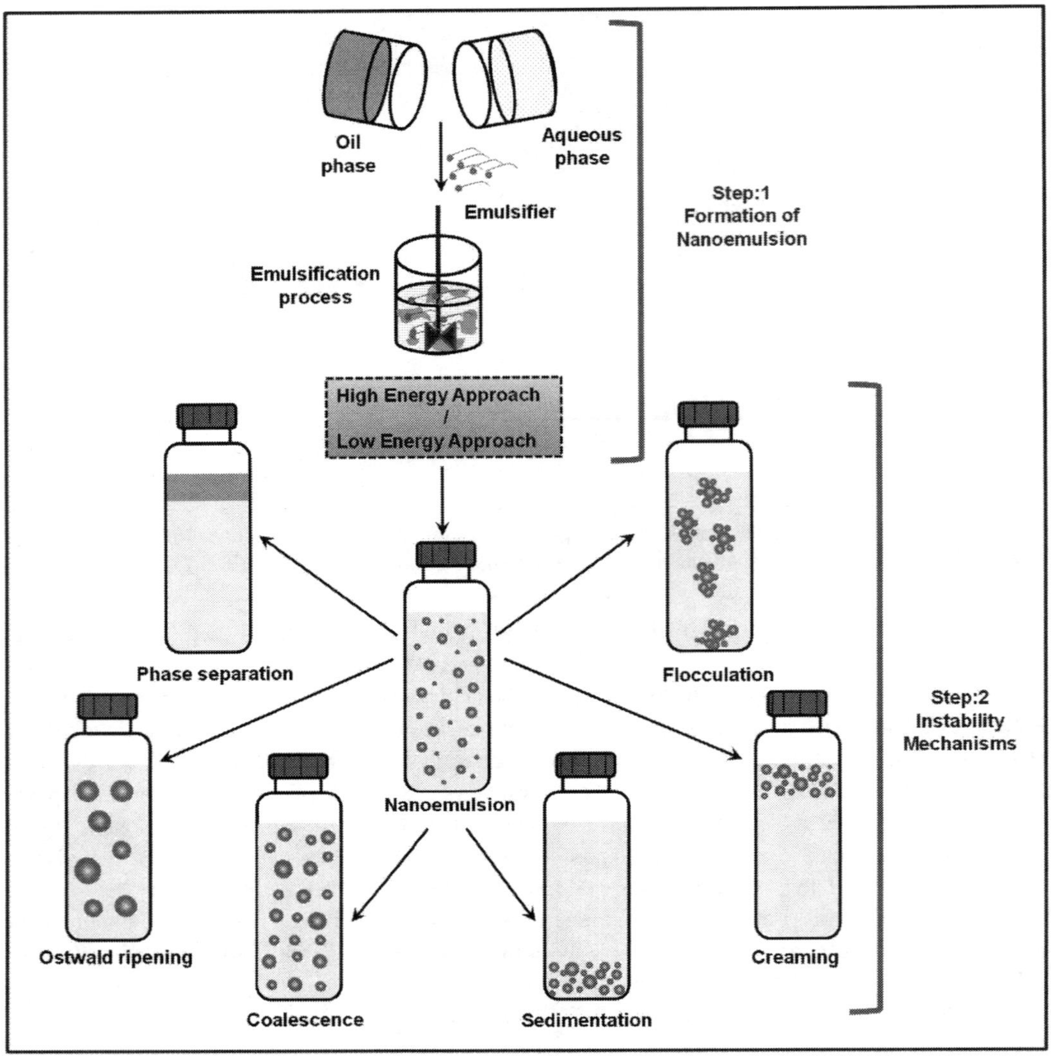

FIGURE 8.13 Step 1 signifies the formation of nanoemulsions via high-energy or low-energy emulsification approaches. Step 2 signifies the breakdown processes of nanoemulsion through different instability mechanisms functioning as creaming, phase separation, coalescence, sedimentation, flocculation, and Ostwald ripening. (Karthik et al., 2017.)

high-pressure homogenization techniques. Also, no flocculation was observed over a storage period of 100 days at different temperatures. Tadros (2004) stated that emulsion flocculation is influenced by van der Waals attractions and it happens due to insufficient repulsive energy between oil droplets. Züge et al. (2013) reported that storing the emulsion at lower temperatures can significantly reduce the molecular movement. On the contrary, sodium caseinate protein stabilized emulsion stored at 28°C and 40°C showed flocculation from the 10th day onwards (Figure 8.15). This is because of the rupture of the interfacial layer during the storage of nanoemulsion at a higher temperatures; further sodium caseinate may not offer enough stability to prevent the oil droplets from flocculation and structural deformation (Karthik and Anandharamakrishnan, 2016a). Jafari et al. (2007) also investigated this with the addition of Tween 20 (surfactant) to Hi-Cap (biopolymer) to form D-limonene nanoemulsion. This showed more flocculation due to surfactant biopolymer interactions and emulsifier displacement. This is because of a mixture of surfactants and biopolymers present in an emulsion, which have contest with each other to adsorb at oil-water interface and

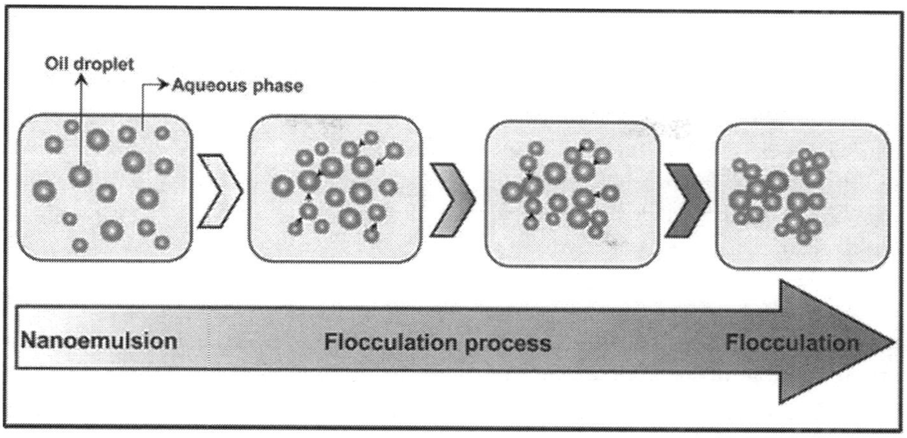

FIGURE 8.14 Graphical representation of flocculation process for oil-in-water nanoemulsion. (Karthik et al., 2017.)

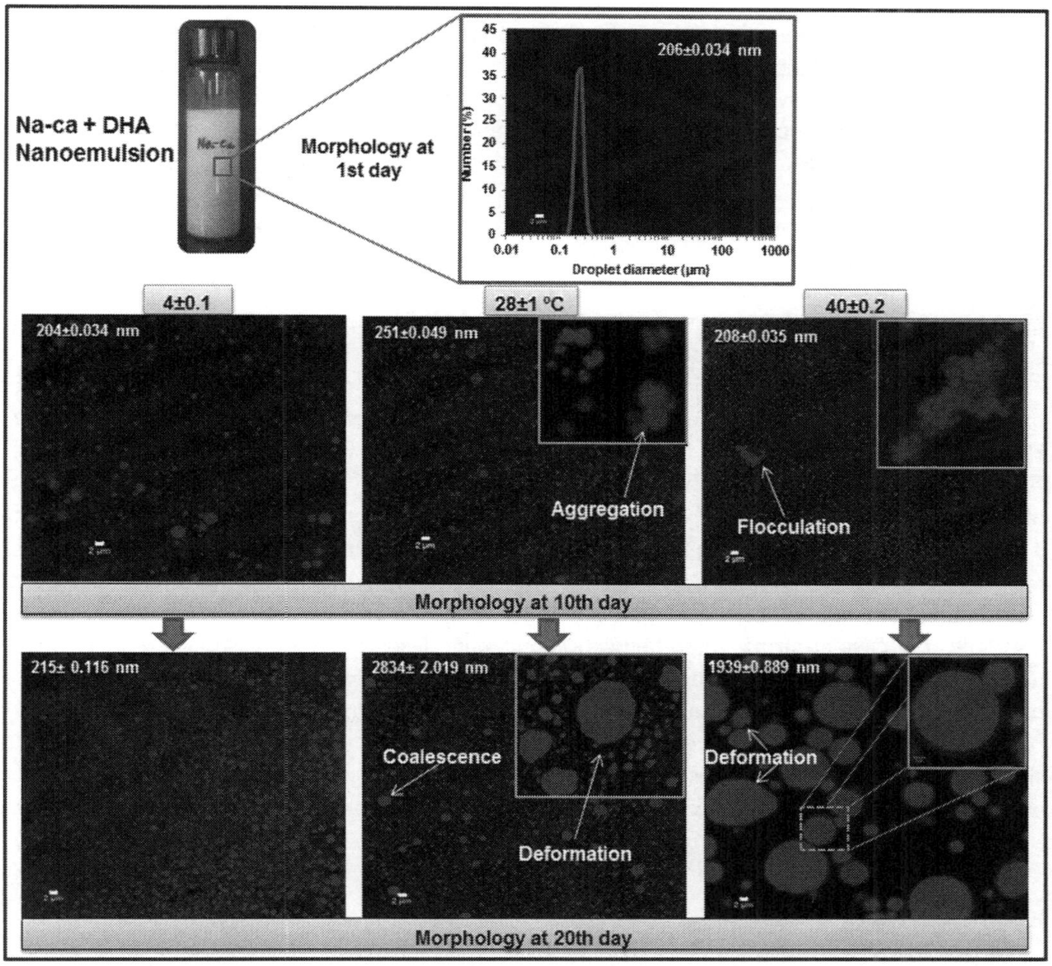

FIGURE 8.15 Effect of different temperatures on the storage stability of DHA nanoemulsions stabilized by sodium caseinate (Na-ca) emulsifier. Flocculation, coalescence, and deformation are mentioned in the micrograph. (Karthik and Anandharamakrishnan, 2016a.)

reduce interfacial tension. The kind of surfactant and concentration used also influences flocculation. Wulff-Pérez et al. (2009) observed flocculation in natural oil nanoemulsions beyond a certain concentration of Pluronic F68 surfactant. Contrarily, Mao et al. (2009) reported no flocculation in Tween20-stabilized β-carotene nanoemulsions due to the use of a lower concentration of surfactant. Qian et al. (2012) observed an increase in mean particle size (78–80 nm) of β-carotene nanoemulsions stabilized with β-lactoglobulin after 15 days of storage. They stated that flocculation was caused by the increased surface hydrophobicity of protein-coated droplets once the globular proteins unfold. Also, there was an increase in the hydrophobic attraction between oil droplets which promoted droplet flocculation.

In general, emulsification techniques and storage temperatures lead to the flocculation of droplets in nanoemulsion. Moreover, the higher surfactant concentration and surfactant–biopolymer interaction resulted in flocculation of the nanoemulsion. Further, an increase in ionic strength can influence the formation of flocculation due to a higher salt concentration that reduces the electrostatic repulsion between oil droplets. Therefore, it is important to optimize the concentration of surfactant during the emulsification process to inhibit flocculation.

8.3.2 Coalescence

Coalescence is a process in which two droplets collide with each other and merge together in the emulsion system. It is the destabilization process over the time evolution of the droplet size which expresses different behaviors from homogeneous growth to strongly heterogeneous growth. Nano-sized emulsions are more stable to coalescence compared to micro-sized emulsions (Capek, 2004); further decrease in particle size is directly related to the increase in coalescence due to higher Brownian motion (Kong and Park, 2011). Tadros et al. (2004) stated that the rate of coalescence is determined by the adsorption ability of the surfactant, which is governed by surfactant surface activity and concentration. In the case of protein-stabilized emulsions, coalescence was slower than creaming and flocculation; however, it was highly increased in the presence of shear stress. Different factors such as salt concentration, pH, composition, formulation, and storage conditions are more relatively influenced by the rate of coalescence in nanoemulsion (Britten and Giroux, 1991).

Recently, DHA nanoemulsion prepared with different emulsifier exhibited coalescence for the protein (sodium caseinate)-stabilized emulsion stored at a higher temperature (see Figure 8.15). However, emulsion stored at lower temperatures did not show any coalescence (Karthik and Anandharamakrishnan, 2016a). Similarly, Rao and McClements (2011) studied the storage stability of sucrose monopalmitate (SMP)-stabilized lemon oil nanoemulsions. It was found that there was a sharp increase in coalescence in the emulsion stored at a higher temperature. On the other hand, nanoemulsification techniques also need to be considered as important tools to prevent coalescence (Karthik and Anandharamakrishnan, 2016b). In addition, Mahdi Jafari et al. (2006) stated that the high rate of collision during microfluidization technique leads to an increase in droplet size with higher coalescence in the nanoemulsion. This is due to the use of energy input beyond a certain level in the emulsification process, resulting in an over-processing of the emulsion. The nature of emulsifiers/surfactants, concentration, and the ratio of oil to surfactant are influenced by coalescence. The stability of the emulsions to coalescence improved when storage temperature was decreased. Further, using a higher concentration of emulsifier during the formation of emulsion reduced the interfacial free energy that creates a mechanical barrier to coalescence.

8.3.3 Creaming

The perception of the creaminess in emulsion is measured in terms of increase in cream layer thickness. Creaming is formed due to the dispersed particles that either settle or float in the lower or upper portion of the emulsion system and it can be more opaque or creamier (Lieberman et al., 1998). A reduction in the average mean droplet size decreases the creaming velocity (Stokes' law) that improves emulsion stability (Desrumaux and Marcand, 2002). Creaming was perceived as a destabilization of the emulsions into an opaque layer at the top (cream) and a marginally turbid or transparent layer at the bottom (serum). The creaming stability of nanoemulsions is monitored by quantifying the height of the cream layer on top (HC) and the height of total emulsion (HE) in the emulsion stored tube or container (Huimin et al., 2014). Creaming stability in terms of creaming index (%) was calculated using the following equation:

$$\text{Creaming Index } (CI\%) = \left(\frac{H_C}{H_E}\right) \times 100 \tag{8.1}$$

Recently, Karthik and Anandharamakrishnan (2016b) investigated the storage stability of DHA nanoemulsions prepared by different emulsification techniques such as high-speed homogenization, high-pressure homogenization, and a combination of both. It was found that the emulsion prepared with a combination of emulsification techniques is highly stable to creaming over storage for 100 days. This is because a decrease in average fat droplet size was achieved during a high-pressure homogenization process, which helps to reduce creaming velocity. Likewise, Karthik and Anandharamakrishnan (2016a) studied the creaming stability of DHA nanoemulsion prepared by different emulsifiers (such as sodium caseinate, soya lecithin and Tween 40) in accelerated storage conditions. It was observed that soya lecithin and Tween 40 had higher creaming stability than sodium caseinate during the storage period and the average mean droplet size did not change. This is due to sodium caseinate not being stable when stored at higher temperatures. Similarly, Bhushani et al. (2016) studied creaming stability by changing the concentration of oil and green tea catechins of nanoemulsion. They observed enhanced stability in the higher concentration of catechins and oil against creaming during storage. This suggested that there is a difference in the viscosity of nanoemulsions. On the other hand, a change in creaming stability was reported when storing at higher temperatures. However, some studies (Donsì et al., 2011; Ahmed et al., 2012; Li et al., 2012) reported no creaming instability in the nanoemulsion over the storage period. The creaming of nanoemulsion decreased with an increase in storage temperature; however, it also depends on the composition and homogenization technique. Nanoemulsions are more stable to creaming when stored at low temperature.

8.3.4 Sedimentation

Sedimentation is one of the destabilization mechanisms of emulsions during long-term storage processes and handling in adverse environmental conditions. As emulsion droplets are denser than the surrounding liquid, they move downward and are referred to as sedimentation. McClements (2005) stated that in oil-in-water emulsion the droplets tend to move upwards (cream), while in water-in-oil emulsion they tend to move downwards (sediment). This is related to the gravitational force over the emulsion oil droplets. In addition, Tadros (2009) stated that smaller droplets lead to less gravitational force with higher Brownian motion that may act against the gravity in the emulsion; further, this helps to prevent sedimentation over the storage period.

Teo et al. (2010) investigated the storage stability of vitamin-E-loaded palm oil esters-based nanoemulsions with a different formulation of Pluronic F-68. The formulation with 2.4% higher emulsifier showed more stability during storage at 45°C. Whereas, nanoemulsion prepared with a lower concentration of emulsifiers resulted in sedimentation. This is due to the smaller size of nanoemulsion obtained while using a high concentration of emulsifier. Likewise, green tea nanoemulsion prepared with lower oil concentration was found to be subject to sedimentation over the storage period (Bhushani et al., 2016). The DHA nanoemulsion prepared at high-speed homogenization stored for 100 days at 40°C showed more sedimentation (Karthik and Anandharamakrishnan, 2016b). In addition, different characteristics of emulsifiers are also influenced by the formation of sedimentation observed by Karthik and Anandharamakrishnan, (2016a). On the contrary, Ahmed et al. (2012) prepared curcumin nanoemulsion containing LCT or MCT as the lipid phase had no sedimentation. This is because of the low water solubility effect of LCT and MCT. All the studies generally reported that emulsification techniques, emulsion system, oil concentration, emulsifier/surfactant concentration, storage time, and temperature can lead to sedimentation in the nanoemulsion.

8.3.5 Phase Separation

The two clear phases with separated boundary formed in the emulsion are called phase separation. This can be a physical indication of nanoemulsion instability which is visually observed. Robins (2000) stated that the emulsion droplets have the ununiformed density of the liquid system that surrounds them and net gravitational forces acts on them and result in phase separation. Dickinson (2003) described that initial instability of emulsion is due to creaming, which ultimately influences macroscopic phase separation into separate visible regions of cream and serum. Phase separation is the undesirable spontaneous separation of two layers in the solution and results in specific morphology with interesting rheological properties. Kinetics of phase separation can be calculated through the rheological behavior of the interconnected emulsion droplet regions. Therefore, changing the polysaccharide ratio in the emulsion system resulted in the formation of a droplet network which prevents or even constrains phase separation (Moschakis et al., 2010). In general, Karthik and Anandharamakrishnan (2013) reported that the nature of emulsifiers, storage time, and temperature influence phase separation in the emulsion system.

Lee et al. (2010) prepared triacylglycerol oil nanoemulsion using WPI emulsifier and compared the nano and conventional emulsions. It was observed that different patterns of phase separation were found near the PI in both types of emulsions. Recently, Bhushani, Karthik, and Anandharamakrishnan (2016) reported that changing the concentration of oil and core material leads to phase separation. Karthik and Anandharamakrishnan (2016b) observed that the emulsification technique is highly influenced by the formation of phase separation in nanoemulsion. They suggested that the use of high-pressure homogenization technique prevents or restricts phase separation. In addition, types of emulsifiers and storage temperatures also influence the formation of phase separation in the nanoemulsion prepared by microfluidization (Karthik and Anandharamakrishnan, 2016a).

8.3.6 Ostwald Ripening

Ostwald ripening is a process in which larger oil droplets are formed in the emulsion system at the expense of small droplets due to its molecular diffusion of oil between droplets in the continuous phase. Because of the Kelvin effect, small droplets have the capacity to increase in oil solubility, unlike larger emulsion droplets, due to the difference in Laplace pressures. The Ostwald ripening is influenced by oil solubility in the continuous phase that can be described by LSW kinetics theory (Lifshitz and Slyozov, 1961; Wooster et al., 2008). Kabalnov (2001) postulated that the rate of Ostwald ripening is proportionate to the interfacial tension. Mean droplet diameter (d) change with respect to time (t), owing to Ostwald ripening of a single component emulsified lipid in the steady-state system, is defined by the following equation (Taylor, 1934):

$$d^3 - d_0^3 = \bar{\omega} = \frac{32}{9}\alpha c D t \tag{8.2}$$

where:

- $\bar{\omega}$ denotes the Ostwald ripening rate
- α (=2γVm/RT)
- γ is the interfacial tension
- V_m is the molar volume of the lipid
- R is the gas constant
- T is the absolute temperature
- c is the water solubility of the lipid
- D is the translational diffusion coefficient of the lipid in the water

Dickinson (2003) stated that Ostwald ripening generally occurs in beverage emulsions, i.e. water-soluble flavor oils, short-chain triglycerides, and essential oils (orange oil, D-limonene, lemon oil, etc.); however, it does not happen in dairy-based emulsions due to the presence of water-insoluble triglyceride oils. Hence, nanoemulsion instability can be prevented by the addition of a ripening inhibitor (non-polar molecule) which is soluble in oil and insoluble in the water phase, such as long-chain triacylglycerol (corn oil, soybean oil, sunflower oil, fish oil, etc.) (McClements and Rao, 2011; Li et al., 2009). Ziani et al. (2011) observed the increase in mean particle diameter, creaming and oiling off in thyme oil nanoemulsions during three days of storage. It was highly prone to Ostwald ripening due to high water solubility, rapid creaming, and low density. However, the addition of more than 75% corn oil (a hydrophobic material with low water solubility) into nanoemulsion prohibited Ostwald ripening. Similarly, Wooster et al. (2008) reported the influence of long-chain triglycerides (i.e. peanut oil) and n-alkane oil on Ostwald ripening of nanoemulsion. It was observed that nanoemulsion prepared with peanut oil (120 nm) was stable for more than three months of storage without Ostwald ripening. Conversely, n-alkane (tributyrin) nanoemulsion showed higher Ostwald ripening rate. Therefore, adding peanut oil into the soluble n-alkane oil (tricaprylin) emulsion made it more stable against Ostwald ripening and thermodynamic property in the emulsion system. The large molar volume of long-chain triglycerides (LCT) oils provides a kinetic barrier to Ostwald ripening.

Rao and McClements (2012) examined the effect of lemon oil in different fold such as 1×, 3×, 5×, and 10× (degree of distillation of oil) based on the formation and stability of nanoemulsions. It was found that the lower fold oils (1×, 3×, and 5×) had highly unstable and susceptible to Ostwald ripening over the storage of nanoemulsion. But the higher fold oil (10×) nanoemulsion showed as stable in Ostwald ripening due to very low water solubility which prevented the emulsion droplet growth. On the contrary, Ahmed et al. (2012) reported that there was no Ostwald ripening in curcumin nanoemulsion prepared with long-chain triacylglycerols (LCT) and short chain triacylglycerols (SCT) using the ratio of 50:50. In addition, this nanoemulsion was found to be stable for particle growth and gravitational separation. This was due to the mixing of highly water-soluble SCT oil into low water-soluble LCT oil during the preparation of the emulsion. This effect is due to the compositional ripening process that responds to Ostwald ripening.

In general, the large molar volume of insoluble oils is more stable towards Ostwald ripening as it can provide a kinetic barrier. Further, the nanoemulsions with high distilled oil were also stable to the formation of Ostwald ripening due to low water solubility. In addition, the nanoemulsions with higher oil (Rao and McClements, 2012) and lower surfactant concentration (Eid et al., 2013) are susceptible to Ostwald ripening.

8.4 APPLICATION OF NANOEMULSIONS

Nanoemulsion techniques are commonly applied to the encapsulation of bioactive compounds in aqueous solutions (Anandharamakrishnan, 2014b; Bhushani et al., 2017). This technique is a primary step for making nanoemulsion into dried or powdered form using drying techniques such as spray drying or freeze drying (Ezhilarasi et al., 2013; Parthasarathi and Anandharamakrishnan, 2016). Based on its application of nanoemulsion, the emulsification technique has been preferred. Nanoemulsion offers great potential to encapsulate a high concentration of oil-soluble nutraceuticals or bioactive compounds. Lipophilic active agents such asβ-carotene, plant sterols, carotenoids, fish oil, and dietary fats can be encapsulated and delivered by O/W emulsion, whereas the W/O emulsion method was used to encapsulate water-soluble active agents such as polyphenols, flavonoids, and minerals (Zuidam and Shimoni, 2010). In the following subsection, delivery system and antibacterial activity of nanoemulsion are discussed due to higher usage in recent years.

8.4.1 Nanoemulsion-Based Delivery Systems

Nanoemulsion-based delivery systems are mainly focused on food and pharmaceutical application due to their important properties of encapsulating, protecting, and delivering poorly water-soluble nutraceuticals and drugs through the oral route (McClements and Li, 2010; Chakraborty et al., 2009). In addition, emulsion-based delivery systems are more appropriate for oral applications due to their formulation from edible components (e.g., oil, water, surfactants, and proteins) and lower operating cost (e.g. mixing and homogenization) (Troncoso et al., 2012). Emulsion-based delivery systems are particularly suitable for oral applications because they can be fabricated from edible components (e.g. oil, water, surfactants, and proteins) using economic processing operations (e.g. mixing and homogenization). In addition, a nanoemulsion is an effective approach for improving bioavailability through controlled delivery and sustaining the physicochemical stability of bioactive compounds.

Encapsulation through nanoemulsion-based bioactive compounds is an effective approach for improving bioavailability through controlled delivery and maintaining physicochemical stability. Lipophilic-based bioactive compounds have very poor aqueous solubility, which gives them poor bioavailability. Thus, nanoemulsions have shown a superior capability to increase the bioavailability of lipophilic compounds such as vitamin-E (Parthasarathi and Anandharamakrishnan, 2016), DHA (Walker et al., 2015), fish oil (Belhaj et al., 2012), curcumin (Yu and Huang, 2012), etc.

8.4.2 Antimicrobial Activity

The use of essential oils as natural antimicrobial compounds in food has attracted growing interest in recent years, to meet the consumer's requirements in terms of food quality and safety. The antimicrobial activities of different essential oils and essential oil components have already been proved on different species of microorganisms (Di Pasqua et al., 2007). Nevertheless, direct incorporation of essential oils in food still encounters technological limitations related to the hydrophobic, reactive, and volatile nature of the bioactive molecules constituting essential oils. Nanoemulsions have a unique class of disinfectant property that is produced through the blending of the immiscible liquid phase into an aqueous phase under high pressure. Moreover, this gives a uniform population of high-energy droplets which is kinetically stable and has the properties of whole milk in appearance and consistency (Lawrence and Rees, 2012). Nanoemulsions can act with biocidal efficacy against bacteria, enveloped viruses, and fungi by disrupting their outer membranes. The mixed immiscible liquids of soybean oil and water correspond to the creation of disinfectants that selectively disrupt membranes of bacteria, fungi in dilutions up to 1:1000, and enveloped virus (Lawrence and Rees, 2012; Lee et al., 2010). Ramalingam et al. (2012) investigated the nanoemulsion of soybean oil prepared by microfluidizer at 20,000 psi at room temperature. Further, the minimal inhibitory concentration (MIC) and minimal bactericidal concentration (MBC) of the prepared nanoemulsion was tested in *S. mutans* (MIC 19683 and MBC, 2187 dilutions), *L. casei* (MIC, 2187 and MBC 729 dilutions), *A. viscosus* (MIC 19683 and MBC 2187 dilutions), *C. albicans* (MIC 2187 and MBC 729 dilutions) and mixed culture (MIC 2187 and MBC 243 dilutions). However, higher MIC was observed for *A. viscosus*and *S. mutans*. Authors concluded that nanoemulsion has improved efficacy against microorganisms and a nanoemulsion mean diameter size of 308 nm showed remarkable inhibitory effects on their growth.

Joe et al. (2012a) have studied the preparation of O/W nanoemulsion, which consists of selected cooking oils such as sunflower, castor, coconut, groundnut, and sesame oils and biosurfactant (surfactin). All the different oils were investigated for the development of possible food-grade nanoemulsion formulations. Surfactin-based sunflower nanoemulsion (designated as AUSN-1) was found to be highly transparent with a particle size of 72.5 nm and survived thermodynamic stability tests with recorded low surface tension, viscosity, and density. AUSN-1 showed excellent antibacterial activity against selected food-borne bacterial species, with more prominent effects on *S. typhi*. Moreover, sporicidal and antifungal activities of AUSN-1 were also found to be high. AUSN-1 also showed the ability to be used as a preservative by its capacity to reduce the total cultivable heterotrophic bacterial and fungal populations in selected food products. In conclusion, this study suggests the possible use of AUSN-1 as a preservative in the food industry. Similarly, Joe et al. (2012b) prepared sunflower oil nanoemulsion using polysorbate 80 surfactant and the mean droplet size was 42.52 ± 0.04 nm. Nanoemulsion treatment could be used for short-term storage of fish

(*Scomberomorus guttatus*) when marketing in areas with poor refrigeration facilities. In conclusion, this work supports the possibility of employing nanoemulsions to prevent spoilage of fish products during commercial transportation where temperature fluctuations are unavoidable.

Sugumar et al. (2013) also studied the antimicrobial activity of eucalyptus oil nanoemulsion through ultrasonication method (10 min at 750 W) using Tween 80 emulsifier. The prepared O/W nanoemulsion was bluish in color, highly transparent, less viscous, and the mean droplet size was found to be 16.2 nm. Moreover, different concentrations such as undiluted to 10-, 100-, and 1000-fold dilutions were used to investigate the antibacterial activity in three different microorganisms, namely *Bacillus cereus, Staphylococcus aureus* (gram-positive), and *Escherichia coli* (gram-negative). All the species confirmed 100% bactericidal activity at 10-fold dilution formulation and these results showed in the following order: *B. cereus* at 0th min, *S. aureus* at 15 min, and *E. coli* at 1 h, respectively. Further, authors concluded that 10-fold dilution of the nanoemulsion exhibited that the cytoplasmic content leakage from the bacterial species was higher for *S. aureus* than in *B. cereus* and *E. coli*. However, the optimized nanoemulsion showed inherent antimicrobial activity and it can act as a self-preservative in the food system.

Donsì et al. (2012) optimized the effect of nanoemulsion delivery systems on the antimicrobial activity of different essential oil components, thus carvacrol, limonene, and cinnamaldehyde were encapsulated in the sunflower oil droplets of nanoemulsions. The nanoemulsion was prepared by high-speed stirring at 24,000 rpm for 5 min followed by high-pressure homogenization technique and stabilized by different emulsifiers such as lecithin, pea proteins, sugar ester, and a combination of Tween 20 and glycerol monooleate. The antimicrobial activity was measured against three different microorganisms: *Escherichia coli, Lactobacillus delbrueckii* and *Saccharomyces cerevisiae*, and mean droplet size was achieved between 100 nm and 200 nm. Moreover, the effect of delivery systems on the antimicrobial activity was correlated to the concentration in the aqueous phase of the active molecules and emulsifier capability. These studies revealed that the prepared nanoemulsion-based on sugar esters or on a combination of Tween 20 and glycerol monooleate showed a considerable increase in bactericidal activity over shorter time scales (2 h). Nevertheless, lecithin or pea proteins emulsifiers involving nanoemulsion result in longer time scale (24 h) antimicrobial activity. Authors reported that the selection of emulsifiers depends on the desired function of the antimicrobials within the food system (Table 8.2).

8.5 CASE STUDY: NANOEMULSIONS OF ESSENTIAL OILS

8.5.1 Case Study I

Docosahexaenoic acid (DHA) is an important omega-3 fatty acid responsible for the development of the human brain, the function of the retina, and heart health; it also has beneficial effects on pregnant and lactating women and their infants. DHA is highly unstable towards environmental factors due to its oxidation and physical instability. Therefore, protecting the functional properties of DHA during storage is highly required. To overcome these problems, Karthik and Anandharamakrishnan (2016b) investigated the physicochemical properties and storage stability of DHA nanoemulsions prepared through different emulsification techniques, such as high-speed homogenization (HSH), high-pressure homogenization (HPH), and their combination (HSH+HPH). The HPH and HSH+HPH DHA nanoemulsions produced lower mean particle size than HSH. It was found that the HPH-involved emulsification process (HPH and HSH+HPH) produces stable DHA nanoemulsions in terms of lower particle size, morphology, and other physical stability over 100 days at different storage temperatures (Figure 8.16). Physical stability includes flocculation, creaming, coalescence, phase separation, sedimentation, and oiling off.

The HPH and HSH+HPH DHA exhibited lower lipid oxidation than the emulsion stored at 4 ± 1°C than other conditions (28 ± 1°C and 40 ± 0.2°C). However, HSH+HPH nanoemulsion exhibited higher thermal transition stability than the other emulsification method. Therefore, amongst the

TABLE 8.2
Application of Nanoemulsion in Antimicrobial Activity

Food Components	Target Organisms	Emulsifier	Equipment Used	Size (nm)	Advantages	References
D-limonene, sunflower oil, and palm oil	*Lactobacillus delbrueckii, Saccharomyces cerevisiae*, and *Escherichia coli*	Soy lecithin, Tween 20 glycerol monooleate, and clear gum co 01	High shear homogenizer and high-pressure homogenizer	75 to 175	To enhance the antimicrobial activity of the encapsulated compounds, to evaluate juice preservation, and to extend the shelf life of juices.	Donsi et al. (2011)
Thyme oil and corn oil	*Zygosaccharomyces bailli, Saccharomyces cerevisiae, Brettanomyces bruxellensis and Brettanomyces naardenensis*	Emulsifiers: Tween80, lauric arginate, sodium dodecyl sulfate	Homogenizer and high-pressure homogenizer	196 andr < 85	They are effective antimicrobial agents for utilization in emulsion-based food and beverage; they are also used to improve the overall antimicrobial efficacy	Ziani et al. (2011)
Soybean oil	*Streptococcus mutans*		Microfluidizer	308	Showed effective antibacterial activity against *S. mutans* and is a useful medication in the prevention of dental caries	Karthikeyan et al. (2012)
Sunflower, castor, coconut, groundnut, and sesame oils	Bacterial activity: *Salmonella typhi, Listeria monocytogenes, and Staphylococcus aureus* Fungicidal activity: *Rhizopus nigricans, Aspergillus niger, and Penicillium sp.* Sporicidal activity: *Bacillus cereus and Bacillus circulans*	Biosurfactant: surfactin	–	72.52 to 875.22	To enhance the antimicrobial activity on food products such as raw chicken, apple juice, milk, and mixed vegetable with reduction in the native cultivable bacterial and fungal populations of these products	Joe et al. (2012a)
Sunflower	Heterotrophic bacteria	Polysorbate 80	–	42.52 ± 0.04	To preserve and increase the shelf life of beverages, fish, and other seafood products.	Joe et al. (2012b)
Eucalyptus oil	*Bacillus cereus, Staphylococcus aureus* (gram-positive), and *Escherichia coli* (gram-negative)	Tween 80	Ultrasonicator	17.1	Antibacterial activity against food-borne pathogens	Sugumar et al. (2013)
Sunflower oil, carvacrol, D-limonene and trans-cinnamaldehyde	*Escherichia coli, Lactobacillus delbrueckii* and *Saccharomyces cerevisiae*	Lecithin, pea proteins, sugar ester, and a combination of Tween 20 and glycerol monooleate	High-speed stirrer and high-pressure homogenizer	100 and 200	Improved bacterial activity and delivery system	Donsi et al. (2012)

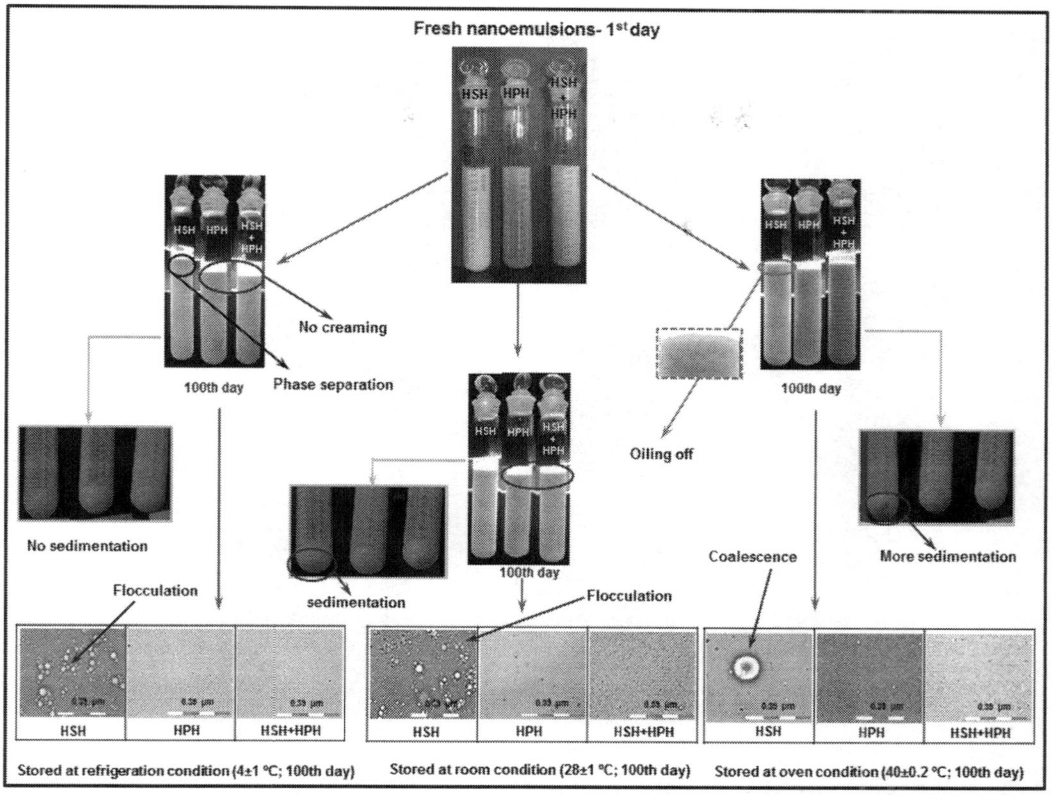

FIGURE 8.16 Physical stability studies of the DHA nanoemulsions during storage for 100 days at different temperature conditions. (Karthik and Anandharamakrishnan, 2016b.)

high-energy techniques, better stability was achieved with the combination of HSH+HPH techniques compared to HPH and HSH. Thus, the formulated DHA nanoemulsion system can be used in food and pharmaceutical applications to enhance the stability and bioavailability of DHA.

8.5.2 Case Study II

Nanoemulsion-based delivery systems are very important to study the digestibility of functional food ingredients under simulated small intestinal fluid (SSIF) conditions. Recently, Karthik et al. (2018) studied the physicochemical fate of DHA nanoemulsions during *in-vitro* small intestine digestion prepared with different emulsifiers such as sodium caseinate, soya lecithin and Tween 40 (Figure 8.17). The DHA nanoemulsions prepared via microfluidization resulted in mean droplet diameter in the range of 148 nm–760 nm. Initially, nanoemulsion using Tween 40 emulsifier showed sustained release of free fatty acids (FFA) (55–60 min) and thereafter it increased rapidly. This initial slow release of % FFA in the Tween 40 nanoemulsion was due to the insignificant effect on the formation of flocculation or coalescence during the *in-vitro* digestion. However, sodium caseinate-based nanoemulsion showed a higher rate of % FFA release for the first 30 min; later the release of % FFA was constant. Interestingly, soya lecithin-based nanoemulsion exhibited an intermediate behavior between those prepared with Tween 40 and sodium caseinate emulsifier. Thus, the overall study confirmed that droplet coalescence is a potential marker for lipid digestibility in DHA nanoemulsions as indicated by its relationship with the free fatty acid (FFA) release. Moreover, this fundamental idea about nanoemulsion during *in-vitro* digestion can be useful in the controlled release of bioactive compounds or drug molecules in the food and pharmaceutical sectors.

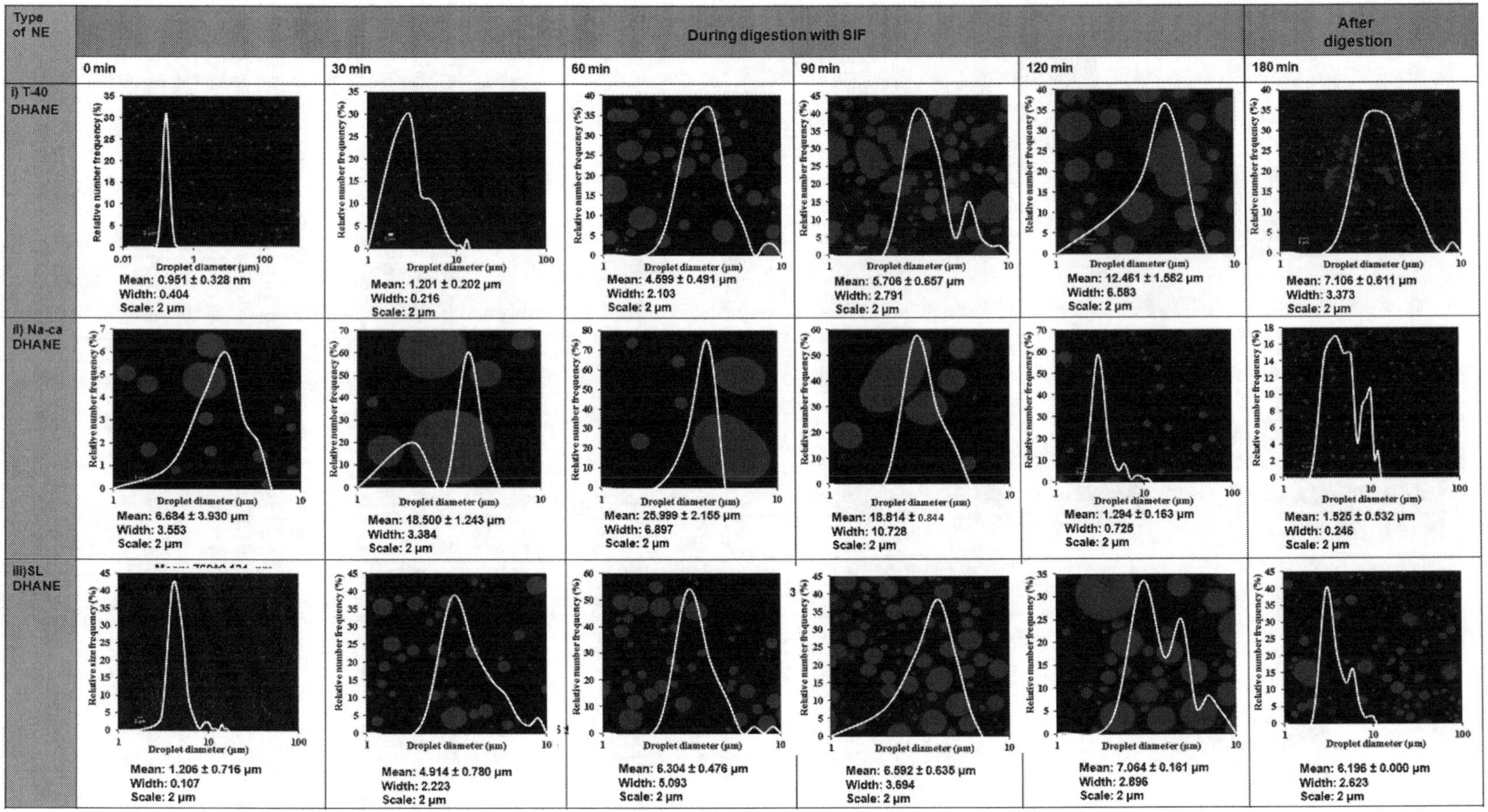

FIGURE 8.17 Confocal images of DHA nanoemulsions during *in-vitro* lipid digestion stabilized by different emulsifiers. (Karthik et al., 2018.)

8.6 FUTURE TRENDS

Nanoemulsion has shown a lot of potential in improving the effectiveness of delivery of food bioactive compounds to the benefit of human health. Presently, different kinds of techniques have been used for the preparation of nanoemulsion with their own merits and demerits. Nanoemulsions can contribute to the development of novel food products due to their specific characteristics and higher stability compared to conventional emulsions. Fortification of highly sensitive food bioactives, especially lipophilic compounds in the different food systems (beverage, milk, fruit juice, bread, health mix, etc.) has become more convenient than the nanoemulsion method. The enhanced stability and unique morphological behavior of nanoemulsion can be useful in the food, agriculture, and pharmaceutical industries, based on their specific requirements.

REFERENCES

Abbas, Shabbar, Khizar Hayat, Eric Karangwa, Mohanad Bashari, and Xiaoming Zhang. 2013. "An overview of ultrasound-assisted food-grade nanoemulsions." *Food Engineering Reviews,* 5 (3):139–157.

Ahmed, Kashif, Yan Li, David Julian McClements, and Hang Xiao. 2012. "Nanoemulsionand emulsion-based delivery systems for curcumin: Encapsulation and release properties." *Food Chemistry,* 132 (2):799–807.

Anandharamakrishnan, C. 2014a. "Liquid-based nanoencapsulation techniques." In *Techniques for Nanoencapsulation of Food Ingredients*: 29–41. New York: Springer.

Anandharamakrishnan, C. 2014b. *Techniques for Nanoencapsulation of Food Ingredients*: New York: Springer.

Anandharamakrishnan, C. and S. Padma Ishwarya. 2015. "Introduction to encapsulation of food ingredients." *Spray Drying Techniques for Food Ingredient Encapsulation*: 37–64. Chicago: John Wiley & Sons.

Anandharamakrishnan, C. and K. Usha Kiran. 2014. "Bionanocomposites and their potential applications in food packaging." *Polymers for Packaging Applications*: 229–262. Boca Raton, FL: CRC Press.

Anton, Nicolas and Thierry F. Vandamme. 2009. "The universality of low-energy nano-emulsification." *International Journal of Pharmaceutics*, 377(1–2):142–147.

Barras, A., A. Mezzetti, A. Richard, S. Lazzaroni, S. Roux, P. Melnyk, D. Betbeder, and N. Monfilliette-Dupont. 2009. "Formulation and characterization of polyphenol-loaded lipid nanocapsules." *International Journal of Pharmaceutics*, 379 (2):270–277.

Belhaj, Nabila, François Dupuis, Elmira Arab-Tehrany, Frédéric M. Denis, Cédric Paris, Isabelle Lartaud, and Michel Linder. 2012. "Formulation, characterization and pharmacokinetic studies of coenzyme Q10 PUFA's nanoemulsions." *European Journal of Pharmaceutical Sciences,* 47 (2):305–312.

Bhushani, Anu and C. Anandharamakrishnan. 2017. "Food-grade nanoemulsions for protection and delivery of nutrients." In *Nanoscience in Food and Agriculture*, 4:99–139. Switzerland: Springer.

Bhushani, J. Anu, P. Karthik, and C.Anandharamakrishnan. 2016. "Nanoemulsion based delivery system for improved bioaccessibility and Caco-2 cell monolayer permeability of green tea catechins." *Food Hydrocolloids,* 56:372–382.

Bhushani, Anu, Udayakumar Harish, and Chinnaswamy Anandharamakrishnan. 2017. "Nanodelivery of nutrients for improved bioavailability." In *Nutrient Delivery*: 369–411. Amsterdam: Elsevier.

Bouchemal, Kawthar, S. Briançon, E. Perrier, and H. Fessi. 2004. "Nano-emulsion formulation using spontaneous emulsification: Solvent, oil and surfactant optimisation." *International Journal of Pharmaceutics,* 280 (1–2):241–251.

Britten, Michel and Hélène J. Giroux. 1991. "Emulsifying properties of whey protein and casein composite blends1." *Journal of Dairy Science,* 74 (10):3318–3325.

Capek, Ignac. 2004. "Degradation of kinetically-stable o/w emulsions." *Advances in Colloid and Interface Science*, 107 (2–3):125–155.

Chakraborty, Subhashis, Dali Shukla, Brahmeshwar Mishra, and Sanjay Singh. 2009. "Lipid–an emerging platform for oral delivery of drugs with poor bioavailability." *European Journal of Pharmaceutics and Biopharmaceutics,* 73 (1):1–15.

Chen, Cen, Wei Yang, Dan-Tong Wang, Chao-Long Chen, Qing-Ye Zhuang, and Xiang-Dong Kong. 2014. "A modified spontaneous emulsification solvent diffusion method for the preparation of curcumin-loaded PLGA nanoparticles with enhanced in vitro anti-tumor activity." *Frontiers of Materials Science,* 8 (4):332–342.

Chuesiang, Piyanan, Ubonrat Siripatrawan, Romanee Sanguandeekul, Lynne McLandsborough, and David Julian McClements.2018. "Optimization of cinnamon oil nanoemulsions using phase inversion temperature method: Impact of oil phase composition and surfactant concentration." *Journal of Colloid and Interface Science,* 514:208–216.

Cook, Edward J and Arthur P. Lagace. 1985. Apparatus for forming emulsions. Google Patents. https://patents.google.com/patent/US4533254A/en.

da Costa Silva, Lívia Maria, Vânia Paula Salvianodos Santos, Andrea Medeiros Salgado, and Karen Signori Pereira. 2013. "Biosensors for contaminants monitoring in food and environment for human and environmental health." In *State of the Art in Biosensors-Environmental and Medical Applications* Ed: Toonika Rinken. London: InTech.

Dalgleish, D.G., S.M. Tosh, and S. West. 1996. "Beyond homogenization: The formation of very small emulsion droplets during the processing of milk by a microfluidizer." *Netherlands Milk and Dairy Journal,* 50 (2):135-148.

Davidov-Pardo, Gabriel and David Julian McClements. 2015. "Nutraceutical delivery systems: Resveratrol encapsulation in grape seed oil nanoemulsions formed by spontaneous emulsification." *Food Chemistry,* 167:205–212.

Delmas, Thomas, Hélene Piraux, Anne-Claude Couffin, Isabelle Texier, Françoise Vinet, Philippe Poulin, Michael E. Cates, and Jérôme Bibette. 2011. "How to prepare and stabilize very small nanoemulsions." *Langmuir,* 27 (5):1683–1692.

Desrumaux, Anne and Julie Marcand. 2002. "Formation of sunflower oil emulsions stabilized by whey proteins with high-pressure homogenization (up to 350 MPa): Effect of pressure on emulsion characteristics." *International Journal of Food Science and Technology,* 37 (3):263–269.

DiPasqua, Rosangela, Gail Betts, Nikki Hoskins, Mike Edwards, Danilo Ercolini, and Gianluigi Mauriello. 2007. "Membrane toxicity of antimicrobial compounds from essential oils." *Journal of Agricultural and Food Chemistry,* 55 (12):4863–4870.

Dickinson, Eric. 2003. "Hydrocolloids at interfaces and the influence on the properties of dispersed systems." *Food Hydrocolloids,* 17 (1):25–39.

Donsì, Francesco, Marianna Annunziata, Mariarenata Sessa, and Giovanna Ferrari. 2011. "Nanoencapsulation of essential oils to enhance their antimicrobial activity in foods." *LWT-Food Science and Technology,* 44 (9):1908–1914.

Donsì, Francesco, Marianna Annunziata, Mariarosaria Vincensi, and Giovanna Ferrari. 2012. "Design of nanoemulsion-based delivery systems of natural antimicrobials: Effect of the emulsifier." *Journal of Biotechnology,* 159 (4):342–350.

Eid, Ahmad Mustafa Masoud, NAE lmarzugi, and Hesham Ali El-Enshasy. 2013. "Preparation and evaluation of olive oil nanoemulsion using sucrose monoester." *International Journal of Pharmacy and Pharmaceutical Sciences,* 5 (Suppl 3):434–440.

Elgammal, Mahmoud, Reinhold Schneider, and Michael Gradzielski. 2015. "Preparation of latex nanoparticles using nanoemulsions obtained by the phase inversion composition (PIC) method and their application in textile printing." *Colloids and Surfaces A: Physicochemical and Engineering Aspects,* 470:70–79.

Esmaeili, Farnaz, Fatemeh Atyabi, and Rassoul Dinarvand. 2007. "Preparation of PLGA nanoparticles using TPGS in the spontaneous emulsification solvent diffusion method." *Journal of Experimental Nanoscience,* 2 (3):183–192.

Ezhilarasi, P.N., P. Karthik, N. Chhanwal, and C. Anandharamakrishnan. 2013. "Nanoencapsulation techniques for food bioactive components: A review." *Food and Bioprocess Technology,* 6 (3):628–647.

Fessi, HPFD, François Puisieux, J. Ph. Devissaguet, Nicolas Ammoury, and Simon Benita. 1989. "Nanocapsule formation by interfacial polymer deposition following solvent displacement." *International Journal of Pharmaceutics,* 55 (1):R1–R4.

Ganachaud, Francois and Joseph L. Katz. 2005. "Nanoparticles and nanocapsules created using the ouzo effect: Spontaneous emulsification as an alternative to ultrasonic and high-shear devices." *Chem Phys Chem,* 6 (2):209–216.

Gulotta, Alessandro, Amir Hossein Saberi, Maria Cristina Nicoli, and David Julian McClements.2014. "Nanoemulsion-based delivery systems for polyunsaturated (ω-3) oils: Formation using a spontaneous emulsification method." *Journal of Agricultural and Food Chemistry,* 62 (7):1720–1725.

Huimin, Xing, Li Lin, Gui Shilin, Walid Elfalleh, He Shenghua, Sheng Qinghai, and Ma Ying. 2014. "Formation, stability, and properties of an algae oil emulsion for application in UHT milk." *Food and Bioprocess Technology,* 7 (2):567–574.

Jafari, Seid Mahdi, Yinghe He, and Bhesh Bhandari. 2007. "Optimization of nano-emulsions production by microfluidization." *European Food Research and Technology,* 225 (5–6):733–741.

Joe, Manoharan Melvin, K. Bradeeba, Rengasamy Parthasarathi, Palanivel Karpagavinaya Sivakumaar, Puneet Singh Chauhan, Sherlyn Tipayno, Abitha Benson, and Tongmin S.A. 2012a. "Development of surfactin based nanoemulsion formulation from selected cooking oils: evaluation for antimicrobial activity against selected food associated microorganisms." *Journal of the Taiwan Institute of Chemical Engineers,* 43 (2):172–180.

Joe, Manoharan Melvin, Puneet Singh Chauhan, K. Bradeeba, Charlotte Shagol, Palanivel Karpagavinaya Sivakumaar, and Tongmin SA. 2012b. "Influence of sunflower oil based nanoemulsion (AUSN-4) on the shelf life and quality of Indo-Pacific king mackerel (Scomberomorus guttatus) steaks stored at 20 C." *Food Control,* 23 (2):564–570.

Kabalnov, Alexey. 2001. "Ostwald ripening and related phenomena." *Journal of Dispersion Science and Technology,* 22 (1):1–12.

Kaltsa, O., I. Gatsi, S. Yanniotis, and I. Mandala. 2014. "Influence of ultrasonication parameters on physical characteristics of olive oil model emulsions containing xanthan." *Food and Bioprocess Technology,* 7 (7):2038–2049.

Kaminsky, V.A., A.V. Vyaz'min, N.N. Kulov, and V.V. Dil'man. 1998. "Marangoni effect in the presence of bulk turbulence." *Chemical Engineering Science,* 53 (19):3347–3353.

Karthik P., and C. Anandharamakrishnan. 2013. "Microencapsulation of docosahexaenoic acid by spray-freeze-drying method and comparison of its stability with spray-drying and freeze-drying methods." *Food and Bioprocess Technology,* 6 (10):2780–2790.

Karthik, P. and C. Anandharamakrishnan. 2016a. "Enhancing omega-3 fatty acids nanoemulsion stability and in-vitro digestibility through emulsifiers." *Journal of Food Engineering,* 187:92–105.

Karthik, P. and C. Anandharamakrishnan. 2016b. "Fabrication of a nutrient delivery system of docosahexaenoic acid nanoemulsions via high energy techniques." *RSC Advances,* 6 (5):3501–3513.

Karthik, P., P.N. Ezhilarasi, and C. Anandharamakrishnan. 2017. "Challenges associated in stability of food grade nanoemulsions." *Critical Reviews in Food Science and Nutrition,* 57 (7):1435–1450.

Karthik, P., S. Padma Iswarya, and C. Anandharamakrishnan. 2018. "Droplet coalescence as a potential marker for physicochemical fate of nanoemulsions during in-vitro small intestine digestion." *Colloids and Surfaces A: Physicochemical and Engineering Aspects,* 553:278–287.

Kentish, S., T.J. Wooster, M. Ashokkumar, S. Balachandran, R. Mawson, and L. Simons. 2008. "The use of ultrasonics for nanoemulsion preparation." *Innovative Food Science and Emerging Technologies,* 9 (2):170–175.

Khademosseini, Ali and Robert Langer. 2006. "Applications in drug delivery and tissue engineering." *Chemical Engineering Progress,* 102 (2):38–42.

Komaiko, Jennifer. 2016. "Optimization of the fabrication, stability, and performance of food grade nanoemulsions with low and high energy methods." Doctoral dissertation, University of Massachusetts, Amherst.

Komaiko, Jennifer and David Julian McClements. 2015. "Low-energy formation of edible nanoemulsions by spontaneous emulsification: Factors influencing particle size." *Journal of Food Engineering,* 146:122–128.

Komaiko, Jennifer and David Julian McClements. 2016. "Formation of food-grade nanoemulsions using low-energy preparation methods: A review of available methods." *Comprehensive Reviews in Food Science and Food Safety,* 15 (2):331–352.

Kong, Ming and Hyun Jin Park. 2011. "Stability investigation of hyaluronic acid based nanoemulsion and its potential as transdermal carrier." *Carbohydrate Polymers,* 83 (3):1303–1310.

Lawrence, M. Jayne and Gareth D. Rees. 2012. "Microemulsion-based media as novel drug delivery systems." *Advanced Drug Delivery Reviews,* 64:175–193.

Lee, Valerie A., Ramalingam Karthikeyan, H. Ralph Rawls, and Bennett T. Amaechi. 2010. "Anti-cariogenic effect of a cetylpyridinium chloride-containing nanoemulsion." *Journal of Dentistry,* 38 (9):742–749.

Leong, T.S.H., T.J. Wooster, S.E. Kentish, and M. Ashokkumar. 2009. "Minimising oil droplet size using ultrasonic emulsification." *Ultrasonics Sonochemistry,* 16 (6):721–727.

Li, Yan, Solene Le Maux, Hang Xiao, and David Julian McClements. 2009. "Emulsion-based delivery systems for tributyrin, a potential colon cancer preventative agent." *Journal of Agricultural and Food Chemistry,* 57 (19):9243–9249.

Li, Yan, Jinkai Zheng, Hang Xiao, and David Julian McClements. 2012. "Nanoemulsion-based delivery systems for poorly water-soluble bioactive compounds: Influence of formulation parameters on polymethoxyflavone crystallization." *Food Hydrocolloids,* 27 (2):517–528.

Li, Yanan, Zijie Zhang, Qipeng Yuan, Hao Liang, and Frank Vriesekoop. 2013. "Process optimization and stability of d-limonene nanoemulsions prepared by catastrophic phase inversion method." *Journal of Food Engineering,* 119 (3):419–424.

Li, Jinglei, In-Cheon Hwang, Xiguang Chen, and Hyun Jin Park. 2016. "Effects of chitosan coating on curcumin loaded nano-emulsion: Study on stability and in vitro digestibility." *Food Hydrocolloids,* 60:138–147.

Liang, Rong, Charles F. Shoemaker, Xiaoqing Yang, Fang Zhong, and Qingrong Huang. 2013. "Stability and bioaccessibility of β-carotene in nanoemulsions stabilized by modified starches." *Journal of Agricultural and Food Chemistry,* 61 (6):1249–1257.

Lieberman, Herbert A., Martin M. Rieger, and Gilbert S. Banker. 1998. *Pharmaceutical Dosage Forms: Disperse Systems*. New York: Marcel Dekker.

Lifshitz, Ilya M. and Vitaly V. Slyozov. 1961. "The kinetics of precipitation from supersaturated solid solutions." *Journal of Physics and Chemistry of Solids*, 19 (1–2):35–50.

Lovelyn, Charles and Anthony A. Attama. 2011. "Current state of nanoemulsions in drug delivery." *Journal of Biomaterials and Nanobiotechnology*, 2 (5):626.

Maa, Yuh-Fun and Chung C. Hsu. 1999. "Performance of sonication and microfluidization for liquid–liquid emulsification." *Pharmaceutical Development and Technology,* 4 (2):233–240.

Mahdi Jafari, Seid, Yinghe He, and Bhesh Bhandari. 2006. "Nano-emulsion production by sonication and microfluidization—a comparison." *International Journal of Food Properties,* 9 (3):475–485.

Mao, Like, Duoxia Xu, Jia Yang, Fang Yuan, Yanxiang Gao, and Jian Zhao. 2009. "Effects of small and large molecule emulsifiers on the characteristics of β-carotene nanoemulsions prepared by high pressure homogenization." *Food Technology and Biotechnology*, 47 (3):336–342.

McClements, David Julian. 2005. "Theoretical analysis of factors affecting the formation and stability of multilayered colloidal dispersions." *Langmuir,* 21:9777–9785.

McClements, David Julian. 2011. "Edible nanoemulsions: Fabrication, properties, and functional performance." *Soft Matter,* 7 (6):2297–2316.

McClements, David Julian. 2015. *Food Emulsions: Principles, Practices, and Techniques*. Boca Raton, FL: CRC press.

McClements, David Julian, and Yan Li. 2010. "Structured emulsion-based delivery systems: Controlling the digestion and release of lipophilic food components." *Advances in Colloid and Interface Science,* 159 (2):213–228.

McClements, David Julian and Jiajia Rao. 2011. "Food-grade nanoemulsions: Formulation, fabrication, properties, performance, biological fate, and potential toxicity." *Critical Reviews in Food Science and Nutrition,* 51 (4):285–330.

Miller, Clarence A. 1988. "Spontaneous emulsification produced by diffusion – A review." *Colloids and Surfaces,* 29 (1):89–102.

Meleson, Kieche, Sara Graves, and Thomas G. Mason. 2004. "Formation of concentrated nanoemulsions by extreme shear." *Soft Materials*, 2 (2–3):109–123.

Mora-Huertas, C.E., H. Fessi, and A. Elaissari. 2011. "Influence of process and formulation parameters on the formation of submicron particles by solvent displacement and emulsification–diffusion methods: Critical comparison." *Advances in Colloid and Interface Science,* 163 (2):90–122.

Moschakis, Thomas, Brent S. Murray, and Costas G. Biliaderis. 2010. "Modifications in stability and structure of whey protein-coated o/w emulsions by interacting chitosan and gum arabic mixed dispersions." *Food Hydrocolloids*, 24 (1):8–17.

Nagavarma, B.V.N., Hemant K.S. Yadav, A. Ayaz, L.S. Vasudha, and H.G. Shivakumar. 2012. "Different techniques for preparation of polymeric nanoparticles–a review." *Asian Journal of Pharmaceutical and Clinical Research,* 5 (3):16–23.

Neethirajan, Suresh and Digvir S. Jayas. 2011. "Nanotechnology for the food and bioprocessing industries." *Food and Bioprocess Technology,* 4 (1):39–47.

Oliveira, Davi Rocha Bernardes, Mariano Michelon, Guilhermede Figueiredo Furtado, Rita Sinigaglia-Coimbra, and Rosiane Lopes Cunha. 2016. "β-Carotene-loaded nanostructured lipid carriers produced by solvent displacement method." *Food Research International,* 90:139–146.

Olson, D.W., C.H. White, and R.L. Richter. 2004. "Effect of pressure and fat content on particle sizes in microfluidized milk." *Journal of Dairy Science,* 87 (10):3217–3223.

Ostrovsky, Mikhail V. and Rafail M. Ostrovsky. 1983. "Dynamic interfacial tension in binary systems and spontaneous pulsation of individual drops by their dissolution." *Journal of Colloid and Interface Science,* 93 (2):392–401.

Pan, Hong, Lijie Yu, Jian Xu, and Dejun Sun. 2014. "Preparation of highly stable concentrated W/O nanoemulsions by PIC method at elevated temperature." *Colloids and Surfaces A: Physicochemical and Engineering Aspects,* 447:97–102.

Parthasarathi, S. and C. Anandharamakrishnan. 2016. "Enhancement of oral bioavailability of vitamin E by spray-freeze drying of whey protein microcapsules." *Food and Bioproducts Processing,* 100:469–476.

Pinnamaneni, S., N.G. Das, and S.K. Das. 2003. "Comparison of oil-in-water emulsions manufactured by microfluidization and homogenization." *Die Pharmazie – An International Journal of Pharmaceutical Sciences*, 58 (8):554–558.

Qian, Cheng and David Julian McClements.2011. "Formation of nanoemulsions stabilized by model food-grade emulsifiers using high-pressure homogenization: Factors affecting particle size." *Food Hydrocolloids,* 25 (5):1000–1008.

Qian, Cheng, Eric Andrew Decker, Hang Xiao, and David Julian McClements. 2012. "Nanoemulsion delivery systems: Influence of carrier oil on β-carotene bioaccessibility." *Food Chemistry,* 135 (3):1440–1447.

Quintanilla-Carvajal, María Ximena, Brenda Hildeliza Camacho-Díaz, Lesvia Sofia Meraz-Torres, José Jorge Chanona-Pérez, Liliana Alamilla-Beltrán, Antonio Jimenéz-Aparicio, and Gustavo F. Gutiérrez-López. 2010. "Nanoencapsulation: Anew trend in food engineering processing." *Food Engineering Reviews,* 2 (1):39–50.

Ramalingam, Karthikeyan, Bennett T. Amaechi, Rawls H. Ralph, and Valerie A. Lee. 2012. "Antimicrobial activity of nanoemulsion on cariogenic planktonic and biofilm organisms." *Archives of Oral Biology,* 57 (1):15–22.

Rao, Jiajia, and David Julian McClements. 2010. "Stabilization of phase inversion temperature nanoemulsions by surfactant displacement." *Journal of Agricultural and Food Chemistry,* 58 (11):7059–7066.

Rao, Jiajia and David Julian McClements. 2012. "Impact of lemon oil composition on formation and stability of model food and beverage emulsions." *Food Chemistry,* 134 (2):749–757.

Robins, Margaret M. 2000. "Emulsions—creaming phenomena." *Current Opinion in Colloid andInterface Science*, 5 (5–6):265–272.

Salvia-Trujillo, Laura, M. Alejandra Rojas-Graü, Robert Soliva-Fortuny, and Olga Martín-Belloso. 2014. "Impact of microfluidization or ultrasound processing on the antimicrobial activity against Escherichia coli of lemongrass oil-loaded nanoemulsions." *Food Control,* 37:292–297.

Sanguansri, Peerasak and Mary Ann Augustin. 2006. "Nanoscale materials development–a food industry perspective." *Trends in Food Science and Technology,* 17 (10):547–556.

Santana-Solano, Jesús, Carla M. Quezada, Sandra Ozuna-Chacón, and José Luis Arauz-Lara. 2012. "Spontaneous emulsification at the water/oil interface." *Colloids and Surfaces A: Physicochemical and Engineering Aspects,* 399:78–82.

Schrade, Anika, Zhihai Cao, Katharina Landfester, and Ulrich Ziener. 2011. "Preparation of raspberry-like nanocapsules by the combination of pickering emulsification and solvent displacement technique." *Langmuir,* 27 (11):6689–6700.

Setya, Sonal, Sushama Talegaonkar, and B.K. Razdan. 2014. "Nanoemulsions: Formulation methods and stability aspects." *World Journal of Pharmacy and Pharmaceutical Sciences,* 3 (2):2214–2228.

Shu, Gaofeng, Nauman Khalid, Tai Boon Tan, Yiguo Zhao, Marcos A. Neves, Isao Kobayashi, and Mitsutoshi Nakajima. 2017. "Comparison of ergocalciferol nanodispersions prepared using modified lecithin and sodium caseinate: Insights of formulation, stability and bioaccessibility." *Journal of Functional Foods,* 38:28–35.

Silva, Hélder Daniel, Miguel Ângelo Cerqueira, and António A.Vicente. 2012. "Nanoemulsions for food applications: Development and characterization." *Food and Bioprocess Technology*, 5 (3):854–867.

Silvestre, M.P.C., E.A. Decker, and D.J. McClements. 1999. "Influence of copper on the stability of whey protein stabilized emulsions." *Food Hydrocolloids,* 13 (5):419–424.

Solans, Conxita, Daniel Morales, and Maria Homs. 2016. "Spontaneous emulsification." *Current Opinion in Colloid andInterface Science,* 22:88–93.

Su, Dan and Qixin Zhong. 2016. "Lemon oil nanoemulsions fabricated with sodium caseinate and Tween 20 using phase inversion temperature method." *Journal of Food Engineering,* 171:214–221.

Sugumar, Saranya, Joyce Nirmala, Vijayalakshmi Ghosh, Haridasan Anjali, Amitava Mukherjee, and Natarajan Chandrasekaran. 2013. "Bio-based nanoemulsion formulation, characterization and antibacterial activity against food-borne pathogens." *Journal of Basic Microbiology*, 53 (8):677–685.

Sugumar, Saranya, Amitava Mukherjee, and Natarajan Chandrasekaran. 2015. "Nanoemulsion formation and characterization by spontaneous emulsification: Investigation of its antibacterial effects on Listeria monocytogenes." *Asian Journal of Pharmaceutics*, 9 (1):23-28.

Surh, Jeonghee, Eric Andrew Decker, and David Julian McClements. 2017. "Utilisation of spontaneous emulsification to fabricate lutein-loaded nanoemulsion-based delivery systems: Factors influencing particle size and colour." *International Journal of Food Science andTechnology*, 52 (6):1408–1416.

Tadros, Tharwat. 2004. "Application of rheology for assessment and prediction of the long-term physical stability of emulsions." *Advances in Colloid and Interface Science,* 108:227–258.

Tadros, T.F. 2009. *Emulsion Science and Technology.* Weinheim: Wiley-VCH Verlag GmbH & Co.

Tan, Tai Boon, Nor Shariffa Yussof, Faridah Abas, Hamed Mirhosseini, Imededdine Arbi Nehdi, and Chin Ping Tan. 2016. "Forming a lutein nanodispersion via solvent displacement method: The effects of processing parameters and emulsifiers with different stabilizing mechanisms." *Food Chemistry,* 194:416–423.

Taylor, Geoffrey Ingram. 1934. "The formation of emulsions in definable fields of flow." *Proceedings of the Royal Society of London A*, 146 (858):501–523.

Teo, Brian Sheng Xian, Mahiran Basri, Mohd Rezuwan Shah Zakaria, Abu Bakar Salleh, Raja Noor Zaliha, Raja Abdul Rahman, and Mohd Basyaruddin Abdul Rahman. 2010. "A potential tocopherol acetate loaded palm oil esters-in-water nanoemulsions for nanocosmeceuticals." *Journal of Nanobiotechnology*, 8 (1):4.

Troncoso, Elizabeth, José Miguel Aguilera, and David Julian McClements. 2012. "Fabrication, characterization and lipase digestibility of food-grade nanoemulsions." *Food Hydrocolloids*, 27 (2):355–363.

Walker, Rebecca, Eric A. Decker, and David Julian McClements. 2015. "Development of food-grade nanoemulsions and emulsions for delivery of omega-3 fatty acids: Opportunities and obstacles in the food industry." *Food and Function*, 6 (1):41–54.

Walstra, Pieter. 1983. "Formation of emulsions." In Paul Belcher (ed.), *Encyclopedia of Emulsion Technology, vol. 1. Basic Theory:* 57–128. New York: Marcel Decker.

Washington, C. 1987. "Emulsion production by microfluidizer." *Lab Equipment Designed,* 85:69–71.

Wooster, Tim J., Matt Golding, and Peerasak Sanguansri. 2008. "Impact of oil type on nanoemulsion formation and Ostwald ripening stability." *Langmuir*, 24 (22):12758–12765.

Wulff-Pérez, M., A. Torcello-Gómez, M.J. Gálvez-Ruíz, and A. Martín-Rodríguez. 2009. "Stability of emulsions for parenteral feeding: Preparation and characterization of o/w nanoemulsions with natural oils and Pluronic f68 as surfactant." *Food Hydrocolloids*, 23 (4):1096–1102.

Yu, Hailong and Qingrong Huang. 2012. "Improving the oral bioavailability of curcumin using novel organogel-based nanoemulsions." *Journal of Agricultural and Food Chemistry,* 60 (21):5373–5379.

Zhang, Jia, Shufang Nie, and Shu Wang. 2013. "Nanoencapsulation enhances epigallocatechin-3-gallate stability and its antiatherogenic bioactivities in macrophages." *Journal of Agricultural and Food Chemistry,* 61 (38):9200–9209.

Zhang, Ruojie, Zipei Zhang, Taha Kumosani, Sawsan Khoja, Khalid Omer Abualnaja, and David Julian McClements. 2016. "Encapsulation of β-carotene in nanoemulsion-based delivery systems formed by spontaneous emulsification: Influence of lipid composition on stability and bioaccessibility." *Food Biophysics,* 11 (2):154–164.

Ziani, Khalid, Yuhua Chang, Lynne McLandsborough, and David Julian McClements.2011. "Influence of surfactant charge on antimicrobial efficacy of surfactant-stabilized thyme oil nanoemulsions." *Journal of Agricultural and Food Chemistry,* 59 (11):6247–6255.

Züge, Luana Carolina Bosmuler, Charles Windson Isidoro Haminiuk, Giselle Maria Maciel, Joana Léa Meira Silveira, and Agnes de Paula Scheer. 2013. "Catastrophic inversion and rheological behavior in soy lecithin and Tween 80 based food emulsions." *Journal of Food Engineering*, 116 (1):72–77.

Zuidam, Nicolaas Jan, and Eyal Shimoni. 2010. "Overview of microencapsulates for use in food products or processes and methods to make them." In N.J. Zuidam and Viktor Nedovic (eds.), *Encapsulation Technologies for Active Food Ingredients and Food Processing*: 3–29. New York: Springer.

9 Electrospraying and Spinning Techniques

Fabrication and Its Potential Applications

Maria Leena, K.S. Yoha, J.A. Moses, and C. Anandharamakrishnan

CONTENTS

9.1 INTRODUCTION

Nanotechnology has emerged as a technological advancement that deals with the manipulation and understanding of materials at atomic and molecular levels, generally with structures less than 100 nm in size. Materials with at least one dimension, and internal or external structure at the nanoscale are considered as nanomaterials. At the nanoscale level, material properties are governed by quantum mechanics, and its physical, chemical, mechanical, and biological properties (such as color, solubility, chemical reactivity, optical properties, conductivity, strength, and toxicity) are quite different from their macroscopic counterparts (Weiss et al., 2006; Pathakoti et al., 2017). Nanotechnology has revolutionized all fields including material science, electronics, tissue engineering, pharmaceutical, cosmetics, and recently food processing and nutraceuticals. Though its entry into food processing is relatively new, it created remarkable impact on almost all stages from farm to fork, including agriculture and food production, improving health benefits of foods, food processing, taste enhancement, food packaging, food safety, etc. (Peters et al., 2014; Peters et al., 2016). Various nanometerials in the form of zero dimensional (e.g. quantum dots), 1D (e.g. nanofibers, nanotubes), 2D (e.g. thin films), and 3D (e.g. nano-patterned bulk structures) materials are widely used to get new functionalities and efficient food processing (Ravichandran, 2010; Duncan, 2014; Fraceto et al., 2016). Among them, nanomaterials in the form of fibers and particles/capsules are of interest in this chapter.

Nanomaterials are an integral part of the natural food system, as the characteristic properties of many foods rely on nanometer-size components. For example, proteins, carbohydrates, and fats, whose basic unit is at the nanoscale level, self-assemble into higher-order structures and form functional compounds (Peters et al., 2016). Recent technological developments lead the way for the development of engineered nanomaterials. These could be finely divided forms of existing ingredients, or completely novel chemical structures. Engineered nanomaterials are synthesized by chemical, physical, and biological routes using numerous inorganic, organic, and hybrid nanomaterials. The electrohydrodynamic process described in this chapter is a kind of physical synthesis method which involves sister technologies called electrospinning and electrospraying to form 1D nanofibers and 3D spheres (fine aerosols), respectively. It provides various advantages including (a) continuous synthesis of nanomaterials with high reproducibility offered by electrostatic driven process; (b) production of high purity nanomaterials; (c) processing at ambient condition (does not involve heat or pressure); (d) possibility to form nanomaterials from wide range of organic/inorganic and hybrid materials; (e) possibility to form diverse architecture including solid spherical particles, core-shell structures, fibers, porous fibers, aligned structures, hollow structures, 3D structures, etc. with simple modification in spinneret or collector design and tuning composition of precursor solution (Zong et al., 2017).

The ambient processing condition of this process makes it ideal for encapsulation of bioactive molecules, enzymes, and live cells that are sensitive to heat, light, and pressure (Bock et al., 2012; Leena et al., 2017; Wen et al., 2017). The high surface to volume ratio of the electrospray nanoparticles and electrospun nanofiber offers more space for interaction with their surrounding medium. Additionally, electrospun nanofibers have interconnected and nanoscale porosity, tunable mechanical and electrical property, extreme strength, etc., which make it ideal for many applications. The electrohydrodynamic process also allows the formation of nanostructures from a wide range of natural polymers (as listed in Figure 9.1) and synthetic polymers such as polyvinyl alcohol (PVA), polyethylene glycol (PEG), polyethylene oxide (PEO), polylactic acid (PLA), poly(lactic-co-glycolic acid) (PLGA), polyvinyl pyrrolidone (PVP), Nylon 6,6, polyurethanes (PU), polyaniline (PANI), etc. This basic nanostructure can be further functionalized with reinforcement of bioactive molecules, nanosized metal, clay, ceramic, minerals, polymeric particles, carbon-based nanomaterials (CNT), etc. Due to these overwhelming advantages, electrosprayed nanoparticles and electrospun nanofibers are used in many applications including controlled bioactive molecule release (Leena et al., 2017), drug delivery, enzyme immobilization, stem cell engineering (Rana et al., 2016, 2017), tissues engineering (Hasan et al., 2014; Pelipenko et al., 2015), sensing (Ding et al., 2009; Aussawasathien et al., 2005), filtration, and food packaging applications. Considering its significant properties and advantages, this chapter essentially focuses on the potential applications of electrospraying and electrospinning in the food processing field.

9.2 ELECTROHYDRODYNAMIC PROCESS

The electrohydrodynamic process is a technique in which the charged liquid is atomized into uniform fine jets or atomized to fine particles using electrohydrodynamic force (Xie et al., 2015;

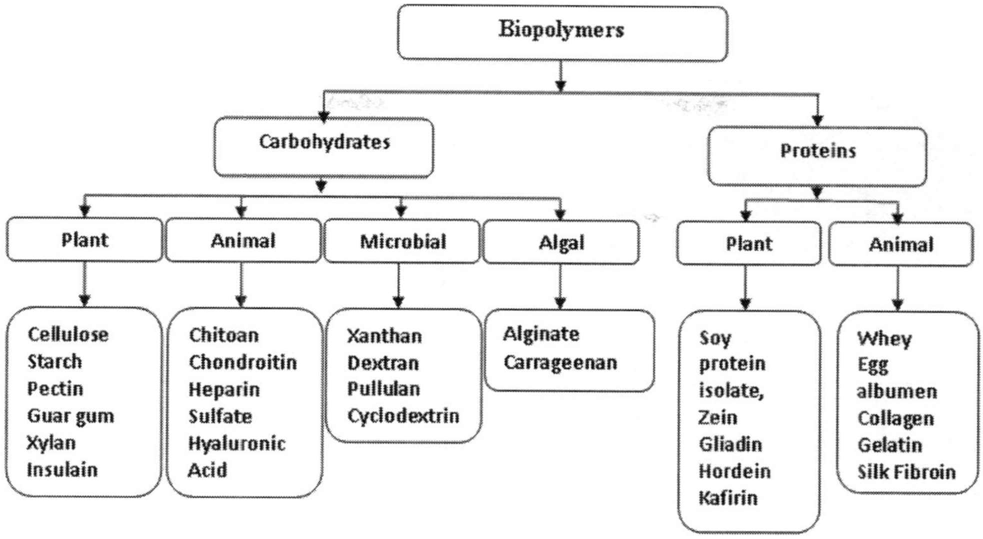

FIGURE 9.1 Choice of biopolymers for nanostructure formation by electrohydrodynamic process.

Hartman et al., 2000; Grace and Marijnissen, 1994). Electrospinning and electrospraying are sister technologies of the electrohydrodynamic process where polymer solution gets spun or sprayed under high potential electric field to obtain 1D nanofibers or 3D nanospheres (fine aerosols), respectively. The typical setup for the electrohydrodynamic process consists of four main components: (1) a syringe pump, (2) spinneret (nozzle), (3) a high voltage source (1–30kV) usually operated in direct current mode, though alternating current mode is also possible (Kessick et al., 2004), and (4) a grounded collector, either flat plate or rotating drum (Figure 9.2a). In this process, the polymer solution loaded in the syringe (reservoir) is fed through the spinneret at a steady and controllable feed rate with the help of a syringe pump. The polymer solution forms the hemispherical surface at the tip of the spinneret. When the high voltage power supply has connected the spinneret, it induces free charges in the polymer solution and creates an electric potential between spinneret and collector. Upon an increase in electric potential beyond critical voltage, the hemispherical surface of the droplet is distorted into a conical shape known as the Taylor cone (Taylor, 1969). Under high electric potential, two major electrostatic forces, namely the electrostatic repulsion of like charges and the Coulombic force of the external electric field, overcome the surface tension and eject the charged polymer towards the grounded collector. When the polymer concentration or the molecular weight is low, electric force atomizes the polymer liquid droplets and causes electrospraying (Figure 9.2b), which results in the formation of solid nanoparticles. On the other hand, when the polymer concentration and the molecular weight are high, the polymer jet is driven towards the collector. During this process, the unevenly distributed charges cause a whipping or bending motion of the jet which leads to elongation of the jet and rapid evaporation of the solvents and thus produces solid nanofibers though a process known as electrospinning (Figure 9.2c) (Yang et al., 2005; Anu Bhushani and Anandharamakrishnan, 2014; Rana et al., 2016)

The main controlling parameters of this process are (i) solution parameter (which includes the molecular weight of a polymer, concentration, solvent or solution conductivity, polymer solubility, viscosity, and surface tension), (ii) atomization process parameters (such as applied voltage, set flow rate, nozzle diameter, distance between nozzle tip to collector, and type of collector), (iii) ambient parameters (such as temperature, pressure, and humidity). This dynamic process offers flexibility in the formation of nanostructures from a wide range of natural and synthetic polymers.

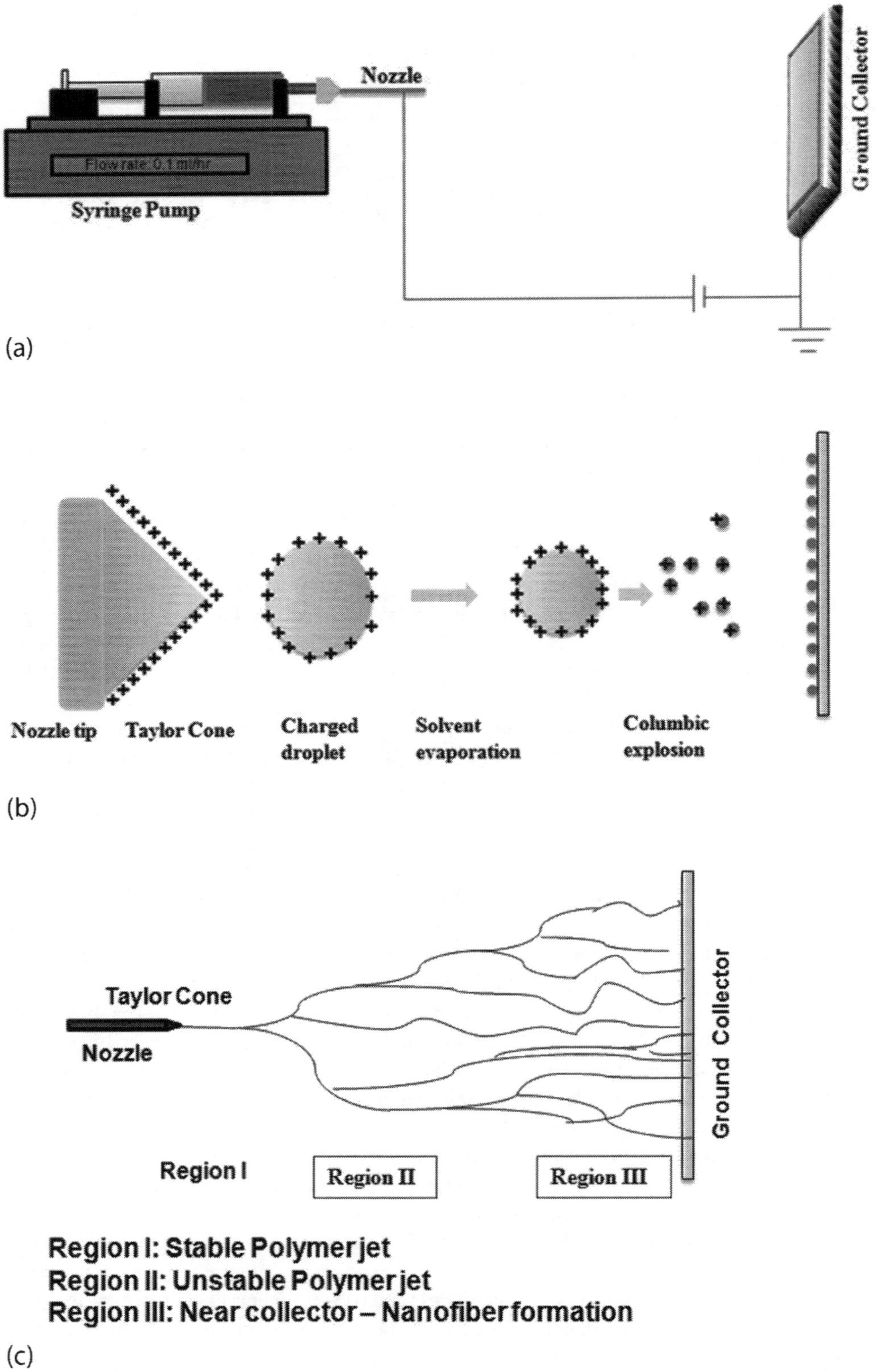

FIGURE 9.2 Electrospinning and electrospraying mechanism (a) electrospinning and electrospraying equipment setup; (b) electrospraying mechanism; (c) electrospinning mechanism.

9.2.1 Theory and Model of Electrospinning and Electrospraying

Over the years many studies were conducted on the electrohydrodynamic process and many models have been proposed. In 1878, Rayleigh found that a charged droplet becomes unstable and results in Coulombic fission of the liquid droplet into smaller droplets. Based on this study, the maximum limit on the surface charge density is defined and termed as Rayleigh limit, beyond which the electrohydrodynamic forces overcome the surface tension forces of the droplets (Rayleigh, 1878). In 1971, the first systemic study of electrospraying was reported by Zeleny (1917). Vonnegut and Neubauer reported the monodispersed aerosol droplets formed by the electrospraying process (Vonnegut and Neubauer, 1952). Taylor proposed the first explanation and theoretical description of the conical shape droplet at the capillary exit by investigating the hydrostatic balance between electrical and surface tension forces (Taylor, 1964). Larrondo and Manley explained polymer melting and jet formation of polymers (nylon 12 and polyethylene) subjected to an electric field (Larrondo and Manley, 1981). In the early 1990s, Doshi and Reneker proposed electrospinning as a method to prepare polymer fibers (Doshi and Reneker, 1993). In later years, various controlling parameters of the electrospinning and electrospraying processes were reported (Cloupeau and Prunet-Foch, 1994; de la Mora and Loscertales, 1994). Several solution parameters were studied and based on a theoretical model of charge transport researchers derived the scaling laws for electrospraying currents, charges, and sizes of the droplets (Gañán-Calvo et al., 1997). This scaling law coincided with the theoretical results.

In electrospraying, two scaling laws can be applied to different solutions with relatively high viscosity and electrical conductivity or low viscosity and electrical conductivity, depending on a dimensionless parameter. In most cases, the dimensionless parameter satisfies equation (9.1):

$$\delta\mu \cdot \delta^{1/3} \sim \left[\frac{\gamma^3 \varepsilon_0^{\,2}}{\mu^3 K^2 Q}\right]^{1/3} \tag{9.1}$$

where:

- Q is the flow rate of liquid
- ε_0 is the permittivity of vacuum
- K is the electrical conductivity
- γ is the surface tension
- μ is the viscosity of the solution

If $\delta\mu \cdot \delta^{1/3} \leq 1$, the current I, and size of the droplets d, in a cone-jet electrospraying mode are:

$$I / I_0 = 6.2\left(Q / (\beta-1)^{1/2} \cdot Q_0\right)^{1/2};$$

$$d / (\beta-1)^{\frac{1}{3}} d_0 = 1.6\left(Q / (\beta-1)^{1/2} Q_0\right)^{1/3} - 1.0 \tag{9.2}$$

If $\delta\mu \cdot \delta^{1/3} \geq 1$, the current I, and size of the droplets d, in a cone-jet electrospraying mode are:

$$I / I_0 = 11.0\left(Q / Q_0\right)^{1/4} - 5.0;\ d / d_0 = 1.2\left(Q / Q_0\right)^{1/2} - 0.3 \tag{9.3}$$

where:

$$I_0 = \left(\gamma^2 \varepsilon_0 / \rho\right)^{1/2}$$

$$Q_0 = \gamma \varepsilon_0 / (\rho K)$$

$$d_0 = \left(\gamma \varepsilon_0^{\,2} \left(\rho K^2\right)\right)^{1/3}$$

where β and ρ are the liquid to vacuum permittivity ratio and the density of the liquid, respectively.

Before 1999, researchers thought that in electrospinning, 1D fibers were generated due to the splitting or splaying of the electrified jet. In 1999, Hartman et al. developed a model to calculate the shape of the cone-jet. This model is also capable of calculating the surface charge density on the cone surface, the magnitude, and direction of the electric field inside and outside the cone-jet (Hartman et al., 1999). Radial charge repulsion triggers the primary jet splitting into multiple sub-jets while the electrified jet fluid accelerates along its trajectory to the collector. However, recent studies and experiments have demonstrated that jet thinning occurs due to bending instability associated with the electrified jet. The electrified jet is the first to form as a straight line in a cone-jet mode referred to as the classical Rayleigh instability, which is axisymmetric with respect to the jet centerline. The second is referred to as an axisymmetric instability, where the varicose structures are visible. The third is the non-axisymmetric, or "whipping" instability, where a non-axisymmetric bending of the fiber occurs. Non-axisymmetric or whipping instability is responsible for the thinning of fibers and is also the key mechanism to reduce the jet diameter from the micrometer to nanometer scale. This instability causes bending and stretching of the jet at very high frequencies (Reneker et al., 2000). The "inverse cone" was caused by a bending instability, not by a splitting of the primary jet into multiple jets (Shin et al., 2001a). Fridrikh et al. derived the equation of motion for the electrified jet and explained the thinning rate of the electrified jet in the bending instability region (Fridrikh et al., 2003).

In the electrospinning process, the precursor solution is held at the end of the nozzle in the form of a droplet by its surface tension. When the voltage is applied to the nozzle the surface tension fluid is overcome by applied electric potential and creates Taylor cone and subsequent jet formation of the precursor solution which undergoes fluid instability and forms as thin fibers in micro to nano diameter range (Reneker et al., 2000; Shin et al., 2001a). When the applied electric field (V) reaches critical voltage (Vc), the electrostatic force overcomes the surface tension and a charged polymer jet is ejected from the tip of the Taylor cone. Taylor(1969) showed that Vc (in kilovolts) is given as:

$$Vc^2 = 4\left(\frac{H^2}{L^2}\right)\left(\ln\frac{2L}{R} - 1.5\right)(0.117\pi R\gamma) \tag{9.4}$$

where:

H is the air-gap distance (cm)
L is the length of the spinneret tube (cm)
R is the radius of the spinneret (cm)
γ is the surface tension of the fluid (dyn/cm)

This jet initiation is followed by jet elongation with or without branching and/or splitting process known as "splaying" (Doshi and Reneker, 1993) followed by solidification of jet into nanofibers. The behavior of the liquid jets during the electrospinning process is well studied by Rayleigh (1882) and Taylor (1969). Ejected jets undergo changes in three regions, namely, (i) near the nozzle as reported by Rutledge and Warner (2001), (ii) far from the nozzle, which was determined by Rutledge and Warner (2001) and Cramariuc et al. (2013), and (iii) at jet terminal (see Figure 9.3). When the surface tension of a fluid has the same value as the surface charge repulsion, that region is considered as jet terminal. The scaling law of the diameter of the jet at the collector shows that flow rate, electric current, and the surface tension of the fluid control the terminal diameter of the whipping jet as reported by Fridrikh et al. (2003).

$$h_n = \left(\frac{6\mu\rho Q^2}{\pi I E_\infty} \right)^{\frac{1}{2}} . z^{-1} [m] \quad (9.5)$$

$$h_x = \left(\frac{\rho Q^3}{2\pi^2 I E_\infty} \right)^{\frac{1}{4}} . z^{\frac{-1}{4}} [m] \quad (9.6)$$

$$h_t = \left[\gamma\varepsilon \left(\frac{Q}{I} \right)^2 \frac{2}{\pi(2\ln\chi - 3)} \right]^{\frac{1}{3}} [m] \quad (9.7)$$

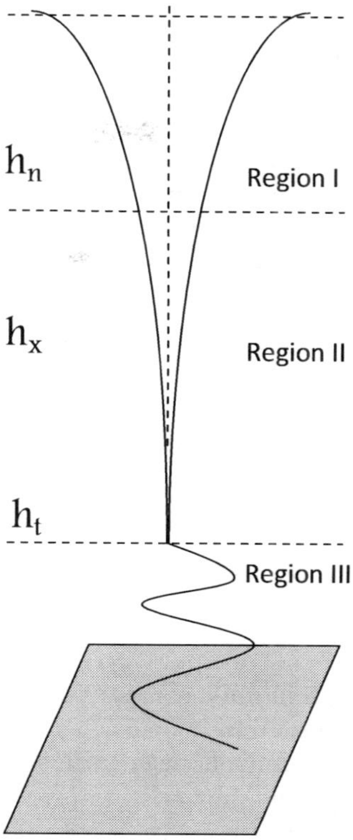

FIGURE 9.3 Schematic jet radius – h_n in region I, h_x in region II, and h_t in region III. (Widartiningsih et al., 2016.)

Explanations and units for the notations are mentioned in the figure.

h_n	Jet radius near the nozzle [m] (Region I)
h_x	Jet radius far from the nozzle [m] (Region II)
h_t	Jet radius near the collector [m] (Region III)
P	Density [kg/m^3]
Q	Flow rate [m^3/s]
Z	Distance from the spinneret to the collector [m]
M	Kinematic viscosity [m^2/s]
E_∞	External electric field [V/m]
I	Electric current [A]
Γ	Surface tension of the fluid [N/m]
E	Dielectric permittivity [F/m]
χ=R/h	Dimensionless whipping instability
R	Radius of the instability jet [m]
H	Fiber diameter [m]
Q/I	Inverse of the net volume density of the charge induced in the fluid at the top electrode [m^3/C]

The final empirical equation of fiber diameter is given as

$$d_f = h_t . c^{\frac{1}{2}} \quad (9.8)$$

where:

d_f fiber diameter collected on the ground collector

C solution concentration

Widartiningsih et al. (2016) predicted that the fiber diameter was not only determined by concentration, but also by other working variable in the electrospinning process. Modifying the above equation by considering the position where the jet start to be unstable (*xc*), the value of h_t in the above equation is substituted by *h*(*xc*)

$$d_f = 6.5373h\left[x_c\right].C^{4.089} \tag{9.9}$$

Here the constants, 6.5373 and 4.089, are a function of other working variables in electrospinning processes.

9.2.2 Controlling Parameters of Electrohydrodynamic Process

9.2.2.1 Solution Parameters

Solution property is one of the essential factors which directly affects the generation of smooth and bead-free electrospun fibers in the electrospinning process. Polymer concentration, molecular weight, solubility, viscosity, surface tension, solvent, and solution conductivity come under solution property. In the electrospinning process, due to the high voltage applied, an electric charge moves through the metallic needle and induces the charge on the polymer droplet which makes the polymer instable at the same time; the reciprocal charge repulsion produces a force that opposes the surface tension, and eventually the polymer solution flows in the electric field direction. When there are increases in an electric field, the spherical droplet will become deformed into a conical shape. This conical polymer droplet is known as Taylor cone, and leads to the formation of ultrafine nanofibers directed towards the metallic collector placed at an optimized distance. If the polymer solution has sufficient cohesive force, the stable charge jet will be able to occur. Sometimes, whipping of the liquid jet occurs due to internal and external charge forces which allow stretching of the polymer solution, resulting in the creation of nanofibers.

9.2.2.2 Molecular Weight of the Polymer

The molecular weight of the polymer has a significant effect on rheological and electrical properties (Bhardwaj and Kundu, 2010). In electrohydrodynamics, polymers are electrosprayed to produce particles as well as electrospun to create fibers, based on the range of molecular chain entanglement regimes. There are three regimes: dilute, semi-dilute unentangled, and semi-dilute entangled, which are determined by the polymer molecular weight and solution concentration. For electrospraying, the semi-dilute entangled regime is suitable where solid particles can be produced (Bock et al., 2012). In terms of the electrospinning, a concentration above the semi-dilute entangled regime is suitable to produce smooth fibers (Gupta et al., 2005). When the molecular weight decreases, the fibers decrease in size, hence the high molecular weight polymers create fibers with a large diameter. High molecular weight compounds with sufficient viscosity have been preferred for electrospinning. Low molecular weight polymers are not recommended to form fibers (Colmenares-roldán et al., 2017). However, non-polymeric materials (e.g. cyclodextrin) have an ability to produce nanofibers through electrospinning (Celebioglu and Uyar, 2011). In this case, there is no more polymer molecular entanglement but a high solution concentration/viscosity plays a role in creating electrospun fibers. And also it has been reported already that the molecular weight of polymers plays a major role in the surface morphology when the solution concentration is reduced. A lower molecular weight leads to the formation of particles with porous nature (Xie et al., 2011). Even at low concentration, the polymer with too high molecular weight favors the formation of microfibers (Zhao et al., 2005). Recently the supra-molecular cross-linking strategy has been considered to enable the electrospinning of low molecular weight polymers, even well below their critical entanglement molecular weight (Xiaoxiao Wang et al., 2016).

9.2.2.3 Concentration and Viscosity

In electrospinning, the fiber diameter is increased with increased polymer concentration. When the concentration of the solution is increased, the critical voltage value is also increased. Based on critical overlap concentration, the polymer solution may either go for electrospinning or electrospraying. If the polymer concentration is lower than critical overlap concentration, the intermolecular entanglement is significantly weak, which limits the viscous effects and results in electrospraying. The study showed that low polymer concentrations between 2 and 4 wt % lead to electrospraying, at intermediate concentrations of 10 wt % and 15 wt % create beads on strings, while at higher concentrations of 20 wt % and 25 wt % electrospinning and fine fibers are obtained (Figure 9.4). At higher concentrations the bead density was reduced and the beads appeared to be more spindle-shaped (Husain et al., 2016).

In electrospraying, viscous effects are less insignificant; surface tension and electrical conductivity are the competing forces. Low viscosity solutions are used for electrospraying, as the concentration is sufficiently low to destabilize the electrified jet that breaks down into small droplets that further solidify through rapid evaporation process and form particles. Whereas, in electrospinning, more viscous solutions with higher polymer concentration are used to form the charged jet. At the initial stage the jet follows a straight path and then undergoes "bending instability" and "whipping motion" which result in the formation of fibers (Shin et al., 2001b). In electrospinning, apart from bead formation, visco-elastic stress does not change in fiber morphology. But when visco-elastic stress competes with the electric field, this promotes tension within fibers, leading to the formation of thin fibers (Thompson et al., 2007).

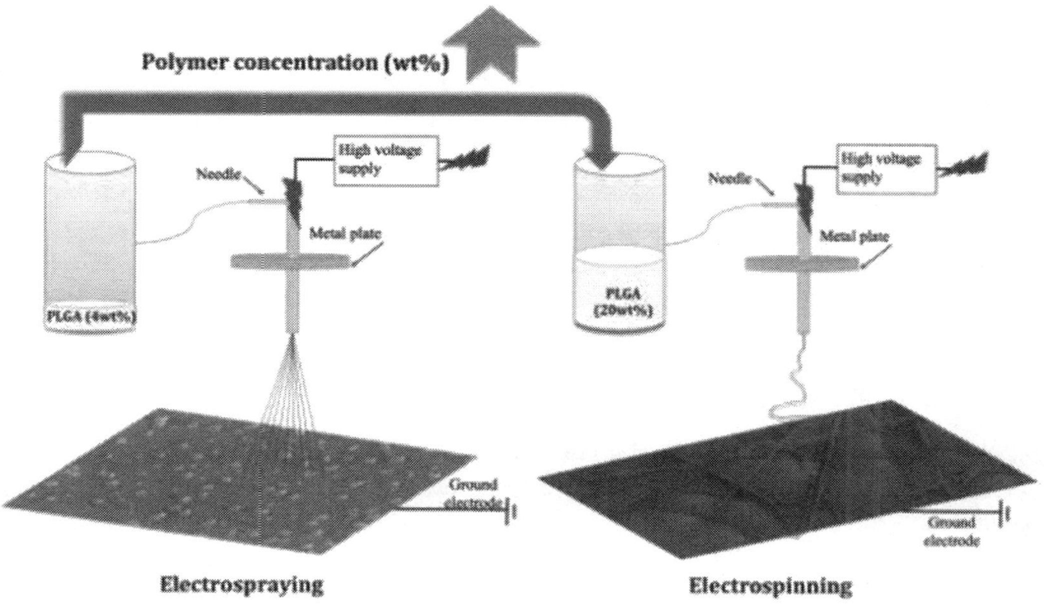

FIGURE 9.4 Effect of polymer concentration. (Husain et al., 2016.)

9.2.2.4 Solvent/Solution Conductivity

The electric conductivity of the liquid is the most important parameter affecting the cone-jet domain of the electrospray. By increasing the electric conductivity of the solution, the cone-jet domain is shifted toward smaller flow rates (Tang and Gomez, 1996). As the liquid electric conductivity increases, the domain shifts toward smaller flow rate which implies that smaller droplets are generated in the spray. To enable the electrospinning to occur, sufficient levels of the dipole moment of the solvent and the solution conductivity are essential. Wet fibers can be created by using the low volatile solvents for electrospinning. If the solvent is too volatile the Taylor cone will solidify and will not be able to form fibers. Low conductivity solutions cannot create fine fibers. Solution conductivity can be modified by introducing additives to the solution such as ionic salts (Chunxue et al., 2005). By adding ionic salts, the conductivity of the solution can be increased and this enables the jet formation of fibers.

9.2.2.5 Surface Tension

In electrospinning, the liquid drop elongates with the increasing electric field and is distorted into a conical shape when the repulsive force is equal to the surface tension of the liquid. Reducing surface tension leads to bead-free fibers. Lower surface tension and boiling point, as well as the increased viscosity of polymer solutions, could provide significant resistance to sustain the elongation of polymer chains and forms the bead free homogeneous fibers. (Huan et al., 2015). Surface tension acts to reduce the surface area of the jet promoting larger particle size (Chakraborty et al., 2009)

Polymer concentration impacts the surface tension of the solution and thus affects the particle size. Hartman and his group developed the relation between particles size and surface tension as (Hartman et al., 2000):

$$d = \alpha \left(\frac{\rho \varepsilon_0 Q^4}{I^2} \right)^{1/6} \tag{9.10}$$

$$I \propto \left(\gamma K Q \right)^{1/2} \tag{9.11}$$

where:

- d is the droplet diameter (m)
- α is a constant
- Q is the liquid flow rate ($m^3 .s^{-1}$)
- ρ is the solution density
- I is the current
- ε_0 is the permittivity of vacuum
- γ is the surface tension of solution in ambient air
- K is the liquid conductivity

These equations indicate that particle size increases with decreasing surface tension. When the applied Coulomb force overcomes the cohesive force of the droplet surface tension, the Taylor cone at the droplet will undergo a breakup into smaller droplets in the micro to nanoscales.

9.2.2.6 Optimization of Process Parameters

Voltage, flow rate, type of collector and the distance between tip and target are the main processing parameters.

9.2.2.6.1 Applied Voltage

Applied voltage directly affects the dynamics of the liquid flow. Changes in the voltage reflect on the shape of the suspending droplets out of the nozzle of the spinneret, on its surface charge, dripping rate, and especially on the structural morphology of the electrospun fibers. The flow of current from a high voltage power supply into a solution via a metallic needle will cause a spherical droplet to deform into a Taylor cone and form ultrafine nanofibers at a critical voltage. This critical value of applied voltage varies from polymer to polymer. The formation of smaller-diameter nanofibers with an increase in the applied voltage is attributed to the stretching of the polymer solution in correlation with the charge repulsion within the polymer jet (Sill and von Recum, 2008). An increase in the applied voltage beyond the critical value causes greater amounts of charge to emerge, making the droplets and the jet accelerate toward the collector faster and resulting in the formation of beads or beaded nanofibers. The effect of applied voltage on the fiber diameters can be explained in terms of the relationships between the three major forces (Coulombic, viscoelastic, and surface tension). Huan et al. (2015) clearly explained that at low applied potentials (e.g. 10 kV), the Coulombic force was not high when compared with that of the surface tension. This resulted in as-spun fibers with large diameters and the presence of large beads along the fibers. At a moderate applied potential (e.g. 15 kV), all three forces were well balanced, resulting in a narrow distribution of the fiber diameters. With a further increase in the applied potential (e.g. 20 kV), the Coulombic force was much greater than the viscoelastic force. This might result in higher possibility of breakage of an over-stretched charged jet during its flight to the target.

In the electrospraying process, applied voltage plays an important role in controlling the diameter of droplets. Applied voltage impacts the surface tension of the polymeric solution, affecting polymer droplet size, as reported by Hartman et al. in the micro-dipping mode of electrospraying (Hartman et al., 2000).

$$d = \alpha\left[\frac{\rho\varepsilon_0 Q^4}{I^2}\right] \tag{9.12}$$

where:

- d is droplet diameter
- α is a constant
- Q is liquid flow rate
- ρ is solution density
- I is current
- ε_0 is permittivity of vacuum

This equation indicates that as the voltage (V=IR) increases, the diameter of the polymer droplet decreases. At the lower applied voltage of 10 kV, the mean diameter of polymer droplet was 0.254 μm, and this decreased to 0.091 μm at a higher applied voltage of 25 kV. At a lower voltage (5 kV) the polymer solution was sprayed as a narrow jet, since low applied voltage resulted in a smaller spray angle (Jadhav et al., 2011). At higher voltage (30 kV) polymer droplets, the high electrohydrodynamic force led to a large spray angle, which resulted in lower polymer aggregation (Jadhav et al., 2012) because when applied voltage increases the polymer droplet diameter decreases (Jadhav et al., 2013).

9.2.2.6.2 Flow Rate

Flow rate is considered as one of the key parameters which influences the fiber diameter distribution, droplet size and initiating shape at the capillary tip, trajectory of the jet, maintenance of Taylor cone, deposition area density, and nanofibers morphology (Supaphol et al., 2005; Baumgarten, 1971; Dalton et al., 2006). At lower flow rates, a small amount of solution was ejected from the capillary tips which led to the formation of small droplet size and made various types of charged jet. The applied electric field draws the charged jet and reduce its diameter. It was observed that at 0.5 mL/hr, the Taylor cone was more stable, which resulted in the narrowest fiber diameter distribution. While considering the evaporation of solvents in the polymer solution, the lower flow rates are more desirable as the solvent will have sufficient time for evaporation (Yuan et al., 2004). On the other hand, when the flow rate is high, a greater amount of solution volume will be ejected from the needle tip and result in the formation of larger droplets. The same applied electric field cannot draw the larger droplets and causes unstable polymeric jet, resulting in electrospraying without sufficient stretching.

The formed Taylor cone was dropped off at the end of the syringe tip, which resulted in wider distribution. Additionally, high flow rate caused an increase in the bending instabilities and the occurrence of more splitting of the polymer jet (Zargham et al., 2012). It is believed that at high flow rates, there is a limitation to the increase in the fiber diameter because the amount of charge and flow rates has not increased simultaneously (Rutledge and Warner, 2001; Zong et al., 2002). In other cases, when the flow rate is high, a larger volume of solution is drawn from the needle tip, which needs a longer time to allow the evaporation of solvents in the polymer solution (Yuan et al., 2004). Without sufficient solvent evaporation, fiber formation will have some defects such as branched, splitting fibers, blobs, and flattened web-like structures. Hence its necessary to control the flow rate appropriately.

9.2.2.6.3 Distance between Tip and Target

Working distance, together with electrostatic potential, will determine the strength of the electric field. The electric field strength varies with the variation of the applied electrostatic potential and the working distance. It has been found that increasing the electrostatic potential and the working distance results in the formation of thinner fibers (Reneker et al., 2000; Silke et al., 2002). In other cases, increasing working distance without increasing electric field results in the formation of thicker fibers (Xinhua Zong et al., 2002). The working distance drastically affects the total flight time available for the fiber formation. However, at too long or short a distance, jet initiation becomes difficult due to the electric field being too strong or weak.

9.2.2.7 Ambient Parameters

9.2.2.7.1 Temperature

Throughput and fiber diameter are temperature dependent – increasing temperature, the throughput and fiber diameter decreases (Cengiz Çallıoğlu et al., 2010). This also affects the morphology of the fibers, which were found to be "beaded" when the solution was at lower temperatures (<40°C), but when the polymer solution was at higher temperatures (>60°C), the fibers became flat. Mit-uppatham et al. (2004) had proven that increasing temperature favors the thinner fiber diameter with polyamide-6 fibers for the inverse relationship between the solution viscosity and temperature. The study by Kim and co-workers found that increasing the local temperature resulted in the solvent evaporating faster (Kim et al., 2005).

9.2.2.7.2 Vapour Pressure

Vapour pressure plays a major role in evaporation of solvents used in electrospinning and electrospraying. Due to the lower vapor pressure, the solvent loss would be faster for high evaporation rate solvents. Low temperature polymer solutions can be optimized by controlling the vapor pressure of the solution.

9.2.2.7.3 Humidity

Low humidity increases the velocity of the solvent evaporation and allows the complete evaporation of the solvent, whereas high humidity leads to incomplete solvent evaporation. Dao et al., (2010) have reported that humidity strongly affects the fiber property: higher humidity leads to non-fibrous formation, however there is no significant effect on fiber diameter. An increase in humidity level leads to the pore formation on the fiber surface. Similar results were reported by Casper et al., (2004), where higher humidity level formed the pores on polystyrene fibers and coalescence of pores upon further increase of humidity level.

9.3 POTENTIAL APPLICATIONS IN FOOD INDUSTRY

9.3.1 Bioactive Molecules Encapsulation

Natural foods are a rich source of bioactive molecules, micronutrients, and antioxidants such as vitamin A, vitamin E, β-carotene, catechins, gallic acid, resveratrol, alpha-tocopherol, iron, manganese, etc. These bioactive compounds provide potential health benefits including inhibition of oxidative stress, diabetic and cardiovascular diseases, cancer, and aging (Herrera and Barbas, 2001). These compounds are highly sensitive to oxygen, water, heat, light, and UV irradiation (Ananingsih et al., 2013). In addition to its low stability, the poor water solubility of the compound affects its bioavailability which limits the health benefits of food products. Hence it is essential to protect these bioactive compounds from undesirable damage by various processing steps, at the same time improving their bioavailability. Conventionally, encapsulation by spray drying, spray chilling, spray freeze drying, liposomal encapsulation, nanoemulsion, molecular inclusion using cyclodextrin, protein precipitation, hydrogel entrapment, and coacervation are used to protect these sensitive food bioactive compounds form moisture, oxidation, heat, or extreme conditions during processing (Anandharamakrishnan, 2014a,b,c and 2015; Anandharamakrishnan and Ishwarya, 2015a,b; Bhushani et al. 2016; Bhushani and Anandharamakrishnan, 2017; Karthik et al., 2017; Karthik and Anandharamakrishnan, 2013; Ezhilarasi et al., 2013; Parthasarathi and Anandharamakrishnan, 2016; Pasrija et al., 2015; Ezhilarasi et al., 2014; Parthasarathi et al., 2016; Hundre et al., 2015; Remondetto et al., 2004; Yuliani et al., 2004). These methods typically produce capsules of several microns to millimeters in size. Materials in nanosize can penetrate cells without disrupting cell membrane and blood-brain barriers and are thus proven to increase bioavailability (Nakamura and Watano, 2018; Zhou et al., 2018). Hence, encapsulation of food bioactive molecules in nanosize emerged to increase its bioavailability (Quintanilla-Carvajal et al., 2010; Parthasarathi et al., 2016). To form nanoencapsulates, techniques such as nanoprecipitation (Khayata et al., 2012), nano spray drying, electrospraying and electrospinning (Bhushani and Anandharamakrishnan, 2014) are in common practice. Among these, electrospraying/spinning is a one step method to produce nanoscale solid materials by dehydrating molecules. This technique can form matrix encapsulation or core-shell encapsulation based on the spinneret used. A potential application for effective delivery of drug molecules in the pharmaceutical field is already proven. A wide range of food bioactive molecules are also encapsulated by this technique and proven to improve the stability and bioavailability of encapsulated molecules.

9.3.1.1 Catechin Encapsulation

Epigallocatechin-gallate (EGCG), the main polyphenol compound present in green tea, is a potential health benefit molecule with antioxidant and antimicrobial properties which is widely used in the nutraceutical industry. Unencapsulated EGCG molecules are more stable in acidic digestive conditions, whereas they lack stability under the higher pH conditions found in the duodenum and get degraded, which affects the bioaccessibility of this potent compound. To improve the bioaccessibility of catechin, Bhushani et al. encapsulated catechin in zein wall material by electrospraying technique and formed monodisperse spherical nanoparticles in powder form with a mean diameter of 157 nm. This nanopowders significantly improved the *in vitro* gastrointestinal stability and Caco-2 cell monolayer permeability of catechins compared to unencapsulated catechins (Bhushani et al., 2017). Further improvement in the protection of the EGCG molecule in the gastric phase and increased availability of EGCG in the intestinal segment (bioaccessability) are achieved by multilayer encaplsulation by electrospraying technique with zein and gelatin wall materials. The antioxidant capacity of multilayer encapsulated EGCG was high in the intestinal phase compared to matrix-encapsulated and free EGCG molecules (Gómez-Mascaraque et al., 2018). Similary, catechin encapsulated in zein was also produced in the form of nanofibers by electrospinng technique. Encapsulation in zein nanofibers was found to increase the stability of EGCG molecules in an aqueous environment. One-day aging of nanofibers at 0% relative humidity (RH) increased the immobilization of EGCG in zein nanofibers (with recovery of > 98%) compared to freshly spun fibers when tested by immersing them in water. Upon aging at elevated RH (75% of RH), progressive degrading of EGCG with aging time (with 66–73% recovery) was observed (Li et al., 2009).

9.3.1.2 Curcumin Encapsulation

Curcumin is an important food bioactive molecule with a wide range of biological properties and potential health benefits. However, its poor solubility in water, rapid degradation properties under neutral or alkaline pH conditions and exposure to light, and poor bioavailability limits its potential benefits in food preservation and nutraceutical application. Gomez et al. (2017) proved that matrix encapsulation of curcumin in gelatin nanoparticles through electrohydrodynamic atomization process could overcome this limitation. Uniform spherical particles with size ranging from a few nanometers to micrometers were formed. Curcumin encapsulation in gelatin nanoparticles increased its water solubility by 38.6-fold and antioxidant activity compared to native curcumin. Additionally, a significant improvement in the antimicrobial property of gelatin-encapsulated curcumin at 0.4 mg/mL concentration was observed, which could not be achieved with native curcumin even at a concentration as high as 10 mg/mL (Gomez-Estaca et al., 2017). Similarly, encapsulation of curcumin in zein nanoparticles achieved solubility of curcumin in an aqueous medium and caused it to gain its food coloring property (Gomez-Estaca et al., 2012). Considerable storage stability of curcumin-encapsulated nanoparticles was achieved over a period of three months under 23°C at 43% relative humidity with no change in morphology of nanoparticles, and without any agglomeration or recrystallization of curcumin. When mixed in semi-skimmed milk (an aqueous food matrix), nanoparticles showed good dispersion and coloring capacity compared to commercial curcumin. The addition of commercial curcumin to milk did not show any significant color change (with the difference value of 4) due to its poor solubility. Whereas adding the same amount of curcumin (0.1 g/100 mL) incorporated into zein nanoparticles showed significant change in color, which was observed based on change in hue angle (from 110.7° to 96.6°), chroma (from 7.8 to 61.7), and lightness (from 87.8 to 83) compared to the milk without additions (Gomez-Estaca et al., 2012). In another study, the free radical scavenging ability of curcumin was maintained by encapsulating it in zein nanofibers. Curcumin-encapsulated zein nanofibers were shown to increase the fluorescence with an increase in curcumin content and could release cucurmin in a sustained manner. This curcumin-loaded nanofibers were proven to support the growth of fibroblast cells and demonstrated their potential for soft tissue engineering (Brahatheeswaran et al., 2012). In addition to protecting curcumin from undesirable conditions, combining it with natural enhancers such as piperine from black pepper will enhance its bioadsorption. In this regard, core-shell nanoparticles with curcumin in zein matrix as core and piperine in chitosan matrix as shell materials is formed by coaxial electrospraying. This curcumin and piperine-loaded zein-chitosan nanoparticles showed more than 50% cytotoxicity effect against neuroblastoma cells, which revealed the potential of developed nanoparticles (Baspinar et al., 2018).

9.3.1.3 Fish Oil Encapsulation

Polyunsaturated fatty acids (PUFA) are primarily omega-3 (α-linolenic acid, ALA; docosahexanoic acid, DHA; eicosapentaenoic acid, EPA) and omega-6 (arachidonic acid, linoleic acid, Y-linolenic acid) fatty acids, essential for maintaining normal growth and development. However, they are prone to oxidation due to the presence of a high degree of unsaturated carbon-carbon double bonds which render them sensitive to oxygen, light, and heat. Their low solubility in many food matrices also limits their benefits. Oxidation stability of fish oil is achieved by micro and nanoencapsulation. For example, fish oil is encapsulated in zein nanofibers by electrospinning which increased its oxidative stability compared to unencapsulated fish oil (Moomand and Lim, 2014). Further improvement in oxidative stability is achieved by coaxial electrospinning (Yang et al. 2017b). However, the core-shell nanofiber structure constrains the release profile of fish oil compared to single nanofiber. Thus, to retain the higher release profile of single nanofibers and at the same time better oxidative stability, ferulic acid, a polyphenolic compound with antioxidant activity, is added along with fish oil in zein and forms as nanofibers (Yang et al. 2017a). The resulting composite nanofibers exhibit better oxidative stability with additional nutritional quality due to the synergic effect of fish oil and ferulic acid.

Similarly, nanoencapsulation of fish oil in carbohydrate matrix was also studied to improve its oxidative stability. As polysaccharides are thermally stable molecules, encapsulating fish oil in polysaccharide matrix will increase its thermal stability and make it suitable for thermal food processing steps (e.g. incorporation in baked products). Fish oil as neat and emulsified oil is encapsulated in pullulan and formed as nano/microfibers by electrospinning. It exhibited higher oxidative stability. Similarly, fish oil was encapsulated in dextran and produced nano/micro particles. Compared to pullulan, matrix dextran exhibited poor interaction fish oil and showed low oxidative stability (García-Moreno et al., 2017). Nanoencapsulation of fish oil with protein and carbohydrate mixture was also studied. A combination of whey protein/pullulan/dextran or whey protein/pullulan/glucose was used to encapsulate fish oil and formed nanoparticles by electrospraying (García-Moreno et al., 2018). Pilot scale facility was also employed to increase the bulk production of nanoparticles by pressurized air-assisted electrospraying. Glucose-based coating offered better oxidative stability compared to dextran-based coating. This difference is attributed to oxygen permeability between both matrices. In addition to its use as protective coating in bioactive molecules, nanoencapsulation was also employed to mask undesirable flavors and tastes of active molecules, and to meet consumer requests for organoleptic quality and functionality of functional foods.

9.3.1.4 β-Carotene Encapsulation

Other food coloring molecules with well known health benefits but limited by poor stability are also encapsulated by an electrohydrodynamic process. For example, light stability of β-carotene was improved by encapsulating it within zein nanofibers by electrospinning. While the β-carotene molecule decades within 20 min upon UV-VIS exposure, nanofiber-encapsulated β-carotene was stable for up to 1 hr of tested time (Fernandez et al., 2009). Another food coloring molecule, anthocyanin, was also encapsulated in chitosan/gelatin matrix by electrospraying. Faster release of anthocyanin from chitosan/gelatin nanoparticles was observed under 3% acetic acid medium (Hern et al., 2017).

9.3.1.5 Other Bioactive Molecules

Similarly, many food bioactives found in natural fruits and vegetables which lack stability were effectively encapsulated by electrohydrodynamic process. For example, quercetin, an important flavonoid found in berries, apples, green tea, etc., was encapsulated in zein nanofibers which offered effective protection along with the sustained release, enhancing its functionality and therapeutic effects (Li et al., 2014a). Similarly, in another study, gallic acid was incorporated in zein nanofibers by electrospinning and formed as functional ingredient delivery system (Ping et al., 2013). Resveratrol, a potential nutraceutical compound found in grapes, red wine, and berries was also protected by nanofiber encapsulation and proven to have therapeutic potential in anticancer, periodontal disease (Zupančič et al., 2015), and bone tissue engineering (Dhanasekaran, 2014) applications. A combination of ferulic acid and resveratrol in core-shell nanofibers was proven to have potential as wound healing application (Poornima and Korrapati, 2016).

In another study, alpha-tocopherol (α-TOC) (biologically active form of vitamin E), which is commonly applied in beverage products and used as supplements in nutraceuticals, is also preserved by encapsulation within nanofibers. Soluble dietary fibers (SDF) extracted from rice bran, which have good oxygen barrier properties, were used in combination with zein to protect α-TOC through the formation of nanofibers and films (Li et al., 2016). Addition of SDF to zein nanofibers protected α-TOC from applied heat, UV irradiation, and storage conditions, and retained (~85 μg/100 μg) α-TOC even after exposure to UV light (253.7 nm, 40 W) for 8 hr at room temperature and (~73 μg/100 μg) after 15 days of storage at room temperature. This is much higher than the earlier reported liposomal carrier system, which retained 65 μg/100 μg of vitamin E, after storage at 25°C for 15 days (Zhao et al., 2011).

Thermal stability of α-linolenic acid (ALA), a hydrophobic molecule, was achieved by encapsulation in zein or gelatin matrix by electrospraying. While unencapsulated, ALA molecules degraded in 400 min upon exposure to 80°C, encapsulated molecules were stable up to 50 hrs. Additionally, multilayer encapsulation by coaxial electrospraying with gelatin/zein structure provided better thermal statbility of the compound due to the lower affinity of ALA with gelatin, forming a better protective layer around ALA (Gómez-Mascaraque et al., 2018).

9.3.2 Live Cell Encapsulation

Probiotics are dietary supplements of live microorganisms which improve the intestinal microbial balance and provide health benefits. Many microorganisms belonging to *Lactobacillus* and *Bifidobacterium* genera are recognized as potential probiotics and considered safe to be used as food products. Increasingly, health-conscious consumers demand the formulation of functionalized food products based on probiotics. Many types of dairy (i.e. yogurts, ice cream, and fermented milk) and non-dairy (i.e. juices, jelly) probiotics functionalized foods have been formulated. However, food processing steps including temperature, stress, oxygen, heat, extreme salinity, and the extreme pH conditions of the human stomach affect the viability of probiotics. Notably, maintaining the viability of probiotics during passage through the gastrointestinal tract and improving their viability in the intestine is the primary target to confer a health benefit to the host.

Encapsulation techniques such as spray drying, freeze drying, coacervation, and cross-linking are in practice to encapsulate probiotics; however, the large porosity of encapsulating matrix may allow leakage of free cells into the product, exposing them to the adverse environment. Hence, recent techniques by electrohydrodynamic process have been employed to entrap the live cells in a suitable solid matrix with size ranging from micro to submicron.

Various wall materials such as proteins like whey protein, soy protein, maltodextrin, gliadin, hordein, zein; polysaccharides such as gelatinized starch, alginate, xanthan, and chitosan, and synthetic polymers such as PVA and PVP have been employed to encapsulate probiotics. For example, probiotic cells were encapsulated in different wall materials such as WPC, maltodextrin, Fibersol, zein, and PVP by electrohydrodynamic atomization. Cell viability was retained up to 600 days, and additionally, WPC coating retained the highest viability of encapsulated *bifidobacteria* (Librán et al., 2017).

In a study by Zaeim et al, encapsulation efficiency of probiotics in acacia gum by electrospraying at 19 kV and freeze drying at –56°C and 100 mTorr was compared. The freeze drying process reduced the viability of probiotics cells to 9.02 CFU/g, whereas electrospraying reduced it to 9.74 CFU/g. The freeze drying process reduced survival efficiency to 89% with 1 log cycle of bacterial viability loss, while electrospraydrying offered survival efficiency as high as 96%. This highlighted the efficacy of the electro spray drying process for probiotics encapsulation (Zaeim et al., 2018). Similarly, various wall materials have been used to encapsulate probiotics and produce solid powders by electrospraying. *Lactobacillus plantarum* was encapsulated in sodium alginate and in a mixture of sodium alginate andcitric pectin matrices by wet electrospraying in $CaCl_2$ solution. Through this process, round and rod-shaped microcapsules were formed. The efficiency of the protective coating to prolong the viability of cells under simulated gastric and intestinal conditions was evaluated. Upon exposure to simulated gastric juice (SGJ) (for 120 min) and simulated intestinal juice (SIJ) (for 120 min), unencapsulated cells lost their viability by 6 and 4.2 log CFU/mL, whereas encapsulated cells decreased only by 2.9 CFU/mL and 2.7 CFU/mL respectively. Upon refrigeration storage at 4°C for 21 days, encapsulated cells showed prolonged survivability of 9 log CFU/mL while, unencapsulated cells showed only 1 log CFU/mL survival (Coghetto et al., 2016).

In another study, single and double layer encapsulation of *Lactobacillus plantarum* was achieved by wet electrospraying. Encapsulation of bacteria in Ca-alginate/chitosan microcapsules as a single layer and double layer matrix was formed. When formed as a single layer protective coating, chitosan molecules were present in close proximity to probiotics cells and adversely affected their viability and showed no protection against low pH environments in stimulated gastric conditions. No viable cells were present after 2 h exposure to SGJ condition. Whereas, when chitosan coated Ca-alginate microcapsules by double-stage procedure, it successfully retained the cells viably inside a micro structure under simulated gastrointestinal conditions. Additionally, delayed inward diffusion of H+ ions and improved cell viability by 1 log cycle compared to free cells was observed (Zaeim et al., 2017).

Similarly, encapsulation of probiotics in nanofibers as the solid matrix was also formed. Heunis et al. (2010) encapsulated *Lactobacillus plantarum* 423 along with its antimicrobial peptide, plantaricin 423 (bacteriocins), in nanofibers by electrospinning. They were effectively encapsulated within polyethylene oxide (PEO 200 000 Da) nanofibers produced by electrospinning. High shear and high voltage conditions of electrospinning process reduced the antimicrobial activity of bacteriocins (plantaricin 423) from 51 200 AU/mL to 25 600 AU/mL against *Lactobacillus sakei* DSM 20017 strain and from 204 800 AU/mL to 51 200 AU/mL against *Enterococcus faecium* HKLHS strain. Even though only 2% of the *L. plantarum* cells remained viable after encapsulation within nanofibers, entrapped cells continued to produce plantaricin 423, which proves the effectiveness of nanofibers for delivery of bacteriocins (Heunis et al., 2010). Fung et al. (2011) used soluble dietary fibers (SDF) obtained from agrowastes such as okra (soybean solid waste), oil palm trunk (OPT), and oil palm frond (OPF) to encapsulate *Lactobacillus acidophilus* and formed nanofiber matrix by blending it with poly(vinyl alcohol) (PVA), an easily spinnable synthetic polymer. Nanofibers with a diameter ranging from 229 to 703 nm were formed and effective encapsulation of probiotic within nanofibers was confirmed through a fluorescence microscope. Even after the high shear stress and high voltage conditions of the electrospinning process, 78–90% survivability of bacteria was retained. Additionally, encapsulated probiotics were viable up to 21 days under refrigeration. The high melting temperature of the PVA (> 180°C) matrix provided good thermal protection to encapsulated probiotics, which indicated the potential of incorporating nanofiber-encapsulated probiotics in heat-processed food products (e.g. bakery products), where they can remain stable at working temperatures of 50–180°C.

In a recent study by Moayyedi et al. (2018), three different wall materials: whey protein isolate (WPI), WPI+ inulin (IN), and WPI+IN+Persian gum (PG) were used to encapsulate *L. rhamnosus* ATCC 7469 through the three different techniques of freeze drying, spray drying, and electrospraying. Their efficacy was then compared. The morphology of the dried particles obtained with these three methods is presented in Figure 9.5. Spray dried particles appeared as capsules with dents/concavities on the surface, freeze drying formed irregular shapes, whereas electrospray drying formed uniform spherical particles. When tested against lysozyme, penicillin G, bile, and NaCl conditions, a significant difference in viability was observed among these three methods. Lower cell damage was observed in freeze dried product, while electrospray drying caused more damage to cells. In addition, coating with WPI+IN+PG wall provided better protection to cells owing to the lower moisture retaining capacity of Persian gum, which prevented excessive drying of the encapsulates. During a storage period of 24 h, significantly higher viability of probiotics cells was obtained in

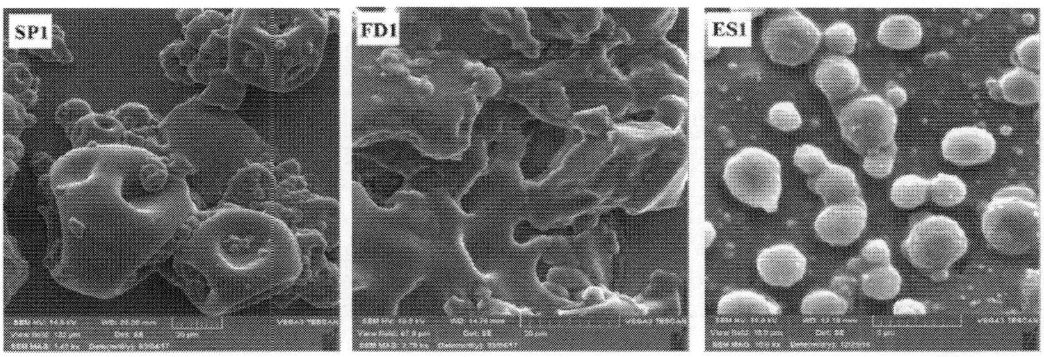

FIGURE 9.5 SEM image of *Lactobacillus rhamnosus* ATCC 7469 cell encapsulated in whey protein isolate by three different techniques: spray drying, freeze drying, and electrospray drying. (Moayyedi et al., 2018.)

freeze dried microcapsules than in spray dried and electrosprayed microcapsules, where viability loss was 10.44–10.78 log CFU/g, 9.43–9.69 log CFU/g, and 8.72–9.44 log CFU/g, respectively (Moayyedi et al., 2018).

9.3.3 Enzyme Encapsulation

Generally, enzymes for industrial application are formulated in liquid or solid form. Typically, enzymes have a shortened shelf life at ambient conditions and require lyophilization for longer storage. Removal of water from the enzyme will restrict the enzyme mobility and limit various physical effects (unfolding, surface adsorption, aggregation) and chemical degradation pathways (covalent bond modification through oxidation and deamidation). Thus, solid state enzyme formulation would be advantageous during processing, storage, and shipping of enzymes. While freeze drying and spray drying are conventionally used to produce solid state enzymes, extreme temperature and pressure conditions during these processes affect the biological activity of the enzymes. In recent years, the electrohydrodynamic process proved its potential to encapsulate enzymes and to protect their activity in the form of solid state materials. For example, an important industrial enzyme, cyclodextrin glucanotransferase (CGTase), which catalyzes the formation of cyclodextrins (CDs) from starch, is formed as fine solid particles by electrospraying. The electrospraying process does not alter any functional group of the CGTase. The further activity of native and electrosprayed enzymes does not vary much. Dry particles collected at longest distance (25 cm) between needle and collector produced fine particles of 75 nm size and exhibited higher activity than particles collected at a shorter distance (10 cm), with 200 nm size (Saallah et al., 2014). These studies showed the importance of particle size in controlling enzyme activity. In another study, horseradish peroxidase (HRP) enzyme was encapsulated within electrospun nanofibers using water soluble polyvinylpyrrolidone (PVP) polymers. The electrospinning process decreased enzyme activity by 20%; however, 40% enzyme activity was retained even after 280 days storage at ambient condition (Dai et al., 2012). This was developed as reagent storage on chip.

In a study by Li et al., lysozyme was effectively immobilized in nanofiber matrix to protect its activity. Lysozyme is a positively charged antibacterial enzyme able to hydrolyze the peptidoglycan layer of the cell wall of some gram-positive bacteria and used as bacterial inhibitor in food packaging food preservation applications. However, it lacks stability. A mixture of lysozyme and rectorite was electrosprayed on negatively charged cellulose acetate nanofibers and immobilized. Rectorite was added to control the release rate of enzymes from the nanofibers. Here the enzyme was entrapped in semi-permeable nanofibers support material, thus immobilizing it while allowing substrates, co-factor and products to pass through the membrane. This process is proposed as a method to overcome issues with recovery and reuse of enzymes and also to increase the long term stability of the enzymes for industrial applications (Li et al., 2014b).

Recently, in a the study by Huang et al. (2017), naringinase (a debittering enzyme) was immobilized on cellulose acetate nanofibers by layer-by-layer self-assembly process. By altering the number of layers, the amount of naringinase in the membrane can be controlled. This enzyme-immobile nanofibers were employed to remove the major bitter compounds, namely naringin and limonin, from grape juice. This nanofiber mat could remove 22.72% of naringin by hydrolysis with naringinase enzyme and 60.71% of limonin by adsorption on cellulose acetate nanofibers from grape juice. Similarly, naringinase-entrapped water soluble PVA electrospun nanofibers were formed to hydrolyze naringin (Nunes et al., 2015). Nanofiber entrapment retained 100% enzyme activity even after 8 days of continuous operation at 25°C. In addition, cross-linking with boronic acids increased the thermal stability of enzyme-entrapped nanofibers and showed a continuous enzymatic reaction for 16 days at 45°C. Thus, immobilization of the enzyme in/on nanofibers increased its tolerance to higher temperature, lower pH, storage, and operational conditions required in food processing applications.

9.3.4 Application in Sensing Field

Agricultural commodities and food products play a critical role in fulfilling the growing population demand. Any contamination in food or agricultural products by adulteration, pathogenic infestation, and spoilage will reduce their nutritive values. Hence monitoring quality, nutritional value, and safety of the food products are most important to ensure food safety globally. This necessitates developments of novel sensors based on faster, simpler, and less expensive detection techniques. Recently, many chemical biosensors have been developed for analyzing food analytes. The sensitivity of these sensors can be increased by increasing the specific surface area of interacting material with the analyte. In this scenario, nanomaterials play an important role in increasing the surface area of a sensing element. In recent years, nanostructures formed by an electrohydrodynamic process in the form of nanofibers are used to develop highly sensitive sensors. High surface area, porosity, 1D electron confinement, flexibility in surface functionalization, tunable mechanical properties, hydrophobicity, and conductivity are based on the polymer, chemical treatment, and processing parameters used make nanofibers ideal for sensing applications.

Towards the quality analysis of food samples, many nanofiber-based sensors have been developed. For example, to check the safety of drinking water, electrospun polyacrylic acid/polyvinyl alcohol (PAA/PVA) nanofibers are integrated as sensing layer in a light addressable potentiometric (LAP) sensor for rapid detection of *Escherichia coli* (*E. coli*) in drinking water. Presence of *E. coli* is monitored by the change in pH of media caused by fermentation of sugar molecule by *E. coli*. High sensitivity with a detection limit of 20 CFU/mL and nernstian response of a 74 mV/pH change was achieved (Mojir et al., 2016). This pH sensitive hydrogels get deprotonated and swell in the basic media, which increases the density of similar charges and mobile counter ions, whereas in the acidic media, hydrogels get protonated and decrease the density of charges and mobile counterions, causing shrinking in the nanofiber gel.

To access the quality of fruit juices, an electrospun polyaniline nanofiber-based taste sensor array was developed to detect citric acid concentration in juice samples. Due to the high sensitivity of nanofibers, this sensor could detect a change in the citric acid concentration at values as low as 2 ppm. This sensor could effectively classify eight different types of orange juices, including four commercial brands of ready-to-drink juice, two commercial brands of artificially reconstituted juices, and freshly squeezed orange juices from sweet lime (*Citrus limettioides Tanaka*) and "Pera" sweet orange (*Citrus sinensis (L.) Osbeck*). It was also able to group similar ones. The possibility of monitoring the aging process as a function of the storage conditions was also explored (Medeiros et al., 2009).

Ascorbic acid is a potential antioxidant with health benefits. This is a sensitive compound which can be easily oxidized by excessive heat, light, and heavy metal ions, and degraded by enzymes and atmospheric oxygen (Danet et al., 2008). Monitoring the ascorbic acid content of beverages is a critical parameter for monitoring quality, safety, and storage stability as well as reducing health risks associated with an excess of ascorbic acid. A nanofiber-based colorimetric sensor was developed for rapid detection of L-ascorbic acid to the naked eye. This colorimetric sensor was developed based on polyaniline (PANI)/polyamide 66 (PA66) electrospun nanofibers (Wen et al., 2015). The reversible non-redox acid/base doping process of PANI was explored for the purpose. Three oxidation states with considerable changes in electrical and optical properties from one state to another make PANI a promising material for sensing applications. Blending PANI with polyamide 66 improves the thermal stability and solubility in water or organic solvents. This nanofiber sensor could detect ascorbic acid concentrations as low as 50 ppb with a visible color change in the sensor strips. Furthermore, RGB (red-green-blue) digital parameters obtained from photographs of the nanofiber strips were automatically read out via an iPhone. This is an incredibly inexpensive system, especially for unskilled home users for visual detection of L- ascorbic acid (L-AA) (Wen et al., 2015). Similarly, highly sensitive detection of L-AA was developed by iridium oxide [IrOx $(0 < x \leq 2)$]/polyvinylpyrrolidone (PVP) electrospun nanofibers which could sense ascorbic acid concentration as low as $<0.4\ \mu M$ (Kim et al., 2014).

To ensure the safety of food products, nanofibers were also widely used for the detection of pesticides and fungicides in vegetables, fruits, and beverage products. For example, Liou et al. (2017) used silver nanopartile-coated cellulose nanofibers as an effective surface-enhanced Raman spectroscopy (SERS) platform to measure the amount of thiabendazole (TBZ) (commonly used pesticide and fungicide) in apples (Liou et al., 2017). Wang et al. developed a highly selective colorimetric nanofiber sensor for visual detection of formaldehyde. Hydroxylamine sulfate and methyl yellow dye were impregnated on a nylon membrane. When the analyte formaldehyde reacts with hydroxylamine in nanofiber matrix, it produces sulfuric acid which causes methyl yellow dye to change its color from yellow to red based on the pH value (Figure 9.6). Thus, it helps visual detection of formaldehyde with a detection limit of 50 ppb (Wang et al., 2012).

Similarly, a highly sensitive colorimetric strip for formaldehyde sensing was developed based on 4-amino-3-penten-2-one (fluoral-p) functionalized electrospun polyacrylonitrile (PAN) (PAN/fluoral-p) nanofibrous membranes which could detect HCOH concentration as low as 40 ppm (Zueqin et al., 2013).

Ebrahim et al. developed an electrochemical-based nanosensor by depositing polyaniline nanofiber and single-walled carbon nanotube on a graphite electrode for the detection of pesticide. This nanofiber-based electrochemical sensor will generate different voltage characteristics based on the presence and concentration of malathion (Ebrahim et al., 2014). Similary, nanofibers were also used to detect nitrite, one of the most widely used salt and preservatives in foods products. N-doped graphene quantum dots-decorated N-doped carbon nanofibers membrane was used for highly sensitive electrochemical detection of nitrite. Incorporation of N-doped graphene quantum dots increases electron transfer rate, and N-doped carbon nanofibers provide higher electrical conductivity and large electroactive area, freeing standing film structures, which makes this composite appropriate as an electrochemical sensor. This quantum dots-based nanofiber sensor displays superior performance for nitrite sensing with low detection limit up to 3 μM, a wide range of linearity, high

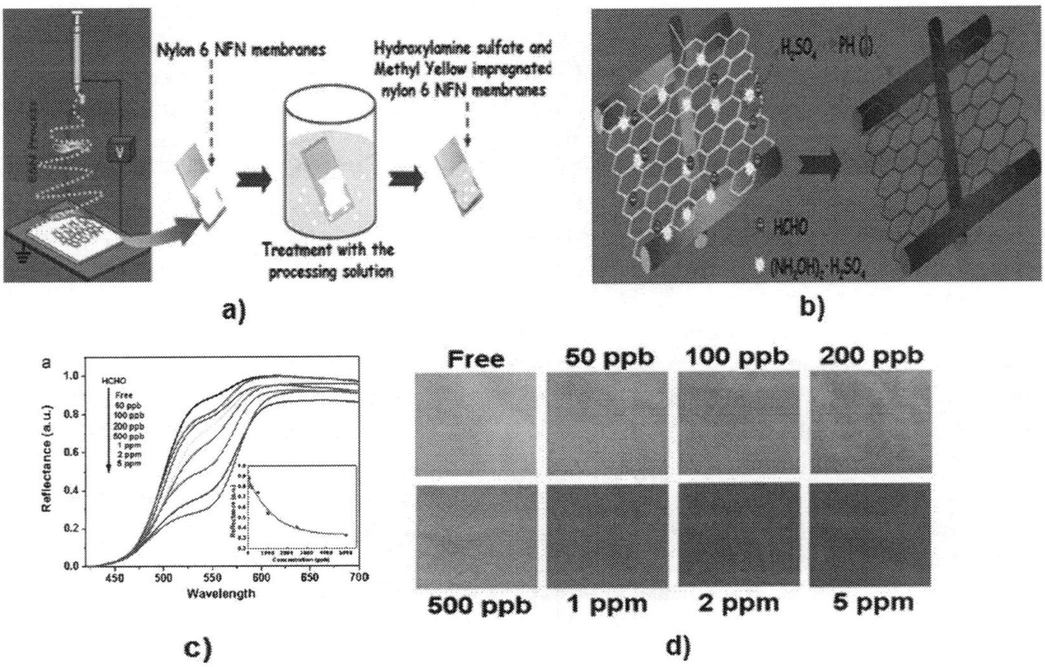

FIGURE 9.6 (a) preparation of the Methyl Yellow-impregnated nylon 6 colorimetric NFN membranes; (b) illustration of the colorimetric detection of formaldehyde based on the nylon 6 NFN membranes; (c) reflectance spectra; and (d) optical colorimetric response of the Methyl Yellow-impregnated nylon 6 NFN sensor strips upon exposure to increasing formaldehyde concentrations. (Wang et al., 2012.)

selectivity, and excellent reproducibility (Li et al., 2017). A nanofiber-based e-tongue for detection of paraoxon in water, specifically in waters of agroproduce washing in post-harvest treatment, was developed. For this, nylon nanofibers were modified with layer-by-layer (LbL) films of conductive polypyrrole and poly(o-ethoxyaniline) polymers that were assembled onto graphite-interdigitated polyethylene terephtalate substrates, and used to detect pesticide in water. The electronic tongue apparatus, assembled to a flow analysis system, was able to distinguish water from different sources, with and without the presence of contaminants, at nanomolar concentrations (Oliveira et al., 2012).

9.3.5 Packaging

Active packaging materials in the form of antimicrobial packaging to eliminate microbial contamination, spoilage, and foodborne pathogens have been developed using antimicrobial agents such as organic, inorganic, and biologically active substances. Among the biologically active substances, essential oils are preferred as eco-friendly substances with powerful antimicrobial activity. However, poor solubility and special flavor which could alter the original sensory property of the foods limit their usage in food processing. Encapsulation by the electrohydrodynamic process is proven to improve solubility and stability as well as masking the undesirable flavors of essential oils. For example, cinnamon essential oil was encapsulated in PVA-based electrospun nanofibers with b-cyclodextrin inclusion complex. The developed nanofibers exhibited excellent antimicrobial activity against *Staphylococcus aureus* and *Escherichia coli*. When tested for effective packaging of strawberry fruits, cinnamon essential oil-loaded nanofibers could extend the shelf life of fruits by more than six days, whereas unpacked fruits and fruits covered with fresh-keeping films were spoiled within the test period (Wen et al., 2016). Similarly, many active packaging nanofiber films loaded with essential oils such as cinnamon (Wen et al., 2016), carvacrol (Tampau et al., 2017), peppermint, thyme, orange, etc., have been developed for food packaging applications. The electrohydrodynamic process has been employed to form an edible coating on food products. For example, electrospraying was employed to coat water-in-oil emulsion (sunflower oil and olive oil were coated on apple slices) on apple slices and chocolate on candy tablets. Emulsion coating by electrospraying was effective in preventing the browning of apple slices but could not control the moisture loss. The coated emulsion penetrated the apple slices leaving a thin film on the surface. Whereas chocolate-based coatings formed a solid coating on the food product, and reduced water vapor flux effectively compared to emulsion-based coatings (Kashif et al., 2013). In another study, an active edible packaging material was developed by encapsulating alpha-tocopherol, an antioxidant, in different wall materials such as whey protein isolate (WPI), zein, and soy protein isolate (SPI) by the electrohydrodynamic process. Alpha-tocopherol-encapsulated SPI and WPI were deposited as nanoparticles on wheat gluten film; where alpha-tocopherol-encapsulated zein was deposited as nanofibers on gluten film (Figure 9.7). Active bilayer packaging materials were developed by depositing these nanostructures on thermoplastic wheat gluten film (Fabra et al., 2016). Alpha-tocopherol release from WPI and SPI-encapsulated nanoparticles was much faster due to faster dissolution of these water-soluble protein matrices, whereas zein wall material provides delayed release of active compounds. So, by selecting the appropriate wall material, we can tune the release profile of the active material. The high melting point of zein provides thermally stable encapsulation around alpha-tocopherol which could preserve its activity even under industrial sterilization process. This bilayer packaging material is a novel active packaging material in food preservation. Thus, the electrohydrodynamic process has proven its potential to increase the shelf life of minimally processed food products.

9.4 CONCLUSION

Electrohydrodynamic processes, namely elelctrospinning and electrospraying, are used to produce two different nanostructures using the same principle and device just by changing the polymer solution properties and processing condition. These versatile techniques allow the creation of 1D

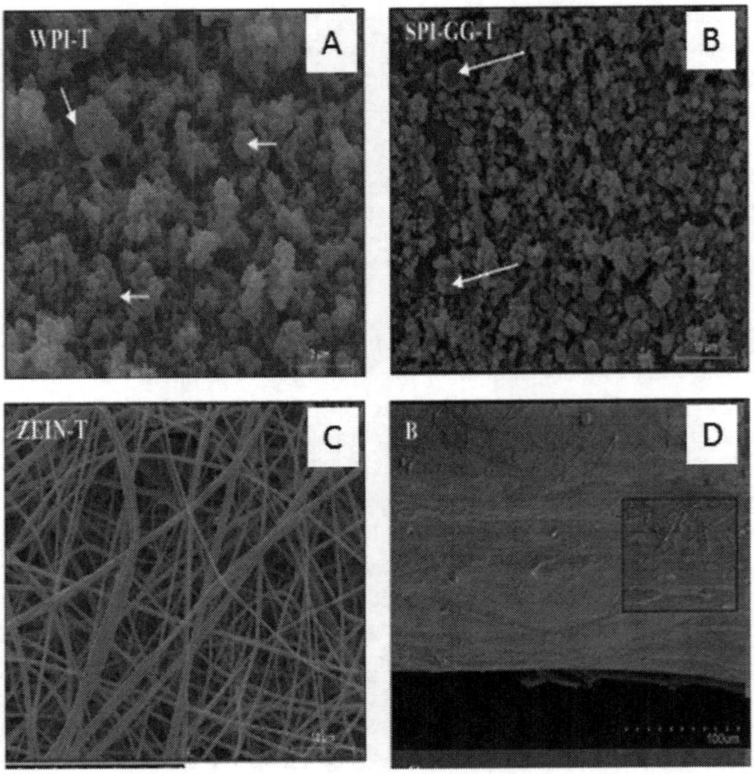

FIGURE 9.7 SEM microphotographs of (a) WPI-alpha-tocopherol capsules; (b) SPI-GG-alpha-tocopherol capsules; (c) zein-alpha-tocopherol fibers; and (d) cryo-fractured section of the developed bilayer with nanofiber coating on wheat gluten film formed by means of the electrohydrodynamic process. (Fabra et al., 2016.)

nanofibers and 3D nanospheres from a wide range of synthetic and natural polymers. The possibility to form nanostructures from natural polymers such as polysaccharides and proteins brought these techniques to a focal point in the food processing field. Ambient processing conditions and dehydration of the polymer in a few seconds make these processes ideal for encapsulation of bioactive molecules which are sensitive to heat, pressure, pH, and light; otherwise, they would be subjected to conventional food processing techniques such as spray drying, freeze drying, extrusion, etc. The potential application of these nanostructures to encapsulate and deliver bioactive/drug molecules in controlled rates is well explored in pharmaceutical and tissue engineering applications. With a relatively recent entry in the food processing field, electrospun nanofibers and electrosprayed nanoparticles have been used to encapsulate food bioactive molecules such as antioxidants, flavonoids and non-flavonoids, live cells, enzymes, essential oils, micronutrients, etc. to provide protective coating against undesirable environmental/processing conditions such as heat, light, pH, etc (Anu Bhushani and Anandharamakrishnan 2014; Ghorani and Tucker 2015). Additionally, these protective coatings were used to mask undesirable flavors and taste of active molecules (like omega-3 fatty acids), and to meet consumers' requests for organoleptic quality and functionality of functional foods. Though there are many claims reporting effective protection of active molecules by micro/nanocoatings, accessing the behavior of encapsulated molecules in real food processing conditions needs to be emphasized (Hern et al., 2017).

Flexibility to reinforce any type of organic and inorganic materials, offers tunable optical, electrical, chemical, mechanical, and biological properties to these nanostructures. Reinforcement of metal nanoparticles such as silver, gold, ZnO nanoparticles, carbon nanotubes (CNT), graphene, graphite, etc. has been used to alter the conductivity of the nanoparticles and nanofibers, and in sensing applications. Silver nanoparticles-based electrospun nanocomposite fibers are widely investigated for effective antimicrobial and protective coating/packaging properties in food products. In addition to high surface to volume ratio, electrospun nanofibers present unique structural properties such as intertwined fibrous structure, high porosity and interconnected pores, improved mechanical properties, etc. Electrospun nanofibers are potent candidates for filtration and clarification of beverages, bioreactor formation through enzyme immobilization, sensing elements, packaging materials, etc. While it presents unique advantages, the low throughput of electrospinning/electrospraying is a major limitation of this process. The synergetic effort of researchers and industrialists brought a solution to this limitation through the development of a multi-nozzle processing system for continuous industrial applications. Recently, an air pressure-assisted electrohydrodynamic process has also been developed to scale up this process. With these credentials, electrosprayed nanoparticles and electrospun nanofibers are proving their potential in the food processing field.

ACKNOWLEDGMENTS

The author Maria Leena acknowledges the Department of Science and Technology (DST), Government of India for the award of Women Scientist Fellowship (WOS-A).

REFERENCES

Anandharamakrishnan, C. 2014a. "Liquid-based nanoencapsulation techniques." In *Techniques for Nanoencapsulation of Food Ingredients*, New York, Springer, Chapter 4, 29–41.

Anandharamakrishnan, C. 2014b. "Techniques for formation of nanoemulsions." In *Techniques for Nanoencapsulation of Food Ingredients*, New York, Springer, Chapter 2, 7–16.

Anandharamakrishnan, C. 2014c. "Bioactive entrapment using Lipid based nanocarrier technology." *Techniques for Nanoencapsulation of Food Ingredients*, New York, Springer, Chapter 3, pp 17–28.

Anandharamakrishnan, C. 2015. "Spray drying for nanoencapsulation of food components." *In Spray Drying Techniques for Food Ingredient Encapsulation*, Chichester, John Wiley & Sons, Chapter 8, pp 180–197. 9.

Anandharamakrishnan, C. and S. Padma Ishwarya. 2015a. "Introduction to encapsulation of food ingredients." In *Spray Drying Techniques for Food Ingredient Encapsulation,* Cambridge, John Wiley & Sons, Chapter 2, pp 37–64.

Anandharamakrishnan, C. and S. Padma Ishwarya. 2015b. "Encapsulation of bioactive ingredients by spray drying." In *Spray Drying Techniques for Food Ingredient Encapsulation,* Chichester, John Wiley & Sons, Chapter 7, pp 156–179. 9.

Ananingsih, V. K., Sharma, A., and Zhou, W. 2013. "Green tea catechins during food processing and storage: A review on stability and detection." *Food Research International*, 50 (2). Elsevier: 469–479. doi:10.1016/J.FOODRES.2011.03.004.

Anh Tuan Dao, Oldrich JIRSAK. 2010. "Effect of cross-linking agent on electrospinning of poly(vinyl) alcohol." 7th International Conference- TEXSCI 2010, September 6–8.

Anu Bhushani, J. and C. Anandharamakrishnan. 2014. "Electrospinning and electrospraying techniques: Potential food based applications." *Trends in Food Science and Technology*, 38 (1). Elsevier Ltd: 21–33. doi:10.1016/j.tifs.2014.03.004.

Aussawasathien, D. J.-H., J.-H Dong, and L. Dai. 2005. "Electrospun polymer nanofiber sensors." *Synthetic Metals,* 154 (1–3). Elsevier: 37–40. doi:10.1016/J.SYNTHMET.2005.07.018.

Baspinar, Y., Üstündas, M-, Bayraktar, O., and Sezgin, C. 2018. "Curcumin and piperine loaded zein-chitosan nanoparticles : Development and in-vitro characterisation." *Saudi Pharmaceutical Journal*, 26 (3): 323–334. doi:10.1016/j.jsps.2018.01.010.

Baumgarten, Peter K. 1971. "Electrostatic spinning of acrylic microfibers." *Journal of Colloid and Interface Science*, 36 (1): 71–79. doi:https://doi.org/10.1016/0021-9797(71)90241-4.

Bhardwaj, Nandana and Subhas Kundu. 2010. "Electrospinning: A fascinating fiber fabrication technique. *Biotechnology Advances*, 28: 325–347.

Bhushani, Anu and C. Anandharamakrishnan. 2017. "Food-grade nanoemulsions for protection and delivery of nutrients." In *Nanoscience in Food and Agriculture*, 4, 99–139. Springer.

Bhushani, J. Anu, P. Karthik, and C. Anandharamakrishnan. 2016. "Nanoemulsion based delivery system for improved bioaccessibility and Caco-2 cell monolayer permeability of green tea catechins." *Food Hydrocolloids,* 56:372–382.

Bhushani, J. Anu, Nawneet Kumar Kurrey, and C. Anandharamakrishnan. 2017. "Nanoencapsulation of green tea catechins by electrospraying technique and its effect on controlled release and in-vitro permeability." *Journal of Food Engineering*, 199: 82–92. doi:10.1016/j.jfoodeng.2016.12.010.

Bock, N., T. R. Dargaville, and M. A. Woodruff. 2012. "Progress in polymer science electrospraying of polymers with therapeutic molecules: State of the art." *Progress in Polymer Science*, 37 (11). Elsevier Ltd: 1510–1551. doi:10.1016/j.progpolymsci.2012.03.002.

Brahatheeswaran, Dhandayuthapani, Anila Mathew, Ravindran Girija Aswathy, Yutaka Nagaoka, K. Venugopal, Yasuhiko Yoshida, Toru Maekawa, and D. Sakthikumar. 2012. "Hybrid fluorescent curcumin loaded zein electrospun nanofibrous scaffold for biomedical applications." *Biomedical Materials*, 7 (4): 045001. doi:10.1088/1748-6041/7/4/045001.

Casper, C.L. 2004. "Controlling surface morphology of electrospun polystyrene fibers: effect of humidity andmolecular weight in the electrospinning process." *Macromolecules*. 37:573–8.

Celebioglu, Asli and Tamer Uyar. 2011. "Electrospinning of polymer-free nanofibers from cyclodextrin inclusion complexes." *Langmuir : The ACS Journal of Surfaces and Colloids*, 27: 6218–26.

Chakraborty, Syandan, I-Chien Liao, Andrew Adler, and Kam W. Leong. 2009. "Electrohydrodynamics: A facile technique to fabricate drug delivery systems." *Advanced Drug Delivery Reviews*, 61 (12): 1043–1054. doi:https://doi.org/10.1016/j.addr.2009.07.013.

Chunxue, Zhang, Yuan Xiaoyan, Wu Lili, Han Yue, and Shenga Jing. 2005. "Study on morphology of electrospun poly(vinyl alcohol) mats." *European Polymer Journal,* 41 (3). Pergamon: 423–432. doi:10.1016/J.EURPOLYMJ.2004.10.027.

Cloupeau, Michel and Bernard Prunet-Foch. 1994. "Electrohydrodynamic spraying functioning modes: A critical review." *Journal of Aerosol Science,* 25 (6): 1021–1036. doi:https://doi.org/10.1016/0021-8502(94)90199-6.

Coghetto, C. C., G. B. Brinques, N. M. Siqueira, J. Pletsch, R. M. D. Soares, and M. A. Z. Ayub. 2016. "Electrospraying microencapsulation of lactobacillus plantarum enhances cell viability under refrigeration storage and simulated gastric and intestinal fluids." *Journal of Functional Foods,* 24: 316–326. doi:10.1016/j.jff.2016.03.036.

Colmenares-Roldán, Gabriel Jaime, Yeixon Quintero-Martínez, Liliana María Agudelo-Gómez, Luis Fernando Rodríguez, and Lina Marcela Hoyos-Palacio. 2017. "Influence of the molecular weight of polymer, solvents and operational condition in the electrospinning of polycaprolactone," 84: 35–45. doi:10.17533/udea.redin.n84a05.

Cramariuc, Bogdan, Radu Cramariuc, Roxana Scarlet, Liliana Rozemarie Manea, Iuliana G. Lupu, and Oana Cramariuc. 2013. "Fiber diameter in electrospinning process." *Journal of Electrostatics,* 71 (3). Elsevier Ltd: 189–198. doi:10.1016/j.elstat.2012.12.018.

Dai, Minhui, Shengquan Jin, and Sam R. Nugen. 2012. "Water-soluble electrospun nanofibers as a method for on-chip reagent storage." *Biosensors*, 2 (4): 388–395. doi:10.3390/bios2040388.

Dalton, Paul, Kristina Klinkhammer, Jochen Salber, Doris Klee, and Martin Möller. 2006. "Direct in vitro electrospinning with polymer melts." *Biomacromolecules*, 7: 686–690.

Danet, Andrei Florin, Aurelia Magdalena Pisoschi, and Slawomir Kalinowski. 2008. "Ascorbic acid determination in commercial fruit juice samples by cyclic voltammetry." *Journal of Automated Methods and Management in Chemistry,* 2008 (Ii). doi:10.1155/2008/937651.

Dhanasekaran, M. 2014. "Polycaprolactone scaffold engineered for sustained release of resveratrol: Therapeutic enhancement in bone tissue engineering,"*International Journal of Nanomedicine*, 9:183–195.

Ding, Bin, Moran Wang, Jianyong Yu, and Gang Sun. 2009. "Gas sensors based on electrospun nanofibers." *Sensors,* 9 (3): 1609–1624. doi:10.3390/s90301609.

Doshi, J. and D.H. Reneker. 1993. "Electrospinning process and applications of electrospun fibers." *Conference Record of the 1993 IEEE Industry Applications Conference Twenty-Eighth IAS Annual Meeting,* 35: 151–160. doi:10.1109/IAS.1993.299067.

Duncan, Timothy V. 2014. "Applications of nanotechnology in food packaging and food safety: Barrier materials, antimicrobials and sensors." *Journal of Colloid and Interface Science,* 363 (1). Elsevier Inc.: 1–24. doi:10.1016/j.jcis.2011.07.017.

Ebrahim, Shaker, Radwa El-Raey, Ahmed Hefnawy, Hesham Ibrahim, Moataz Soliman, and Tarek M. Abdel-Fattah. 2014. "Electrochemical sensor based on polyaniline nanofibers/single wall carbon nanotubes composite for detection of malathion." *Synthetic Metals,* 190. Elsevier B.V.: 13–19. doi:10.1016/j.synthmet.2014.01.021.

Ezhilarasi, P.N., P. Karthik, N. Chhanwal, and C. Anandharamakrishnan. 2013. "Nanoencapsulation techniques for food bioactive components: A review." *Food and Bioprocess Technology*, 6 (3):628–647.

Ezhilarasi, Perumal Natarajan, Dasappa Indrani, Bhabani Sankar Jena, and Chinnaswamy Anandharamakrishnan. 2014. "Microencapsulation of *Garcinia* fruit extract by spray drying and its effect on bread quality." *Journal of the Science of Food and Agriculture*, 94 (6). John Wiley & Sons, Ltd: 1116–1123. doi:10.1002/jsfa.6378.

Fabra, María José, Amparo López-Rubio, and Jose M. Lagaron. 2016. "Use of the electrohydrodynamic process to develop active/bioactive bilayer films for food packaging applications." *Food Hydrocolloids*, 55: 11–18. doi:10.1016/j.foodhyd.2015.10.026.

Fernandez, Avelina, Sergio Torres-giner, and Jose Maria Lagaron. 2009. "Novel route to stabilization of bioactive antioxidants by encapsulation in electrospun fibers of zein prolamine," 23. Elsevier Ltd: 1427–1432. doi:10.1016/j.foodhyd.2008.10.011.

Fraceto, Leonardo F., Renato Grillo, Gerson A. de Medeiros, Viviana Scognamiglio, Giuseppina Rea, and Cecilia Bartolucci. 2016. "Nanotechnology in agriculture: Which innovation potential does it have?" *Frontiers in Environmental Science,* 4 (March). Frontiers: 20. doi:10.3389/fenvs.2016.00020.

Fridrikh, Sergey V., Jian H. Yu, Michael P. Brenner, and Gregory C. Rutledge. 2003. "Controlling the fiber diameter during electrospinning." *Physical Review Letters*, 90 (14): 144502. doi:10.1103/PhysRevLett.90.144502.

Fung, Wai Yee, Kay Hay Yuen, and Min Tze Liong. 2011. "Agrowaste-based nanofibers as a probiotic encapsulant: Fabrication and characterization." *Journal of Agricultural and Food Chemistry*, 59 (15): 8140–8147. doi:10.1021/jf2009342.

Gañán-Calvo, A. M., J. Dávila, and A. Barrero. 1997. "Current and droplet size in the electrospraying of liquids. Scaling laws." *Journal of Aerosol Science*, 28 (2): 249–275. doi:https://doi.org/10.1016/S0021-8502(96)00433-8.

García-Moreno, Pedro J., Necla Özdemir, Karen Stephansen, Ramona V. Mateiu, Yolanda Echegoyen, Jose M. Lagaron, Ioannis S. Chronakis, and Charlotte Jacobsen. 2017. "Development of carbohydrate-based nano-microstructures loaded with fish oil by using electrohydrodynamic processing." *Food Hydrocolloids,* 69 (August): 273–285. doi:10.1016/j.foodhyd.2017.02.013.

García-Moreno, Pedro J., Andres Pelayo, Sen Yu, María Busolo, Jose M. Lagaron, Ioannis S. Chronakis, and Charlotte Jacobsen. 2018. "Physicochemical characterization and oxidative stability of Fish oil-loaded electrosprayed capsules: Combined use of whey protein and carbohydrates as wall materials." *Journal of Food Engineering,* 231. Elsevier Ltd: 42–53. doi:10.1016/j.jfoodeng.2018.03.005.

Ghorani, Behrouz and Nick Tucker. 2015. "Fundamentals of electrospinning as a novel delivery vehicle for bioactive compounds in food nanotechnology." *Food Hydrocolloids*, 51. Elsevier Ltd: 227–240. doi:10.1016/j.foodhyd.2015.05.024.

Gomez-Estaca, J., M. P. Balaguer, R. Gavara, and P. Hernandez-Munoz. 2012. "Formation of zein nanoparticles by electrohydrodynamic atomization: Effect of the main processing variables and suitability for encapsulating the food coloring and active ingredient curcumin." *Food Hydrocolloids*, 28 (1): 82–91. doi:10.1016/j.foodhyd.2011.11.013.

Gomez-Estaca, J., M.P. Balaguer, G. Lopez-Carballo, R. Gavara, and P. Hernandez-Munoz. 2017. "Improving antioxidant and antimicrobial properties of curcumin by means of encapsulation in gelatin through electrohydrodynamic atomization." *Food Hydrocolloids*, 70: 313–320. doi:10.1016/j.foodhyd.2017.04.019.

Gómez-Mascaraque, Laura G., Fran Tordera, Maria Jose Fabra, and Marta Martínez-Sanz. 2018. "Coaxial electrospraying of biopolymers as a strategy to improve protection of bioactive food ingredients." *Innovative Food Science and Emerging Technologies*, February. Elsevier: 0–1. doi:10.1016/j.ifset.2018.03.023.

Grace, J.M. and J. C. M. Marijnissen. 1994. "A review of liquid atomization by electrical means." *Journal of Aerosol Science*, 25 (6): 1005–1019. doi:10.1016/0021-8502(94)90198-8.

Gupta, Pankaj, Casey Elkins, Timothy E. Long, and Garth L. Wilkes. 2005. "Electrospinning of linear homopolymers of poly(methyl methacrylate): Exploring relationships between fiber formation, viscosity, molecular weight and concentration in a good solvent." *Polymer,* 46 (13): 4799–4810. doi:10.1016/j.polymer.2005.04.021.

Hartman, R. P. A., D. J. Brunner, D. M. A. Camelot, J. C. M. Marijnissen, and B. Scarlett. 1999. "Electrohydrodynamic atomization in the cone-jet mode physical modeling of the liquid cone and jet." *Journal of Aerosol Science,* 30 (7): 823–849. doi:10.1016/S0021-8502(99)00033-6.

Hartman, R. P. A., D. J. Brunner, D. M. A. Camelot, J. C. M. Marijnissen, and B. Scarlett. 2000. "Jet break up in electrohydrodynamic atomization in the cone jet mode." *Journal of Aerosol Science*, 31 (1): 65–95. doi:10.1016/S0021-8502(99)00034-8.

Hasan, Anwarul, Adnan Memic, Nasim Annabi, Monowar Hossain, Arghya Paul, Mehmet R. Dokmeci, Fariba Dehghani, and Ali Khademhosseini. 2014. "Electrospun scaffolds for tissue engineering of vascular grafts." *Acta Biomaterialia*, 10 (1). Acta Materialia Inc.: 11–25. doi:10.1016/j.actbio.2013.08.022.

Hern, Marai, Susana Fiszman, Mathieu Tenon, and Nicolas Feuill. 2017. "Impact of microencapsulation within electrosprayed proteins on the formulation of green tea extract-enriched biscuits." *LWT - Food Science and Technology,* 81: 77–86. doi:10.1016/j.lwt.2017.03.041.

Herrera, E. and C. Barbas. 2001. "Vitamin E: Action, metabolism and perspectives." *Journal of Physiology and Biochemistry,* 57 (2): 43–56. http://www.ncbi.nlm.nih.gov/pubmed/11579997.

Heunis, T. D. J., M. Botes, and L. M. T. Dicks. 2010. "Encapsulation of lactobacillus plantarum 423 and its bacteriocin in nanofibers." *Probiotics and Antimicrobial Proteins,* 2 (1): 46–51. doi:10.1007/s12602-009-9024-9.

Huan, Siqi, Guoxiang Liu, Guangping Han, Wanli Cheng, Zongying Fu, Qinglin Wu, and Qingwen Wang. 2015. "Effect of experimental parameters on morphological, mechanical and hydrophobic properties of electrospun polystyrene fibers." 2718–2734. doi:10.3390/ma8052718.

Huang, Weijuan, Yingfei Zhan, Xiaowen Shi, Jiajia Chen, Hongbing Deng, and Yumin Du. 2017. "Controllable immobilization of naringinase on electrospun cellulose acetate nanofibers and their application to juice debittering." *International Journal of Biological Macromolecules,* 98. Elsevier B.V.: 630–636. doi:10.1016/j.ijbiomac.2017.02.018.

Hundre, Swetank Y., P. Karthik, and C. Anandharamakrishnan. 2015. "Effect of whey protein isolate and B-cyclodextrin wall systems on stability of microencapsulated vanillin by spray-freeze drying method." *Food Chemistry*, 174: 16–24. doi:10.1016/j.foodchem.2014.11.016.

Husain, O., W. Lau, M. Edirisinghe, and M. Parhizkar. 2016. "Investigating the particle to fibre transition threshold during electrohydrodynamic atomization of a polymer solution." *Materials Science and Engineering C.* Elsevier B.V. doi:10.1016/j.msec.2016.03.076.

Jadhav, Amit, Li Jing Wang, Carl Lawrence, and Rajiv Padhye. 2011. "Study of electrospraying characteristics of polymer solution coating on textile substrate." *Advanced Materials Research*, 332: 710–715.

Jadhav, Amit, Lijing Wang, and Rajiv Padhye. 2012. "Effect of applied voltage on polymer aggregation in electrospraying of thermoplastic polymer" 537: 2522–2525. doi:10.4028/www.scientific.net/AMR.535-537.2522.

Jadhav, Amit, Lijing Wang, and Rajiv Padhye. 2013. "Influence of applied voltage on droplet size distribution in electrospraying of thermoplastic polyurethane." *International Journal of Materials, Mechanics and Manufacturing*, 1 (3): 287–289. doi:10.7763/IJMMM.2013.V1.61.

Karthik, P. and C. Anandharamakrishnan. 2013. "Microencapsulation of docosahexaenoic acid by spray-freeze-drying method and comparison of its stability with spray-drying and freeze-drying methods." *Food and Bioprocess Technology,* 6 (10):2780–2790.

Karthik, P., P.N. Ezhilarasi, and C. Anandharamakrishnan. 2017. "Challenges associated in stability of food grade nanoemulsions." *Critical Reviews in Food Science and Nutrition,* 57 (7):1435–1450.

Kashif, Muhammad, Iqbal Khan, Hulya Cakmak, Ş. Tavman, Maarten Schutyser, and Karin Schro. 2013. "Anti-browning and barrier properties of edible coatings prepared with electrospraying." *Innovative Food Science and Emerging Technologies.* doi:10.1016/j.ifset.2013.10.006.

Khayata N., W. Abdelwahed, M. F. Chehna, C. Charcosset, and H. Fessi. 2012. "Stability study and lyophilization of vitamin E-loaded nanocapsules prepared by membrane contactor." *International Journal of Pharmaceutics*, 439 (1–2). Elsevier: 254–259. doi:10.1016/J.IJPHARM.2012.09.032.

Kim, Choo-Won, Margaret Frey, Manuel Marquez, and Yong Joo. 2005. "Preparation of submicron-scale, electrospun cellulose fibers via direct dissolution." *Journal of Polymer Science Part B: Polymer Physics*, 43: 1673–1683.

Kim, Su Jin, Yu Lim Kim, Areum Yu, Jiyoung Lee, Sang Cheol Lee, Chongmok Lee, Myung Hwa Kim, and Youngmi Lee. 2014. "Electrospun iridium oxide nanofibers for direct selective electrochemical detection of ascorbic acid." *Sensors and Actuators, B: Chemical,* 196. Elsevier B.V.: 480–488. doi:10.1016/j.snb.2014.02.032.

la Mora, J. Fernández de, and I. G. Loscertales. 1994. "The current emitted by highly conducting Taylor cones." *Journal of Fluid Mechanics,* 260 (special issue): 155–184. doi:10.1017/S0022112094003472.

Larrondo, L. and R. St. John Manley. 1981. "Electrostatic fiber spinning from polymer melts. III. Electrostatic deformation of a pendant drop of polymer melt." *Journal of Polymer Science: Polymer Physics Edition*, 19 (6): 933–940. doi:10.1002/pol.1981.180190603.

Leena, M., A. Barade, D. Rana, C. Dhand, Seeram Ramakrishna, and Murugam Ramalingam. 2017. "Nanofiber composites in biomolecular delivery." In *Nanofiber Composites for Biomedical Applications*, 225–252. Elsevier Ltd. doi:10.1016/B978-0-08-100173-8.00009-0.

Li, Y., L. T. Lim, and Y. Kakuda. 2009. "Electrospun zein fibers as carriers to stabilize (–) -epigallocatechin gallate." *Food Chemistry,* 74 (3): 233–240. doi:10.1111/j.1750-3841.2009.01093.x.

Li, Wei, Xueyong Li, Qun Wang, Yijun Pan, Ting Wang, Hanqing Wang, Rong Song, and Hongbing Deng. 2014a. "Antibacterial activity of nanofibrous mats coated with lysozyme-layered silicate composites via electrospraying." *Carbohydrate Polymers,* 99. Elsevier Ltd.: 218–225. doi:10.1016/j.carbpol.2013.07.055.

Li, Xiao Yan, Chen Jie Shi, Deng Guang Yu, Yao Zu Liao, and Xia Wang. 2014b. "Electrospun quercetin-loaded zein nanoribbons." *Bio-Medical Materials and Engineering,* 24 (6): 2015–2023. doi:10.3233/BME-141011.

Li, Juan, Arranee Chotiko, Damien A. Narcisse, and Subramaniam Sathivel. 2016. "Evaluation of alpha-tocopherol stability in soluble dietary fiber based nanofiber" 68: 485–490. doi:10.1016/j.lwt.2015.12.042.

Li, Libo, Dong Liu, Kun Wang, Hanping Mao, and Tianyan You. 2017. "Quantitative detection of nitrite with N-doped graphene quantum dots decorated N-doped carbon nanofibers composite-based electrochemical sensor." *Sensors and Actuators, B: Chemical,* 252. Elsevier B.V.: 17–23. doi:10.1016/j.snb.2017.05.155.

Librán, C. M., S. Castro, and J. M. Lagaron. 2017. "Encapsulation by electrospray coating atomization of probiotic strains." *Innovative Food Science and Emerging Technologies,* 39. Elsevier Ltd: 216–222. doi:10.1016/j.ifset.2016.12.013.

Liou, Polly, Francois Xavier Nayigiziki, Fanbin Kong, Azlin Mustapha, and Mengshi Lin. 2017. "Cellulose nanofibers coated with silver nanoparticles as a SERS platform for detection of pesticides in apples." *Carbohydrate Polymers,* 157 (February): 643–650. doi:10.1016/j.carbpol.2016.10.031.

Medeiros, Eliton S., Rinaldo Gregório, Rodrigo A. Martinez, and Luiz H. C. Mattoso. 2009. "A taste sensor array based on polyaniline nanofibers for orange juice quality assessment," February. doi:10.1166/sl.2009.1005.

Mit-uppatham, Chidchanok, Manit Nithitanakul, and Pitt Supaphol. 2004. "Ultrafine electrospun polyamide-6 fibers: Effect of solution conditions on morphology and average fiber diameter." *Macromolecular Chemistry and Physics,* 205: 2327–2338.

Moayyedi, Mahsa, Mohammad Hadi, Amir Hossein, Elhami Rad, Esmaeil Ziaee, Mohammad Hossein, Haddad Khodaparast, and Mohammad-taghi Golmakani. 2018. "Effect of drying methods (electrospraying, freeze drying and spray drying) on survival and viability of microencapsulated lactobacillus rhamnosus ATCC 7469." *Journal of Functional Foods,* 40 (November 2017). Elsevier: 391–399. doi:10.1016/j.jff.2017.11.016.

Mojir, Parmiss, Keren Jiang, Ghazaleh Haghighat, Mahtab Hassanpourfard, Hashem Etayash, Selvaraj Naicker, and Thomas Thundat. 2016. "The detection of Escherichia Coli (E. Coli) with the pH sensitive hydrogel nanofiber-light addressable potentiometric sensor (NF-LAPS)." *Sensors and Actuators: B. Chemical,* 226. Elsevier B.V.: 176–183. doi:10.1016/j.snb.2015.11.135.

Moomand, Khalid and Loong-Tak Lim. 2014. "Oxidative stability of encapsulated fish oil in electrospun zein fibres." *Food Research International,* 62. Elsevier Ltd: 523–532. doi:10.1016/j.foodres.2014.03.054.

Nakamura, Hideya and Satoru Watano. 2018. "Direct permeation of nanoparticles across cell membrane: A review." *KONA Powder and Particle Journal,* 2018 (35): 49–65. doi:10.14356/kona.2018011.

Nunes, Mário A. P, Samuel Martins, M. Emilia Rosa, Pedro M. P. Gois, Pedro C. B. Fernandes, and Maria H. L. Ribeiro. 2015. "Improved thermostable polyvinyl alcohol electrospun nanofibers with entangled naringinase used in a novel mini-packed bed reactor." *Bioresource Technology,* 213. Elsevier Ltd: 208–215. doi:10.1016/j.biortech.2016.03.058.

Oliveira, Juliano E., Vanessa P. Scagion, Viviane Grassi, Daniel S. Correa, and Luiz H. C. Mattoso. 2012. "Modification of electrospun nylon nanofibers using layer-by-layer films for application in flow injection electronic tongue: Detection of paraoxon pesticide in corn crop." *Sensors and Actuators: B. Chemical,* 171–172. Elsevier B.V.: 249–255. doi:10.1016/j.snb.2012.03.056.

Parthasarathi, S. and C. Anandharamakrishnan. 2016. "Enhancement of oral bioavailability of vitamin E by spray-freeze drying of whey protein microcapsules." *Food and Bioproducts Processing,* 100 (A): 469–476. doi:10.1016/j.fbp.2016.09.004.

Parthasarathi, S., S.P. Muthukumar, and C. Anandharamakrishnan. 2016. "The influence of droplet size on the stability, in vivo digestion, and oral bioavailability of vitamin E emulsions." *Food and Function,* 7 (5):2294–2302.

Pasrija, D., P. N. Ezhilarasi, D Indrani, and C Anandharamakrishnan. 2015. "Microencapsulation of green tea polyphenols and its effect on incorporated bread quality." *LWT - Food Science and Technology,* 64 (1): 289–296. doi:10.1016/j.lwt.2015.05.054.

Pathakoti, Kavitha, Manjunath Manubolu, and Huey Min Hwang. 2017. "Nanostructures: Current uses and future applications in food science." *Journal of Food and Drug Analysis,* 25 (2). Elsevier Ltd: 245–253. doi:10.1016/j.jfda.2017.02.004.

Pelipenko, J., P. Kocbek, and J. Kristl. 2015. "Critical attributes of nanofibers: Preparation, drug loading, and tissue regeneration." *International Journal of Pharmaceutics,* 484 (1–2). Elsevier B.V.: 57–74. doi:10.1016/j.ijpharm.2015.02.043.

Peters, Ruud, Puck Brandhoff, Stefan Weigel, Hans Marvin, Hans Bouwmeester, Karin Aschberger, Hubert Rauscher, et al. 2014. "Inventory of nanotechnology applications in the agricultural, feed and food sector." *EFSA Supporting Publications*, 11 (7). doi:10.2903/sp.efsa.2014.EN-621.

Peters, Ruud J. B., Hans Bouwmeester, Stefania Gottardo, Valeria Amenta, Maria Arena, Puck Brandhoff, Hans J. P. Marvin, et al. 2016. "Nanomaterials for products and application in agriculture, feed and food." *Trends in Food Science and Technology*, 54. Elsevier Ltd: 155–164. doi:10.1016/j.tifs.2016.06.008.

Ping, Yun, Sudip Ray, Jianyong Jin, Marija Gizdavic-nikolaidis, Michel K. Nieuwoudt, Dongyan Liu, and Siew Young. 2013. "Encapsulation of food grade antioxidant in natural biopolymer by electrospinning technique: A physicochemical study based on zein – gallic acid system." *Food Chemistry,* 136 (2). Elsevier Ltd: 1013–1021. doi:10.1016/j.foodchem.2012.09.010.

Poornima, Balan and Purna Sai Korrapati. 2016. "Fabrication of chitosan-polycaprolactone composite nanofibrous scaffold for simultaneous delivery of ferulic acid and resveratrol." *Carbohydrate Polymers,* Elsevier Ltd. doi:10.1016/j.carbpol.2016.11.056.

Quintanilla-Carvajal, María Ximena, Brenda Hildeliza Camacho-Díaz, Lesvia Sofia Meraz-Torres, José Jorge Chanona-Pérez, Liliana Alamilla-Beltrán, Antonio Jimenéz-Aparicio, and Gustavo F. Gutiérrez-López. 2010. "Nanoencapsulation: A new trend in food engineering processing." *Food Engineering Reviews*, 2 (1): 39–50. doi:10.1007/s12393-009-9012-6.

Rana, Deepti, Maria Leena, Manasseh Nithyananth, Renu Pasricha, Geetha Manivasagam, and Murugan Ramalingam. 2016. "Control of stem cell fate and function by polymer nanofibers." *Journal of Nanoscience and Nanotechnology*, 16 (9): 9015–9021. doi:10.1166/jnn.2016.12735.

Rana, Deepti, Minal Thacker, Maria Leena, and Murugan Ramalingam. 2017. "Induced pluripotent stem cells in scaffold-based tissue engineering." In *Tissue Engineering for Artificial Organs*, 111–142. Weinheim, Germany: Wiley-VCH Verlag GmbH & Co. KGaA. doi:10.1002/9783527689934.ch4.

Ravichandran, R. 2010. "Nanotechnology applications in food and food processing: Innovative green approaches, opportunities and uncertainties for global market." *International Journal of Green Nanotechnology: Physics and Chemistry*, 1: 72–96. doi:10.1080/19430871003684440.

Rayleigh, Lord. 1878. "On the instability of jets." *Proceedings of the London Mathematical Society*, S1–10 (1): 4–13. doi:10.1112/plms/s1-10.1.4.

Rayleigh, Lord. 1882. "On the equilibrium of liquid conducting masses charged with electricity." *Philosophical Magazine Series, 5*. Vol. 14. doi:10.1080/14786448208628425.

Remondetto, Gabriel E., Erick Beyssac, and Muriel Subirade. 2004. "Iron availability from whey protein hydrogels: An in vitro study." *Journal of Agricultural and Food Chemistry,* 52 (26). American Chemical Society: 8137–8143. doi:10.1021/JF040286H.

Reneker, Darrell H., Alexander L. Yarin, Hao Fong, and Sureeporn Koombhongse. 2000. "Bending instability of electrically charged liquid jets of polymer solutions in electrospinning." *Journal of Applied Physics,* 87 (9): 4531–4547. doi:10.1063/1.373532.

Rutledge, G. C., Li, Y., Fridrikh, S., Warner, S. B., Kalayci, V. E., & Patra, P. 2001. "Electrostatic spinning and properties of ultrafine fibers." *National Textile Center Annual Report*, (M98-D01), 1-10.

Saallah, Suryani, M. Nazli Naim, Mohd Noriznan, Noor Fitrah, Abu Bakar, Masao Gen, and I. Wuled Lenggoro. 2014. "Enzyme and microbial technology transformation of cyclodextrin glucanotransferase (CGTase) from aqueous suspension to fine solid particles via electrospraying." *Enzyme and Microbial Technology,* 64–65. Elsevier Inc.: 52–59. doi:10.1016/j.enzmictec.2014.06.002.

Shin, Y. M., M. M. Hohman, M. P. Brenner, and G. C. Rutledge. 2001a. "Experimental characterization of electrospinning: The electrically forced jet and instabilities." *Polymer,* 42 (25): 9955–9967. doi:https://doi.org/10.1016/S0032-3861(01)00540-7.

Shin, Y. M., M. M. Hohman, M. P. Brenner, and G. C. Rutledge. 2001b. "Electrospinning: A whipping fluid jet generates submicron polymer fibers." *Applied Physics Letters,* 78 (8): 1149–1151. doi:10.1063/1.1345798.

Silke, Megelski, Stephens Jean S., Chase D. Bruce, and F. Rabolt John. 2002. "Micro- and nanostructured surface morphology on electrospun polymer fibers." *Macromolecules*, 35 (22). American Chemical Society: 8456–8466. doi:10.1021/MA020444A.

Sill, Travis J. and Horst A. von Recum. 2008. "Electrospinning: Applications in drug delivery and tissue engineering." *Biomaterials,* 29 (13): 1989–2006. doi:https://doi.org/10.1016/j.biomaterials.2008.01.011.

Supaphol, Pitt, Chidchanok Mit-uppatham, and Manit Nithitanakul. 2005. "Ultrafine electrospun polyamide-6 fibers: Effects of solvent system and emitting electrode polarity on morphology and average fiber diameter." *Macromolecular Materials and Engineering*, 290: 933–942.

Tampau, Alina and Amparo Chiralt. 2017. "Carvacrol encapsulation in starch or PCL based matrices by electrospinning." *Journal of Food Engineering*. doi:10.1016/j.jfoodeng.2017.07.005.

Tang, Keqi and Alessandro Gomez. 1996. "Monodisperse electrosprays of low electric conductivity liquids in the cone-jet mode." *Journal of Colloid and Interface Science*, 184 (2): 500–511. doi:https://doi.org/10.1006/jcis.1996.0645.

Taylor,G. 1964. "Disintegration of water drops in an electric field." *Proceedings of the Royal Society of London A: Mathematical, Physical and Engineering Sciences,* 280 (1382). The Royal Society: 383–397. doi:10.1098/rspa.1964.0151.

Taylor, G. 1969. "Electrically driven jets." *Proceedings of the Royal Society A: Mathematical, Physical and Engineering Sciences*, 313 (1515): 453–475. doi:10.1098/rspa.1969.0205.

Thompson, C. J., George Chase, A. L. Yarin, and D. H. Reneker. 2007. "Effects of parameters on nanofiber diameter determined from electrospinning model." *Polymer,* 48: 6913–6922.

Vonnegut, Bernard and Raymond L. Neubauer. 1952. "Production of monodisperse liquid particles by electrical atomization." *Journal of Colloid Science*, 7 (6): 616–622. doi:https://doi.org/10.1016/0095-8522(52)90043-3.

Wang, Xianfeng, Yang Si, Jialin Wang, Bin Ding, Jianyong Yu, and Salem S. Al-deyab. 2012. "A facile and highly sensitive colorimetric sensor for the detection of formaldehyde based on electro-spinning/netting nano-fiber/nets." *Sensors and Actuators: B. Chemical*, 163 (1). Elsevier B.V.: 186–193. doi:10.1016/j.snb.2012.01.033.

Wang, Xiaoxiao, Christian Pellerin, and C. Geraldine Bazuin. 2016. "Enhancing the electrospinnability of low molecular weight polymers using small Effective cross-linkers." doi:10.1021/acs.macromol.5b02670.

Weiss, Jochen, Paul Takhistov, and D. Julian McClements. 2006. "Functional materials in food nanotechnology." *Journal of Food Science*, 71 (9): 107–116. doi:10.1111/j.1750-3841.2006.00195.x.

Wen, Yanan, Yan Li, Yang Si, Xueqin Wang, Faxue Li, Jianyong Yu, and Bin Ding. 2015. "Ready-to-use strip for L-ascorbic acid visual detection based on polyaniline/polyamide 66 nano-fibers/nets membranes." *Talanta*, 144. Elsevier: 1146–1154. doi:10.1016/j.talanta.2015.07.086.

Wen, Peng, Ding-he Zhu, Hong Wu, Min-hua Zong, and Yi-ru Jing. 2016. "Encapsulation of cinnamon essential oil in electrospun nano fibrous film for active food packaging," 59: 366–376. doi:10.1016/j.foodcont.2015.06.005.

Wen, Peng, Yan Wen, Min Hua Zong, Robert J. Linhardt, and Hong Wu. 2017. "Encapsulation of bioactive compound in electrospun fibers and its potential application." *Journal of Agricultural and Food Chemistry*, 65 (42): 9161–9179. doi:10.1021/acs.jafc.7b02956.

Widartiningsih, P.M., F. Iskandar, M. M. Munir, and S. Viridi. 2016. "Predicting jet radius in electrospinning by superpositioning exponential functions." *Journal of Physics: Conference Series*, 739 (1). doi:10.1088/1742-6596/739/1/012097.

Xie, Huafei, Zhengyu Jia, Jinhui Huang, and Cuirong Zhang. 2011. "Preparation of low molecular weight chitosan by complex enzymes hydrolysis." *International Journal of Chemistry,* 3(2): 180–186. doi:10.5539/ijc.v3n2p180.

Xie, J., Jiang, J., Davoodi, P., Srinivasan, M. P., & Wang, C.-H. (2015). Electrohydrodynamic atomization: A two-decade effort to produce and process micro-/nanoparticulate materials. *Chemical Engineering Science*, 125, 32–57. http://doi.org/10.1016/j.ces.2014.08.061.

Zong, Xinhua, Kwangsok Kim, Dufei Fang, Shaofeng Ran, Benjamin S. Hsiao, and Benjamin Chu. 2002. "Structure and process relationship of electrospun bioabsorbable nanofiber membranes." *Polymer,* 43 (16). Elsevier: 4403–4412. doi:10.1016/S0032-3861(02)00275-6.

Yang, F., R. Murugan, S. Wang, and S. Ramakrishna. 2005. "Electrospinning of nano/micro scale poly(L-lactic acid) aligned fibers and their potential in neural tissue engineering." *Biomaterials,* 26 (15): 2603–2610. doi:10.1016/j.biomaterials.2004.06.051.

Yang, Huan, Kun Feng, Peng Wen, Min Hua Zong, Wen Yong Lou, and Hong Wu. 2017a. "Enhancing oxidative stability of encapsulated fish oil by incorporation of ferulic acid into electrospun zein mat." *LWT - Food Science and Technology,* 84. Elsevier Ltd: 82–90. doi:10.1016/j.lwt.2017.05.045.

Yang, Huan, Peng Wen, Kun Feng, Min H. Zong, Wen Y. Lou, and Hong Wu. 2017b. "Encapsulation of fish oil in a coaxial electrospun nanofibrous mat and its properties." *RSC Advances,* 7 (24). Royal Society of Chemistry: 14939–14946. doi:10.1039/C7RA00051K.

Yuan, Xiaoyan, Yuanyuan Zhang, Cunhai Dong, and Jing Sheng. 2004. "Morphology of ultrafine polysulfone fibers prepared by electrospinning." *Polymer International,* 53 (11). John Wiley & Sons, Ltd.: 1704–1710. doi:10.1002/pi.1538.

Yuliani, Sri, Bhesh Bhandari, Rulande Rutgers, and Bruce D'Arcy. 2004. "Application of microencapsulated flavor to extrusion product." *Food Reviews International,* 20 (2). Taylor & Francis Group: 163–185. doi:10.1081/FRI-120037159.

Zaeim, Davood, Mahboobe Sarabi-jamab, Behrouz Ghorani, Rassoul Kadkhodaee, and R. Hans Tromp. 2017. "Electrospray assisted fabrication of hydrogel microcapsules by single- and double-stage procedures for encapsulation of probiotics." *Food and Bioproducts Processing*, Institution of Chemical Engineers. doi:10.1016/j.fbp.2017.01.004.

Zaeim, D., M. Sarabi-Jamab, B. Ghorani, R. Kadkhodaee, and R. H. Tromp. 2018. "Electrospray-assisted drying of live probiotics in acacia gum microparticles matrix." *Carbohydrate Polymers*, 183. Elsevier Ltd.: 183–191. doi:10.1016/j.carbpol.2017.12.001.

Zargham, Shamim, Saeed Bazgir, Amir Tavakoli, Abo Saied Rashidi, and Rogheih Damerchely. 2012. "The effect of flow rate on morphology and deposition area of electrospun nylon 6 nanofiber" 7 (4): 42–49.

Zeleny, John. 1917. "Instability of electrified liquid surfaces." *Physical Review,* 10 (1). American Physical Society: 1–6. doi:10.1103/PhysRev.10.1.

Zhao, Y. Y., Q. B. Yang, X. F. Lu, C. Wang, and Y. Wei. 2005. "Study on correlation of morphology of electrospun products of polyacrylamide with ultrahigh molecular weight." *Journal of Polymer Science Part B: Polymer Physics*, 43: 2190–2195.

Zhao, L., H. Xiong, H. Peng, Q. Wang, D. Han, C. Bai YuZhen Liu et al. 2011. "PEG-coated lyophilized proliposomes: Preparation, characterizations and in vitro release evaluation of vitamin E." *European Food Research and Technology*, 232 (4): 647–654.

Zhou, Yiqun, Zhili Peng, Elif S. Seven, and Roger M. Leblanc. 2018. "Crossing the blood-brain barrier with nanoparticles." *Journal of Controlled Release*, 270. Elsevier B.V: 290–303. doi:10.1016/j.jconrel.2017.12.015.

Zong, Huixin, Xin Xia, Yongri Liang, Songyuan Dai, Ahmed Alsaedi, Tasawar Hayat, Fantai Kong, and Jia Hong Pan. 2017. "Designing function-oriented artificial nanomaterials and membranes via electrospinning and electrospraying techniques." *Materials Science and Engineering C,* Elsevier B.V. doi:10.1016/j.msec.2017.11.007.

Zueqin, Wangab, C. Huaping Wanga, Bin Ding, Yang Si, A. Xue Mao, A. Yan Li, and C. Jianyong Yu. 2013. "Colorimetric sensor strips for formaldehyde assay utilizing fluoral-P decorated polyacrylonitrile nanofibrous membranes." *Analust*, 138: 5129–5136. doi:10.1039/c3an00812f.

Zupančič, Š., S. Baumgartner, Z. Lavrič, M. Petelin, and Julijana Kristl. 2015. "Local delivery of resveratrol using polycaprolactone nanofibers for treatment of periodontal disease." *Journal of Drug Delivery Science and Technology*. Elsevier Ltd. doi:10.1016/j.jddst.2015.07.009.

10 Nano-Delivery System for Food Bioactives

Das Trishitman and C. Anandharamakrishnan

CONTENTS

10.1 INTRODUCTION

A nano-delivery system is defined as a controlled, engineered release of bioactive materials which have been entrapped inside of a shell. Nano-delivery systems are designed to allow delivery of food bioactives to a particular or targeted site. Polyphenols, probiotics, micronutrients, enzymes, nutraceuticals, polyunsaturated fatty acids, and proteins are the main food bioactives which can be added to improve and increase the value addition of foods (Rajam and Anandharamakrishnan, 2015; Karthik and Anandharamakrishnan, 2016; Karthik et al., 2017; Anandharamakrishnan, 2014; Bhushani et al., 2016; Bhushani and Anandharamakrishnan, 2017; Bhushani et al., 2017a,b). An efficient targeted delivery system allows a systematic, relatively higher, and more efficient dose to be released to a particular site. Nano-delivery of these food bioactives is used to cure diseases and improve health benefits (Jiménez-Colmenero, 2013).

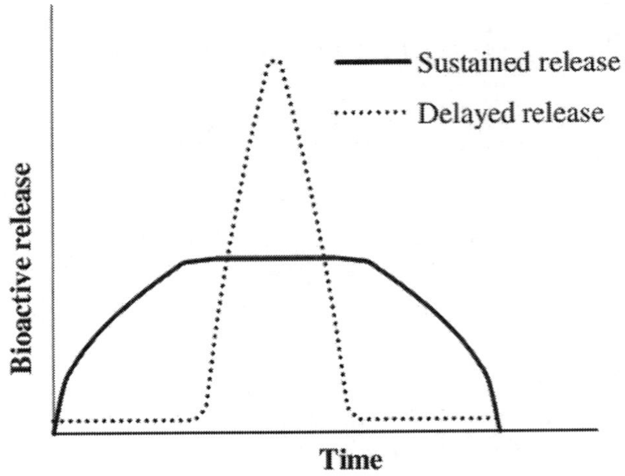

FIGURE 10.1 Models of bioactive release from nanocarrier systems. (Fathi et al., 2012.)

Generally, two mechanisms control the release of bioactives in the delivery system (Lakkis, 2007; Fathi et al., 2012):

1. Delayed release, which delays the release of bioactive compounds from a definite lag time up to a point when/where their release is allowed and is no longer impeded. This release pattern could include the release of probiotics in the intestine, color in beverages, and flavor in ready meals.
2. Sustained release, which provides a continuous release of bioactives at the same rate to a targeted site.

An example of this is the release of sweeteners in chewing gum, and the release of drugs (Figure 10.1). In the delivery system, the release of bioactives depends/varies according to the core material (encapsulates) and the wall material (encapsulating material). Control release also varies with the amount and type of wall materials. Nano-delivery systems are classified into two major group: liquid and solid. Liquid systems are of three types: nanoemulsion, nanoliposomes, nano polymersomes, and solid systems distinguished into lipid nanoparticles, polymeric nanoparticles, and nanocrystals. Lipid nanoparticles are composed of solid lipid nanoparticles (SLNs) and nanostructured lipid carriers (NLCs) (Figure 10.2).

10.2 SOLID LIPID NANOPARTICLES (SLNS)

Solid lipid nanoparticles (SLNs) are nano-sized particles made from solid lipids at room temperature and stabilized by surfactants. SLNs are the same as nanoemulsions with the exception of slower digestion as the lipids are in solid phase. These lipids are highly purified triglycerides, complex glycerides, a mixture of both, or even waxes (Wissing et al., 2004). SLNs are the change of the liquid lipids (oils) of the emulsion into lipids which are solid at room temperature as well as in the body. The use of SLNs in the delivery system is an attractive and innovative idea where the release of bioactives is controlled as compared to oil emulsions (Wolfgang Mehnert and Mäder, 2001; McClements and Li, 2010).

10.2.1 Advantages and Disadvantages of SLNs

Advantages:

- High flexibility of controlled targeting and release.
- Increased stability of bioactives.

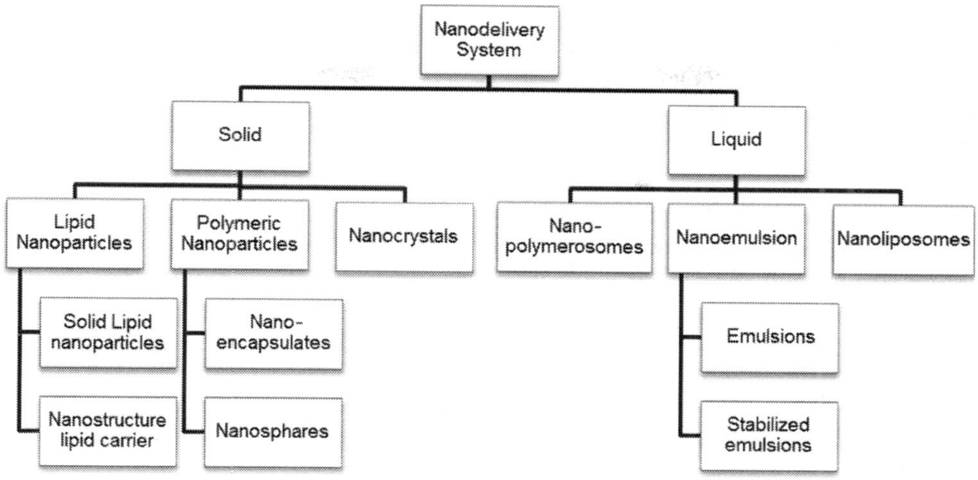

FIGURE 10.2 Classification of nano-delivery system.

- Improvement in the bioavailability of bioactives.
- High encapsulation efficiency.
- Feasibility in the incorporation of both lipophilic and hydrophilic nutraceuticals.
- High concentration of bioactives is possible.
- Scale-up in production can be done.
- Avoidance of organic solvents.

Disadvantages:

- Loading capacity is low.
- Relatively high water content may create expulsion of core materials during storage.
- Use of more complex lipids (mono-, di-, triglycerides, and different chain lengths) is more sensitive to higher loading.

The loading of bioactives in conventional SLNs is poor and it is limited to approximately 25% with respect to lipids, and up to 50% for special actives such as ubidecarenone (Schwarz et al., 1994; Kirsten Westesen et al., 1993; Bunjes et al., 1996; Müller et al., 2000; Westesen et al., 1997). The loading of bioactives depends on their solubility in the lipid matrix, the structure of the lipid matrix, and the polymorphic state of the lipid matrix, so incorporated bioactives are located between the fatty acid chains, between lipid layers, and in crystal imperfections. A highly ordered crystal lattice cannot accommodate a large number of drugs in the lipid matrix, so the use of more complex lipids is more sensible for loading higher amounts of drugs (Mehnert et al., 1997). The conversion of high ordered lipids to SLNs is also the reason for bioactive expulsion during storage. Synthesis of SLNs for food bioactives includes mainly three methods, such as high shear homogenization, high-pressure homogenization, and solvent emulsification and evaporation.

10.2.2 High-Shear Homogenization

High-shear homogenization is a dispersing technique which was initially used for the production of solid lipid nanodispersions (Mehnert and Mäder, 2001). This method is widespread as it is easy to handle. However, dispersion quality is poor and often compromised by the presence of microparticles. Different process parameters control the production of SLNs in the high-shear homogenizer, including emulsification time, stirring rate, and cooling conditions. The characterization processes of SLNs are performed based on the particle size and zeta potential, which are controlled by the process parameters of high shear homogenization. In most SLNs in high-shear homogenization, the average particle size obtained is in the ranges of 100 to 200 nm, using stirring rate of 20,000 to 25,000 rpm for 8–10 min and controlled cooling with a stirring rate of 5000 rpm (Mader and Mehnert, 2004).

10.2.3 High-Pressure Homogenization

SLNs can be produced by high-pressure homogenization, like to parenteral O/W emulsions. High-pressure homogenization is a well-established technique and pharmaceutical industries are using it on a large scale. The production lines are well-equipped and temperature controlled for parenteral emulsions, as an increase in temperature may affect the production (Müller et al., 2000).

Two basic production techniques are adapted for SLNs by high-pressure homogenization, (i) hot homogenization and (ii) cold homogenization technique. Both the cases include, disperse, or dissolve bioactives in the melt bulk lipid to form SLNs (Figure 10.3).

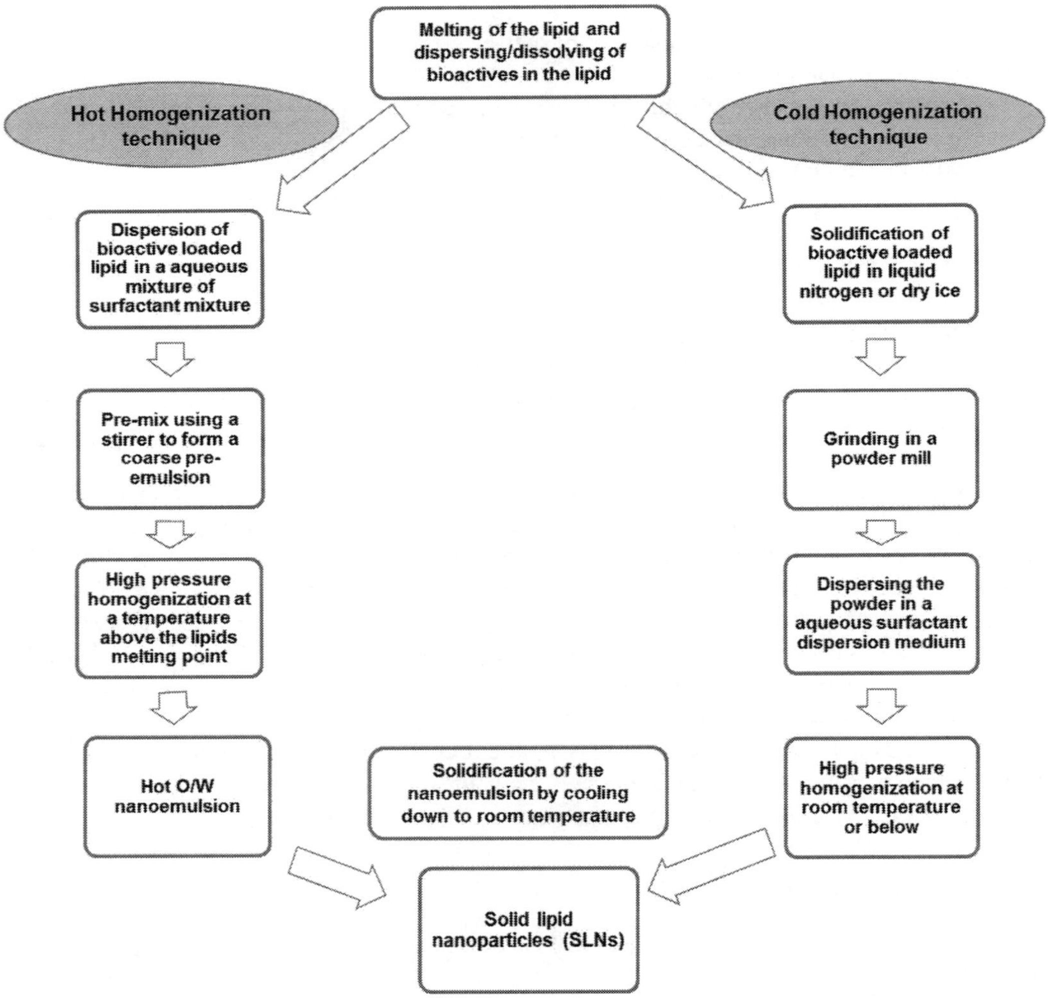

FIGURE 10.3 Schematic flow diagram of hot and cold homogenization technique for production of SLNs.

10.2.3.1 Hot Homogenization

Hot homogenization is carried out at a higher temperature than the melting point of the lipids to form a homogenized emulsion. A pre-emulsion aqueous mixture of loaded bioactives and emulsifier at the same temperature as a melted lipid is obtained using a high shear homogenizer. The temperature and homogenization of the aqueous pre-mixture affect the encapsulation efficiency, quality, and size of the final product. Higher temperature may lead to lower particle size due to the decrease in viscosity of the inner phase (Lander et al., 2000). However, high temperature increases the degradation of bioactives are still needed to keep the temperature above the melting temperature of the lipid. The high-pressure homogenization process is been carried out at 500–1500 bar for 3–5 cycles. During the process, the temperature may increase up to 10°C for 500 bar (Benita and Böhm, 1998). Increase in homogenization pressure and cycles may result in an increase of particle size due to the coalescence effects which occur as a result of the high kinetic energy of the particles (Kirsten Westesen and Siekmann, 1997).

The initial product of hot homogenization is a nanoemulsion due to the liquid state of the lipid at a higher temperature than its melting point. Solidification of nanoparticles is obtained by cooling down the samples to room temperature or below. The presence of emulsifiers and size particles at nano form retards the crystallization of lipids and keeps the sample as a supercooled melt after several months of storage (Benita, 1998) (Figure 10.4).

10.2.3.2 Cold Homogenization

The cold homogenization technique is most suitable for heat-labile bioactives and hydrophilic compounds. The main difference between this technique and the one described above is that cold homogenization is carried out with solid lipids and bioactives, in a high-pressure milling suspension. A high temperature and regulation control are needed to keep the lipids in an unmolten state due to the rapid increase of temperature during the homogenization process. Cold homogenization is been adopted over hot homogenization to overcome three major problems:

1. Degradation of drug/bioactives during the increase in temperature
2. Distribution of bioactives in the aqueous phase during homogenization
3. Complexity in the crystallization process of nanoemulsions (Wolfgang Mehnert and Mäder, 2001)

The primary step of cold homogenization is the same as in hot homogenization as bioactives are dissolved or dispersed in the molten lipid and make it bulk. The following step involves rapid cooling of the mixture by dry ice or liquid nitrogen. Rapid cooling of the disperse bioactives with lipid provides a homogenous matrix. The solid mixture of lipid-containing bioactives is milled to microparticles. Milling of the solid mix by ball or mortar milling gives typical size particles in the range 50–100 μm. The solid milled microparticles are dispersed in an aqueous surfactant dispersion medium. The suspension of the aqueous solution and disperse bioactives is subjected to high-pressure homogenization at or below room temperature. The homogenization conditions are generally

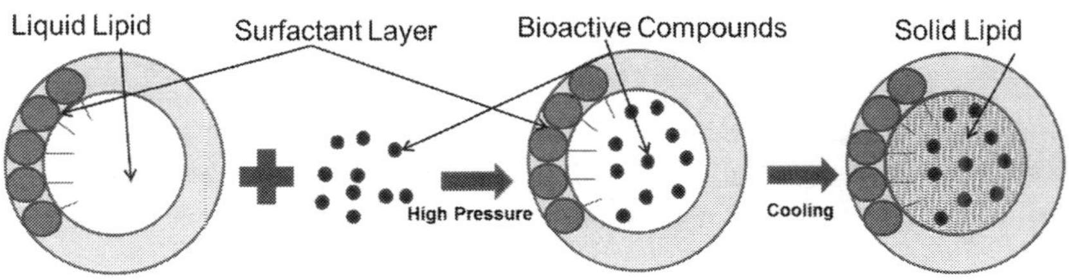

FIGURE 10.4 Structural formation of solid lipid nanoparticles (SLNs).

5 cycles at 500 bar (Wissing et al., 2004) or 10 cycles at 1500 bar (Zur Mühlen et al., 1998). In cold homogenization, a larger particle size has been observed compared to hot homogenization. Cold homogenization minimizes the exposure of bioactives to higher temperatures but does not avoid the melting of the lipid and bioactive mixture during the primary step. The cold homogenization process provides the desired crystal structure as the crystallization process is controllable due to the rapid cooling of the mixture (Tamjidi et al., 2013).

10.2.4 Solvent Emulsification and Evaporation

In solvent emulsification and evaporation, lipid and bioactive compounds are dissolved in a water-immiscible organic solvent (e.g. methylene chloride, toluene, chloroform) with a low boiling point. The mixture is then emulsified in an aqueous phase before proceeding for further evaporation (Wissing et al., 2004; Tamjidi et al., 2013). Evaporation of the organic solvent under reduced pressure (40–60 mbar) (Wolfgang Mehnert and Mäder, 2001) leaves the lipid and bioactive compounds in nanoparticle form. An important benefit of this method is the reduction in the thermal exposure of the sample, which makes it suitable for heat sensitive compounds. The mean particle size ranges from approximately 30 to 100 nm depending upon the lipid load, emulsifier type, and production conditions. Solvent residue in the final product is the main problem of this technique. Also, the dispersion of bioactive compounds is generally quite low, because of the limited solubility of the lipid in the organic solvents or the use of diluted organic solvents in the procedure. The typical concentration of the final SLN dispersion ranges around 0.1 g/l and the particle concentration of SLNs can be increased by means of ultra-filtration or evaporation.

SLNs size often varies depending upon the surfactant concentration; if the concentration is too high or low, the surfactant forms a cluster. The application of ionic surfactants usually forms smaller size particles, and their stability is increased by using more than one surfactant (Wolfgang Mehnert and Mäder, 2001).

10.3 NANOSTRUCTURED LIPID CARRIERS (NLCs)

Nanostructured lipid carriers (NLCs) are partially crystallized lipid particles as well as solid lipids with a mean radius of less than 100 nanometers which are dispersed/dissolved in an aqueous phase of emulsifier(s) as a potential delivery system for bioactive compounds. NLCs are useful in nutraceutical delivery systems with high drug loading, encapsulation, solubility, and storage stability. NLCs have some advantages over other colloidal carrier systems as their encapsulation efficiency, drug loading, and stability are higher. NLCs may increase consumer acceptability, functionality, and safety in the food system, and improve nutritional value by adding bioactives and providing controlled release of encapsulated material due to their high loading capacity, encapsulation efficiency, and stability.

The process of full crystallization or recrystallization of lipids reduces the solubility of drugs and leads to expulsion of drugs during long storage time. The majority the drugs have higher solubility in liquid lipids rather than solid lipids. The purpose of NLCs is to increase the availability of drugs or bioactive compounds and the stability of the encapsulated material. NLCs are modified SLNs where both solid (fat) and liquid (oil) lipids are present in the lipidic phase at ambient temperature (Müller et al., 2002b).

In the late 1990s, Muller et al. developed NLCs or oil-loaded SLNs to overcome the limitations of SLNs with a less ordered crystalline structure or an amorphous solid structure of lipid droplets. As controlled release is greater in liquid oil, the purpose of NLCs is to increase the drug loading capacity in the core of the solid encapsulated lipids (Varshosaz et al., 2010). The polymorphic transition and crystallinity index are low in lipid blend NLCs, and along with this the spherical shape of the particles may account for an improvement in properties, thus showing an advantage over SLNs and other lipid carriers (Müller et al., 2002b; Jores et al., 2004). Therefore, NLCs may increase

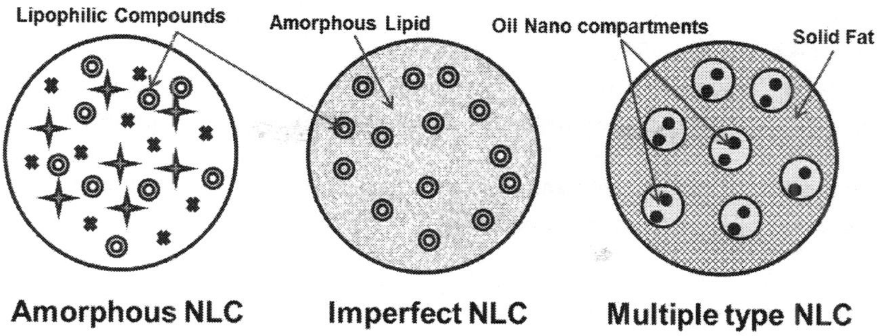

FIGURE 10.5 Types of NLCs.

encapsulation efficiency, drug loading capacity, and stability, thus improving chemical stability, bioavailability, and controlled release of functional lipophilic compounds in food. NLCs immobilize the bioactive compounds in solid lipid matrix systems and the outer lipid matrix protects the core compounds and works as a physical barrier from damaging factors in the aqueous phase (Pardeike et al., 2009; Teeranachaideekul et al., 2007a) (Figure 10.5).

10.3.1 NLCs Ingredients

NLCs are a nano-sized carrier system derived from O/W nanoemulsions where the main ingredients are a lipid, surface active agent (surfactant), and water. In the NLC carrier system, the major portion of the O/W emulsion is being replaced by a solid lipid matrix at body temperature (Jenning et al., 2000; Jores et al., 2004). To prepare the blend for the particle matrix with solid lipid and liquid lipid (oils), a ratio of 70:30 up to a ratio of 99.9:0.1 is preferable (Tamjidi et al., 2013). SLNs are composed with dispersed solid lipid of 0.1% to 30% (w/w) in an aqueous mixture, whereas in NLCs the total solid content can be up to 95% (Mueller et al., 2000; Pardeike et al., 2009). The higher the solid content, the higher the drug load; this is an important advantage of NLCs over SLNs. As the drug load increases, the requirement of NLCs in comparison to SLNs to deliver drugs decreases, which makes it more cost-effective.

10.3.2 Lipids

A successful production of NLCs for suitable physical and chemical stability needs a proper selection of an appropriate lipidic blend. The lipid blend has a significant impact on the chemical stability of the bioactive compounds inside the lipid matrix (Mitri et al., 2011; Müller et al., 2002b). The following parameters should be considered before choosing a suitable lipid blend:

1. The solubility of bioactive compounds is essential in the lipid matrices, before attempting incorporation of molecules into nanocarriers, because drug loading capacity, encapsulation efficiency, and usefulness of NLCs are influenced by it (Kasongo et al., 2011; Jaspart et al., 2005; Müller et al., 2000; Müller et al., 1995). The solubility of various actives in different liquid lipids, surfactants, co-surfactants, solubilizers or a mixture of these can be improved by heating or by aqueous phase emulsions (Müller et al., 2000).
2. The liquid and solid lipid molecules should be principally incompatible as much as possible; this means oil molecules should not take part in the formation of the solid crystalline matrix and the crystal solid lipid should not be dissolved in the liquid lipid. Also it is necessary that the solid and liquid lipids used to form NLCs be miscible at the specific concentrations (Kasongo et al., 2011; Müller et al., 2002a); this means that macroscopic phase separation should not occur at a temperature below the melting point of the fat, the bulk should be homogenous, and the liquid lipids should be within the core of the fat matrix.

3. The lipid phase should be protected from any chemical degradation such as oxidation and lipolysis (Tamjidi et al., 2013).
4. When the oil phase has low viscosity and/or lower interfacial tension then it is easy to produce very fine O/W emulsions. So the lipids should be capable of producing particles in nanoscale (Walstra, 2002).
5. The lipids should not produce any toxic substances during NLC preparation and should have an acceptable toxicological profile.

Generally, lipids used to prepare NLCs are mono-, di-, and triacylglycerols, fatty acids, and waxes.

10.3.2.1 Liquid Lipids

The most used liquid lipids for production of NLCs are medium-chain triglycerides (MCTs) and oleic acid. Caprylic (C8:0; 50–80%) and capric (C10:0; 20–50%) fatty acids with small amounts of caproic (C6:0; ≤ 2%), lauric (C12:0; ≤ 3%), and myristic (C14:0; ≤ 1%) fatty acids are the main constituent of MCTs. MCTs solidify at ≈0°C and have low viscosity (28–32 mPa.s at 20°C) and also count as an emulsifier because of their uniqueness as saturated lipids (Rowe et al., 2012). MCTs are usually produced by modification and fractionation of natural oils, such as coconut or palm kernel oils. MCTs are food grade, odorless oils, and have faster digestibility and higher stability against oxidation than any other long-chain triglycerides (e.g. corn oil) (Porter et al., 2007; Hippalgaonkar et al., 2010). The US Food and Drug Administration (FDA) approved MCTs as generally recognized as safe (GRAS) substances that can be applied directly to many functional food items (e.g. as emulsifier, vehicle, solvent, or release agent).

Odorless oleic acid [*cis*-9-octadenoic acid; C18:1 (9)] is usually obtained by the hydrolysis of many natural fats from animal and vegetable sources (e.g. olive oil) followed by separation of the liquid fatty acids. The European Pharmacopeia (Ph. Eur.) documented the specifications of oleic acid including assay for varying amounts of saturated and unsaturated fatty acids. It has been widely used in food and pharmaceutical formulations as an emulsifier as it has low viscosity (~26 mPa.s at 25°C) and solidifies at ~4°C (Rowe et al., 2012). Also, it has health-promoting values which are documented by Lopez-Huertas (2010). Oleic acid and other fatty acids are quite prone to oxidation compared to other triglycerides. However, free fatty acids are more surface active than triacylglycerols and thus superior for building up at the oil-water interface, which increases their susceptibility to oxidation (Coupland et al., 1996). The encapsulated sensitive bioactive compounds inside the NLCs may be damaged due to the formation of free radicals during the oxidation process. FDA listed the fatty acids derived from edible sources (e.g. capric, caprylic, lauric, oleic, palmitic, myristic, and stearic) and included in permitted food additives for direct addition to food (21CFR172.860).

NLCs can also be produced using natural edible oils such as sunflower oil (Gramdorf et al., 2008), corn oil, and soybean oil (Table 10.1). These natural edible oils are "friendly labeled" and some of them may contain a natural antioxidant which leads to effective protection of drugs from oxidative degradation. These oils are a cheaper alternative to MCTs and oleic acid for food manufacturers. However, these oils are not very surface active, have high viscosity, a high degree of saturation, and possibly a poor dissolving agent for some bioactive compounds. Therefore, utilizing these oils for NLC production may lead to poor encapsulation efficiency, disruption of droplets during size reduction, and oxidative instability.

Other natural substances such as vitamin E and squalene are also been used in NLC formulation as liquid lipid (Table 10.1). Although tocopherols and squalene have health-promoting effects on humans, as an ingredient of NLCs they are prone to oxidation and somewhat expensive.

10.3.2.2 Solid Lipids

Glyceryl behenate, glyceryl monosterate/monosterin, glycerol palmitosterate, cetyl palmitate, and stearic acid (Table 10.1) are the most used solid lipids for NLC preparation. These fats are commercially available and have also been approved as GRAS for direct addition to food. Many of these fats not only work as a vehicle but also act as surface active and, moreover, form a lattice. Glyceryl

TABLE 10.1
Commonly Used Lipids and Emulsifiers for Preparation of NLCs

Liquid Lipid	References
Soybean oil	Chinsriwongkul et al. (2012) and Zhang et al. (2011)
Medium-chain triglycerides (MCT)/Caprylic and capric triglycerides	Cirri et al. (2012), Esposito et al. (2012), Fathi et al. (2013), Chen-yu et al. (2012), Sf et al. (2011), Han et al. (2008), Hu et al. (2005), Jia et al. (2010), Jores et al. (2004), ZhenLi et al. (2009), Li et al. (2010), Lin et al. (2007), Luo et al. (2011), Mitri et al. (2011), Nayak et al. (2010), Obeidat et al. (2010), Puglia et al. (2012), Sanad et al. (2010), Saupe et al. (2006), Saupe et al. (2005), Varshosaz et al. (2012b), Teeranachaideekul et al. (2007a), Xia et al. (2007), Zhang et al. (2010), Zhuang et al. (2010)
Oleic acid	Chinsriwongkul et al. (2012), Fathi et al. (2013), Hu et al. (2005), Kuo and Chung (2011), Liu et al. (2011), Li et al. (2010), Pardeike et al. (2011), Sanad et al. (2010), Tiwari and Pathak (2011), Varshosaz et al. (2012c), Varshosaz et al. (2012b), Yuan et al. (2007)
α-tocopherol/Vitamin E	Tsai et al. (2012) and Souto and Müller (2005)
Corn oil	Liu and Wu (2010), Mitri et al. (2011), and Nguyen et al. (2012)
Squalene	Chen et al. (2010), Fang et al. (2008), and Araújo et al. (2011)
Solid Lipid	
Stearic acid	Fathi et al. (2013), Chen-yu et al. (2012), Hu et al. (2005), Kuo and Chung, (2011), Fang et al. (2008), and Yuan et al. (2007)
Carnauba wax (Cera carnauba)	Mitri et al. (2011) and Xia et al. (2007)
Glyceryl tristearate/Tristearin	Ali et al. (2010), Esposito et al. (2012), Sf et al. (2011), and Nayak et al. (2010)
Glycerol monostearate	Fathi et al. (2013), Chen-yu et al. (2012), Han et al. (2008), Hu et al. (2006), Jia et al. (2010), Lacatusu et al. (2012), Li et al. (2010), Liu et al. (2011), Müller et al. (2002), Nayak et al. (2010), Sanad et al. (2010), Teeranachaideekul et al. (2007a), Tiwari and Pathak, (2011), Varshosaz et al. (2012b), Yuan et al. (2007), Zhang et al. (2011), Zhang et al. (2010), Zheng et al. (2013), and Zhuang et al. (2010)
Cetyl palmitate	Ali et al. (2010), Chinsriwongkul et al. (2012), Eskandari et al. (2011), Müller et al. (2002), Obeidat et al. (2010), Saupe et al. (2005), Saupe et al. (2006), Teeranachaideekul et al. (2007a), Varshosaz et al. (2010), Varshosaz et al. (2012a), and Xia et al. (2007)
Glycerol monocaprate	Li et al. (2009) and Lin et al. (2007)
Glyceryl palmitostearate	Ali et al. (2010), Araújo et al. (2011), Chen et al. (2010), Doktorovová et al. (2010), Fang et al. (2008), Kasongo et al. (2011), Li and Ge, (2012), Liu and Wu (2010), Pardeike et al. (2011), and Puglia et al. (2012)
Glycerol behenate	Ali et al. (2010), Cirri et al. (2012), Fathi et al. (2013), Jores et al. (2004), Kuo and Chung, (2011), Luo et al. (2011), Müller et al., (2002), Souto and Müller, (2005), Varshosaz et al. (2010), Xia et al. (2007), and Zhuang et al. (2010)
Propylene glycol monostearate	Gramdorf et al. (2008) and Hentschel et al. (2008)
Emulsifier	
Tween 80	Chinsriwongkul et al. (2012), Doktorovová et al. (2010), Fang et al. (2008), Fathi et al. (2013), Gramdorf et al. (2008), Han et al. (2008), Hentschel et al. (2008), Hung et al. (2011), Kuo and Chung, (2011), Lacatusu et al. (2012), Li and Ge, (2012), Li et al. (2009), Lin et al. (2007), Luo et al. (2011), Nayak et al. (2010), Nguyen et al. (2012), Puglia et al. (2012), Teeranachaideekul et al. (2007a), Varshosaz et al. (2010), Varshosaz et al. (2012), Varshosaz et al. (2012c), and Zhang et al. (2010)
Lecithin	Chen et al. (2010), Chinsriwongkul et al. (2012), Doktorovová et al. (2010), Eskandari et al. (2011), Fang et al. (2008), Chen-yu et al. (2012), Han et al. (2008), Jia et al. (2010), Kuo and Chung, (2011), Lacatusu et al. (2012), Li and Ge (2012), Liu et al. (2012), Nguyen et al. (2012) and Tsai et al. (2012), Varshosaz et al. (2010), Varshosaz et al. (2012), Varshosaz et al. (2012c), Varshosaz et al. (2012), Zhang et al. (2011), Zheng et al. (2013) and Zhuang et al. (2010)

(*Continued*)

TABLE 10.1 (CONTINUED)
Commonly Used Lipids and Emulsifiers for Preparation of NLCs

Emulsifier	References
Poloxamer 188	Ali et al. (2010), Araújo et al. (2011), Chen et al. (2010), Cirri et al. (2012), Eskandari et al. (2011), Esposito et al. (2012), Fang et al. (2008), Han et al. (2008), Hu et al. (2006), Jia et al. (2010), Jores et al. (2004), Li et al. (2010), Liu and Wu, (2010), Liu et al. (2011), Nayak et al. (2010), Puglia et al. (2012), Souto and Müller, (2005), Tsai et al. (2012), Varshosaz et al. (2010), Varshosaz et al., (2012), Zhang et al. (2010), and Zhuang et al. (2010)
Myverol 18–04 K	Chen et al. (2010), Fang et al. (2008), Liu and Wu, (2010), and Nayak et al. (2010)
Polyglyceryl-3-methylglucose distearate	Obeidat et al. (2010), Saupe et al. (2006), Saupe et al. (2005), Teeranachaideekul et al. (2007a), and Teeranachaideekul et al. (2007b)
Sodium dodecyl sulfate (SDS)	Hu et al. (2005) and (2006)
Sodium deoxycholate (SDC)	Han et al. (2008), Li et al. (2010), and Souto and Müller (2005)
Tween 20	Lacatusu et al. (2012), Varshosaz et al. (2012c), Yuan et al. (2007), and Zheng et al. (2013)

Source: Tamjidi et al. (2013).

behenate is mainly used in food, oral pharmaceutical, and cosmetics formulations, and Europe has accepted it as a food additive. Glyceryl behenate, which consists of diacylglycerols of behenic acid (C22:0) together with mono- and triacylglycerols (Tamjidi et al., 2013) has high encapsulation efficiency and stability (Castro et al., 2007). There may be much disfigurement in the crystal lattice caused by high encapsulation efficiency which leads to higher drug entrapment.

Mono- and diacylglycerols are the main constituents of glycerol monostearate. There have been differences in the specifications of glycerol monostearate by Ph. Eur. and United States Pharmacopeia (USP). Glycerol monostearate contains 40% monostearoylglycerol based on the report of Ph. Eur., whereas USP reported it consists of no less than 90% monoacylglycerols of saturated fatty acids, mainly monostearoylgycerols and glyceryls monopalmitate. As glycerol monostearate is a non-toxic and non-irritant material, it is been widely used in cosmetics, foods, and pharmaceutical formulations as a non-ionic emulsifier, stabilizer, emollient, and plasticizer. Glyceryl palmitostearate is a good solid lipid for sustained release applications. It consists of mono-, di-, and triacylglycerols of palmitic (C16:0) and stearic (C18:0) fatty acids. Cetyl esters waxes are primarily mixtures of esters of saturated fatty alcohols (C14–C18) and saturated fatty acids (C14–C18) as described by USP and are not permitted in food applications. However, naturally occurring waxes (carnauba, candelilla, and bees) are approved as GRAS for direct addition to food and are consequently used for food application as solid lipid in the formulation of NLCs.

Stearic acid is a principal constituent of most hydrogenated fat and consists of an endogenous long chain of saturated fatty acids; it is also the principal component of lipids in plant and animal sources. It has better biocompatibility and lower toxicity than synthesized acids. It is accepted as a food additive in Europe.

NLCs also include tristearin, glyceryl monocaprate, and propylene glycol as a solid lipid for formulation. FDA included tristearin and propylene glycol monostearate on the list of direct food additives (21CFR172.811 & 21CFR172.856), while glyceryl monocaprate is included in the indirect list of food additives (21CFR176.180).

10.3.3 Influence of Surfactants on NLCs

Surfactants play an important role in the formation and characterization of NLCs. Polysorbate 80, Tween 80, Poloxamer 188, lecithin, and sodium deoxycholate (SDC) are the most frequently used surfactants for the preparation of NLCs (Tamjidi et al., 2013; Han et al., 2008). Surfactants are able to absorb rapidly to droplet surfaces and reduce the interfacial tension in high-energy methods (high-pressure homogenization) (McClements and Rao, 2011).

Changes in NLCs composition can impact on particle size, degree of crystallization, chemical composition, and polymorphism. A decrease in the particle size can be achieved by the addition of more than one type of surfactant in NLC formulation. Ionic surfactant on the surface of the NLCs aids electrostatic repulsion between particles and steric hindrance to agglomeration by bulky non-ionic surfactant surface constituents (Karn-Orachai et al., 2014). However, non-ionic surfactants are less toxic and have lower irritation potential compared to ionic ones (McClements and Rao, 2011).

The addition of two or more surfactant types to NLC production may lead to producing different properties, especially in their chemical composition and crystallinity. As the combination of two ionic surfactants produces repelling forces, crystalline balance and non-ionic surfactants are required in order to regulate and balance crystallization. Two surfactant-based NLCs show lower crystallinity and a more homogenous mixture of oil and solid lipid than a three surfactant system (Karn-Orachai et al., 2014). An appropriate concentration of surfactant is required to cover the surface well and thus cause a reduction in the aggregation of particles and produce stable NLCs. Surfactants not only give effective stability to crystalline structure but also prevent recrystallization during storage.

A proper surfactant and its amount helps to keep the physical state of the core liquid lipids in NLCs. Thus, drug molecules can find their appropriate position in the lipid matrix, which improves the drug loading and encapsulation efficiency (Han et al., 2008). Whereas, on the other hand, surfactants protect drugs from defusing to the solid lipid matrix during storage.

10.4 TARGETED DELIVERY USING THE NANOPARTICULATE SYSTEM

Targeted delivery of drug/bioactives by the nanoparticulate system can be achieved actively or passively. Active targeted delivery can be achieved by conjugating the therapeutic drug or carrier system to a tissue- or cell-specific ligand. Passive targeting is achieved by integrating the therapeutic agent into a macromolecule or nanoparticle that passively reaches the targeted organ.

The development of the nanoparticulate system was affected to increase bioavailability and absorption; control/sustain, release or solubilize drugs; improve targeted delivery, and overcome biological barriers. Among all the different nanoparticulate systems, SLN and NLC are advantageous as they can: (i) encapsulate various types of drugs, (ii) control and extend drug release, (iii) extend blood circulation time, and (iv) enhance permeability and retention of the drug during delivery to increase the treatment effect. The release profile of any drug or bioactive compound is a function of the lipid matrix, surfactant concentration, and process parameters (e.g. temperature).

Many proteins and peptides are being developed but they are unable to cross cell membranes and have a short half-life. SLNs and NLCs are the best carriers for therapeutic drugs and protect drugs from degradation. transporting them into the cell interior. Hydrophilic drugs are also a candidate for delivery by SLNs and NLCs (Joshi and Müller, 2009).

The drugs inside the lipid carrier are protected against any enzymatic degradation (i.e. nucleases and proteases). Therefore, it is advantageous to use SLNs and NLCs for targeted drug delivery as they are small and the drug will be protected inside the lipid matrix, which can extravasate through the endothelium inflammatory sites, epithelium (e.g. intestinal tract and liver), and tumors, or penetrate microcapillaries (Singh and Lillard, 2009).

Govern delivery routes of nanoparticulate system:

a) Parental
b) Oral
c) Rectal
d) Nasal
e) Respiratory

f) Topical
g) Ocular

Different therapeutic applications of the nanoparticulate system for targeted delivery:

1. Lipid nanoparticle for treatment of cancer
2. Liver-targeting lipid nanoparticles
3. Lipid nanoparticles for transfection
4. Lipid nanoparticles for targeting the central nervous system
5. Cardiovascular treatment using lipid nanoparticles
6. Treatment of parasitic diseases with lipid nanoparticles
7. Lipid nanoparticles for treatment of rheumatoid arthritis

The potential of the lipid nanoparticulate system for the parenteral delivery of various therapeutic agents/drugs has been successfully established for different applications. However, there are no injectable SLNs or NLCs in the market to date, so more studies and progress are needed on the targeted delivery of the nanoparticulate system.

10.5 MODELING BIOACTIVE RELEASE OF THE NANOSCALE DELIVERY SYSTEM

Mathematical modeling plays a great role in the bioactive release of the control delivery system. It explains the mechanism of bioactive release, provides an optimization scenario for the carrier system, and helps avoid excessive experiments. Bioactive release is mainly controlled by one or a combination of three different mechanisms, i.e. (i) diffusion, (ii) erosion, (iii) fragmentation, and (iv) swelling (Lesmes and McClements, 2009). All mechanisms other than diffusion are more likely to occur in hydrophilic (carbohydrate- and protein-based) carriers. Mathematical models related to the diffusion mechanism are mostly responsible for lipid-based carrier systems (SLNs and NLCs), (Fathi et al., 2012).

The drug release in the loaded core model is membrane-controlled and is governed by Fick's law of diffusion as lipid surround the drug as a membrane (Müller et al., 2000). Fick's first law (Equation 10.1) assumes constant diffusion and boundaries. The concentration is diffusion coefficient-dependent and can be written as Fick's second law (Equation 10.2), which can be solved both analytically or numerically (Fathi et al., 2012).

$$\frac{\partial C}{\partial t} = D\frac{\partial^2 C}{\partial Z^2} \tag{10.1}$$

$$\frac{\partial C}{\partial t} = \frac{\partial}{\partial Z}\left[D(C)\frac{\partial C}{\partial Z}\right] \tag{10.2}$$

where:

C is the concentration of the bioactive compound
t is time
Z is the position in the carrier
D is the diffusion coefficient of the bioactive material

The release phenomenon of bioactive compounds from spherical particles is described by (Higuchi, 1963) with a square root of time model (Equation 10.3).

$$M_t = \sqrt{DC_m\left(2C_t - 2C_s\right)t} \tag{10.3}$$

In this equation C_m is the solubility of bioactives in the encapsulated matrix, C_t, is the initial bioactive concentration, and C_s is the solubility of bioactives in the sink phase. Higuchi describes the release of dispersed solute from a solid matrix of spherical-shaped particles with no swelling or erosion or fragmentation and written as Equation 10.4.

$$M_t = 4r^2\left[\sqrt{2(C_0 - C_s)DC_s t} - \frac{4C_s D}{9r}\left(\frac{C_s}{2C_0 - C_s} - 3\right)t\right] \tag{10.4}$$

where C_0 is initial bioactive concentration. Large values of release time (t) made the second term significant and, thus, the particle size became important in encapsulation.

The Higuchi model can be written in another form (Equation 10.5).

$$R = k_H\sqrt{t} \tag{10.5}$$

$$\text{where, } k_H = \left[2DC_s\varepsilon\left(C_m - 0.5C_s\varepsilon\right)\right]^{0.5} \tag{10.6}$$

R, defines the amount of bioactive released in time t

ε is the porosity

k_H is the Higuchi constant

The above Equation (10.5) can also be expressed as Equation 10.7.

$$\frac{M_t}{M_\infty} = k_H\sqrt{t} \tag{10.7}$$

When this was investigated (Haidar et al., 2008), results showed these models have a high ability to investigate the release rate. Also, it has been found that there are two distinct regions in the release mechanism, (i) the initial phase (shell release) and (ii) the terminal phase (core release).

Kumari et al. (2010) described the kinetic analysis of bioactive release from nanoparticles by monoexponential and biexponential equations.

$$C = C_0 e^{-kt} \tag{10.8}$$

$$C = Ae^{-\alpha t} + Be^{-\beta t} \tag{10.9}$$

where:

C is the concentration of drug remaining in the nanocapsule at time t

A (for diffusion control system)

B (for erosion control system) are system characteristic constants

α (for burst phase)

β (for sustain phase) are rate constants that can be obtained from semilog-arithmetic plots

All these equations are being used in non-bioactive compound release mechanisms. In spite of the importance of bioactive release modeling in the nano-delivery system, not much research has been done on this.

10.5.1 Challenges and Limitations in Bioactive Release Modeling

Bioactive release modeling has not been used widely in industry and research due to its limited predictability and accuracy. All the present models are empirically based and mechanism-based models; it is hard to mimic reality because of biological, chemical, and delivery processes. High-quality experimental research and different parameters data are required to calibrate, validate, and analyze the errors in mechanistic models. Each model has limitations and is unable to predict the release of bioactives across length along with time. The inclusion of bioactive release across the length of the capsules with time and linking different models is difficult, but it will provide great accuracy to overcome limitations.

10.6 RELEASE OF BIOACTIVES FROM NANOCAPSULES

Bioactive incorporating models are crucial to the release pattern. The release of bioactives from nanocapsules is an important factor for targeted delivery. It is related to the composition and production method of nanocapsules. Various parameters such as composition of formulation (e.g. surfactant type, surfactant concentration, surfactant mixture, amount of drug, lipid, and drug proportion), production methods, and conditions (such as time, production temperature, sterilization, equipment, and cooling) control the particle size of SLNs which affects the release rate of drugs. In NLCs, the oil content of the particles dissolves the drugs and controls the release characteristics with high loading capacity. The imperfect type and amorphous type of NLCs provides more flexibility and desired prolonged release of drugs (Müller et al., 2002a). In case of SNLs and NLCs, bioactives are set inside the solid lipid system which extends the mobility of bioactives dispersion.

A good understanding and knowledge of bioactive release mechanisms are needed to control rate and time of bioactive release. The release rate of bioactive compounds depends on: (i) their solubility, (ii) desorption of the surface-bound or absorbed compound, (iii) diffusion through the nanoparticle matrix, (iv) nanocapsule erosion or degradation, (v) a combination of erosion and diffusion.

Mostly, bioactives release rate depends on both diffusion and biodegradation of matrix material. Bioactive release can be modified to some extent by the choice of shell material. The initial release in SLNs is mainly due to the burst effect, which exists in the first five minutes in the drug-enriched shell of the nanocapsule model (Üner and Yener, 2007). The burst effect in nanocapsules is a result of the bursting of the outer layer due to the large surface area of drug deposition on the particle (e.g. SLN) surface (Müller et al., 1995; zur Mühlen et al., 1998). The burst release can be reduced by increasing particle size, or through the type of surfactant used and its concentration. Concentrations of surfactant, interaction with the outer shell, and structure of nanocapsules are important factors for minimal burst and prolong the release of bioactives. Prolonged release of drugs can be achieved in NLCs with the lower melting point of nanocapsule lipid, which is dependent upon the interaction between drug and lipid molecule, surfactant and lipid molecules, and solubility of the drugs in liquid and solid lipid matrix (zur Mühlen et al., 1998).

When a drug is uniformly distributed, the diffusion of the drug is faster than matrix erosion, and then the mechanism of release is mostly controlled by a diffusion process. When rapid initial release or burst happens, this is mainly attributed to weakly bound drugs with the shell matrixes (Kumari et al., 2010).

10.7 CASE STUDY: DELIVERY OF RESVERATROL (RSV) THROUGH SLNS

Resveratrol (3,5,4-trihydroxy-trans-stilbene) is a stilbenoid, a natural polyphenolic compound produced by a wide variety of plant sources. It has interesting functions and works as antioxidant, anticarcinogenic, anti-inflammatory, anti-aging, cardioprotector, neuroprotector, cell cycle inhibitor, and obesity and diabetics preventive. The food sources of resveratrol include the skin of grapes, blueberries, raspberries, mulberries, lingonberry, and senna. The richest natural source of resveratrol is dark grapes and consequently the wines that are produced from these grapes.

10.7.1 Dermal Delivery of Resveratrol

Gokce et al. (2012) studied the potential of SLNs in delivering resveratrol by dermal application. SLNs were produced by high shear homogenization method and delivered dermally as SLNs is advantageous in control release of bioactives. Compritol C888 surfactant was used for the lipid phase and a 60:1 lipid-to-drug ratio was maintained. The dermal delivery was examined using rat abdominal skin in an *ex vivo* study of Franz diffusion cell method. The *ex vivo* studies showed that resveratrol alone was unable to permeate skin tissue; however, resveratrol SLNs permitted the penetration and accumulation into the skin. 50 µM of resveratrol was sufficient to decrease the ROS (radical oxygen species) concentration, which is a contributor to skin pathogens.

10.7.2 Oral Delivery of Resveratrol

Oral delivery of resveratrol towards the enhancement of supplements and nutraceuticals through SLNs was investigated in terms of stability, compatibility, and intestinal permeability (Neves et al., 2016a). The formed SLNs gave a mean diameter of 160–190 nm, a high zeta potential of 30 mV and entrapment efficiency of 80%, which suggested suitability for oral delivery. It has been shown that oral absorption is enhanced by resveratrol SLNs as intestinal permeability is significantly increased by fed-state intestinal juice.

10.7.3 Brain-Targeted Delivery of Resveratrol

Resveratrol as a neuroprotective compound has beneficial effects. The potential of colloidal carriers such as SLNs in delivering resveratrol to the brain was checked by Teskac and Kristl (2010). The effect of SLNs when loaded (SLN-RSV) and/or empty was compared with RSV solution on the internalization, growth, morphology, metabolic activity, and genetic material of keratinocytes. Resveratrol solubility, stability, and intracellular delivery were increased by loading into a size below 180 nm SLN. SLN-RSV moves through the cell membrane and distributes through the cytosol without inducing cytotoxicity. The cytotoxic effect was reduced compared to RSV solution by using SLN-RSV. The loading of RSV in SLNs has potential benefits for preventing skin cancer, as it was effective in decreasing cell multiplication by inhibiting the growth of keratinocytes via cell cycle modulation.

In another study, Neves et al. (2016b) functionalized resveratrol loaded in SLNs with apolipoprotein E and showed that low-density lipoprotein (LDL) receptor identified it on the blood-brain barrier (BBB). SLN loaded with resveratrol, of less than 200 nm diameter and around -13 mV zeta potential and encapsulation efficiency of 90%, has no cytotoxicity (up to 50 µM) after *in vitro* studies of 4h incubation. A release kinetics study showed only 5%–10% of the release of resveratrol after 15 minutes of incubation in simulated body fluid (SBF). And *in vitro* cellular studies, when performed in the surroundings of hCMEC/D3 (immortalized human cerebral microvascular endothelial) cells, showed a 1.8 higher fold of resveratrol permeability when encapsulated in apolipoprotein E SLNs compared with non-functionalized SLNs. The study shows a promising strategy for the treatment of brain neurological disorders by delivering RVS through SLNs. However, the majority of drugs are restricted by the BBB, which obstacles the entry of drugs into the brain, restricting many therapies for several central nervous system diseases.

10.8 FUTURE TRENDS

Nano-delivery of pharmaceutical drugs is already an established system; more studies need to adopt nano lipid carriers for food bioactives. The future of lipid-based nanocarriers is mainly dependent on two things: (i) development of a suitable technique for large-scale production and increase encapsulation efficiency, and (ii) novel concepts and approaches to accomplish the on-demand release of different food bioactives.

Overcoming the stress caused by bioactive compound structures and developing the lipid matrix of a multicomponent system will allow the introduction of different functionalities at the same time. The development of a multicomponent system with different functionalities in the same delivery system which can be absorbed within the gastrointestinal tract and overcome the blood-brain barrier is the future. Furthermore, the development of multilayer systems is required to increase the functionality of food bioactives in the different delivery systems along with multicomponents. The delivery of bioactives with different carrier systems through different routes (e.g. parental, oral, nasal, rectal, respiratory, topical, ocular) help us increase the functionality of the bioactive system.

Future trends in nano-delivery systems should focus more on developing a mechanical and computational model system to understand the release mechanism by considering all existing/additional factors from novel mathematical models, in order to speed up the process and reduce existing human and animal trials. Lastly, the selection of a carrier system should be in such a way that lipids can encapsulate the bioactives as well as taking part in improving human health (e.g. unsaturated fats).

REFERENCES

Ali, Hazem, Khalid El-Sayed, Paul W. Sylvester, and Sami Nazzal. 2010. "Molecular Interaction and Localization of Tocotrienol-Rich Fraction (TRF) within the matrices of lipid nanoparticles: Evidence Studies by Differential Scanning Calorimetry (DSC) and Proton Nuclear Magnetic Resonance Spectroscopy (1H NMR)." *Colloids and Surfaces B: Biointerfaces,* 77 (2): 286–297.

Anandharamakrishnan, C. 2014. "Techniques for formation of nanoemulsions." In *Techniques for Nanoencapsulation of Food Ingredients*, 7–16. New York: Springer.

Araújo, Joana, Sasha Nikolic, Maria A. Egea, Eliana B. Souto, and Maria L. Garcia. 2011. "Nanostructured lipid carriers for triamcinolone acetonide delivery to the posterior segment of the eye." *Colloids and Surfaces. B, Biointerfaces*, 88 (1): 150–157.

Benita, Simon. 1998. *Submicron Emulsions in Drug Targeting and Delivery.* Vol. 9. Boca Raton: CRC Press.

Bhushani, Anu and C. Anandharamakrishnan. 2017. "Food-grade nanoemulsions for protection and delivery of nutrients." In *Nanoscience in Food and Agriculture*, 4, eds. S. Ranjan, N. Dasgupta, and E. Lichtfouse, 99–139. New York: Springer.

Bhushani, J. Anu, P. Karthik, and C. Anandharamakrishnan. 2016. "Nanoemulsion based delivery system for improved bioaccessibility and Caco-2 cell monolayer permeability of green tea catechins." *Food Hydrocolloids,* 56:372–382.

Bhushani, Anu, Udayakumar Harish, and Chinnaswamy Anandharamakrishnan. 2017a. "Nanodelivery of nutrients for improved bioavailability." In *Nutrient Delivery*, ed. A. M. Grumezescu, 369–411. Cambridge: Elsevier.

Bhushani, J. Anu, Nawneet Kumar Kurrey, and C. Anandharamakrishnan. 2017b. "Nanoencapsulation of green tea catechins by electrospraying technique and its effect on controlled release and in-vitro permeability." *Journal of Food Engineering,* 199:82–92.

Bunjes, Heike, Kirsten Westesen, and Michel H. J. Koch. 1996. "Crystallization tendency and polymorphic transitions in triglyceride nanoparticles." *International Journal of Pharmaceutics,* 129 (1–2): 159–173.

Castro, Gisele A., Rodrigo L. Oréfice, José M. C. Vilela, Margareth S. Andrade, and Lucas A. M. Ferreira. 2007. "Development of a new solid lipid nanoparticle formulation containing retinoic acid for topical treatment of acne." *Journal of Microencapsulation,* 24 (5): 395–407.

Chen, Chih-Chieh, Tung-Hu Tsai, Zih-Rou Huang, and Jia-You Fang. 2010. "Effects of lipophilic emulsifiers on the oral administration of lovastatin from nanostructured lipid carriers: Physicochemical characterization and pharmacokinetics." *European Journal of Pharmaceutics and Biopharmaceutics,* 74 (3): 474–482.

Chen-yu, Guo, Yang Chun-fen, Li Qi-lu, Tan Qi, Xi Yan-wei, Liu Wei-na, and Zhai Guang-xi. 2012. "Development of a quercetin-loaded nanostructured lipid carrier formulation for topical delivery." *International Journal of Pharmaceutics*, 430 (1–2): 292–298.

Chinsriwongkul, Akhayacatra, Ponwanit Chareanputtakhun, Tanasait Ngawhirunpat, Theerasak Rojanarata, Warisada Sila-on, Uracha Ruktanonchai, and Praneet Opanasopit. 2012. "Nanostructured Lipid Carriers (NLC) for parenteral delivery of an anticancer drug." *AAPS PharmSciTech*, 13 (1): 150–158.

Cirri, M., M. Bragagni, N. Mennini, and P. Mura. 2012. "Development of a new delivery system consisting in 'drug – in cyclodextrin – in nanostructured lipid carriers' for ketoprofen topical delivery." *European Journal of Pharmaceutics and Biopharmaceutics*, 80 (1): 46–53.

Coupland, J. N., Z. Zhu, H. Wan, D. J. McClements, W. W. Nawar, and P. Chinachoti. 1996. "Droplet composition affects the rate of oxidation of emulsified ethyl linoleate." *Journal of the American Oil Chemists' Society*, 73 (6): 795.

Doktorovová, Slavomíra, Joana Araújo, Marisa L. Garcia, Erik Rakovský, and Eliana B. Souto. 2010. "Formulating Fluticasone Propionate in Novel PEG-Containing Nanostructured Lipid Carriers (PEG-NLC)." *Colloids and Surfaces B: Biointerfaces,* 75 (2): 538–542.

Eskandari, Sharareh, Jaleh Varshosaz, Mohsen Minaiyan, and Majid Tabbakhian. 2011. "Brain delivery of valproic acid via intranasal administration of nanostructured lipid carriers: In vivo pharmacodynamic studies using rat electroshock model." *International Journal of Nanomedicine*, 6: 363–371.

Esposito, Elisabetta, Paolo Mariani, Laura Ravani, Catia Contado, Mattia Volta, Simone Bido, Markus Drechsler, et al. 2012. "Nanoparticulate lipid dispersions for bromocriptine delivery: Characterization and in vivo study." *European Journal of Pharmaceutics and Biopharmaceutics,* 80 (2): 306–314.

Fang, Jia-You, Chia-Lang Fang, Chi-Hsien Liu, and Yu-Han Su. 2008. "Lipid nanoparticles as vehicles for topical psoralen delivery: Solid Lipid Nanoparticles (SLN) versus Nanostructured Lipid Carriers (NLC)." *European Journal of Pharmaceutics and Biopharmaceutics*, 70 (2): 633–640.

Fathi, Milad, M. R. Mozafari, and M. Mohebbi. 2012. "Nanoencapsulation of food ingredients using lipid based delivery systems." *Trends in Food Science and Technology,* 23 (1): 13–27.

Fathi, Milad, Jaleh Varshosaz, Mohebbat Mohebbi, and Fakhri Shahidi. 2013. "Hesperetin-loaded solid lipid nanoparticles and nanostructure lipid carriers for food fortification: Preparation, characterization, and modeling." *Food and Bioprocess Technology*, 6 (6): 1464–1475.

Gokce, Evren H., Emrah Korkmaz, Eleonora Dellera, Giuseppina Sandri, M. Cristina Bonferoni, and Ozgen Ozer. 2012. "Resveratrol-loaded solid lipid nanoparticles versus nanostructured lipid carriers: Evaluation of antioxidant potential for dermal applications." *International Journal of Nanomedicine,* 7: 1841–1850.

Gramdorf, Sönke, Stephanie Hermann, Anke Hentschel, Katrin Schrader, Rainer H. Müller, Mont Kumpugdee-Vollrath, and Matthias Kraume. 2008. "Crystallized miniemulsions: Influence of operating parameters during high-pressure homogenization on size and shape of particles." *Colloids and Surfaces A: Physicochemical and Engineering Aspects*, Frontiers in Formulation Science (Formula V - 2007) Selected papers from the 5th International Conference on Formulation Technology, November 2007, Potsdam, Germany, 331 (1–2): 108–113.

Haidar, Ziyad S., Reggie C. Hamdy, and Maryam Tabrizian. 2008. "Protein release kinetics for core–shell hybrid nanoparticles based on the layer-by-layer assembly of alginate and chitosan on liposomes." *Biomaterials*, 29 (9): 1207–1215.

Han, Fei, Sanming Li, Ran Yin, Hongzhuo Liu, and Lu Xu. 2008. "Effect of surfactants on the formation and characterization of a new type of colloidal drug delivery system: Nanostructured lipid carriers." *Colloids and Surfaces A: Physicochemical and Engineering Aspects,* 315 (1–3): 210–216.

Hentschel, A., S. Gramdorf, R. H. Müller, and T. Kurz. 2008. "Beta-carotene-loaded nanostructured lipid carriers." *Journal of Food Science*, 73 (2): N1–N6.

Higuchi, T. 1963. "Mechanism of sustained-action medication. theoretical analysis of rate of release of solid drugs dispersed in solid matrices." *Journal of Pharmaceutical Sciences*, 52 (12): 1145–1149.

Hippalgaonkar, Ketan, Soumyajit Majumdar, and Viral Kansara. 2010. "Injectable lipid emulsions-advancements, opportunities and challenges." *AAPS PharmSciTech*, 11 (4): 1526–1540.

Hu, Fu-Qiang, Sai-Ping Jiang, Yong-Zhong Du, Hong Yuan, Yi-Qing Ye, and Su Zeng. 2005. "Preparation and characterization of stearic acid nanostructured lipid carriers by solvent diffusion method in an aqueous system." *Colloids and Surfaces B: Biointerfaces*, 45 (3–4): 167–173.

Hu, Fu-Qiang, Sai-Ping Jiang, Yong-Zhong Du, Hong Yuan, Yi-Qing Ye, and Su Zeng. 2006. "Preparation and characteristics of monostearin nanostructured lipid carriers." *International Journal of Pharmaceutics*, 314 (1): 83–89.

Hung, Loo Chew, Mahiran Basri, Bimo A. Tejo, Rosnah Ismail, Harrison Lau Lik Nang, Hazimah Abu Hassan, and Choo Yuen May. 2011. "An improved method for the preparations of nanostructured lipid carriers containing heat-sensitive bioactives." *Colloids and Surfaces. B, Biointerfaces,* 87 (1): 180–186.

Jaspart, Séverine, Géraldine Piel, Luc Delattre, and Brigitte Evrard. 2005. "Solid lipid microparticles: Formulation, preparation, characterisation, drug release and applications." *Expert Opinion on Drug Delivery,* 2 (1): 75–87.

Jenning, Volkhard, Karsten Mäder, and Sven H. Gohla. 2000. "Solid Lipid Nanoparticles (SLN™) based on binary mixtures of liquid and solid lipids: A 1H-NMR study." *International Journal of Pharmaceutics,* 205 (1–2): 15–21.

Jia, Lejiao, Dianrui Zhang, Zhenyu Li, Cunxian Duan, Yancai Wang, Feifei Feng, Feihu Wang, Yue Liu, and Qiang Zhang. 2010. "Nanostructured lipid carriers for parenteral delivery of silybin: Biodistribution and pharmacokinetic studies." *Colloids and Surfaces B: Biointerfaces*, 80 (2): 213–218.

Jiménez-Colmenero, Francisco. 2013. “Potential applications of multiple emulsions in the development of healthy and functional foods.” *Food Research International*, 52 (1): 64–74.

Jores, Katja, Wolfgang Mehnert, Markus Drechsler, Heike Bunjes, Christoph Johann, and Karsten Mäder. 2004. “Investigations on the Structure of Solid Lipid Nanoparticles (SLN) and oil-loaded solid lipid nanoparticles by photon correlation spectroscopy, field-flow fractionation and transmission electron microscopy.” *Journal of Controlled Release*, 95 (2): 217–227.

Joshi, Medha D. and Rainer H. Müller. 2009. “Lipid nanoparticles for parenteral delivery of actives.” *European Journal of Pharmaceutics and Biopharmaceutics,* 71 (2): 161–172.

Karn-Orachai, Kullavadee, Siwaporn Meejoo Smith, Sarunya Phunpee, Alongkot Treethong, Satit Puttipipatkhachorn, Sirapat Pratontep, and Uracha Rungsardthong Ruktanonchai. 2014. “The effect of surfactant composition on the chemical and structural properties of nanostructured lipid carriers.” *Journal of Microencapsulation*, 31 (6): 609–618.

Karthik, P. and C. Anandharamakrishnan. 2016. “Enhancing omega-3 fatty acids nanoemulsion stability and in-vitro digestibility through emulsifiers.” *Journal of Food Engineering,* 187:92–105.

Karthik, P., P. N. Ezhilarasi, and C. Anandharamakrishnan. 2017. “Challenges associated in stability of food grade nanoemulsions.” *Critical Reviews in Food Science and Nutrition,* 57 (7):1435–1450.

Kasongo, Kasongo Wa, Jana Pardeike, Rainer H. Müller, and Roderick B. Walker. 2011. “Selection and characterization of suitable lipid excipients for use in the manufacture of didanosine-loaded solid lipid nanoparticles and nanostructured lipid carriers.” *Journal of Pharmaceutical Sciences*, 100 (12): 5185–5196.

Kumari, Avnesh, Sudesh Kumar Yadav, and Subhash C. Yadav. 2010. “Biodegradable polymeric nanoparticles based drug delivery systems.” *Colloids and Surfaces B: Biointerfaces,* 75 (1): 1–18.

Kuo, Yung-Chih and Jui-Fang Chung. 2011. “Physicochemical properties of nevirapine-loaded solid lipid nanoparticles and nanostructured lipid carriers.” *Colloids and Surfaces B: Biointerfaces*, 83 (2): 299–306.

Lacatusu, Loana, Nicoleta Badea, Oprea Ovidiu, Dionezie Bojin, and Aurelia Meghea. 2012. “Highly antioxidant carotene-lipid nanocarriers: Synthesis and antibacterial activity.” *Journal of Nanoparticle Research*, 14 (6): 1–16.

Lakkis, Jamileh M. 2007. “Introduction.” In *Encapsulation and Controlled Release Technologies in Food Systems*, edited by Jamileh M. Lakkis, 1–11. Hoboken:Blackwell Publishing.

Lander, R., W. Manger, M. Scouloudis, A. Ku, C. Davis, and A. Lee. 2000. “Gaulin homogenization: A mechanistic study.” *Biotechnology Progress*, 16 (1): 80–85.

Lesmes, Uri and David Julian McClements. 2009. “Structure–function relationships to guide rational design and fabrication of particulate food delivery systems.” *Trends in Food Science and Technology*, 20 (10): 448–457.

Li, Bei and Zhi-Qiang Ge. 2012. “Nanostructured lipid carriers improve skin permeation and chemical stability of idebenone.” *AAPS PharmSciTech,* 13 (1): 276–283.

Li, Zhen, Xiaohong Lin, Li Yu, Xinwei Li, Fei Geng, and Liqiang Zheng. 2009. “Effects of chloramphenicol on the characterization of solid lipid nanoparticles and nanostructured lipid carriers.” *Journal of Dispersion Science and Technology,* 30 (7): 1008–1014.

Li, F., Y. Weng, L. Wang, H. He, J. Yang, and X. Tang 2010. “The efficacy and safety of bufadienolides-loaded nanostructured lipid carriers.” *International Journal of Pharmaceutics*, 393 (1–2): 203–211.

Lin, Xiaohong, Xinwei Li, Li Qiang Zheng, Li Yu, Qiqing Zhang, and Weichang Liu. 2007. “Preparation and characterization of monocaprate nanostructured lipid carriers.” *Colloids and Surfaces A: Physicochemical and Engineering Aspects*, *Engineering Particle Technology*, 311 (1–3): 106–111.

Liu, Chi-Hsien and Chao-Ting Wu. 2010. “Optimization of nanostructured lipid carriers for lutein delivery.” *Colloids and Surfaces A: Physicochemical and Engineering Aspects*, 353 (2–3): 149–156.

Liu, Donghua, Zhihong Liu, Lili Wang, Cai Zhang, and Na Zhang. 2011. “Nanostructured lipid carriers as novel carrier for parenteral delivery of docetaxel.” *Colloids and Surfaces B: Biointerfaces*, 85 (2): 262–269.

Liu, G. Y., J. M. Wang, and Q. Xia. 2012. “Application of nanostructured lipid carrier in food for the improved bioavailability.” *European Food Research and Technology,* 234 (3): 391–398.

Lopez-Huertas, Eduardo. 2010. “Health effects of oleic acid and long chain omega-3 fatty acids (EPA and DHA) enriched milks. A review of intervention studies.” *Pharmacological Research*, 61 (3): 200–207.

Luo, Qiuhua, Junming Zhao, Xiangrong Zhang, and Weisan Pan. 2011. “Nanostructured Lipid Carrier (NLC) coated with chitosan oligosaccharides and its potential use in ocular drug delivery system.” *International Journal of Pharmaceutics*, 403 (1–2): 185–191.

Mäder, Karsten and Wolfgang Mehnert. 2004. “Solid lipid nanoparticles ? Concepts, procedures, and physicochemical aspects.” In *Lipospheres in Drug Targets and Delivery*, ed. C. Nastruzzi. 1–22. Boca Raton: CRC Press.

McClements, David Julian, and Yan Li. 2010. "Structured emulsion-based delivery systems: Controlling the digestion and release of lipophilic food components." *Advances in Colloid and Interface Science*, 159 (2): 213–228.

McClements, David Julian and Jiajia Rao. 2011. "Food-grade nanoemulsions: Formulation, fabrication, properties, performance, biological fate, and potential toxicity." *Critical Reviews in Food Science and Nutrition,* 51 (4): 285–330.

Mehnert, Wolfgang and Karsten Mäder. 2001. "Solid lipid nanoparticles: Production, characterization and applications." *Advanced Drug Delivery Reviews,* 47 (2): 165–196.

Mehnert, W., A. Zur Mühlen, A. Dingler, H. Weyhers, and R. H. Müller. 1997. "Solid lipid nanoparticles - ein neuartiger Wirkstoff-Carrier für Kosmetika und Pharmazeutika 2. Mitteilung: Wirkstoff-Inkorporation, Freisetzung und Sterilisierbarkeit." *Pharmazeutische Industrie,* 59 (6): 511–514.

Mitri, Khalil, Ranjita Shegokar, Sven Gohla, Cecilia Anselmi, and Rainer H. Müller. 2011. "Lipid nanocarriers for dermal delivery of lutein: Preparation, characterization, stability and performance." *International Journal of Pharmaceutics,* 414 (1–2): 267–275.

Mueller, Rainer, Karsten Maeder, Andreas Lippacher, and Volkhard Jenning. 2000. Solid-liquid (semi-solid) lipid particles (nano-Complement-Carrier-NCC) and processes for the preparation of highly concentrated lipid particles. DE19945203 A1, filed September 21, 1999, and issued December 21, 2000. http://www.google.com/patents/DE19945203A1.

Mühlen, Annette zur, Cora Schwarz, and Wolfgang Mehnert. 1998. "Solid Lipid Nanoparticles (SLN) for controlled drug delivery–drug release and release mechanism." *European Journal of Pharmaceutics and Biopharmaceutics*, 45 (2): 149–155.

Müller, R. H., W. Mehnert, J. S. Lucks, C. Schwarz, A. Zur Mühlen, H. Meyhers, C. Freitas, and D. Rühl. 1995. "Solid Lipid Nanoparticles (SLN) : An alternative colloidal carrier system for controlled drug delivery." *European Journal of Pharmaceutics and Biopharmaceutics*, 41 (1): 62–69.

Muller, Rainer H., Benita, Simon and Bernhard H. L. Böhm. 1998. *Emulsions and Nanosuspensions for the Formulation of Poorly Soluble Drugs.* Boca Raton:CRC Press.

Müller, Rainer H., Karsten Mäder, and Sven Gohla. 2000. "Solid Lipid Nanoparticles (SLN) for controlled drug delivery – a review of the state of the art." *European Journal of Pharmaceutics and Biopharmaceutics,* 50 (1): 161–177.

Müller, R. H, M. Radtke, and S. A. Wissing. 2002a. "Nanostructured lipid matrices for improved microencapsulation of drugs." *International Journal of Pharmaceutics*, 242 (1–2): 121–128.

Müller, R. H., M. Radtke, and S. A. Wissing. 2002b. "Solid Lipid Nanoparticles (SLN) and Nanostructured Lipid Carriers (NLC) in cosmetic and dermatological preparations." *Advanced Drug Delivery Reviews*, Human skin: the Medium of Touch, 54, Supplement (November): S131–S155.

Nayak, Aditya P., Waree Tiyaboonchai, Swati Patankar, Basavaraj Madhusudhan, and Eliana B. Souto. 2010. "Curcuminoids-loaded lipid nanoparticles: Novel approach towards malaria treatment." *Colloids and Surfaces. B, Biointerfaces*, 81 (1): 263–273.

Neves, Ana Rute, Susana Martins, Marcela A. Segundo, and Salette Reis. 2016a. "Nanoscale delivery of resveratrol towards enhancement of supplements and nutraceuticals." *Nutrients,* 8 (3): 131.

Neves, Ana Rute, Joana Fontes Queiroz, and Salette Reis. 2016b. "Brain-targeted delivery of resveratrol using solid lipid nanoparticles functionalized with apolipoprotein E." *Journal of Nanobiotechnology*, 14: 27.

Nguyen, H. M., I. C. Hwang, J. W. Park, and H. J. Park. 2012. "Enhanced payload and photo-protection for pesticides using nanostructured lipid carriers with corn oil as liquid lipid." *Journal of Microencapsulation*, 29 (6): 596–604.

Obeidat, Wasfy M., Kay Schwabe, Rainer H. Müller, and Cornelia M. Keck. 2010. "Preservation of Nanostructured Lipid Carriers (NLC)." *European Journal of Pharmaceutics and Biopharmaceutics*, 76 (1): 56–67.

Pardeike, Jana, Aiman Hommoss, and Rainer H. Müller. 2009. "Lipid Nanoparticles (SLN, NLC) in cosmetic and pharmaceutical dermal products." *International Journal of Pharmaceutics,* 366 (1–2): 170–184.

Pardeike, J., S. Weber, T. Haber, J. Wagner, H. P. Zarfl, H. Plank, and A. Zimmer. 2011. "Development of an itraconazole-loaded Nanostructured Lipid Carrier (NLC) formulation for pulmonary application." *International Journal of Pharmaceutics*, 419 (1–2): 329–338.

Porter, Christopher J. H., Natalie L. Trevaskis, and William N. Charman. 2007. "Lipids and lipid-based formulations: Optimizing the oral delivery of lipophilic drugs." *Nature Reviews. Drug Discovery*, 6 (3): 231–248.

Puglia, Carmelo, Giuseppina Frasca, Teresa Musumeci, Luisa Rizza, Giovanni Puglisi, Francesco Bonina, and Santina Chiechio. 2012. "Curcumin loaded NLC induces histone hypoacetylation in the CNS after intraperitoneal administration in mice." *European Journal of Pharmaceutics and Biopharmaceutics,* 81 (2): 288–293.

Rajam, R. and C. Anandharamakrishnan. 2015. "Spray freeze drying method for microencapsulation of Lactobacillus plantarum." *Journal of Food Engineering,* 166:95–103.

Rowe, Raymond C., Paul J. Sheskey, Walter G. Cook, and Marian E. Fenton. 2012. *Handbook of Pharmaceutical Excipients*. London:Pharmaceutical Press.

Sanad, Rania A., Nevine Shawky Abdelmalak, Tahany S. Elbayoomy, and Alia A. Badawi. 2010. "Formulation of a novel oxybenzone-loaded Nanostructured Lipid Carriers (NLCs)." *AAPS PharmSciTech,* 11 (4): 1684–1694.

Saupe, Anne, Sylvia A. Wissing, Andreas Lenk, Corinna Schmidt, and Rainer H. Müller. 2005. "Solid Lipid Nanoparticles (SLN) and Nanostructured Lipid Carriers (NLC) – structural investigations on two different carrier systems." *Bio-Medical Materials and Engineering* ,15 (5): 393–402.

Saupe, Anne, Keith C. Gordon, and Thomas Rades. 2006. "Structural investigations on nanoemulsions, solid lipid nanoparticles and nanostructured lipid carriers by cryo-field emission scanning electron microscopy and raman spectroscopy." *International Journal of Pharmaceutics,* 314 (1): 56–62.

Schwarz, C., W. Mehnert, J. S. Lucks, and R. H. Müller. 1994. "Solid Lipid Nanoparticles (SLN) for controlled drug delivery. I. production, characterization and sterilization." *Journal of Controlled Release*, 30 (1): 83–96.

Sf, Haag, M. Chen, D. Peters, C.M. Keck, B. Taskoparan, A. Fahr, C. Teutloff, et al. 2011. "Nanostructured lipid carriers as nitroxide depot system measured by electron paramagnetic resonance spectroscopy." *International Journal of Pharmaceutics*, 421 (2): 364–369.

Singh, Rajesh and James W. Lillard. 2009. "Nanoparticle-based targeted drug delivery." *Experimental and Molecular Pathology*, 86 (3): 215–223.

Souto, E. B. and R. H. Müller. 2005. "SLN and NLC for topical delivery of ketoconazole." *Journal of Microencapsulation*, 22 (5): 501–510.

Tamjidi, Fardin, Mohammad Shahedi, Jaleh Varshosaz, and Ali Nasirpour. 2013. "Nanostructured Lipid Carriers (NLC): A potential delivery system for bioactive food molecules." *Innovative Food Science and Emerging Technologies*, 19 (July): 29–43.

Teeranachaideekul, Veerawat, Rainer Helmut Müller, and Varaporn Buraphacheep Junyaprasert. 2007a. "Encapsulation of Ascorbyl Palmitate in Nanostructured Lipid Carriers (NLC)—Effects of Formulation Parameters on Physicochemical Stability." *International Journal of Pharmaceutics,* 340 (1–2): 198–206.

Teeranachaideekul, Veerawat, Eliana B. Souto, Varaporn B. Junyaprasert, and Rainer H. Müller. 2007b. "Cetyl palmitate-based NLC for topical delivery of coenzyme Q10 – development, physicochemical characterization and in vitro release studies." *European Journal of Pharmaceutics and Biopharmaceutics,* 67 (1): 141–148.

Teskac, K. and J. Kristl. 2010. "The evidence for solid lipid nanoparticles mediated cell uptake of resveratrol." *International Journal of Pharmaceutics*, 390 (1): 61–69.

Tiwari, Radheshyam and Kamla Pathak. 2011. "Nanostructured lipid carrier versus solid lipid nanoparticles of simvastatin: Comparative analysis of characteristics, pharmacokinetics and tissue uptake." *International Journal of Pharmaceutics*, 415 (1–2): 232–243.

Tsai, Ming-Jun, Pao-Chu Wu, Yaw-Bin Huang, Jui-Sheng Chang, Chin-Lin Lin, Yi-Hung Tsai, and Jia-You Fang. 2012. "Baicalein loaded in tocol nanostructured lipid carriers (tocol NLCs) for enhanced stability and brain targeting." *International Journal of Pharmaceutics,* 423 (2): 461–470.

Üner, Melike and Gülgün Yener. 2007. "Importance of solid lipid nanoparticles (SLN) in various administration routes and future perspectives." *International Journal of Nanomedicine*, 2 (3): 289–300.

Varshosaz, Jaleh, Sharareh Eskandari, and Majid Tabakhian. 2010. "Production and optimization of valproic acid nanostructured lipid carriers by the Taguchi design." *Pharmaceutical Development and Technology*, 15 (1): 89–96.

Varshosaz, Jaleh, Sharareh Eskandari, and Majid Tabbakhian. 2012a. "Freeze-drying of nanostructure lipid carriers by different carbohydrate polymers used as cryoprotectants." *Carbohydrate Polymers*, 88 (4): 1157–1163.

Varshosaz, Jaleh, Farshid Hassanzadeh, Hojjat Sadeghi, and Mostafa Khadem. 2012b. "Galactosylated nanostructured lipid carriers for delivery of 5-FU to hepatocellular carcinoma." *Journal of Liposome Research*, 22 (3): 224–236.

Varshosaz, J., F. Hassanzadeh, H. Sadeghi, and S. Andalib 2012c. "Synthesis of octadecylamine-retinoic acid conjugate for enhanced cytotoxic effects of 5-FU using LDL targeted nanostructured lipid carriers." *European Journal of Medicinal Chemistry*, 54 (August): 429–438.

Walstra, Pieter. 2002. *Physical Chemistry of Foods.*Boca Raton: CRC Press.

Westesen, K., H. Bunjes, and M. H. J. Koch. 1997. "Physicochemical characterization of lipid nanoparticles and evaluation of their drug loading capacity and sustained release potential." *Journal of Controlled Release*, 48 (2–3): 223–236.

Westesen, Kirsten and Britta Siekmann. 1997. "Investigation of the gel formation of phospholipid-stabilized solid lipid nanoparticles." *International Journal of Pharmaceutics*, 151 (1): 35–45.

Westesen, Kirsten, Britta Siekmann, and Michel H. J. Koch. 1993. "Investigations on the physical state of lipid nanoparticles by synchrotron radiation X-ray diffraction." *International Journal of Pharmaceutics,* 93 (1–3): 189–199.

Wissing, S. A., O. Kayser, and R. H. Müller. 2004. "Solid lipid nanoparticles for parenteral drug delivery." *Advanced Drug Delivery Reviews*, 56 (9): 1257–1272.

Xia, Q., A. Saupe, R. H. Müller, and E. B. Souto. 2007. "Nanostructured lipid carriers as novel carrier for sunscreen formulations." *International Journal of Cosmetic Science*, 29 (6): 473–482.

Yuan, Hong, Lei-Lei Wang, Yong-Zhong Du, Jian You, Fu-Qiang Hu, and Su Zeng. 2007. "Preparation and characteristics of nanostructured lipid carriers for control-releasing progesterone by melt-emulsification." *Colloids and Surfaces. B, Biointerfaces*, 60 (2): 174–179.

Zhang, Xiaoyun, Jianping Liu, Hua Qiao, Huan Liu, Jingman Ni, Wenli Zhang, and Yanbin Shi. 2010. "Formulation optimization of dihydroartemisinin nanostructured lipid carrier using response surface methodology." *Powder Technology,* 197 (1–2): 120–128.

Zhang, Tiecheng, Jianian Chen, Yi Zhang, Qi Shen, and Weisan Pan. 2011. "Characterization and evaluation of nanostructured lipid carrier as a vehicle for oral delivery of etoposide." *European Journal of Pharmaceutical Sciences: Official Journal of the European Federation for Pharmaceutical Sciences,* 43 (3): 174–179.

Zheng, Kai, Aihua Zou, Xiaomin Yang, Fang Liu, Qiang Xia, Ruqiang Ye, and Bozhong Mu. 2013. "The effect of polymer–surfactant emulsifying agent on the formation and stability of α-lipoic acid loaded nanostructured lipid carriers (NLC)." *Food Hydrocolloids,* 32 (1): 72–78.

Zhuang, Chun-Yang, Ning Li, Mi Wang, Xiao-Ning Zhang, Wei-San Pan, Jun-Jie Peng, Yu-Sheng Pan, and Xin Tang. 2010. "Preparation and characterization of vinpocetine loaded nanostructured lipid carriers (NLC) for improved oral bioavailability." *International Journal of Pharmaceutics,* 394 (1–2): 179–185.

11 Stability and Viability of Food Nanoparticles

S. Parthasarathi and C. Anandharamakrishnan

CONTENTS

11.1 INTRODUCTION

There has been growing interest in the utilization of different types of nanoparticles in the food and pharmaceutical industries for encapsulation, protection, and release of bioactive ingredients (Anandharamakrishnan, 2014a; Anandharamakrishnan, 2015; Anandharamakrishnan and Ishwarya, 2015a,b; Bhushani and Anandharamakrishnan, 2017; Ezhilarasi et al., 2013). The acceptability of food product is achieved by the quality and retention of nutritional content after processing, distribution, and storage until it reaches the consumer's hand. The physiological benefits of nutraceuticals can be achieved only if (1) the product of bioactive compounds is present in the required concentration, and (2) the bioactive compound is stable and available for absorption in the intestine. If either of these conditions is not met by product processing or during storage, then the nutraceutical loses its beneficial effects. Physical and chemical stability of nanoparticles is a major consideration during processing and bulk production nanoparticles. Stability of nanoparticles can be improved with the use of small molecule excipients. For instance, addition of lactose, sucrose or trehalose during the drying process prevent the aggregation of nanoparticles (Joye et al., 2015).

Colloidal dispersions play a pivotal role in food, nutraceutical, and pharmaceutical industries for the delivery of bioactive compounds (Anandharamakrishnan, 2014b; Bhushani, Harish, and Anandharamakrishnan, 2017; Karthik and Anandharamakrishnan, 2016a; Karthik and Anandharamakrishnan, 2016b). Colloidal dispersions contain the dispersed phase which remains kinetically stable for a long-time scale, for months or even years, and is considered as a stable dispersion. The dispersed phase contains mostly bioactive compounds, such as carotene, vitamin E, DHA, etc. Such colloidal dispersion may be stable for thermodynamic or for kinetic reasons. In the latter case, typified by electrostatic stabilization, the system is said to be thermodynamically metastable (Napper, 1983). Utilization of many lipid-soluble bioactive compounds is limited because of their poor water solubility and chemical instability. These lipophilic bioactive compounds should be dissolved in oils or in other suitable matrices for possible utilization.

Nanotechnology has become one of the most promising and interesting research fields during the past few decades. Tan and Nakajima (2005) and Chu et al. (2007)reported that the nanodelivery of lipophilic bioactive compounds and natural antioxidants enhanced dissolution rate and saturation solubility. A nanoparticle delivery system uses nanocarriers to encapsulate and then deliver bioactive compounds. These nanocarriers either enhance the absorption of the bioactive compound in the digestive tract by endocytosis or remain in the biological activities circulating through the body with specific targeting (Muller et al., 2000; Liu et al., 2016).

Many bioactive compounds are highly prone to chemical degradation during food processing and storage due to various environmental effects including thermal stress (Chen et al., 2014). Protection of encapsulated bioactive compounds from the external biological environment is one of the major benefits of employing nanoparticle delivery systems. Stability of the bioactive compound is a critical issue that arises during storage, processing, or shipping (Shah et al., 2015). However, stability issues are trivial in dry nanoparticles and this chapter is focused mainly on the stability of the nanoparticle dispersion.

11.2 STABILITY MECHANISMS

The major concern during the formulation of nanoparticles is to create sufficiently small particles with long kinetic stability. Nanoparticles are prone to agglomeration because the attractive van der Waal force and/or the driving force tend to minimize the total surface energy. Hence, the presence of repulsive force is necessary to prevent the agglomeration of nanoparticles. Two terms are commonly used to represent repulsive forces: (i) electrostatic stabilization and (ii) steric stabilization (Chow and Gonsalves, 1996). The kinetic stability of the nanoemulsion can be improved by controlling its particle size, or by incorporating stabilizers such as emulsifiers, texture modifiers, weighting agents, or ripening retarders (McClements and Rao, 2011b).

11.2.1 Electrostatic Stabilization

Electrostatic repulsion results from the interaction between the electric double layers of the particle surface. Due to this phenomenon, an unequal charge distribution always exists between the particle surface and the continuous phase. However, electrostatic stabilization is sensitive to electrolyte concentration, because a small change in electrolyte concentration may destroy the electric double layer and result in particle agglomeration (Chow and Gonsalves, 1996).

When a nanoparticle is suspended in a solution/emulsion system, it often exhibits charge on its surface due to asymmetrical bonds of surface atoms. Ions with opposite charge are attracted towards the surface to balance the charge and form a layer called "Stern layer." For instance, when a negatively charged nanoparticle is suspended in water, the positively charged hydrogen part of the water molecules binds to the surface and forms a Stern layer. Outside the Stern layer, another layer

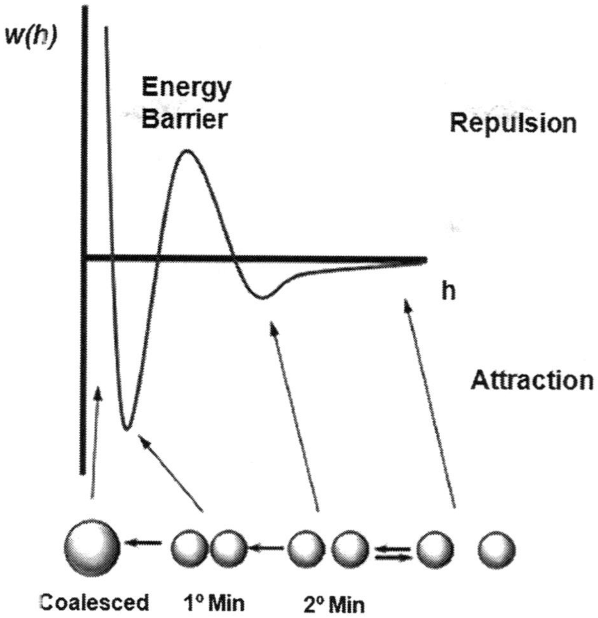

FIGURE 11.1 The DLVO theory. The total interaction between two particles is a combination of electrostatic repulsion and van der Waals attraction. (Piorkowski and McClements, 2014.)

of negative ions will be formed which is called "diffusion layer." So, the combination of Stern layer and diffusion layer is called "double layer."

This ionic double layer offers electrostatic stabilization to the system and keeps the nanoparticles from aggregation by exceeding the van der Waals interactions between nanoparticles. The behavior of nanoparticles in a solution is the result of the combination of two forces, i.e. van del Waals attraction and electrostatic repulsion with respect to distance asdescribed by the Derjaguin-Landau-Verwey-Overbeek (DLVO) theory (Figure 11.1). The DLVO theory was first developed by Derjaguin and Landau (1941), and later by Verwey and Overbeek (1948). According to the DLVO theory, the total potential energy of interaction, $\Phi_T(h)$, is the sum of attractive van der Waals (Φ_{vdW}) and repulsive electrostatic interactions ($\Phi_{electrostatic}$). So, the thumb rule on nanoparticle stability is that if attractive interaction dominates, nanoparticles tend to aggregates, and if repulsive interactions dominate, nanoparticles remain individual (McClements, 2004). Gambinossi et al. (2015) included other contributions such as hydration ($\Phi_{hydration}$) and osmotic forces ($\Phi_{osmotic}$) in the interaction potential:

$$\phi(h) = \phi_{vdw}(h) + \phi_{electrostatic}(h) + \phi_{hydration}(h) + \phi_{osmotic}(h) \tag{11.1}$$

Electrostatic potential can be calculated by the following two equations, which depend on the κR_H value (Piorkowski and McClements, 2014):

$$\phi_{electrostatic}(h) = 2\pi R_H \varepsilon_r \varepsilon_0 \Psi_0^2 ln\left[1 + e^{-\kappa h}\right];\ \kappa R_H > 5 \tag{11.2}$$

$$\phi_{electrostatic}(h) = 4\pi R_H^2 \varepsilon_r \varepsilon_0 Y^2 \left(\frac{k_B T}{e}\right) \frac{e^{-\kappa h}}{h + 2R_H};\ \kappa R_H < 5 \tag{11.3}$$

where:

ε_r is the relative permittivity of the medium
ε_0 is the permittivity of free space
$\Psi(0)$ is the surface potential

R_H is the particle size
κ^{-1} and Y are the Debye length calculated from the following expression:

$$\kappa^{-1} = \sqrt{\frac{\varepsilon_0 \varepsilon_R \kappa_B T}{1000 e^2 N_A \left(2C_{\text{NaCl}}\right)}} \tag{11.4}$$

$$Y = \frac{8 tanh\left(\varepsilon_0 \varepsilon_R \psi_0 / 4k_B T\right)}{1+\sqrt{1-2\kappa R_H / \left(\kappa R_H + 1\right)^2 tanh^2\left(\varepsilon_0 \varepsilon_R \psi_0 / 4k_B T\right)}} \tag{11.5}$$

where:
k_B is the Boltzmann constant
T is the absolute temperature
e is the electronic charge

At room temperature, κ^{-1} for aqueous solutions can be represented as: $k^{-1} = 0.304 / \sqrt{I}$, where I is the ionic strength of 1:1 electrolytes (e.g. NaCl) in moles per liter. Rajagopalan and Hiemenz (1997) reported that multivalent counterions such as Ca^{2+}, Fe^{2+}, and Fe^{3+} are much more effective at screening electrostatic interactions than monovalent counterions (e.g. Na^+, Cl^-, and K^+). Hence, a lesser concentration of multivalent counterions screens more charges than monovalent counterions (McClements, 2004). The distribution of counterions in the droplet interior is governed by the Poisson-Boltzmann equation (Feldman et al., 2001):

$$\Delta\Psi = \frac{-4\pi e c(r)}{\varepsilon_w} = \frac{-4\pi e c_0}{\varepsilon_w} e - \frac{e\left(\Psi(r) - \Psi(0)\right)}{k} \tag{11.6}$$

where:
Ψ is the electrostatic potential
ε_w is the dielectric permittivity of the water core
c_0 is the counterion density
$\Psi(0)$ is the surface potential

Here, the reference point is chosen at the center (r=0) of the spherical particle.

11.2.2 Steric Stabilization

Electrostatic repulsion depends on the z-potential (charge) and ionic composition of the dispersion medium, whereas steric repulsion depends on the thickness and chemistry of the interfacial layer. Steric stabilization arises from the hairy layers of emulsifiers, typically present in protein or polysaccharide interfacial layer. These surfactant molecules can adsorb onto the nanoparticle surfaces, and their lyophilic chains will extend into the solvent. This solvent-chain interaction creates a free energy in the system and forms an energy barrier surrounding the nanoparticle. Unlike electrostatic stabilization, steric stabilization is effective in both aqueous and non-aqueous media, and less sensitive to impurities or trace additives (Chow and Gonsalves, 1996).

Steric interactions are extremely strong and operate between all types of colloidal particles (McClements, 2015). When two colloidal particles approach each other at a distance of less than twice the thickness of the adsorbed polymer layer, then the interaction of two layers takes place. The degree of stabilization depends on the energy change that occurs during the interaction of adsorbed layers. The Gibbs free energy change ΔG during the interaction can be expressed as:

$$\Delta G = \Delta H - T\Delta S \tag{11.7}$$

where:

- ΔG is the change in Gibbs free energy
- ΔH is the change in enthalpy
- ΔS is the change in entropy
- T is the temperature

If Gibbs free energy change (ΔG) is negative upon the overlap of the adsorbed layer, then the colloidal particles undergo flocculation or coagulation and for a stable nanoparticle, the Gibbs free energy change (ΔG) should be positive. When the interfacial layer of two colloidal particles overlaps, it generates a strong repulsive force because of the entropy effects. Steric repulsion can be described by the following equation (McClements, 2015):

$$w_s(h) = \left(2\frac{\delta}{h}\right)^{12} kT \tag{11.8}$$

where:

- δ is the thickness of the interfacial layer of the particles
- k is the Boltzmann's constant (1.381×10^{-23} J K^{-1})
- T is the absolute temperature (Table 11.1)

11.2.2.1 Advantages of Steric Stabilization

Steric stabilization is widely exploited industrially because it offers several distinct advantages over electrostatic stabilization (Guzey and McClements, 2006).

Relative insensitivity to the presence of electrolytes: For 1:1 electrolytes, a charge-stabilized dispersion will not be stable and will coagulate when the electrolyte concentration exceeds the 10^{-1} limit. Polymer chains have no such dramatic sensitivity and sterically stabilized colloidal nanoemulsions are relatively insensitive to the electrolyte concentration.

Equal efficacy in both aqueous and non-aqueous dispersion media: Steric stabilization is effective in both aqueous and non-aqueous media. Hence, steric stabilization is usually a preferred method of stabilization in non-aqueous dispersion media.

Equal efficacy at both high and low solid contents: Unlike electrostatic stabilization, steric stabilization is equally effective at both high and low volume fractions of the dispersed phase.

TABLE 11.1
Comparison between Steric and Electrostatic Stabilization Mechanisms

Steric Stabilization	Electrostatic Stabilization
Insensitive to pH	pH-dependent: when emulsifier loses charge, it leads to aggregation
Insensitive to electrolytes	Aggregation at high electrolyte concentrations (above critical flocculation concentration)
High emulsifier concentration required to cover the droplet surface	Low or high emulsifier concentration required to cover droplet surface
Flocculation is weak and commonly reversible	Flocculation is stronger and often irreversible
Good freeze-thaw stability	Poor freeze-thaw stability
Equally effective for both aqueous and non-aqueous dispersion	Usually effective in aqueous dispersions

Source: Kuroiwa et al. (2015).

Reversibility of flocculation: Flocculation of sterically stabilized dispersions (stabilization of moieties induced by the addition of nonsolvent) can be reversed spontaneously by dilution of nonsolvent concentration to a suitably low value. On the other hand, flocculation of electrostatically stabilized dispersions is irreversible to subsequent dilution due to the coagulation of charge-stabilized particles. The reason for this behavior is that sterically stabilized dispersion is thermodynamically stable while electrostatic stabilized dispersion is only thermodynamically metastable.

11.3 DESTABILIZATION MECHANISM

The lifespan of nanoparticles in the emulsion system may vary considerably from one system to another, based on the nature of surfactants and on the nature of both phases and their volume ratio. Timescale may vary from a few seconds to years. The thermodynamic instability of the emulsion system is caused by interfacial tension, which ranges from 40 mN/m (hydrocarbons in water system) to 1 mN/m (high amount of surfactant and/or high amount of water or oil in both phases). The interface area can be represented by the following equation (Bergenstahl, 2015):

$$A = \frac{[6\phi]}{d} \tag{11.9}$$

where:

A is the interfacial area (m_2/m^3)

Φ is the volume fraction dispersed phase; d is the particle diameter (m)

Interfacial tension is a major driving force for instability. For instance, an oil-in-water emulsion system (10% oil) with a particle size of 10 μm possesses a surface area of around 600 m^2/liter; however, the interfacial energy is just 6 J/liter.

Emulsions are thermodynamically unfavorable systems, which tend to break down due to different physicochemical mechanisms, including gravitational separation, flocculation, coalescence, and Ostwald ripening.

11.3.1 Gravitational Separation

Gravitational separation is one of the most common forms of instability that occur in the emulsion. Gravitational separation may be either creaming or sedimentation, depending on the relative densities of the dispersed and continuous phases. Creaming is the upward movement of particles due to their density when this is lower than that of the surrounding liquid, whereas sedimentation is the downward movement of particles due to their density when this is higher than that of the surrounding liquid. In some cases, the lipid particles may be surrounded by thick and dense biopolymer coatings, which may cause either creaming or sedimentation.

The velocity at which a particle moves upward in nanoemulsion due to gravitational separation is given by Stokes' law:

$$v_{\text{Stokes}} = \frac{-2gr^2(\rho_2 - \rho_1)}{9\eta_1} \tag{11.10}$$

where:

v_{Stokes} is the creaming velocity

r is the particle radius

g is the acceleration due to gravity

ρ is the density

η_1 is the shear viscosity of the continuous phase

η_2 is the shear viscosity of the dispersed phase

Stokes' law is valid only for spherical particles. In a real emulsion, a modification to this equation is necessary to define r^2 (Binks, 1998),

$$r^2 = \frac{\sum_i n_i r_i^5}{\sum_i n_i r_i^3} \tag{11.11}$$

where n is the number of particles of radius r_i

The above Stokes' law equation indicates that creaming can be inhibited by reducing the particle size or the difference in density between the continuous phase and the dispersed phase or by increasing the viscosity of the continuous phase. Reducing the particle size is possible by using any of the high-energy emulsification techniques (high-pressure homogenization, sonication, microfludization). Kentish et al. (2008) formulated flax seed oil nanoemulsion of average particle size <70 nm using food grade non-ionic surfactant, Tween 40. Authors related the sonication parameters with the emulsion particle size and found that the particle size decreases whenincreasing the sonication time, power level and, emulsifier concentration. Another strategy to reduce sedimentation is decreasing the particle size or increasing viscosity. Although the density of lipid particles (dispersed phase) cannot be changed, the density of the dispersion medium can be altered by the addition of density modifiers, such as mannitol and sorbitol.

11.3.2 Flocculation

Flocculation is the process during which two or more droplets come together to form an aggregate without losing their individual identity, i.e. the droplets do not merge into one droplet and have separate interfacial films. The flocculation rate can be calculated from the product of the frequency factor (how often each particle encounters each other) and a probability factor (time duration upon contact). The frequency factor can be calculated for the case of Brownian motion (perikinetic flocculation) or under shear flow (orthokinetic flocculation). The probability factor depends on the interaction energy, the energy required to bring the droplets from infinity to a specified distance apart. Three terms are normally considered for calculating interaction energy (Binks, 1998):

- van der Waals attraction (depends on droplet diameter and Hamaker constant)
- electrostatic repulsion (depends on surface potential, droplet diameter, and ionic strength of the continuous phase)
- steric repulsion

Flocculation occurs in the emulsions either by increasing the attractive forces or decreasing the repulsive forces between the droplets. The interaction between the two droplets can be described by the sum of van der Waals (ω_{VDV}), electrostatic (ω_E), steric (ω_s), and hydrophobic (ω_H) interactions (Guzey and McClements, 2006; McClements, 2015; McClements and Jiajia Rao, 2011):

$$\omega(h) = \omega_{VDV}(h) + \omega_E(h) + \omega_S(h) + \omega_H(h) \tag{11.12}$$

The van der Waals and hydrophobic interactions are attractive, whereas the steric and electrostatic interactions are usually repulsive. Depending on the types of emulsifiers used in the emulsion and the solution conditions (e.g., pH, ion type and concentration, and functional ingredients), depletion flocculation or bridging flocculation can occur.

11.3.3 Coalescence

Coalescence is an irreversible process in which two or more emulsion droplets fuse/combine together and form a single larger droplet. The driving force for coalescence is the surface or film fluctuations, which result in a close approach of the emulsion droplets whereby the van der Waals forces are strong and prevent their separation. For coalescence to occur, the film of the continuous phase separating the droplets must become sufficiently thin that film rupture is a likely possibility. For two large emulsion droplets (around 1 millimeter) approaching each other, when the distance between them is less than 1 um, their interfaces interact and begin to deform, and at the final stage, a parallel thin film is formed. The main factors that determine the stability of emulsion against coalescence are the rate of thinning and rupture of this thin film (Binks, 1998).

11.3.4 Ostwald Ripening

The process whereby the mean droplet size of the emulsion increases due to the diffusion of oil molecules from small to large droplets through intervening fluid is called "Ostwald ripening" (Napper, 1983). It is a widely observed phenomenon in O/W emulsions, in which the oil droplets diffuse through the water phase and join with the larger oil droplets. Over a period, the population of larger droplets increases and leads to phase separation. The rate of increase in droplet size can be described by the following equation (McClements, 2017; McClements and Rao, 2011b):

$$d(t)^3 - d(0)^3 = \omega t = \frac{32}{9}\alpha S_\infty Dt \tag{11.13}$$

$$\alpha = 2\nu V_m / RT \tag{11.14}$$

where:

- d(t) is the number mean droplet diameter at time t
- d_0 is the initial number mean droplet diameter
- ω is the Ostwald ripening rate
- S_α is the watersolubility of the oil phase in the aqueous phase
- D is the translational diffusion coefficient
- V_m is the molar volume of oil
- γ is the oil-water interfacial tension
- R is the gas constant
- T is the absolute temperature

Laplace pressure is the internal overpressure among the emulsion droplets due to interfacial tension.

$$\Pi_{\text{Laplace}} = \frac{4.\gamma}{d} \tag{11.15}$$

where y is the interfacial tension between oil and water. Stability of nanoemulsion to Ostwald ripening usually gets worse as the emulsion droplets get smaller because the driving force (Laplace pressure) increases with a decrease in droplets' size (Kabalnov, 2001).

11.4 STABILITY OF NANOPARTICLES IN FOOD

Nanoformulations are aimed at maintaining the stability of bioactive compounds in the food system throughout the anticipated shelf life. The aim of a researcher is to maintain the stability of the nanoparticles and to prevent instability mechanisms including coalescence, flocculation, gravitational separation, and Ostwald ripening. Nanoemulsions become unstable due to different environmental conditions, such as pH, ionic strength, thermal processing, freeze-thaw cycle, drying, and mechanical agitation during their manufacture, storage, transport, and utilization.

11.4.1 Thermal Processing

Nanoformulated food particles may undergo thermal processing during production, storage, or utilization, e.g. pasteurization, sterilization, or cooking. Hence, researchers are always concernedaboutmaintaining stability during thermal treatments. During thermal processing, food emulsion may undergo flocculation or coalescence because many emulsifiers fail to prevent droplet aggregation during thermal processing.

11.4.1.1 Experimental Protocol to Evaluate the Thermal Stability of Nanoparticles

Kabalnov and Shchukin (1992) obtained thermal stability by extruding the eugenol-loaded chitosan nanoparticle into the model plastic, i.e. thermoplastic flour (TPF). TPF is a mixed flour containing cassava flour/rice flour/waxy rice flour (50:30:20 w/w/w) and glycerol (30% w/w). The remaining eugenol content was estimated in the TPF after the extrusion process (a thermal process that involves the forcing of molten material through a die at 155°C). Thermogravimetry studies (TGA) have been used to study the temperature at which maximum degradation occurs. Saberi et al. (2013) dried cellulose nanoparticles at 105°C for 24 h before thermal analysis. Then TGA was carried out at the rate of 10°C/min under nitrogen atmosphere in the range of 30C–900°C. Estimating particle size and bioactive content is a widely accepted protocol to evaluate the thermal stability of nanoparticles. Woranuch and Yoksan (2013) heated the beta-carotene-loaded lipid nanoparticles to 35, 55, 75, or 95°C and estimated the thermal stability by measuring mean particle size and beta-carotene content.

Sharma and Varma (2014) developed an approach to overcome denaturation of particles during thermal processing. Nanoemulsions were formed using a non-ionic surfactant (Brij 30) with relatively low PIT (phase inversion temperature) and diluted the resulting nanoemulsions in a solution containing another surfactant (SDS or Tween 80) with no or high PIT. The second surfactant was aimed at replacing the part of the original surfactant and thereby increase the repulsive interaction between the droplets. Gomes et al. (2017) investigated the thermal stability of model protein (lysozyme) in the presence and absence of Fe_2O_3 magnetic nanoparticles. Hen egg white lysozyme is used in the pharmaceutical and food industries. It has bactericidal and bacteriostatic properties. The isoelectric points of lysozyme and Fe_2O_3 were 11.2 and 5.1 respectively. This makes a strong electrostatic interaction between the negatively charged Fe_2O_3 magnetic nanoparticle and the positively charged lysozyme. The influences of Fe_2O_3 magnetic nanoparticles on the thermal properties of lysozyme are listed in Table 11.2. Results suggested that increasing the nano Fe_2O_3 concentration increases the thermal stability of lysozyme. Rao and McClements (2010) also reported higher stability against temperature, pH, and salt concentration for gliadin nanoparticles. Thermal stability of nanoparticles was improved by altering the solution pH, which increases the electrostatic repulsion between particles.

TABLE 11.2
Thermodynamic Data of Lysozyme at Various Nano Fe_2O_3Concentrations

Fe_2O_3 (mg/ml)	T_m (k)	$\Delta H°_m$ (J/mol)	$\Delta S°_m$ (J/mol)
0.015	326.5	55,423.33	169.75
0.018	330.0	56,446.5	171.05
0.023	334.5	68,508.9	204.81

Source: Shareghi et al. (2015).

Joye et al. (2015) reported DSC thermograms of beta-carotene-loaded lipid nanoparticles of cupuacu butter. The melting peak of the lipid nanoparticles was detected at a lower temperature (12–13°C) than that of the melting point (32°C) of bulk cupuacu butter. The relation between particle size and melting temperature can be explained by the Gibbs-Thomson equation:

$$ln\frac{T}{T_0} = \frac{2\gamma_{sl}V_s}{r\Delta H_{fus}} \tag{11.16}$$

where:

T is the melting temperature of nanoparticles with radius r
T_0 is the bulk temperature of the lipid
γ_{sl} is the solid-liquid interfacial tension
V_s is the specific volume of the solid
ΔH_{fus} is the fusion of heat of the pure solid

According to the Gibbs-Thomson equation, the smaller the particle size, the higher the decrease in melting temperature. There was no significant change in particle size even when the nanoparticles weresubjected to extreme heat stress at 95°C.

11.4.2 pH and Ionic Strength

In practical applications, the pH and ionic strength of the surrounding system may vary, and that may impact on the stability of nanoencapsulated nutrients. The resistance of nanodelivery systems to these effects has to be characterized before making them into the final product. The stability of nanoparticles against pH and ionic strength can be characterized by testing their properties with a wide range of pH and mineral concentrations (McClements, 2015). For example, food and pharmaceutical products are formulated with a range of different pH values, as they have to experience the range of pH values within the gastrointestinal tract (Shareghi et al., 2015). Hence it is important to understand the influence of pH and ionic strength on the stability of the delivery system.

11.4.2.1 Experimental Protocol to Evaluate Stability against pH and Ionic Strength

The effects of pH on the stability of nanoparticles are evaluated by adjusting the pH of the samples to a wide range of pH from 2 to 12 with 1 M NaOH and/or 1 M HCl. Similarly, for ionic strength, the samples are dispersed in NaCl solution (0 to 1 M). Then, they are stored overnight at room temperature and will be analyzed for hydrodynamic diameter, zeta potential, and bioactive contents.

Crystallinity is an important parameter which determines the aqueous stability and release pattern of the solid lipid nanoparticles (SLN) suspension. Gomes et al. (2017) reported melting characteristics and relative crystallinity of SLNs samples prepared at different electrolyte concentrations (see Table 11.3). Relative crystallinity gradually decreases when electrolyte concentration increases and reaches the minimum relative crystallinity for the electrolyte concentration of 100 mM. Increasing the concentration of electrolytes suppresses the transition of liquid lipid droplets into the crystal.

Hu and McClements (2014) studied the influence of pH (2.0 to 8.0) and ionic strength (0 to 300 mM NaCl concentration) on the stability of zein nanoparticles. Colloidal dispersions were stable to sedimentation and had a relatively small particle size ($d < 100$ nm) at pH 2.0 to 4.5 and pH 7.0 to 8.0. However, at pH 5.0 to 6.5, a milky white appearance of the colloidal dispersion was observed with an increase in particle size of $100 \text{ nm} < d < 200$ nm. Further, at pH 5.5 to 6.0, there was sedimentation of zein particles and a tremendous increase in mean particle size of $d > 1000$ nm. Hu and McClements (2014) reported the isoelectric point of zein is around pH 6.2, but the authors observed zero charges at pH 5.8 for the uncoated zein nanoparticles. The possible mechanism for destabilization at pH 5.0 to 6.5 is due to the rate of surfactant adsorption to the nanoparticle surface, lower

TABLE 11.3
Effect of Electrolyte Concentration on the Melting Temperature and Crystallinity of Solid Lipid Nanoparticles

NaCl Concentration (mM)	Melting Temperature (°C)	Relative Crystallinity (%)
0	64.9	63.0 ± 0.2
5	65.0	61.7 ± 0.4
10	65.1	54.2 ± 0.9
20	65.0	53.5 ± 0.9
100	65.1	47.6 ± 0.1

Source: Choi et al. (2014).

than that of the nanoparticle aggregation. Similarly, in the NaCl solution, counterions (Na^+ and Cl^-) partly neutralize the surface charges thereby weakening the electrostatic repulsion. Thus, in the presence of NaCl (>100 mM), the zein nanoparticles precipitate and sediment at the bottom of the test tubes. Choi et al. (2014) reported stable nanoparticles at 200 mM NaCl for anionic pectin-coated gliadin nanoparticles.

Electrostatic repulsion plays a major role in stabilizing nanoparticles against aggregation. Recently, Li et al. (2017) reported increasing particle size of starch nanoparticles with increasing electrolyte concentration (>200 mM), due to the decrease in net charge. The electrical charge on the particles generates an electrostatic repulsion, which should be strong enough to overcome van der Waals and hydrophobic interactions. The net charge of the starch nanoparticles (StNP) decreases from 10.39 mV to 4.4 mV with an increase in salt concentration from 200 mM to 500 mM. However, gum arabic and sodium alginate-templated StNP are more stable in the NaCl solutions than the pure StNP. The increase in stability is due to the stronger electrostatic repulsion provided by the polysaccharides, i.e. gum arabic and sodium alginate. This study revealed that the coating of food gums on the surface of nanoparticles could increase steric repulsion and maintain stability.

11.5 STABILITY EVALUATION

Four basic ingredients are required to make lipid nanoparticles: two immiscible phases (oil and water), energy to create the oil-water interface of the emulsion droplets, and surfactants to stabilize the oil-water interface. These four parameters have to be optimized to control key properties such as droplet volume fraction, droplet size distribution, and the nature of the stabilizing layer around the droplets (Bergenstahl, 2015). These emulsifiers and emulsification techniques play a predominant role in maintaining the stability of nanoemulsion against environmental stresses such as pH, ionic strength, heating, cooling, or long-term storage. Based on energy input, emulsification processes can be classified as high-energy (high-pressure homogenization, sonication, microfluidizer) and low-energy processes (membrane emulsification, spontaneous emulsification, solvent displacement method, phase inversion method). Further, emulsifiers/surfactants can be classified according to their electrical characteristics: ionic, non-ionic, and zwitterionic.

In view of evaluating the stability of lipid nanoparticles, a wide variety of analytical techniques and experimental protocols have been used by the researchers. The most important characteristics that need to be established are the net electrical charge of the droplets (zeta potential), thickness of the interface (film rheometry, X-ray reflectometry), microstructure (scanning electron microscopy, SEM; atomic force microscopy, AFM; transmission electron microscopy, TEM, and confocal laser scanning microscopy, CLSM), and the long-term stability of the emulsion, e.g. to droplet aggregation, creaming, and Ostwald ripening (single particle light scattering, SPLS; dynamic light scattering, DLS).

11.5.1 Zeta Potential

Lipid nanoparticles in the system often have an electrical charge due to the adsorption of ionized emulsifiers, mineral ions, or biopolymers to their surfaces (McClements,2015). The sign and magnitude of the electrical charge have a significant impact on its stability, such as aggregation stability, interaction with other compounds, etc. These electrical properties are characterized in terms of surface-charge density (σ), surface potential (Ψ_0), and/or ζ–potential (ζ) (Hunter, 1986). The science that deals with the surface charge of the nanoparticles is called electroacoustics. This works based on two principles: (1) colloid vibration potential (or current) (CVP or CVC), which measures surface charge in kilohertz range (only zeta potential, not the particle size distribution), and (2) electrokinetic sonic amplitude (ESA), which measures surface charge in megahertz range (determining zeta potential and particle size distribution simultaneously). In the early 80s, engineers of Matec Applied Sciences first recognized the ESA effect and patented the instrument ESA-8000 (Oja et al., 1985). Later, subsequent researchers developed aninstrument based on the ESA theory to estimate the zeta potential and particle size distribution simultaneously (Hunter, 2001).

In colloidal science, zeta potential is the difference between the dispersed phase and the continuous phase. The magnitude of 30 mV (+ve or −ve) can be considered as the arbitrary value that separates low-charged surfaces from highly charged surfaces (Shukla and Cheryan, 2001). The stability of nanoparticles in the colloidal dispersions can be determined using the zeta potential value, which indicates the degree of repulsion between adjacent, similarly charged particles in the dispersion. If the zeta potential value is high, there will be high repulsion between individual particles, and it will resist aggregation. On the other hand, low zeta potential value represents a weak repulsive force between individual particles and the system tends to flocculate/coagulate. So, the rule of thumb is, a zeta potential higher than $\pm$30 mV indicates stability and a zeta potential lower than $\pm$30 mV indicates the instability of the nanoparticles in the colloidal dispersion (ASTM, 1985). Shukla and Cheryan (2001) reported a higher stability and narrow particle size distribution for the nanoemulsion formulation with higher zeta potential value.

Zeta potential value depends on the type of emulsifiers used for the formulation of the emulsion. In principle, emulsions stabilized by non-ionic surfactants (like Tweens and Spans) should not have electrical charge, but practically they often have a significant negative charge due to the presence of free fatty acids or any other ionic impurities formed during storage (Preetz et al., 2010; Silva et al., 2012). Emulsion-stabilized anionic surfactants possess significant negative charge and those stabilized by cationic surfactants have a net positive charge (Gong et al., 2012). Hur et al. (2009) stabilized the soybean oil-in-water emulsion using four different emulsifiers viz., lecithin, Tween 20, sodium caseinate, and whey protein and evaluated the change in zeta potential during each stage of digestion (Figure 11.2). An electrical charge was negative for the emulsions stabilized with lecithin, Tween 20, and sodium caseinate throughout digestion, while whey protein showed a positive charge in the stomach stage. Mun et al. (2005) stabilized curcumin nanoemulsion using three different emulsifiers, including Tween 20 (nonionic), SDS (sodium dodecyl sulfate, anionic emulsifier) and DTAB (dodecyl trimethyl ammonium bromide, cationic emulsifier). The zeta potential of the fresh emulsion was -1.03 ± 0.64 mV (Tween 20-stabilized nanoemulsion), -92.86 ± 5.89 mV (SDS-stabilized nanoemulsion), and 89.42 ± 7.18 mV (DTAB-stabilized nanoemulsion). No significant change in electrical charges was observed for the Tween 20- and DTAB-stabilized nanoemulsions during gastric digestion. Later, McClements and Rao (2011a) developed three emulsions (small, d = 180 nm; medium, d = 520 nm; large, d = 14 μm) stabilized with a non-ionic surfactant and studied the electrical charges during each stage of digestion. Electrical charges for the initial emulsions were −5 mV, which is due to the presence of a non-ionic surfactant. During the first stage of digestion, at simulated oral digestion, an association of mucin to the lipid surface brings the electrical charge to−10 mV. But during gastric digestion, the electrical charge of the particles became less negative, −2 mV, due to the low-acid environment and high ionic strength of the gastric fluids, because low pH enhances the protonation of free fatty acids (−COOH) and reduces the net electrical charge of

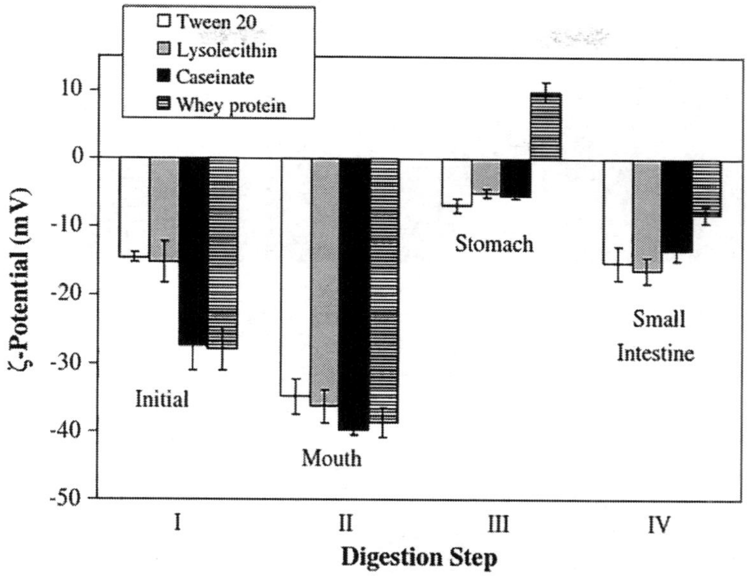

FIGURE 11.2 ξ-potentials during simulated *in vitro* digestion of soybean oil-in-water emulsions stabilized by different emulsifiers (Tween 20, lysolecithin, caseinate, whey protein). (Joye et al., 2015.)

the droplets. In the next stage of digestion (intestinal digestion), electrical charge increased rapidly (−40 to −50 mV) due to the presence of anionic constituents, such as bile salts, phospholipids, and free fatty acids. This study revealed that the electrical characteristics of the nanoemulsion during digestion purely depend on the emulsifier and not on the particle size of the nanoemulsion.

11.5.2 Microstructure

Microscopy can be used as a direct imaging technique to evaluate the structural changes of the lipid nanoparticles.

11.5.2.1 Transmission Electron Microscopy (TEM)

Transmission electron microscopy (TEM) is used to evaluate the three-dimensional structure and composition of the sample (Myhra and Riviere, 2012). TEM is based on the signal from electrons transmitted through the sample: part of the electron beam is absorbed or scattered by the sample and the rest is transmitted. The transmittance of the electron beam depends on the electron density; the lower the electron density, the greater the transmittance of electrons, and thisproduces a darker image. The magnification factor for TEM is in the range of 10^2 to 10^6 times, which enables to characterize very fine structures and interactions.

Pinheiro et al. (2013) analyzed the structural changes of milk fat globules during *in vivo* digestion and reported that pasteurization of milk decreases the extent of lipolysis and homogenization increases the lipolysis. Fu et al. (2013) studied the distribution of silica nanoparticles in the liver, spleen, and small intestine. Silica nanoparticle (particle size = 110 nm) suspension in sterile 5% glucose was administered using different routes including intravenous, hypodermic, and intramuscular injection and oral administration. For TEM observation, tissues of liver and spleen and intestinal segments of duodenum, jejunum, and ileum were excised at 24 h and seven days after administration of silica nanoparticles. Tissues were immediately fixed in 3% glutaraldehyde solution overnight. After 24 h of oral administration, nanoparticles were observed in the liver but not found seven days post-exposure (Figure 11.3). Upon administration of nanoparticles by hypodermic and intramuscular routes, nanoparticles were not found in liver and spleen. Authors revealed thatoral

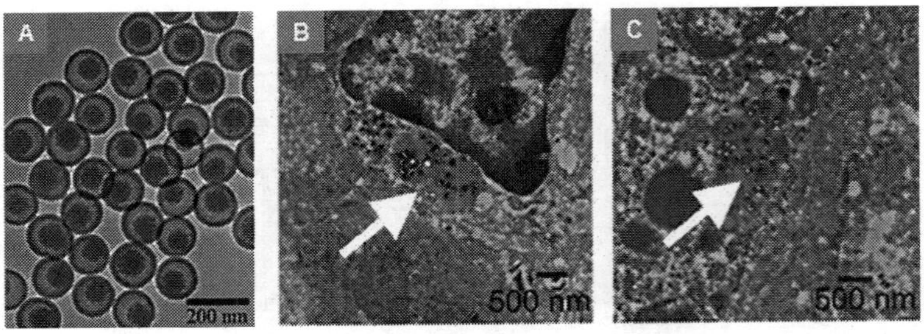

FIGURE 11.3 (a) TEM images of silica nanoparticles. TEM images of the liver after 24 h administration: (b) intravenous administration of silica; (c) oral administration of silica nanoparticles. (Hur et al., 2009.)

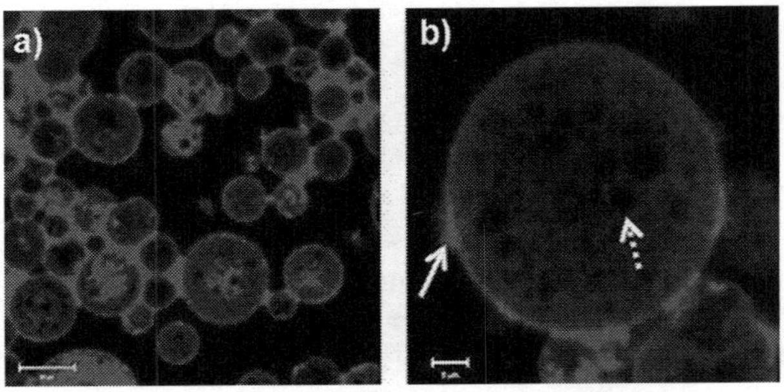

FIGURE 11.4 (a) Confocal micrograph of oil powders; (b) internal structure of single granule containing cross-linked emulsion droplets. The solid arrow represents the protein shell of the granule and the dotted arrows represent protein layers protecting individual oil droplets within a powder granule. (Sarkar et al., 2016.)

and intravenous administration of mesoporous silica nanoparticles is relatively safe for possible biomedical applications.

11.5.2.2 Confocal Laser Scanning Microscopy

Confocal microscopy is frequently used in conjunction with fluorescent or radioactive labels for lipid nanoparticles. Confocal microscopy has been extensively used by researchers to understand the microstructural changes of lipid nanoparticles during *in vitro*/*in vivo* digestion (Zou et al., 2015). Fu et al. (2013) studied the stability of emulsion droplets during gastric digestion using confocal laser scanning microscopy. Two types of dyes were used for staining; Nile red for oil phase (argon laser with an excitation line at 488 nm) and fast green for protein (He–Ne laser with an excitation line at 633 nm). Confocal micrograph revealed the cohesive cross-linked whey protein film (stained using rhodamine 6G dye) at the oil-water interface (Gallier et al., 2013) (Figure 11.4).

Li et al. (2012) and Ozturk et al. (2015) evaluated the microstructural analysis of curcumin-excipient emulsions (small, d = 180 nm; medium, d = 520 nm; large, d = 14 μm) after exposure to thermal treatment (30°C and 100°C) and simulated gastrointestinal digestion. There was no observable change in microstructure observed after incubation at 30°C. However, coalescence was observed for the samples incubated at 100°C. Also, an appreciable increase in particle size was observed during the gastrointestinal tract passage. These microstructural changes in the simulated intestinal conditions were due to aggregation of lipid droplets and formation of mixed micelles and insoluble calcium soaps. In a recent study, Karthik and Anandharamakrishnan (2018) proposed a methodology to investigate the rheological transformations of nanoemulsions in response to the simulated conditions of small intestinal mechanical forces.

11.5.3 Particle Size Distribution

Particle size distribution of lipid nanoparticles is widely used for evaluating the stability of nanoparticles. Many variables are involved in maintaining the particle size of the system. In high-energy methods, the particle size of lipid nanoparticles mainly depends on the type of mechanical device used, intensity and duration of the energy input, emulsifier concentration, interfacial tension in the oil–water phase, and relative viscosities of the dispersed and continuous phases. In the case of the low-energy method, particle size depends on the system composition (surfactant type, pH of the system) and preparation method (including the order of addition, stirring speed and temperature) (Zou et al., 2015).

A variety of techniques are performed to measure the particle size distribution. The light scattering method (photon correlation spectroscopy, laser diffraction method) has been accepted by researchers because of its speed and easy operational procedures. Photon correlation spectroscopy is commonly referred to as dynamic light scattering (DLS), the most widely used technique to determine particle size in emulsion systems. This technique is able to detect particles in the size range of 3–3,000 nm and it ascertains the particle size distribution in a small amount of sample. However, DLS is not a suitable technique for systems containing particles in the micrometer range and irregular structure (DLS assumes all particles are spherical). Laser diffraction (LD), another laser light scattering technique, is commonly used in combination with DLS. This technique detects particles in the size range of 20 nm–2,000 μm (Guo et al., 2014).

Many researchers evaluated the stability of nanoemulsion after each stage of digestion (oral, gastric, and intestinal digestion). The particle size of the emulsion during each stage of digestion depends on the carrier oil type and emulsifier used to stabilize the emulsion. Sarkar et al. (2016) evaluated the influence of carrier oil on the physical stability of nanoemulsion. Nanoemulsions were prepared on two types of carrier oils, (1) digestible oil (including medium-chain triglyceride (MCT), corn oil, and fish oil), and (2) non-digestible oil (orange oil and mineral oil). Lipid digestion may promote a reduction in particle size, because some of the oil phases may be removed during digestion. However, lipid digestion may also induce droplet aggregation due to change in interfacial and core characteristics, and interaction of droplets with intestinal fluids, such as mineral ions, bile salts, or proteins.

Zou et al. (2015) developed three emulsions (small, d=240 nm; medium, d=570 nm; large, d=16 μm) stabilized with a non-ionic surfactant (Tween 80) and studied the stability of curcumin emulsion during each stage of digestion (Figure 11.5). The mean particle diameter of emulsion samples was relatively stable and maintained monomodal distribution after exposure to oral and stomach digestion. Droplets were coated with Tween 80 which stabilized them by steric repulsion

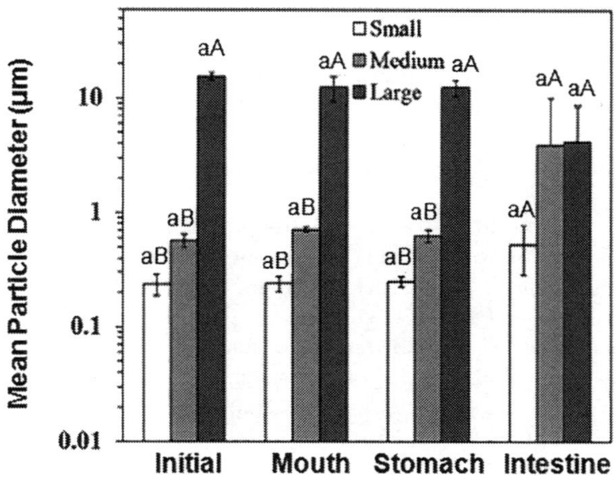

FIGURE 11.5 Mean particle diameter of the emulsion during each stage of digestion. (Zho et al., 2015.)

and provided stability against changes in pH, ionic strength, and protease activity during the digestion process.

Parthasarathi et al. (2016) reported the stability of nanoemulsion (277 nm) and conventional emulsion (1.28 μm) upon storage under different conditions (4, 25, and 40°C). Nanoemulsion was stable with an insignificant increase in mean particle diameter after one month of storage: the mean particle size of nanoemulsion ranged from 277 ± 0.02 to 307 ± 0.023 nm (at 4°C), 277 ± 0.02 to 292 ± 0.018 nm (at 25°C) and 277 ± 0.02 to 289 ± 0.003 nm (at 40°C). On the other hand, conventional emulsion showed a decrease in mean particle diameter, which may be due to the solubilization mechanism (movement of oil molecules from the emulsion droplets into the surfactant micelles).

11.6 CONCLUSION AND FUTURE TRENDS

Bringing food bioactive compounds into nano size is a recent trend in food and nutraceutical industries due to the unique inherent properties of food nanoparticles. The stability of nanoparticles in the food system remains a challenging task during food processing. Because various factors govern the stability of nanoparticles, including dispersion medium (aqueous vs. non-aqueous), dosage form (nanoemulsions vs. dry powders), delivery route (oral vs. transdermal), formulation technique (bottom-up vs. top-down), and nature of the bioactive compounds. This chapter discussed stability mechanisms (electrostatic stabilization, steric stabilization), destabilization mechanisms (gravitational separation, flocculation, coalescence, Ostwald ripening), and evaluation techniques (particle size, zetapotential, microstructural change). Most of the studies have been carried out using comparatively simple solution chemistry and system composition, where stability is controlled by pH, ionic strength, and type of ions (monovalent or multivalent). Future research should focus on determining the types and properties of nanoparticle degradation and understand the potential environmental influence on stability. The experimental investigation should be complex enough to understand the real-world problem and not based on simple assumptions. Finally, the development of more sensitive and advanced analytical equipment is mandatory for inline monitoring of nanoparticle stability during process flow.

REFERENCES

Anandharamakrishnan, C.2014a. *Techniques for Nanoencapsulation offood Ingredients*. New York: Springer.

Anandharamakrishnan, C.2014b. "Liquid-based nanoencapsulation techniques." In *Techniques for Nanoencapsulation of Food Ingredients*, 29–41. New York: Springer.

Anandharamakrishnan, C. 2015. *Spray Drying Techniques for Food Ingredient Encapsulation*. Chichester: John Wiley & Sons.

Anandharamakrishnan, C. and S. Padma Ishwarya. 2015a. "Introduction to encapsulation of food ingredients." *Spray Drying Techniques for Food Ingredient Encapsulation*: 37–64. Chichester: John Wiley & Sons.

Anandharamakrishnan, C. and S. Padma Ishwarya. 2015b. "Spray drying for nanoencapsulation of food components." *Spray Drying Techniques for Food Ingredient Encapsulation*: 180–197. Chichester: John Wiley & Sons.

ASTM. 1985. *Zeta potential of colloids in water and waste water. American Society for Testing and Materials*, D 4187–4182.

Bergenstahl, B. 2015. "Emulsion formation and instability." In: *Engineering Aspects of Food Emulsification and Homogenization*. Ed: M. Rayner and P. Dejmek. Boca Raton, FL CRC Press.

Bhushani, A. and C. Anandharamakrishnan. 2017. "Food-grade nanoemulsions for protection and delivery of nutrients." In *Nanoscience in Food and Agriculture*, 4, 99–139. Switzerland: Springer.

Bhushani, A., Udayakumar H., and C. Anandharamakrishnan. 2017. "Nanodelivery of nutrients for improved bioavailability." In *Nutrient Delivery*, 369–411. London: Elsevier.

Binks, B.P. 1998. "Emulsions-recent advances in understanding." In: *Modern Aspects of Emulsion Science*, Cambridge, UK: The Royal Society of Chemistry.

Chen, L., G. Bai, R. Yang, J. Zang, T. Zhou, and G. Zhao. 2014. "Encapsulation of β-carotene within ferritin nanocages greatly increases its water-solubility and thermal stability." *Food Chemistry*, 149 (April): 307–312. doi:10.1016/j.foodchem.2013.10.115.

Choi, K.-O., N.P. Aditya, and Sanghoon K.2014. "Effect of aqueous pH and electrolyte concentration on structure, stability and flow behavior of non-ionic surfactant based solid lipid nanoparticles." *Food Chemistry,* 147 (March): 239–244. doi:10.1016/j.foodchem.2013.09.095.

Chow, G.M. and Gonsalves, K.E. 1996. "Particle synthesis by chemical routes."In: *Nanomaterials: Synthesis, Properties and Applications.* Ed: A.S. Edelstein and R.C. Cammarata. London: The Institute of Physics.

Chu, B.-S., S. Ichikawa, S. Kanafusa, and M. Nakajima. 2007. "Preparation of protein-stabilized β-carotene nanodispersions by emulsification–evaporation method." *Journal of the American Oil Chemists' Society*, 84 (11): 1053–1062. doi:10.1007/s11746-007-1132-7.

Derjaguin, B. and L. Landau. 1941. "The theory of stability of highly charged lyophobic sols and coalescence of highly charged particles in electrolyte solutions." *Acta Physicochim. URSS*, 14 (633–52): 58.

Ezhilarasi, P.N., P. Karthik, N. Chhanwal, and C. Anandharamakrishnan. 2013. "Nanoencapsulation techniques for food bioactive components: Areview." *Food and Bioprocess Technology*, no. 6 (3): 628–647.

Feldman, Y., T. Skodvin, and J. Sjöblom2001. "Dielectric spectroscopy on emulsion and related colloidal systems – a review. In: *Encyclopedic Handbook of Emulsion Technology.* Ed: J. Sjöblom. New York: Marcel Dekker, Inc.

Fu, C., T. Liu, L. Li, H. Liu, D. Chen, and F. Tang. 2013. "The absorption, distribution, excretion and toxicity of mesoporous silica nanoparticles in mice following different exposure routes." *Biomaterials*, 34 (10): 2565–2575. doi:10.1016/j.biomaterials.2012.12.043.

Gallier, S., X. Q. Zhu, S. M. Rutherfurd, A. Ye, P. J. Moughan, and H. Singh. 2013. "In vivo digestion of bovine milk fat globules: Effect of processing and interfacial structural changes. II. upper digestive tract digestion." *Food Chemistry,* 141 (3): 3215–3223. doi:10.1016/j.foodchem.2013.06.019.

Gambinossi, F., S. E. Mylon, and J. K. Ferri. 2015. "Aggregation kinetics and colloidal stability of functionalized nanoparticles." *Advances in Colloid and Interface Science,* 222 (August): 332–349. doi:10.1016/j.cis.2014.07.015.

Gomes, G. V. L., M. R. Sola, L. F. P. Marostegan, C. G. Jange, C. P. S. Cazado, A. C. Pinheiro, A. A. Vicente, and S. C. Pinho. 2017. "Physico-chemical stability and in vitro digestibility of beta-carotene-loaded lipid nanoparticles of cupuacu butter (*Theobroma grandiflorum*) produced by the phase inversion temperature (PIT) method." *Journal of Food Engineering*, 192: 93–102. doi:10.1016/j.jfoodeng.2016.08.001.

Gong, Y., Y. Wu, C. Zheng, L. Fan, F. Xiong, and J. Zhu. 2012. "An excellent delivery system for improving the oral bioavailability of natural vitamin E in rats." *AAPS PharmSciTech,* 13 (3). Springer US: 961–66. doi:10.1208/s12249-012-9819-y.

Guo, Q., A. Ye, M. Lad, D. Dalgleish, H. Singh, N. M. Rigby, K. Cross, N. Wellner, and A. R. Mackie. 2014. "Effect of gel structure on the gastric digestion of whey protein emulsion gels." *Soft Matter*, 10 (8). The Royal Society of Chemistry: 1214. doi:10.1039/c3sm52758a.

Guzey, D. and D. J. McClements. 2006. "Formation, stability and properties of multilayer emulsions for application in the food industry." *Advances in Colloid and Interface Science,* 128–130 (December): 227–248. doi:10.1016/j.cis.2006.11.021.

Hu, K. and D. J. McClements. 2014. "Fabrication of surfactant-stabilized zein nanoparticles: A pH modulated antisolvent precipitation method." *Food Research International,*64: 329–335. doi:10.1016/j.foodres.2014.07.004.

Hunter, R. J. 1986. *Foundations of Colloid Science.* Oxford University Press, Oxford, UK.

Hunter, R. J., 2001. "Electroacoustic characterization of emulsions." In *Encyclopedic Handbook of Emulsion Technology.* Boca Raton, FL: CRC Press.

Hur, S. J., E. A. Decker, and D. J. McClements. 2009. "Influence of initial emulsifier type on microstructural changes occurring in emulsified lipids during in vitro digestion." *Food Chemistry*, 114 (1): 253–262. doi:10.1016/j.foodchem.2008.09.069.

Joye, I. J., V. A. Nelis, and D. J. McClements. 2015. "Gliadin-based nanoparticles: Fabrication and stability of food-grade colloidal delivery systems." *Food Hydrocolloids,* 44: 86–93. doi:10.1016/j.foodhyd.2014.09.008.

Kabalnov, A. 2001. "Ostwald ripening and related phenomena." *Journal of Dispersion Science and Technology,* 22 (1). Taylor & Francis Group: 1–12. doi:10.1081/DIS-100102675.

Kabalnov, A. S. and E. D. Shchukin. 1992. "Ostwald ripening theory: Applications to fluorocarbon emulsion stability." *Advances in Colloid and Interface Science,* 38 (March): 69–97. doi:10.1016/0001-8686(92)80043-W.

Karthik, P. and C. Anandharamakrishnan. 2016a. "Enhancing omega-3 fatty acids nanoemulsion stability and in-vitro digestibility through emulsifiers." *Journal of Food Engineering,* 187: 92–105.

Karthik, P. and C. Anandharamakrishnan. 2016b. "Fabrication of a nutrient delivery system of docosahexaenoic acid nanoemulsions via high energy techniques." *RSC Advances*, 6 (5): 3501–3513.

Karthik, P., S. Padma Ishwarya, and C. Anandharamakrishnan. 2018. "Droplet coalescence as a potential marker for physicochemical fate of nanoemulsions during in-vitro small intestine digestion." *Colloids and Surfaces A: Physicochemical and Engineering Aspects* 553:278-287.

Kentish, S., T. J. Wooster, M. Ashokkumar, S. Balachandran, R. Mawson, and L. Simons. 2008. "The use of ultrasonics for nanoemulsion preparation." *Innovative Food Science andEmerging Technologies,* 9 (2): 170–175. doi:10.1016/j.ifset.2007.07.005.

Kuroiwa, T., I. Kobayashi, A. Mey Chuah, M. Nakajima, and S. Ichikawa. 2015. "Formulation and stabilization of nano-/microdispersion systems using naturally occurring edible polyelectrolytes by electrostatic deposition and complexation." *Advances in Colloid and Interface Science,* 226 (December): 86–100. doi:10.1016/j.cis.2015.09.003.

Li, Y., J. Kim, Y. Park, D. J. McClements, D. J. McClements, A. Astrup, J. H. Park, et al. 2012. "Modulation of lipid digestibility using structured emulsion-based delivery systems: Comparison of in vivo and in vitro measurements." *Food and Function,* 3 (5). The Royal Society of Chemistry: 528. doi:10.1039/c2fo10273k.

Li, X., S. Ge, J. Yang, R. Chang, C. Liang, L. Xiong, M. Zhao, M. Li, and Q. Sun. 2017. "Synthesis and study the properties of StNPs/gum nanoparticles for Salvianolic acid B-Oral delivery system." *Food Chemistry,* 229 (August): 111–119. doi:10.1016/j.foodchem.2017.02.059.

Liu, C., S. Ge, J. Yang, Y. Xu, M. Zhao, L. Xiong, and Q. Sun. 2016. "Adsorption mechanism of polyphenols onto starch nanoparticles and enhanced antioxidant activity under adverse conditions." *Journal of Functional Foods,* 26 (October): 632–644. doi:10.1016/j.jff.2016.08.036.

McClements. 2004. *Food Emulsions: Principles, Practice, and Techniques.* CRC Series in Contemporary Food Science, 2nd ed. Boca Raton, FL: CRC Press.

McClements, D. J. 2014. *Nanoparticle-and Microparticle-Based Delivery Systems: Encapsulation, Protection and Release of Active Compounds.* Boca Raton, FL: CRC Press.

McClements, D. J. 2015. *Food Emulsions: Principles, Practices, and Techniques.* Boca Raton, FL: CRC Press.

McClements, D. J. 2017. *Food Emulsions: Principles, Practice, and Techniques.* CRC Series in Contemporary Food Science. Boca Raton, FL: CRC Press.

McClements, D. J. and J. Rao. 2011. "Food-grade nanoemulsions: Formulation, fabrication, properties, performance, biological fate, and potential toxicity." *Critical Reviews in Food Science and Nutrition,* 51 (May 2015): 285–330. doi:10.1080/10408398.2011.559558.

Muller, R. H., K. Mader, and S. Gohla. 2000. "Solid lipid nanoparticles (SLN) for controlled drug delivery – a review of the state of the art." *European Journal of Pharmaceutics and Biopharmaceutics,* 50 (1): 161–177. doi:10.1016/S0939-6411(00)00087-4.

Mun, S., E. A. Decker, and D. J. McClements. 2005. "Influence of droplet characteristics on the formation of oil-in-water emulsions stabilized by surfactant–chitosan layers." *Langmuir,* 21 (14). American Chemical Society: 6228–6234. doi:10.1021/la050502w.

Myhra, S. and J.C. Riviere2012. Electron–optical imaging of nanostructures ((HR)TEM, STEM, and SEM). In: *Characterization of Nanostructures,* ed. S.Myhra and J.C.Riviere. Boca Raton, FL: CRC Press.

Napper, D. H. 1983. "*Polymeric Stabilization of Colloidal Dispersions.*" London: Academic Press.

Oja, T., D. Cannon, and G. L. Petersen. 1985. USPatent 4 497 208.

Ozturk, B., S. Argin, M. Ozilgen, and D. J. McClements. 2015. "Nanoemulsion delivery systems for oil-soluble vitamins: Influence of carrier oil type on lipid digestion and vitamin D3 bioaccessibility." *Food Chemistry,* 187: 499–506. doi:10.1016/j.foodchem.2015.04.065.

Parthasarathi, S., S. P. Muthukumar, C. Anandharamakrishnan, A. Alayoubi, S. Nazzal, P. W. Sylvester, A. Kaddoumi, et al. 2016. "The influence of droplet size on the stability, in vivo digestion, and oral bioavailability of vitamin E emulsions." *Food Function,* 7 (5). The Royal Society of Chemistry: 2294–2302. doi:10.1039/C5FO01517K.

Pinheiro, A. C., M. Lad, H. D. Silva, M. A. Coimbra, M. Boland, A. A. Vicente, V. J. Morris, and P. J. Wilde. 2013. "Unravelling the behaviour of curcumin nanoemulsions during in vitro digestion: Effect of the surface charge." *Soft Matter,* 9 (11). The Royal Society of Chemistry: 3147. doi:10.1039/c3sm27527b.

Piorkowski, D. T. and D. J. McClements.2014. "Beverage emulsions: Recent developments in formulation, production, and applications." *Food Hydrocolloids,* 42 (December): 5–41. doi:10.1016/j.foodhyd.2013.07.009.

Preetz, C., A. Hauser, G. Hause, A. Kramer, and K. Mäder. 2010. "Application of atomic force microscopy and ultrasonic resonator technology on nanoscale: Distinction of nanoemulsions from nanocapsules." *European Journal of Pharmaceutical Sciences,* 39 (1): 141–151. doi:10.1016/j.ejps.2009.11.009.

Rajagopalan, R. and P. C. Hiemenz. 1997. *Principles of Colloid and Surface Chemistry,* 3rd Ed. New York: Marcel Dekker, Inc. http://www.eng.uc.edu/~gbeaucag/Classes/Properties/HiemenzRajagopalanDLS.pdf.

Rao, J. and D. J. McClements.2010. "Stabilization of phase inversion temperature nanoemulsions by surfactant displacement." *Journal of Agricultural and Food Chemistry*, 58 (11). American Chemical Society: 7059–7066. doi:10.1021/jf100990r.

Saberi, A. H., Y. F., and D. J. McClements. 2013. "Fabrication of vitamin e-enriched nanoemulsions: Factors affecting particle size using spontaneous emulsification." *Journal of Colloid and Interface Science,* 391: 95–102. doi:10.1016/j.jcis.2012.08.069.

Sarkar, A., J. Arfsten, P.-A. Golay, S. Acquistapace, and E. Heinrich. 2016. "Microstructure and long-term stability of spray dried emulsions with ultra-high oil content." *Food Hydrocolloids*, 52: 857–867. doi:10.1016/j.foodhyd.2015.09.003.

Shareghi, B., S. Farhadian, N. Zamani, M. Salavati-Niasari, H. Moshtaghi, and S. Gholamrezaei. 2015. "Investigation the activity and stability of lysozyme on presence of magnetic nanoparticles." *Journal of Industrial and Engineering Chemistry*, 21: 862–867. doi:10.1016/j.jiec.2014.04.024.

Sharma, P. R. and A. J. Varma. 2014. "Thermal stability of cellulose and their nanoparticles: Effect of incremental increases in carboxyl and aldehyde groups." *Carbohydrate Polymers,* 114: 339–343. doi:10.1016/j.carbpol.2014.08.032.

Shah, R., D. Eldridge, E. Palombo, and L. Harding. 2015. *Lipid Nanoparticles: Production, Characterization* and *Stability*. New York: Springer International Publishing.

Shukla, R. and M. Cheryan. 2001. "Zein: The industrial protein from corn." *Industrial Crops and Products*,13 (3): 171–192. doi:10.1016/S0926-6690(00)00064-9.

Silva, H. D., M. Â. Cerqueira, and A. A. Vicente. 2012. "Nanoemulsions for food applications: Development and characterization." *Food and Bioprocess Technology*, 5 (3): 854–867. doi:10.1007/s11947-011-0683-7.

Tan, C. and M. Nakajima. 2005. "β-carotene nanodispersions: Preparation, characterization and stability evaluation." *Food Chemistry*, 92 (4): 661–671. doi:10.1016/j.foodchem.2004.08.044.

Verwey, E. J. W. and J. T. G. Overbeek.1948.*Theory of the Stability of Lyophobic Colloids*.Elsevier Publishing Company, New York.

Woranuch, S. and R. Yoksan. 2013. "Eugenol-loaded chitosan nanoparticles: I. thermal stability improvement of eugenol through encapsulation." *Carbohydrate Polymers*, 96 (2): 578–585. doi:10.1016/j.carbpol.2012.08.117.

Zou, L., B. Zheng, W. Liu, C. Liu, H. Xiao, and D. J. McClements.2015. "Enhancing nutraceutical bioavailability using excipient emulsions: influence of lipid droplet size on solubility and bioaccessibility of powdered curcumin." *Journal of Functional Foods*, 15: 72–83. doi:10.1016/j.jff.2015.02.044.

12 Biological Fate of Nanoparticles

S. Parthasarathi and C. Anandharamakrishnan

CONTENTS

12.1 INTRODUCTION

Advances in nanoscience and nanotechnology resulted in the development of numerous consumer product applications and finally reached a transformation stage from lab scale research to consumer's hand. The application of nanotechnology to the food sector could improve texture, taste, sensory attributes, stability, and new product development. Nano-based food bioactive compounds can be effectively employed to improve water solubility, thermal stability, physicochemical stability, bioaccessibility, and bioavailability (Parthasarathi and Anandharamakrishnan, 2016; Ezhilarasi et al., 2013 and 2016; Karthik et al., 2017; Karthik and Anandharamakrishnan, 2016b; Bhushani et al., 2017; Anandharamakrishnan, 2014a,b). Although a number of nanomaterial-based products have been developed, research on their potential effects on the environment and human health continues (Vance et al., 2015). To document nanotechnology-based products, the Woodrow Wilson International Center for Scholars and the Project on Emerging Nanotechnology created the Nanotechnology Consumer Product Inventory (CPI) in 2005, with 54 nanotechnology-based products. Now, CPI has identified nearly 1800 nano-based products from 1600 manufacturers in different categories, such as home appliances (battery, laundry, and clothing care), automotive (lubricants, watercraft), computers and electronics (computer hardware, mobile devices), food and beverages, health and fitness, etc. Vance et al. (2015) made a detailed study of the claimed nanoparticles from the CPI listed products. Nearly 47% of products claimed the presence of at least one nanomaterial and 62 products advertised the presence of more than one nanomaterial (see Figure 12.1).

There is an urgent need to study the health impact of nanoparticles since humans have already been exposed to nanoparticles through various scenarios (nano-based consumer products, workplace) and routes (respiratory, skin, gastrointestinal tract). CPI reported nearly 680 metal-based nanoparticles, which may be dangerous for two reasons: the catalytic activity of metal nanoparticles interferes with numerous intracellular biochemical processes and with intracellular free metal ion homeostasis, which is essential for cell metabolism (Karakoti et al., 2010; Llop et al., 2014). Knowledge of physicochemical properties and toxicity of nanomaterials will be used to design the fabrication and selection of nanomaterials to ensure safety during human exposure.

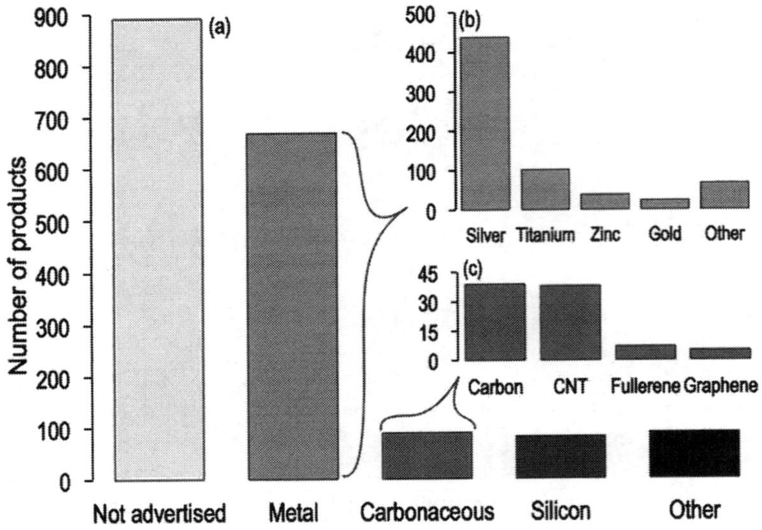

FIGURE 12.1 Composition of nanomaterials in nano-based products (a) claimed composition of nanomaterials listed in the CPI – not advertised, (b) claimed elemental composition of nanomaterials listed in the metals category, and (c) claimed carbonaceous nanomaterials. (Vance et al., 2015.)

12.2 INTERACTION OF NANOPARTICLES WITH PLASMA PROTEINS

Nanoparticles possess huge surface to volume ratios and exhibit different chemical, physical, and biological properties to those of their counterparts, which may lead to adverse effects on human health. Interaction of nanoparticles with proteins plays a crucial role in their biomedical behavior. Proteins compose 75% of the dry weight in the human body and more than 90% of the dry weight of blood plasma. Quite often nanoparticles are administered to patients through intravenous route, and upon exposure to the blood, they immediately encounter a complex mixture of ions, small molecules, proteins, and cells (Lane et al., 2015). When nanoparticles enter or are exposed to biological fluids (like blood, plasma or interstitial fluid), several proteins (apolipoproteins, fibronectin, vitronectin, etc.) adhere to the surface of nanoparticles. This leads to the formation of "protein corona" and endows the nanoparticles with new biological identities (Rahman et al., 2013). These newly formed biological identities are different from their originals, with respect to their chemical composition, shape, size, surface charge, hydrophobicity, as well as other physicochemical properties. The surface charge of the nanoparticle plays an important role in dictating biological fate. Thorek and Tsourkas (2008) and Slowing et al. (2006) reported that nanoparticles with positive surface charge are taken up by the cells at a faster rate than neutral or negatively charged nanoparticles because the cell membrane possesses a slight negative charge which induces electrostatic attraction between the cell membrane and positively charged nanoparticles. Therefore, understanding the mechanism behind the nanoparticle-protein interaction is very important and it could lead to meaningful guidelines in both the cellular internalization pathway and corresponding biological responses to the nanoparticles.

The governing factors involved in the formation of protein corona are: (1) protein-nanoparticle binding affinities and (2) protein-protein interactions. Proteins that adsorb on the surface of the nanoparticle with high affinity are known as "hard corona," which consists of tightly bound proteins that do not desorb easily. Similarly, proteins that adsorb with low affinity are called "soft corona," which consists of loosely bound proteins. Further, the thickness of the protein corona depends on many factors such as protein concentration, particle size, and surface properties of the particle. A hard protein corona consists of 100 proteins and it is highly dynamic, changing its composition over time. Protein composition includes albumin, immunoglobulin G (IgG), fibrinogen, and apolipoproteins in the corona (Aggarwal et al., 2009).

Tenzer et al. (2011) investigated the long-lived hard corona on monodispersed amorphous silica nanoparticles in the size range of 20, 30, and 100 nm. With the aid of liquid chromatography, mass spectrometry, one- and two-dimensional gel electrophoresis, and immunoblotting, compositions of the protein corona were analyzed both qualitatively and quantitatively. Authors identified 125 binding proteins and observed enrichment of specific lipoproteins as well as proteins involved in coagulation and the complement pathways. In contrast, immunoglobulins and acute phase response proteins displayed a lower affinity for the particles. Electrostatic and steric forces dominated the interactions between the nanoparticle (TiO_2 anatase) and protein (Vergaro et al., 2015).

Verderio et al. (2014) described the structure of a protein corona by five parameters:

(1) Thickness and density, which determines the overall size of the nanomaterial
(2) Identity and quantity, which affects the array of possible biological interactions according to their binding strengths
(3) Arrangement and orientation, which determine the accessibility of potential binding and/or catalytic domains
(4) Conformation, which influences the activity of a protein and its interaction with other molecules
(5) Affinity, which regulates the biophysical interactions or translocation to a new biological compartment

Altogether, these parameters define the interaction of a nanomaterial within a specific biological environment. Once the nanoparticle enters the body, a corona can be developed due to the high frequency of collisions between the protein and nanoparticles. The frequency of collisions ($f_{collision}$) between a nanoparticle and a protein in the blood can be described as,

$$f_{\text{collision}} = \frac{\text{RT}}{1500\eta} \cdot \frac{\left(r_{\text{NP}} + r_P\right)^2}{r_{\text{NP}} \cdot r_P} \cdot c_P \tag{12.1}$$

where:

- R is the gas constant
- T is the temperature
- η is the viscosity of blood
- r_{NP} is the radius of the nanoparticle
- r_P is the radius of the protein
- C_p is the concentration of protein

The frequency of collision ($f_{collision}$) is usually in the order of 10^6 s^{-1} or 1 μs. So after administration of nanoparticle, it meets its first protein in tenths of a microsecond and after each passing second, administered nanoparticles experience millions of collisions (Lane et al., 2015). But the formation of the corona after these collisions depends on the balance between the adsorption and desorption rate of the proteins on the particle surface and on how strongly the protein is bound to the surface. Adsorption and desorption rate are proportional to their corresponding rate constants (k_{ads} and k_{des}). Adsorption rate constant (k_{ads}) depends on the frequency of collision between the proteins and nanoparticles. Similarly, desorption rate constant (k_{des}) depends on the binding strength of the protein to the surface. Adsorption and desorption rate constants can be described as,

$$k_{\text{ads}} \propto f_{\text{collision}} \exp\left(-\frac{E_{\text{ads}}}{\text{RT}}\right) \tag{12.2}$$

$$k_{\text{des}} \propto \exp\left(-\frac{E_{\text{des}}}{\text{RT}}\right) \tag{12.3}$$

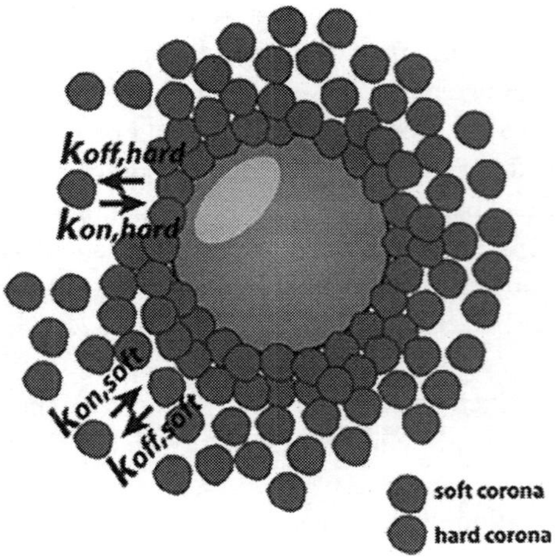

FIGURE 12.2 Schematic representation of protein corona. Hard corona composed of strongly adsorbed proteins and soft corona composed of proteins that associated with hard corona via protein-protein interaction. (Ge et al., 2015.)

Protein binding affinity for the nanoparticle surface is described in terms of equilibrium dissociation constant (K_d),

$$K_d = \frac{k_{\text{des}}}{k_{\text{ads}}} = \exp\left(\frac{\Delta G_{\text{abs}}}{\text{RT}}\right) \quad (12.4)$$

The mechanism of protein absorption is mostly regulated by changes in Gibbs free energy:

$$\Delta G = \Delta H_{\text{ads}} - T\Delta G_{\text{ads}} \leq 0 \quad (12.5)$$

where:

ΔG_{ads} is the free energy during adsorption
ΔH_{ads} is the enthalpy during adsorption
ΔS_{ads} is the entropy during adsorption
T is the temperature

For weak protein interaction, ΔG_{ads} and dissociation constant (K_d) would be largely positive, indicating rapid desorption of protein from nanoparticles. Similarly, for strong protein interactions, ΔG_{ads} and dissociation constant (K_d) would be negative (Figures 12.2 and 12.3).

12.3 PROTEIN CORONA – ANALYTICAL APPROACHES

Interaction of proteins with nanoparticles in biological fluids dramatically changes their physicochemical and biological identities. More than 3,700 proteins at different concentrations compete for binding to the surface of the nanoparticle. A conventional technique to analyze the evaluation of the corona is to incubate the nanoparticles with a complex mixture of proteins for a certain period of time (10 minutes to a few hours) and wash the unbound proteins using ultracentrifugation, column chromatography, or density gradient purification(Rahman et al., 2013).

Interaction of nanoparticles with proteins can be studied using the analytical techniques described in Table 12.1. The most frequently used technique to study the nanoparticle-corona interaction is

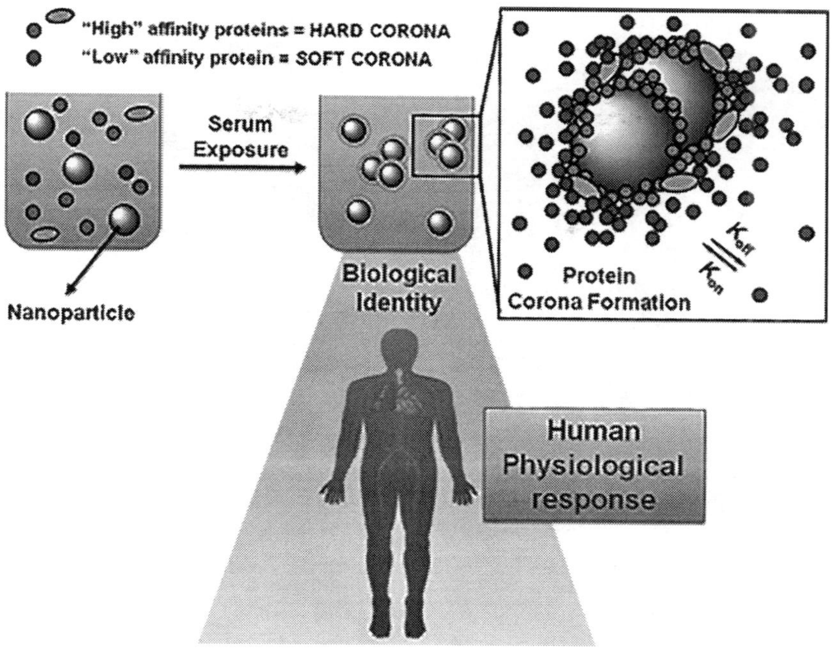

FIGURE 12.3 The formation of a protein corona occurs when a nanomaterial is exposed to physiological environment. (Verderio et al., 2014.)

TABLE 12.1
Analytical Techniques to Study the Parameters Influencing the Structure and the Composition of Protein Corona

Corona Parameter	Impact on the Nano-Systems	Techniques
Thickness and density	Influence on the hydrodynamic size of the nanomaterial	DLS, DCS, SEC, TEM
Identity and quantity	Influence on the array of biological interactions	PAGE, LC-MS/MS
Conformation	Influence on the activity of a protein and itsinteraction	CD, fluorescence quenching, computational simulation
Affinity	Influence on the biophysical interactions ortranslocation to a new physiological compartment	SEC, SPR, ITC

Source: Rahman et al. (2013).

mass spectrometry (MS)-based proteomics (Figure 12.4). First, nanoparticles should be incubated with the protein, which results in protein adsorption with the nanoparticle surface. Protein concentration may affect the amount and identity of proteins adsorbed on the presented NP surface (Figure 12.4a). Then centrifugation is employed to wash the unbound proteins from the NP-protein pellets and repeated washing results in isolation of the "hard protein corona" (Figure 12.4b). Isolation of NP-protein corona should be achieved through denaturing agents such as Laemmli buffer (sodium dodecyl sulfate and 2-mercaptoethanol), which facilitates the overall desorption of proteins (Figure 12.4c), gel electrophoresis for separating the desorbed proteins (Figure 12.4d), and identification using mass spectrometric methods (Figure 12.4e).

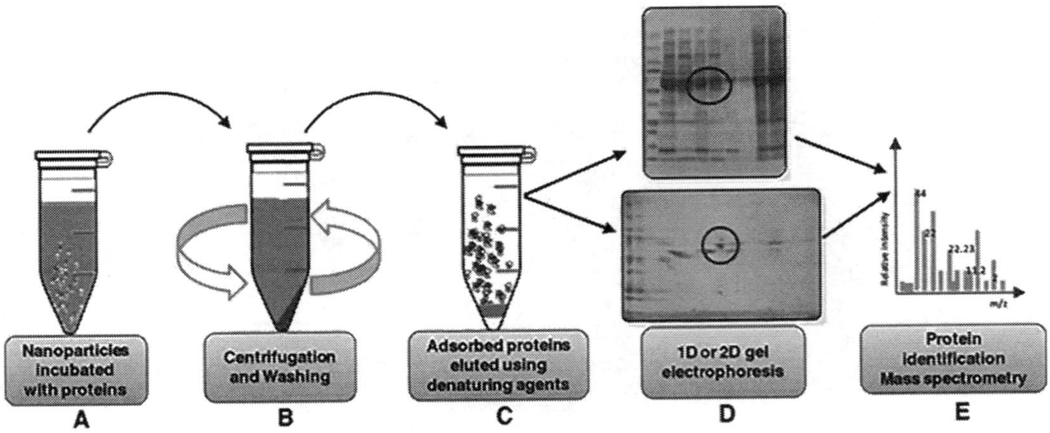

FIGURE 12.4 Schematic representation of the commonly used strategy to isolate and identify surface adsorbed proteins, when nanoparticles interact with complex protein mixtures. (Saptarshi et al., 2013.)

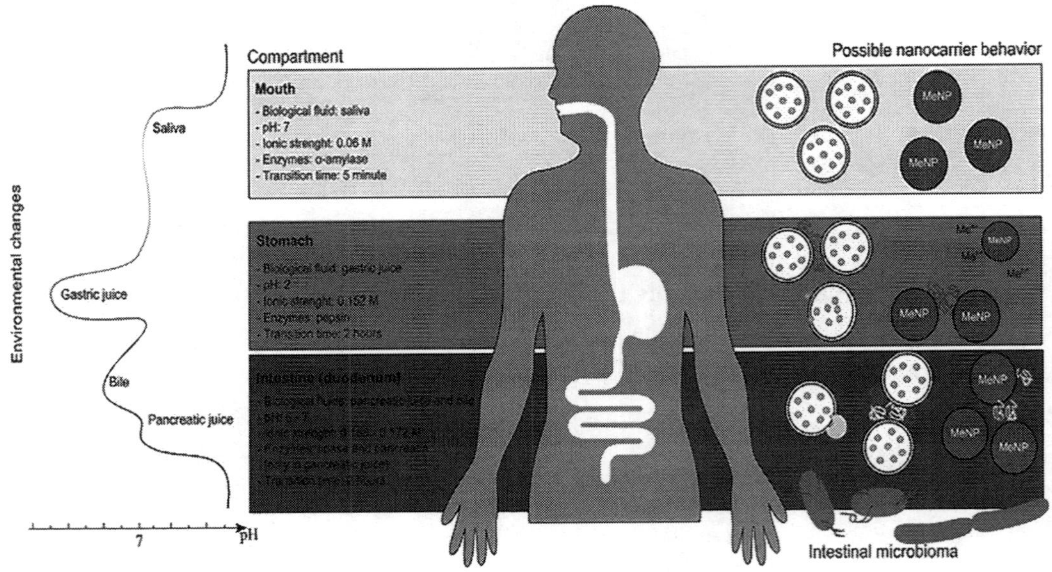

FIGURE 12.5 Gastrointestinal digestion and nanoparticlebehaviors. (Ciappellano et al., 2016.)

12.4 FATE OF ORALLY ADMINISTERED NANOPARTICLES

The oral route of administration is considered to be the most common and accessible way of administrating nutraceutical formulations. The food digestion process is carried out in the gastrointestinal tract and it is composed of several sections which differ in their chemical and enzymatic compositions and pH (Figure 12.5). Mastication or chewing is the first step in food digestion, which reduces particle size and creates the food bolus. The oral cavity (mouth) is followed by the pharynx and esophagus, and the food bolus is conveyed to the stomach by a peristalsis process. The stomach is divided into four major regions: fundus, body, antrum, and pylorus, and it has three main motor functions: storage, mixing, and emptying. The proximal part of the stomach is fundus and body, which act as a reservoir for undigested food materials (Kong and Singh, 2008). The proximal part initiates contact between bolus and gastric juice. Gastric juice is mainly composed of enzymes, pepsin, and lipase, hydrochloric acid leading to a regular fall in pH (approximately from pH 6–5 to 1.5), promoting digestion. The distal part of the stomach is antrum and pylorus, responsible for grinding, mixing, and sieving solid food materials, and acts as a pump for gastric emptying by propelling actions. Gastric emptying conveys the food bolus into the small intestine, where absorption of nutrients takes place at more neutral or slightly basic pH values (Malhaire et al., 2016).

The intestinal compartment is divided into three parts: duodenum, a short section receiving digestive secretions from the pancreas and liver, and two longer ones, the jejunum and ileum. The small intestinal segment is responsible for the breakdown of macromolecules and absorption of water and nutrients. In the duodenum, acidic chyme is neutralized by sodium bicarbonate ($NaHCO_3$) and the secretion of pancreatic enzymes (containing proteases, amylases, and lipases) and other digestive enzymes helps in the breakdown of food constituents. Bile secretion (produced by the liver) plays a crucial role in lipid digestion by emulsifying dietary fats into small droplets promoting pancreatic lipase activity. The segmentation and peristalsis activities of the intestinal segment allow chyme to mix and propel the digesta through the small intestine. Water and nutrients are absorbed by villus enterocytes via different mechanisms: simple diffusion, facilitated diffusion, or active transport (Guerra et al., 2012).

Nutraceutical bioactive compounds will encounter a series of barriers in the gastrointestinal tract before reaching the capillaries in the subepithelial tissue (Figure 12.6). Gastrointestinal barriers include gastric and intestinal enzymes, mucus barrier, tight junctions blocking paracellular passage, intestinal epithelial cells lining the gastrointestinal tract, and subepithelial tissue (Lundquist and Artursson, 2016).

For an effective oral delivery, the bioactive compound must effectively pass through the digestive tract, permeate through the mucus layer, transverse across the intestinal epithelium, enter the portal vein, and finally reach the peripheral circulation (Chen et al., 2011). One of the most promising strategies to achieve sufficient oral bioavailability is to formulate the bioactive compounds with the use of nanoparticle carriers. Nanocarriers protect the encapsulated bioactive compounds in the digestive tract, facilitate cellular uptake, and provide controlled release at the targeted site (des Rieux et al., 2006; Jiang et al., 2015). A thorough understanding of the fate of nanoparticles following oral administration is essential in order to develop technologies to enhance oral delivery, facilitate targeting, and minimize toxicity because the impact of nanocarriers on the pharmacokinetics (absorption, distribution, metabolism, and elimination) is significantly different from that of conventional formulations.

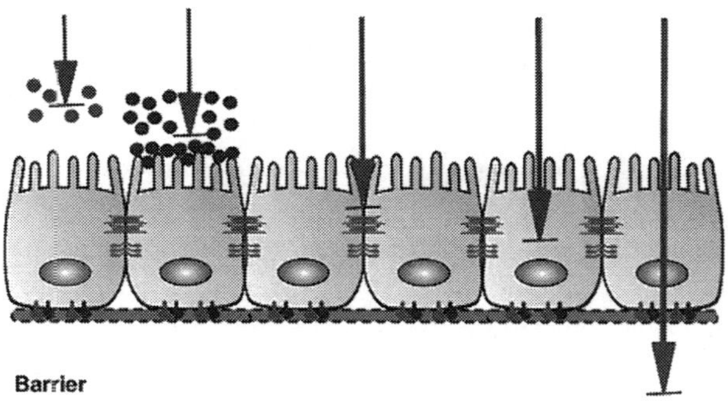

Barrier

Intestinal enzymes	Mucus	Tight junctions	Epithelial cells	Sub epithelium

Penetraton strategy

Inhibitors	Mucus penetrating nanoparticles	Permeation enhancers	Cell penetrating peptides	Small effects in vivo
Nanoparticles		MLCP inhibition	Transcytosis	

FIGURE 12.6 Gastrointestinal barriers. (Lundquist and Artursson, 2016.)

12.4.1 Stability in the Gastrointestinal Tract

Nanoencapsulation is the most synergetic approach to improve the oral bioavailability of poorly water-soluble bioactive compounds. Nanoencapsulation technology has attracted wide attention in the food and pharmaceutical industries for various applications, such as protection of bioactivity and controlled release for improving bioavailability. Emulsion-based delivery systems are most suitable for encapsulating and delivering lipophilic bioactive compounds, such as carotenoids (Salvia-Trujillo et al., 2013; Qian et al., 2012), curcumin (Pinheiro, Coimbra, and Vicente, 2016; Sari et al., 2015), DHA (Karthik and Anandharamakrishnan, 2016a; Karthik et al., 2018), quercetin (Pool et al., 2013), vitamin E (Parthasarathi and Anandharamakrishnan, 2016; Parthasarathi et al. (2016), and vitamin D (Ozturk et al., 2015). These emulsion systems can simply be prepared by solubilizing the lipophilic bioactive compounds in the oil phase and then homogenizing them with an aqueous phase containing a water-soluble emulsifier. The stability of nanoparticles and release of bioactive compounds in the gastrointestinal tract depend on their composition and physicochemical properties, such as concentration, particle-size distribution, electrical charge, and interfacial characteristics. The lipophilic core in the emulsion droplets may comprise one or more non-polar components, such as triacylglycerols, diacylglycerols, monoacylglycerols, flavor oils, essential oils, mineral oils, fat substitutes, weighting agents, oil-soluble vitamins, and nutraceuticals (such as carotenoids, phytosterols, and co-enzyme Q). In the gastrointestinal tract, some of these lipophilic components are typically digested (e.g. triacylglycerols and diacylglycerols), some pass through the GI tract without absorption (e.g. mineral oils and fat substitutes), and some of the lipophilic compounds are absorbed in the GI tract without digestion (e.g. carotenoids and phytosterols) (McClements and Rao, 2011).

Lipid digestion is an interfacial phenomenon involving adsorption of lipase to the surface of the lipid droplets. The specific surface area of the lipids in an emulsion can be described by the following expression (McClements and Xiao, 2012):

$$A_N = \frac{6\varphi}{d_{32}} \tag{12.6}$$

where:

A_N is the surface area of lipids exposed to the aqueous phase per unit volume of emulsion
ϕ is the volume fraction of dispersed phase
d_{32} is the surface weighed mean particle diameter

From this expression, one can understand that lipid digestion rate depends on droplet size. The decrease in droplet size increases the lipid digestion rate, because of the increased surface area of lipid-exposed to lipase molecules.

Qian et al. (2012) analyzed the impact of carrier oil (long-chain triglycerides, medium-chain triglycerides, orange oil) on nanoemulsion stability during in vitro digestion. The particle diameter of the nanoemulsion remained relatively small after 10 min incubation in artificial saliva (pH 7.0) and 2 h incubation in simulated gastric fluid (pH 2.5). However, after incubation in the simulated intestinal fluid, the nanoemulsion of all three preparations exhibited a large increase in particle diameter (Figure 12.7). The nanoemulsions were stabilized by non-ionic surfactant (Tween 20) which is a high surface-active molecule with a relatively large hydrophilic head group (i.e. three polyoxyethylene chains). Simulated oral and gastric fluids did not contain any highly surface-active ingredients, which did not affect the emulsion droplet size. In the simulated intestinal fluid, the mean particle diameter of the nanoemulsion increased from 170 to 920 nm, indicating that nanoemulsion droplets were unstable due to droplet aggregation. Salvia-Trujillo et al. (2013) observed the stability of nanoemulsion droplets as they passed through the simulated gastrointestinal conditions by confocal microscopy and light scattering experiments. The particle size distribution of nanoemulsion in the simulated GI conditions indicated that a large fraction of droplets had fairly similar sizes to the initial emulsions. But the microscopy technique indicated that the largest particles were present in

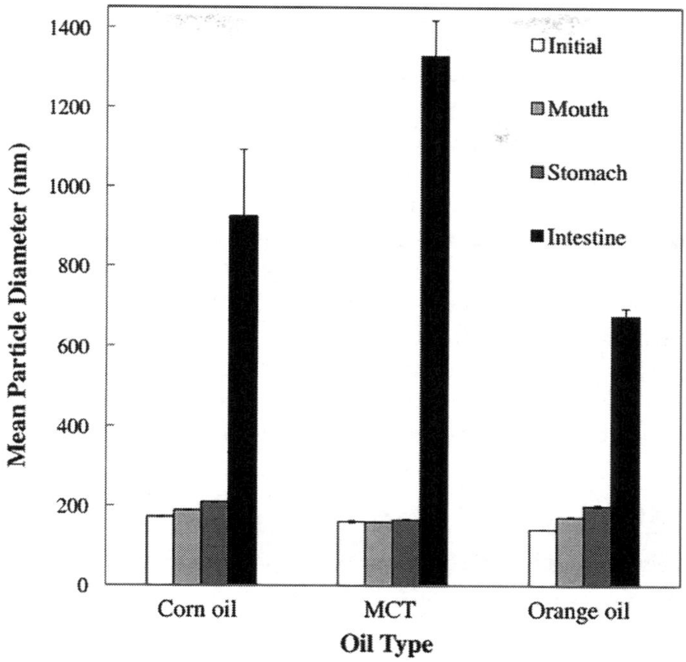

FIGURE 12.7 Influence of carrier oil type and in vitro digestion on the mean diameter of the particles in oil-in-water nanoemulsions initially stabilized by Tween 20. (Qian et al., 2012.)

the oral conditions, whereas with the aid of light scattering data, large particles were observed in the small intestinal condition. This contradiction may be due to the artifacts associated with the analytical methods. Light scattering experiments require high dilution prior to analysis which may have changed the particle size of the samples. Also, the authors suggested the importance of using a combination of different analytical techniques to determine the stability of nanoparticles in gastrointestinal conditions.

Change in particle size during GI conditions may be due to a number of reasons, such as lipid digestion, droplet aggregation, mixed micelle formation, or generation of insoluble sediments (such as calcium soaps). Apart from the above-mentioned reasons, the stability of nanoparticle in the GI tract also depends on the magnitude of the electrical charge (ζ-potential) of the droplets. Nanoemulsion droplets often have an electrical charge, as their surfaces are adsorbed with an ionic emulsifier, mineral ions, or biopolymers. In principle, nanoemulsions stabilized by non-ionic surfactants (e.g. Tweens and Spans) should not possess any electrical charge. But practically these nanoemulsions often have a significant negative charge due to the presence of free fatty acids or other ionic impurities formed during production or storage (Mun et al., 2005). Emulsions stabilized by cationic surfactants possess a positive charge and those stabilized by anionic surfactants possess a negative charge (Asker et al., 2008; McClements and Xiao, 2012). Ozturk et al. (2015) reported the change in interfacial characteristics (ζ-potential) during the GI transit of nanoemulsion stabilized by natural surfactant (quillaja saponin). The freshly prepared nanoemulsions were highly negatively charged with −65 to −70 mV due to the presence of adsorbed surfactant layer. In the mouth stage, an appreciable decrease in the magnitude of the negative charge was observed due to the adsorption of mucin molecules to the surface of the droplet. In the simulated gastric conditions, the magnitude of the negative charge was decreased further, as the surfactant (quillaja saponin) loses its negative charge under highly acidic conditions (Ozturk et al., 2014; Yang et al., 2013). But in the intestinal condition, an appreciable increase in the magnitude of the negative charge was observed due to the increase in pH and adsorption of anionic phospholipids and bile salts to the droplet surfaces.

Various in vitro GI models and simulated GI conditions have been used to unravel the interesting structural and chemical changes of nanoemulsified lipids within the GI tract (Parthasarathi et al., 2018). However, many in vitro models have failed to simulate the complex physicochemical and physiological events that occur in the GI tract (Parthasarathi et al., 2016). It is undeniable that in vivo animal experiments are more time-consuming and costly and have ethical constraints. Armand et al. (1999) examined the stability of differently sized emulsions (fine, 0.7 μm and coarse emulsion, 10 μm) in eight healthy male volunteers (subject) of 24 ± 1.4 years of age. Each subject was intubated with a single-lumen nasogastric tube and a single-lumen nasoduodenal tube. Before ingestion of emulsion samples, gastric juice and duodenal fluid were removed by manual aspiration. Then emulsion was intubated intragastrically over a 10-min period. Half an hour after feeding the stomach and duodenal contents were aspirated for analysis. During digestion, both emulsions showed an increase in droplet size. Fine emulsion showed a tenfold increase in diameter from 0.7 μm to 6–8 μm, compared to that of the coarse emulsion. Li et al. (2012) also reported strong evidence of droplet flocculation and coalescence in the emulsion samples collected from a rat's stomach.

12.4.2 Mucoadhesion

Bioadhesion is a phenomenon by which two materials, one of which should be biological in nature, are held together by interfacial force for an extended period of time. If the adhesive attachment is to mucus or a mucous membrane, then the phenomenon is referred to as mucoadhesion (Smart, 2005; Gu et al., 1988). Mucous membranes are also called mucosae, and represent the moist surface of various body cavities, such as the gastrointestinal wall and respiratory tract. The molecular weight of mucins varies from 500 kDa to 20 MDa and the thickness of the mucus layer differs on different mucosal surfaces, 50 to 450 μm in the stomach, and >1 μm in the oral cavity. The composition of mucus gels includes mucin glycoproteins (usually termed mucins), lipids, inorganic salts, and water, the latter accounting for more than 95% of mucus weight, providing a highly hydrated system (Marriott and Gregory, 1990). Most of the mucins are negatively charged due to the presence of sialic acids and ester sulfates, and fully ionized at pH > 2.6 (Sogias et al., 2008).

In the early 80s, Prof. Joseph R. Robinson, University of Wisconsin, revealed a new strategy to prolong the residence time of drugs on ocular surfaces by mucoadhesion (Bernkop-Schnürch, 2005). If the formulations are developed with mucoadhesive polymers, then the residence time can be prolonged significantly, and provide sustained release at the target site. Interaction with mucous membrane can be explained by two concepts, (i) mucoadhesion and (ii) mucodiffusion. Mucoadhesion is due to high affinity between the nanoparticles and the mucus. Different processes are involved in mucoadhesion, such as hydrogen bonding, macromolecule entanglement, electrostatic, hydrophobic, or van der Waals interactions. Mucodiffusion is due to diffusion of a particle through the mucus layer from the luminal part to the apical site of the epithelial cells. Different theories have been suggested to explain mucoadhesion (Table 12.2).

Mucoadhesive polymers: The mucoadhesivity of dosage forms depends on the ability of polymers to stick to mucosal membranes. Mucins tend to form larger aggregates through hydrophobic interactions between non-polar groups, hydrogen bonding between sugar units, and disulfide linkages between cysteine residues (Sogias et al., 2008). Mucoadhesive polymers should be biocompatible, non-toxic, and economically favorable and should possess sufficient flexibility to penetrate the mucus network (Shojaei and Li, 1997). Mucoadhesive polymers can be classified based on their origin (natural, synthetic, semi-synthetic polymers), kind of mucosa being applied (e.g. buccal/ocular), chemical structure, and functionality (e.g.anionic, cationic, non-ionic, amphoteric polymers). Park and Robinson (1984) classified mucoadhesive polymers into three broad categories:

(1) Polymers that become sticky when placed in aqueous media and owe their bioadhesion to stickiness

TABLE 12.2
Theories of Mucoadhesion

Theory	Short Description
Electronic theory	Adhesion is established due to the electrostatic attraction between negatively charged mucin and positively charged materials
Adsorption theory	Adhesive interactions are related to the establishment of hydrogen and van der Waals bonding; hydrophobic effects and chemisorption may also contribute
Wetting theory	Adhesion is related to the ability of a mucoadhesive (when in liquid form) to spread over the mucus layer
Diffusion theory	Considers that adhesion is established by the interpenetration of macromolecular mucoadhesives (either polymeric or other) with mucin fibers, as driven by a concentration gradient differential
Fracture theory	Relates adhesion with the force required for interfacial detachment of two previously joint solid surfaces
Mechanical theory	Adhesion is dependent on the roughness of two different surfaces and the available area for interaction

Source: Sosnik et al. (2014).

(2) Polymers that adhere through non-specific, non-covalent interactions that are primarily electrostatic in nature (although hydrogen and hydrophobic bonding may be significant)
(3) Polymers that bind to specific receptor sites on the cell surface

Chitosan: Chitosan is a natural cationic polysaccharide derived from chitin by partially deacetylating its acetamido groups with strong alkaline solutions (Sogias et al., 2008). Chitosan is a polycation, with aqueous solubility below a pH of 5.8 due to the progressive protonation of pendant amine moieties. Sosnik et al. (2014) reported the strong mucoadhesive property of chitosan is due to the ability to establish ionic, hydrogen, and hydrophobic bonds with the negatively charged mucin. Chitosan-based nanoparticles emerge as a promising vehicle for oral delivery of therapeutic molecules, proteins, and nutrition. The safety of chitosan has been demonstrated in both animal models and humans (Roy et al., 1999; Maezaki et al., 1993; Kanauchi et al., 1994). Chitosan has emerged as a fundamental mucoadhesive biomaterial and has been used for a broad range of administration routes such as oral (Li et al., 2016; Jana et al., 2013), ocular (Costa et al., 2015), pulmonary (Jafarinejad et al., 2012; Sinsuebpol et al., 2013), and topical (Calderón et al., 2013).

Shin et al. (2013) prolonged the absorption of curcumin in the gastrointestinal tract using chitosan mucoadhesive polymer. The mucoadhesive properties of curcumin nanoliposomes (Cur-NLs) and chitosan-coated curcumin nanoliposomes (CS-Cur-NLs) were studied using invitro mucin adsorption studies. A nearly twofold increase in mucin adsorption was observed for the CS-Cur-NLs compared to Cur-NLs, suggesting CS-Cur-NLs stay at the target site longer than Cur-NLs. Mucin proteins secreted from epithelial cells possess negative charge that create electrostatic interactions with positively charged chitosans. Ramalingam et al. (2016) increased the bioavailability of curcumin by chitosan-coated solid lipid nanoparticles (SLN). The absorption of curcumin was studied for curcumin suspension and chitosan-coated curcumin solid lipid nanoparticles. Curcumin suspension resulted in a sharp Cmax at 0.53 h after oral administration, whereas SLN formulation resulted in 1.25 h. Results suggested that the degree of exposure to the SLN formulation showed a sevenfold increase in bioavailability. Improvements in oral bioavailability demonstrate chitosan's mucoadhesive and high adsorption properties, protection of SLN from enzymatic degradation, and sustained release properties.

Yuan et al. (2013) evaluated the mucus penetrating ability of poly(ethylene glycol) (PEG) for improving oral delivery of bioactive compounds across gastrointestinal mucus. PEG is FDA-approved, non-toxic, nonimmunogenic, nonantigenic and highly soluble in water (Kouchakzadeh et al., 2010). PEGylated solid lipid nanoparticles were studied for cellular uptake, transport across the Caco-2/HT29 coculture cell monolayers, intestinal absorption, and pharmacokinetics, and compared with that SLN. Transport across Caco-2 cell monolayer showed decreased permeation ability of pSLN compared to that of SLN. However, the permeation ability of pSLN increased through mucus-secreting Caco-2/HT29 coculture cell monolayer model, which indicates the impact of the mucus layer on the delivery of nanoformulations. Caco-2/HT29 coculture cell monolayer simulates the intestinal epithelium and is usually applied for evaluating the effect of the mucus gel layer on penetration (Mahler et al., 2009). The influence of the mucus penetrating effect of pSLN nanocarrier was further verified with an in vivo bioavailability study using a hydrophobic model drug, doxorubicin (DOX). The pharmacological effects of the DOX-HCl solution, SLN/DOX, and pSLN/DOX were evaluated on male SD rats. Maximum plasma concentration (C_{max}) for a DOX-HCl solution and SLN/DOX showed at 1 and 1.5 h, respectively. Peak concentration for pSLN/DOX reached 0.5 and 4 h, representing the quick uptake from the gastrointestinal tract into systemic circulation.

12.5 BIOLOGICAL BARRIER

An endless number of potential injuring agents such as viruses, bacteria, dust, and physical agents surround humans; however, their access to the human organs are strongly limited by sophisticated protective systems called biological barriers (Pietroiusti et al., 2013). These biological barriers are designed to prevent undesired access to sensitive organs. Without such a barrier, for instance, our lungs, our bloodstream, and subsequently all our organs would be exposed to a multitude of dangerous inhalation scenarios (Rocks and Dawson, 2014). There are three primary anatomical barriers including the epithelia of the skin, the gastrointestinal tract, and the respiratory system. In unfortunate conditions, if any agents overcome these primary barriers, secondary barriers protect the strategic sites. Secondary barriers include the blood-brain barrier (BBB) for protecting the brain, the blood-testis barrier (BTB), which protects the male reproductive apparatus, and the placenta for protecting the developing embryo (Pietroiusti et al., 2013). The BBB is a physical and physiological barrier regulating the passage of external biomolecules from the systemic circulation into the brain parenchyma.

Nanoparticles of size ranging from less than 100 nm are similar to physiological molecules and might be allowed to cross the internal barriers. The penetration capacity of nanoparticles across the biological barriers mainly depends on the particle size and surface functionalization. Extensive research has been initiated on this basis for the diagnosis and treatment of diseases in internal barrier-protected organs (Kievit and Zhang, 2011). However, nanoparticles produced for industrial and commercial purposes pose growing concerns about their unintentional access to these highly protected organs, when these nanoparticles come in contact with workers (Shvedova et al., 2013). Sharma et al. (2009) reported that administration of engineered nanoparticles such as Cu, Ag, and Al of 50–60 nm size, alter the BBB and induce brain pathology in adult rats. Neurotoxic experimentation with rats indicates that young and elderly age groups may be more vulnerable to nanoparticles-induced neurotoxicity.

On the other hand, a wide range of drugs including genes need to enter the cells to function properly; however, cell membranes are not permeable to them. So understanding these biological barriers is necessary for fabricating engineered nanoparticles with the ability to bypass these barriers and reach the target cells. Two main mechanisms are involved in the crossing of internal barriers: (1) passive and active transport, and (2) endocytosis. In passive diffusion, concentration gradient across the barrier and solute governs the passage of solutes between cells (i.e. paracellular) or across cells (i.e. transcellular). In the case of passive facilitated diffusion, transport is facilitated by influx transporters and the passage through the barrier depends on size, concentration gradient, hydrophobicity, polarity, and protein binding. In the case of active transport, the passage is mediated by influx transporters and it takes place against concentration gradient with energy expenditure. Endocytosis is another mechanism followed by nanoparticles for crossing internal barriers. Endocytosis of nanoparticles involves engulfment in membrane invagination and subsequent inclusion into the cytoplasm of the cell (Pietroiusti et al., 2013).

12.6 CONCLUSION AND FUTURE PERSPECTIVES

In this chapter, we discussed the biological fate and toxicity of nanomaterials in the environment. Nanotechnology has gained public interest due to the commercial development of nanomaterial-based products in many areas of human endeavors, from electronics to public health. Any particle of less than 10 μm has the tendency to reach the alveolar spaces of the lungs. Nanoparticles of size ranging from less than 100 nm are similar to physiological molecules and might be allowed to cross internal barriers. Many nanomaterial-based products were developed from gold and silver nanoparticles and reports demonstrated the accumulation of these nanoparticles in the liver. Human exposure to nanomaterials is inevitable, due to which nanotoxicology research is gaining attention. Many reports claim potential benefits of nanotechnology on one side and potential risks of nanotechnology on the other side. Research on nanofabrication and nanotoxicology should have a common platform and ensure the safety of nanomaterial-based products. Further, there is a lack of in vivo and in vitro toxicity information that might allow correlations between the findings to be made.

REFERENCES

Aggarwal, P., J. B. Hall, C. B. McLeland, M. A. Dobrovolskaia, and S. E. McNeil. 2009. "Nanoparticle interaction with plasma proteins as it relates to particle biodistribution, biocompatibility and therapeutic efficacy." *Advanced Drug Delivery Reviews*, 61 (6):428–437.

Anandharamakrishnan, C. 2014a. "Liquid-based nanoencapsulation techniques." In *Techniques for Nanoencapsulation of Food Ingredients*, 29–41. New York: Springer.

Anandharamakrishnan, C. 2014b. *Techniques for Nanoencapsulation of Food Ingredients*. New York: Springer.

Armand, M., B. Pasquier, M. André, P. Borel, M. Senft, J. Peyrot, J. Salducci, H. Portugal, V. Jaussan, and D. Lairon. 1999. "Digestion and absorption of 2 fat emulsions with different droplet sizes in the human digestive tract." *The American Journal of Clinical Nutrition*, 70 (6):1096–1106.

Asker, D., J. Weiss, and D. J. McClements. 2008. "Analysis of the interactions of a cationic surfactant (lauric arginate) with an anionic biopolymer (pectin): Isothermal titration calorimetry, light scattering, and microelectrophoresis." *Langmuir*, 25 (1):116–122.

Bernkop-Schnürch, A. 2005. "Mucoadhesive polymers: Strategies, achievements and future challenges." *Advanced Drug Delivery Reviews*, 57:1553-1555.

Bhushani, J. A., N. K. Kurrey, and C. Anandharamakrishnan. 2017. "Nanoencapsulation of green tea catechins by electrospraying technique and its effect on controlled release and in-vitro permeability." *Journal of Food Engineering*,199:82–92.

Calderón, L., R. Harris, M. Cordoba-Diaz, M. Elorza, B. Elorza, J. Lenoir, E. Adriaens, J. P. Remon, A. Heras, D. Cordoba-Diaz. 2013. "Nano and microparticulate chitosan-based systems for antiviral topical delivery." *European Journal of Pharmaceutical Sciences*, 48 (1–2):216–222.

Chen, M.-C., K. Sonaje, K.-J. Chen, and H.-W. Sung. 2011. "A review of the prospects for polymeric nanoparticle platforms in oral insulin delivery." *Biomaterials*, 32 (36):9826–9838.

Ciappellano, S. G., E. Tedesco, M. Venturini, and F. Benetti. 2016. "In vitro toxicity assessment of oral nanocarriers." *Advanced Drug Delivery Reviews,* 106:381–401.

Costa, J. R., N. C. Silva, B. Sarmento, and M. Pintado. 2015. "Potential chitosan-coated alginate nanoparticles for ocular delivery of daptomycin." *European Journal of Clinical Microbiology andInfectious Diseases*, 34 (6):1255–1262.

CPI. http://www.nanotechproject.org/cpi/.

des Rieux, A., V. Fievez, M. Garinot, Y.-J. Schneider, and V. Préat. 2006. "Nanoparticles as potential oral delivery systems of proteins and vaccines: Amechanistic approach." *Journal of Controlled Release*, 116 (1):1–27.

Ezhilarasi, P. N., P. Karthik, N. Chhanwal, and C. Anandharamakrishnan. 2013. "Nanoencapsulation techniques for food bioactive components: A review." *Food and Bioprocess Technology*, 6 (3):628–647.

Ezhilarasi, P. N., S. P. Muthukumar, and C. Anandharamakrishnan. 2016. "Solid lipid nanoparticle enhances bioavailability of hydroxycitric acid compared to a microparticle delivery system." *RSC Advances*, 6 (59):53784–53793.

Ge, C., J. Tian, Y. Zhao, C. Chen, R. Zhou, and Z. Chai. 2015. "Towards understanding of nanoparticle–protein corona." *Archives of Toxicology*, 89 (4):519–539.

Gu, J. M., J. R. Robinson, and S. H. Leung. 1988. "Binding of acrylic polymers to mucin/epithelial surfaces: structure-property relationships." *Critical Reviews in Therapeutic Drug Carrier Systems*, 5 (1):21–67.

Guerra, A., L. Etienne-Mesmin, V. Livrelli, S. Denis, S. Blanquet-Diot, and M. Alric. 2012. "Relevance and challenges in modeling human gastric and small intestinal digestion." *Trends in Biotechnology*, 30 (11):591–600.

Jafarinejad, S., K. Gilani, E. Moazeni, M. Ghazi-Khansari, A. Rouholamini Najafabadi, and N. Mohajel. 2012. "Development of chitosan-based nanoparticles for pulmonary delivery of itraconazole as dry powder formulation." *Powder Technology*, 222:65–70.

Jana, S., N. Maji, A. Kumar Nayak, K. Kumar Sen, and S. Kumar Basu. 2013. "Development of chitosan-based nanoparticles through inter-polymeric complexation for oral drug delivery." *Carbohydrate Polymers*, 98 (1):870–876.

Jiang, S., C. Peng Teng, W. C. Puah, M. Wasser, K. Y. Win, and M.-Y. Han. 2015. "Oral administration and selective uptake of polymeric nanoparticles in Drosophila larvae as an in vivo model." *ACS Biomaterials Science and Engineering*, 1 (11):1077–1084.

Kanauchi, O., K. Deuchi, Y. Imasato, and E. Kobayashi. 1994. "Increasing effect of a chitosan and ascorbic acid mixture on fecal dietary fat excretion." *Bioscience, Biotechnology, and Biochemistry*, 58 (9):1617–1620.

Karakoti, A., S. Singh, J. M. Dowding, S. Seal, and W. T. Self. 2010. "Redox-active radical scavenging nanomaterials." *Chemical Society Reviews*, 39 (11):4422–4432.

Karthik, P. and C. Anandharamakrishnan. 2016a. "Enhancing omega-3 fatty acids nanoemulsion stability and in-vitro digestibility through emulsifiers." *Journal of Food Engineering*, 187:92–105.

Karthik, P. and C. Anandharamakrishnan. 2016b. "Fabrication of a nutrient delivery system of docosahexaenoic acid nanoemulsions via high energy techniques." *RSC Advances*, 6 (5):3501–3513.

Karthik, P., P. N. Ezhilarasi, and C. Anandharamakrishnan. 2017. "Challenges associated in stability of food grade nanoemulsions." *Critical Reviews in Food Science and Nutrition*, 57 (7):1435–1450.

Karthik, P., S. Padma Iswarya, and C. Anandharamakrishnan. 2018. "Droplet coalescence as a potential marker for physicochemical fate of nanoemulsions during in-vitro small intestine digestion." *Colloids and Surfaces A: Physicochemical and Engineering Aspects*, 553:278-287.

Kievit, F. M., and M. Zhang. 2011. "Cancer nanotheranostics: Improving imaging and therapy by targeted delivery across biological barriers." *Advanced Materials*, 23 (36):H217-H247.

Kong, F. and R. P. Singh. 2008. "Disintegration of solid foods in human stomach." *Journal of Food Science*, 73 (5):R67-R80.

Kouchakzadeh, H., S. A. Shojaosadati, A. Maghsoudi, and E. V. Farahani. 2010. "Optimization of PEGylation conditions for BSA nanoparticles using response surface methodology." *Aaps Pharmscitech*, 11 (3): 1206–1211.

Lane, L. A., X. Qian, A. M. Smith, and S. Nie. 2015. "Physical chemistry of nanomedicine: Understanding the complex behaviors of nanoparticles in vivo." *Annual Review of Physical Chemistry*, 66:521–547.

Li, Y., J. Kim, Y. Park, and D. J. McClements. 2012. "Modulation of lipid digestibility using structured emulsion-based delivery systems: Comparison of in vivo and in vitro measurements." *Food and Function*, 3 (5):528–536.

Li, J., C. Jiang, X. Lang, M. Kong, X. Cheng, Y. Liu, C. Feng, and X. Chen. 2016. "Multilayer sodium alginate beads with porous core containing chitosan based nanoparticles for oral delivery of anticancer drug." *International Journal of Biological Macromolecules*, 85:1–8.

Llop, J., I. Estrela-Lopis, R. F. Ziolo, A. González, J. Fleddermann, M. Dorn, V. Gomez-Vallejo, R. Simon-Vazquez, E. Donath, and Z. Mao. 2014. "Uptake, biological fate, and toxicity of metal oxide nanoparticles." *Particle and Particle Systems Characterization*, 31 (1):24–35.

Lundquist, P. and P. Artursson. 2016. "Oral absorption of peptides and nanoparticles across the human intestine: Opportunities, limitations and studies in human tissues." *Advanced Drug Delivery Reviews*,106:256–276.

Maezaki, Y., K. Tsuji, Y. Nakagawa, Y. Kawai, M. Akimoto, T. Tsugita, W. Takekawa, A. Terada, H. Hara, and T. Mitsuoka. 1993. "Hypocholesterolemic effect of chitosan in adult males." *Bioscience, Biotechnology, and Biochemistry*, 57 (9):1439–1444.

Mahler, G. J., M. L. Shuler, and R. P. Glahn. 2009. "Characterization of Caco-2 and HT29-MTX cocultures in an in vitro digestion/cell culture model used to predict iron bioavailability." *The Journal of Nutritional Biochemistry*, 20 (7):494–502.

Malhaire, H., J.-C. Gimel, E. Roger, J.-P. Benoît, and F. Lagarce. 2016. "How to design the surface of peptide-loaded nanoparticles for efficient oral bioavailability?"*Advanced Drug Delivery Reviews*, 106:320–336.

Marriott, C. and N. P. Gregory. 1990. "Mucus physiology and pathology." *Bioadhesive Drug DeliverySystems*:1–24.

McClements, D. J. and J. Rao. 2011. "Food-grade nanoemulsions: Formulation, fabrication, properties, performance, biological fate, and potential toxicity." *Critical Reviews in Food Science and Nutrition*, 51 (4):285–330.

McClements, D. J., and H. Xiao. 2012. "Potential biological fate of ingested nanoemulsions: Influence of particle characteristics." *Food and Function*, 3 (3):202–220.

Mun, S., E. A. Decker, and D. J. McClements. 2005. "Influence of droplet characteristics on the formation of oil-in-water emulsions stabilized by surfactant–chitosan layers." *Langmuir*, 21 (14):6228–6234.

Ozturk, B., S. Argin, M. Ozilgen, and D. J. McClements. 2014. "Formation and stabilization of nanoemulsion-based vitamin E delivery systems using natural surfactants: Quillaja saponin and lecithin." *Journal of Food Engineering,*142:57–63.

Ozturk, B., S. Argin, M. Ozilgen, and D. J. McClements. 2015. "Nanoemulsion delivery systems for oil-soluble vitamins: Influence of carrier oil type on lipid digestion and vitamin D3 bioaccessibility." *Food Chemistry,* 187:499–506.

Park, K. and J. R. Robinson. 1984. "Bioadhesive polymers as platforms for oral-controlled drug delivery: Method to study bioadhesion." *International Journal of Pharmaceutics*,19 (2):107–127.

Parthasarathi, S. and C. Anandharamakrishnan. 2016. "Enhancement of oral bioavailability of vitamin E by spray-freeze drying of whey protein microcapsules." *Food and Bioproducts Processing,*100:469–476.

Parthasarathi, S., S. P. Muthukumar, and C. Anandharamakrishnan. 2016. "The influence of droplet size on the stability, in vivo digestion, and oral bioavailability of vitamin E emulsions." *Food and Function*, 7 (5):2294–2302.

Parthasarathi, S., J. A. Bhushani, and C. Anandharamakrishnan. 2018. "Engineered small intestinal system as an alternative to in-situ intestinal permeability model." *Journal of Food Engineering,* 222:110–114.

Pietroiusti, A., L. Campagnolo, and B. Fadeel. 2013. "Interactions of engineered nanoparticles with organs protected by internal biological barriers." *Small*, 9 (9–10):1557–1572.

Pinheiro, A. C., M. A. Coimbra, and A. A. Vicente. 2016. "In vitro behaviour of curcumin nanoemulsions stabilized by biopolymer emulsifiers–effect of interfacial composition." *Food Hydrocolloids*, 52:460–467.

Pool, H., S. Mendoza, H. Xiao, and D. J. McClements.2013. "Encapsulation and release of hydrophobic bioactive components in nanoemulsion-based delivery systems: Impact of physical form on quercetin bioaccessibility." *Food and Function*, 4 (1):162–174.

Qian, C., E. A. Decker, H. Xiao, and D. J. McClements. 2012. "Nanoemulsion delivery systems: Influence of carrier oil on β-carotene bioaccessibility." *Food Chemistry*, 135 (3):1440–1447.

Rahman, M., S. Laurent, N. Tawil, L. Yahia, and L. Mahmoudi. 2013. *Protein-Nanoparticles Interactions: The Bio-Nano Interface*. Berlin, Springer.

Ramalingam, P., S. WooYoo, and Y. TagKo. 2016. "Nanodelivery systems based on mucoadhesive polymer coated solid lipid nanoparticles to improve the oral intake of food curcumin." *Food Research International,* 84:113–119.

Rocks, L. and K. A. Dawson. 2014. "The interaction between nanoparticles and biological barriers." *European Journal of Nanomedicine*, 6 (3):121–122.

Roy, K., H.-Q. Mao, S.-K. Huang, and K. W. Leong. 1999. "Oral gene delivery with chitosan–DNA nanoparticles generates immunologic protection in a murine model of peanut allergy." *Nature Medicine*, 5 (4):387.

Salvia-Trujillo, L., C. Qian, O. Martín-Belloso, and D. J. McClements. 2013. "Influence of particle size on lipid digestion and β-carotene bioaccessibility in emulsions and nanoemulsions." *Food Chemistry*, 141 (2):1472–1480.

Saptarshi, S. R., A. Duschl, and A. L. Lopata. 2013. "Interaction of nanoparticles with proteins: Relation to bio-reactivity of the nanoparticle." *Journal of Nanobiotechnology*, 11 (1):26.

Sari, T. P., B. Mann, R. Kumar, R. R. B. Singh, R. Sharma, M. Bhardwaj, and S. Athira. 2015. "Preparation and characterization of nanoemulsion encapsulating curcumin." *Food Hydrocolloids,* 43:540–546.

Sharma, H. S., S. F. Ali, S. M. Hussain, J. J. Schlager, and A. Sharma. 2009. "Influence of engineered nanoparticles from metals on the blood-brain barrier permeability, cerebral blood flow, brain edema and neurotoxicity. An experimental study in the rat and mice using biochemical and morphological approaches." *Journal of Nanoscience and Nanotechnology*, 9 (8):5055–5072.

Shin, Gye Hwa, Seoung KyunChung, Jun TaeKim, Hee JoungJoung, and Hyun JinPark. 2013. "Preparation of chitosan-coated nanoliposomes for improving the mucoadhesive property of curcumin using the ethanol injection method." *Journal of Agricultural and Food Chemistry*, 61 (46):11119–11126.

Shojaei, A. H. and X. Li.1997. "Mechanisms of buccal mucoadhesion of novel copolymers of acrylic acid and polyethylene glycol monomethylether monomethacrylate." *Journal of Controlled Release*, 47 (2):151–161.

Shvedova, A. A., N. Yanamala, E. R. Kisin, A. V. Tkach, A. R. Murray, A. Hubbs, M. M. Chirila, P. Keohavong, L. P. Sycheva, and V. E. Kagan. 2013. "Long-term effects of carbon containing engineered nanomaterials and asbestos in the lung: One year postexposure comparisons." *American Journal of Physiology-Lung Cellular and Molecular Physiology*, 306 (2):L170-L182.

Sinsuebpol, C., J. Chatchawalsaisin, and P. Kulvanich. 2013. "Preparation and in vivo absorption evaluation of spray dried powders containing salmon calcitonin loaded chitosan nanoparticles for pulmonary delivery." *Drug Design, Development and Therapy*, 7:861.

Slowing, I., B. G. Trewyn, and V. S.-Y. Lin. 2006. "Effect of surface functionalization of MCM-41-type mesoporous silica nanoparticles on the endocytosis by human cancer cells." *Journal of the American Chemical Society*, 128 (46):14792–14793.

Smart, J. D. 2005. "The basics and underlying mechanisms of mucoadhesion." *Advanced Drug Delivery Reviews*, 57 (11):1556–1568.

Sogias, I. A., A. C. Williams, and V. V. Khutoryanskiy. 2008. "Why is chitosan mucoadhesive?"*Biomacrom olecules*, 9 (7):1837–1842.

Sosnik, A., J. Neves, and B. Sarmento. 2014. "Mucoadhesive polymers in the design of nano-drug delivery systems for administration by non-parenteral routes: Areview." *Progress in Polymer Science*, 39 (12):2030–2075.

Tenzer, S., D. Docter, S. Rosfa, A. Wlodarski, J. Kuharev, A. Rekik, S. K. Knauer, C. Bantz, T. Nawroth, and C. Bier. 2011. "Nanoparticle size is a critical physicochemical determinant of the human blood plasma corona: a comprehensive quantitative proteomic analysis." *ACS Nano*, 5 (9):7155–7167.

Thorek, D. L. J. and A. Tsourkas. 2008. "Size, charge and concentration dependent uptake of iron oxide particles by non-phagocytic cells." *Biomaterials*, 29 (26):3583–3590.

Vance, M. E., T. Kuiken, E. P. Vejerano, S. P. McGinnis, M. F. Hochella Jr., D. Rejeski, and M. S. Hull. 2015. "Nanotechnology in the real world: Redeveloping the nanomaterial consumer products inventory." *Beilstein Journal of Nanotechnology*, 6:1769.

Verderio, P., S. Avvakumova, G. Alessio, M. Bellini, M. Colombo, E. Galbiati, S. Mazzucchelli, J. Peñaranda-Avila, B. Santini, and D. Prosperi. 2014. "Delivering colloidal nanoparticles to mammalian cells: Anano–bio interface perspective." *Advanced Healthcare Materials*, 3 (7):957–976.

Vergaro, V., C. Carlucci, M. Cascione, C. Lorusso, F. Conciauro, B. F. Scremin, P. M. Congedo, G. Cannazza, C. Citti, and G. Ciccarella. 2015. "Interaction between human serum albumin and different anatase TiO_2 nanoparticles: A nano-bio interface study." *Nanomaterials and Nanotechnology*, 5:30.

Yang, Y., M. E. Leser, A. A. Sher, and D. J. McClements. 2013. "Formation and stability of emulsions using a natural small molecule surfactant: Quillaja saponin (Q-Naturale®)." *Food Hydrocolloids*, 30 (2):589–596.

Yuan, H., C.-Y. Chen, G.-H. Chai, Y.-Z. Du, and F.-Q. Hu. 2013. "Improved transport and absorption through gastrointestinal tract by PEGylated solid lipid nanoparticles." *Molecular Pharmaceutics*, 10 (5):1865–1873.

13 Nanocomposite for Food Packaging

Vimala Bharathi S.K., B. Rohini, J.A. Moses, and C. Anandharamakrishnan

CONTENTS

13.1 INTRODUCTION

The packaging of food products plays an important role, and absence of it can lead to fast deterioration of quality and safety of food products, which paves the way to massive commercial losses of valuable foodstuffs. The packaging industry consumes more than 42% of the polymers produced; among which, the share of food packaging industry is almost 50% of it (Rhim et al., 2013; Silvestre et al., 2011). Food packaging constitutes the largest demanding application of long-lived petrochemical-based polymeric product due to its flexibility, mechanical property, lightweight, low density, and economicity. It should withstand harsh conditions during transport, distribution, marketing, and storage (Anandharamakrishnan and Usha Kiran, 2014). During the last decades, there has been a strong motivation for replacing polymeric, non-biodegradable packaging material with biodegradable, eco-friendly polymers. Recently, the European Union has taken the initiative to reduce plastic litter in oceans, especially single-use plastics. As an alternative, the EU has started promoting the usage of bio-based polymers. Also, the proposal requires producers to cover the waste management and clean-up cost (EU, 2018).

Biopolymers are materials obtained from renewable sources that are degradable without leaving lethal remains in the environment. This includes polymers such as starch, cellulose, chitosan, and other proteins of plant and animal origin. Though these biopolymers have been explored for decades, they could not replace the non-degradable polymers in practical application due to prime limitations such as poor mechanical properties and relative sensitivity towards moisture content due to their hydrophilic nature (Lluis et al., 2006). The boundaries of the biopolymer packaging material can be tailored to the application of nanotechnology.

Briefly, nanotechnology deals with the handling, preparation, characterization, and application of materials with at least one dimension in the nanometer range (10^{-9}) (Bawa et al., 2005). Because of their larger surface to volume and aspect ratios, nanoparticles exhibit different physical, chemical, mechanical, thermal, electrical, and optical properties compared to those of micro and macro scale particles. In food applications, nanotechnology can be used for enhancing taste, nutrient availability, preservation, storage, processing, and packaging (Bhushani, Kurrey, and Anandharamakrishnan, 2017; Kathiravan et al., 2017; Anandharamakrishnan, 2014a,b,c; Anandharamakrishnan and Ishwarya, 2015a,b; Bhushani and Anandharamakrishnan, 2017; Bhushani et al., 2017; Ezhilarasi et al., 2013). Recent researches focus on the application of nanotechnology in food packaging application.

This chapter explains the different biopolymer materials and reinforced nanomaterials with a view to make them feasible material for food packaging application. An overview of researches conducted in those areas is also provided for better understanding. Further, the vital roles of nanoreinforcements that are essential for food packaging application are also given in brief with the ways to predict them.

13.2 BIOPOLYMER-BASED NANOCOMPOSITES

13.2.1 Chitosan-Based Nanocomposites

Chitosan is manufactured primarily from crustaceans (crab and crayfish) because a significant amount of the crustacean exoskeleton is available as a by-product of food processing. Chitosan is a partially deacetylated derivative of chitin, which is the second most abundant natural biopolymer next to cellulose. Structurally chitosan is a heteropolysaccharide consisting of linear β-1,4-linked GlcN and GlcNAc units (Figure 13.1). During deacetylation, complex conditions yield a block-wise distribution of acetyl groups in chitosan, whereas homogeneous conditions yield a random distribution of acetyl groups. Due to its rigid and specific crystalline structures, obtained through intra- and intermolecular hydrogen bonding, chitosan can exist in different polymorphic forms, whose properties vary considerably (Harish Prashanth and Tharanathan, 2007).

Unique properties such as biodegradability, non-toxicity, biocompatibility, film-forming property, and similar mechanical properties of medium-chain commercial polymers have made chitosan one of the most commonly explored biopolymers in food packaging industries (Butler et al., 1996). Moreover, chitosan has strong antimicrobial properties against fungi, bacteria, and viruses, and its biocide properties have been known for many years (Fernandez-Saiz et al., 2008; Di Pierro et al., 2011). Various types of chitosan nanocomposites have been synthesized with different kinds

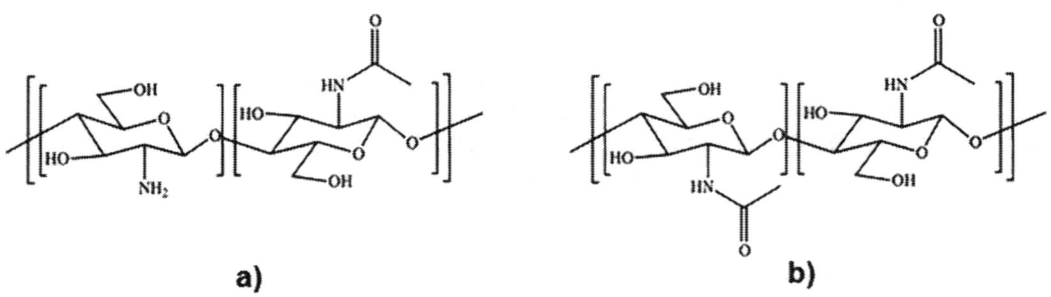

FIGURE 13.1 Chemical structure of (a) chitosan and (b) chitin. (Croisier and Jérôme, 2013.)

of nanoparticles (Dehnad et al., 2014; Di Pierro et al., 2011). Xu et al. (2006) proved that the addition of clay into the chitosan matrix resulted in the improved tensile strength of the film with a reduced elongation at break. Similarly, Mujeeb Rahman et al. (2016) carried out one-pot synthesis of chitosan nanocomposite film with nano-ZnO as nanofiller and observed that tensile strength was enhanced with the additional property of antimicrobial inhibitory effects. Moreover, the presence of additional materials in nanocomposites can have an effect on the composite as well. For instance, nanocomposites of chitosan were prepared with varying levels of MMT-Na and encapsulated α-tocopherol and it was found that addition of α-tocopherol showed enhanced water vapor permeability as well as improved intercalation of MMT-Na, resulting in a more hydrophilic surface (Dias et al., 2014). Contradictorily, the addition of neem oil to chitosan-ZnO nanocomposite improved the physical, barrier, and mechanical properties of the film and the resultant film exhibited positive results on cut pieces of carrot (Sanuja et al., 2015).

The addition of nanoreinforcements into the chitosan films has an effect not only on the property of the film itself but also on the material packed inside. A study was carried out by Youssef et al. (2016) on improving the shelf-life of Egyptian soft white cheese using chitosan, carboxymethyl cellulose, and nano-ZnO film. The packaging film exhibited excellent mechanical and thermal properties compared to chitosan/carboxymethyl cellulose blend. The prepared nanocomposites displayed good antimicrobial activity against gram-positive (*Staphylococcus aureus*) and gram-negative (*Pseudomonas aeruginosin*, *Escherichia coli*) bacteria and fungi (*Candida Albicans*). Nanocrystalline cellulose (NCC)-reinforced chitosan-based biodegradable films were prepared by solution casting (Khan et al., 2012). The NCC content in the films varied from 1% to 10% (dry basis). The results showed that 5% of NCC improved the tensile property of the film and also water vapor permeability. Swelling studies revealed a decrease in water uptake of the NCC-reinforced chitosan films. A study on the effect of plasticizers (oleic acid and glycerol) on the chitosan- Na-MMT nanocomposite film was performed. It was found that MMT was able to increase the plasticizing effect acting as plasticizers carrier at higher concentration (Vlacha et al., 2016). The addition of a nanoclay (Cloisite®30B) was studied to improve the water vapor permeability and the mechanical properties of chitosan films, and ethyl-Na-dodecanoyl-L-arginine hydrochloride (LAE) was added as an antimicrobial agent. The films obtained with the highest concentrations of LAE (1 g/L) and/or nanoclay (10 g/L) improved the mechanical properties (tensile and puncture strengths, and elongation at break) and decreased the water solubility and water vapor transmission rate. Besides, this concentration of nanoclay increased the opacity of the films, which is of potential interest for the protection of food from light (Rubilar et al., 2016). Montmorillonite (MMT) nanoclay and rosemary essential oil (REO) were incorporated into chitosan film to improve its physical and mechanical properties as well as antimicrobial and antioxidant behavior. The antimicrobial properties of the films were also enhanced by REO incorporation in 1.5% v/v. This indicates a good interaction between chitosan and MMT could improve water vapor permeability, water sensitiveness, and mechanical properties of the chitosan films (Ngah et al., 2011). The problems of moisture sensitivity and higher permeability of chitosan film are the major bottlenecks to overcome, especially for food packaging application, where the material comes in contact with food with high moisture content. Thus, incorporating appropriate nanofillers at an appropriate concentration can help chitosan to find its potential commercial application in the food packaging sector.

13.2.2 Starch-Based Nanocomposites

Starch is a renewable and biodegradable polymer produced by many plants as a source of stored energy. Starch is derived from maize (82%), wheat (8%), potatoes (5%), and cassava (5%) (Angellier et al., 2004). Structurally, starch is composed of (1) granule (120–500 nm), (2) growth rings (20–50 nm), (3) amorphous and crystalline lamellae containing amylopectin and amylose chains (0.1–1 nm). Starch granules are composed of two glucosidic macromolecules: amylose (linear molecule of glucose units linked by α-D-glycoside bonds, slightly branched by α-linkages) and

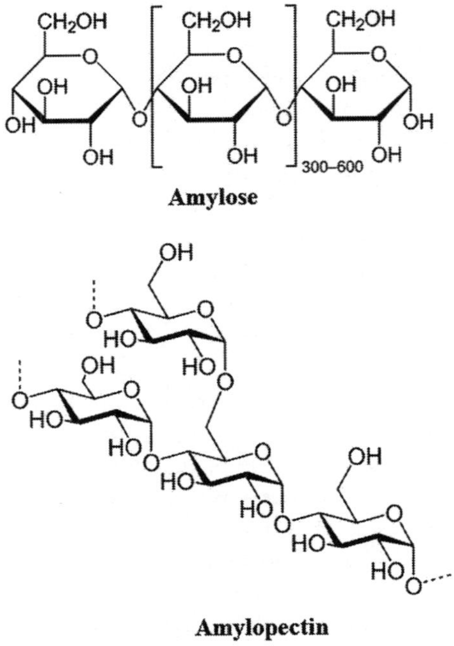

FIGURE 13.2 Chemical structure of amylose and amylopectin. (Zhou et al., 2010.)

amylopectin (polymer of short branches of α-D-(1–4) glycopyranose linked by α-D-(1–6)-glycosidic linkages) (Le Corre and Angellier-Coussy, 2014). Figure 13.2 illustrates the chemical structure of amylose and amylopectin.

Despite being a polysaccharide with high moisture permeability and poor mechanical properties, starch exhibits a good oxygen barrier property (Byun and Kim, 2014; Mali et al., 2004; Mali et al., 2004). The properties of starch-based film depend on amylose content, type of plasticizer added, film thickness, method of preparation, moisture content, etc. (Mali et al., 2004; Visakh and Yu, 2016). For instance, with an increase in amylose content, there would be an increase in viscosity and tensile strength, whereas water absorption decreases (Mani and Bhattacharya, 1998; Xie et al., 2009). Forssell et al. (2002) revealed that the barrier property of starch is better when moisture content is less than 15%.

Starch nanocrystals: Starch nanocrystals are widely used in food industries as emulsifiers (Ge et al., 2017), nanocomposites (Costa et al., 2017), bioplastic materials (Moro et al., 2017) and so on. Starch nanocrystals refer to the crystalline part of the starch obtained by disruption of the semicrystalline structure of starch by acid (normally sulfuric acid) hydrolysis at controlled environment including temperature, acid and starch concentration, hydrolysis duration and stirring speed. Structural properties of the starch crystals depend on the starch source; temperature and acid treatment (see Table 13.1).

In a recent study, Hao et al. (2018) adopted enzymatic (glucoamylase) pretreatment followed by acid hydrolysis for starch nanocrystals. The glucoamylase pretreatment created pores by breaking the alpha-1,4 and alpha-1,6 links of the polymer (Jurado et al., 2004). These pores at the surface enhance the diffusion of acid during hydrolysis, and a few researchers further increased the acid hydrolysis by coupling glucoamylase and ultrasound (Zhu, 2015; Wu et al., 2011). Starch nanocrystals are frequently preferred for their reinforcement property. Wang et al. (2017) developed biodegradable starch-based composites with jute fibers as a reinforcement material. The degradation period of the composite film was increased due to the addition of jute and showed higher mechanical strength, which can be applied to various applications. To enhance the water resistance and tensile properties of starch-based packaging material, Cao et al. (2008) prepared starch-flax

TABLE 13.1
Effects of Different Extraction Methods and Starch Sources on Starch Nanocrystal Size

Starch Source	Concentration of Acid for Hydrolysis	Temperature	Duration	Structural Parameters[a]
Waxy corn starch	3.16 mol/L H_2SO_4	40°C	5 days	T = 6–8 nm; L = 70–100 nm
Corn starch	2.87 mol/L H_2SO_4	45°C	7 days	L = 50 nm
Pea starch	3.16 mol/L H_2SO_4	40°C	5 days	L = 60–150 nm; W = 15–30 nm
Potato starch	3.16 mol/L H_2SO_4	40°C	5 days	L = 40–70 nm; W = 10–20 nm

Source: Lin et al. (2011).
[a] Structural parameters are length (L); width (W), and thickness (T).

cellulose nanocrystal-based nanocomposite films and showed that with an increase in nanocrystals (100–500 nm length and 10–30 nm diameter) concentration up to 30% (w/w), the mechanical properties and water resistance increased.

13.2.3 Poly(Lactic Acid)-Based Nanocomposites

Poly(lactic acid) (PLA) belongs to the family of aliphatic polyesters commonly made from α-hydroxy acids, which include polyglycolic acid or polymandelic acid (Ahmed and Varshney, 2011). Lactic acid (2-hydroxy propionic acid) is the simplest hydroxyl acid with an asymmetric carbon atom and it exists in two optically active configurations, the L(+) and D(−) isomers. The monomeric units of lactic acid obtained from the fermentation of corn, beet sugar, cane sugar, etc., undergo either ring-opening polymerization or condensation polymerization, leading to the formation of PLA (see Figure 13.3). During the initiation of polymerization, lactic acid monomers are oligomerized and catalytically dimerized to produce cyclic lactic acid monomers which are further subjected to ring-opening polymerization to produce high molecular weight PLA.

The properties of PLA depend upon the ratio of distribution of two isomeric forms of lactic acid (L-lactic acid and D-lactic acid) and the molecular weight of the polymer. PLA homopolymers are a pseudo-plastic and non-Newtonian fluid having glass-transition and melting temperature of about 55°C and 175°C respectively, which is lower than PET and PS and require processing temperature around 185°C–190°C. High molecular weight PLA is transparent, glossy, and stiff, which is similar to polystyrene. The mechanical properties such as tensile strength and elongation percentage are similar to PS but lower than PET under the same test conditions. The barrier to O_2 and CO_2 of PLA is similar to PET but lower than PS. Surprisingly, though PLA is polar in nature, water permeability coefficient does not vary with relative humidity. PLA is more soluble in alcohol, nitrogen compounds, and anhydride compounds but poor interaction with water, aromatic hydrocarbons, ketone, esters, and sulfur compounds. Application of PLA in food packaging applications is increasing because of their good mechanical properties, ease of processing, and lower cost compared to those of other biodegradable polymers (Parker et al., 2011; Sin et al., 2012). PLA films exhibit inferior property in terms of thermal stability, brittleness, barrier properties, and hydrophobic nature (Haafiz et al., 2013; Sin et al., 2012).

Chang et al. (2003) studied the interaction between the PLA matrix and two different organoclays such as C16-Mica and C16-MMT and found that small amounts of added organoclay improved the thermo-mechanical properties of PLA nanocomposite films. This showed that C16-Mica was effective and it was believed that the interfacial interaction between PLA and organoclay strengthened the film.

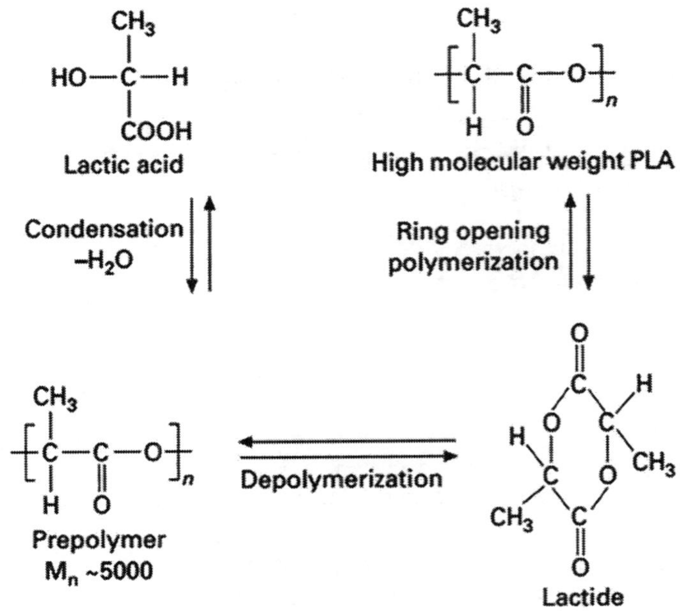

FIGURE 13.3 Schematic representation of PLA synthesis. (Mochizuki, 2009.)

The incorporation of TiO_2 nanoparticle in PLA polymers has been reported to improve transparency and achieve photo-degradability. It was also found that surface modification of TiO_2 nanoparticle improved its dispersion in the polymer matrix without any aggregation, which was key for improving the optical properties of PLA-TiO_2 nanocomposite film (Nakayama and Hayashi, 2007).

Busolo et al. (2010) prepared PLA-based nanocomposite film from nanoclay and also incorporated nano-silver particles to improve its antimicrobial properties. The developed films were found to possess strong antimicrobial activity against gram-negative *Salmonella* sp. with minimum inhibitory concentration and minimum bactericidal concentration below 1 mg per 10 ml. Rhim et al. (2013) found that tensile, water vapor barrier and antimicrobial properties were improved when PLA polymer was blended with different nanoclays. It was also stated that a suitable type of nanoclay and its optimum concentration imparts good mechanical and barrier properties to PLA nanocomposite films (Auras et al., 2004).

The new Cargill Dow LLC polylactide plant in Blair, NB, USA, began to minimize the cost of production of PLA and increase its demand. Mitsui Chemicals are conducting PLA business development. Danone and McDonalds' companies in Europe are already using PLA as packaging materials for yogurt and cutleries. Also, in countries such as Japan, Europe, and the United States, PLA are used for packaging fresh fruits and vegetables for short shelf life.

13.2.4 Cellulose-Based Nanocomposites

Cellulose is the most abundant linear polysaccharide made of repeating monomers of anhydrous glucose linked via a β-1,4 glycosidic bond called cellobiose (Clarke, 1997; Scott, 1999). Depending on the source of the cellulose, the degree of polymerization varies from 700 to 27,000 (Hon, 1989). Cellulose is found in wood, plant materials (cotton, jute, bagasse, straw, potato tubers, etc.) animal source (tunicate), algae (red, green, yellow-green, etc.), and bacteria (*Gluconacetobacter xylinus*) (Moon et al., 2011). Among these, cotton seed possesses the maximum purity, in the range of 90%–99% (Hon et al., 1989). There are different cellulose particles that are mainly made from wood and plant fibers, but those particles differ in method of preparation, characteristics, and aspect ratio (Table 13.2).

TABLE 13.2
Types, Method of Preparation and Characteristics of Cellulose

Type of Particle	Method of Preparation	Characteristics	Aspect Ratio
Microcrystalline cellulose (MCC)	Acid (most commonly sulfuric acid) hydrolysis, followed by back-neutralization with alkali and spray drying	High cellulose content and crystallinity	~1
Micro-fibrillated cellulose (MFC)	Mechanical refining	• Very high cellulose content (~100%) • Presence of amorphous and crystalline region	100–150
Nano-fibrillated cellulose (NFC)	Mechanical refining (finer particle than MFC)	• Very high cellulose content (~100%) • Presence of amorphous and crystalline region • High strength	>100
Cellulose nanocrystals (CNC)	Acid hydrolysis	• Very high crystallinity • High thermal stability	50–60

Source: Moon et al. (2011) and Siro and Plackett (2010).

13.2.5 Protein-Based Nanocomposites

13.2.5.1 Gluten

Gluten is a storage protein that consists of prolamin and glutelin. Alcohol solubilizes prolamin; whereas dilute acids or bases solubilize glutelin (Georgiev and Dekova, 2011). The major sources of gluten are wheat, rye, barley, triticale, and oats. Wheat gluten *(Triticum aestivum)* is the most commonly used biopolymer and is employed in the form of films and edible coatings. Wheat gluten is composed of gliadin (a prolamin protein) and glutenin (a glutelin protein), which are responsible for viscous properties, and strength and elasticity respectively (Khatkar et al., 1995). Hence, the properties of film vary with gliadin and glutenin ratio. Furthermore, wheat gluten films exhibit lower permeability compared to other films (Telis, 2012).

Rafieian et al. (2014) used carboxylated cellulose nanocrystals (C.CNC) as a nanofiller material for developing gluten-based nanocomposite packaging film. The C_6 hydroxyls of cellulose were converted to carboxyl groups by oxidization reaction which induces a repulsive force between the cellulose nanocrystals, thus ensuring the complete dispersion of CNCs within the gluten matrix. They observed an improved tensile strength for nanocomposite when compared with gluten film without any filler material. They also observed reduced water vapor permeability and decreased breaking elongation for nanocomposite films. El-wakil et al. (2015) used TiO_2 nanoparticle along with CNCs to coat commercial unbleached kraft paper sheets. When three layers of the nanocomposite coating were given, the breaking strength and burst index were increased by 56% and 53% respectively. The coated paper also exhibited an improved antimicrobial property against *Saccharomyces cervisiae*, gram-negative bacteria *Escherichia coli* and gram-positive bacteria *Staphylococcus aureus*.

13.2.5.2 Gelatin

Gelatin is a water-soluble protein obtained from partial hydrolysis of an insoluble structural protein, collagen. Collagen is found abundantly in fibrous tissues of vertebrate and invertebrate animals such as skin, tendons, ligaments, and bones. Gelatin is mainly obtained from pig, cow, buffalo, and fish scales (Ramos et al., 2016; Williams and Phillips, 2012). Gelatin films exhibit higher strength, flexibility, and gas barrier properties (Shankar et al., 2016). These properties depend on collagen characteristics and processing method. Acid-treated gelatin possesses a higher isoelectric point and lowers hydroxyl groups with residues of aspartate and glutamate compared to those of alkaline-treated gelatin (Ramos, 2011; Ramos et al., 2016).

Several researchers have evaluated the potential application of gelatin as a biopolymer for nanocomposite preparation (Martucci et al., 2007; Rouhi et al., 2013). Kanmani and Rhim (2014) developed silver and organoclay reinforced gelatin nanocomposite film and proved that the films exhibited stronger antimicrobial and tensile properties and stated that the gelatin-based nanocomposites help to improve the quality of the food. Similarly, Chen et al. (2018) prepared chitin nanofiber/gelatin-based nanocomposite and proved that nanofiber and gelatin exhibited good compatibility and showed improved mechanical properties.

13.2.5.3 Soy Protein

Soy protein is a complex macromolecule consisting of 58% polar and reactive amino acid residues and 38% non-polar and non-reactive amino acid residues. Though soy protein films exhibit poor mechanical properties, they have great potential as food packaging material because of their high oxygen barrier property (Gupta and Nayak, 2014; Park et al., 2000). Furthermore, owing to their hydrophilicity, soy protein films exhibit low water vapor barrier properties. So, in general, these films are used in blends with other biodegradable polymers with high hydrophobic properties such as lipids to enhance their moisture barrier properties (Gupta and Nayak, 2014).

Echeverría et al. (2018) developed an active nanocomposite packaging material based on soy protein isolate-MMT and clove essential oil for the preservation of muscle fillets of bluefin tuna. This packaging material was able to maintain freshness up to 17 days at 2°C without any microbial growth. Also, the study stated that addition of MMT, besides improving the mechanical property, prolonged the antimicrobial activity by favoring the release of the antimicrobial agent. To improve the tensile strength of soy protein isolate, a nanocomposite material was prepared with microcrystalline cellulose and metal nanoclusters. The study proved that the tensile properties and water resistance capacity of soy protein improved with the nanoreinforcements (Li et al., 2017).

13.3 NANOREINFORCEMENTS FOR FOOD PACKAGING

Nanoreinforcements are nanomaterials dispersed in a continuous polymer phase. They can be divided based on source, dimensions, or geometry. Based on their source, nanoreinforcements can be either organic or inorganic (Ramos et al., 2017). Organic nanoreinforcements include polysaccharides (starch, chitosan, chitin, and cellulose), antimicrobials (nisin), and carbon nanofillers (fullerenes, graphene, and carbon nanotubes). Inorganic nanoreinforcements include clays (montmorillonite, mica, sepiolite, kaolinite, halloysite, and laponite) (Sanchez et al., 2005), metals (gold, silver, copper, iron, and palladium), metal oxides (silica, zinc oxide, titanium dioxide, magnesium oxide, silver oxide, and cupric oxide), and others (calcium carbonate and polyhedral oligomeric silsesquioxane) (Saba et al., 2014). Based on number of dimensions, nanoreinforcements can be divided into three categories: nanolayers (one nano-sized dimension), nanotubes/nanowhiskers (two nano-sized dimensions), and nanoparticles (three nano-sized dimensions) (Murray et al., 2015); whereas based on geometry, nanoreinforcements can be layered, spherical, or acicular particles (Chivrac et al., 2009). Nanoreinforcements are often used to improve the properties of biopolymers (Mishnaevsky et al., 2016).

It is clear from the illustrations (Table 13.3) that the property of the nanocomposite depends on the amount of filler used and the added filler can be a limiting factor when the amount added is increased beyond a threshold value. For instance, when bacterial cellulose nanocrystals were added to gelatin in the range of 1%–5% (w/w), the tensile strength values were found to increase gradually up to 4% with a maximum value of 108.6 MPa, whereas neat gelatin showed only 83.7 MPa. In this case, 4% is the threshold value. When nanocrystals were added beyond that, say, 5%, tensile

TABLE 13.3
Effect of Nanoreinforcement Type and Amount on the Properties of Nanocomposite

Matrix	Filler	Amount of Filler Added	Property	Change	Percent Change (%)	References
Pectin	Crystalline nanocellulose	5% (w/w)	Glass transition temperature	Increase	1.50	Chaichi et al. (2017)
			Melting temperature	Increase	0.96	
			Melting enthalpy	Increase	20.34	
			Moisture absorption	Decrease	5.74	
		7% (w/w)	Glass transition temperature	Increase	1.36	
			Melting temperature	Decrease	1.26	
			Melting enthalpy	Increase	15.15	
			Moisture absorption	Increase	10.32	
Chitosan incorporated with 5% *Carum copticum* essential oil	Cellulose nanofibers	4% (w/w)	Tensile strength	Increase	8.46	Jahed et al. (2017)
			Water vapor permeability	Decrease	26.44	
			Antioxidant activity	Decrease	25.24	
	Ligno-cellulose nanofibers	4% (w/w)	Tensile strength	Increase	27.83	
			Water vapor permeability	Decrease	29.32	
			Antioxidant activity	Decrease	37.12	
Chitosan	Sodium montmorillonite (MMT)	3%(w/w)	Water solubility	Decrease	30.00	Ghelejlu et al., (2016)
			Moisture content	Decrease	26.81	
			Water vapor permeability	Decrease	39.57	
Starch	Halloysite	0.1% (w/v)	Young modulus	Increase	137.87	Meira et al. (2016)
			Tensile strength	Increase	81.41	
			Elongation at break	Decrease	30.00	
			Transparency value	Increase	48.00	
			Water vapor permeability	Decrease	21.81	
Potato starch	Sodium montmorillonite	3% (w/w)	UV-C transmittance (240 nm)	Decrease	71.96	Oleyaei et al. (2016)
			UV-B transmittance (300 nm)	Decrease	30.94	
			UV-A transmittance (360 nm)	Decrease	9.56	
			Visible light transmittance	Decrease	1.73	

(Continued)

TABLE 13.3 (CONTINUED)
Effect of Nanoreinforcement Type and Amount on the Properties of Nanocomposite

Matrix	Filler	Amount of Filler Added	Property	Change	Percent Change (%)	References
			Opacity	Increase	34.37	
			Color difference	Decrease	0.41	
			Glass transition temperature	Increase	21.76	
			Melting temperature	Increase	16.02	
			Water vapor permeability	Decrease	41.01	
	Sodium montmorillonite and nano-TiO_2	3% MMT (w/w) and 2% TiO_2 (w/w)	UV-C transmittance (240 nm)	Decrease	99.95	
			UV-B transmittance (300 nm)	Decrease	99.97	
			UV-A transmittance (360 nm)	Decrease	99.93	
			Visible light transmittance	Decrease	93.69	
			Opacity	Increase	2015.62	
			Color difference	Increase	119.33	
			Glass transition temperature	Increase	106.48	
			Melting temperature	Increase	29.02	
			Water vapor permeability	Decrease	69.29	
	Sodium montmorillonite	5% (w/w)	UV-C transmittance (240 nm)	Decrease	82.49	
			UV-B transmittance (300 nm)	Decrease	38.40	
			UV-A transmittance (360 nm)	Decrease	11.14	
			Visible light transmittance	Decrease	2.96	
			Opacity	Increase	62.5	
			Color difference	Increase	1.89	
			Glass transition temperature	Increase	50.46	
			Melting temperature	Increase	20.03	
			Water vapor permeability	Decrease	50.81	

(Continued)

TABLE 13.3 (CONTINUED)
Effect of Nanoreinforcement Type and Amount on the Properties of Nanocomposite

Matrix	Filler	Amount of Filler Added	Property	Change	Percent Change (%)	References
	Sodium montmorillonite and nano-TiO_2	5% MMT (w/w) and 2% TiO_2(w/w)	UV-C transmittance (240 nm)	Decrease	99.95	
			UV-B transmittance (300 nm)	Decrease	99.97	
			UV-A transmittance (360 nm)	Decrease	99.88	
			Visible light transmittance	Decrease	95.38	
			Opacity	Increase	2093.75	
			Color difference	Increase	121.00	
			Glass transition temperature	Increase	114.35	
			Melting temperature	Increase	25.24	
			Water vapor permeability	Decrease	59.89	
Poly(3-hydroxybutyrate-co-18 mol%- 3-hydroxyvalerate) (PHBV18)/poly(3-hydroxybutyrate-co-3 mol%- 3-hydroxyvalerate) (PHBV3)	Silver Nanoparticles	0.04% (w/w)	Melting enthalpy	Increase	1.06	Castro-Mayorga et al. (2015)
			Degree of crystallinity	Increase	0.99	
			Tensile strength	Decrease	11.76	
			Elongation at break	Decrease	14.14	
			Water uptake	Decrease	4.72	
			Water vapor permeability (100% RH)	Decrease	36.21	
			Oxygen permeability (80% RH)	Decrease	59.64	
Amaranth flour	Montmorillonite	9.4% (w/w)	Puncture force	Increase	100.00	Diéguez et al. (2015)
			Puncture deformation	Decrease	10.18	
			Tensile strength	Increase	33.33	
			Percent elongation	Decrease	49.58	
			Young's modulus	Increase	86.30	
			Solubility	Decrease	6.64	
			Water vapor permeability	Decrease	96.27	

(Continued)

TABLE 13.3 (CONTINUED)
Effect of Nanoreinforcement Type and Amount on the Properties of Nanocomposite

Matrix	Filler	Amount of Filler Added	Property	Change	Percent Change (%)	References
			Moisture content	Increase	3.38	
Wheat gluten	Cellulose nanocrystals	12.5% (w/w)	Contact angle	Increase	74.01	El-wakil et al. (2015)
			Elongation at break	Decrease	88.20	
			Tensile strength	Increase	70.17	
			Water vapor permeability at 100% Rh	Decrease	37.42	
	7.5% (w/w) Cellulose nanocrystals and TiO_2	1% (w/w)	Contact angle	Increase	40.07	
			Tensile strength	Increase	19.57	
			Water vapor uptake at 98% Rh	Decrease	40.80	
			Glass transition temperature	Increase	93.55	
Cellulose acetate	Natural montmorillonite (Na-MMT)	5%(w/w)	Glass transition temperature	Increase	2.07	Ferfera-Harrar and Dairi (2015)
			Melting temperature	Decrease	0.85	
	Na-MMT with plasticizer triethyl citrate	5%(w/w)	Glass transition temperature	Decrease	47.66	
			Melting temperature	Decrease	26.38	
	Organomodified MMT with gelatin (Ge-MMT)	5%(w/w)	Glass transition temperature	Increase	2.59	
			Melting temperature	Decrease	0.85	
	Ge-MMT with plasticizer triethyl citrate	5%(w/w)	Glass transition temperature	Decrease	47.15	
			Melting temperature	Decrease	26.80	
	Organomodified MMT with chitosan (Cs-MMT)	5%(w/w)	Glass transition temperature	Increase	1.03	
			Melting temperature	Decrease	0.85	
	Cs-MMT with plasticizer triethyl citrate	5%(w/w)	Glass transition temperature	Decrease	47.66	
			Melting temperature	Decrease	26.38	
			Contact angle (indicator of hydrophilic/hydrophobic property)	Increase	113.12	
			Degradation temperature	Decrease	2.32	

(Continued)

TABLE 13.3 (CONTINUED)
Effect of Nanoreinforcement Type and Amount on the Properties of Nanocomposite

Matrix	Filler	Amount of Filler Added	Property		Change	Percent Change (%)	References
Fish gelatin	Chitosan nanoparticles	8% (w/w)	Moisture content		Decrease	31.34	Hosseini et al. (2015)
Cellulose acetate	Copper montmorillonite modified	5% (w/w)	Crystallinity		Decrease	60.00	Bruna et al. (2014)
			Opacity		Increase	500.00	
			Tensile modulus		Decrease	3.82	
Chitosan	Nanocellulose	2% (w/w)	Melting temperature		Decrease	1.27	Dehnad et al. (2014)
			Glass transition temperature		Decrease	2.53	
Starch	Oil palm empty fruit bunch Cellulose Nanofiber	2% (w/w)	Tensile strength		Increase	27.86	Salehudin et al. (2014)
			Percent elongation		Decrease	33.51	
			Young's modulus		Increase	92.35	
		10% (w/w)	Tensile strength		Decrease	20.76	
			Percent elongation		Decrease	60.33	
			Young's modulus		Increase	59.17	
Soy protein	Montmorillonite	10% (w/w)	Rheological characteristics of film-forming solution	Consistency index	Increase	570.91	Echeverria et al. (2014)
				Flow behavior	Decrease	21.28	
				Apparent viscosity at shear rate of 60 s^{-1}	Increase	242.23	
				Apparent viscosity at shear rate of 300 s^{-1}	Increase	174.01	
				Apparent viscosity at shear rate of 500 s^{-1}	Increase	153.51	
			Moisture content		Decrease	22.03	
			Solubility		Decrease	34.99	
			Water vapor permeability		Decrease	73.33	
			Effective diffusion		Decrease	78.14	

(Continued)

TABLE 13.3 (CONTINUED)
Effect of Nanoreinforcement Type and Amount on the Properties of Nanocomposite

Matrix	Filler	Amount of Filler Added	Property		Change	Percent Change (%)	References
			Opacity		Decrease	47.82	
Kefiran–whey protein isolate	Montmorillonite	5% (w/w)	Whiteness index		Decrease	6.95	Zolfi et al. (2014)
			Roughness parameters	Average roughness	Increase	55.67	
				Root mean square (RMS)	Increase	53.57	
				RMS gradient	Increase	108.54	
	Nano-TiO_2	5% (w/w)	Whiteness index		Increase	13.98	
			Roughness parameters	Average roughness	Increase	43.47	
				Root mean square (RMS)	Increase	40.15	
				RMS gradient	Increase	92.68	
Alginate	Montmorillonite	5% (w/w)	Water solubility		Decrease	38.2	Abdollahi et al. (2013)
			Contact angle		Decrease	11.10	
			Water vapor permeability		Decrease	20	
			Melting temperature		Increase	18.74	
			Opacity		Increase	19.57	
	Cellulose nanoparticles	5% (w/w)	Water solubility		Decrease	22.06	
			Contact angle		Increase	83.54	
			Water vapor permeability		Decrease	18	
			Melting temperature		Decrease	1.99	
			Opacity		Increase	17.53	
Alginate (conc 1% w/v)	Montmorillonite	3% (w/w)	Tensile strength		Increase	1.05	Alboofetileh et al. (2013)
			Elongation at break		Decrease	24.68	
			Opacity		Increase	29.89	
			Moisture content		Decrease	14.14	

(Continued)

TABLE 13.3 (CONTINUED)
Effect of Nanoreinforcement Type and Amount on the Properties of Nanocomposite

Matrix	Filler	Amount of Filler Added	Property	Change	Percent Change (%)	References
		5% (w/w)	Tensile strength	Decrease	4.21	
			Elongation at break	Decrease	21.86	
			Opacity	Increase	19.56	
			Moisture content	Decrease	2.40	
Alginate (conc 1.5% w/v)		3% (w/w)	Tensile strength	Increase	13.33	
			Elongation at break	Decrease	18.02	
			Opacity	Increase	24.43	
			Moisture content	Decrease	21.80	
		5% (w/w)	Tensile strength	Decrease	10.78	
			Elongation at break	Decrease	37.17	
			Opacity	Increase	24.96	
			Moisture content	Decrease	14.39	
Poly (lactic acid)	Silver nanoparticles	1% (w/w)	Oxygen transmission rate	Decrease	21.64	Fortunati et al. (2013)
			Water vapor permeabillity	Decrease	3.85	
	Cellulose nanocrystals and silver nanoparticles	1% CNC (w/w) and 1% Ag (w/w)	Oxygen transmission rate	Decrease	45.90	
			Water vapor permeabillity	Decrease	36.54	
		5% CNC (w/w) and 1% Ag (w/w)	Oxygen transmission rate	Decrease	41.64	
			Water vapor permeability	Decrease	47.11	
	Surfactant modified cellulose nanocrystals and silver nanoparticles	1% s-CNC (w/w) and 1% Ag (w/w)	Oxygen transmission rate	Decrease	53.44	
			Water vapor permeability	Decrease	58.65	

(Continued)

TABLE 13.3 (CONTINUED)
Effect of Nanoreinforcement Type and Amount on the Properties of Nanocomposite

Matrix	Filler	Amount of Filler Added	Property	Change	Percent Change (%)	References
		5% s-CNC (w/w) and 1% Ag (w/w)	Oxygen transmission rate	Decrease	58.69	
			Water vapor permeability	Decrease	58.65	
Poly(lactic acid)	Microcrystalline cellulose	1% (w/w)	Onset decomposition temperature	Increase	14.61	Haafiz et al. (2013)
			10% decomposition temperature	Increase	6.82	
			50% decomposition temperature	Increase	1.77	
			Residual weight % at 800° C	Increase	0.86	
		5% (w/w)	Onset decomposition temperature	Increase	7.04	
			10% decomposition temperature	Increase	3.53	
			50% decomposition temperature	Increase	1.77	
			Residual weight % at 800° C	Increase	2.40	
Poly(lactic acid)	Silver nanoparticles	1% (w/w)	Yield stress	Decrease	42.59	Fortunati et al. (2012)
			Strain	Decrease	4.54	
			Tensile strength	Decrease	48.84	
			Elongation at break	Decrease	87.78	
			Elastic modulus	Increase	5.00	
			Glass transition temperature	Decrease	5.08	
			Cold crystallization temperature	Decrease	2.29	
			Cold crystallization enthalpy	Increase	33.33	
			Melting temperature	Decrease	3.61	
			Melting enthalpy	Increase	30.43	
			Crystallinity	Increase	30.43	
	Cellulose nanocrystals	5% (w/w)	Yield stress	Decrease	27.78	
			Strain	Increase	4.54	
			Tensile strength	Decrease	34.19	
			Elongation at break	Decrease	60.00	

(Continued)

TABLE 13.3 (CONTINUED)
Effect of Nanoreinforcement Type and Amount on the Properties of Nanocomposite

Matrix	Filler	Amount of Filler Added	Property	Change	Percent Change (%)	References
			Elastic modulus	Increase	22.08	
			Glass transition temperature	Decrease	7.46	
			Cold crystallization temperature	Decrease	3.77	
			Cold crystallization enthalpy	Increase	102.67	
			Melting temperature	Decrease	3.67	
			Melting enthalpy	Increase	59.62	
			Crystallinity	Increase	28.26	
	Cellulose nanocrystals and silver nanoparticles	5% CNC (w/w) and 1% Ag (w/w)	Yield stress	Decrease	11.11	
			Strain	Decrease	4.54	
			Tensile strength	Decrease	32.56	
			Elongation at break	Decrease	70.00	
			Elastic modulus	Increase	12.50	
			Glass transition temperature	Decrease	7.28	
			Cold crystallization temperature	Decrease	4.75	
			Cold crystallization enthalpy	Increase	88.00	
			Melting temperature	Decrease	4.06	
			Melting enthalpy	Increase	22.36	
			Crystallinity	Decrease	30.43	
	Surfactant modified cellulose nanocrystals	5% w/w	Strain	Decrease	45.45	
			Tensile strength	Increase	7.21	
			Elongation at break	Decrease	80.00	
			Elastic modulus	Increase	83.33	
			Glass transition temperature	Decrease	8.13	

(Continued)

TABLE 13.3 (CONTINUED)
Effect of Nanoreinforcement Type and Amount on the Properties of Nanocomposite

Matrix	Filler	Amount of Filler Added	Property	Change	Percent Change (%)	References
			Cold crystallization temperature	Decrease	9.59	
			Cold crystallization enthalpy	Increase	192.00	
			Melting temperature	Decrease	4.25	
			Melting enthalpy	Increase	137.26	
			Crystallinity	Increase	101.08	
	Surfactant modified cellulose nanocrystals and silver nanoparticles	5% s-CNC (w/w) and 1% Ag (w/w)	Strain	Decrease	13.64	
			Tensile strength	Decrease	5.12	
			Elongation at break	Decrease	33.33	
			Elastic modulus	Increase	25.00	
			Glass transition temperature	Decrease	8.30	
			Cold crystallization temperature	Decrease	7.86	
			Cold crystallization enthalpy	Increase	132.00	
			Melting temperature	Decrease	3.41	
			Melting enthalpy	Increase	104.96	
			Crystallinity	Increase	92.39	
Gelatin	Bacterial (*Gluconacetobacterxylinus*) cellulose nanocrystals	4% (w/w)	Tensile strength	Increase	30.12	George and Siddaramaiah (2012)
			Elastic modulus	Increase	7.35	
			Glass transition temperature	Increase	4.33	
			Degradation temperature	Increase	5.03	

strength value decreased to 89.8 MPa (George and Siddaramaiah, 2012). The properties of nanoreinforcements used depend on shape (Kwon et al., 2016; Pal et al., 2007), size (Guz et al., 2017), type (Shankar et al., 2015), flexibility (Kwon et al., 2016), and aspect ratio (Corcinone et al., 2007)

13.3.1 Antimicrobial Properties

Nanoreinforcements can also be used to improve the antimicrobial properties of the film and, thus, increase the shelf life of the product. Moreover, antimicrobial agents in the nano form when reinforced into the polymer matrix possess improved activity owing to their high surface to volume ratio (Llorens et al., 2012). The selection of a suitable antimicrobial agent is based on its effect on the food product and the microorganism associated.

Several studies have been conducted on nanoreinforcement of antimicrobials into the polymer matrix both in individual and combined form (Jin and Gurtler, 2011; Zohri et al., 2013). To improve the antimicrobial properties, essential oils are also incorporated with other reinforcements. The addition of essential oils may affect the mechanical, physical, and barrier properties of the film. For instance, addition of plai essential oil in hydroxypropyl methylcellulose decreased water activity by 20% and increased water vapor permeability and oxygen permeability by 38% and 40% respectively (Klangmuang and Sothornvit, 2016). Apart from the antimicrobials discussed above, some inorganic nanofillers such as zinc oxide (Mallakpour and Behranvand, 2016), copper oxide (Oun and Rhim, 2017), silver (Franci et al., 2015), and copper (Chatterjee et al., 2014) also possess antimicrobial properties.

13.3.1.1 Titanium Di-Oxide (TiO_2)-Based Nanocomposite

TiO_2 possesses a property of high refractive index and is resistant to discoloration. It has the tendency to give opacity to papers, coatings, plastics, films, etc. The photocatalytic reaction of TiO_2 leads to the generation of hydroxyl radicals (_OH) and reactive oxygen species (ROS) on the TiO_2 surface, resulting in the oxidation of polyunsaturated phospholipids present in cell membranes of microorganisms. As a consequence, the microorganism is inactivated. The exhibited photocatalytic property is stronger in nano form than that of the micro-sized particle. Many researchers (Babaei-Ghazvini et al., 2018; Oleyaei et al., 2016; Vejdan et al., 2016) have reported on blending TiO_2 nanoparticles with suitable packaging material to increase its efficiency. Whey protein film was blended with TiO_2 nanoparticles to obtain nanocomposite food package and the improved mechanical strength of the film was observed. The variation of TiO_2 concentration had an effect on the tensile strength and water barrier properties of the film (Zhou et al., 2009). Long et al. (2014) studied the effects of UV exposure time, nano-TiO_2 concentrations and initial bacteria populations on nano-TiO_2 photocatalytic disinfection against gram-negative (*Salmonella typhimurium)* and gram-positive (*Listeria monocytogenes)* pathogen bacteria, the two most typical pathogens found in meat products, and found that TiO_2-based packaging material can be effectively used to prevent microbial contamination of meat-based products. Xing et al. (2012) observed the antibacterial activity of the TiO_2-polyethylene film against *Escherichia coli* and *Staphylococcus aureus* and found that the nanocomposite film exhibited more effective antibacterial activity for *Staphylococcus aureus*.

13.3.1.2 Silver (Ag)-Based Nanocomposite

Ag nanoparticles possess unique properties such as thermal conductivity, catalytic activity, and chemical stability; they also exhibit antimicrobial and antifungal properties which find their application in various fields such as textiles, medical applications, consumer products, etc. (Akter et al., 2018; González et al., 2015). De Moura et al. (2012) prepared nanocomposite film by blending hydroxypropyl methylcellulose (HPMC) with Ag nanoparticles and found that the developed nanocomposite film had excellent mechanical property and barrier property; it also exhibited antimicrobial property against *E. coli* culture. It is also revealed that the Ag nanoparticle, besides acting as an antimicrobial agent, increases the mechanical properties of the film as well. This strengthens the point that the Ag nanoparticle finds its potential application in the field of food packaging. Nanostructured LDPE/Ag_2O film bags developed by Zhou et al. (2011) have decreased microbial spoilage in apple slices. The quality of apple slices stored at 5°C in an LDPE/Ag_2O bag was acceptable after 12 days, while those packaged in conventional LDPE bags showed visible deterioration after six days. In a few cases, two or more antimicrobial reinforcements are also combined to produce the synergistic effect of microbial inactivation. For instance, LDPE (low-density polyethylene) nanocomposite packaging materials containing Ag and ZnO were found to inactivate *Lactobacillus plantarum* in orange juice at 4°C (Emamifar et al., 2011). Also, the Ag nanoparticle was found to possess higher antimicrobial activity in *Lactobacillus plantarum* compared with ZnO nanoparticles, especially for longer storage times. Similarly, orange juice packed in nanocomposites of LDPE filled with a powder containing TiO_2 and nanosilver was found to have longer shelf life without any deterioration (Emamifar et al., 2010). Zhou et al. (2012) prepared nanosilver/gelatin/carboxymethyl chitosan hydrogel by a green and simple fabrication method, i.e. radiation-induced reduction and cross-linking. The hydrogels exhibited an antibacterial effect on *Escherichia coli*.

13.3.1.3 Zinc Oxide (ZnO)-Based Nanocomposite

Zinc oxide (ZnO) is an inorganic compound widely used in various applications. ZnO has been regarded as a generally recognized as safe (GRAS) material by the Food and Drug Administration and is used as food additive. Among other known metals for antimicrobial activity, ZnO has been proved to be highly toxic against various microorganisms (Espitia et al., 2012; Stoimenov et al., 2002). Li et al. (2009) worked on ZnO-PVC for inactivation of foodborne pathogens and this proved to be effective against *E. coli* and *S. aureus*. Akbar and Anal (2014) used ZnO nanoparticles against two pathogens, *Salmonella typhimurium* and *Staphylococcus aureus,* and were found highly effective against both of them. ZnO nanoparticles loaded film of calcium alginate was prepared for active packaging application against the foodborne pathogens present in ready-to-eat poultry meat.

13.3.2 Barrier Properties

Barrier property is one of the key parameters in selecting the packaging material for food packaging application. Barrier properties denote permeability of gases such as O_2, CO_2 and N_2, water vapor, and aroma compounds which determine the quality of the food. For instance, permeability to oxygen facilitates degradation through mechanisms such as corrosive phenomena, oxidations, and modification of organoleptic properties (López et al., 2015; Russo et al., 2006). Knowledge of the diffusion/permeation behaviors of these molecules through the packaging material is necessary to design novel packaging material. Diffusion of gas/water vapor across the packaging film is influenced by the following factors (Siracusa, 2012):

- film structure
- film permeability properties to a specific substance
- physical properties of the film such as thickness and area
- conditions of the packaging material and the environment (temperature)
- difference in pressure and concentration gradient across the film

Theoretically, the permeation of gas across the packing film is described by the diffusion model: Henry's law and Fick's law relate the permeation rate with the area and thickness of the packaging film. Figure 13.4 describes the permeation mechanism of gas across the film in terms of thickness (l), pressure (p) and concentration (c).

The permeation of gas/vapor across the film (one-dimensional diffusion) can be described by Fick's first law:

$$J = -D \cdot \Delta C \tag{13.1}$$

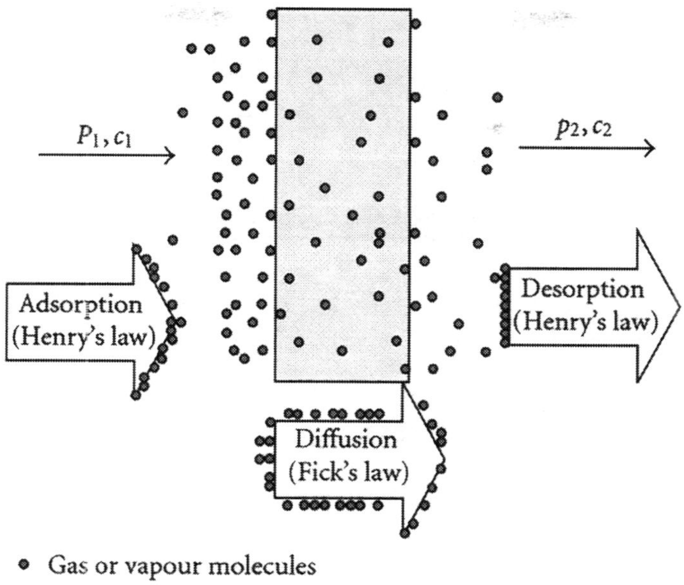

FIGURE 13.4 Mechanism of gas/vapor permeation across the packaging film. (Siracusa, 2012.)

where:

- J is the diffusion flux (mol cm^{-2} s^{-1})
- D is the diffusion coefficient (cm^2/s)
- ΔC is the concentration gradient (mol/cm^3)

If the diffusion is in steady state, then the equilibrium relationship between the concentration of sorbed gas at the interfaces and partial pressure will be considered through Henry's law (Li and Henley, 1964):

$$C = kp \tag{13.2}$$

where:

- C is the solubility of a gas at a fixed temperature in a particular solvent
- k is Henry's law constant
- p is the partial pressure of the gas

Bio-based packaging materials such as starch, cellulose, chitin, and biodegradable synthetic polymers such as polycaprolactone and polylactic acid are getting attention in recent days. However, the application of these natural polymers in the packaging industry is limited due to their poor barrier properties and weak mechanical properties (Rhim, 2007). Incorporation of nanoreinforcements in the polymeric film was found to increase the mechanical and barrier properties, which lead to the development of nanocomposite packaging material (Khan et al., 2012). Nanocomposite technology using nanofillers such as multiwalled carbon nanotubes (Paszkiewicz et al., 2017), nanoclays (Staroszczyk et al., 2016), and chitin nanoparticles (Sahraee et al., 2017) has been reported as an effective way to improve thermal stability and enhance mechanical and barrier properties. Similarly, the addition of cellulose nanofibers to the starch nanocomposites (reinforced thermoplastic starch nanocomposite) was found to increase the barrier properties (water vapor permeability) of the composite (Babaee et al., 2015).

Hosseini et al. (2015) reported excellent barrier properties on chitosan nanoparticle-reinforced bionanocomposite films against UV light and induction of lipid oxidation in the food system. In a

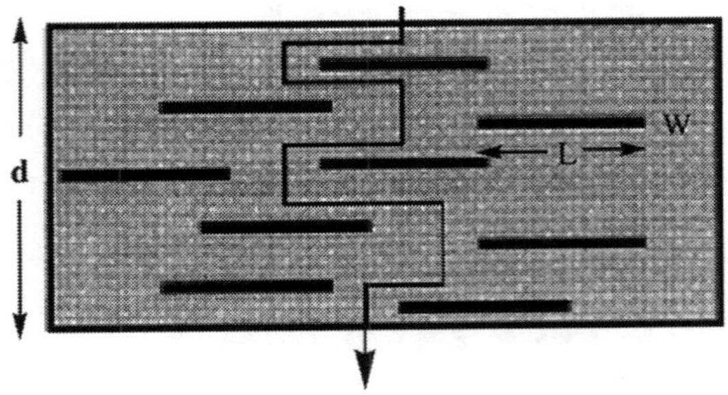

FIGURE 13.5 Tortuous diffusive pathway in nanocomposites. (Adame and Beall, 2009.)

recent study, Babaei-Ghazvini et al. (2018) reported starch/kefiran/ZnO nanocomposite films for UV-protective property and reduced water vapor permeability. The developed films were better suited for food packaging application, not only because of their moisture barrier property but also due to their ability to protect against UV light and transparency in the visible range. It was reported that the moisture barrier property of the film was due to the formation of a hydrogen bond between the nanoparticle and the polymer matrix.

In general, incorporation of nanoreinforcements enhances the barrier properties due to the formation of tortuous diffusive pathway (Azeredo, 2009) (Figure 13.5). In food packaging application, nanoclays, especially montmorillonite, are widely explored due to their nano-sized thickness and macro-sized length. Hence, migration into the food system is prevented. Avella et al. (2005) developed biodegradable starch-MMT nanocomposite and proved that migration of clay into contact food material and stimulants was within the regulatory limit of the European Commision directive. In a polymer matrix, the layered silicates are distributed in tactoid (immiscible clay and polymer structures), intercalated (penetration of polymer into the clay with ordered matrix), or exfoliated (penetration of polymer matrix into the delaminated clay with randomly dispersed matrix) form (Figure 13.6). Among these, the exfoliated structure is proved to improve barrier properties significantly (Adame and Beall, 2009; Alexandre and Dubois, 2000).

Liu et al. (2016) prepared hydroxypropyl starch/Na-MMT with different concentrations of MMT and found that water vapor permeability was lower at the threshold MMT concentration of 3%, above which water vapor permeability increased due to the exfoliation of Na-MMT at 3%. Similarly, Saha et al. (2016) found that methylcellulose and pectin films, when incorporated with 3% MMT, showed enhanced barrier properties. Chen et al. (2015) developed hemicelluloses-MMT-based hybrid films with polyvinyl alcohol and chitin nanowhiskers as fillers. The study found that the barrier properties of the hemicelulose-MMT-PVA hybrid films were significantly better compared to those of plain hemicelluloses-MMT films.

13.3.2.1 Prediction of Barrier Properties

The improved barrier properties of biopolymers with the addition of silicate layered clays has been initially explained by Nielsen (1967). It is based on the tortuous pathway that a permeant follows in a polymer matrix and according to the theory, the path length increases as a function of aspect ratio and volume fraction. Moreover, the theory assumes complete and uniform dispersion of clay platelets parallel to the polymer matrix and hence the maximum possible permeability achieved can only be predicted through this theory. Since deviations from the assumptions occur in the experimental condition, the predicted value is greater than the obtained experimental permeability value. Later, Bharadwaj (2001) modified the model and studied the effect of length, concentration, orientation, and level of aggregation of the layered silicates on the relative permeability. The study concluded that the longer silicates can effectively reduce the relative permeability of the polymer.

Beall (2000) proposed a new "conceptual model" that considers the polymer-clay interface as well as the tortuous pathway for the influence of clay on the barrier properties of the polymer.

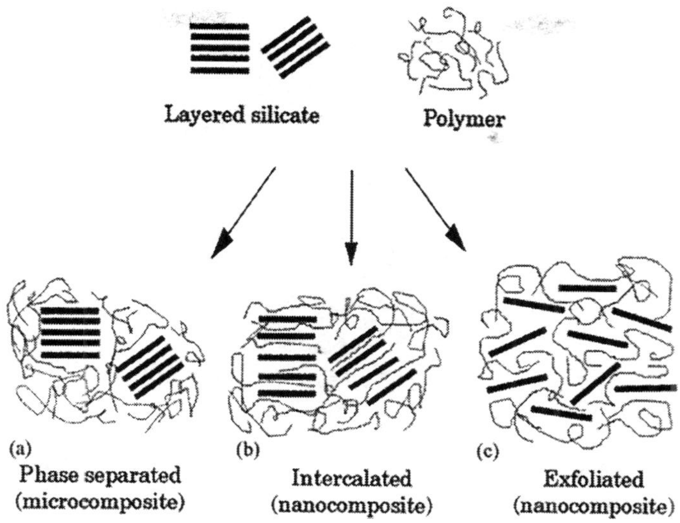

FIGURE 13.6 Schematic representation of various composites obtained from interaction of polymers and layered silicates. (Alexandre and Dubois, 2000.)

The model describes surface modifier region, constrained, and unconstrained polymer region around the clay platelets. According to the model, the surface modifier region of the polymer is the area that binds with the clay. The unconstrained polymer is not taken into consideration as it is not affected by the clay; whereas the constrained polymer region is the region (50–100 nm) that is in direct contact with the clay. As a consequence, the free volume of the polymer is reduced and hence the diffusion coefficient. This eventually improves the barrier property of the film. Thus, the "conceptual model" introduces a correction factor to the Nielsen model (Adame and Beall, 2009). Statler and Gupta (2007) proved that the significant difference between the experimental and predicted relative permeability value was because of not considering the structural orientation of the clay silicates. Through finite element model, the study proved that platelet-platelet overlap can lead to improved moisture barrier property as predicted by the mathematical equations. The experimental barrier properties of epoxy/zirconium phosphate as a function of exfoliation level and aspect ratio were correlated with several models such as Nielsen, Cussler (regular array and random array), Gusev-Lusti, Fredrickson-Bicerano, and Bharadwaj. Among the predicted values of different models, the Gusev-Lusti model and the Nielsen model were found to correlate with the experimental readings (Sun et al., 2008). Moreover, the proposed theories were only considering the 2D structures of nanoclay. Minelli et al. (2008) developed a numerical algorithm based on the finite volume method and found that considering 2D structures overpredicts the barrier properties, whereas 3D structures can be suitably correlated with the experimental data.

13.4 COMMERCIAL ASPECTS OF FOOD PACKAGING

Nanocomposites have attracted significant research interest due to their potential application in various industrial sectors including transportation, defense systems, cosmetics, pharmaceutical delivery system, energy storage, food packaging, etc. The market for nanocomposites is growing rapidly and is expected to reach above \$5,100 million by 2020 and \$7,475 million by 2022. Clay-based nanocomposites segment was reported as a global player in the nanocomposite market in 2017 (Plastemart, 2016).

Biodegradable and bio-based plastic materials have the potential to move the packaging industry from conventional packing to the next level of intelligent packaging and biodegradation packaging. Recently, the European Commission released its communication "A European Strategy for Plastics

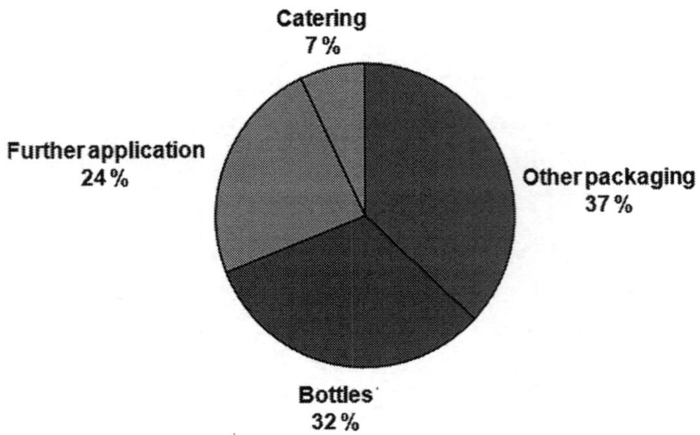

FIGURE 13.7 Global market segments of bio-plastics. (European Bioplastics, https://www.european-bioplastics.org.)

in a Circular Economy" to ensure all plastic packaging material in the European Union market should be reusable or recycled by 2030 and making the European plastics system more resource efficient. The application of bioplastics in different packing sectors is given in Figure 13.7. In 2007, the global market share of starch (bioplastics) was 50% followed by polylactic acid (PLA). Global opportunity analysis and industry forecast reported that the starch-based bioplastic market is expected to reach USD 561 million by 2023 with the compound annual growth rate of 3.7% from 2017 to 2023.

Currently, bioplastics represent about 1% of the about 320 million tonnes of plastic produced annually. However, the demand and market for bioplastic are rising and the global bioplastic production is predicted to reach 2.44 million tonnes in 2022.

13.5 CONCLUSION AND FUTURE TRENDS

The increasing scientific and industrial interest on the applications of nanomaterials in the packaging sector led to the development of numerous reinforcement materials. To protect food products from external factors and to maintain the quality and safety of the food packed inside, it is necessary to satisfy the mechanical and barrier properties of nanocomposite films. Though the nanocomposite is the outcome of the recent research with validated results, there is still a question on safety issues with the possibility of diffusion/migration of nanomaterials from the packaging material to the food. There is limited scientific data available on the migration of nanoparticles from the packaging material. Future research should be focused on the potential toxicity and improvement of water resistance characteristics of the nanocomposite materials as well as environmental safety.

REFERENCES

Abdollahi, Mehdi, Mehdi Alboofetileh, Masoud Rezaei, and Rabi Behrooz. 2013. "Comparing physico-mechanical and thermal properties of alginate nanocomposite films reinforced with organic and/or inorganic nanofillers." *Food Hydrocolloids,* 32(2):416–424.

Adame, Daniel and Gary W. Beall. 2009. "Applied clay science direct measurement of the constrained polymer region in polyamide/clay nanocomposites and the implications for gas diffusion." *Applied Clay Science*, 42(3–4):545–552.

Ahmed, Jasim and Sunil K. Varshney. 2011. "Polylactides-chemistry, properties and green packaging technology: A review." *International Journal of Food Properties,* 14(1):37–58.

Akbar, Ali and Anil Kumar Anal. 2014. "Zinc oxide nanoparticles loaded active packaging, a challenge study against salmonella typhimurium and staphylococcus aureus in ready-to-eat poultry meat." *Food Control*, 38(1):88–95.

Akter, Mahmuda et al. 2018. "A systematic review on silver nanoparticles-induced cytotoxicity: Physicochemical properties and perspectives." *Journal of Advanced Research*, 9:1–16.

Alboofetileh, Mehdi, Masoud Rezaei, Hedayat Hosseini, and Mehdi Abdollahi. 2013. "Effect of montmorillonite clay and biopolymer concentration on the physical and mechanical properties of alginate nanocomposite films." *Journal of Food Engineering*, 117(1):26–33.

Alexandre, Michael and Philippe Dubois. 2000. "Polymer-layered silicate nanocomposites: Preparation, properties and uses of a new class of materials." *Materials Science and Engineering R*, 28(March):1–63.

Anandharamakrishnan, Chinnaswamy. 2014a. "Electrospraying and electrospinning techniques for nanoencapsulation." In *Techniques for Nanoencapsulation of Food Ingredients*, 43–49. New York: Springer.

Anandharamakrishnan, Chinnaswamy. 2014b. "Liquid-based nanoencapsulation techniques." In *Techniques for Nanoencapsulation of Food Ingredients*, 29–41. New York: Springer.

Anandharamakrishnan, Chinnaswamy. 2014c. "Techniques for formation of nanoemulsions." In *Techniques For Nanoencapsulation Of Food Ingredients*, 7–16. New York: Springer.

Anandharamakrishnan, Chinnaswamy. and Kolli Usha Kiran. 2014. "Bionanocomposites and their potential applications in food packaging." In *Polymers for Packaging Applications edited by S.* Alavi et al., 229-262. Boca Raton: CRC Press.

Anandharamakrishnan, Chinnaswamy and S. Padma Ishwarya. 2015a. "Introduction to encapsulation of food ingredients." *Spray Drying Techniques for Food Ingredient Encapsulation,* 37–64. Chichester: Wiley.

Anandharamakrishnan, Chinnaswamy and S. Padma Ishwarya. 2015b. "Spray drying for nanoencapsulation of food components". *Spray Drying Techniques for Food Ingredient Encapsulation,* 180–197. Chichester: Wiley.

Angellier, Hélène, Luc Choisnard, Sonia Molina-Boisseau, Patrick Ozil, and Alain Dufresne. 2004. "Optimization of the preparation of aqueous suspensions of waxy maize starch nanocrystals using a response surface methodology." *Biomacromolecules*, 5(4):1545–1551.

Auras, Rafael, Bruce Harte, and Susan Selke. 2004. "An overview of polylactides as packaging materials." *Macromolecular Bioscience,* 4(9):835–864.

Avella, Maurizio et al. 2005. "Biodegradable starch/clay nanocomposite films for food packaging applications." *Food Chemistry*, 93(3):467–474.

Babaee, Mehran, Mehdi Jonoobi, Yahya Hamzeh, and Alireza Ashori. 2015. "Biodegradability and mechanical properties of reinforced starch nanocomposites using cellulose nanofibers." *Carbohydrate Polymers,* 132:1–8.

Babaei-Ghazvini, Amin, Iman Shahabi-Ghahfarrokhi, and Vahid Goudarzi. 2018. "Preparation of UV-protective starch/Kefiran/ZnO nanocomposite as a packaging film: characterization." *Food Packaging and Shelf Life,* 16(December 2017):103–111.

Bawa, Raj, T. S. R. Bawa, Stephen B. Maebius, Ted Flynn, and Chiming Wei. 2005. "Protecting new ideas and inventions in nanomedicine with patents." *Nanomedicine: Nanotechnology, Biology and Medicine.* 1(2):150–158.

Beall, G. W. 2000. "New conceptual model for interpreting nanocomposite behavior." In *Polymer Clay Nanocomposites*, edited by T. J. Pinnavaia and G. W. Beall, 267–279. Hoboken: John Wiley & Sons, Ltd.

Bharadwaj, R. K. 2001. "Modeling the barrier properties of polymer-layered silicate nanocomposites." *Macromolecules*, 34:9189–9192.

Bhushani, Anu and Chinnaswamy Anandharamakrishnan. 2017. "Food-grade nanoemulsions for protection and delivery of nutrients." In *Nanoscience in Food and Agriculture Vol 24*, edited by S. Ranjan, N. Dasgupta, and E. Lichtfouse, 99–139. Switzerland: Springer.

Bhushani, J. Anu, Nawneet Kumar Kurrey, and Chinnaswamy Anandharamakrishnan. 2017a. "Nanoencapsulation of green tea catechins by electrospraying technique and its effect on controlled release and in-vitro permeability." *Journal of Food Engineering*, 199:82–92.

Bhushani, Anu, Udayakumar Harish, and Chinnaswamy Anandharamakrishnan. 2017b. "Nanodelivery of nutrients for improved bioavailability." In *Nutrient Delivery*, edited by A. M. Grumezescu, 369–411. Amsterdam: Elsevier.

Bruna, Julio, Maria Jose Galotto, A. Guarda, Francisco Rodríguez. 2014. "A novel polymer based on MtCu2+/cellulose acetate with antimicrobial activity." *Carbohydrate Polymers*, 102:317–323.

Busolo, Maria A., Patricia Fernandez, Maria J. Ocio, and Jose M. Lagaron. 2010. "Novel silver-based nanoclay as an antimicrobial in polylactic acid food packaging coatings." *Food Additives and Contaminants - Part A Chemistry, Analysis, Control, Exposure and Risk Assessment,* 27(11):1617–1626.

Butler, B. L., P. J. Vergano, R. F. Testin, J. M. Bunn, and J. L. Wiles. 1996. "Mechanical and barrier properties of edible chitosan films as affected by composition and storage." *Journal of Food Science*, 61(5):953–596.

Byun, Youngjae and Young Teck Kim. 2014. "Bioplastics for food packaging: Chemistry and physics." In *Innovations in Food Packaging*, edited by J. H. Han, 353–368, Amsterdam: Elsevier.

Cao, X., Y. Chen, P. R. Chang, A. D. Muir, and G. Falk. 2008. "Starch-based nanocomposites reinforced with flax cellulose nanocrystals." *Express Polymer Letters*, 2(7):502–510.

Castro-Mayorga, J. L., M. J. Fabra, and J. M. Lagaron. 2015. "Stabilized nanosilver based antimicrobial poly(3-hydroxybutyrate-co-3-hydroxyvalerate) nanocomposites of interest in active food packaging." *Innovative Food Science and Emerging Technologies,* 33:524–533.

Chaichi, Maryam, Maryam Hashemi, Fojan Badii, and Abdorreza Mohammadi. 2017. "Preparation and characterization of a novel bionanocomposite edible film based on pectin and crystalline nanocellulose." *Carbohydrate Polymers,* 157:167–175.

Chang, Jin-hae, Yeong Uk An, and Gil Soo Sur. 2003. "Poly(lactic acid) nanocomposites with various organoclays. I. Thermomechanical properties, morphology, and gas permeability." *Journal of Polymer Science: Part B: Polymer Physics*, 41(1):94–103.

Chatterjee, Arijit Kumar, Ruchira Chakraborty, and Tarakdas Basu. 2014. "Mechanism of antibacterial activity of copper nanoparticles." *Nanotechnology,* 25(13): 135101.

Chen, Ge-gu et al. 2015. "Hemicelluloses/montmorillonite hybrid films with improved mechanical and barrier properties." *Scientific Reports,* 5: 16405.

Chen, Chuchu et al. 2018. "High strength gelatin-based nanocomposites reinforced by surface-deacetylated chitin nanofiber networks." *Carbohydrate Polymers*, 195:387–392.

Chivrac, Frederic, Eric Pollet, and Luc Averous. 2009. "Progress in nano-biocomposites based on polysaccharides and nanoclays." *Materials Science and Engineering R,* 67(December):1–17.

Clarke, Anthony J. 1997. *Biodegradation of Cellulose: Enzymology and Biotechnology*. Lancaster: Technomic Pub. Co.

Corcinone, Carola Esposito, Fabrizio Freuli, and Alfonso Maffezzoli. 2007. "The aspect ratio of epoxy matrix nanocomposites reinforced with graphene stacks." *Polymer Engineering and Science*, 47:21–25.

Costa, Élia Karina de Carvalho, Carolina Oliveira de Souza, Jania Betânia Alves da Silva, and Janice Izabel Druzian. 2017. "Hydrolysis of part of cassava starch into nanocrystals leads to increased reinforcement of nanocomposite films." *Journal of Applied Polymer Science*, 134(41):1–9.

Croisier, Florence and Christine Jérôme. 2013. "Chitosan-based biomaterials for tissue engineering." *European Polymer Journal*, 49(4):780–792.

de Azeredo, Henriette M. C. 2009. "Nanocomposites for food packaging applications." *Food Research International*, 42(9):1240–1253.

De Moura, Márcia R., Luiz H. C. Mattoso, and Valtencir Zucolotto. 2012. "Development of cellulose-based bactericidal nanocomposites containing silver nanoparticles and their use as active food packaging." *Journal of Food Engineering*, 109(3):520–524.

Dehnad, Danial, Habibollah Mirzaei, Zahra Emam-Djomeh, Seid-Mahdi Jafari, and Saeed Dadashi. 2014. "Thermal and antimicrobial properties of chitosan – nanocellulose films for extending shelf life of ground meat." *Carbohydrate Polymers,* (June 2015):148–154.

Dias, Marali Vilela et al. 2014. "Development of chitosan/montmorillonite nanocomposites with encapsulated α-tocopherol." *Food Chemistry,* 165:323–329.

Diéguez, María Cristina Villamán, Franciele Maria Pelissari, Paulo José, and Florencia Cecilia Menegalli. 2015. "Effect of process conditions on the production of nanocomposite films based on amaranth flour and montmorillonite." *LWT - Food Science and Technology*, 6(1):70–79.

Echeverria, Ignacio, Patricia Eisenberg, and Adriana N. Mauri. 2014. "Nanocomposites films based on soy proteins and montmorillonite processed by casting." *Journal of Membrane Science,* 449(June 2017):15–26.

Echeverría, Ignacio, María Elvira López-Caballero, María Carmen Gómez-Guillén, Adriana Noemi Mauri, and María Pilar Montero. 2018. "Active nanocomposite films based on soy proteins-montmorillonite-clove essential oil for the preservation of refrigerated bluefin tuna (thunnus thynnus) fillets." *International Journal of Food Microbiology,* 266:142–149.

El-wakil, Nahla A., Enas A. Hassan, Ragab E. Abou-zeid, and Alain Dufresne. 2015. "Development of wheat gluten/nanocellulose/titanium dioxide nanocomposites for active food packaging." *Carbohydrate Polymers,* 124:337–346.

Emamifar, Aryou, Mahdi Kadivar, Mohammad Shahedi, and Sabihe Soleimanian-Zad. 2010. "Evaluation of nanocomposite packaging containing Ag and ZnO on shelf life of fresh orange juice." *Innovative Food Science and Emerging Technologies*, 11(4):742–748.

Emamifar, Aryou, Mahdi Kadivar, Mohammad Shahedi, and Sabihe Soleimanian-zad. 2011. "Effect of nanocomposite packaging containing Ag and ZnO on inactivation of lactobacillus plantarum in orange juice." *Food Control,* 22(3–4):408.

Espitia, Paula Judith Perez et al. 2012. "Zinc oxide nanoparticles: Synthesis, antimicrobial activity and food packaging applications." *Food and Bioprocess Technology*, 5(5):1447–1464.

EU (European Commission). 2018. *Proposal for a Directive of the European Parliament and of the Council on the Reduction of the Impact of Certain Plastic Products on the Environment.*

Ezhilarasi, P. N., P. Karthik, N. Chhanwal, and C. Anandharamakrishnan. 2013. "Nanoencapsulation techniques for food bioactive components: A review." *Food and Bioprocess Technology,* 6(3):628–647.

Ferfera-Harrar, Hafida and Nassima Dairi. 2015. "Green nanocomposite films based on cellulose acetate and biopolymer-modified nanoclays: Studies on morphology and properties." *Iranian Polymer Journal,* (November 2014):917–931.

Fernandez-Saiz, P., Jose Maria Lagaron, P. Hernandez-Muñoz, and M. J. Ocio. 2008. "Characterization of antimicrobial properties on the growth of S. aureus of novel renewable blends of gliadins and chitosan of interest in food packaging and coating applications." *International Journal of Food Microbiology*, 124(1):13–20.

Forssell, P., Ritta Partanen, M. Lahelin, and P. Myllarinen. 2002. "Oxygen permeability of amylose and amylopectin films." *Carbohydrate Polymers*, 47(2):125–129.

Fortunati, Elena, Ilaria Armentano, Qi Zhou, and Jose M. Kenny. 2012. "Multifunctional bionanocomposite films of poly(lactic acid), cellulose nanocrystals and silver nanoparticles." *Carbohydrate Polymers*, 87(2):1596–1605.

Fortunati, Elena, M. Peltzer, Ilario Armentano, A. Jiménez, Jose M. Kenny. 2013. "Combined effects of cellulose nanocrystals and silver nanoparticles on the barrier and migration properties of PLA nano-biocomposites." *Journal of Food Engineering*, 118(1):117–124.

Franci, Gianluigi, et al. 2015. "*Silver nanoparticles as potential antibacterial agents.*" *Molecules* 20(5): 8856–8874.

Ge, Shengju et al. 2017. "Characterizations of pickering emulsions stabilized by starch nanoparticles: Influence of starch variety and particle size." *Food Chemistry*, 234:339–347.

George, Johnsy and Siddaramaiah. 2012. "High Performance Edible Nanocomposite Films Containing Bacterial Cellulose Nanocrystals." *Carbohydrate Polymers* 87(3):2031–2037.

Georgiev, S. and Tereza Dekova. 2011. "Gluten: A natural biopolymer." In *Biopolymers: Biomedical and Environmental Applications*, edited by S. Kalia and L. Avérous, 377–401, Hoboken: John Wiley & Sons, Ltd.

Ghelejlu, Sara Beigzadeh, Mohsen Esmaiili, and Hadi Almasi. 2016. "Characterization of chitosan–nanoclay bionanocomposite active films containing milk thistle extract." *International Journal of Biological Macromolecules*, 86:613–621.

González, Agustín and Cecilia I. Alvarez Igarzabal. 2015. "Nanocrystal-reinforced soy protein films and their application as active packaging." *Food Hydrocolloids,* 43:777–784.

Gupta, Pratima and Kush Kumar Nayak. 2014. "Characteristics of protein-based biopolymer and its application." *Polymer Engineering and Science,* 47:21–25.

Guz, Lucas, L. Famá, R. Candal, and Silvia Goyanes. 2017. "Size effect of ZnO nanorods on physicochemical properties of plasticized starch composites." *Carbohydrate Polymers*, 157:1611–1619.

Haafiz, M. K. Mohamad, Azman Hassan, Zainoha Zakaria, I.M. Inuwa, M.D. Saiful Islam, and Mohammad Jawaid. 2013. "Properties of polylactic acid composites reinforced with oil palm biomass microcrystalline cellulose." *Carbohydrate Polymers*, 98(1):139–145.

Hao, Yacheng, Yun Chen, Qian Li, and Qunyu Gao. 2018. "Preparation of starch nanocrystals through enzymatic pretreatment from waxy potato starch." *Carbohydrate Polymers,* 184:171–177.

Hon, David N. S. 1989. "Cellulosic adhesives." In *Adhesives from Renewable Resources*, edited by A. H. Hemingway Richard, W. Conner and S. J. Branham, 289–304, Washington, DC: American Chemical Society.

Hosseini, Seyed Fakhreddin, Masoud Rezaei, Mojgan Zandi, and Farhid Farahmandghavi. 2015a. "Preparation and characterization of chitosan nanoparticles-loaded fish gelatin-based edible films." *Journal of Food Process Engineering*: 39 (5): 521-530.

Hosseini, Seyed Fakhreddin, Masoud Rezaei, Mojgan Zandi, and Farhid Farahmandghavi. 2015b. "Fabrication of bio-nanocomposite films based on fish gelatin reinforced with chitosan nanoparticles." *Food Hydrocolloids*, 44:172–182.

Jahed, Eisa, Mohammad Alizadeh, Hadi Almasi, and Reza Hasanzadeh. 2017. "Physicochemical properties of carum copticum essential oil loaded chitosan films containing organic nanoreinforcements." *Carbohydrate Polymers,* 164:325–338.

Jin, T. and Joshua B. Gurtler. 2011. "Inactivation of salmonella in liquid egg albumen by antimicrobial bottle coatings infused with allyl isothiocyanate, nisin and zinc oxide nanoparticles." *Journal of Applied Microbiology,* 110(3):704–712.

Jurado, Encarnación, Fernando Camacho, German Luzón, and Jose Vicaria. 2004. "Kinetic models of activity for β-galactosidases: Influence of pH, ionic concentration and temperature." *Enzyme and Microbial Technology,* 34 (1):33–40.

Kanmani, Paulraj and Jong Whan Rhim. 2014. "Physical, mechanical and antimicrobial properties of gelatin based active nanocomposite films containing AgNPs and nanoclay." *Food Hydrocolloids,* 35:644–652.

Kathiravan, T., J. A. Moses, and Chinnaswamy Anandharamakrishnan. 2017. "Nano food fortification." *Everyman's Science,* LI(4):171–175.

Karthik, P., P.N. Ezhilarasi, and Chinnaswamy Anandharamakrishnan. 2017. "Challenges associated in stability of food grade nanoemulsions." *Critical Reviews in Food Science and Nutrition,* 57(7):1435–1450.

Khan, Avik et al. 2012. "Mechanical and barrier properties of nanocrystalline cellulose reinforced chitosan-based nanocomposite films." *Carbohydrate Polymers*, 90(4):1601–1608.

Khatkar, Bhupendar, Alan Bell, and J. D. Schofield. 1995. "The dynamic rheological properties of glutens and gluten sub-fractions from wheats of good and poor bread making quality." *Journal of Cereal Science*, 22(1):29–44.

Klangmuang, Ploy and Rungsinee Sothornvit. 2016. "Barrier properties, mechanical properties and antimicrobial activity of hydroxypropyl methylcellulose-based nanocomposite films incorporated with thai essential oils." *Food Hydrocolloids,* 61:609–616.

Kwon, Seulki, Hyun Woo Cho, Gyemin Gwon, Heesuk Kim, and Bong June Sung. 2016. "Effects of shape and flexibility of conductive fillers in nanocomposites on percolating network formation and electrical conductivity." *Physical Review E*, 93(3):1–9.

Le Corre, Déborah and Hélène Angellier-Coussy. 2014. "Preparation and application of starch nanoparticles for nanocomposites: A review." *Reactive and Functional Polymers,* 85:97–120.

Li, N. N. and E. J. Henley. 1964. "Permeation of gases through polyethylene films at elevated pressures." *A.I.Ch.E. Journal*, 10(5):666–670.

Li, Xihong, Yage Xing, Yunhong Jiang, Yulong Ding, and Weili Li. 2009. "Antimicrobial activities of ZnO powder-coated PVC film to inactivate food pathogens." *International Journal of Food Science and Technology,* 44(11):2161–2168.

Li, Kuang, Shicun Jin, Hui Chen, Jing He, and Jianzhang Li. 2017. "A high-performance soy protein isolate-based nanocomposite film modified with microcrystalline cellulose and Cu and Zn nanoclusters." *Polymers*, 9(5):1–12.

Lin, Ning, Jin Huang, Peter R. Chang, Debbie P. Anderson, and Jiahui Yu. 2011. "Preparation, modification, and application of starch nanocrystals in nanomaterials: A review." *Journal of Nanomaterials,* 2011:20.

Liu, Siyuan et al. 2016. "Effect of film multi-scale structure on the water vapor permeability in hydroxypropyl starch (HPS)/Na-MMT nanocomposites." *Carbohydrate Polymers*, 154:186–193.

Llorens, Amparo, Elsa Lloret, Pierre A. Picouet, Raul Trbojevich, and Avelina Fernandez. 2012. "Metallic-based micro and nanocomposites in food contact materials and active food packaging." *Trends in Food Science and Technology*, 24(1):19–29.

Lluis, Cabedo, Jose Luis Feijoo, Maria Pilar Villanueva, Jose Maria Lagaron, and Enrique Gimenez. 2006. "Optimization of biodegradable nanocomposites based on APLA/PCL blends for food packaging applications." *Macromolecular Symposia,* 233:191–197.

Long, Men et al. 2014. "Performance and mechanism of standard nano-TiO_2(P-25) in photocatalytic disinfection of foodborne microorganisms – salmonella typhimurium and listeria monocytogenes." *Food Control*, 39(1):68–74.

López, Olivia V., Luciana A. Castillo, M. Alejandra García, Marcelo A. Villar, and Silvia E. Barbosa. 2015. "Food packaging bags based on thermoplastic corn starch reinforced with talc nanoparticles." *Food Hydrocolloids*, 43:18–24.

Mali, Suzana et al. 2004. "Barrier, mechanical and optical properties of plasticized yam starch films." *Carbohydrate Polymers,* 56:129–135.

Mallakpour, Shadpour, and Vajiheh Behranvand. 2016. "Nanocomposites based on biosafe nano ZnO and different polymeric matrixes for antibacterial, optical, thermal and mechanical applications." *European Polmer Journal,* 84: 377–403.

Mani, Ramaswamy and Mrinal Bhattacharya. 1998. "Properties of injection moulded starch/synthetic polymer blends-III. Effect of amylopectin to amylose ratio in starch." *European Polymer Journal,* 34(10):1467–1475.

Martucci, J. F., A. Vázquez, R. A. Ruseckaite. 2007. "Nanocomposites based on gelatin and montmorillonite: Morphological and thermal studies." *Journal of Thermal Analysis and Calorimetry,* 89(1):117–122.

Meira, Stela Maris Meister, Gislene Zehetmeyer, Júlia Orlandini Werner, and Adriano Brandelli. 2016. "A novel active packaging material based on starch-halloysite nanocomposites incorporating antimicrobial peptides." *Food Hydrocolloids*, 63:560–5671.

Minelli, Matteo, Marco Giacinti Baschetti, Ferruccio Doghieri, and Environment Technology. 2008. "A 3-D Approach to model diffusion in randomly distributed nanocomposite." *The 2008 Annual Meeting,* 1–8.

Mishnaevsky, Leon, Alessandro Pontefisso, and Gaoming Dai. 2016. "Hierarchical nanoreinforced composites: Computational analysis of damage mechanisms." *IOP Conference Series: Materials Science and Engineering*, 139(1):1–6.

Mochizuki, Masatsugu. 2009. "Synthesis, properties and structure of polylactic acid fibres." In *Handbook of Textile Fiber Structure: Vol. 1: Fundamentals and Manufactured Polymer Fibers,* edited by S. J. Eichhorn, J. W. S. Hearle, M. Jaffe and T. Kikutani. Cambridge: Woodhead Publishing Limited.

Moon, Robert J., Ashlie Martini, John Nairn, John Simonsen, and Jeff Youngblood. 2011. *Cellulose Nanomaterials Review: Structure, Properties and Nanocomposites*. Vol. 40.

Moro, Thaisa M. A., José L. R. Ascheri, Juan A. R. Ortiz, Carlos W. P. Carvalho, and Arturo Meléndez-Arévalo. 2017. "Bioplastics of native starches reinforced with passion fruit peel." *Food and Bioprocess Technology*, 10(10):1798–1808.

Mujeeb Rahman, P., V. M. Abdul Mujeeb, K. Muraleedharan, and Steni K. Thomas. 2016. "Chitosan/nano ZnO composite films: Enhanced mechanical, antimicrobial and dielectric properties." *Arabian Journal of Chemistry,* 11(1):120–127.

Murray, Kieran A., John A. Killion, Ian Major, and Luke M. Geever. 2015. "Thermal degradation of bio-nanocomposites." In *Thermal Degradation of Polymer Blends, Composites and Nanocomposites*, edited by P. M. Visakh and Y. Arao, 221–245. Cham: Springer.

Nakayama, Norio and Toyoharu Hayashi. 2007. "Preparation and characterization of poly(L-lactic acid)/TiO_2 nanoparticle nanocomposite films with high transparency and efficient photodegradability." *Polymer Degradation and Stability*, 92(7):1255–1264.

Ngah, W. S. Wan, L. C. Teong, and M. A. K. M. Hanafiah. 2011. "Adsorption of dyes and heavy metal ions by chitosan composites: A review." *Carbohydrate Polymers*, 83(4):1446–1456.

Nielsen, Lawrence E. 1967. "Models for the permeability of filled polymer systems." *Journal of Macromolecular Science: Part A-Chemistry*, A1(5):929–942.

Oleyaei, Seyed Amir, Hadi Almasi, Babak Ghanbarzadeh, and Ali Akbar Moayedi. 2016. "Synergistic reinforcing effect of TiO_2 and montmorillonite on potato starch nanocomposite films: thermal, mechanical and barrier properties." *Carbohydrate Polymers,* 152:253–262.

Oun, Ahmed A., and Jong-Whan Rhim. 2017. "Carrageenan-based hydrogels and films: Effect of ZnO and CuO nanoparticles on the physical, mechanical, and antimicrobial properties." *Food Hydrocolloids* 67: 45–53.

Pal, Sukdeb, Yu Kyung Tak, and Joon Myong Song. 2007. "Does the antibacterial activity of silver nanoparticles depend on the shape of the nanoparticle? A study of the gram-negative bacterium *Escherichia coli*." *Applied and Environmental Microbiology*, 290(42):1712–1720.

Park, S. K., D. H. Bae, and K. C. Rhee. 2000. "Soy protein biopolymers cross-linked with glutaraldehyde." *Journal of the American Oil Chemists' Society,* 77(8):879–884.

Parker, Kate, Jean-Philippe Garancher, Samir Shah, Stephanie Weal, and Fernyhough Alan. 2011. "Polylactic acid (PLA) foams for packaging application." In *Handbook of Bioplastics; Biocomposites Engineering Applications*, edited by S. Pilla, 161–176. Hoboken: Wiley.

Paszkiewicz, Sandra et al. 2017. "Comparative study on the properties of poly(trimethylene terephthalate)-based nanocomposites containing multi-walled carbon (MWCNT) and tungsten disulfide (INT-WS2) nanotubes." *Polymers for Advanced Technologies,* 28(6):645-657.

Pierro Di, Prospero, Angela Sorrentino, Loredana Mariniello, Concetta Valeria L. Giosafatto, and Raffaele Porta. 2011. "Chitosan/whey protein film as active coating to extend ricotta cheese shelf-life." *LWT - Food Science and Technology*, 44(10):2324–2327.

Plastemart. 2016. "Packaging sector – the highest consumer of polymer nanocomposites drives global demand." Retrieved June 19, 2018 (http://www.plastemart.com/plastic-technical-articles/Demand-from-construction-automotive-in-APAC-drives-global-growth-of-masterbatches/2314).

Prashanth, Harish, and R. N. Tharanathan. 2007. "Chitin/chitosan: modifications and their unlimited application potential – an overview." *Trends in Food Science and Technology,* 18(3):117–131.

Rafieian, Fatemeh, Mohammad Shahedi, Javad Keramat, and John Simonsen. 2014. "Mechanical, thermal and barrier properties of nano-biocomposite based on gluten and carboxylated cellulose nanocrystals." *Industrial Crops and Products*, 53:282–288.

Ramos, Betina Giehl Zanetti. 2011. "Biopolymers employed in drug delivery." In *Biopolymers : Biomedical and Environmental Applications*, edited by S. Kalia and L. Avérous, 559–574. Hoboken: Wiley.

Ramos, Marina, Arantzazu Valdés, Ana Beltrán, and María Garrigós. 2016. "Gelatin-based films and coatings for food packaging applications." *Coatings,* 6(4):41.

Ramos, Marina, Alfonso Jiménez, and María Carmen Garrigós. 2017. "Active nanocomposites in food contact materials." In *Nanoscience in Food and Agriculture*, edited by S. Ranjan, N. Dasgupta, and E. Lichtfouse, 1–44. Cham: Springer.

Rhim, Jong-Whan. 2007. "Natural biopolymer-based nanocomposite films for packaging." *Critical Reviews in Food Science and Nutrition,* 4(47):411–433.

Rhim, Jong-Whan, Hwanman Park, and Chang-Sik Ha. 2013. "Bio-nanocomposites for food packaging applications." *Progress in Polymer Science,* 38(10–11):1629–1652.

Rouhi, Jalal, Shahrom Mahmud, Nima Naderi, C. H. Raymond Ooi, and Mohamad Rusop Mahmood. 2013. "Physical properties of fish gelatin-based bio-nanocomposite films incorporated with ZnO nanorods." *Nanoscale Research Letters*, 8(1):1–6.

Rubilar, Javiera F., Davinia Candia, Angel Cobos, Olga Díaz, and Franco Pedreschi. 2016. "Effect of nanoclay and ethyl-Nα-dodecanoyl-l-arginate hydrochloride (LAE) on physico-mechanical properties of chitosan films." *LWT – Food Science and Technology*, 72:206–214.

Russo, G. M., G. P. Simon, and L. Incarnato. 2006. "Correlation between rheological, mechanical, and barrier properties in new copolyamide-based nanocomposite films." *Macromolecules*, 3855–3864.

Saba, Naheed, Paridah Tahir, and Mohammad Jawaid. 2014. "A review on potentiality of nano filler/natural fiber filled polymer hybrid composites." *Polymers*, 6:2247–2273.

Saha, Nayan Ranjan et al. 2016. "Studies on methylcellulose/pectin/montmorillonite nanocomposite films and their application possibilities." *Carbohydrate Polymers*, 136:1218–1227.

Sahraee, Samar, Babak Ghanbarzadeh, Jafar M. Milani, and Hamed Hamishehkar. 2017. "Development of gelatin bionanocomposite films containing chitin and ZnO nanoparticles." *Food and Bioprocess Technology,* 10(8):1441–1453.

Salehudin, Harfiz, E. Salleh, Ida Idayu Muhamad, and S. N. H. Mamat. 2014. "Starch-based biofilm reinforced with empty fruit bunch cellulose nanofibre." *Materials Research Innovations,* 18(Supplement 6): 322–325.

Sanchez, Clement, Beatriz Julia, Philippe Belleville, and Michael Popall. 2005. "Applications of hybrid organic – inorganic nanocomposites." *Journal of Material Chemistry,* 15:3559–3592.

Sanuja, S., A. Agalya, and M. J. Umapathy. 2015. "Synthesis and characterization of zinc oxide-neem oil-chitosan bionanocomposite for food packaging application." *International Journal of Biological Macromolecules*, 74:76–84.

Scott, Gerald. 1999. *Polymers and the Environment*. Cambridge: Royal Society of Chemistry.

Shankar, Shiv, Jeevan Prasad Reddy, Jong Whan Rhim, and Hee Yun Kim. 2015. "Preparation, characterization, and antimicrobial activity of chitin nanofibrils reinforced carrageenan nanocomposite films." *Carbohydrate Polymers,* 117:468–475.

Shankar, S., L. Jaiswal, and J. W. Rhim. 2016. "Gelatin-based nanocomposite films: potential use in antimicrobial active packaging." In *Antimicrobial Food Packaging,* edited by Jorge Barros-Velazquez. Amsterdam: Elsevier.

Silvestre, Clara, Donatella Duraccio, and Sossio Cimmino. 2011. "Food packaging based on polymer nanomaterials." *Progress in Polymer Science,* 36(12):1766–1782.

Sin, Lee Tin., Abdul Razak. Rahmat, and Wan Azian Wan Abdul. Rahman. 2012. *Polylactic Acid: PLA Biopolymer Technology and Applications*. Waltham: Elsevier.

Siracusa, Valentina. 2012. "Food packaging permeability behaviour: A report." Article ID 302029. *International Journal of Polymer Science*, 2012:1–11.

Siró, István, and David Plackett. 2010. "Microfibrillated cellulose and new nanocomposite materials: a review." *Cellulose* 17(3): 459–494.

Staroszczyk, Hanna, Edyta Malinowska-Pańczyk, Karolina Gottfried, and Ilona Kołodziejska. 2016. "Fish gelatin-nanoclay films. Part I: Effect of a kind of nanoclays and glycerol concentration on mechanical and water barrier properties of nanocomposites." *Journal of Food Processing and Preservation,* 41(5):1–7.

Statler, David L. and Rakesh K. Gupta. 2007. "A finite element analysis of the influence of morphology on barrier properties of polymer-clay nanocomposites." *Proceedings of the COMSOL Conference, Boston*.

Stoimenov, Peter K., Rosalyn L. Klinger, George L. Marchin, and Kenneth J. Klabunde. 2002. "Metal oxide nanoparticles as bactericidal agents." *Langmuir*, 18(17):6679–6686.

Sun, Luyi, W. Boo, A. Clearfield, H. Sue, and H. Q. Pham. 2008. "Barrier properties of model epoxy nanocomposites." *Journal of Membrane Science*, 318:129–136.

Telis, Vania Regina Nicoletti, ed. 2012. "Films and coatings produced from biopolymers and composites." In. *Biopolymer Engineering in Food Processing*, edited by V. R. N. Telis. Boca Raton: CRC Press.

Vejdan, Akbar, Seyed Mahdi Ojagh, Afshin Adeli, and Mehdi Abdollahi. 2016. "Effect of TiO_2 nanoparticles on the physico-mechanical and ultraviolet light barrier properties of fish gelatin/agar bilayer film." *LWT - Food Science and Technology,* 71:88–95.

Visakh, P. M. and Long Yu, eds. 2016. *Starch-Based Blends, Composites and Nanocomposites*. Cambridge: The Royal Society of Chemistry.

Vlacha, Maria et al. 2016. "On the efficiency of oleic acid as plasticizer of chitosan/clay nanocomposites and its role on thermo-mechanical, barrier and antimicrobial properties – comparison with glycerol." *Food Hydrocolloids*, 57:10–19.

Wang, Peilei et al. 2017. "Large-scale preparation of jute fiber reinforced starch-based composites with high mechanical strength and optimized biodegradability." *Starch-Starke,* Nov 1.

Williams, Peter A. and Glyn O. Phillips. 2012. *Gums and Stabilisers for the Food Industry*, p. 16. Cambridge: RSC Publishing.

Wu, Yun, Xianfeng Du, Honghua Ge, and Zhen Lv. 2011. "Preparation of microporous starch by glucoamylase and ultrasound." *Starch-Stärke,* 63 (4):217–225.

Xie, Fengwei et al. 2009. "Rheological properties of starches with different amylose/amylopectin ratios." *Journal of Cereal Science,* 49(3):371–377.

Xing, Yage et al. 2012. "Effect of TiO_2 nanoparticles on the antibacterial and physical properties of polyethylene-based film." *Progress in Organic Coatings,* 73(2–3):219–224.

Xu, Yixiang, Xi Ren, and Milford A. Hanna. 2006. "Chitosan/clay nanocomposite film preparation and characterization." *Journal of Applied Polymer Science*, 99(4):1684–1691.

Youssef, Ahmed M., Samah M. El-Sayed, Hoda S. El-Sayed, Heba H. Salama, and Alain Dufresne. 2016. "Enhancement of egyptian soft white cheese shelf life using a novel chitosan/carboxymethyl cellulose/zinc oxide bionanocomposite film." *Carbohydrate Polymers,* 151:9–19.

Zhou, J. J., S. Y. Wang, and S. Gunasekaran. 2009. "Preparation and characterization of whey protein film incorporated with TiO_2 nanoparticles." *Journal of Food Science,* 74(7):50–56.

Zhou, Marvellous, Collins Kimbeng, Serge Edme, Anna Hale, Ryan Viator, and Gillian Eggleston. 2010. "Sustainability of low starch concentrations in sugarcane through short-term optimized amylase and long-term breeding strategies". In *Sustainability of the Sugar and Sugar–Ethanol Industries*, edited by G. Eggleston, 229–250. Washington: ACS Publications.

Zhou, Ling, Sining Lv, Guiping He, Qiang He, and Bi Shi. 2011. "Effect of PE/Ag_2O nano-packaging on the quality of apple slices." *Journal of Food Quality*, 34(3):171–176.

Zhou, Ying et al. 2012. "Radiation synthesis and characterization of nanosilver/gelatin/carboxymethyl chitosan hydrogel." *Radiation Physics and Chemistry*, 81(5):553–560.

Zhu, Fan. 2015. "Impact of ultrasound on structure, physicochemical properties, modifications, and applications of starch." *Trends in Food Science and Technology,* 43 (1):1–17.

Zohri, Maryam et al. 2013. "Nisin-loaded chitosan/alginate nanoparticles: A hopeful hybrid biopreservative." *Journal of Food Safety,* 33:40–49.

Zolfi, Mohsen, Faramarz Khodaiyan, Mohammad Mousavi, and Maryam Hashemi. 2014. "The improvement of characteristics of biodegradable films made from kefiran – whey protein by nanoparticle incorporation." *Carbohydrate Polymers*, 109(January):118–125.

Section 3

Diagnostics and Characterization

14 Nanosensors for Food Contaminant Detection

Heera Jayan, L. Bhavani Devi, and C. Anandharamakrishnan

CONTENTS

14.1 INTRODUCTION

Food is a vehicle for the interaction of the human being with the environment, and it is the primary source of all nutrients to the human body. It can also act as a source of human exposure to infectious, toxic, and pharmacological agents (Stewart, 1987). Further, the analysis of food composition not only involves nutrients but also other contaminants that possess the potential to cause serious health problems. The analysis of nutrient contents in food gives an insight into the quantity of each component in food. Based on the amount of nutrients present, the quantity of food that should be consumed to meet the daily requirement of each nutrient can be fixed. Similarly, the detection of hazardous components in food allows us to determine whether the food is safe for consumption. There are fixed maximum limits for hazardous compounds that can be present in each food. If the quantity exceeds the maximum limit, then the food is considered unsafe.

According to the Food and Agriculture Organization of the United Nations (FAO), "Food safety is about handling, storing and preparing food to prevent infection and help to make sure that our food keeps enough nutrients for us to have a healthy diet"(FAO, 2018). Consumption of contaminated food causes serious diseases and malnutrition, particularly affecting infants, young children, and elder people. To ensure the health of consumers, the safety of food is a major concern for the food industry. Maintaining the safety and quality of food products throughout the processing steps is important to produce a good quality product.

14.2 CONTAMINANTS IN FOOD

Food contamination is the presence of harmful substances in a food product that causes hazardous health effects in the human body. Globalization and international agribusiness caused food contamination to spread around the world and damaged the food supply chain system. For instance, hospitalization of more than 200 persons and six deaths were reported due to the consumption of *Salmonella*-contaminated cucumber imported from Mexico in 2015 (CDC, 2018). According to the Centers for Disease Control and Prevention (CDC), 128,000 people are hospitalized and 3,000 deaths take place annually due to food poisoning in the United States (Ingelfinger, 2008). In China, the major outbreak was "China's milk scandal." It was reported that the melamine-tainted infant formula led to kidney stones, and even the death of infants (Zhang et al., 2009).

Food contamination occurs due to the presence of physical, chemical or biological hazards in the food product during storage, processing or transportation. Food contamination not only causes illness but also reduces the absorption of certain nutrients by the body. Food contamination can be divided into physical, chemical, and biological contaminants. These contaminants lead to serious diseases like kidney stones, cancer, liver failure, and malnutrition (WHO et al., 1987). The microbial contaminations cause allergy and infections, and it also affects many organ systems based on the type of microorganism or toxin present in food. It has been proved that mycotoxins present in food cause hepatic, gastrointestinal, and carcinogenic diseases (Fung and Clark, 2004). In contrast, the pesticide residues in food material generally affect neurodevelopment and growth (Eskenazi et al., 1999).

14.2.1 PHYSICAL CONTAMINATION

Physical contaminants are potentially harmful extraneous matter which is not normally present in food (Corlett, 1991). These contaminants are generally visible and include glass pieces, wood chips, stones, insects and other pests, insulation, and bone fragments. Physical contamination is the most reported complaint because it creates immediate effects after consumption. The effects include choking, injury, and pain in the stomach, and may require surgery to remove the contaminant. The presence of bone fragments, or any unwanted hazardous materials such as metals and plastics in boneless poultry products, is a major concern of poultry processors (Chen et al., 2003). Physical contaminants can be identified and controlled using various methods. Metal detectors can

TABLE 14.1
Commonly Occurring Physical Contaminants

Physical Contaminants	Sources
Glass	Light bulbs, glass containers, glass part of machinery
Metal	Fragments from equipment, broken veterinary needles, fragments of prior processing procedures.
Plastics	The material used for packaging, fragments of utensils, fragments from equipment
Stones	Improper handling during harvesting of crops
Wood	Fragments of wooden pallets used for handling of products, raw materials
Bone	Poultry products
Natural components of food	Hard outer coatings of food products like shell of nuts

be employed to identify the presence of metal fragments in packed products. Bone fragments can be detected using X-ray technology. X-ray imaging technology detects the residual bone fragments in the production of boneless chicken and also addresses the thickness of products. Above all this, the removal of foreign objects from plant premises are of paramount importance (WHO et al., 1987). Some of the major physical contaminants and their sources are given in Table 14.1.

14.2.2 Chemical Contamination

Chemical contamination can be divided into three categories (Rhodehamel et al., 1992):

- Naturally occurring chemicals
- Intentionally added chemicals
- Unintentionally added chemicals

Chemical contaminants are extremely harmful if they are not properly controlled and maintained during processing and storage. Pesticide residues on fruits and vegetables cannot be avoided but can be reduced by processing properly. All the chemical contaminants can't be completely removed from the product, but there is a permissible limit for every chemical. If the contaminants are above the permissible limit, it causes a harmful effect in the human body. Some of the commonly occurring chemical contaminants are listed in Table 14.2.

Naturally occurring chemicals involves plant, animal, or micro-organism derived substances which are naturally present in the food material. Histamine is a toxin, which is produced by spoilage bacteria in fish products. The consumption of histamine leads to skin irritation, localized inflammation, and hypertension (Arnold and Brown, 1978). Aflatoxin is a major toxin found in certain food products including corn and nuts. It is produced by *Aspergillus flavus* and *Aspergillus parasiticus,* which cause liver, kidney, and spleen enlargement, and fatty liver syndrome (Eaton and Groopman, 2013). Similarly, certain hazardous compounds are formed during the processing of food products due to chemical reactions. During thermal treatment of certain food products containing asparagine, a potential cancer-causing agent called acrylamide is released (Stadler et al., 2002).

Intentionally added chemicals include those added during processing to aid financial advantage by manufacturers. They include preservatives, colorants, artificial sweeteners, etc. More than 2,500 chemical substances can be added to improve the properties of food products (Branen et al., 2001). Some of the chemical substances are permitted in a food product in a certain safe level. If the amount is above the safe limit, the food is regarded as contaminated. Sibutramine (Sibutramine hydrochloride monohydrate), a synthetic slimming drug, is added to herbal slimming foods and supplements to improve the claimed activity of the product (Liang et al., 2006). Additional intake

TABLE 14.2
Commonly Occurring Chemical Contaminants

S.No	Type of Contaminants	Examples	References
1	Pesticide residues	Profenofos, endosulfan, deltamethrin, malathion, propiaconazole, chlorpyriphos, hexaconazole, DDT, lindane, methamidophos, parathion	Kaushik et al. (2008)
2	Veterinary drug residues	Nitrofurans, quinolones, sulfonamides and diaminopyrimidine potentiators, tetracyclines, benzamides, carbanilides, nitroimidazoles	Botsoglou et al. (2000)
3	Environmental contaminants	Heavy metals, polychlorinated biphenyls (PCBs), polychlorinated dibenzodioxins, polycyclic aromatic hydrocarbons	Ghaedi et al. (2008); Khan et al. (1976); Larsen (2006) and Wenzl et al. (2006)
4	Processing contaminants	Acrylamide, nitrosamines, 3-monochloropropane-1,2 diol (3-MCPD), furan, ethyl carbamate	Capuano and Fogliano (2011); Bulushi et al. (2009); Bakhiya et al. (2011); Crews and Castle (2007); and Battaglia et al. (1990)
5	Migrants from packaging materials	Styrene, laurolactam, limonene, 1-octene, bisphenol A, diisopropylnaphthlene, triacetin, tri-n-butylacetyl citrate, triclosan	García et al. (2006)
6	Toxins	Avidin, heam agglutanins, saponins phytates, cynogenic glycosidases, trypsin inhibitors	Farzana (2017)
7	Adulterants	Lead chromate, lead tetraoxide, melamine, sundan dye, metanin yellow, formaldehyde	Moore et al. (2012)

of this drug leads to heart attack and strokes in human beings. For this reason, the U.S. Food and Drug Administration (FDA) prohibits the use of sibutramine in herbal slimming foods like green tea and green coffee. Some unscrupulous milk processors add hydrogen peroxide to milk products to reduce the microbial load, so as to give it a longer shelf life. Hydrogen peroxide causes cardiac arrhythmia and irritation in the mucosa of the gastrointestinal tract. Prolonged consumption leads to decreased life expectancy in humans (Gilliland et al., 1969).

Chemicals which are present in food products without being intentionally added to food are called unintentionally added chemicals. For example, components from the packaging material leaking out and leaving residue in food products. Bisphenol A is one of the major residues present in packaged drinking water (Krishnan et al., 1993). The consumption of bisphenol A causes neurobehavioral problems, such as attention deficit hyperactivity disorder (ADHD) and autism, obesity and type 2 diabetes, and hormonally mediated cancers, such as prostate and breast cancers (Richter et al., 2007). Residues of sanitizers, which are used to clean the processing equipment, may get incorporated into food. These chemicals are unintentionally added chemicals in food products. Such chemicals do not have an immediate effect on the human body, but prolonged consumption and higher amounts lead to serious health effects. Pesticides are known to be universal contaminants in water. The pesticides that are used in agricultural land leach out and travel into sand through percolation. These pesticides finally reach the groundwater. Some of the pesticides that found in groundwater include acetanilide, diphenyl ether, nitroanilines, triazole, and cyclodienes (Boyd-Boland et al., 1996).

14.2.3 Biological Contamination

Biological contamination occurs due to the presence of undesirable micro-organisms in food products. Generally, physical and chemical contaminants do not alter the nature of food products, but micro-organisms change the properties of the food itself. The undesirable changes that occur in

TABLE 14.3
Commonly Occurring Biological Contaminants

Contaminants	Associated Disease
Clostridium botulinum	Neurotoxicity; shortness of breath, weakness or paralysis, blurred vision; loss of motor capabilities
Salmonella spp	Diarrhea, fever and abdominal cramps
Listeria monocytogens	Fever, muscle contractions, diarrhea, meningitis, convulsions, and death
Vibrio parahaemolyticus	Abdominal cramps, nausea, vomiting, headache, and diarrhea (with occasional blood and mucus in stools) and fever

food products due to the action of micro-organisms are known as spoilage. The food spoils due to the presence of spoilage micro-organisms, and some of them are harmful to the human body when consumed. Biological contaminations are further divided into bacterial, viral, and parasitic (protozoa and worms) contamination. When sufficient numbers of a pathogenic micro-organism are consumed, tissues react to the presence of micro-organisms, which create food infections. Some of the major microbial contaminants and the diseases associated with them are given in Table 14.3 (Rhodehamel, 1992).

14.3 CONVENTIONAL METHODS FOR DETECTION OF CONTAMINANTS

Detection of food contamination is necessary to maintain the safety of food products, and also to control the global trading of contaminated products (Viswanathan et al., 2009). The quality of food products is ensured using various chemical and microbiological analyses. The technique involved in analyzing the contamination depends upon the nature and quantity of contaminants present in the sample. Some techniques for the detection of chemical contaminants include NMR spectroscopy (Charlton et al., 2008), High-Performance Liquid Chromatography (HPLC) (Nerın et al., 2002), and Fourier Transform Infrared Spectroscopy (FTIR) (Galvis-Sánchez et al., 2007). These techniques involve laborious sample preparation procedures and sophisticated instruments. Conventional techniques for the detection of microbial contamination in food include culture, microscopic, immunological, and molecular methods (López-Campos et al., 2012). These methods require long incubation time, time-consuming culture preparation, and isolation techniques (Kaittanis et al., 2010).

Qualitative and quantitative analysis of food contaminants and residues can be done using gas chromatography-mass spectrometry (GC-MS). This technique can be utilized to detect contaminants including bisphenol migrated from packaging (Viñas et al., 2010), pesticides in honey (Rissato et al., 2007), melamine in milk (Zhu et al., 2009) and mycotoxins (Tanaka et al., 2000). Chemical contaminants can also be detected using a more advanced technique like immuno-microarray. This technique is an alternative because it consumes less of the sample, and the analytical time is reduced compared to GC–MS. Immuno-microarray can be employed to detect contaminants including mycotoxins, biotoxins, pharmaceutical residues, and pesticide residues (Zhang et al., 2012). FTIR is used for detection of chemicals contaminants, including various mycotoxins (Greene et al., 1992) and melamine (Mauer et al., 2009) in food products down to parts per billion (ppb) levels. Currently, liquid chromatography-mass spectrometry (LCMS) is the method used for detection of dyes in food products by the FDA. LCMS was employed for the detection of Sudan I in chili powder and chili-containing food products with the detection limit of 7 μg/kg (Mazzetti et al., 2004). Acrylamide, organophosphorus pesticides, and azimsuluron were also determined using LCMS with detection limits of 0.2 ng/ml, 3 ng/L, and 0.1 μg/kg respectively (Ono et al., 2006; Cappiello et al., 2002; Pareja et al., 2011). LCMS requires skilled labor and time-consuming sample preparation. Raman spectroscopy is an alternative technique to detect microorganisms (Yang and Irudayaraj, 2003), pesticides (Skoulika et al., 2000), and adulterants (Lopez-Diez et al., 2003).

14.4 NEED FOR SENSOR

Even though all the conventional methods provide good selectivity and sensitivity against all contaminants, they possess several disadvantages. The methods involve time-consuming steps, which are very complicated to perform. Equipment used in conventional methods is expensive and requires a skilled person to operate. None of the methods can be used for rapid or field detection of contaminants. Microbial examination requires days to analyze the presence of certain pathogens in food. All these disadvantages limit the application of conventional technique for the extensive applications. In order to avoid these limitations, various sensors have been developed to provide on-site, cost-effective, and real-time detection of food contaminants.

14.4.1 Biosensor

Biosensors are integrated device which provides selective quantitative and semi-quantitative analytical information using a biological recognition element (Barceló, 2006). Enzymes, antibodies, DNA, or microorganisms can be used as a bio-recognition element. It also contains a transducer that converts the signal produced by the recognition element, which is a measure of receptor-analyte interaction process (Perumal and Hashim, 2014). The common components of the biosensor are given in Figure 14.1 (Vargas-Bernal et al., 2012). The key advantages of using biosensors for the detection of contaminants include the ease of portability, working on-site, and the ability to measure contaminants in complex food matrices with minimal sample preparation. Biosensors use optical, amperometric, potentiometric, electrochemical, impedimetric, and piezoelectric transduction mechanisms for the detection of food contaminants. Batra et al. (2014) developed an amperometric biosensor containing enzyme glutamate oxidase as a recognition element for monosodium glutamate (MSG) detection in food products. Biosensors can also used for the detection of heavy metals in food products. Syshchyk et al. (2015) developed a photoluminescence biosensor based on porous silicon for the detection of glucose, urea, and heavy metals using enzyme as a recognition element. Enzyme can also be used as a biorecognition element for the detection of pesticides residues in food products. Arduini et al. (2006) developed a biosensor in which the enzyme specific to pesticides are immobilized on Prussian Blue-modified screen-printed electrodes. The signal generated during the binding of the pesticide with the enzyme is measured using cyclic voltammetry. Enzyme-based electrochemical biosensors can also be prepared for the detection of aflatoxin B1 in food samples. Soldatkin et al. (2013) developed acetylcholinesterase-based conductometric biosensors for the detection of aflatoxin B1 with a detection limit of 1 nM.

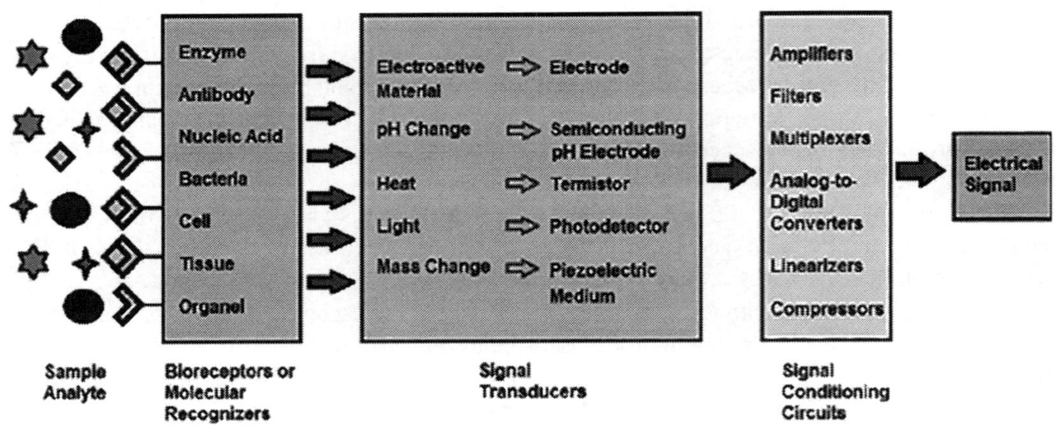

FIGURE 14.1 Operating principle of biosensor with different components. (Vargas-Bernal et al., 2012.)

14.4.2 Nanosensor

Nanotechnology has been touted as the next revolution in food and agricultural industries. Nanotechnology has been revolutionizing the entire food system from production to processing, storage, and development of innovative materials, products, and applications (Anandharamakrishnan, 2014; Bhushani and Anandharamakrishnan, 2017; Ezhilarasi et al., 2013; Karthik et al., 2017). Besides, the introduction of nanomaterials (NMs) into the sensing field improved the sensitivity and selectivity of sensors. Nanoparticles (NPs) have at least one dimension in the range of 1–100 nm, and they exhibit unique mechanical, chemical, and electro-optical properties different from their bulk material. These improved properties are due to the large surface area to volume ratio of NPs (Alagarasi, 2011). The major advantage of using NMs in sensor includes improved performance in terms of sensitivity and lower detection limit. In addition, the surface of NMs is highly specific for ligands, and higher units of immobilization of bioreceptors can be done. Common NPs included as recognition or transduction elements in sensors include gold nanoparticle, silver nanoparticle, quantum dot, carbon nanotube, Titanium dioxide (TiO_2) nanoparticles, and magnetic nanoparticle (Huang et al., 2017). Each NMs behaves differently, and it offers a specific property to sensors. Gold and silver NPs are generally employed in the development of colorimetric detection of contaminants in food. When the light of a particular wavelength interacts with the electrons on the surface of gold NPs, oscillation of conduction band electrons occurs. This phenomenon is known as surface plasmon resonance, and it is demonstrated in Figure 14.2. The wavelength (frequency) with which the interaction occurs depends mainly on the characteristics of gold NPs like size, shape, and agglomeration (Willets et al., 2007). The surface of gold/silver NP can be easily coated with a functionalizing agent that is specific to the analyte. In the presence of an analyte, aggregation of silver/gold NP occurs, which leads to a change in the color of the NP. The shift in peak can be directly correlated with the concentration of analyte present.

Quantum dot and fluorescent semiconductor NMs are also frequently used in the field of the sensor. Quantum dot exhibits large absorption and narrow emission spectrum. This property depends on the size of the quantum dot, as the band gap of the semiconductor varies with size (Eychmüller et al., 1993). This unique property of the quantum dot makes it an important material in the sensing field for the development of a fluorescent sensor with a suitable bioreceptor (Zhang et al., 2005). The unique optical property of carbon nanotubes makes them an ideal material as an electrochemical transducer in developing a biosensor for the detection of contaminants (Cai et al., 2003; Lin et al., 2004). Recently, the use of new NMs derived from their original forms by chemical modifications, such as oxidation (Liu et al., 2015), sulfonation (Zhang et al., 2014), and fluorination (Nair et al., 2010) have gained attention. Zhang et al. (2014) prepared a novel sulfonated graphene/polypyrrole for solid phase microextraction of terpenes. They found that the use of sulfonated graphene/polypyrrole has increased the extraction yield of terpenes from fennel and star anise. Also, fluorinated silica NPs have been prepared and used to deposit a coating on aluminum alloy surfaces which resulted in the superhydrophobic surfaces. Thus it has been proven that the chemically modified form of NP exhibit more functionality than their original form.

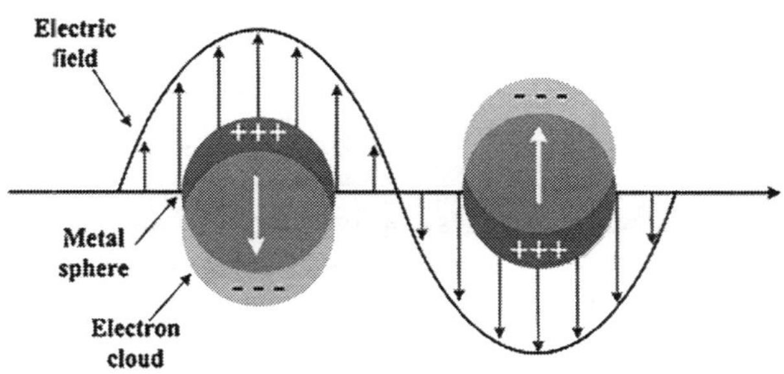

FIGURE 14.2 Illustration of localized surface plasmon resonance of metal nanoparticle. (Sun et al., 2016.)

This chapter reviews various types of nanosensors used for the detection of food contamination. It also discusses the recent application of nanosensors for effective monitoring of food contamination to ensure food safety and quality. The working mechanism of various sensors for the detection of hazardous contaminants is discussed in detail.

14.5 NANOMATERIALS IN SENSORS

With the advent of nanotechnology in the past few decades, a myriad of new materials and devices have been brought into application in various fields. Beginning with energy and health-related applications, NMs play a significant role in everyday consumer products. NMs are defined as materials with at least one external dimension in the size range from approximately 1–100 nanometers. They possess distinctive optical, electrical, catalytic, and magnetic properties compared to their bulk material due to the quantum effect stemming from the large surface area-to-volume ratio. These properties offer enviable and essential characteristics for biological detection, which makes NMs an ideal candidate for the biosensor development. These materials possess unusual target-binding characteristics due to their unique physical and chemical properties leading to diverse signal transduction modes. In addition, due to the evolution in the synthetic advances of NMs with variant size, shape, and composition, surface modification enabled them to conjugate with various biomolecules, which aided the development of nanosensors with the desired properties.

Due to their large surface area, strong adsorption capacity, high surface reactivity, and catalytic efficiency, NMs play different roles in the design of sensors (Andreescu et al., 2009; Halvorson and Vikesland, 2010; Lee et al., 2010; Wang, 2005; Lide Zhang and Fang, 2010). In particular, NM-based sensors have gained considerable attention in the field of analytical chemistry for the development of various types of simple and cost-effective sensors for the detection of diverse analytes (Elghanian et al., 1997; Kalluri et al., 2009; Wang et al., 2004; Zhang et al., 2008). For example, colorimetric sensors make use of color change in nanoparticles (AuNPs and AgNPs) for the detection of environmental pollutants. NMs, such as carbon nanotubes (CNTs) or silicon nanowires, which can be readily functionalized, can be utilized for the development of nanosensors for ultrasensitive detection of pollutants. AuNPs, CNTs, and graphene possessing large surface area and strong adsorption capacity, can be used to bind with the target and amplify the signal for detection. In a word, compared with the traditional sensors, NM-based sensors exhibit many advantages, such as small size, fast response time, high sensitivity, and selectivity.

14.5.1 Methods of Detection Using Nanoparticles

Nanotechnology provides a potential impact on a various aspect of food science, from the growth of food to packaging, and finally consumption. Owing to their small size, NPs are regarded as promising materials in the food industry. In addition, nanotechnology possesses broad application in food safety, nanoparticulate delivery systems, and biosecurity (Figure 14.3).

Nanosized particles in foods can either occur naturally or engineered NMs derived from food materials can be added intentionally or may be tailored using materials that are not endogenous to foods. The application of NMs in the food industry is extended to nutritional additives, flavorings, colorings, and antibacterial coatings for packaging, provided preventive measures are taken prior to their use. Though nano-based food products are rare, nano-enhanced barriers keep food fresher, and vitamins encapsulated with NMs improves the bioavailability. Apart from developing food products, nanotechnology plays a vital role in a majority of new developments in food sensors through integration with other sciences and technologies. Nanosensors enhance the safety and security of food products by detecting pathogens and contaminants that emerge during food processing, transportation, and storage. In combination with the developing analytical techniques, such as

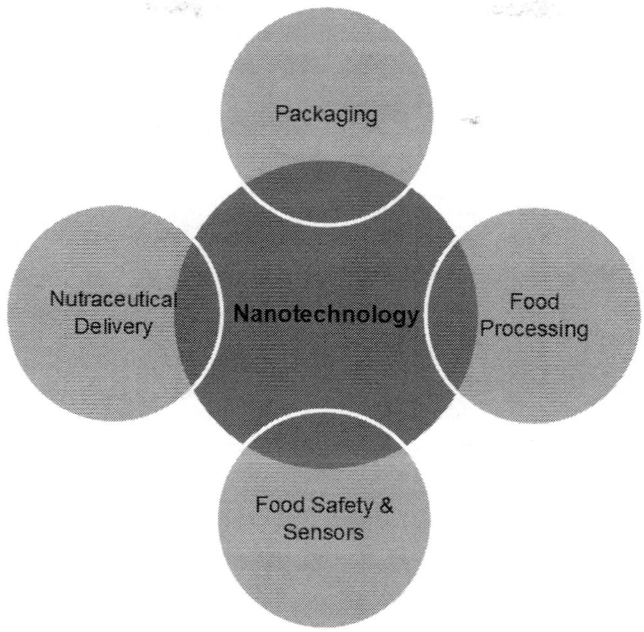

FIGURE 14.3 Application of nanotechnology in the food industry. (Chen et al., 2006.)

optical, surface-enhanced Raman scattering (SERS), and electrochemical techniques, nano-based sensors have significantly enlarged to scrutinize the food safety.

14.5.2 Optical Sensors

At the nanoscale, the property of the materials differs considerably from that of bulk. For example, silver nanoparticles appear yellow, which is different from the bulk material, due to the influence of quantum confinement. In particular, the electrons are densely packed, with increasing energy difference between the highest valence band and the lowest conduction band. As a result, when the electrons return to the ground state, the highest energy is released, as the energy required to excite the particles also increases. This effect is reflected as a color shift in the property of NPs in terms of optical characteristics known as Surface Plasmon Resonance (SPR). The principle of the SPR is based on the measurement of the angle of reflectance generated by the Plasmon waves on a dielectric interface. The change in the reflectance angle is based on the refractive index of the surface (i.e. binding of the target to the bioaffinity agent).

Optical sensors utilize visible or UV light for the purpose of analysis. The sensor material is represented by a wavelength selectable light source, which interacts with the analytes and a light detector (Figure 14.4).

The detection is based on the different techniques (e.g. refractive index, scattering, diffraction, absorbance, reflectance, photoluminescence, chemiluminescence, etc.), which cover the various regions of the electromagnetic spectrum that allow the measurement of multiple properties (e.g. intensity of light, lifetime, polarization, etc.) (Grundler, 2007; Jerónimo et al., 2007). However, here the discussion is on the optical sensor that relies on the absorbance-based technique.

Colorimetry is the quantitative measurement of absorbance spectra, which is one of the well-known analytical techniques with simple, easy, efficient, and sensitive naked-eye detection and identification of a wide range of chemical substrates. In general, each optical sensor must contain both an active center that interacts strongly with the desired analytes and an intense chromophore

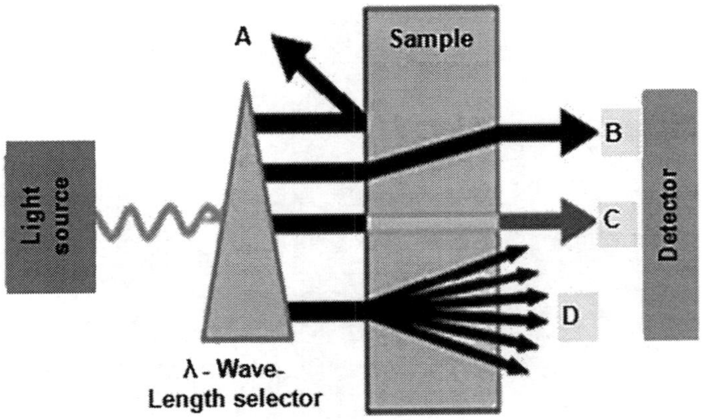

FIGURE 14.4 General arrangement of spectroscopic measurements: (A) light reflection, (B) light refraction; (C) light absorption, and (D) fluorescent emission. (Jerónimo et al., 2007.)

that is strongly coupled to that active center. It is the intermolecular interactions of analytes with the active center that leads to the colorimetric signal in a chemoresponsive manner.

14.5.3 Electrochemical Sensors

Amperometric measurement monitors the change in current due to the oxidation/reduction reactions with respect to the applied potential at the working electrode surface. A higher selectivity is achieved based on the redox potential by the separation of electroactive species using potential dynamic techniques such as cyclic voltammetry, differential pulse voltammetry, and square wave voltammetry. In addition, cleaning of the electrode surface minimizes biofouling by sample constituents or by-products of the electrochemical reaction. Potentiometric sensors measure the change in potential between the working and reference electrodes, which in turn depends on the mobility, in addition to the activity of ions. Though a wide range of sensors has been designed in the case of potentiometric sensors, ion-selective electrodes and gas sensors are considered to be major potentiometric sensors (Śliwińska et al., 2014). The difference in electrical resistance between two electrodes is measured in conductometric sensors. NMs that are employed in biosensor devices modify the conductivity with respect to the changes in the electrode surface, i.e. binding of the target to bioaffinity agent.

Electrochemical sensors have gained considerable attention in the field of analytical chemistry owing to their specific advantages, such as being easy-to-operate, economical, sensitive, portable, and simple-to-construct, by the integration of recognition elements with electronic elements. Based on the electrochemical change over electrode surface, these sensors play a vital role in a wide variety of applications for detecting targets. Amperometric, electrochemical impedance, electrochemical luminescence, and photoelectrochemical sensors are the most important electrochemical sensors used for the detection. In combination with nanotechnology, NMs on electrode surfaces enhance the sensitivity and selectivity of electrochemical sensors. It is evident that the electrode materials have a significant role in developing electrochemical sensors for the detection of target molecules, which are based on diverging analytical theory. In addition to the electrode materials, the synergic effects of NMs not only hasten signal transduction by enhancing catalytic activity, conductivity, and biocompatibility but also led to the development of a highly sensitive biosensor by the amplification of biorecognition events with the application of unambiguously tailored signal tags. The development of high-performance functional electrode materials using nanotechnology, coupled with various electrochemical techniques provided widespread application of electrochemical devices. For example, Ge et al. (2014) demonstrated the application of NM-enhanced paper-based biosensors, which includes lateral flow test-strips and paper microfluidic devices. In particular, the sensitivity and specificity of a paper-based biosensor for the diagnosis of disease are enhanced by the utilization of two different kinds of nanoparticles. Therefore, detailed information on the electrode engineering, strategies for electrochemical signal amplification, and novel electroanalytical techniques are employed for the miniaturization and integration of the sensors, which results in the development of enhanced electrochemical nanosensors.

SERS based sensors: Surface plasmon resonance (SPR) is a unique technique for in-situ detection of various biological and chemical analytes (Bedford et al., 2012; Hoa et al., 2007; Homola et al., 1999; Prasad, 2003; Shalabney and Abdulhalim, 2011). SPR sensors possess distinctive abilities to quantify and characterize low molecular weight molecules, which enabled them to use as a powerful tool in a wide range of applications such as theranostics, pharmaceutics, food safety, environmental monitoring, and homeland security (Situ et al., 2010; Szunerits et al., 2013). In principle, SPR sensors are optical refractometers that evaluate the change in the refractive index of the medium at the SPR sensing surfaces, which is based on the simple optical phenomenon. This phenomenon involves the excitation of the collective coherent oscillations of free electrons in the conduction band of a metal by the interactive electromagnetic field at a metal/dielectric interface (Ho and Lam, 2003; Ho et al., 2006), and these created charge density oscillations called surface plasmon polaritons (SPPs). The SPPs then exponentially decay into their surrounding medium by forming an electric field with a penetration depth in a range of hundreds of nanometers. As a consequence, this transient field is highly sensitive towards the change in the refractive index of the surrounding medium, which determines the characteristics (e.g. angle, wavelength, phase, etc.) of the incident beam for SPR excitation. NMs such as noble metal NPs (AuNPs, AgNPs), CNTs, and graphene-based materials are employed for the enhancement of sensitivity of SPR sensors in the various fields ranging from medicine to the environment.

Noble metal NPs, also called plasmonic NPs, are the most commonly used NMs for ameliorating the performance of SPR sensors. Since they possess strong absorption in the visible and near-infrared wavelength regions, a large electric field will be excited on the surface of NPs due to LSPR (Localized Surface Plasmon Resonance). It is evident that the larger the electric field generated on the surface of NPs, the more sensitive the SPR sensor which is related to the change in its surrounding medium. Furthermore, functionalization of the NPs' surface enhances the surface coverage and target binding on the surface of the NPs (Zeng et al., 2011).

In recent times, graphene-modified SPR substrates have gained considerable attention owing to their ability to induce SPR signal compared to that of bare metal films in the SPR substrates (Wu et al., 2010a; Choi et al., 2011). Such an enhancement in the SPR signal is attributed to the charge transfer from graphene to the surface of the metal film due to the difference in their work function (Giovannetti et al., 2008). In addition, strong pi-stacking in the graphene molecules allows the selective binding of aromatics molecules over the SPR substrates, which opens up a new avenue for the detection of peptides, DNA, RNA, and siRNA molecules (Akca et al., 2011). In this chapter, we have highlighted the current state-of-the-art that describes NMs-enhanced surface plasmon resonance sensors for the detection of food contaminants in various food products.

14.6 DETECTION OF CHEMICAL FOOD CONTAMINANTS

In recent decades, in spite of the efforts made to ensure the food safety of traditional products, many toxological, microbial, and carcinogenic substances prevail in the products tailored for the utilization of society. Though human can sense the quality of food whether it is good or bad, analytical testing provides more reliable information about the quality and safety of food. There is ongoing demand for the development of sensors for the detection of food safety on-site. Nanotechnology provides efficient devices to assess food safety, which is accurate, specific, sensitive, low-cost, battery-free, and reliable. Herein, we review recent achievements of nano-based sensors in combination with other techniques, such as a field-effect transistor (FET), fluorescence, electrochemistry, surface-enhanced Raman scattering (SERS), and colorimetric, for the detection of food contaminants such as pesticides, veterinary drugs, food additives, organic pollutants, and toxic metals.

14.6.1 Pesticides

Pesticides as the name indicates, are used to eliminate pests that interrupt plant growth and to protect them from disease. Though these pesticides are helpful for the agricultural sector, their persistent nature and the presence of their residues in the environment plays a major role in its toxicity (Chaudhry et al., 2002). Different categories are involved in pesticide classification, depending on the nature of the target. Insecticides kill or prevent the growth of insects. Herbicides control or destroy plants. Fungicides control or destroy fungi. Acaricides control or destroy mites. Nematicides control or destroy nematodes. Owing to their high water solubility, extensive use in agriculture, and toxicity, it has become mandatory to monitor the trace contamination level of these pesticides in water and in foods consumed by humans in day-to-day life. Though traditional techniques like gas/liquid chromatography (GC/LC)–mass spectroscopy (–MS) (Balinova et al., 2007; Sicbaldi et al., 1997; Stan and Linkerhägner, 1996) are used for the detection of pesticides, their lengthy work-up and expensive equipment limit their application for in-situ analysis. Nanosensors have been developed to overcome these limitations of conventional sensors with multifunctional advantageous properties like small size, high sensitivity and selectivity, low detection limits, and fast responses (Liu et al., 2008).

Though organophosphorus (OP) compounds and carbamates are extensively used pesticides in the agricultural field for their effectiveness in insect eradication and low persistence in the field, they inhibit the activity of acetylcholinesterase (AChE), an important enzyme involved in the neurotransmitter. A highly sensitive assay with dual readouts namely, colorimetric and fluorimetric methods, has been developed based on Rhodamine B (RB)-AuNPs for the detection of OP and carbamates pesticides. The detection mechanism was based on the aggregation of AuNPs, which in turn depends on the inhibition of enzyme AChE activity by pesticides. In brief, the color of the RB-AuNPs changed from red to blue due to the addition of acetylthiocholine and AchE, which simultaneously recovered the fluorescence of RB that was initially quenched by AuNPs (Liu et al., 2012). A simple FRET of CdTe QDs based on ligand replacement was used for the detection of organophosphorothioate pesticides. Dithizone on the surface of CdTe QDs strongly quenched the fluorescence, whereas, in the presence of organophosphorothioate pesticides, the fluorescence was turned on due to the ligand replacement (diozone) by hydrolyzate of organophosphorothioate (Zhang et al., 2010). A ratiometric nanosensor has been developed using carbon quantum dots coated with vitamin B12 (VitB12), which exhibits dual emission for the detection of carbofuran extracted from soy sauce. Upon excitation at 355 nm, a typical blue emission (417 nm) and red emission (550 nm) was observed due to the coating effect of VitB12. The complexation of carbofuran with VitB12 lead to the quenching of the red emission leading to the ratiometric detection of pesticide with good selectivity (Campos et al., 2017). The luminescence property of up-conversion nanoparticles has been made use of for the ratiometric detection of organophosphorus pesticides, mimic diethyl chlorophosphate (DCP), or OPs dimethoate. Tween-20 modified $NaYF_4$: Yb^{3+}, Tm^{3+}@$NaYF_4$ core-shell UCNPs, and an oxime-based pesticide probe HODN were combined to develop a nanosensor that is highly sensitive at detecting pesticides. The major advantage of this nanosensor is that the upconversion fluorescence is observed in the far IR region, which is not influenced by any interference (Wang et al., 2016). In recent decades, quantum dot sensors have gained considerable attention due to their exceptional optical properties and high quantum yield compared to that of conventional fluorophores. Complete reviews of their applications in the detection of various pesticides in different media and substrates have been done (Nsibande and Forbes, 2016).

An electrochemical sensor with carbon nanotube modified electrode was used to detect triazophos insecticide. The electrode was prepared by the deposition of AuNPs on a CNT modified GC electrode followed by electropolymerization of o-hydroxyphenol at the AuNP/CNT/GC electrode surface in the presence of triazophos as a template. The as-prepared electrode was highly selective for the detection of triazophos in the presence of other pesticides and can be applied for the rapid detection of triazophos in post-harvest vegetables (Li et al., 2012). An impedimetric immunosensor with graphene as the substrate has been employed for the specific detection of parathion in the presence of other pesticides. The validation of this biosensor has been done in real samples like tomatoes and carrots for the detection of parathion (Mehta et al., 2016). AChE based electrochemical sensor for the detection of OP pesticides have developed using AuNPs/SiNWs nanocomplex. It was demonstrated that the nanocomplex not only improved the enzymatic activity of AChE but also assist electron communication between the enzyme and the electrode with high sensitivity and fast response (Su et al., 2008). In another report, AChE biosensor with chitosan-prussian blue-multiwall CNT-hollow gold nanospheres (CS-PB-MWNTs-HGNs) film has been developed for the detection of malathion, chlorpyrifos, monocrotophos, and carbofuran. The incorporation of CNT and HGNs enhanced the electron transfer, thereby promoting the electrochemical response for the detection of pesticides (Zhai et al., 2013).

SERS active tape has been proposed for the rapid in-situ detection of pesticides in apple and vegetables. SERS tape was prepared using AuNPs, whose strong signal was able to distinguish different pesticide residues such as parathion-methyl, thiram, and chlorpyrifos in the real samples such as green vegetables, cucumbers, oranges, and apples. The process involves direct extraction of pesticide residue in fruits and vegetables by a paste and peel-off approach (Chen et al., 2016). Fabrication of nanocomposite using cellulose nanofiber (CNF) coated with AgNPs has been used as SERS substrate for the detection of thiabendazole. CNF retarded the aggregation of AgNPs thereby enhancing the sensitivity of the detection of pesticide (Liou et al., 2017). AgNP monolayer has been fabricated as a SERS substrate, which aids the detection of thiabendazole pesticide in apples and various food products. In this study, 4-aminothiophenol was used as an effective Raman indicator (Li Zhang, 2013). SERS based sensor for the detection of organochlorine pesticides aldrin, endosulfan, lindane, and dieldrin have been proposed by the functionalization of metal surface using thiols as SERS substrate. Surface modification and organization of thiols into multilayers led to the development of sensitive hot spots with a large number of binding sites, which serve as host for the pesticide (Kubackova et al., 2015).

14.6.2 Veterinary Drugs

"Drug" is defined as any substance or combination of substances used for treating or preventing diseases. Such drugs improve the health, condition, and productivity of animals, which ensures food and environmental safety. It appears not to be a problem after oral administration, the route where consumers will be exposed to residues of the drug. But these residues are pharmacologically active, which remains in food derived from animals, and is considered to be toxic if it exceeds the maximum residual limit (MRL) as prescribed by Council Regulation (EEC) No. 2377/90, 1990. Care should be taken to keep residues within the MRL through the use of licensed animal medicines so as to avoid any food contamination. Therefore, testing methods and regulatory bodies are essential in order to check whether residues are within the specified limit in foods. Though traditional methods can detect the presence of residues, their quantification can be accurately measured using nanosensors by overcoming the drawback of traditional techniques like HPLC, GC–MS, etc.

A simple electrochemical sensor has been developed for the detection of antibiotic chloramphenicol in real samples using magnetite nanostructures stabilized with carboxymethyl cellulose (Fe_3O_4-CMC) and decorated with AuNPs (Fe_3O_4-CMC@Au). This sensor indicates high accuracy, with a recovery value of around 97% (Jakubec et al., 2016). Molecularly imprinted silica nanospheres embedded CdTe quantum dots (CdTe@SiO_2@MIP) were used for the detection of chloramphenicol (CAP). The fluorescence of prepared CdTe@SiO_2@MIP was remarkably quenched by

CAP via electron transfer mechanism. The developed nanosensor detects CAP in the linear range of 40–500 μg L^{-1} with a detection limit of 5.0 μg L^{-1}. Therefore, the above-proposed fluorescence-based sensor possesses good selectivity and is utilized for the detection of CAP in spiked human and bovine serum and milk samples (Amjadi et al., 2015). A label-free nanosensor was used for the specific detection of the antibiotic enrofloxacin (ENRO) using multi-branched gold-silica-molecularly imprinted polymer (bAu@mSiO_2@MIP) core-shell NPs by surface-enhanced Raman scattering. The prepared core-shell NPs possess specific recognition ability towards ENRO with the detection of 1.5 nM. The proposed SERS based nanosensor affords insignificant cross-reactivity towards other potential interferents (Carrasco et al., 2016).

Erythromycin imprinted nanoparticles coupled with SPR act as an excellent optical fiber sensor for the detection of ERY, and possess additional advantages, such as the capability of online monitoring and remote sensing. The sensor operates at the range of 1.62×10^{-3} to 100 μM. The linear range for ERY detection was found to be 0.1 to 5 μM. The maximum sensitivity of the sensor was observed to be 205 nm/μM, whereas the detection limit and quantification limit of the sensor were found to be 1.62×10^{-3} μM and 6.14×10^{-3} μM respectively. In addition, ERY imprinted NPs which serve as a highly selective recognition medium also detects the concentration of ERY in milk and honey (Shrivastav et al., 2017). An RNA aptamer-based gold nanoparticle (AuNP) sensor was developed for the fluorometric detection of Neomycin B in milk. This aptasensor detects Neomycin based on the self-assembly of the RNA aptamer/Neomycin B complex followed by fluorescence quenching by AuNPs. The above-proposed biosensor detects Neomycin in milk samples in the linear range of 0.1 to 10 μM, with a detection limit of 0.01 μM (Ling et al., 2016). As it is evident that semiconductor quantum dots serve as excellent sensitive probes for fluorescence sensing, a fluorescent assay was proposed for the detection of oxytetracycline hydrochloride (OTC) using functionalized cadmium telluride quantum dots (CdTe QDs). The detection was based on the fluorescence quenching of CdTe QDs by OTC, which may be due to effective electron transfer from OTC to QDs. CdTe QDs sensor exhibited excellent performance for the detection of OTC in the linear range of $1.34 \sim 13.4 \times 10^{-5}$ mol/L with a limit of detection of 3.08×10^{-7} mol/L (Gao et al., 2012).

Tetracycline (TC) is the most widely used veterinary drug for preventing intestinal infection and growth promotion, and excessive use may result in drug residuals. This residual TC could reach humans via the food chain and would cause harm to human health. TC was detected by using a fluorescence probe via the photoinduced electron transfer (PET) fluorescence quenching mechanism, which was constructed using a hybrid quantum dot/mesoporous silica/molecularly imprinted polymer (QD/MS/MIP). This molecular imprinting technology detects TC in the linear range of 50 to 1000 ng mL^{-1}, with a limit of detection of 15 ng mL^{-1}, and the corresponding recoveries were found to be 90.2–97.2% with relative standard deviations of 2.2–5.7% (Zhang and Chen et al., 2016). A simple enzymic penicillin biosensor with layer-by-layer (LbL) films of single-graphene nanosheets (SGN)-hematein/ILs/penicillinase was demonstrated for the detection of penicillin. The immobilized penicillinase catalyzes the hydrolysis of penicillin to penicilloic acid, where H^+ is liberated and monitored amperometrically with hematein as a pH indicator. Therefore, the linear range of detection for penicillin using SGN-hematein/ILs/penicillinase biosensor was from 1.25×10^{-13} M to 7.5×10^{-3} M and a relatively low detection limit of 10^{-13} M (0.04 ppt). Moreover, the proposed method was applied for the detection of penicillin in milk, which exhibits acceptable reproducibility, stability, good specificity, and high sensitivity (Wu et al., 2014). Sulfamethazine (SMZ), one of the most frequently used sulfonamides (SA) in veterinary clinics, was detected using a chemiluminescence resonance energy transfer (CRET) technique by employing hapten functionalized quantum dots (QDs) in a competitive immunoassay. Amphiphilic polymers were used for the synthesis of core/multishell QDs, which was further functionalized using hapten 4-(4-aminophenyl sulfonamido)butanoic acid. The above described CRET based immunoassay detects SMZ with LOD of 9 pg mL^{-1}, which was found to be greater than two orders of magnitude than a heterogeneous enzyme-linked immunosorbent assay and four orders of magnitude better than a homogeneous fluorescence polarization immunoassay (Ma et al., 2016).

14.6.3 Food Additives

According to the Codex Alimentarius Commission 1995b (Rees and Watson 2000), a food additive is defined as:

> Any substance not normally consumed as a food by itself and not normally used as a typical ingredient of the food, whether or not it has nutritive value, the intentional addition of which to food for a technological (including organoleptic) purpose ….

With the advent of processed foods, the use of food additives such as nutrients, preservatives, flavors and flavor enhancers, texture modifiers, and colorants have become prevalent. Therefore, it is necessary to prevent health risks due to the inclusion of these food additives in the human diet by monitoring their levels in the food system.

Benzoic acid and benzoates are generally used as preservatives in drinks and other food products to prevent degradation caused by micro-organisms during storage. However, they are harmful if not used in small quantities (minimal usage limit) (Mota et al., 2003; Tfouni et al., 2002). Electrochemical sensors based on Copper germanate ($CuGeO_3$) nanowire was proposed for the detection of benzoic acid (BA). The electrochemical peaks (anodic peak potential) of $CuGeO_3$ nanowire modified GCE increases with the addition of BA concentration, thereby detecting BA in samples. The current electrochemical sensor detects BA in the linear range of 0.01–2 mM with LOD of 0.91 μM at a signal-to-noise ratio of three and exhibits good reproducibility and stability (Cai et al., 2013). Copper vanadate nanobelts modified GCE was used for the detection of BA based on a pair of semi-reversible electrochemical peaks whose anodic peak potential is located at 0.13 and −0.06 V. Furthermore, the intensities of electrochemical peaks have a linear relationship with the concentration of BA, and detect BA with good reproducibility and stability in the linear range of 0.001–2 mM, with detection limits of 0.61 and 1.21 μM for cvp1 and cvp2, respectively (Lin et al., 2016). A novel glutamate electrochemical biosensor was developed based on nano-porous pseudo carbon paste electrode (Nano-PPCPE) as the working electrode, coupled with L-glutamate oxidase, catalase, and bovine serum albumin, which was followed by cross-linking with glutaraldehyde. Nano-PPCPE possesses excellent selectivity and sensitivity with a linear range of detection from 5×10^{-7} M to 1×10^{-5} M, with the detection limit of 25×10^{-7} M (Deng et al., 2013).

Food colorants play a major role in the food industry due to colors being attractive to consumers. Though the utilization of synthetic food colorants possesses several advantages like high stability, lower cost, and color equivalence, their content must be monitored, since they can potentially cause harm to human beings. Tartrazine is one such synthetic organic food dye, which has applications in bakery products, dairy products, candies, and beverages. A sensitive colorimetric method was developed for the determination of tartazine in food samples using silver nanoparticles (AgNPs). This method depends on the catalytic effect of AgNPs, which catalyzes the oxidation of tartazine by potassium iodate in the acetate buffer medium. A spectrophotometric method was applied to measure the change in absorbance at 420 nm, which monitors the reaction and detect tartazine in the linear range of 0.7–360 ng/mL, with a detection limit of 0.3 ng/mL. The proposed method was applied for the detection of tartazine in lemon and papaya flavored gelatin and in fruit syrup (Sahraei et al., 2013). A fluorescent carbon dot was applied for the detection of tartazine in the food sample using the static quenching process. The decrease in fluorescence intensity was correlated with tartazine concentration in the linear range of 0.25 to 32.50 μM, and the detection limit was calculated to be 73 nM. As a real-time application, the current method was used for the detection of tartazine in food samples collected from local markets, demonstrating its successful application in routine food analysis (Xu et al., 2015).

The above-mentioned methods utilize the fluorescent principle for the detection of organic pollutants. Electrochemical sensors for the detection of some of the organic pollutants are presented herein. AgNPs-graphene oxide modified glassy carbon electrode (AgNPs@GO/GCE) was used for the detection of Sudan I in chili powder using the electrochemical method. AgNPs@GO/GCE detected Sudan I by the reduction of the azo group followed by electrochemical oxidation of –OH group, which was present in the dye molecule. Furthermore, the amperometric method was employed for the quantitative detection of Sudan I in food products, whose reduction potential was fixed at −0.77 V vs. Ag/AgCl. The sensitivity and detection limit of the amperometric method was found to be 6.83 μA mM^{-1} and 11.4×10^{-7} M L^{-1}. The present method can detect Sudan I in red chili powder in the linear range of 3.90×10^{-6} to 3.19×10^{-5} M L^{-1} (Prabakaran and Pandian, 2015). The voltammetric method was developed for the detection of Sunset Yellow (SY) and Tartazine (Tz) using a carbon-paste electrode (CPE) which is chemically modified with gold nanoparticles (nAu) fabricated and used for the determination of Sunset Yellow (SY) and Tartrazine (Tz). Two well-resolved anodic peaks revealed the oxidation signals of these colorants, which increased with concentration. The detection limit was found to be 3.0×10^{-8} and 2.0×10^{-9} mol L^{-1} for SY and Tz, using this method, which possesses high sensitivity, selectivity, and reproducibility (Ghoreishi et al., 2012).

14.6.4 Organic Pollutants

For the past two decades, organic contaminants have gained much attention in the food industry and environmental pollution. The significance of contaminants with respect to human health is based on persistence, accumulation and toxic effect. The significance of contaminants with regard to the health of humans is normally based on the following criteria: persistence; bioaccumulation and toxic effect. In the case of food safety, consumers concern on food contaminants increased due to the utilization of polychlorinated biphenyls (PCB)-oil for the production of feed. Aquatic organisms are also susceptible to contamination of these substances that have entered the environment. Therefore, it is essential to screen the levels of these organic contaminants in food materials. The following section deals with the detection of some of the important organic contaminants like crystal violet, malachite green, bisphenol A, benzo[a] pyrene using nanosensor.

SERS based method was proposed for the detection of Crystal Violet (CV) and Malachite Green (MG), which were found to be significant contaminants in seafood products. For the detection, fractal-like gold nanostructure was prepared via the self-assembly process and utilized for the detection of dyes using surface-enhanced Raman spectroscopy (SERS). SERS differentiates both the dyes very effectively by corresponding Raman signals with the lowest detection limit of ~0.2 ppb level and the enhancement factor was found to be of $\sim4\times10^7$ (He et al., 2008). Three-dimensional hybrid nanostructure with AgNPs, TiO_2 nanorods was constructed for the detection of MG by SERS. The present hybrid substrate possesses good SERS activity due to the formation of high-density hot spots among the AgNPs. The results indicate that the hybrid structure reported serving as an excellent optical sensor for screening MG with the detection limit of 1×10^{-12} M (Tan et al., 2012). Besides, Qiu et al. (2016) proposed a portable system of SERS detection of malachite green with the detection limit of 50 pM, with high specificity against potential interferents. Here, nanoporous gold disk functionalized with stabilized guanine-Quadraplex moieties used as SERS substrate. With the increase in the concentration of MG from 50 pM to 5.0 μM, the SERS intensity increased, followed by the saturation beyond 5.0 μM, which is an indication of slower binding of MG molecules on nanoporous disk.

Aptamer-based gold nanoprobe was developed for the specific detection of PCBs by utilizing AuNPs and choosing single-stranded DNA aptamers that specifically bind to PCB. A highly sensitive aptamer-based fluorescent assay was created, which detect PCB in the wide linear range of 0.1 to 100 ng/ml. In addition, the present aptamer-based gold nanoprobe exhibits specificity towards 3,3',4,4'-tetrachlorobiphenyl (PCB77) compared with other few common PCB77 structural analogs (Xu et al., 2012). Quantification of bisphenol A (BPA) in food samples was performed by a voltammetric method using ZnO/CNT nanocomposite with 1,3 – dipropylimidazolium bromide as a binder in the carbon matrix. BPA can be determined in the presence of Sudan I, wherein peak current of BPA 8.7 times higher than that of sudan I and peak potential of BPA (0.47 V) and Sudan I (0.70 V) were well separated. The current voltammetric method detected BPA in the linear range of 0.002–700 μmol L^{-1} with the detection limit of 9.0 nmol L^{-1} (Najafi et al., 2014). The aptamer-based fluorescent method was proposed for the detection of BPA based on the specific recognition of aptamers and the inner filter effect of AuNPs on the fluorescence of CdTeQDs. Aggregation of AuNPs could be controlled by recognition between BPA and aptamers, which influence IFE of CdTeQDs. A good linear correlation was displayed in the concentration range of10 to 80 ng/mL with the detection limit of 1.86 ng/mL, which was found to be much lower than that of Maximum Residual Limits of BPA in drinking water. Furthermore, this method showed excellent selectivity over other structural analogs such as tetrabromobisphenol A, p-tert-butylphenol, and 4,4-bisphenol due to the specific recognition between BPA and aptamers. The results obtained using this method for the detection of BPA in drinking water was comparable with that of the HPLC technique. Besides, the higher sensitivity of the fluorescent method was attributed to the fluorescent probe CdTeQds and IFE of AuNPs (Li et al., 2016).

CdTe modified TiO_2 nanotubes (NT) was utilized for the rapid detection of benzo(a)pyrene (BaP) based on the FRET between CdTeQDs and PAHs. Since PAHs exhibit adsorption and fluorescence emission around 364/410 nm, the phenomenon FRET occurs with CdTeQDs as donors and PAHs as acceptor. The higher sensitivity was monitored in BaP due to the presence of a larger number of rings present in BaP. The fluorescent sensor exhibits good linear behavior to the proposed sensor to the logarithm of BaP concentration in the range of 400 nM to 40 pM, with a detection limit of 15 pM, which was found to be much closer to the quality criteria (15.1 pM) in drinking water set by U.S. Environment Protection. Furthermore, calibration with a standard sensor can be avoided due to the quantification of PAHs based on the relative fluorescence intensity at 410–370 nm and therefore, the sensor can be directly applied for the detection of BaP neglecting sensor-to-sensor variation. The real-time application of this method was tested by analyzing BaP in River water in the concentration range of 0.045 to 2.847 ng/L (Yang et al., 2010).

14.6.5 Toxic Metal

Contamination of the environment and food system by toxic heavy metals are common nowadays. The presence of these metals creates a serious impact on human health and ecosystems, thus the detection of heavy metal has high importance in maintaining the safety of the food. Common heavy metal ions considered as environmental pollutants are mercury ion (Hg^{2+}), copper (Cu^{2+}), and lead (Pb^{2+}). Amongst, few nanosensors that have been developed for the detection of toxic metal ions are discussed below.

14.6.6 Mercury Ions

Recently, oligonucleotides have been employed in the detection of mercury ions. Mercury ions specifically interact with thiamine which results in the formation of thiamine- Hg^{2+} -thiamine complex (Lee et al., 2008). This specific interaction was utilized for the development of high sensitivity sensor for the detection of mercury ions in aqueous media. Ye et al. (2008) used T- Hg^{2+}-T complex formation ability of Hg^{2+} for the detection of the same. They used AuNPs enhancement approach to decrease the limit of detection to 0.2 ppb. In this sensor, AuNPs and fluorescence dye FAM (6-fluorescein-CE phosphoramidite) was functionalized with complementary DNA, which will only hybridize in the presence of mercury ions. This hybridization resulted in an increase in fluorescence polarization value of the solution which is measured using fluorescent spectroscopy.

Li et al. (2011) used both quantum dot and AuNPs for developing nano-metal surface energy transfer (NEST) sensor for Hg^{2+} detection in water samples. This special mechanism has several advantages over conventional FRET sensors including high sensitivity, better reproducibility, and reduced photobleaching. They have used quantum dot/DNA/AuNPs ensemble in which both QD and AuNPs have been hybridized with DNA. In presence of mercury ion, DNA hybridization takes place which results in surface energy transfer from QD to the gold nanoparticle. This mechanism results in quenching of fluorescent activity of QD linearly to the concentration of mercury. The sensor exhibited high selectivity in buffer samples but the response was also observed for other metal ions also. Recently, Lu et al. (2017) used CH3NH3PbBr3 perovskite quantum dots (QDs) for detection of ultra-trace amount of mercury ions. The mechanism of detection was based on the quenching of fluorescence of developed QD by mercury ion. The sensor was able to detect mercury ion up to 24.87 ppt level.

14.6.7 Iron Ions

Simple colorimetric detection of iron ions can be done using functionalized AuNPs. Salimi et al. (2018) used methylene blue as a capping agent for sodium borohydrate reduced AuNPs. Methylene blue (MB) imparts a positive charge to AuNPs which helps in the formation of mono-dispersed MB-AuNPs. AuNPs aggregate in the presence of Fe (II) ion, because it disrupts the electrostatic repulsion between gold nanoparticles. Even though the sensor gives a limit of detection of 11.21 nM, compromise in selectivity was a major drawback because the response was observed in other metal ions also. Selectivity can be improved by functionalization with pyrophosphate (Wu et al., 2011). Pyrophosphate selectively binds to the ferric ion to form a ferric pyrophosphate $Fe_4(P_2O7)_3$ complex, which results in aggregation and color change that can be detected using a UV spectrometer or the naked eye. Thatai et al. (2014) demonstrated the application of gold nanorods for the detection of Fe (III) ions in aqueous media. Gold nanorods were prepared using cetyl tri-methylammonium bromide (CTAB) capped seed and used for the detection of Fe (III) ion. The Fe (III) ion will selectively bind on the surface of gold nanorods which will result in a shift in surface plasmon absorbance to blue, due to the aggregation of gold nanorods. This sensor can be employed to detect the presence of Fe (III) ion as low as 100 ppb.

14.6.8 Lead and Copper Ions

The colorimetric nanosensors can be developed by functionalizing gold or silver NPs for the detection of lead ions. The functionalizing agent can be DNA or non-DNA based. AuNPs can be functionalized with gallic acid (Yoosaf et al., 2007) or peptide (Joseph et al., 2008) for the selective binding of lead ions with a limit of detection of 5.0 mM and 242 nM respectively. However, these sensors were not efficient in real complex matrices. Instead, the functionalization of AuNPs with aptamer showed improved detection in pond and lake water. Also, the use of aptamer decreased the limit to detection to as low as 250 nM (Liu et al., 2014). Further, Ma et al. (2018) used aptamer for the specific binding of Cu^{2+} and silver staining technology. A solid-phase SiO_2 support has been used for the functionalization of aptamers, and silver was used for the staining of bound lead ions, which allows visual detection and semi-quantitative measurement of lead ions present in the sample. Further, DNAzyme can also be used to decrease the detection limit to 0.2 and 0.5 nM for Pb^{2+} and Cu^{2+} respectively (Wu et al., 2010b). In this study, the quantum dots (QDs) were protected with ZnS and silica shell to provide stability in aqueous solution. Further, the surface silanizated QDs were functionalized with DNAzyme which breaks down in presence of metal ions, resulting in increased emission from QDs.

Recently, fluorescent carbon dots have been introduced in the sensing field to overcome the drawbacks of conventional semiconductor quantum dot. The carbon dots can be functionalized with polyamine (Dong et al., 2012) or N-(2-aminoethyl)-N,N,N-tris(pyridine-2-ylmethyl) ethane-1,2-diamine (Qu et al., 2012) to detect Cu^{2+} in cellular imaging. Liu et al. (2014) developed a ratiometric fluorescence sensor based on dual-emission fluorescent nanoparticles for the detection of Cu^{2+} ions. The prepared Rhodamine B-doped silica NPs were coated with carbon dot and exhibited two characteristic absorption wavelengths at 467 and 585 nm. The intensity of absorption reduces according to the concentration of Cu^{2+} ions present in the sample because of the presence of ethylenediamine groups on the surface of CDs which selectively bind to Cu^{2+} ions. The sensor was tested in tap and river water with the detection limit as low as 35.2 nM. Tian et al. (2018) also used carbon dot for the detection of Cu^{2+} and Mg^{2+} ions in water as a measure of the hardness of water. They used hexametaphosphate-capped Au NPs and carbon dot for the development of hybrid nanosensors, which exhibit dual emission fluorescent spectra at 439 nm and 608 nm, attributed to carbon dot hexametaphosphate-capped AuNPs respectively. Hexametaphosphate chelate with Cu^{2+} and Mg^{2+} ions, which resulted in increased and decreased fluorescent intensity at 439 nm and 608 nm respectively.

14.7 DETECTION OF MICROBIAL CONTAMINATION

According to the World Health Organization (WHO) almost 1 in 10 people are affected by foodborne diseases every year, and 420,000 die as a result of consumption of contaminated foods. The major agents for the contamination of foods include norovirus, *Campylobacter*, *Salmonella*, pathogenic *E. coli*, *Taenia solium*, and *aflatoxins* (WHO, 2015). The consumption of food containing pathogenic micro-organisms or associated toxins causes health problems in humans. Thus, it is important to analyze the sample for assuring the microbial safety of the food product. Conventional techniques used for the detection of pathogenic microorganisms in food include culture methods, microscopic methods, immunological methods, and molecular methods (López-Campos et al., 2012). These methods require long incubation times and time-consuming culture preparation and isolation techniques (Kaittanis et al., 2010). Similarly, various toxins produced by bacteria and fungus affect tissues, which will lead to organ damage in the human body. Besides, these toxins remain in the food even after the death of corresponding micro-organism (Inbaraj and Chen, 2016). Thus, it is necessary to prevent intoxication by detecting the presence of toxins in food. Common methods employed for the detection of toxins include mass spectrometry (Boyer et al., 2011), high-performance liquid chromatography (Al-Hazmi, 2010), and biosensors (Lian et al., 2010). This involves sophisticated equipment and laborious sample purifications techniques. In order to provide real-time detection of pathogens and toxins in food, various sensors are employed, which ensures immediate detection and prevention. Introduction of nanoscale materials into the sensing field improves the sensitivity and reduces the detection time of sensors (Jain, 2005; Rosi and Mirkin, 2005). Some of the nanosensors employed for the detection of pathogenic microorganism and toxins are discussed below.

14.7.1 Nanosensors for the Detection of Pathogenic Bacteria

14.7.1.1 *Escherichia coli*

Escherichia coli (*E.* coli) are present in human and animal intestines and they are generally not harmful. But the serotype *E. coli* O157: H7 is observed to cause acute diarrhea and gastroenteritis when consumed. The contamination occurs mainly through drinking water. It also produces verotoxin, similar to Shiga toxin (*Shigella dysenteriae*), which induces renal failure (Scheutz et al., 2011).

Magnetic and optical sensors have been developed for the detection of *E. coli*, but both exhibit their own limitations. Magnetic NPs provides high sensitivity and rapid detection in a low population of bacterial cells, but quantitative detection is less accurate in a high concentration of bacterial cells. On the other hand, fluorescent detection shows sensitivity in high population, but it is less sensitive in low populations. In order to overcome these limitations, both techniques (magnetic and fluorescent) were combined to produce a magneto-fluorescent nanosensor, which allows specific detection of *E. coli* O157: H7, regardless of the source of contamination (Banerjee et al., 2016).

Maurer et al. (2012) developed a multi-array sensor with a target-specific ribonucleic acid (RNA) strand, which was immobilized on AuNPs. The sensor contains five layers as shown in Figure 14.5. CNT was deposited on the graphite layer, which was coated with a thin layer of silica. AuNPs were directly synthesized on the surface of CNT to provide an adsorption location for the RNA sequence. *E. coli* showed a higher affinity for RNA coated AuNPs, and different pathogens can be detected by changing the RNA sequence specific to targeted cells.

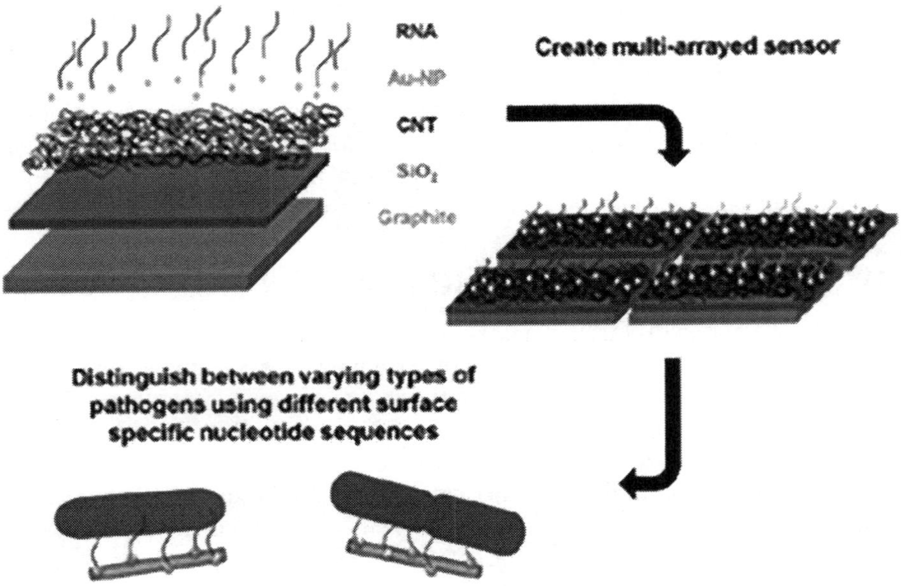

FIGURE 14.5 Schematic representation of layers of the multi-array sensor.

Liu et al. (2014) developed a fluorescence sensor based on lysozyme functionalized fluorescence gold nanoclusters (AuNCs) because of its non-cytotoxic and multifunctional characteristics. Lysozyme specifically binds to *E. coli*, which enhanced the red fluorescence corresponding to the population of microbes in the sample. They have also found that bovine serum albumin (BSA) stabilized AuNCs do not produce red fluorescence, but the lysozyme functionalized AuNCs exhibited a limit of detection of 2×10^4 CFU/mL.

Another attempt has been made to produce a nanocomposite film for the detection of *E. coli*. They developed ZnO-rGO nanocomposite film by sol-gel method and the surface microstructure of the film has been modified with gamma radiation. The irradiated nanocomposite film was used to detect *E. coli* by measuring the change in conductivity of the film. They proved that the irradiation process improved the sensitivity of the sensor from 1.04×10^{-3} to 4.62×10^{-3} CFU/ml compared to un-radiated films (Azmy et al., 2017).

14.7.1.2 *Salmonella* spp

Salmonella organisms may be considered as a single species with ~2500 serotypes (e.g. *Salmonella typhimurium*, *Salmonella enteritidis*, and *Salmonella paratyphi*). Recent outbreaks of *Salmonella* infection are associated with eggs, cheese, dry cereal, ice cream premix, and some fresh vegetables. The disease associated with *Salmonella* are typhoid, gastroenteritis, salmonellosis, diarrhea, nausea, pneumonia, meningitis, and osteomyelitis (Inbaraj and Chen, 2016).

Electrochemical biosensors have been employed for the detection of microorganism for a long period of time. The performance of an electrochemical biosensor can be improved by the inclusion of NPs. CNTs are commonly employed for modification of glassy carbon electrodes. CNT provides a high surface area for immobilization of antibodies, and it also allows electrochemical measurement. Jain et al. (2012) modified a glassy carbon electrode with an antibody-functionalized single-walled carbon nanotube (SWNT) for the detection of *Salmonella enterica*. Antibody-antigen (*Salmonella*) reaction changed the resistance and impedance, which was measured using cyclic voltammetry device. Glassy carbon electrodes can also be modified with ethylene diamine (through a carbon-nitrogen bond) and self-assembled AuNPs (Yang et al., 2009). The antibody bound to AuNPs forms an immuno-complex with Salmonella and leads to a change in capacitance. The capacitance change was linear to the presence of *Salmonella* in the sample, measured by electrochemical impedance spectroscopy (EIS) technique. The detection limit was found to be 1.0×10^2 CFU/mL.

Joo et al. (2012) combined immunomagnetic separation and optical sensing techniques for the detection of *Salmonella* in milk samples. *Salmonella* bacteria which were bound to the antibody-immobilized magnetic nanoparticle (MNPs) can be separated from the sample by applying an external magnetic field. This *Salmonella*-MNPs forms a bond with the TiO_2 nanocrystal, which would change the absorption of light, thus it can be detected through an optical method. The limit of detection of *Salmonella* in milk through this sensor was 100 CFU/mL. The separated *Salmonella*-MNPs can be reacted with biotin-labeled anti-*Salmonella* antibody followed by Streptavidin-conjugated quantum dots (Yang and Li, 2005). The streptavidin-biotin interaction changes the fluorescence, which was directly correlated with the population of *Salmonella* present in the sample.

All the aforementioned sensors are based on antibodies as a recognition element. Molecular imprinted polymer (MIP) as a recognition element possess several advantages over antibodies, such as long-term stability, chemical inertness, and insolubility in water and most organic solvents (Cai et al., 2010). Percin et al. (2017) used a molecular imprinting technique to create a pattern on the surface plasmon resonance (SPR) sensor as a binding zone for *Salmonella paratyphi,* with a limit of detection of 1.4×10^6 CFU/mL.

A simple calorimetric sensor, containing thiolated oligonucleotides functionalized AuNPs can be employed for the detection of *Salmonella* in milk samples. The presence of *Salmonella* causes the aggregation of AuNPs due to binding of oligonucleotides to the targeted organism, leading to a change in color of the solution (Fu et al., 2013). The non-thiol probe can also be used for the functionalization of AuNPs for the detection of *Salmonella,* which involves electrostatic reciprocal action (Verma et al., 2015).

14.7.1.3 *Listeria* spp

Listeria monocytogenes is a foodborne pathogen that causes listeriosis and are present in environmental mediums such as waters, sediment, and seafood. *Listeria* contamination commonly occurs in meat, poultry, and seafood products (Jemmi et al., 2006).

Noiseux et al. (2010) developed a sensor containing fluorescent and magnetic NPs with specific antibodies. Antibody-coated AuNPs binds to *listeria* in food samples and allows separation by applying an external magnetic field. The fluorescent NPs conjugated with antibody provided amplification of intensity of fluorescence, which was detected by the cell reader developed by Noiseux et al. (2010). The measured fluorescence allowed quantification of *listeria* in the given sample.

Aptasensors are more specific for the determination of pathogenic micro-organisms, and they are commonly used as a biorecognition element. Ding et al. (2014) developed label-free potentiometric aptasensor for the detection of *Listeria monocytogenes.* The aptamer binds to protein internalin A on the surface of the *listeria* cell, which will prevent aptamer from electrostatically interacting with protamine, an indicator for the transduction of potential signal (Ohk et al., 2010). This signal can be measured using a polycation-sensitive membrane electrode, and the limit of detection was found to be 10 CFU/mL.

Recently, superparamagnetic iron oxide NPs have drawn attention in the sensing field because of their high specificity against the target analyte. Upon target interaction, the NPs alternate between dispersed and clustered states, which exhibit a change in spin-spin relaxation time. Based on this principle, superparamagnetic NPs detect the micro-organism. Superparamagnetic NPs conjugated with an antibody specific to *Listeria monocytogenes* were employed as a probe for the detection of contamination (Zhao et al., 2017). The change in relaxation time can be directly correlated with the concentration of bacterial cells in the food sample.

14.7.1.4 *Pseudomonas* spp

Pseudomonas aeruginosa (*P. aeruginosa*) is a foodborne pathogen that causes urinary tract infections, pneumonia, and lung diseases like cystic fibrosis (Stover et al., 2000). *P. aeruginosa* survives in low temperatures and are present even in chilled meats. Jia et al. (2017) developed an aptasensor consisting of superparamagnetic iron oxide NPs conjugated with the specific aptamer. When the aptamer binds to the micro-organism, the spin-spin relaxation time will increase. This signal can be measured to quantify the population of *P. aeruginosa* in the given sample.

14.7.2 Nanosensors for the Detection of Toxins Produced by Microorganisms

14.7.2.1 Aflatoxin

Among the various types of aflatoxins, a secondary metabolite produced by *Aspergillus flavus* and *Aspergillus parasiticus*, aflatoxin B1 is highly toxic and commonly found in food.

Xu et al. (2012) developed an optical sensor with gold nanorods for the quantification of aflatoxin B1 in food. Gold nanorods possess more stability in ionic solutions than AuNPs, which make them suitable for the detection of the toxin with great linearity. The sensor contained gold nanorod-aflatoxin B1-Bovine serum albumin (BSA) conjugates and antibodies mixed together in a buffer solution. The target aflatoxin molecules compete with the conjugate molecule for the antibody. Each BSA requires 8 to 12 aflatoxin B1, and it leads to aggregation when the amount of aflatoxin is not sufficient. The aggregation depends upon the concentration of the aflatoxin in the solution. The sensor has a limit of detection of 0.04 ppb.

Microcantilever can be used for detecting the presence of aflatoxin in food materials. Zhou et al. (2016) used stress microcantilever for nanomechanical label-free detection of aflatoxin in food and feed materials. A sulfhydrylated antibody specific to aflatoxin was immobilized on the surface of a gold microcantilever to ensure the stability and sensitivity of the immunosensor. When the antibody binds to aflatoxin, the microcantilever bends toward the gold surface. The stress created on the gold surface can be measured using a light detector and source, which gives the amount of aflatoxin in the food materials. The limit of detection of this sensor was found to be 0.03 ng/mL.

14.7.2.2 Cholera Toxin

The cholera toxin is produced by the bacterium *Vibrio cholerae*. The toxin is present in contaminated food and water, which causes watery diarrhea when consumed. The cholera toxin binds to the GM1 ganglioside receptor on the surface of mucosal cells when it reaches the small intestine. So GM1 ganglioside can be used as a receptor in the development of a sensor for the detection of *Vibrio cholerae* (Tark et al., 2010). Ganglioside-incorporated nanodiscs can be used to functionalize microcantilever and the changes in the vertical deflection are measured through position sensitive detector (PSD). A stable receptor layer has been created without compromising the bioactivity of the membrane protein by the use of nanodiscs for the immobilization of GM1 ganglioside on the microcantilever surface.

Zhang et al. (2016) also used GM1 ganglioside as a receptor, which is bound to plasmonic nanoparticles for the development of plasmon coupling enhanced Raman scattering sensor for the detection of the cholera toxin. AuNPs have been synthesized with Raman dye and a bilayer phospholipid coating has been given to provide multiple sites for the binding GM1 molecule. The GM1 act as a specific receptor, and it will also help to form a sandwiched GM1–CT–GM1 interaction that minimizes the distance between the PCERS nanoparticles due to the smaller size, which leads to enhanced plasmonic coupling. The sensor showed high sensitivity against cholera toxin with a limit of detection of 0.3 pg/mL.

14.7.2.3 Shiga Toxin

Shiga toxin, produced by *E. coli* enters into the human body through contaminated water and food, which results in diarrhea, hemorrhagic colitis, and hemolytic-uremic syndrome (HUS). Shiga toxin belongs to the “AB” family of protein toxins, and consists of one enzymatically active part (A) and five identical subunits (B), which bind to glycolipid receptors (Gal-α1,4-Gal disaccharides) on the cell surface (Sandvig, 2001). Li et al. (2017) used Gal-α(1→4)-Gal-β(1→4)-GlcNAc, which was present in pigeon ovalbumin (POA) as a recognition element for the detection of Shiga toxin. They synthesized pigeon egg white protein-immobilized AuNPs, which showed a limit of detection of 40 pM. Unlike another sensor probe where functionalizations take place after the synthesis of the nanoparticle, this method involves synthesis and functionalization in a single step.

14.7.2.4 Staphylococcal Enterotoxin

Staphylococcal enterotoxins B (SEB) are toxic substances produced by different strains of *Staphylococcus aureus,* which causes toxic shock syndrome (TSS) (Balaban and Rasooly, 2000). Many strains of *Staphylococcus aureus* produce one or more serotypes of specific exotoxins, which results in food poisoning and staphylococcal scarlet fever (Jarraud et al., 1999). Rong-Hwa et al. (2010) developed a strip assay for the real-time detection of Staphylococcal enterotoxin B (SEB) containing a double-antibody sandwich format on a porous nitrocellulose membrane. SEB reacts with polyclonal antibody coated AuNPs and then reacts with the fixed antibody on the membrane. This reaction will lead to the formation of the red line on the device. The intensity of the red line will be proportional to the concentration of SEB in the sample. This assay allows detection of up to 1 ng/mL and it can be improved to 10 pg/mL with silver enhancement.

Zhou et al. (2016) used the formation of AuNPs as a signal generation step to detect SEB. Initially, the SEB binding aptamer and the probe G were hybridized, and in the presence of SEB, the G-quadruplex forming sequence was produced. This results in the consumption of hydrogen peroxide, which leads to the formation of aggregated ill-defined AuNPs (blue colored solution). The absence of SEB can be indicated by the generation of dispersed spherical gold nanoparticle (red color solution). The sensor allows a detection limit of 1 pg/mL.

14.7.3 Multiplex Detection of Toxins

Several sensors have been developed for simultaneous detection of multiple toxins. Foubert et al. (2017) also used a CdSeS/ZnS quantum dot as a label for a lateral flow immunoassay (LFIA) system for the detection of mycotoxins like deoxynivalenol (DON), zearalenone (ZEN) and T_2/HT_2-toxin (T_2/HT_2) in cereal matrices. They showed that colloidal AuNPs can also be used as a label with the same antibody. However, the quantum dot-based LFIA consumed fewer immunoreagents, and it also showed more sensitivity.

Nagatsuka et al. (2013) developed a localized surface plasmon resonance (LSPR) sensor for the detection of ricin, Shiga toxin, and cholera toxin. Glycosyl ceramides (β-lactoside, globosyl trisaccharide (Gb3), or GM1 pentasaccharide) were used as a recognition element and were bound to AuNPs to visually detect the response. They proved that the attached glycosyl molecule and the size of AuNPs affect the response. For the detection of ricin, Shiga toxin, and cholera toxin, the highest LSPR was obtained when the size of nanoparticle was 20 nm, 40 nm, and 40 nm respectively. The limit of detection of the developed sensor for various toxins was found to be 30 ng/mL for ricin, 10 ng/mL for Shiga toxin, and 20 ng/mL for cholera toxin.

14.8 CONCLUSION AND FUTURE PERSPECTIVES

Nanomaterials (NMs)-based sensors for the detection of food contaminants have been discussed with plenty of available literature. It is evident that NMs play a vital role in the development of analytical methods for the detection of food contaminants. They have demonstrated their superior capabilities over conventional materials by improving the sensitivity and selectivity of detection, with the reduction in sample consumption and analytical time. In addition, they can be integrated with portable or miniaturized devices for specific applications, owing to their biocompatibility, stability, and selective catalytic activity, which resulted in the development of SPR nanosensors, electrochemical sensors, SERS-based sensors, and bionanosensors.

Though recent advancements in the synthesis of NMs offer a variety of materials, their production and the commercialization of low-cost NMs limits their application in the detection of food safety issues. Therefore, the development of low-cost NM synthesis, integration, sensor fabrication, and the miniaturization of testing devices and techniques may provide an alternative to the current analytical disputes in tailoring sensor devices.

Although the distinct properties of NMs possess various advantages towards bio-application, they are lacking in their real-world application. Due to the continuous increase in the release of numerous contaminants to the environment, the selectivity and sensitivity of the sensing platforms remain a challenge. The complex nature of the released chemicals has a tendency to combine, which leads to the evolution of new products. Therefore, the collaboration between multidisciplinary sciences and engineering sciences may offer a solution to meet these challenges by facilitating the design of sensors with superior capabilities. It has been anticipated that the rapid development in the fabrication of NMs will assist in screening for widespread contaminants in the environment and food in the near future.

REFERENCES

Akca, S., Foroughi, A., Frochtzwajg, D., and Postma, H. W. C. 2011. Competing interactions in DNA assembly on graphene. *PLOS ONE* 6(4): 18442.

Al-Hazmi, N. A. 2010. Determination of patulin and ochratoxin a using HPLC in apple juice samples in Saudi Arabia. *Saudi Journal of Biological Sciences* 17(4): 353–359.

Alagarasi, A. 2011. *Introduction to nanomaterials*. National Center for Environmental Research. https://nccr.iitm.ac.in/2011.pdf (accessed May 2018).

Amjadi, M., Jalili, R., and Manzoori Jamshid, L. 2015. A sensitive fluorescent nanosensor for chloramphenicol based on molecularly imprinted polymer-capped CdTe quantum dots. *Luminescence* 31(3): 633–639.

Anandharamakrishnan, C. 2014. Liquid-based nanoencapsulation techniques. In *Techniques for Nanoencapsulation of Food Ingredients*, ed. Anandharamakrishnan, C., 29–41. New York: Springer.

Andreescu, S., Njagi, J., Ispas, C., and Ravalli, M. T. 2009. JEM Spotlight: Applications of advanced nanomaterials for environmental monitoring. *Journal of Environmental Monitoring* 11(1): 27–40.

Arduini, F., Ricci, F., Tuta, C. S., Moscone, D., Amine, A., and Palleschi, G. 2006. Detection of carbamic and organophosphorus pesticides in water samples using a cholinesterase biosensor based on Prussian Blue-modified screen-printed electrode. *Analytica Chimica Acta* 580: 155–162.

Arnold, S. H. and Brown, W. D. 1978. Histamine toxicity from fish products. In *Advances in Food Research*, ed. Chichester, C.O., Vol 24, 113–154. New York: Academic Press.

Azmy, N. A. N., Ahmad, A. A., Arsad, N., Idris, S., Mohmad, A. R., and Abdul Hamid, A. 2017. Enhancement of ZnO-rGO nanocomposite thin films by gamma radiation for *E. coli* sensor. *Applied Surface Science* 392: 1134–1143.

Bakhiya, N., Abraham, K., Gürtler, R., Appel, K. E., and Lampen, A. 2011. Toxicological assessment of 3-chloropropane-1,2-diol and glycidol fatty acid esters in food. *Molecular Nutrition and Food Research* 55(4): 509–521.

Balaban, N. and Rasooly, A. 2000. Staphylococcal enterotoxins. *International Journal of Food Microbiology* 61(1): 1–10.

Balinova, A., Mladenova, R., and Shtereva, D. 2007. Solid-phase extraction on sorbents of different retention mechanisms followed by determination by gas chromatography–mass spectrometric and gas chromatography–electron capture detection of pesticide residues in crops. *Journal of Chromatography A* 1150(1): 136–144.

Banerjee, T., Sulthana, S., Shelby, T., et al. 2016. Multiparametric magneto-fluorescent nanosensors for the ultrasensitive detection of Escherichia coli O157:H7. *The Journal of Infectious Diseases* 2(10): 667–673.

Batra, B., Kumari, S., and Pundir, C. S. 2014. Construction of glutamate biosensor based on covalent immobilization of glutmate oxidase on polypyrrole nanoparticles/polyaniline modified gold electrode. *Enzyme and Microbial Technology* 57: 69–77.

Battaglia, R., Conacher, H. B. S., and Page, B. D. 1990. Ethyl carbamate (urethane) in alcoholic beverages and foods: A review. *Food Additives and Contaminants* 7(4): 477–496.

Bedford, E., Spadavecchia, J., Pradier, C. M., and Gu Frank, X. 2012. Surface plasmon resonance biosensors incorporating gold nanoparticles. *Macromolecular Bioscience* 12(6): 724–739.

Bhushani, J. A., N. Kumar Kurrey, and C. Anandharamakrishnan. 2017. Nanoencapsulation of green tea catechins by electrospraying technique and its effect on controlled release and in-vitro permeability. *Journal of Food Engineering* 199: 82–92.

Botsoglou, N. A. and Fletouris, D. J. 2000. *Drug Residues in Foods: Pharmacology, Food Safety, and Analysis*: Boca Raton, FL: CRC Press.

Boyd-Boland, A. A., Magdic, S., and Pawliszyn, J. B. 1996. Simultaneous determination of 60 pesticides in water using solid-phase microextraction and gas chromatography–mass spectrometry. *The Analyst* 121:7, 929–937.

Boyer, A. E., Gallegos-Candela, M., Lins, R. C., Kuklenyik, Z. et al. 2011. Quantitative mass spectrometry for bacterial protein toxins – A sensitive, specific, high-throughput tool for detection and diagnosis. *Molecules* 16(3): 2391–2413.

Branen, A. L., Davidson, P. M., Salminen, S., and Thorngate III, J. H. 2001. *Food Additives*, CRC Press. http://ariefm.lecture.ub.ac.id/files/2012/10/A._Larry_Branen_P._Michael_Davidson_Seppo_SalmiBookFi.org-FOOD-ADDITIVES.pdf (accessed May 2018).

Bulushi, I. A., Poole, S., Deeth, H. C., and Dykes, G. A. 2009. Biogenic amines in fish: Roles in intoxication, spoilage, and nitrosamine formation—a review. *Critical Reviews in Food Science and Nutrition* 49(4): 369–377.

Cai, H., Cao, X., Jiang, Y., He, P., and Fang, Y. 2003. Carbon nanotube-enhanced electrochemical DNA biosensor for DNA hybridization detection. *Analytical and Bioanalytical Chemistry* 375(2): 287–293.

Cai, D., Ren, L., Zhao, H., Xu, C., Zhang, L., Yu, Y., Wang, H., Lan, Y., Roberts, M. F., Chuang, J. H., Naughton, M. J., Ren, Z., and Chiles, T. C. 2010. A molecular-imprint nanosensor for ultrasensitive detection of proteins. *Nature Nanotechnology* 5(8): 597–601.

Cai, Z. Y., Pei, L. Z., Xie, Y. K., Fan, C. G., and Fu, D. G. 2013. Electrochemical determination of benzoic acid using CuGeO 3 nanowire modified glassy carbon electrode. *Measurement Science and Technology* 24(9): 095701.

Campos, B. B., Contreras-Cáceres, R., Bandosz, T. J., Jiménez-Jiménez, J., Rodríguez-Castellón, E., da Silva, J. C. G. E., and Algarra, M. 2017. Carbon dots coated with vitamin B12 as selective ratiometric nanosensor for phenolic carbofuran. *Sensors and Actuators B: Chemical* 239: 553–561.

Cappiello, A., Famiglini, G., Palma, P., and Mangani, F. 2002. Trace level determination of organophosphorus pesticides in water with the new direct-electron ionization LC/MS interface. *Analytical Chemistry* 74(14): 3547–3554.

Capuano, E. and Fogliano, V. 2011. Acrylamide and 5-hydroxymethylfurfural (HMF): A review on metabolism, toxicity, occurrence in food and mitigation strategies. *LWT – Food Science and Technology* 44(4): 793–810.

Carrasco, S., Benito-Peña, E., Navarro-Villoslada, F., Langer, J., Sanz-Ortiz, M. N., Reguera, J., Liz-Marzán, L. M., and Moreno-Bondi, M. C. 2016. Multibranched gold–mesoporous silica nanoparticles coated with a molecularly imprinted polymer for label-free antibiotic surface-enhanced Raman scattering analysis. *Chemistry of Materials* 28(21): 7947–7954.

CDC. 2018. List of selected multistate foodborne outbreak investigations. Washington, DC: Center for Disease Control. https://www.cdc.gov/foodsafety/outbreaks/multistate-outbreaks/outbreaks-list.html (accessed May 2018).

Charlton, A. J., Robb, P., Donarski, J. A., and Godward, J. 2008. Non-targeted detection of chemical contamination in carbonated soft drinks using NMR spectroscopy, variable selection and chemometrics. *Analytica Chimica Acta* 618(2): 196–203.

Chaudhry, Q., Schröder, P., Werck-Reichhart, D., Grajek, W., and Marecik, R. 2002. Prospects and limitations of phytoremediation for the removal of persistent pesticides in the environment. *Environmental Science and Pollution, Research* 9(1): 4.

Chen, H., Weiss, J., and Shahidi, F. 2006. Nanotechnology in Nutraceuticals and Functional Foods, *Food technology.*

Chen, X., Jing, H., Tao, Y., Carr, L., and Wheaton, F. 2003. Real-time detection of physical hazards in de-bonded poultry using high-resolution X-ray imaging. *ASAE International Annual Meeting of the American Society of Agricultural and Biological Engineers, 2003.*

Chen, J., Huang, Y., Kannan, P., Zhang, L., Lin, Z., Zhang, J., Chen, T., and Guo, L. 2016. Flexible and adhesive surface enhance raman scattering active tape for rapid detection of pesticide residues in fruits and vegetables. *Analytical Chemistry* 88(4): 2149–2155.

Choi, S. H., Kim, Y. L., and Byun, K. M. 2011. Graphene-on-silver substrates for sensitive surface plasmon resonance imaging biosensors. *Optics Express* 19(2): 458–466.

Corlett, D. A. and Stier, R. F. 1991. Risk assessment within the HACCP system. *Food Control* 2(2): 71–72.

Crews, C. and Castle, L. 2007. A review of the occurrence, formation and analysis of furan in heat-processed foods. *Trends in Food Science and Technology* 18(7): 365–372.

Deng, Y., Wang, W., Zhang, L., Lu, Z., Li, S., and Xu, L. 2013. Preparation and electrochemical behavior of L-glutamate electrochemical biosensor. *Journal of Biomedical Nanotechnology* 9(2): 318–321.

Ding, J., Lei, J., Ma, X., Gong, J., and Qin, W. 2014. Potentiometric aptasensing of Listeria monocytogenes using protamine as an indicator. *Analytical Chemistry* 86(19): 9412–9416.

Dong, Y., Wang, R., Li, G., Chen, C., Chi, Y., and Chen, G. 2012. Polyamine-functionalized carbon quantum dots as fluorescent probes for selective and sensitive detection of copper ions. *Analytical Chemistry* 84(14): 6220–6224.

Eaton, D. L. and Groopman, J. D. 2013. *The Toxicology of Aflatoxins: Human Health, Veterinary, and Agricultural Significance.* St Louis, MO: Elsevier.

Elghanian, R., Storhoff, J. J., Mucic, R. C., Letsinger, R. L., and Mirkin, C. A. 1997. Selective colorimetric detection of polynucleotides based on the distance-dependent optical properties of gold nanoparticles. *Science* 277(5329): 1078.

Eskenazi, B., Bradman, A., and Castorina, R. 1999. Exposures of children to organophosphate pesticides and their potential adverse health effects. *Environmental Health Perspectives,* 107(Suppl 3): 409–419.

Ezhilarasi, P.N., Karthik, P., Chhanwal, N., and Anandharamakrishnan, N. 2013. Nanoencapsulation techniques for food bioactive components: a review. *Food and Bioprocess Technology* 6 (3):628–647.

Health and Safety Executive. 2006. *Review of the Adequacy of Current Regulatory Regimes to Secure Effective Regulation of Nanoparticles Created by Nanotechnology: The Regulations Covered by HSE.* London: The Stationery Office.

Eychmüller, A., Mews, A., and Weller, H. 1993. A quantum dot quantum well: CdS/HgS/CdS. *Chemical Physics Letters* 208(1–2): 59–62.

FAO. 2018. Retrieved from www.fao.org/docrep/008/a0104e/a0104e08.html

Farzana, E. W. 2017. *Antinutritional factors in foods.* https://www.biotecharticles.com/Applications-Article/Antinutritional-Factors-In-Foods-4216.html (accessed June 2018).

Foubert, A., Beloglazova, N. V., and De Saeger, S. 2017. Comparative study of colloidal gold and quantum dots as labels for multiplex screening tests for multi-mycotoxin detection. *Analytica Chimica Acta* 955: 48–57.

Fu, Z., Zhou, X. and Xing, D. 2013. Rapid colorimetric gene-sensing of food pathogenic bacteria using biomodification-free gold nanoparticle. *Sensors and Actuators B: Chemical* 182: 633–641.

Fung, F. and Clark, R. F. 2004. Health effects of mycotoxins: A toxicological overview. *Journal of Toxicology: Clinical Toxicology* 42(2): 217–234.

Galvis-Sánchez, A., Barros, A., and Delgadillo, I. 2007. FTIR-ATR infrared spectroscopy for the detection of ochratoxin A in dried vine fruit. *Food Additives and Contaminants* 24(11): 1299–1305.

Gao, C., Liu, Z., Chen, J., and Yan, Z. 2012. A novel fluorescent assay for oxytetracycline hydrochloride based on fluorescence quenching of water-soluble CdTe nanocrystals. *Luminescence* 28(3): 378–383.

García, R. S., Silva, A. S., Cooper, I., Franz, R., and Losada, P. P. 2006. Revision of analytical strategies to evaluate different migrants from food packaging materials. *Trends in Food Science and Technology* 17(7): 354–366.

Ge, X., Asiri, A. M., Du, D., Wen, W., Wang, S., and Lin, Y. 2014. Nanomaterial-enhanced paper-based biosensors. *TrAC Trends in Analytical Chemistry* 58: 31–39.

Ghaedi, M., Shokrollahi, A., Kianfar, A. H., Mirsadeghi, A. S., Pourfarokhi, A., and Soylak, M. 2008. The determination of some heavy metals in food samples by flame atomic absorption spectrometry after their separation-preconcentration on bis salicyl aldehyde, 1,3 propan diimine (BSPDI) loaded on activated carbon. *Journal of Hazardous Materials* 154(1–3): 128–134.

Ghoreishi, S. M., Behpour, M., and Golestaneh, M. 2012. Simultaneous determination of Sunset Yellow and Tartrazine in soft drinks using gold nanoparticles carbon paste electrode. *Food Chemistry* 132(1): 637–641.

Gilliland, S. E. 1969. Enzymatic determination of residual hydrogen peroxide in milk. *Journal of Dairy Science* 52(3): 321–324.

Giovannetti, G. A. K. P. A., Khomyakov, P. A., Brocks, G., Karpan, V. V., Van den Brink, J., and Kelly, P. J. 2008. Doping graphene with metal contacts. *Physical Review Letters* 101(2).

Greene, R. V., Gordon, S. H., Jackson, M. A., Bennett, G. A., McClelland, J. F., and Jones, R. W. 1992. Detection of fungal contamination in corn: Potential of FTIR-PAS and-DRS. *Journal of Agricultural and Food Chemistry* 40(7): 1144–1149.

Gründler, P. 2007. *Chemical Sensors: An Introduction for Scientists and Engineers.* Berlin: Springer Science & Business Media.

Halvorson, R. A. and Vikesland, P. J. 2010. Surface-Enhanced Raman Spectroscopy (SERS) for environmental analyses. *Environmental Science and Technology* 44(20): 7749–7755.

He, L., Kim, N.-J., Li, H., Hu, Z., and Lin, M. 2008. Use of a fractal-like gold nanostructure in surface-enhanced raman spectroscopy for detection of selected food contaminants. *Journal of Agricultural and Food Chemistry* 56(21): 9843–9847.

Ho, H. P. and Lam, W. W. 2003. Application of differential phase measurement technique to surface plasmon resonance sensors. *Sensors and Actuators B: Chemical* 96(3): 554–559.

Ho, H. P., Law, W. C., Wu, S. Y., Liu, X. H., Wong, S. P., Lin, C., and Kong, S. K. 2006. Phase-sensitive surface plasmon resonance biosensor using the photoelastic modulation technique. *Sensors and Actuators B: Chemical* 114(1): 80–84.

Hoa, X. D., Kirk, A. G., and Tabrizian, M. 2007. Towards integrated and sensitive surface plasmon resonance biosensors: A review of recent progress. *Biosensors and Bioelectronics* 23(2): 151–160.

Homola, J., Yee, S. S., and Gauglitz, G. 1999. Surface plasmon resonance sensors: review. *Sensors and Actuators B: Chemical* 54(1): 3–15.

Huang, X., Liu, Q., Yao, S., and Jiang, G. 2017. Recent progress in the application of nanomaterials in the analysis of emerging chemical contaminants. *Analytical Methods* 9(19): 2768–2783.

Inbaraj, S.B. and Chen, B. H. 2016. Nanomaterial-based sensors for detection of foodborne bacterial pathogens and toxins as well as pork adulteration in meat products. *Journal of Food and Drug Analysis* 24(1): 15–28.

Ingelfinger, J. R. 2008. Melamine and the global implications of food contamination. New *England Journal of Medicine* 359(26): 2745–2748.

Jain, K. K. 2005. Nanotechnology in clinical laboratory diagnostics. *Clinica Chimica Acta* 358(1–2): 37–54.

Jain, S., Davis, V. A., Pillai, S. R., Pillai, S., Singh, S. R., Horn, D. W., and Kumar Ram, M. 2012. Development of an antibody functionalized carbon nanotube biosensor for foodborne bacterial pathogens enzymatic fuel cells view project role of TGF β and regulatory T cells in prostate cancer article deep desulfurization of diesel fuel by selective adsorption over Ni/Al2O3 and Ni/ZSM-5 extrudates view project development of an antibody functionalized carbon nanotube biosensor for foodborne bacterial pathogens. *Journal of Biosensors and Bioelectronics* S11:002.

Jakubec, P., Urbanová, V., Medříková, Z., and Zbořil, R. 2016. Advanced sensing of antibiotics with magnetic gold nanocomposite: Electrochemical detection of chloramphenicol. *Chemistry – A European Journal* 22(40): 14279–14284.

Jarraud, S., Cozon, G., Vandenesch, F., Bes, M., Etienne, J., and Lina, G. 1999. Involvement of enterotoxins G and I in staphylococcal toxic shock syndrome and staphylococcal scarlet fever. *Journal of Clinical Microbiology* 37(8): 2446–2449.

Jemmi, T. and Stephan, R. 2006. Listeria monocytogenes: Food-borne pathogen and hygiene indicator. *Revue Scientifique et Technique* 25(2): 571–580.

Jerónimo, P. C. A., Araújo, A. N., and Conceição B.S.M. Montenegro, M. 2007. Optical sensors and biosensors based on sol–gel films. *Talanta* 72(1): 13–27.

Jia, F., Xu, L., Yan, W., Wu, W., Yu, Q., Tian, X., Dai, R., and Li, X. 2017. A magnetic relaxation switch aptasensor for the rapid detection of Pseudomonas aeruginosa using superparamagnetic nanoparticles. *Microchimica Acta* 184(5): 1539–1545.

Joo, J., Yim, C., Kwon, D., Lee, J., Shin, H. H., Cha, H. J., and Jeon, S. 2012. A facile and sensitive detection of pathogenic bacteria using magnetic nanoparticles and optical nanocrystal probes. *Analyst* 137(16): 3609–3612.

Kaittanis, C., Santra, S., and Perez, J. M. 2010. Emerging nanotechnology-based strategies for the identification of microbial pathogenesis. *Advanced Drug Delivery Reviews* 62(4–5): 408–423.

Kalluri, J. R., Arbneshi, T., Afrin Khan, S., Neely, A., Candice, P., Varisli, B., Washington, M., McAfee, S., Robinson, B., Banerjee, S., Singh, A. K., Senapati, D., and Ray, P. C. 2009. Use of gold nanoparticles in a simple colorimetric and ultrasensitive dynamic light scattering assay: selective detection of arsenic in groundwater. *Angewandte Chemie International Edition* 48(51): 9668–9671.

Karthik, P., Ezhilarasi, P. N., and Anandharamakrishnan, C. 2017. Challenges associated in stability of food grade nanoemulsions. *Critical Reviews in Food Science and Nutrition* 57(7): 1435–1450.

Kaushik, G., Satya, S., and Naik, S. N. 2008. Food processing a tool to pesticide residue dissipation – A review. *Food Research International* 42: 26–40.

Khan, M. A., Rao, R. M., Novak, A. F., and Murthy, M. R. V. 1976. Polychlorinated biphenyls (Pcbs) in food. *C R C Critical Reviews in Food Science and Nutrition* 7(2): 103–145.

Krishnan, V., Permuth, F., and Alto, P. 1993. Bisphenol-A: an estrogenic substance is released from polycarbonate flasks during autoclaving. *Endocrinology, 132*(6), 2279–2286.

Kubackova, J., Fabriciova, G., Miskovsky, P., Jancura, D., and Sanchez-Cortes, S. 2015. Sensitive Surface-Enhanced Raman Spectroscopy (SERS) detection of organochlorine pesticides by Alkyl Dithiol-functionalized metal nanoparticles-induced plasmonic hot spots. *Analytical Chemistry* 87(1): 663–669.

Larsen, J. C. 2006. Risk assessments of polychlorinated dibenzo-p-dioxins, polychlorinated dibenzofurans, and dioxin-like polychlorinated biphenyls in food. *Molecular Nutrition and Food Research* 50(10): 885–896.

Lee, J.-S., Ulmann, P. A., Han, M. S., and Mirkin, C. A. 2008. A DNA–gold nanoparticle-based colorimetric competition assay for the detection of cysteine. *Nano Letters* 8(2): 529–533.

Lee, J., Mahendra, S., and Alvarez, P. J. J. 2010. Nanomaterials in the construction industry: A review of their applications and environmental health and safety considerations. *ACS Nano* 4(7): 3580–3590.

Li, M., Wang, Q., Shi, X., Hornak, L. A., and Wu, N. 2011. Detection of Mercury(II) by Quantum Dot/DNA/gold nanoparticle ensemble based nanosensor via nanometal surface energy transfer. *Analytical Chemistry* 83(18): 7061–7065.

Li, H., Xie, C., Li, S., and Xu, K. 2012. Electropolymerized molecular imprinting on gold nanoparticle-carbon nanotube modified electrode for electrochemical detection of triazophos. *Colloids and Surfaces B: Biointerfaces* 89: 175–181.

Li, Y., Xu, J., Wang, L., Huang, Y., Guo, J., Cao, X., Shen, F., Luo, Y., and Sun, C. 2016. Aptamer-based fluorescent detection of bisphenol A using nonconjugated gold nanoparticles and CdTe quantum dots. *Sensors and Actuators B: Chemical* 222: 815–822.

Li, C.-H., Bai, Y.-L., Selvaprakash, K., Mong, K.-K. T., and Chen, Y.-C. 2017. Selective detection of shiga-like toxin 1 from complex samples using pigeon ovalbumin functionalized gold nanoparticles as affinity probes. *Journal of Agricultural and Food Chemistry* 65(21): 4359–4365.

Lian, W., Wu, D., Lim, D. V., and Jin, S. 2010. Sensitive detection of multiplex toxins using antibody microarray. *Analytical Biochemistry* 401(2): 271–279.

Liang, Q., Qu, J., Luo, G., and Wang, Y. 2006. Rapid and reliable determination of illegal adulterant in herbal medicines and dietary supplements by LC/MS/MS. *Journal of Pharmaceutical and Biomedical Analysis* 40(2): 305–311.

Lin, Y., Lu, F., and Wang, J. 2004. Disposable carbon nanotube modified screen-printed biosensor for amperometric detection of organophosphorus pesticides and nerve agents. *Electroanalysis* 16(12): 145–149.

Lin, N., Pei, L., Wei, T., Liu, H., and Cai, Z. 2016. Electrochemical sensor based on glassy carbon electrode modified with copper vanadate nanobelts for detection of benzoic acid. *IET Science, Measurement and Technology* 10(4): 247–252.

Ling, K., Jiang, H., Zhang, L., Li, Y., Yang, L., Qiu, C., and Li, F.-R. 2016. A self-assembling RNA aptamer-based nanoparticle sensor for fluorometric detection of Neomycin B in milk. *Analytical and Bioanalytical Chemistry* 408(13): 3593–3600.

Liou, P., Nayigiziki, F. X., Kong, F., Mustapha, A., and Lin, M. 2017. Cellulose nanofibers coated with silver nanoparticles as a SERS platform for detection of pesticides in apples. *Carbohydrate Polymers* 157: 643–650.

Liu, S., Yuan, L., Yue, X., Zheng, Z., and Tang, Z. 2008. Recent advances in nanosensors for organophosphate pesticide detection. *Advanced Powder Technology* 19(5): 419–441.

Liu, D., Chen, W., Wei, J., Li, X., Wang, Z., and Jiang, X. 2012. A highly sensitive, dual-readout assay based on gold nanoparticles for organophosphorus and carbamate pesticides. *Analytical Chemistry* 84(9): 4185–4191.

Liu, X., Zhang, N., Bing, T., and Shangguan, D. 2014. Carbon dots based dual-emission silica nanoparticles as a ratiometric nanosensor for Cu^{2+}. *Analytical Chemistry* 86(5): 2289–2296.

Liu, J., Lu, L., Xu, S., and Wang, L. 2015. One-pot synthesis of gold nanoclusters with bright red fluorescence and good biorecognition Abilities for visualization fluorescence enhancement detection of E. coli. *Talanta* 134: 54–59.

Lu Sun et al., 2016. *Enhanced Molecular Spectroscopy via Localized Surface Plasmon Resonance*. pg. 3. United Kingdom: Intech.

López-Diez, E. C., Bianchi, G., and Goodacre, R. 2003. Rapid quantitative assessment of the adulteration of virgin olive oils with hazelnut oils using Raman spectroscopy and chemometrics. *Journal of Agricultural and Food Chemistry* 51(21): 6145–6150.

López-Campos, G., Martinez-Suarez, J., Aguado-Urda, M., and Lopez-Alonso, V. 2012. *Microarray detection and characterization of bacterial foodborne pathogens*. Springer Verlag: Springer US.

Lu, L.-Q., Tan, T., Tian, X.-K., Li, Y., and Deng, P. 2017. Visual and sensitive fluorescent sensing for ultratrace mercury ions by perovskite quantum dots. *Analytica Chimica Acta* 986: 109–114.

Ma, M., Wen, K., Beier, R. C., Eremin, S. A., Li, C., Zhang, S., Shen, J., and Wang, Z. 2016. Chemiluminescence resonance energy transfer competitive immunoassay employing hapten-functionalized quantum dots for the detection of sulfamethazine. *ACS Applied Materials and Interfaces* 8(28) 17745–17750.

Ma, L.-H., Wang, H.-B., Fang, B.-Y., Tan, F., Cao, Y.-C., and Zhao, Y.-D. 2018. Visual detection of trace lead ion based on aptamer and silver staining nano-metal composite. *Colloids and Surfaces B: Biointerfaces* 162: 415–419.

Mauer, L. J., Chernyshova, A. A., Hiatt, A., Deering, A., and Davis, R. 2009. Melamine detection in infant formula powder using near- and mid-infrared spectroscopy. *Journal of Agricultural and Food Chemistry* 57(10): 3974–3980.

Maurer, E. I., Comfort, K. K., Hussain, S. M., Schlager, J. J., and Mukhopadhyay, S. M. 2012. Novel platform development using an assembly of carbon nanotube, nanogold and immobilized RNA capture element towards rapid, selective sensing of bacteria. *Sensors* 12(6): 8135–8144.

Mazzetti, M., Fascioli, R., Mazzoncini, I., Spinelli, G., Morelli, I., and Bertoli, A. 2004. Determination of 1-phenylazo-2-naphthol (Sudan I) in chilli powder and in chilli-containing food products by GPC clean-up and HPLC with LC/MS confirmation. *Food Additives and Contaminants* 21(10): 935–941.

Mehta, J., Vinayak, P., Tuteja, S. K., Chhabra, V. A., Bhardwaj, N., Paul, A. K., Kim, K.-H., and Deep, A. 2016. Graphene modified screen printed immunosensor for highly sensitive detection of parathion. *Biosensors and Bioelectronics* 83: 339–346.

Moore, J. C., Spink, J., and Lipp, M. 2012. Development and application of a database of food ingredient fraud and economically motivated Adulteration from 1980 to 2010. *Journal of Food Science* 77(4): 118–126.

Mota, F. J. M., Ferreira, I. M. P. L. V. O., Cunha, S. C., Beatriz, M., and Oliveira, P. P. 2003. Optimisation of extraction procedures for analysis of benzoic and sorbic acids in foodstuffs. *Food Chemistry* 82(3): 469–473.

Nagatsuka T., Uzawa, H., Sato, K., Kondo, S., Izumi, M., Yokoyama, K., Ohsawa, I., Seto, Y., Neri, P., Mori, H., Nishida, Y. 2013. Localized surface plasmon resonance detection of biological toxins using cell surface oligosaccharides on glyco chips. *ACS Applied Materials and Interfaces* 5(10): 4173–4180.

Nair, R.R., Ren, W., Jalil, R., Riaz, I., Kravets, V. G., Britnell, L., Blake, P., Schedin, F., Mayorov, A. S., Yuan, S., Katsnelson, M. I. 2010. Fluorographene: A two-dimensional counterpart of Teflon. *Small* 6(24): 2877–2884.

Najafi, M., Khalilzadeh, M. A., and Karimi-Maleh, H. 2014. A new strategy for determination of bisphenol A in the presence of Sudan I using a ZnO/CNTs/ionic liquid paste electrode in food samples. *Food Chemistry* 158: 125–131.

Neŕin, C., Philo, M. R., Salafranca, J., and Castle, L. 2002. Determination of bisphenol-type contaminants from food packaging materials in aqueous foods by solid-phase microextraction–high-performance liquid chromatography. *Journal of Chromatography A* 963(1–2): 375–380.

Noiseux, I., Bouchard, J. P., Gallant, P., Bourqui, P., Cao, H., Vernon, M., and Mermut, O. 2010. Detection of food-borne pathogens by nanoparticle technology coupled to a low-cost cell reader. In *Frontiers in Pathogen Detection: From Nanosensors to Systems*, ed. Fauchet, P. M. 7553: 75530 (March). International Society for Optics and Photonics.

Nsibande, S. A., and Forbes, P. B. C. 2016. Fluorescence detection of pesticides using quantum dot materials–A review. *Analytica Chimica Acta* 945: 9–22.

Ohk, S. H., Koo, O. K., Sen, T., Yamamoto, C. M., and Bhunia, A. K. 2010. Antibody-aptamer functionalized fibre-optic biosensor for specific detection of Listeria monocytogenes from food. *Journal of Applied Microbiology* 109(3): 808–817.

Ono, Y., Yamagami, T., Nishina, T., and Tobino, T. 2006. Pesticide multiresidue analysis of 303 compounds using supercritical fluid extraction. Analytical sciences: *The International Journal of the Japan Society for Analytical Chemistry* 22(11): 1473–1476.

Pareja, L., Cesio, V., Heinzen, H., and Fernández-Alba, A. R. 2011. Evaluation of various QuEChERS based methods for the analysis of herbicides and other commonly used pesticides in polished rice by LC--MS/MS. *Talanta* 83(5): 1613–1622.

Perçin, I., Idil, N., Bakhshpour, M., Yılmaz, E., Mattiasson, B., and Denizli, A. 2017. Microcontact imprinted plasmonic nanosensors: Powerful tools in the detection of salmonella paratyphi. *Sensors* 17(6), 1375.

Perumal, V. and Hashim, U. 2014. Advances in biosensors: Principle, architecture and applications. *Journal of Applied Biomedicine,* 12(1): 1–15.

Prabakaran, E. and Pandian, K. 2015. Amperometric detection of Sudan I in red chili powder samples using Ag nanoparticles decorated graphene oxide modified glassy carbon electrode. *Food Chemistry* 166: 198–205.

Prasad, P. N. 2003. Bioimaging: principles and techniques. In *Introduction to Biophotonics*, ed. Prasad, P. N., 203–254. Hoboken, NJ: Wiley.

Qiu, S., Zhao, F., Zenasni, O., Li, J., and Shih, W.-C. 2016. Nanoporous gold disks functionalized with stabilized G-Quadruplex moieties for sensing small molecules. *ACS Applied Materials and Interfaces* 8(44): 29968–29976.

Qu, Q., Zhu, A., Shao, X., Shi, G., and Tian, Y. 2012. Development of a carbon quantum dots-based fluorescent Cu^{2+} probe suitable for living cell imaging. *Chemical Communications* 48(44): 5473–5475.

Rees, N. and Watson, D. 2000. *International Standards for Food Safety*. Gaithersburg, MD: Aspen Publishers.

Rhodehamel, E. J. 1992. Overview of biological, chemical, and physical hazards. In Pierson M.D., Corlett D.A. (eds) *HACCP* 8–28. Boston, MA: Springer.

Richter, C. A., Birnbaum, L. S., Farabollini, F., Newbold, R. R., Rubin, B. S., Talsness, C. E., Vandenbergh, J. G., Walser-Kuntz, D. R., and vom Saal, F. S. 2007. In vivo effects of bisphenol a in laboratory rodent studies. *Reproductive Toxicology (Elmsford, N.Y.)* 24(2): 199–224.

Rissato, S. R., Galhiane, M. S., de Almeida, M. V., Gerenutti, M., and Apon, B. M. 2007. Multiresidue determination of pesticides in honey samples by gas chromatography–mass spectrometry and application in environmental contamination. *Food Chemistry 101*(4): 1719–1726.

Rodriguez-Mozaz, S., Lopez de Alda, M. J., and Barceló, D. 2006. Biosensors as useful tools for environmental analysis and monitoring. *Analytical and Bioanalytical Chemistry* 386: 1025–1041.

Rong-Hwa, S., Shiao-Shek, T., Der-Jiang, C., and Yao-Wen, H. 2010. Gold nanoparticle-based lateral flow assay for detection of staphylococcal enterotoxin B. *Food Chemistry* 118(2): 462–466.

Rosi, N. L. and Mirkin, C. A. 2005. Nanostructures in biodiagnostics. *Chemical Reviews* 105(4): 1547–1562.

Sahraei, R., Farmany, A., and Mortazavi, S. S. 2013. A nanosilver-based spectrophotometry method for sensitive determination of tartrazine in food samples. *Food Chemistry* 138(2): 1239–1242.

Salimi, F., Kiani, M., Karami, C., and Taher, M. A. 2018. Colorimetric sensor of detection of Cr (III) and Fe (II) ions in aqueous solutions using gold nanoparticles modified with methylene blue. *Optik* 158: 813–825.

Sandvig, K. 2001. Shiga toxins. *Toxicon* 39(11): 1629–1635.

Scheutz, F., Nielsen, E. M., Frimodt-Møller, J., Boisen, N., Morabito, S., Tozzoli, R., Nataro, J. P., and Caprioli, A. 2011. Characteristics of the enteroaggregative Shiga toxin/verotoxin-producing Escherichia coli O104: H4 strain causing the outbreak of haemolytic uraemic syndrome in Germany, May to June 2011. *Eurosurveillance* 16(24): 19889.

Shalabney, A. and Abdulhalim, I. 2011. Sensitivity-enhancement methods for surface plasmon sensors. *Laser and Photonics Reviews*, 5(4): 571–606.

Shrivastav, A. M., Usha, S. P., and Gupta, B. D. 2017. Highly sensitive and selective erythromycin nanosensor employing fiber optic SPR/ERY imprinted nanostructure: application in milk and honey. *Biosensors and Bioelectronics* 90: 516–524.

Sicbaldi, F., Sarra, A., Mutti, D., and Bo, P. F. 1997. Use of gas-liquid chromatography with electron-capture and thermionic-sensitive detection for the quantitation and identification of pesticide residues. *Journal of Chromatography A* 765(1): 13–22.

Situ, C., Mooney, M. H., Elliott, C. T., and Buijs, J. 2010. Advances in surface plasmon resonance biosensor technology towards high-throughput, food-safety analysis. *TrAC Trends in Analytical Chemistry* 29(11): 1305–1315.

Skoulika, S. G., Georgiou, C. A., and Polissiou, M. G. 2000. FT-Raman spectroscopy Analytical tool for routine analysis of diazinon pesticide formulations. *Talanta* 51(3): 599–604.

Śliwińska, M., Wiśniewska, P., Dymerski, T., Namieśnik, J., and Wardencki, W. 2014. Food analysis using artificial senses. *Journal of Agricultural and Food Chemistry* 62(7): 1423–1448.

Slocik, J. M., Zabinski, J. S., Phillips, D. M., and Naik, R. R. 2008. Colorimetric response of peptide-functionalized gold nanoparticles to metal ions. *Small* 4(5): 548–551.

Soldatkin, O. O., Burdak, O. S., Sergeyeva, T. A., Arkhypova, V. M., Dzyadevych, S. V., and Soldatkin, A. P. 2013. Acetylcholinesterase-based conductometric biosensor for determination of aflatoxin B1. *Sensors and Actuators B: Chemical* 188: 999–1003.

Stadler, R. H., Blank, I., Varga, N., Robert, F., Hau, J., Guy, P. A., and Riediker, S. 2002. Food chemistry: Acrylamide from Maillard reaction products. *Nature 419*(6906): 449.133.

Stan, H.-J. and Linkerhägner, M. 1996. Pesticide residue analysis in foodstuffs applying capillary gas chromatography with atomic emission detection State-of-the-art use of modified multimethod S19 of the Deutsche Forschungsgemeinschaft and automated large-volume injection with programmed-temperature vaporization and solvent venting. *Journal of Chromatography A* 750(1): 369–390.

Stewart, K. K. 1987. On reasons for a new journal. *Journal of Food Composition and Analysis* 1(1): 1–2.

Stover, C. K., Pham, X. Q., Erwin, A. L., et al. 2000. Complete genome sequence of Pseudomonas aeruginosa PAO1, an opportunistic pathogen. *Nature*, 406(6799): 959–964.

Su, S., He, Y., Zhang, M., Yang, K., Song, S., Zhang, X., Fan, C., and Lee, S.-T. 2008. High-sensitivity pesticide detection via silicon nanowires-supported acetylcholinesterase-based electrochemical sensors. *Applied Physics Letters* 93(2): 023113.

Sun, L., Chen, P., and Lin, L. 2016. Enhanced molecular spectroscopy via localized surface plasmon resonance. In *Applications of Molecular Spectroscopy to Current Research in the Chemical and Biological Sciences,* ed. Stauffer, M. London InTech. https://www.intechopen.com/books/applications-of-molecular-spectroscopy-to-current-research-in-the-chemical-and-biological-sciences/enhanced-molecular-spectroscopy-via-localized-surface-plasmon-resonance (accessed May 2018).

Syshchyk, O., Skryshevsky, V. A., Soldatkin, O. O., and Soldatkin, A. P. 2015. Enzyme biosensor systems based on porous silicon photoluminescence for detection of glucose, urea and heavy metals. *Biosensors and Bioelectronics* 66: 89–94.

Szunerits, S., Maalouli, N., Wijaya, E., Vilcot, J.-P., and Boukherroub, R. 2013. Recent advances in the development of graphene-based surface plasmon resonance (SPR) interfaces. *Analytical and Bioanalytical Chemistry* 405(5): 1435–1443.

Tan, E.-Z., Yin, P.-G., You, T.-t., Wang, H., and Guo, L. 2012. Three dimensional design of large-scale TiO_2 nanorods scaffold decorated by silver nanoparticles as SERS sensor for ultrasensitive malachite green detection. *ACS Applied Materials and Interfaces* 4(7): 3432–3437.

Tanaka, T., Yoneda, A., Inoue, S., Sugiura, Y., and Ueno, Y. 2000. Simultaneous determination of trichothecene mycotoxins and zearalenone in cereals by gas chromatography–mass spectrometry. *Journal of Chromatography A* 882(1–2): 23–28.

Tark, S.-H., Das, A., Sligar, S., and Dravid, V. P. 2010. Nanomechanical detection of cholera toxin using microcantilevers functionalized with ganglioside nanodiscs. *Nanotechnology* 21: 435502.

Tfouni, S. A. V. and Toledo, M. C. F. 2002. Determination of benzoic and sorbic acids in Brazilian food. *Food Control* 13(2): 117–123.

Thatai, S., Khurana, P., Prasad, S., and Kumar, D. 2014. A new way in nanosensors: Gold nanorods for sensing of Fe(III) ions in aqueous media. *Microchemical Journal* 113: 77–82.

Tian, X., Wang, J., Li, Y., Yang, C., Lu, L., and Nie, Y. 2018. Sensitive determination of hardness and fluoride in ground water by a hybrid nanosensor based on aggregation induced FRET on and off mechanism. *Sensors and Actuators B: Chemical* 262: 522–530.

Vargas-Bernal, R., Rodriguez-Miranda, E., and Herrera-Prez, G. 2012. Evolution and expectations of enzymatic biosensors for pesticides. In *Pesticides*, ed. Soundararajan, R.P. London: InTech,.

Verma, M. S., Rogowski, J. L., Jones, L., and Gu, F. X. 2015. Colorimetric biosensing of pathogens using gold nanoparticles. *Biotechnology Advances* 33(6): 666–680.

Viñas, P., Campillo, N., Martínez-Castillo, N., and Hernández-Córdoba, M. 2010. Comparison of two derivatization-based methods for solid-phase microextraction-gas chromatography-mass spectrometric determination of bisphenol A, bisphenol S and biphenol migrated from food cans. *Analytical and Bioanalytical Chemistry* 397(1): 115–125.

Viswanathan, S., Radecka, H., and Radecki, J. 2009. Electrochemical biosensors for food analysis. *Monatshefte für Chemie-Chemical Monthly* 140(8): 891.

Wang, J. 2005. Nanomaterial-based electrochemical biosensors. *Analyst* 130(4): 421–426.

Wang, J., Liu, G., and Jan, M. R. 2004. ultrasensitive electrical biosensing of proteins and DNA: Carbon-nanotube derived amplification of the recognition and transduction events. *Journal of the American Chemical Society* 126(10) 3010–3011.

Wang, S., Wang, X., Chen, X., Cao, X., Cao, J., Xiong, X., and Zeng, W. 2016. A novel upconversion luminescence turn-on nanosensor for ratiometric detection of organophosphorus pesticides. *RSC Advances* 6(52): 46317–46324.

Wenzl, T., Simon, R., Anklam, E., and Kleiner, J. 2006. Analytical methods for polycyclic aromatic hydrocarbons (PAHs) in food and the environment needed for new food legislation in the European Union. *TrAC Trends in Analytical Chemistry* 25(7): 716–725.

Willets, K. A., and Van Duyne, R. P. 2007. Localized surface plasmon resonance spectroscopy and sensing. *Annual Review of Physical Chemistry* 58: 267–297.

World Health Organization, International Programme on Chemical Safety, Joint FAO/WHO Expert Committee on Food Additives, WHO Task Group on Updating the Principles for the Safety Assessment of Food Additives and Contaminants in Food, International Labour Organization. et al. 1987. Principles for the safety assessment of food additives and contaminants in food/published under the joint sponsorship of the United Nations Environment Programme, the International Labour Organisation, and the World Health Organization in collaboration with the Food and Agriculture Organization of the United Nations. Geneva: World Health Organization. http://www.who.int/iris/handle/10665/37578 (accessed June 2018).

Wu, C.-S., Khaing Oo, M. K., and Fan, X. 2010a. Highly sensitive multiplexed heavy metal detection using quantum-dot-labeled DNAzymes. *ACS Nano* 4(10): 5897–5904.

Wu, L., Chu, H. S., Koh, W. S., and Li, E. P. 2010b. Highly sensitive graphene biosensors based on surface plasmon resonance. *Optics Express* 18(14): 14395–14400.

Wu, S.-P., Chen, Y.-P., and Sung, Y.-M. 2011. Colorimetric detection of Fe3+ ions using pyrophosphate functionalized gold nanoparticles. *Analyst* 136(9): 1887–1891.

Wu, Y., Tang, L., Huang, L., Han, Z., Wang, J., and Pan, H. 2014. A low detection limit penicillin biosensor based on single graphene nanosheets preadsorbed with hematein/ionic liquids/penicillinase. *Materials Science and Engineering* C 39: 92–99.

Xiarong Zhou, S. W., Hong Liu, Xiaoping Wu, Qingchuan Zhang. 2016. Label-free detection of aflatoxin B1 using a nanomechanical sensor. *International Journal of Mechanical Engineering and Applications* 4(2): 81–87.

Xu, S., Yuan, H., Chen, S., Xu, A., Wang, J., and Wu, L. 2012. Selection of DNA aptamers against polychlorinated biphenyls as potential biorecognition elements for environmental analysis. *Analytical Biochemistry* 423(2): 195–201.

Xu, H., Yang, X., Li, G., Zhao, C., and Liao, X. 2015. Green synthesis of fluorescent carbon dots for selective detection of tartrazine in food samples. *Journal of Agricultural and Food Chemistry* 63(30): 6707–6714.

Yang, H. and Irudayaraj, J. 2003. Rapid detection of foodborne microorganisms on food surface using Fourier transform Raman spectroscopy. *Journal of Molecular Structure* 646(1–3): 35–43.

Yang, L., and Li, Y. 2005. Quantum dots as fluorescent labels for quantitative detection of Salmonella typhimurium in chicken carcass wash water. *Journal of Food Protection* 68(6): 1241–1245.

Yang, G.-J., Huang, J.-L., Meng, W.-J., Shen, M., and Jiao, X.-A. 2009. A reusable capacitive immunosensor for detection of Salmonella spp. based on grafted ethylene diamine and self-assembled gold nanoparticle monolayers. *Analytica Chimica Acta* 647(2): 159–166.

Yang, L., Chen, B., Luo, S., Li, J., Liu, R., and Cai, Q. 2010. Sensitive detection of polycyclic aromatic hydrocarbons using CdTe Quantum Dot-Modified TiO2 nanotube array through fluorescence resonance energy transfer. *Environmental Science and Technology* 44(20): 7884–7889.

Ye, B. C., and Yin, B. C. 2008. Highly sensitive detection of mercury(II) ions by fluorescence polarization enhanced by gold nanoparticles. *Angewandte Chemie International Edition* 47(44): 8386–8389.

Yoosaf, K., Ipe, B. I., Suresh, C. H., and Thomas, K. G. 2007. In situ synthesis of metal nanoparticles and selective naked-eye detection of lead ions from aqueous media. *The Journal of Physical Chemistry C* 111(34): 12839–12847.

Zeng, S., Yong, K.-T., Roy, I., Dinh, X.-Q., Yu, X., and Luan, F. 2011. A review on functionalized gold nanoparticles for biosensing applications. *Plasmonics* 6(3): 491.

Zhai, C., Sun, X., Zhao, W., Gong, Z., and Wang, X. 2013. Acetylcholinesterase biosensor based on chitosan/prussian blue/multiwall carbon nanotubes/hollow gold nanospheres nanocomposite film by one-stepelectrodeposition. *Biosensors and Bioelectronics* 42: 124–130.

Zhang, L. 2013. Self-assembly Ag nanoparticle monolayer film as SERS Substrate for pesticide detection. *Applied Surface Science* 270: 292–294.

Zhang, L., and Chen, L. 2016. Fluorescence probe based on hybrid mesoporous silica/quantum dot/molecularly imprinted polymer for detection of tetracycline. *ACS Applied Materials and Interfaces* 8(25): 16248–16256.

Zhang, L., and Fang, M. 2010. Nanomaterials in pollution trace detection and environmental improvement. *Nano Today* 5(2): 128–142.

Zhang, C. Y., Yeh, H. C., Kuroki, M. T., and Wang, T. H. 2005. Single-quantum-dot-based DNA nanosensor. *Nature Materials 4*(11): 826.

Zhang, J., Wang, L., Pan, D., Song, S., Boey Freddy, Y. C., Zhang, H., and Fan, C. 2008. Visual cocaine detection with gold nanoparticles and rationally engineered aptamer structures. *Small* 4(8): 1196–1200.

Zhang, L., Wu, L.-L., Wang, Y.-P., Liu, A.-M., Zou, C.-C., and Zhao, Z.-Y. 2009. Melamine-contaminated milk products induced urinary tract calculi in children. *World Journal of Pediatrics*, 5(1): 31–35.

Zhang, K., Mei, Q., Guan, G., Liu, B., Wang, S., and Zhang, Z. 2010. ligand replacement-induced fluorescence switch of quantum dots for ultrasensitive detection of organophosphorothioate pesticides. *Analytical Chemistry* 82(22): 9579–9586.

Zhang, Z., Li, P., Hu, X., Zhang, Q., Ding, X., and Zhang, W. 2012. Microarray technology for major chemical contaminants analysis in food: current status and prospects. *Sensors* 12(7): 9234–9252.

Zhang, C., Zhang, Z., and Li, G. 2014. Preparation of sulfonated graphene/polypyrrole solid-phase microextraction coating by in situ electrochemical polymerization for analysis of trace terpenes. *Journal of Chromatography A* 1346: 8–15.

Zhang, C. H., Liu, L. W., Liang, P., Tang, L. J., Yu, R. Q., and Jiang, J. H. 2016. Plasmon coupling enhanced raman scattering nanobeacon for single-step, ultrasensitive detection of cholera toxin. *Analytical Chemistry* 88(15): 7447–7452.

Zhao, Y., Li, Y., Jiang, K., Wang, J., White, W. L., Yang, S., and Lu, J. 2017. Rapid detection of Listeria monocytogenes in food by biofunctionalized magnetic nanoparticle based on nuclear magnetic resonance. *Food Control* 71: 110–116.

Zhou, D., Xie, G., Cao, X., Chen, X., Zhang, X., and Chen, H. 2016. Colorimetric determination of staphylococcal enterotoxin B via DNAzyme-guided growth of gold nanoparticles. *Microchimica Acta* 183(10): 2753–2760.

Zhu, X., Wang, S., Liu, Q. I., Xu, Q., Xu, S., and Chen, H. 2009. Determination of residues of cyromazine and its metabolite, melamine, in animal-derived food by gas chromatography-mass spectrometry with derivatization. *Journal of Agricultural and Food Chemistry* 57(23):11075–11080.

15 Biosensors for Food Component Analysis

Praveena Bhatt, Monali Mukherjee, and Uchangi Satyaprasad Akshath

CONTENTS

15.1 INTRODUCTION

In the past three decades, biosensor research has experienced explosive growth and has generated interest among researchers for its applicability in various fields. A biosensor is an analytical device which converts a biological response into a quantifiable and processable signal (Eggins, 1996). It is comprised of a biological or biologically derived sensing element, also called a biorecognition element (e.g. enzymes, antibodies, micro-organisms, DNA, RNA, etc.) either integrated with or in intimate contact with a physicochemical transducer (e.g. electrochemical, optical, thermometric, or piezoelectric) (see Figure 15.1). In essence, the transducer coupled to the biological recognition element converts or transforms the specific recognition between the biomolecule and the analyte (biological signal) into an electronic one (Eggins, 1996).

Biosensors find wide application in various fields, such as food and agriculture, food processing, environment, medicine, therapeutics, diagnostics, etc. Biosensor development is a multi- and interdisciplinary field that is currently one of the most active areas of research in analytical chemistry. The reason for this is that biosensors typically eliminate the need for sample preparation, unlike conventional analytical techniques that require elaborate sample preparation steps and skilled manpower for analysis. The performance of biosensors when evaluated based on selectivity, sensitivity, limit of detection (LOD), linear and dynamic ranges, reproducibility or precision of the response, and response in varied matrices also superior (Kissinger, 2005). Other parameters that are often compared include the sensor's response time, operational and storage stability, ease of use, portability, etc.

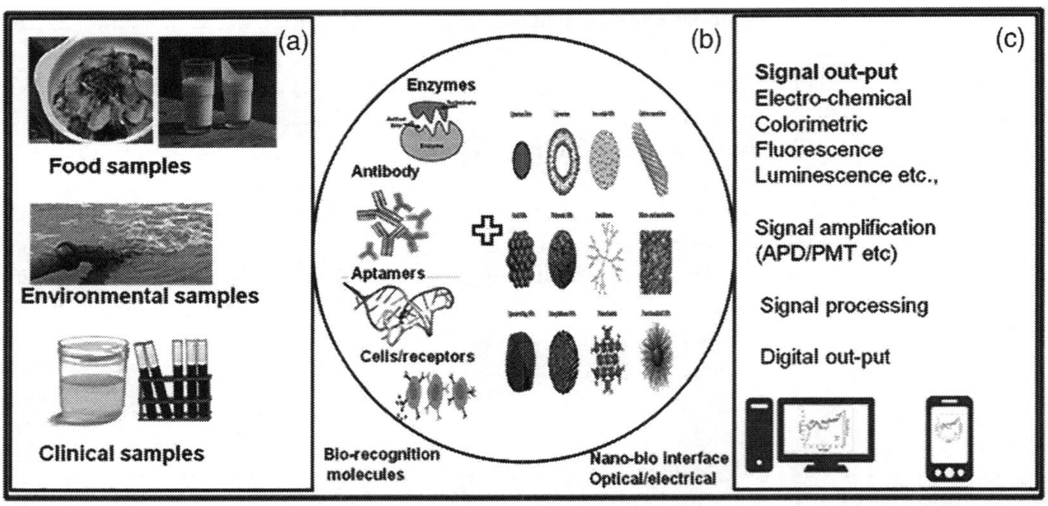

FIGURE 15.1 The components of a biosensor: (a) samples, (b) biorecognition elements and signal amplifiers, (c) signal transducers. (Kissinger, 2005; Wang, 2008.)

From an application perspective, for a successful biosensor design, it is desirable that the sensing platform must be reusable. In several cases, such as in medical diagnostics, food, environment, and national defense applications, the sensor needs to be capable of continuously monitoring the analyte online and real-time (Ronkainen et al., 2010). However, disposable, single-use biosensors also find wide application in point-of-care personalized sensing, such as in blood glucose monitoring by diabetic patients.

The present chapter focuses on biosensors in the area of food component analysis. In the field of "food sensors," the food industry requires suitable analytical methods for analysis of molecules towards two major objectives: (1) for quality control and contamination monitoring along the whole chain process from raw materials to the consumers, and (2) for monitoring and process control of important analytes, such as sugars, alcohols, amino acids, flavors, sweeteners, and other products during processing (Table 15.1).

TABLE 15.1
Major Broad Areas for Biosensor Development in Food Analysis

	Food Composition	
Health Labeling	**Compositional Labeling**	**Aesthetic Labeling**
Cholestrol	Glucose (dairy products)	Flavor compounds (wine, tea)
Polyphenols	Lactose (dairy products)	
Vitamins	Sucrose	
Glucosinolates	Organic acids	
	Process Control	
Fermentation process indicators	**Freshness and storage indicators**	
Glucose	Lactic acid	
Fructose	Malic acid	
Lactose	Acetic acid	
Sucrose	Ethanol	
Sorbitol	Methanol	
	Food Safety	
Contaminants	**Adulterants**	
Pesticides and herbicides Fertilizers	Urea	
Heavy metals	Formaldehyde	
Antibiotics	Melamine	
Pathogens and toxins		

15.2 PREREQUISITES OF A BIOSENSOR

Designing a stable and robust biosensing system requires several prerequisites which need to be adopted for their commercial and operational viability. Some important ones are:

1. Selectivity: It is imperative that a biosensor device should be highly selective for the target analyte and cross-reactivity with structural analogs should be minimal.
2. Sensitivity: The biosensor device should be sensitive and able to measure the analyte at concentrations of interest in the samples used for analysis. It should also avoid any pre-cleaning or pre-treatment steps as far as possible during real sample analysis.
3. The linearity of response: The linear response range of the system should cover the concentration range over which the target analyte is to be measured.
4. Reproducibility of signal response: Reproducible signals independent of user and place are an important prerequisite.
5. Quick response time and recovery time: The biosensor device response should be quick enough that real-time monitoring of the target analyte can be done efficiently. Also, the recovery time should be small for reusability of the biosensor system.
6. Stability and operating life: Stability of a bio-sensing system partially depends on the biological molecule being used in the sensor while the operating life depends on the instrument or device used for detection.

15.3 CLASSIFICATION OF BIOSENSORS

In general, parts comprising a typical biosensor are: (a) bio-receptors which bind specifically to an analyte; (b) the interface platform where the biological reaction takes place; (c) the transduction element that picks up the signal of the biological event and converts it to electronic output; (d) software that converts electrical signals to digital ones; and (e) an interface to the human operator.

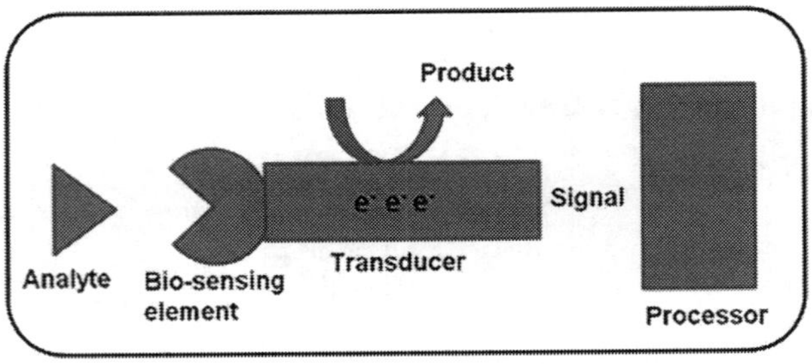

FIGURE 15.2 Scheme of an electrochemical biosensor. (Wang, 2008.)

Based on the components that make up the biosensor, they are divided into:

1. *Based on the transduction element*: electrochemical, optical, piezoelectric, and thermal/calorimetric biosensors (Kissinger, 2005).
2. *Based on the biorecognition element*: enzyme biosensors, immunosensors, DNA sensors, etc.

15.3.1 Biosensors Based on the Transduction Element

15.3.1.1 Electrochemical Biosensors

Among the various types of biosensors, electrochemical biosensors are a class of the most widespread, numerous, and successfully commercialized devices of bio-molecular electronics (Dzyadevych et al., 2008; Fang et al., 2003; Wang, 2008). These devices are very well known for their simple, inexpensive, and yet accurate and sensitive platform for bio-sensing.

Electrochemical biosensors combine the sensitivity of electrochemical transducers with the specific biorecognition process, producing an electrical output as a function of target analyte concentration (Figure 15.2). In general, electrochemical biosensors can be divided into two main categories based on the nature of the recognition process; biocatalytic devices and affinity sensors (Wang, 2008). The first is based on a "catalytic event," while the latter depends on a "binding event" to generate electric signals (Wilson, 2005).

Electrochemical biosensors depend entirely on surface chemistry and do not depend, unlike other conventional techniques, on chromophores or fluorophores which often interfere in detection. The reaction which is being monitored in an electrochemical biosensor typically generates a measurable current (amperometry) or a measurable potential or charge accumulation (potentiometry), or alters the conductive properties of the medium between electrodes (conductometry) (Wang, 2008). The three types of electrochemical biosensors based on the above output are: (a) voltammetric/amperometric biosensors; (b) potentiometric biosensors; (c) conductometric biosensors.

15.3.1.1.1 Voltammetry/Amperometry

Voltammetric and amperometric techniques work on the principle of applying a potential to a working (or indicator) electrode versus a reference electrode and measuring the current (D'Orazio, 2003). The current generated is a result of electrolysis induced by electrochemical reduction or oxidation at the working electrode. Voltammetry is used for those techniques in which the potential is scanned over a set potential range. The current response is usually a peak or a plateau that is proportional to the concentration of the analyte. Many voltammetric methods are known, including sweep voltammetry, cyclic voltammetry, hydrodynamic voltammetry, differential pulse voltammetry, square-wave voltammetry, AC voltammetry, polarography, and stripping voltammetry (Wang, 2008). These methods have a wide dynamic range and are useful for low-level quantitation. In recent years, the application of voltammetry has significantly increased, due to the ease of its application.

In amperometry, changes in current generated by the electrochemical oxidation or reduction are monitored directly with time while a constant potential is maintained at the working electrode with respect to a reference electrode (D'costa et al., 1986; Wang, 2006a,b). It is the absence of a scanning potential that distinguishes amperometry from voltammetry. Amperometric biosensors have additional selectivity, in that the oxidation or reduction potential used for detection is characteristic of the analyte species. Amperometric detection is commonly used with biocatalytic and affinity sensors because of its simplicity and low LOD. Another advantage is that fixed potential during amperometric detection results in a negligible charging current (the current needed to apply the potential to the system), which minimizes the background signal that adversely affects the limit of detection. Of all types of electrochemical methods, amperometric methods occupy top priority for designing hand-held or on-site devices due to the reasons discussed above.

15.3.1.1.2 Potentiometry

Potentiometric sensors are based on measuring the potential of an electrochemical cell while drawing negligible current. The commonly used design of sensing in potentiometry is a pH electrode and ion selective electrodes for ions such as K^+, Ca^{2+}, and Na^+ (Wang, 2006a,b, Grieshaber et al., 2008; Kissinger and Heineman, 1996; Bartlett, 2008). The principle in potentiometry is that sensors use an electrochemical cell with two reference electrodes to measure the potential across a membrane that selectively reacts with the charged ion of interest. For biosensing, the chemical sensors are coated with a biological element, such as an enzyme or antibody or aptamers, that catalyzes a reaction to form an ion which can be detected using specific electrode. Application of potentiometric biosensors is limited by their relatively high cost, sometimes poor performance, and the limited availability of suitable ion-selective electrodes, although they offer a higher level of specificity. Research efforts are now prioritizing reducing the manufacturing cost of potentiometric biosensors.

15.3.1.1.3 Conductometry

In this system, changes in electrical conductivity of the sample solution are monitored in the course of a chemical reaction. Generally, conductometric biosensors include enzymes whose charged products result in ionic strength changes, and thus increase the conductivity as a function of target analyte concentration (Wang, 2006a).

15.3.1.2 Optical Biosensors

Optical biosensors exploit light absorption, fluorescence, luminescence, reflectance, Raman scattering, refractive index, and other such optical techniques for measuring a chemical or biological species (Kissinger, 2005). The most recent work on optical biosensors has concentrated on the evanescent wave technology.

15.3.1.2.1 Evanescent Wave Fiber Optic

When light propagates through a fiber optic on the basis of total internal reflection (TIR), a thin electromagnetic field (the "evanescent wave") generated decays exponentially with the distance from the interface, with a typical penetration depth of up to several hundred nanometers (Dzyadevych et al., 2008). In a biosensor, the evanescent wave is used to excite fluorescent molecules bound to the surface of the waveguide. These evanescent field-based optic devices are well-suited for especially affinity-based biosensors.

15.3.1.2.2 SPR/SPR-Raman

Surface Plasmon Resonance (SPR) is a phenomenon that relies on the excitation of electron plasma (surface plasmon) of a thin metallic layer on the surface of the waveguide. In brief, when p-polarized light, under conditions of total internal reflection (TIR), strikes an electrically conducting metal surface, the energy of photons are directly absorbed by the free electrons of the thin metallic layer (gold or silver) and converted into an evanescent field at a certain angle of incident light (Fang et al., 2003).

The biorecognition element is immobilized on the metal surface, and minute changes in the refractive index during the biorecognition event are captured and produced as an output signal.

15.3.1.2.3 Interferometry

Bio-layer interferometry (BLI) is an optical, label-free technology for measuring biomolecular interactions where the interference pattern of white light reflected from two surfaces, viz. an immobilized analyte on the biosensor tip and an internal reference layer is analyzed. The binding of an analyte present in the solution to a ligand immobilized on the surface of the biosensor tip increases the optical thickness, resulting in a wavelength shift corresponding to the thickness change in the biolayer (Lou et al., 2016). The BLI technique provides measuring of real-time interaction along with monitoring of binding specificity, association, and dissociation rates of molecules during the interaction.

15.3.1.2.4 Luminescence

Luminescence is a mechanism of photon emission either from biological or chemical entities upon interaction with specific molecules. It is a widely used technique in optical biosensors. They are broadly classified into two types: (1) bioluminescence and (2) chemiluminescence.

1. Bioluminescence

   Bioluminescence is a biological phenomenon in which interaction of ATP with the enzyme luciferase produces photons, which can be quantified using a luminometer. These sensors use luminescent micro-organisms or their related genes (Narsaiah et al., 2012). Luminescence is measured before and after exposure to molecules, especially toxic compounds, and the percentage of inhibition of photon emission is calculated. The increasing toxicity of the sample results in greater light loss from the test suspension of luminescent bacteria, or its gene expression. This system combines the advantages of biotoxicity testing and instrumental precision (Ranjan et al., 2012).
2. Chemiluminescence

   Chemiluminescence is a chemical phenomenon in which luminol is oxidized by peroxides in the presence of catalysts to emit photons (Figure 15.3). Chemiluminescence (CL)-based analytical methods are known to be highly sensitive, rapid, and easy to miniaturize, and provide the possibility for robust and inexpensive instrument devices (Sang et al., 2016). Other advantages of CL techniques include: no requirement of an excitation source like fluorescence and phosphorescence; no monochromators (often not even a filter); and no use of radioactive or hazardous chemicals. Luminol is widely used as a chemiluminescent reagent, but peroxyoxalate systems for formaldehyde detection are also reported (Akshath et al., 2012a).

15.3.1.3 Piezoelectric Biosensors

Piezoelectric biosensors are devices that are able to generate and transmit acoustic waves in a frequency dependent manner. These sensors are based on the relationship between the resonant

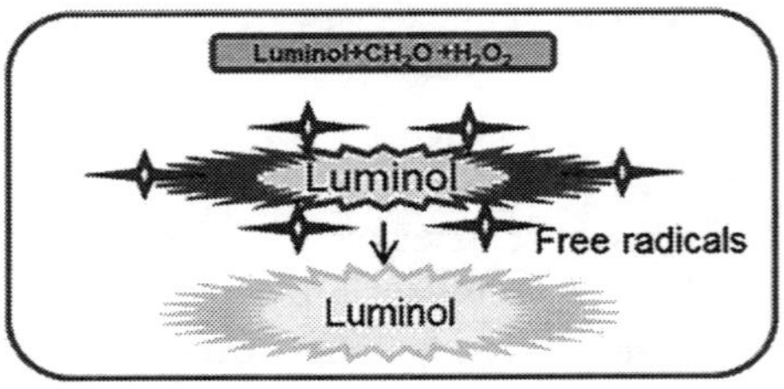

FIGURE 15.3 Luminol reaction for formaldehyde detection. (Akshath et al., 2012.)

frequency of an oscillating piezoelectric crystal and the mass deposited on the crystal surface (Ronkainen et al., 2010).

15.3.1.4 Thermometric Biosensors

Thermometric biosensors are also sometimes referred to as calorimetric biosensors. These sensors are based on biological reactions that are associated with the generation of heat. The reaction takes place in a small reactor. When the substrate enters the reactor, it gets converted to a product, and heat is generated. The temperature difference between the substrate and the product is sensed and produced as an output signal. These sensors have been widely used in the detection of cholesterol, uric acid, penicillin, etc. (Wang, 2008).

15.3.1.5 Nanoparticles as Enhancers of Transduction

Nanomaterials and nanoparticles prepared from metals, semiconductors, carbon, or polymeric substances have been widely employed to enhance the transducing ability of biosensors and consequently improve their sensitivity of detection. The amalgamation of nanotechnology with biological systems is made feasible by combining the intrinsic properties of nanoparticles and their ability to immobilize on specific ligands, such as oligonucleotides, proteins, and other biorecognition surfaces (Vinayaka and Thakur, 2011). Among the various nanomaterials used, zinc oxide, gold nanoparticles (GNPs), quantum dots, carbon nanotubes (CNTs), and graphene have received considerable interest in the development of sensitive biosensors.

15.3.1.5.1 Zinc Oxide

Among nanomaterials, ZnO has offered a wide range of applications for improving the sensitivity of transduction. Due to its wide band gap (3.37 eV), semiconducting, piezoelectric, and pyroelectric properties, it has been widely used in optics, optoelectronics, sensors, and actuators. Nano-structured ZnO possess a high surface area, are non-toxic, biocompatible, and have appreciable chemical stability. Its high electron communication features make it a great potential material for device design. There are many reports on the use of nano-structured ZnO for amperometric sensing (Zhao et al., 2010).

15.3.1.5.2 Gold Nanoparticles

Gold nanoparticles provide a stable immobilization for biomolecules and also a good platform for retaining their bioactivity as a result of their large surface to volume ratio. GNPs can act as "electron wires" in many bio-electrochemical reactions. In an electrochemical system, the electron transfer between the redox-protein and the electrode surface is the key parameter for biomolecular recognition. However, in most of the oxido-reductase enzymes, the active centers are surrounded by insulating protein shells, and therefore resist electron transfer between the electrode and the active center. Due to this fact, compounds such as ferrocene derivatives (Amine et al., 1993), quinones (Ikeda et al., 1985), etc., were used as mediators to shuttle electrons. It was in 1996 that Natan's group demonstrated the direct electron transfer between the electrode and the protein complex by GNPs (Brown et al., 1996). Consequently, several biosensors have been designed based on the unique opto-physical property of GNPs which today occupy center stage as effective transduction enhancers in biosensor development.

The other emerging area where gold nanoparticles continue to play a pivotal role is "plasmonics," a subfield of nano-photonics which empower controlling and manipulating light at nanoscale dimensions. GNPs are employed in plasmonics for their property of collective oscillations of conduction band electrons ("plasmons") in response to external electromagnetic radiation (Hu et al., 2006). One example is that of surface plasmon resonance (SPR), where the interaction between an electromagnetic wave and the conduction electrons in a metal (gold), is used for probing and characterizing physicochemical changes of thin films on the metal surface (Li et al., 2010b; Liu and Lu, 2004). Upon binding of target molecules to the surface of gold-coated films, a variation in the dielectric constant causes a change in the reflection of laser light from the metal-liquid surface which is read as an output of target concentration (Li et al., 2010a; Liu and Lu, 2004).

GNPs are therefore used in many biosensors as efficient transducers for both *in vitro* and *in vivo* detection of molecules.

15.3.1.5.3 Quantum Dots

QDs are colloidal nano-crystalline semiconductors possessing unique spectral properties due to quantum confinement effects (He et al., 2011). Properties such as high emission quantum yield, sharp emission spectra, broad absorption spectra, photostability, and tunable emission frequencies make them a superior tool for QD-based biosensors (Frasco and Chaniotakis, 2009). In comparison to organic dyes, QDs possess distinguishing spectral cross-talk, and thus can be selectively used for Forster resonance energy transfer (FRET)-based applications. FRET is a distance-dependent dipole–dipole interaction phenomenon of non-radiative transfer of excited-state energy from an excited donor molecule to an acceptor molecule (Lakowicz, 2006; Ren et al., 2010; Chen et al., 2005). These unique properties of QDs suggest that the replacement of organic fluorophores in FRET studies with QDs could lead to an experimental setup, which would be ultrasensitive, economical, and easy to configure, as well as providing simultaneous measurements of different macromolecular systems (Xiao et al., 2003; Freeman et al., 2009).

15.3.1.5.4 Graphene

Graphene is a monolayer of carbon atoms which are tightly packed into a two-dimensional honeycomb lattice. The team headed by Andre Geim and Konstantin Novoselov, discovered the exfoliation method and the electronic properties of graphene, for which they were awarded the 2010 Nobel Prize in Physics (Geim and Novoselov, 2007). Due to graphene's delocalized network of electrons and its unique opt-electrochemical properties like high surface area, ease of functionalities interrelation, and fast electron transfer, it has attracted enormous interest in bio-sensing and bio-engineering applications. The derivatives of graphene, like graphene oxide (GO) and graphene quantum dots (GQDs) are now the leading composite materials in the biomedical and diagnostic field. Due to its high dispersibility in solvents, graphene can be deposited in a wide range of conductive materials for electrochemical sensor design (Ma et al., 2013).

15.3.1.5.5 Carbon Nanotubes (CNTs)

Carbon nanotubes (CNTs) are cylinders built of either one or more layers of graphene with desired ends. These layers are termed as a single wall or multiwall layers (SWNTs/MWNTs) with perfect hexagonal lattice and all carbons bonded within these hexagonal lattices (De Volder et al., 2013). CNTs length often range from less than 100 nm to several centimeters, which bridges the molecular and macroscopic scales. Recent interest and advancement in CNTs as components of biosensors and bio-medical devices is guided by the compatible chemical interaction of CNTs with a wide range of biomolecules, like DNA or proteins, enabling their use in biosensors. It is also reported that CNTs have a high electrocatalytic effect and a fast electron-transfer rate leading to the design of nanoelectrode based label-free biosensors (Justino et al., 2013).

15.3.2 Biosensors Based on the Biorecognition Element

15.3.2.1 Enzyme Biosensors

Enzymes are substrate-specific biological molecules that catalyze specific chemical reactions. Exploiting this phenomenon, enzyme-based biosensors have emerged as a valuable technique for qualitative and quantitative analysis of a number of target analytes in food processing, agriculture, medicine, pharmaceuticals, and food safety and monitoring, etc. (Ispas et al., 2012). The major reason for their widespread application is their use for on-site detection, and significant benefits over classical analytical methods including high sensitivity, selectivity, no (or less) clean-up of samples, etc.

Most classical electrochemical enzyme biosensors are based on oxidoreductase enzyme bio-catalytic events, coupled with amperometric detection mediated by an electron transfer mechanism (Algar et al., 2010; Costa-Fernández, 2006). The most successful application of enzyme biosensors commercially has been the self-monitoring blood glucose biosensor (Accu Check, One Touch) (Ispas et al., 2012).

Akshath et al. (2012b) used a NAD^+ dependent formaldehyde dehydrogenase enzyme to develop a biosensing platform for the ultrasensitive detection of formaldehyde (HCHO). The spectral cross-talk of NAD^+/NADH with QDs was used to monitor the target analyte levels. It was proposed that CdTe QD may undergo dipolar interaction with NADH as a result of broad spectral absorption due to multiple excitonic states resulting from quantum confinement effects. NAD^+ acts as an electron acceptor and gets reduced to NADH during the dehydrogenase reaction. Since NADH from the enzymatic reaction is a fluorescent compound, CdTe QD was used as a "plug-in" for NAD^+ co-factor which can route the fluorescent energy from the formed NADH (Figure 15.4). It was possible to detect formaldehyde in the range 1000 ng/mL–0.01 ng/mL with a limit of detection (LOD) at 0.01 ng/mL using the proposed method.

15.3.2.2 Immuno-Biosensors

Antibodies have become an indispensable tool in various biosensors due to their unique ability to recognize and bind other molecules (antigens) or specific structures onto them (Ronkainen et al., 2010). Immunosensors have been regarded as the gold-standard technique in several biosensing fields especially in environmental monitoring and clinical diagnostics. An immunosensor makes use of the highly specific interaction of the two binding sites of an antibody with a particular target which is detected by the transducer (e.g. optical or electronic). Therefore, it provides a highly repeatable and highly specific reaction format. The advantages of immunosensors, apart from specificity, also include the requirement of small sample volumes, low detection limits, little or no sample preparation, reduced use of chemicals, ease of automation etc. Immunosensors are better than other immunological methods (e.g. ELISA formats) in terms of regeneration and binding properties of the sensing surface, which is critical for the successful reuse of the same sensor surface and the accuracy of detection results (D'costa et al., 1986). There are three general formats of an immunosensor: (1) competitive assay, (2) uncompetitive assay, and (3) sandwich assay.

In a previous work by one of the authors of this chapter (Padmaja et al., 2014; Figure 15.5), an immunosensor was designed to assess bovine mastitis by quantifying leukotoxin M/F′-PV (LukM/F′-PV), a potent neutrophil targeting a beta-barrel pore-forming toxin secreted by bovine strains of *S. aureus*. Initially, polyclonal antibodies to the recombinant LukF (rLukF) component of LukM/F′-PV were raised in chicken. They were functionalized to gold nanoparticles, and an immunoassay based on nanosurface energy transfer (NSET) was designed. In brief, rLukF was incubated with functionalized GNPs, and fluorescein isothiocyanate (FITC)-labeled secondary antibodies were added.

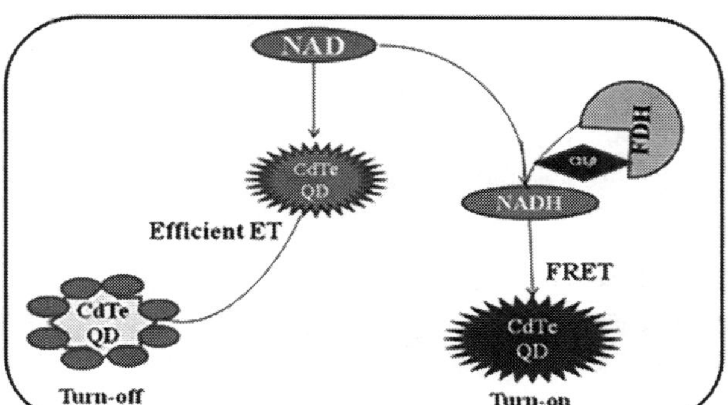

FIGURE 15.4 Enzyme-based FRET reaction. (Akshath et al., 2012.)

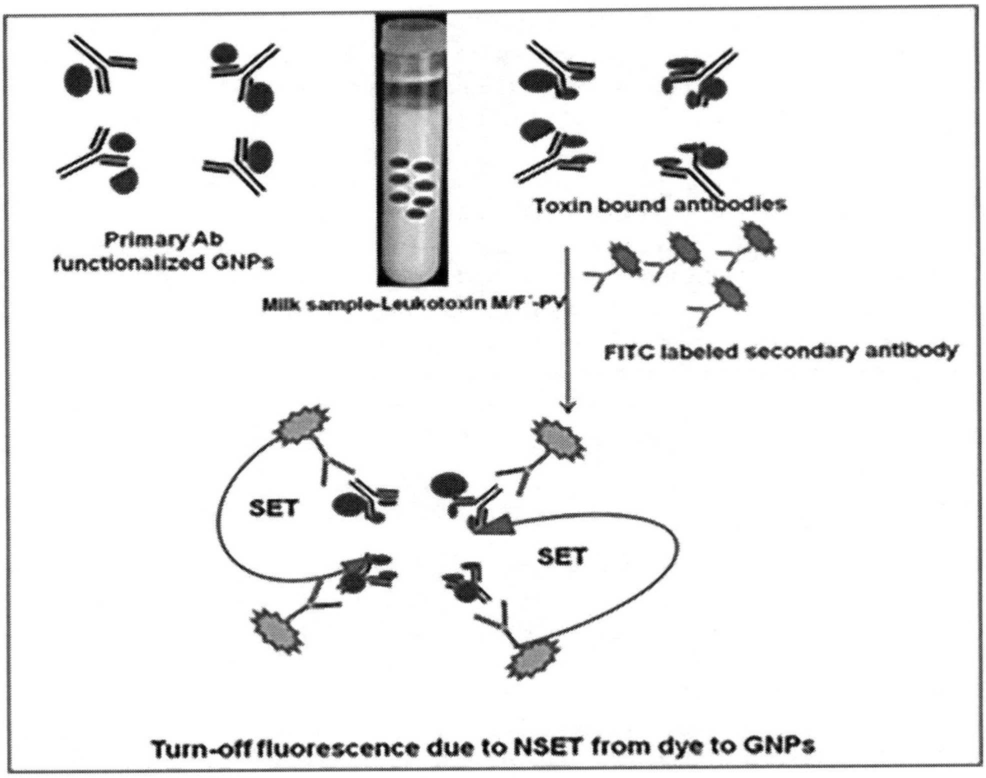

FIGURE 15.5 The NSET assay for detection of LukF in milk samples, an indicator of bovine subclinical mastitis. (Padmaja et al., 2014.)

The simultaneous fluorescence quenching was monitored as a function of toxin concentration. The method enabled detection of leukotoxin in the range of 100–0.1 ng ml^{-1} (LOD of 0.1 ng ml^{-1} and $R^2 = 0.9908$). The suitability of the assay to detect toxins in spiked and field samples was also reported.

15.3.2.3 Receptor-Based Biosensors

Receptors are natural "small molecules" that are impregnated in the cellular membrane that specifically binds to their target analytes resulting in metabolic changes. These act by either opening ion-channels, producing secondary messenger systems, or activating enzymes (Taylor et al., 1988). In biosensing, the binding of the target analyte to the receptor is made use of by capturing the change in conformation of the receptor, which is translated into an optical/electrochemical signal (Wei et al., 2012b; Rogers, 2000). In a complete contrast to Abs that bind tightly with their complementary Ag, receptors transmit signals upon ligand binding as an output signal. Receptors, unlike Abs, are never specific for a single analyte, but are common for a group of target analytes. Therefore, receptor-based biosensors are usually group-specific affinity sensors and may not be suitable for biosensor application where high specificity is required.

15.3.2.4 Nucleic Acid-Based Biosensors

Affinity biosensors based on DNA, gold nanoparticle, quantum dots, and surface plasmon resonance have been widely used in the last two decades for the detection of various analytes. Aptamers are short, structured, single-stranded sequences of DNA, RNA, or peptides that can bind with high affinity, specificity, and selectivity to a wide range of targets, from small chemicals like metal ions, amino acids, toxins, ATP, enzymes, oligosaccharides, growth factors, antibodies, viral proteins or

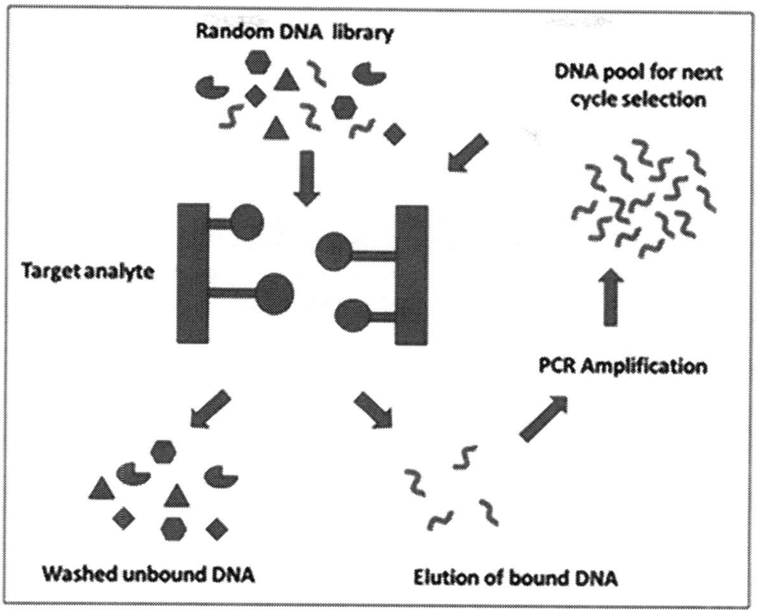

FIGURE 15.6 Scheme for aptamer generation through SELEX (Systematic Evolution of Ligand by EXponential enrichment). (Tuerk and Gold, 1990.)

their components, cells, bacteria, biological cofactors, transcription cofactors, trans-membrane proteins, and other small organic molecules (Liu et al., 2009). Aptamers are generated by a selection process known as SELEX (Systematic Evolution of Ligands by Exponential enrichment) which was first reported in 1990 by two research groups (Tuerk and Gold, 1990; Ellington and Szostak, 1990). The ability of aptamer to fold into three-dimensional structures facilitates the interaction with target molecules with high affinity constants (Ellington and Szostak, 1990). This property of aptamer has been used in a wide variety of sensing application including detection based on fluorescence intensity, energy transfer, electrochemistry, or color change (Figure 15.6).

Aptamer-based colorimetric detection for mercury ions (Hg^{2+}) using unmodified gold nanoparticles as a colorimetric probe exhibited that bare gold nanoparticles interact differently with short single-strand DNA and double-stranded DNA (Li et al., 2009). Another aptamer-based sensor using unmodified AuNPs for protein detection depicted the folding of aptamer into G-quadruplex/duplex structure upon thrombin binding. This colorimetric sensor provided an LOD as low as 0.83 nM with prospects of colorimetric biosensing of broad range of analytes (Wei et al., 2007). Quantum dots (QDs) have also been used in aptamer-based affinity biosensing. In one study, semiconductor QDs conjugated molecular beacons were designed, where fluorescent ZnS-capped cadmium–selenium QDs served as fluorophore. Use of QDs enhanced the lifetime during bioimaging as compared to organic molecular beacon implicating application for *in vivo* imaging (Kim et al., 2004).

15.3.2.5 Others

Until recently, only enzymes, antibodies, and subcellular components have primarily been used as bio-recognition elements of a sensor. But the unstable nature of these biomolecules has affected their wide application in the field of biosensing. Microbial cells offer several advantages over enzymes and antibodies, where various micro-organisms occurring in the natural environment can be stored in culture collections, allowing their availability for a specified purpose. Another important advantage of the use of whole cell as a biosensor is to analyze a component with multiple enzymes involved in a process. Uses of reporter genes fused with their responsive promoters for detection of analytes were employed to develop whole-cell biosensors. The use of the bacterial luciferase gene for developing a bacterial biosensor against naphthalene was first reported by King and group (King et al., 1990). The enzyme catalyzed the oxidation of long chain fatty acid and reduced FMN producing the corresponding fatty acid along with FMN in the presence of oxygen resulted in bioluminescence. Among the five genes, lux C, D, E, A, and B, located in the lux operon, the expression of C, D, and E were required to form fatty aldehydes through the synthesis of fatty acid reductases. Billard and DuBow, 1998). A lab-on-chip whole cell sensor with a micro-opto-electro-mechanical modulator and solid-state photo-based detector with the aim of toxicity detection was fabricated. Analysis of isopropyl beta D thiogalactopyranoside where the sensor was exposed to different concentration of analyte in the lower mM levels, showed potential for implementation in determining environmental toxicity (Elman et al., 2008). Molecularly imprinted polymers (MIPs) are synthetic materials that provide high affinity binding to their specific targets. The principle underlying molecular imprinting is the self-assembly of functional and cross-linking monomers around a target analyte followed by polymerization to form a solid material. In a study, DNA aptamers adsorbed through electropolymerization onto methylene green monomers imprinted with DNA were used for the detection of thrombin. The sensor was coupled to an electrochemical impedance spectroscopy and a potentiometry transducer which provided sensitive detection of thrombin to a LOD of 0.5 nM (Evtugyn et al., 2009).

15.4 FOOD COMPONENT ANALYSIS USING BIOSENSORS

Food is a highly complex substrate, and analysis of food is imperative from a consumers' point of view. Although freshly prepared food is subjected to analysis by the sensory abilities of the person consuming it, harvested and processed food which is prepared and packed for later consumption, and/or is subjected to transportation for consumption elsewhere, has to be analyzed for its quality and safety. Compliance with legislation for food safety outlined by government agencies and food control and monitoring departments is also one of the prime areas which requires biosensor development.

The rationale for food analysis is therefore to check whether the food product retains its nutritional quality, fulfills the requirement of its intended use (in case of fortification etc.), justifies the claims on its label, is stable and consumable for the intended storage period, meets the regulatory guidelines, and, most importantly, is safe for consumption, being devoid of contaminants, microbial pathogens, and adulterants. In addition, the food industry may also require sensors to evaluate the performance or process parameters during the manufacturing of the food product. To this effect therefore, biosensors for food component analysis can be divided into two broad categories (Table 15.1):

1. Biosensors for analysis of metabolites/substrates/nutrients, etc., for food composition and process control
2. Biosensors for detection of contaminants (chemical and biological)/adulterants, etc.) for food safety

15.4.1 Biosensors for Analysis of Food Composition and Process Control

The food industry needs reliable, rapid, and robust biosensors, not only for analysis of the composition of the final food product, but also during the time of food processing (process control). In fermentation processes, for example, methods are required to analyze sugars like glucose, fructose, lactose, sorbitol, sucrose, etc. for better process efficiency and economy (Serna-Cock and Perenguez-Verdugo, 2011). Also, the recent demand for high-quality food products, along with accurate labeling to facilitate customer assurance and acceptance, has also gained high priority. With demand for health-based foods, emphasis is now being given to labeling the products for their nutrient, nutraceuticals, and functional content (Scognamiglio et al., 2014).

This segment discusses some of the biosensors which have been developed for the analysis of food components, and some which are used in the industry for process control (Table 15.2).

TABLE 15.2
Biosensors Reported for Different Food Components

Food Component	Detection Method	LOD	References
	Sugars		
Glucose	Amperometric biosensor based on layer-by-layer films of microperoxidase-11 and liposome-encapsulated glucose oxidase	8.6 uM	Graça et al. (2014)
	Stainless-steel pins based electrochemical biosensing platforms	0.05 mM	Rama et al. (2017)
Lactose	Conductometric biosensor for dairy products	0.05 uM	Nguyen-Boisse et al. (2013)
	Nanoparticle based third generation biosensor based on cellobiose dehydrogenase	3.5 mM	Bollella et al. (2017)
Sorbitol	Sorbitol dehydrogenase enzyme entrapped Ta_2O_5 nanoflowers assembly coupled with fiber optic SPR	3.6 ng/mL	Kant et al. (2016)
	Sorbitol dehydrogenase immobilized in a sol–gel carbon nanotubes–poly(methylene green) composite	0.11 mM	Wang et al. (2013b)
Ethanol	Conducting polymers containing polypeptide and ferrocene side chains	0.28 mM	Kesik et al. (2014)
	Palladium Modified Graphene Nanocomposite Glassy Carbon Electrode	2 mM	Kumar et al. (2017)
	Organic acids		
Lactate	Amperometric biosensor based on microbial lactate oxidase	0.25 µM	Dagar and Pundir (2017)
Malic acid	Electrochemical biosensor based on high enzymatic activity preservation of malate dehydrogenase immobilized in a Langmuir–Blodgett film	10 mM	Gür et al. (2016)
Acetic acid and butyric acid	Plasmonic-based colorimetric and spectroscopic discrimination of acetic and butyric acids produced by different types of Escherichia coli and formation of gold nanoparticles	1 mM	La et al. (2016)
	Vitamins		
Vitamin B12	Glutathione coated silver nanoparticles as a colorimetric probe	50 nM	Rajamanikandan and Ilanchelian (2016)
VitaminB_1 (thiamine)	Fluorescent sensor using coconut water derived luminescent carbon dots	280 nM	Purbia and Paria (2016)
Folic acid	PVA-based nanobiosensor by fluorescence quenching	42.29 ng/mL	Chakravarty et al. (2016)
Cholesterol	Carbon dot–hemoglobin complex-based biosensor	56 uM	Bui and Park (2016)
	Electrochemical sensor based on cholesterol oxidase and MoS_2-AuNPs modified glassy carbon electrode	0.26 uM	Lin et al. (2016)
Acrylamide	Colorimetric detection in potato chips based onnucleophile-initiated thiol–ene Michael addition	28.6 nM	Hu et al. (2016)
	Electrochemical biosensor based on single-stranded DNA modified gold electrode	8.1 nM	Huang et al. (2016)
	Pesticide and insecticides		
DDT	Chemiluminescence based dipstick immunoassay	0.05 ng/mL	Baker et al. (2012)
Organophosphorus	Voltammetric acetylcholinesterase-based biosensor	0.058 ppm 0.194 ppm	Raghu et al. (2014)

(*Continued*)

TABLE 15.2 (CONTINUED)
Biosensors Reported for Different Food Components

Food Component	Detection Method	LOD	References
Heavy metals	Horseradish peroxidase immobilized into maize tassel-multiwalled composite based amperometric biosensor	2.5 μg/L 4.2 μg/L	Moyo et al. (2014)
	Antibiotics		
Chloramphenicol	MIP-Canadian penny-based surface enhanced raman spectroscopy (SERS)	0.1 ppm	Gao et al. (2014)
Streptomycin	Electrichemical aptasensor based on arch shaped aptamer-complementary strand conjugate	11.4 nM	Danesh et al. (2016)
Kanamycin	Aptamer mediated turn-off/on nanozyme	1.49 nM	Sharma et al. (2014)
	Pathogens and Toxins		
Mycotoxin	Aptamer-based dipstick assay	0.1 ng/mL	Shim et al. (2014)
E. coli O157:H7	Lectin-carbohydrate interaction-based SPR biosensor	3×10^3 CFU/mL	Wang et al. (2013a)
Listeriamonocytogenes	Graphene modified carbon ionic liquid electrode with dendritic nanogold-based electrochemical DNA sensor	2.9×10^{-13} mol/L	Sun et al. (2012)
Salmonella	Graphene oxide and GNP modified glassy carbon electrode aptamer biosensor	3 CFU/mL	Ma et al. (2014)
Staphylococcal enterotoxin B	Photonic crystal lab on a chip immunosensor	35 aM	Han et al. (2013)
Urea	Potentiometric biosensor-based urease electrodeposited on polyaniline film	10^{-6} mol/L	Lakard et al. (2011)
Formaldehyde	BODIPY fluorescent turn on	165 nM	Song et al. (2012)
Melamine	MIP-SERS	0.005 mmol/L	Hu et al. (2015)

15.4.1.1 Sugars

Fifty-four years after Clark and Lyons (1962) proposed the concept of glucose enzyme electrodes, numerous advancements have taken place in sugar biosensors for application in industrial process monitoring, as well as medical diagnostics. Biosensors for common sugars like glucose, fructose, sucrose, sorbitol, mannitol, etc. have been reported extensively and are standard procedures followed today in automated industrial processes (Wang, 2001). The most successful use of biosensor application in sugar analysis is the commercially available sensors for self-monitoring of blood glucose levels (Accu Check, One Touch) (Heller and Feldman, 2008).

Most classical electrochemical enzyme biosensors are based on oxidoreductase enzyme biocatalytic events coupled with amperometric detection mediated by electron transfer mechanism (Schuhmann, 2002). However, in these types of first-generation sensors, non-specific oxidation of other food molecules was of major concern (uric acid, ascorbic acid etc). Consequently, extensive efforts were made to minimize non-specific oxidation by the use of nanoparticles and electroactive material. There are many reports on the use of nanostructured ZnO for amperometric sensing. Cao et al., 2008 designed a hydrogen peroxide biosensor using waxberry-like ZnO microstructures consisting of nanorods (8–10 nm) (Xia et al., 2008). Weber et al. obtained ZnO nanowires with a typical length of 0.5–2 μm and a diameter of 40–120 nm, which were grown on the substrate with an array of ZnO nanowires (Weber et al., 2008).

In 2013, Lang et al. designed a solid/nanoporous gold/cobalt oxide hybrid structure for electrochemical non-enzymatic glucose biosensing with sensitivity up to 12.5 mA mM^{-1} cm^{-2} and a short response time of less than 1 s, with an ultralow detection limit of 5 nM (Lang et al., 2013).

For detection of lactose in food processing, enzymes like beta-galactosidase and glucose oxidase are most commonly used in conjunction with nanoparticles. A novel method for lactose determination in milk was developed based on oxidation of lactose by cellobiose dehydrogenase (CDH) from *Phanerochaete chrysosporium*. The enzyme was immobilized in a reactor. A highly reproducible linear response for lactose was obtained between 0.05 mM and 30 mM (Yakovleva et al., 2012).

An integrated biosensor for detection of sucrose, fructose, and glucose was designed using a self-assembled monolayer modified gold disc electrode co-immobilized with invertase and fructose dehydrogenase enzymes (Campuzano et al., 2004). Another example of biosensors for sugar analysis in food processing is the detection of sorbitol in food samples. Hu et al. (2004) reported molecular imprinted biosensors for detection of sorbitol in food samples with LOD of 1 mM (Feng et al., 2004).

15.4.1.2 Alcohol

To assess the food freshness and quality of alcoholic beverages, ethanol and methanol have been used as indicators in the food processing industries. Classically, they are quantified by colorimetric, chromatographic, flow injection analysis and enzyme-spectrophotometric methods (Altman and Bland, 1994). However, these methods are expensive and cannot be used in quality control systems where time is an important factor. To overcome these drawbacks, biosensing has been used extensively for alcohol monitoring in food. To this end, enzymatic biosensors for the detection of alcohol are mainly based on two common enzymes, viz. alcohol dehydrogenase (ADH) and alcohol oxidase, and to a lesser extent, a third enzyme, catalase. Boujtita and Murr (1995) designed an amperometric ethanol sensor by co-immobilization of ADH and bovine serum albumin (BSA) in carbon paste containing entrapped NAD. The sensor showed rapid response time and high sensitivity for ethanol in food samples. A novel voltammetric ethanol biosensor was constructed using ADH immobilized on the surface of a glassy carbon electrode modified by cellulose acetate (CA) bonded to toluidine blue O (TBO). A good detection limit, calculated as three times the signal-to-noise ratio, was obtained (1 uM LOD) with sensor performance stable even with food samples (Boujtita and Murr, 1995).

Apart from enzyme-based sensors, a microbial biosensor for detection of ethanol was developed using *Methylobacterium organophilium* immobilized eggshell membrane and an oxygen electrode (Wen et al., 2013). The proposed microbial sensor was found to be stable and responsive in food matrices like wine and beer with good response and recovery (0.025 mM LOD). Recently, in designing a colorimetric biosensor using gold nanoparticle-enzyme hybrid systems, Willner et al. (2005) reported the use of NADH mediated growth of Au nanoparticles (NPs) for detection of ethanol using alcohol dehydrogenase enzyme (Xiao et al., 2005). The reaction mixture consisted of ascorbic acid, gold chloride salt, and cetyltrimethylammonium bromide which led to the formation of shaped NP structures consisting of dipods, tripods, and tetrapods in the presence of NADH which formed from ethanol dehydrogenase reaction. However, application of these systems in complex food matrices remains questionable due to interfering molecules in food. To date, alcohol dehydrogenase-based electrochemical biosensors are the preferred choice, due to their application in complex food matrices.

15.4.1.3 Organic Acids

Organic acids are present in wide numbers of fruits and vegetables. They are used extensively as acidulants in the manufacture of beverages and juices. The prime organic acids used in the food industries include lactic, malic, and acetic acid. The content of these organic acids influences stability, nutrition, acceptability, and storage capability. Lactate levels serve as an indicator during fermentation processes and provide information about freshness, stability, and storage quality of several food products. Lactate biosensors have been designed using enzyme lactate oxidase or horseradish peroxidase modified lactate oxidase (Mazzei et al., 2007). Multiplex detection of lactic acid along with malic acid has also been performed using co-immobilization of multiple enzymes on self-assembled electrodes (Shkotova et al., 2008; Gamella et al., 2010). Recently, an amperometric biosensor based on the chemical reduction of platinum nanoparticles on graphitized carbon nanofiber surfaces was used to detect lactate levels in wine and cider (Loaiza et al., 2015).

Malic acid present in fruits and vegetables is an important parameter for characterizing wine quality. It is also used as an acidulant in food products. Enzyme-based biosensors have played a significant role in malate determination. Enzymatic biosensors using malate quinone oxidoreductase have been fabricated based on tetrathiafulvalene and AuNP modified electrodes (Molinero-Abad et al., 2014). On-site monitoring of malate during malolactic fermentation of red wine was achieved by a bienzymatic biosensor using malate dehydrogenase and diaphorase (Giménez-Gómez et al., 2016).

Acetic acid, another key component of food industry, is produced during the fermentation process of wine, vinegar, and soy sauce, and is also present as a final component of these products. Enzyme biosensors have been used for acetic acid detection. These include acetate kinase, pyruvate kinase, and lactate dehydrogenase tri-enzyme modified electrodes for acetate detection (Mieliauskiene et al., 2006). Recently, molecularly imprinted polymers have been used, where polypyrrole polymers were used for the electrochemical detection of acetic acid (Xie et al., 2010).

15.4.1.4 Vitamins

Biosensors for vitamin detection are important for the purpose of labeling food products for their content of vitamin present. Vitamin detection is classically performed by HPLC after extraction from its source. However, of late, biosensors have replaced these classical methods, due to their sensitivity and rapid detection capabilities. Although not commercially used at present, vitamin biosensor research has seen great strides in recent years. Ovalle et al. (2015) proposed a sensitive amperometric microbial biosensor for detection of vitamin B_{12} in food samples which is fast, economical, reliable, and can compete with the existing conventional methods (Ovalle et al., 2015). The authors exploited the advantage of the bacterial strain *Tetrasphaera duodecadis* that oxidizes vitamin B_{12} concomitantly utilizing oxygen. Amperometric whole cell biosensor based on *Saccharomyces cerevisiae* immobilized in gelatin was developed for selective and sensitive determination of vitamin B_1 (thiamine). When the vitamin B_1 solution was added to the biosensor system, authors found an increase in respiration activity and a decrease in the dissolved oxygen concentration which was used to monitor vitamin B_1 concentration in vitamin tablets. For multiplexed detection of water-soluble vitamins B_2, B_{12}, folic acid, biotin, and pantothenic acid used to supplement infant formula samples, a biosensor-based technology was developed by Biacore AB. The authors showed that the biosensor-based technology of vitamin analysis is a sensitive, reliable, and realistic alternative to other methods (Gao et al., 2008).

Recently, folate biosensors have been developed as versatile tools for targeting cancer cells and therapeutics. Folic acid analysis in real food samples was performed using molecular imprinted polymers (Hussain et al., 2013). An ultrasensitive probe was developed using gold–silver nanoparticle (AuNP–AgNP) hetero-dimers with SERS substrate, based on antigen–antibody immunoreactions. An LOD of 0.86 pg/mL was achievable using this sensor (Wu et al., 2016).

15.4.1.5 Cholesterol

Much like vitamins, cholesterol sensors are important from the point of view of labeling of food products, especially in the health sector. Cholesterol is an important component in food samples like egg yolk, goat meat, dairy products, chicken, beef meat, etc. Today, enzymatic biosensors based on cholesterol oxidase have practically replaced the chemical methods based on the Libermann–Burchard reaction. To fabricate cholesterol biosensors for food analysis, cholesterol oxidase (ChOx) is most commonly used as the biosensing element (Arya et al., 2008). However, cholesterol (total cholesterol) biosensors in food samples are very limited.

Runu Chakraborty et al. (2007), developed a cholesterol biosensor by co-immobilization of cholesterol esterase (ChEt) and cholesterol oxidase (ChOX) on oxygen electrode with the gold cathode and Ag/AgCl anode for detection of cholesterol in food samples (Basu et al., 2007). In another study, a bi-enzyme cell containing immobilized cholesterol esterase (CE) and cholesterol oxidase was developed for the detection of free cholesterol. Authors were able to detect cholesterol oleate and cholesterol in a linear measuring range of 0.1–0.5 mM respectively in lard, butter, and pasta samples (Adányi and Váradi, 2003). Recently Lei Ma et al. (2016) designed a sensitive and selective chemiluminescence (CL) sensor based on the peroxidase-like activity of copper nanoclusters for the detection of cholesterol in food samples with a detection limit of 1.5 μM (Xu et al., 2016).

15.4.1.6 Acrylamide

Acrylamide is a chemical compound produced during thermal processing of carbohydrate-rich foods, at temperatures above 120°C. It is said to be formed during the Maillard reaction between asparagine and reducing sugars such as glucose. The highest amount of acrylamide is generated during frying of potatoes, roasting of cocoa beans and coffee beans, thermal processing of cereals, baking of bread and cakes, and roasting of meat. Acrylamide can cause damage to the central and peripheral nervous system in humans, along with genotoxic and mutagenic effects in animals. Therefore, a reliable method for easy and rapid detection of this compound is necessary to assess its human exposure (Pedreschi et al., 2010).

Methods employing the immobilization of hemoglobin on the surface of electrodes acting as receptors of acrylamide have been used to design a novel voltammetric biosensor in food samples. Based on the adduct formation of acrylamide with hemoglobin and its interaction with α-NH_2 group of N-terminal valine in hemoglobin, a carbon paste electrode modified with hemoglobin containing four prosthetic groups of heme Fe(III) was fabricated. The electrode displayed a reversible reduction/oxidation process of Hb-Fe(III)/Hb-Fe(II) where the detection was achieved through the decrease in peak current due to hemoglobin Fe(III) reduction. The biosensor achieved an LOD of 1.2×10^{-10} M with successful validation in food products like potato chips (Stobiecka et al., 2007). Another electrochemical biosensor for acrylamide detection was fabricated based on covalent immobilization of hemoglobin (Hb) onto a composite of carbon nanotube/copper nanoparticle/polyaniline. The pencil graphite electrodeposited composite showed a rapid response time of less than 2 s, with a wide linear detection range from 5 nM to 75 mM. The enzyme electrode had an LOD of 0.2 nM with the possibility of usage over 120 times for a period of 100 days when stored at 4°C (Batra et al., 2013).

15.5 BIOSENSORS FOR ASSESSMENT OF FOOD SAFETY

15.5.1 Detection of Food Contaminants

Currently, one of the greatest challenges in food safety is the control of risks associated with both chemical and microbial contaminants. Among the most prominent chemical contaminants that raise huge public concern are low levels of pesticides, heavy metals, and antibiotics. Foodborne pathogens and toxins are also one of the most life-threatening contaminants in the food industry. The major chemical and biological contaminants and research on biosensors developed for their analysis are discussed below.

15.5.1.1 Pesticides

The term pesticide includes a wide range of compounds like insecticides, fungicides, herbicides, rodenticides, nematicides, plant growth regulators, and others. Pesticides are added to improve crop productivity, but when they reach the food chain and enter into food products, they are hazardous to animal and human health. There is an urgent need for continuous monitoring of even extremely low pesticide levels in food and water to ensure food safety, since they are known to have bioaccumulative properties.

The most common biosensors for pesticide determination use inhibition of acetylcholinesterase (AChE) activity by pesticides as a screening test for the evaluation of the presence of these compounds in food matrices. The inhibition can be evaluated using several platforms, including screen-printed electrodes for amperometric or voltammetric measurements (Mascini and Palchetti, 2001). Detection of carbaryl and parathion methyl in complex food matrices like eggs, bovine meat, milk, and honey was reported using an electrochemical enzyme inhibition assay based on screen-printed electrodes (Del Carlo et al., 2004). Disposable screen-printed electrode immobilized enzyme-sensors where inhibition activity of the pure enzyme (AChE) (Hildebrandt et al., 2008) from engineered variants of *Drosophila melanogaster* (Bachmann et al., 2000) and *Nippostrongylus brasiliensis* (Zhang et al., 2005) were analyzed for the presence of carbamate and organophosphorus (OP) residues. Photothermal (Pogačnik and Franko, 1999) biosensors based on acetylcholinesterase activity inhibition have also been reported (Pogačnik and Franko, 2003). In a recent paper, where the fluorescent sensor (PFS) based on the aggregation-induced emission effect of tetraphenylethylene (TPE) and the addition reaction capability of maleimide for rapid naked-eye detection of AChE activity, and consequently OPs, was reported, OPs up to 0.5 ng/mL could be detected using the simple paper-based sensor (Chang et al., 2016).

Another common biosensor used in pesticide analysis includes immunodetection. Immunodetection using the nanomechanical response of a microcantilever was used for detection of DDT. The biosensor achieved subnanomolar sensitivity without the use of any fluorescent labeling, providing a low-cost, portable, and high throughput analysis of environmental contaminants (Alvarez et al., 2003). A portable biosensor for on-site application was developed using GNP-based dipstick competitive immunoassay. This visual detection-based immunosensor provided preliminary and qualitative analysis which (the authors claimed) could be extended to other classes of organochlorine pesticides, depending upon specific antibody availability (Lisa et al., 2009).

An enzyme-based potentiometric biosensor has been reported using dehydrohalogenases which release chloride ions as a result of dehalogenation. The ion-selective detection system provided a complete analysis within 10 to 15 minutes for DDT and HCH, two very widely present organochlorine pesticides in food samples (Murthy et al., 2010). A novel fluorimetric sensing platform was constructed for quantitative detection of OP compounds via tyrosinase (TYR) enzyme-controlled quenching of gold nanoclusters (AuNCs). The method was based on the mechanism that oxidation of dopamine catalyzed by TYR produces dopaminechrome, which efficiently quenches the fluorescence of AuNCs. However, in the presence of OPs, the activity of TYR is inhibited, resulting in the fluorescence recovery of AuNCs. The proposed platform was demonstrated to enable rapid detection for OPs up to 0.1 ng mL^1 (Yan et al., 2016).

An advanced SPR sensor showing three times more sensitivity than colorimetric ELISA was used to monitor DDT in water samples where rapid and real-time detection was achieved within 20 minutes of analyte injection. The sensor surface provided regeneration up to 270 cycles along with the sensitivity of the assay (Mauriz et al., 2007).

15.5.1.2 Heavy Metals

Certain heavy metals (e.g. Fe^{2+}, Cu^{2+}, Mn^{2+}, Zn^{2+}) in small quantities are required for nutritional needs in human health. However, heavy metals tend to form complexes, especially with biological ligands containing nitrogen, sulfur, and oxygen, resulting in changes in the molecular structure of proteins, breaking of hydrogen bonds, or inhibition of enzymes. These interactions, among others, may cause toxicological and carcinogenic effects, such as those affecting the central nervous system (Hg^{2+}, Pb^{2+}, As^{3+}); the kidneys or liver (Cu^{2+}, Cd^{2+}, Hg^{2+}, Pb^{2+}); or skin, bones, or teeth (Ni^{2+}, Cu^{2+}, Cd^{2+}, Cr^{3+}) (Hamilton et al., 1998; Vallee and Ulmer, 1972). An important pathway for the contamination of food by heavy metals is through their uptake by crops irrigated with contaminated wastewater (Partanen et al., 1991).

Recent reports on detection of heavy metals are mostly receptor-based biorecognition elements. A capacitance biosensor based on the concept of synthetic phytochelatins as biorecognition elements was used to detect Hg^{2+}, Cd^{2+}, Pb^{2+}, Cu^{2+}, and Zn^{2+} ions. The sensor provided broad selectivity along with the sensitivity of all heavy metals where regeneration with EDTA offered stability for over 15 days (Sharma et al., 2007). Plant metal binding proteins having an affinity for specific metal ions like AgNt84–6 protein for Ni^{2+}, Zn^{2+}, Co^{2+}, Cu^{2+}, Cd^{2+}, and Hg^{2+} (Bontidean et al., 2003) and metallothionein protein for Cd^{2+} and Zn^{2+} (Cherian et al., 2003) have also been used to develop biosensors based on electrochemical transduction.

More recently, several authors have introduced the concept of a whole-cell biosensor (WCB) as a very useful alternative to classical biosensors for the sensitive detection of heavy metals. A WCB uses the whole prokaryotic or eukaryotic cell as a single reporter, incorporating both bioreceptor and transducer elements into the same cell (Adam et al., 2005). A WCB using *S. cerevisiae* cells and GFP as the reporter protein was developed to detect copper ions (Gutiérrez et al., 2015). The system was based on the ability of transcriptional activator protein Ace1 to control the expression of the reporter gene GFP. The GFPuv protein acted as a reporter of Cu^{2+} ion concentration, the expression of which was governed by an Ace1 protein that activated the CUP1 promoter located upstream of the GFPuv gene in the plasmid pYEX-GFPuv. The biosensor was able to detect Cu^{2+} at concentrations as low as $5X10^{-7}$ M. In another study, the stability and storability of a whole cell biosensor was improved by using poly(2-hydroxyethyl methacrylate) (pHEMA). The biosensor was developed by entrapping the cyanobacterium *Anabaena torulosa* on a cellulose membrane, followed by applying a layer of pHEMA with biosensing in a well format. This sensor was able to detect Cu, Cd, and Pb at concentrations of 1.410, 0.250, and 0.500 µg/L respectively. The authors demonstrated enhanced stability and storability of the biosensor with the addition of pHEMA over the microbial cells. A conductometric biosensor using immobilized *Chlorella vulgaris* microalgae as bioreceptors was used as a bi-enzymatic biosensor for the detection of heavy metals and pesticide. Alkaline phosphatase and acetylcholinesterase inhibition were used for the detection of heavy metal and pesticides respectively. The biosensor was appreciably sensitive for Cd^{2+} and Zn^{2+}, with limits of detection (LOD) of 10 ppb for a 30 minute long exposure (Shetty et al., 2004).

Among enzyme biosensors for heavy metals, a novel concept of the tri-enzymatic conductometric biosensor was proposed for detection of Hg^{2+} and Ag^{+} ions. The three enzymes (invertase, mutarotase, glucose oxidase) immobilized on the transducer surface showed a cascade of enzymatic changes in medium conductivity and demonstrated the best sensitivity towards Hg^{2+} and Ag^{+} (Chouteau et al., 2005). A possibility of the biosensor reactivation, either by EDTA solution after inhibition with silver ions or by cysteine solution after inhibition with mercury ions, was demonstrated (Soldatkin et al., 2012).

An optical biosensor for on-site detection of heavy metals using structure-switching DNA offered a simple, rapid, portable sensing with minimum sample extraction procedures (Wu et al., 2010). Novel aspects of colorimetric biosensing using label-free gold nanoparticles (Au NPs) and alkanethiols were used to sense Hg^{2+}, Ag^{+}, and Pb^{2+} ions (Long et al., 2013; Hun et al., 2010). More recently, a dual channel sensor for the detection and discrimination of heavy metal ions based on colorimetric and fluorescence response of AuNP-DNA conjugates was reported. A highly sensitive discrimination of metal ion targets, with the detection limit as low as 50 nM with 100% identification accuracy of nine different heavy metal ions, was obtained. This technique was based on the exclusion of DNA and AuNPs in the presence of metal ions, with the amount of FAM-labeled DNA adsorbed on AuNP surface increasing, resulting in a fluorescence quenching effect (Tan et al., 2016).

15.5.1.3 Antibiotics

In recent years, the use of several antibiotics as growth promoters in stock farming has led to the emergence of an array of multi-drug-resistant human pathogens in food products. The most common antibiotics entering the food chain include chloramphenicol, streptomycin, kanamycin, streptogramin, and tetracycline.

A three-dimensional polystyrene (PS)-quartz crystal microbalance (QCM)-based immunosensor was fabricated for the biosensing of chloramphenicol (CAP), where the PS membrane and fibers were deposited on the QCM electrode via electrospinning. The 3-mercaptopropionic acid functionalized antibody against CAP showed the best results with good sensitivity and selectivity (5–200 ppb) with a response time per sample of 2–3 s (Sun et al., 2011). A novel fluorescent aptasensor for

CAP was developed where the aptamer-magnetic nanoparticle conjugate served both as recognition and concentration element in combination with upconversion nanoparticles as signal enhancers. The assay was validated in contaminated milk samples and compared with commercial ELISA. This method provided a sensitive detection range of 0.01–1 ng/mL with an LOD of 0.01 ng/mL (Wu et al., 2015).

Streptomycin (STR), an antibiotic obtained from *Streptomyces grisens*, has been extensively administered in the veterinary field for the treatment of animals against gram-negative bacterial infections. Over-usage of STR in livestock has resulted in the presence of its residue in animal-derived foods leading to side effects like nephrotoxicity and ototoxicity (Granja et al., 2009; De Oliveira et al., 2009). An electrochemical-based redox-active magnetic molecularly imprinted polymer (mMIP) was used for the detection of STR in milk and honey samples. The electrochemical response was based on the bioelectrocatalytic reaction of enzymes resulting in signal amplification. The assay was based on a competitive format where the analyte (mMIP nanosphere) competed with glucose oxidase-labeled STR with the addition of free STR. This method exhibited a detection range of 0.05–20 ng/mL with a 10 pg/mL LOD (Liu B et al. 2013). In a recent report, the colorimetric and fluorescence quenching properties of aptasensors were employed for the detection of STR in food samples. The aptasensing was based on the aggregation of AuNPs by salt in the absence of STR, resulting in a color change from red to blue with strong fluorescence emission. The colorimetric and fluorescence quenching aptasensor showed an LOD of 73.1 and 47.6 nM respectively, along with successful detection in milk and serum samples (Emrani et al., 2016).

A label-free immunosensor for kanamycin detection was fabricated using silver hybridized ferroferric oxide nanoparticles and thionine mixed graphene sheets. Thionine served as the electron transfer medium. Characterization of kanamycin recognition was done using cyclic voltammetry and square wave voltammetry, which provided a linear detection range from 0.05–16 ng/mL, an LOD of 15 pg/mL, and a response time of three minutes for evaluation in pork meat samples (Yu et al., 2013). Another label-free amperometric immunosensor was proposed for kanamycin detection contrived on graphene sheet (GS)-nafion(Nf)/thionine(TH)/platinum (Pt) nanoparticles modified glassy carbon electrode. The presence of GS and Pt enhanced the electron transfer mediator activity of the sensor and provided an LOD of 5.74 pg/mL, along with the relevance of detection in animal-derived food (Wei et al., 2012a).

Multiplexed biosensors offer huge advantages in biosensing since more than one analyte can be detected simultaneously, thereby reducing the time for detection as well as associated costs. A single assembly for the detection of more than one antibiotic was fabricated by Weber's group (2005), based on the specific response of antibiotic sensor protein with DNA fragments of a particular class of antibiotics. The assay was based on an ELISA-type format, where the binding of the protein to the DNA resulted in an immune-based colorimetric signal (Weber et al., 2005). Similarly, a multiplexed dipstick assay (tetracycline, streptogramin, and macrolide) was also developed by a biosensor group, based on the release of bacterial transcriptional regulators and their specific binding to these antibiotics. This method promised a 40-fold lower detection beyond the permissible threshold value (Link et al., 2007).

15.5.1.4 Microbial Pathogens and Toxins

Foodborne illness (which includes both "food infection" and "food intoxication") is a constant threat to human health. According to a WHO report (2015), approximately 600 million cases of food illness are reported with roughly 420,000 deaths per year across the globe. The most common causative agents of foodborne illness include mycotoxins, exotoxins, and enterotoxins from *E. coli* O157:H7, some strains of *Staphylococcus aureus*, *Shigella* spp., *Bacillus anthracis*, *Campylobacter jejuni*, *Clostridium perfringens*, *Clostridium botulinum* (produces a powerful paralytic toxin botulinum), *Salmonella* spp., *Listeria monocytogenes*, *Vibrio cholera*, *Yersinia enterocolitica*, and *Coxiella burnetii*. The risk groups include infants, young children, pregnant people, elderly and sick people, and immune-compromised people (Arora et al., 2011). Therefore, rapid and sensitive detection of foodborne pathogens and toxins are highly desirable for addressing biosafety concerns.

The typical biosensing platform for detection of mycotoxins such as aflatoxin (AFB), ochratoxin A (OTA), patulin, etc., use ELISA or immunosensing for their detection. Many kits are commercially available based on these platforms. The recent developments in biosensors for detection of these toxins include the use of nucleic acid sensors or aptamers. Some examples include aptamer-based electrochemical sensor in conjunction with gold nanoparticles (Kuang et al., 2010); BoNT aptamer combining enzymatic amplification (Tok et al., 2008); paramagnetic microparticle beads functionalized OTA specific aptamer (Bonel et al., 2011); and fluorescence resonance energy transfer of labeled aptamers applied for detection of staphylococcal enterotoxin B (SEB) (Wu et al., 2013).

E. coli O157:H7 strain, known as enterohemorrhagic *E. coli* (EHEC), is a foodborne pathogen which produces exotoxins known as verotoxins (also termed Shiga-like toxins). A label-free DNA aptamer-based impedance biosensor was developed for the detection of *E. coli* outer membrane proteins (OMPs). This aptasensor had a dynamic detection range of 1×10^{-7} to 2×10^{-6} M and showed selectivity towards the target in the presence of other contaminants in the sample, and also at low pH conditions. This "signal-on" electrochemical aptasensor seems to be a promising approach for fabricating a biochip for easy detection of *E. coli* in water samples (Queirós et al., 2013).

Listeria monocytogenes is found as a contaminant in several raw and processed foods like milk and dairy products, meat and meat products, and seafood. They are the major cause of concern in products that are stored and delivered in the cold chain. An antibody–aptamer functionalized fiber-optic biosensor was developed for the specific detection of *L. monocytogenes* from ready to eat food products. The aptamer-fiber-optic sensor showed a detection limit of 10^3 CFU/mL and could successfully distinguish pathogenic *L. monocytogenes* from non-pathogenic species or other pathogenic species of *E. coli* or *Salmonella* (Ohk et al., 2010). A sensitive and selective optical biosensor was developed using magnetic nanobeads and quantum dots (QDs) to detect this food pathogen. Magnetic nanobeads were used to separate and concentrate the target bacteria while quantum dots (QDs) served as fluorescent markers. This biosensor could detect *L. monocytogenes* in a range of 10^0–10^7 cfu/mL, with an LOD of 2–3 cfu/mL, and a total detection time of about 1.5 hours (Wang et al., 2007). A quartz crystal microbalance immunosensor was developed with a detection limit of up to 1×10^7 cells/mL, and one that was reusable more than 10 times without detectable loss in activity (Vaughan et al., 2001).

The detection of *Salmonella* and other pathogens in complex sample matrices such as feces, foods, and environmental samples is challenging. However, of late several rapid and sensitive biosensors have been developed for *Salmonella* detection (Helms et al., 2005; Joshi et al., 2009). A surface plasmon resonance-based biosensor was used to detect *Salmonella*, where anti-*Salmonella* antibodies immobilized to the biosensor surface were allowed to bind injected bacteria, followed by a pulse with soluble anti-*Salmonella* immunoglobulins to intensify the signal. The limit of detection of this biosensor was 1.7×10^3 CFU per test, comprising an equicellular mixture of *Salmonella enteritidis* and *Salmonella typhimurium* (Bokken et al., 2003). Very recently, a phage-based magnetoelastic biosensor was used to detect *Salmonella typhimurium* on fresh tomato surfaces where the magnetoelastic resonator platform was coated with filamentous E2 phage to bind with *S. typhimurium*. The tomato surface was spiked with suspensions containing *S. typhimurium* in a range from 5×10^1 to 5×10^8 CFU/mL. The results showed that the biosensor could detect *Salmonella* at 5×10^2 CFU/mL and higher, with a confidence limit of 80% or higher. The study demonstrated the direct detection of foodborne pathogen on fresh produce (Li et al., 2010a).

15.6 BIOSENSORS FOR FOOD ADULTERANTS

An adulterant is a term for the presence of a substance in food which is not declared as content or is not legally allowed. Detection of adulterants, some of them potent carcinogens or mutagens, is an important area of food biosensor research.

15.6.1 Urea

Urea is commonly is added to milk as an adulterant to increase its fat and solid content. It is relatively cheap and easily available. Urea intake is known to cause indigestion, acidity, ulcers, cancers, and kidney malfunctions if consumed in large quantities. Therefore, urea estimation in milk and other food products is of great significance. In a urea biosensor, urease is the bio-component which catalyzes the hydrolysis of urea to ammonium and bicarbonate ions.

$$NH_2CONH_2 + 3H_2O \rightarrow 2NH_4^+ + OH^- + HCO_3^-$$

Potentiometric urea biosensor using electrosynthesized polymer poly-o-phenylenediamine (PPD) showed a linear response with urea concentration in the range 10 μM to 1 mM, due to the buffering capability of PPD film (Chirizzi and Malitesta, 2011). NH_4^+ ion sensitive electrode was used as a transducer in double matrix membrane (DMM) technology where the system prevents interference of Na^+ and K^+ ions towards urea estimation. The biosensor had a response time of few seconds with a detection limit of 2.5×10^{-5} mol/L (Trivedi et al., 2009). A biosensor with urease immobilized on chitosan membranes by adsorption followed by cross-linking with dilute aqueous glutaraldehyde solutions, showed the best response characteristic in the range of 10^{-4} to 10^{-2} M, including a very long lifetime (more than two months) (Magalhães and Machado, 1998)

A portable urea biosensor with a "switch on" of fluorescence allowing the detection of the substrate at concentration levels of 10^{-9} M was developed (Nikoleli et al., 2010). A manometric assay has been used for online measurement of urea in milk during the milking process. The detection was based on the hydrolysis of urea with urease leading to the release of ammonium and carbonate ions which, when enclosed in a sealed cell with citric acid, produced CO_2 which was subsequently measured by observing the change in pressure. The change in pressure was then correlated to the amount of urea present in the milk sample. The assay was inexpensive, simple, accurate, robust, and completed a measurement cycle in about five minutes (Jenkins and Delwiche, 2002).

15.6.2 Formaldehyde

Formaldehyde (HCHO) is reported to be a potent neurotoxic and carcinogenic agent (IARC Monograph, 2006). Although banned for use in food preservation, HCHO continues to be added to products like fish, milk, noodles, wine, etc. for extension of storage life of the products (Akshath and Bhatt, 2016a). A fluorescence turn-on probe using GNPs, CTAB, and fluorescein dye was developed by the author's group recently. Extensive fluorescence quenching was observed upon interaction of fluorescein with the growth solution that consisted of CTAB, Au^{3+}, and fluorescein dye. However, the addition of NADH led to a dose-dependent fluorescence turn-on (Figure 15.7). This behavior

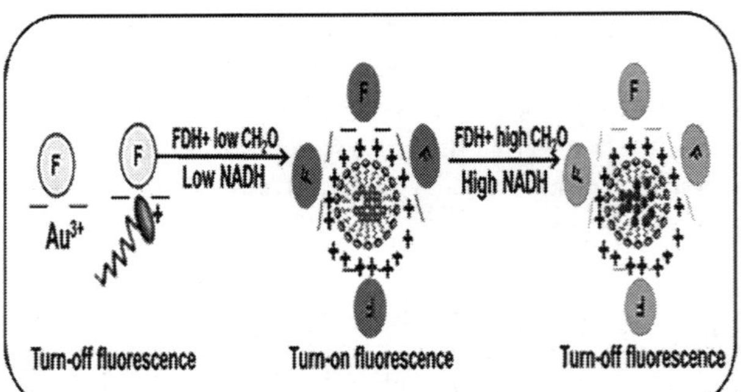

FIGURE 15.7 Fluorescence "turn on" probe using CTAB-GNP-FITC dye and HCHO dehydrogenase. (Akshath and Bhatt, 2016.)

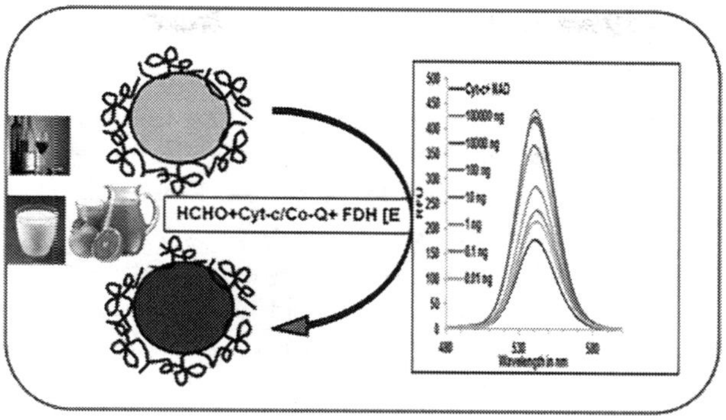

FIGURE 15.8 Non-classical co-factor-based turn on bio-sensing of HCHO in real samples. (Akshath and Bhatt, 2016.)

was successfully employed for ultrasensitive detection of formaldehyde (0.01 pg–300 pg/mL) as a function of NADH using the formaldehyde dehydrogenase enzyme.

In an attempt to investigate whether NADH can be replaced with non-classical cofactors, tunneling of redox enzymes was carried out to develop a probe for HCHO detection. Tunneling of enzyme activity was achieved using redox cofactors, namely oxidized cytochrome-c (Cyt-c) and Co-enzyme-Q (Co-q) immobilized on quantum dots (QDs) which acted as a bio-probe for NAD^+-dependent dehydrogenase catalyzed reaction (Figure 15.8). Electron transfer from substrate to non-native electron acceptors could differentially modify photoluminescence properties of CdTe QDs, and the phenomenon was successfully used for the detection of HCHO in the range of 0.01 ng/mL–100000 ng/mL using both QD-Ox-Cyt-c and QD-Ox-Co-Q. Further, the probe was found to be stable and sensitive (LOD of 0.001 ng/mL) in real samples like milk, wine, and fruit juice (Akshath and Bhatt, 2016b).

15.6.3 Melamine

Melamine is also used as an adulterant to inflate protein levels in food products artificially due to its high nitrogen content. Although it has low acute oral toxicity, excessive intake of melamine is reported to result in the formation of kidney stones (Brown et al., 2007).

A surface plasmon resonance-based optical biosensor was developed for the analysis of melamine in adulterated infant formula and liquid milk samples. A structural mimic of melamine was used to raise polyclonal antibody which was immobilized on the surface of the biosensor chip. The sensitivity of the sensor, in terms of IC_{50} value of 67.9 ng/mL was achieved with an LOD of 0.5 μg/mL in both infant formula as well as a liquid milk sample. The optical sensor did not show any cross-reactivity with the by-products of melamine manufacture, but little cross-reactivity was observed with insecticide cryomazine, for which melamine is a metabolite (Fodey et al., 2011). An indirect competitive immunoassay based on a planar waveguide fluorescence immunosensor was fabricated for the detection and quantification of melamine in milk products. The immunosenor offered an LOD of 6.6 μg/mL, with a quantifiable detection range of 26.6–517.5 μg/mL. Recovery of melamine from spiked milk samples was in the range of 89.8–103.2%. The BSA-melamine immobilized chip offered regeneration for more than 100 assays with 20 minutes of response for each cycle (Guo et al., 2014). On similar lines, a fiber-optic evanescent wave-based immunosensor was designed with a reusable immunosurface for more than 300 assay cycles and a response time of 15 minutes per reaction cycle. The fiber-optic immunosensor achieved an LOD of 5.14 μg/L, with a detectable range from 12.62 to 284.18 μg/L. Validation in milk, yoghurt, and baby formula, with recoveries in the range of 86.1%–106.5%, showed good prospects for its application in the detection of contaminants in dairy industry (Hao et al., 2014).

15.7 COMMERCIALLY AVAILABLE FOOD SENSORS AND FUTURE DIRECTIONS

Since the inception of enzyme electrode by Updike and Hicks (for glucose detection) in 1967, the field of biosensors has progressed in leaps and bounds in almost all fields of science and engineering. It is interesting to note that after its introduction, the first commercial glucose biosensor was not launched until 1987, two decades later, by MediSense Inc. Since then, dramatic advancements in technology and research have led to numerous biosensing devices, and their application in environment, health care, bio-medical, and defense systems (Turner, 2013).

The application of biosensors for food analysis has only recently been gaining attention. Despite a very high number of publications in the field of biosensors and bioelectronics, transfer of these technologies to companies or commercialization is extremely low. There are multiple reasons for this:

(1) Each product has its own analytical requirement and market for individual sensors is lower.
(2) Compared to medical diagnostics, high specificity for food sample analysis may not always be required.
(3) Instability of biosensors in food samples is quite common, as the matrix is highly complex and many times non-specific interactions occur, and may block the sensor surface.
(4) In the food industry, many processes have been standardized over the years and replacing them with biosensors for process monitoring is not always welcome.
(5) Biosensor technology is not always cost-effective.

Nevertheless, the growing demand in the food and agriculture sector has opened new vistas for biosensor market establishment (Antonacci et al., 2016). With changing times, there seems to be an increase in demand for biosensors in food diagnostics. It is estimated that by 2020, the food biosensors market could reach US$16.1 billion. Table 15.3, presents the list of commercial food sensors and companies manufacturing them (Bahadır and Sezgintürk, 2015; Kokkali and van Delft, 2014; Valderrama et al., 2016; Adley, 2014). An extensive review on commercially available sensors was recently published by Scognamiglio et al. (2016; Antonacci et al., 2016), which provides up-to-date details on commercial sensors.

To date, biosensors for food analysis mainly focus on sugar detection, gas sensors, pH sensors, ATP-based microbial load sensors, humidity detection, etc. There is still a large area in the field of food biosensors which needs to be explored. With emerging foodborne pathogens and mutations of existing ones, food biosensors especially for food safety will play a pivotal role. The focus of biosensor research in food analysis should also be in the development of easy-to-use sensors packaged into hand-held devices or paper sensors as simple as pH strips. These sensors should be able to cater to consumers even at individual levels to ensure food safety.

In particular, the gap between research leads in the laboratory to product development needs to be abridged (Valderrama et al., 2016; Adley, 2014; Scognamiglio et al., 2014). It is also extremely necessary that the system developed should be robust and should cater to a wide variety of food matrices like meat, fish, dairy, cereals, fruit, vegetables, bakery products, beverages, confectionary, etc. Many times biosensors developed in the laboratory do not perform well in the field due to several reasons. For example, recently the Supreme Court of Bangladesh recalled a formaldehyde detection kit from market, declaring it unfit for detection (Fresh Plaza, 2014). Future developments of such devices and kits should consider and overcome the limitations associated with real sample analysis and field detection.

Also, the food industry demands that the sensors be accurate in comparison with traditional techniques and have broad applicability in different sample conditions. Considering this, to date, few successful hand-held devices have been commercialized. With increased R&D in nanotechnology, future research incorporating nanoparticles as transduction amplifiers can prove to be important assets for biosensor development. Future research should also focus on minimizing the sample preparation, extraction, purification, processing time, etc. for food analysis, similar to the biosensing field in healthcare and diagnostics, where biosensors have graduated from large and cumbersome glucometers to simple and hassle-free wearable tattoo-based glucose sensors (Bandodkar et al., 2014). Research and development in food biosensors needs to come a long way to establish it as a strong market with a demand-driven product).

TABLE 15.3
Commercially Available Food Biosensors

Company	Products Related	Link
DiagnoSwiss	Development and commercialization of biochip technologies for food, pharmaceuticals healthcare, etc	http://www.diagnoswiss.com/
NeogenCorporation	Products are focused for the detection of pathogens in raw ingredients, finished food products. Test kits to detect foodborne bacteria, mycotoxins, allergens, genetic modifications, drug residues, plant diseases, etc are developed.	http://foodsafety.neogen.com/en/
Roche Diagnostics AG	Biosensors used for analyzing blood samples	http://www.roche-diagnostics.ch/
IBA GmbH	Sucrose, glucose, alcohol, etc	http://www.iba-lifesciences.com/product-shop.html
Hygiena	Freshness meter: Degradation products of ATP	http://www.hygiena.com/other-monitoring-systems-home.html
Biosentec	Lactate biosensors	http://www.biosentec.fr/en/products.html
Sensolytics	Glucose, lactate, sucrose, ethanol, glutamate	http://www.sensolytics.de/en/products/microelectrodes.html
Rapid Biosensors	Micro-organisms and toxic substances	http://www.rapidbiosensor.com/
Biosensores S.L.	Online monitoring of the concentration of total micro-organisms in water samples in 15 min	http://www.biosensores.com/EN/empresa.php
Texas Instruments Inc.	Sensors for gas and pH determination in food industry	http://www.ti.com/lsds/ti/applications/industrial/medical/overview.page
Integrated Genetics	DNA probes for detection of microbial contamination	http://www.integratedgenetics.com/test-menu
Molecular Devices Corporation	Sensors and components for microbial determination	https://www.moleculardevices.com/
Nova Biomedical	Glucose, L-lactate, L-glutamate, L-glutamine, alcohol, sucrose, methanol, ammonia	http://www.novabio.us/
Novas Insignia Technologies	Gas e modified atmosphere	http://insigniatechnologies.com/
Massachusetts Institute of Technology	Detection of *E. coli* O157:H7 in lettuce (Canary)	http://www.atrp.gatech.edu/pt16–3/16-3_p1.html
Universitat Autonoma de Barcelona in collaboration with CSIC	Detection of atrazine traces	http://www.uab.cat/PDF/PDF_1179901814052_en.pdf
Research International	Proteins, toxins, virus, bacteria, spores, and fungi (simultaneous analysis)	http://www.resrchintl.com/
Yellow Springs Instruments	Glucose, sucrose, lactose, L-lactate, galactose, L-glutamate, Ethanol, H2O2, starch, glutamine, choline	https://www.ysi.com/
Biacore AB	Vitamins, chemical veterinary residues, and mycotoxins	https://www.biacore.com/lifesciences/index.html

REFERENCES

Adam, V., Petrlova, J., Potesil, D., Zehnalek, J., Sures, B., Trnkova, L., Jelen, F., and Kizek, R. 2005. Study of Metallothionein modified electrode surface behavior in the presence of heavy metals ions-biosensor. *Electroanalysis* 17(18): 1649–1657.

Adányi, N. and Váradi, M. 2003. Development of organic phase amperometric biosensor for measuring cholesterol in food samples. *European Food Research and Technology* 218(1): 99–104.

Adley, C.C. 2014. Past, present and future of sensors in food production. *Foods* 3(3): 491–510.

Akshath, U.S. and Bhatt, P. 2016a. Gold nanoparticle synthesis coupled to fluorescence turn-on for sensitive detection of formaldehyde using formaldehyde dehydrogenase. *RSC Advances* 6(60): 54777–54784.

Akshath, U.S. and Bhatt, P. 2016b. Tunneling of redox enzymes to design nano-probes for monitoring NAD+dependent bio-catalytic activity. *Biosensors and Bioelectronics* 85: 240–246.

Akshath, U.S., Selvakumar, L.S. and Thakur, M.S. 2012a. Detection of formaldehyde in food samples by enhanced chemiluminescence. *Analytical Methods* 4(3): 699–704.

Akshath, U.S., Vinayaka, A.C. and Thakur, M.S. 2012b. Quantum dots as nano plug-ins for efficient NADH resonance energy routing. *Biosensors and Bioelectronics* 38(1): 411–415.

Algar, W.R., Tavares, A.J. and Krull, U.J. 2010. Beyond labels: A review of the application of quantum dots as integrated components of assays, bioprobes, and biosensors utilizing optical transduction. *Analytica Chimica Acta* 673(1): 1–25.

Altman, D.G. and Bland, J.M. 1994. Diagnostic tests. 1: Sensitivity and specificity. *BMJ: British Medical Journal* 308(6943):1552.

Alvarez, M., Calle, A., Tamayo, J., Lechuga, L.M., Abad, A. and Montoya, A. 2003. Development of nanomechanical biosensors for detection of the pesticide DDT. *Biosensors and Bioelectronics* 18(5): 649–653.

Amine, A., Kauffmann, J.M., Guilbault, G.G. and Bacha, S. 1993. Characterization of mixed enzyme-mediator-carbon paste electrodes. *Analytical Letters* 26(7):1281–1299.

Antonacci, A., Arduini, F., Moscone, D., Palleschi, G. and Scognamiglio, V. 2016. Commercially Available (Bio) sensors in the Agrifood Sector. *Comprehensive Analytical Chemistry,* 74:315–340.

Arora, P., Sindhu, A., Dilbaghi, N. and Chaudhury, A. 2011. Biosensors as innovative tools for the detection of food borne pathogens. *Biosensors and Bioelectronics* 28(1):1–12.

Arya, S.K., Datta, M. and Malhotra, B.D. 2008. Recent advances in cholesterol biosensor. *Biosensors and Bioelectronics* 23(7): 1083–1100.

Bachmann, T.T., Leca, B., Vilatte, F., Marty, J.L., Fournier, D. and Schmid, R.D. 2000. Improved multianalyte detection of organophosphates and carbamates with disposable multielectrode biosensors using recombinant mutants of Drosophila acetylcholinesterase and artificial neural networks. *Biosensors and Bioelectronics* 15(3): 193–201.

Bahadır, E.B. and Sezgintürk, M.K. 2015. Applications of commercial biosensors in clinical, food, environmental, and biothreat/biowarfare analyses. *Analytical Biochemistry* 478: 107–120.

Baker, S., Vinayaka, A.C., Manonmani, H.K. and Thakur, M.S. 2012. Development of dipstick-based immuno-chemiluminescence techniques for the rapid detection of dichlorodiphenyltrichloroethane. *Luminescence* 27(6): 524–529.

Bandodkar, A.J., Jia, W., Yardımcı, C., Wang, X., Ramirez, J. and Wang, J. 2014. Tattoo-based noninvasive glucose monitoring: A proof-of-concept study. *Analytical Chemistry* 87(1): 394–398.

Bartlett, P.N. 2008. *Bioelectrochemistry: Fundamentals, Experimental Techniques and Applications.* Chichester, UK: John Wiley & Sons.

Basu, A.K., Chattopadhyay, P., Roychoudhuri, U. and Chakraborty, R. 2007. Development of cholesterol biosensor based on immobilized cholesterol esterase and cholesterol oxidase on oxygen electrode for the determination of total cholesterol in food samples. *Bioelectrochemistry* 70(2): 375–379.

Batra, B., Lata, S., Sharma, M. and Pundir, C.S. 2013. An acrylamide biosensor based on immobilization of hemoglobin onto multiwalled carbon nanotube/copper nanoparticles/polyaniline hybrid film. *Analytical Biochemistry* 433(2): 210–217.

Billard, P. and DuBow, M.S. 1998. Bioluminescence-based assays for detection and characterization of bacteria and chemicals in clinical laboratories. *Clinical Biochemistry* 31(1): 1–14.

Bokken, G.C., Corbee, R.J., Van Knapen, F. and Bergwerff, A.A. 2003. Immunochemical detection of Salmonella group B, D and E using an optical surface plasmon resonance biosensor. *FEMS Microbiology Letters* 222(1): 75–82.

Bollella, Paolo, Christopher Schulz, Gabriele Favero, Franco Mazzei, Roland Ludwig, Lo Gorton, and Riccarda Antiochia. 2017. Green synthesis and characterization of gold and silver nanoparticles and their application for development of a third generation lactose biosensor. *Electroanalysis* 29(1): 77–86.

Bonel, L., Vidal, J.C., Duato, P. and Castillo, J.R. 2011. An electrochemical competitive biosensor for ochratoxin A based on a DNA biotinylated aptamer. *Biosensors and Bioelectronics* 26(7): 3254–3259.

Bontidean, I., Ahlqvist, J., Mulchandani, A., Chen, W., Bae, W., Mehra, R.K., Mortari, A. and Csöregi, E. 2003. Novel synthetic phytochelatin-based capacitive biosensor for heavy metal ion detection. *Biosensors and Bioelectronics* 18(5): 547–553.

Boujtita, M. and Murr, N. 1995. Biosensors for analysis of ethanol in foods. *Journal of Food Science* 60(1): 201–204.

Brown, K.R., Fox, A.P. and Natan, M.J. 1996. Morphology-dependent electrochemistry of cytochrome c at Au colloid-modified SnO_2 electrodes. *Journal of the American Chemical Society* 118(5): 154–1157.

Brown, C.A., Jeong, K.S., Poppenga, R.H., Puschner, B., Miller, D.M., Ellis, A.E., Kang, K.I., Sum, S., Cistola, A.M. and Brown, S.A. 2007. Outbreaks of renal failure associated with melamine and cyanuric acid in dogs and cats in 2004 and 2007. *Journal of Veterinary Diagnostic Investigation* 19(5): 525–531.

Bui, T.T. and Park, S.Y. 2016. A carbon dot–hemoglobin complex-based biosensor for cholesterol detection. *Green Chemistry* 18(15): 4245–4253.

Campuzano, S., Loaiza, Ó.A., Pedrero, M., de Villena, F.J.M., and Pingarrón, J.M. 2004. An integrated bienzyme glucose oxidase–fructose dehydrogenase–tetrathiafulvalene-3-mercaptopropionic acid–gold electrode for the simultaneous determination of glucose and fructose. *Bioelectrochemistry* 63(1): 199–206.

Chakravarty, S., Dutta, P., Kalita, S. and Sarma, N.S. 2016. PVA-based nanobiosensor for ultrasensitive detection of folic acid by fluorescence quenching. *Sensors and Actuators B: Chemical* 232: 243–250.

Chang, J., Li, H., Hou, T. and Li, F. 2016. Paper-based fluorescent sensor for rapid naked-eye detection of acetylcholinesterase activity and organophosphorus pesticides with high sensitivity and selectivity. *Biosensors and Bioelectronics* 86: 971–977.

Chen, Q., Ma, Q., Wan, Y., Su, X., Lin, Z. and Jin, Q. 2005. Studies on fluorescence resonance energy transfer between dyes and water-soluble quantum dots. *Luminescence* 20(4–5): 251–255.

Cherian, S., Gupta, R.K., Mullin, B.C. and Thundat, T. 2003. Detection of heavy metal ions using protein-functionalized microcantilever sensors. *Biosensors and Bioelectronics* 19(5): 411–416.

Chirizzi, D. and Malitesta, C. 2011. Potentiometric urea biosensor based on urease immobilized by an electrosynthesized poly (o-phenylenediamine) film with buffering capability. *Sensors and Actuators B: Chemical* 157(1): 211–215.

Chouteau, C., Dzyadevych, S., Durrieu, C. and Chovelon, J.M. 2005. A bi-enzymatic whole cell conductometric biosensor for heavy metal ions and pesticides detection in water samples. *Biosensors and Bioelectronics* 21(2): 273–281.

Clark Jr, L.C. and Lyons, C. 1962. Electrode systems for continuous monitoring in cardiovascular surgery. *Annals of the New York Academy of sciences* 102(1): 29–45.

Costa-Fernández, J.M., Pereiro, R. and Sanz-Medel, A. 2006. The use of luminescent quantum dots for optical sensing. *TrAC Trends in Analytical Chemistry* 25(3): 207–218.

D'costa, E.J., Higgins, I.J. and Turner, A.P.F. 1986. Quinoprotein glucose dehydrogenase and its application in an amperometric glucose sensor. *Biosensors* 2(2): 71–87.

D'Orazio, P. 2003. Biosensors in clinical chemistry. *Clinica Chimica Acta* 334(1): 41–69.

Dagar, K. and Pundir, C.S. 2017. An improved amperometric L-lactate biosensor based on covalent immobilization of microbial lactate oxidase onto carboxylated multiwalled carbon nanotubes/copper nanoparticles/polyaniline modified pencil graphite electrode. *Enzyme and Microbial Technology* 96:177–186.

Danesh, N.M., Ramezani, M., Emrani, A.S., Abnous, K. and Taghdisi, S.M. 2016. A novel electrochemical aptasensor based on arch-shape structure of aptamer-complimentary strand conjugate and exonuclease I for sensitive detection of streptomycin. *Biosensors and Bioelectronics* 75:123–128.

de Oliveira, R.C., Paschoal, J.A.R., Sismotto, M., da Silva Airoldi, F.P. and Reyes, F.G.R. 2009. Development and validation of an LC-APCI-MS-MS analytical method for the determination of streptomycin and dihydrostreptomycin residues in milk. *Journal of Chromatographic Science* 47(9): 756–761.

De Volder, M.F., Tawfick, S.H., Baughman, R.H. and Hart, A.J. 2013. Carbon nanotubes: Present and future commercial applications. *Science* 339(6119): 535–539.

Del Carlo, M., Mascini, M., Pepe, A., Diletti, G. and Compagnone, D. 2004. Screening of food samples for carbamate and organophosphate pesticides using an electrochemical bioassay. *Food Chemistry* 84(4): 651–656.

Dzyadevych, S.V., Arkhypova, V.N., Soldatkin, A.P., El'skaya, A.V., Martelet, C. and Jaffrezic-Renault, N. 2008. Amperometric enzyme biosensors: Past, present and future. *IRBM* 29(2): 171–180.

Eggins, B.R. 1996. *Biosensors: An Introduction*. Springer-Verlag.

Ellington, A.D. and Szostak, J.W. 1990. In vitro selection of RNA molecules that bind specific ligands. *Nature* 346(6287): 818–822.

Elman, N.M., Ben-Yoav, H., Sternheim, M., Rosen, R., Krylov, S. and Shacham-Diamand, Y. 2008. Towards toxicity detection using a lab-on-chip based on the integration of MOEMS and whole-cell sensors. *Biosensors and Bioelectronics* 23(11): 1631–1636.

Emrani, A.S., Danesh, N.M., Lavaee, P., Ramezani, M., Abnous, K. and Taghdisi, S.M. 2016. Colorimetric and fluorescence quenching aptasensors for detection of streptomycin in blood serum and milk based on double-stranded DNA and gold nanoparticles. *Food Chemistry* 190: 115–121.

Evtugyn, G., Porfireva, A., Ivanov, A., Konovalova, O. and Hianik, T. 2009. Molecularly imprinted polymerized methylene green as a platform for electrochemical sensing of aptamer–thrombin interactions. *Electroanalysis* 21(11): 1272–1277.

Fang, A., Ng, H.T. and Li, S.F.Y. 2003. A high-performance glucose biosensor based on monomolecular layer of glucose oxidase covalently immobilised on indium–tin oxide surface. *Biosensors and Bioelectronics* 19(1):43–49.

Feng, L., Liu, Y., Tan, Y. and Hu, J. 2004. Biosensor for the determination of sorbitol based on molecularly imprinted electrosynthesized polymers. *Biosensors and Bioelectronics* 19(11): 1513–1519.

Fodey, T.L., Thompson, C.S., Traynor, I.M., Haughey, S.A., Kennedy, D.G. and Crooks, S.R. 2011. Development of an optical biosensor-based immunoassay to screen infant formula milk samples for adulteration with melamine. *Analytical chemistry* 83(12): 5012–5016.

Frasco, M.F. and Chaniotakis, N. 2009. Semiconductor quantum dots in chemical sensors and biosensors. *Sensors* 9(9): 7266–7286.

Freeman, R., Finder, T., Bahshi, L. and Willner, I. 2009. β-Cyclodextrin-modified CdSe/ZnS quantum dots for sensing and chiroselective analysis. *Nano Letters* 9(5): 2073–2076.

Gamella, M., Campuzano, S., Conzuelo, F., Curiel, J.A., Muñoz, R., Reviejo, A.J. and Pingarrón, J.M. 2010. Integrated multienzyme electrochemical biosensors for monitoring malolactic fermentation in wines. *Talanta* 81(3): 925–933.

Gao, Y., Guo, F., Gokavi, S., Chow, A., Sheng, Q. and Guo, M. 2008. Quantification of water-soluble vitamins in milk-based infant formulae using biosensor-based assays. *Food Chemistry* 110(3): 769–776.

Gao, F., Feng, S., Chen, Z., Li-Chan, E.C., Grant, E. and Lu, X. 2014. Detection and quantification of chloramphenicol in milk and honey using molecularly imprinted polymers: Canadian penny-based SERS nano-biosensor. *Journal of Food Science* 79(12): N2542–N2549.

Geim, A.K. and Novoselov, K.S. 2007. The rise of graphene. *Nature Materials* 6(3): 183–191.

Giménez-Gómez, P., Gutiérrez-Capitán, M., Capdevila, F., Puig-Pujol, A., Fernández-Sánchez, C. and Jiménez-Jorquera, C. 2016. Robust L-Malate bienzymatic biosensor to enable the on-site monitoring of malolactic fermentation of red wines. *Analytica Chimica Acta* 954: 105–113.

Graça, J.S., de Oliveira, R.F., de Moraes, M.L. and Ferreira, M. 2014. Amperometric glucose biosensor based on layer-by-layer films of microperoxidase-11 and liposome-encapsulated glucose oxidase. *Bioelectrochemistry* 96: 37–42.

Granja, R.H., Niño, A.M.M., Zucchetti, R.A., Niño, R.E.M., Patel, R. and Salerno, A.G. 2009. Determination of streptomycin residues in honey by liquid chromatography–tandem mass spectrometry. *Analytica Chimica Acta* 637(1): 64–67.

Grieshaber, D., MacKenzie, R., Voeroes, J. and Reimhult, E. 2008. Electrochemical biosensors-sensor principles and architectures. *Sensors* 8(3): 1400–1458.

Guo, H., Zhou, X., Zhang, Y., Song, B., Liu, L., Zhang, J. and Shi, H. 2014. Highly sensitive and rapid detection of melamine in milk products by planar waveguide fluorescence immunosensor (PWFI). *Sensors and Actuators B: Chemical* 194: 114–119.

Gür, B., Işık, M., Kıranşan, K.D., Alanyalıoğlu, M., Beydemir, Ş., and Meral, K. 2016. High enzymatic activity preservation of malate dehydrogenase immobilized in a Langmuir–Blodgett film and its electrochemical biosensor application for malic acid detection. *RSC Advances* 6(83): 79792–79797.

Gutiérrez, J.C., Amaro, F. and Martín-González, A. 2015. Heavy metal whole-cell biosensors using eukaryotic microorganisms: An updated critical review. *Frontiers in Microbiology* 6:48.

Hamilton, J.W., Kaltreider, R.C., Bajenova, O.V., Ihnat, M.A., McCaffrey, J., Turpie, B.W., Rowell, E.E., Oh, J., Nemeth, M.J., Pesce, C.A. and Lariviere, J.P. 1998. Molecular basis for effects of carcinogenic heavy metals on inducible gene expression. *Environmental Health Perspectives* 106: 1005.

Han, J.H., Kim, H.J., Sudheendra, L., Gee, S.J., Hammock, B.D. and Kennedy, I.M. 2013. Photonic crystal lab-on-a-chip for detecting staphylococcal enterotoxin B at low attomolar concentration. *Analytical Chemistry* 85(6): 3104–3109.

Hao, X.J., Zhou, X.H., Zhang, Y., Liu, L.H., Long, F., Song, L. and Shi, H.C. 2014. Melamine detection in dairy products by using a reusable evanescent wave fiber-optic biosensor. *Sensors and Actuators B: Chemical* 204: 682–687.

He, Y., Zhang, S., Zhang, X., Baloda, M., Gurung, A.S., Xu, H., Zhang, X., and Liu, G. 2011. Ultrasensitive nucleic acid biosensor based on enzyme–gold nanoparticle dual label and lateral flow strip biosensor. *Biosensors and Bioelectronics* 26(5): 2018–2024.

Heller, A. and Feldman, B. 2008. Electrochemical glucose sensors and their applications in diabetes management. *Chemical Reviews* 108(7): 2482–2505.

Helms, M., Ethelberg, S. and Molbak, K. 2005. International *Salmonella typhimurium* DT104 infections, 1992–2001. *Emerging Infectious Diseases* 11(6): 859–867.

Hildebrandt, A., Bragos, R., Lacorte, S. and Marty, J.L. 2008. Performance of a portable biosensor for the analysis of organophosphorus and carbamate insecticides in water and food. *Sensors and Actuators B: Chemical* 133(1): 195–201.

Fresh Plaza. 2014. India: Formalin detection kit found unfit to test fruits. http://www.freshplaza.com/article/127601/India-Formalin-detection-kit-found-unfit-to-test-fruits.

Hu, M., Chen, J., Li, Z.Y., Au, L., Hartland, G.V., Li, X., Marquez, M. and Xia, Y. 2006. Gold nanostructures: Engineering their plasmonic properties for biomedical applications. *Chemical Society Reviews* 35(11): 1084–1094.

Hu, Y., Feng, S., Gao, F., Li-Chan, E.C., Grant, E. and Lu, X. 2015. Detection of melamine in milk using molecularly imprinted polymers–surface enhanced Raman spectroscopy. *Food Chemistry* 176: 123–129.

Hu, Q., Fu, Y., Xu, X., Qiao, Z., Wang, R., Zhang, Y. and Li, Y. 2016. A colorimetric detection of acrylamide in potato chips based on nucleophile-initiated thiol–ene Michael addition. *Analyst* 141(3): 1136–1143.

Hung, Y.L., Hsiung, T.M., Chen, Y.Y., Huang, Y.F. and Huang, C.C. 2010. Colorimetric detection of heavy metal ions using label-free gold nanoparticles and alkanethiols. *The Journal of Physical Chemistry C* 114(39): 16329–16334.

Huang, S., Lu, S., Huang, C., Sheng, J., Zhang, L., Su, W. and Xiao, Q. 2016. An electrochemical biosensor based on single-stranded DNA modified gold electrode for acrylamide determination. *Sensors and Actuators B: Chemical* 224: 22–30.

Hussain, M., Iqbal, N. and Lieberzeit, P.A. 2013. Acidic and basic polymers for molecularly imprinted folic acid sensors—QCM studies with thin films and nanoparticles. *Sensors and Actuators B: Chemical* 176: 1090–1095.

IARC Monograph. 2006. IARC monograph on the evaluation of carcinogenic risks to human. Volume 88. https://monographs.iarc.fr/wp-content/uploads/2018/06/mono88.pdf.

Ikeda, T., Hamada, H., Miki, K. and Senda, M. 1985. Glucose oxidase-immobilized benzoquinone-carbon paste electrode as a glucose senso. *Agricultural and Biological Chemistry* 49(2): 541–543.

Ispas, C.R., Crivat, G. and Andreescu, S. 2012. Review: Recent developments in enzyme-based biosensors for biomedical analysis. *Analytical Letters* 45(2–3): 168–186.

Jenkins, D.M. and Delwiche, M.J. 2002. Manometric biosensor for on-line measurement of milk urea. *Biosensors and Bioelectronics* 17(6): 557–563.

Joshi, R., Janagama, H., Dwivedi, H.P., Kumar, T.S., Jaykus, L.A., Schefers, J. and Sreevatsan, S. 2009. Selection, characterization, and application of DNA aptamers for the capture and detection of Salmonella enterica serovars. *Molecular and Cellular Probes* 23(1): 20–28.

Justino, C.I., Rocha-Santos, T.A. and Duarte, A.C. 2013. Advances in point-of-care technologies with biosensors based on carbon nanotubes. *TrAC Trends in Analytical Chemistry* 45: 24–36.

Kant, R., Tabassum, R. and Gupta, B.D. 2016. A highly sensitive and distinctly selective d-sorbitol biosensor using SDH enzyme entrapped Ta_2 O_5 nanoflowers assembly coupled with fiber optic SPR. *Sensors and Actuators B: Chemical* 242: 810–817.

Kesik, M., Akbulut, H., Söylemez, S., Cevher, Ş.C., Hızalan, G., Udum, Y.A., Endo, T., Yamada, S., Çırpan, A., Yağcı, Y. and Toppare, L. 2014. Synthesis and characterization of conducting polymers containing polypeptide and ferrocene side chains as ethanol biosensors. *Polymer Chemistry* 5(21): 6295–6306.

Kim, J.H., Morikis, D. and Ozkan, M. 2004. Adaptation of inorganic quantum dots for stable molecular beacons. *Sensors and Actuators B: Chemical* 102(2): 315–319.

King, J.M.H., DiGrazia, P.M., Applegate, B., Burlage, R., Sanseverino, J., Dunbar, P., Larimer, F. and Sayler, G.A. 1990. Rapid, sensitive bioluminescent reporter technology for naphthalene exposure and biodegradation. *Science(Washington)* 249(4970): 778–781.

Kissinger, P.T. 2005. Biosensors-a perspective. *Biosensors and Bioelectronics* 20(12): 2512–2516.

Kissinger, P. and Heineman, W.R. eds. 1996. *Laboratory Techniques in Electroanalytical Chemistry, Revised and Expanded.* New York: Marcel Dekker.

Kokkali, V. and van Delft, W. 2014. Overview of commercially available bioassays for assessing chemical toxicity in aqueous samples. *TrAC Trends in Analytical Chemistry* 61: 133–155.

Kuang, H., Chen, W., Xu, D., Xu, L., Zhu, Y., Liu, L., Chu, H., Peng, C., Xu, C. and Zhu, S. 2010. Fabricated aptamer-based electrochemical "signal-off" sensor of ochratoxin A. *Biosensors and Bioelectronics* 26(2): 710–716.

Kumar, M.A., Patnaik, S.G., Lakshminarayanan, V., and Ramamurthy, S.S. 2017. Elecrochemical determination of ethanol by a palladium modified graphene nanocomposite glassy carbon electrode. *Analytical Letters*. 50(2): 350–363.

La, J.A., Lim, S., Park, H.J., Heo, M.J., Sang, B.I., Oh, M.K. and Cho, E.C. 2016. Plasmonic-based colorimetric and spectroscopic discrimination of acetic and butyric acids produced by different types of Escherichia coli through the different assembly structures formation of gold nanoparticles. *Analytica Chimica Acta* 933: 196–206.

Lakard, B., Magnin, D., Deschaume, O., Vanlancker, G., Glinel, K., Demoustier-Champagne, S., Nysten, B., Jonas, A.M., Bertrand, P. and Yunus, S. 2011. Urea potentiometric enzymatic biosensor based on charged biopolymers and electrodeposited polyaniline. *Biosensors and Bioelectronics* 26(10): 4139–4145.

Lakowicz, J.R. 2006. Plasmonics in biology and plasmon-controlled fluorescence. *Plasmonics* 1(1): 5–33.

Lang, X.Y., Fu, H.Y., Hou, C., Han, G.F., Yang, P., Liu, Y.B. and Jiang, Q. 2013. Nanoporous gold supported cobalt oxide microelectrodes as high-performance electrochemical biosensors. *Nature Communications* 4: 2169.

Li, L., Li, B., Qi, Y. and Jin, Y. 2009. Label-free aptamer-based colorimetric detection of mercury ions in aqueous media using unmodified gold nanoparticles as colorimetric probe. *Analytical and Bioanalytical Chemistry* 393(8): 2051–2057.

Li, S., Li, Y., Chen, H., Horikawa, S., Shen, W., Simonian, A. and Chin, B.A. 2010a. Direct detection of Salmonella typhimurium on fresh produce using phage-based magnetoelastic biosensors. *Biosensors and Bioelectronics* 26(4): 1313–1319.

Li, Y., Schluesener, H.J. and Xu, S. 2010b. Gold nanoparticle-based biosensors. *Gold Bulletin* 43(1): 29–41.

Lin, X., Ni, Y. and Kokot, S. 2016. Electrochemical cholesterol sensor based on cholesterol oxidase and MoS 2-AuNPs modified glassy carbon electrode. *Sensors and Actuators B: Chemical* 233: 100–106.

Link, N., Weber, W. and Fussenegger, M. 2007. A novel generic dipstick-based technology for rapid and precise detection of tetracycline, streptogramin and macrolide antibiotics in food samples. *Journal of Biotechnology* 128(3): 668–680.

Lisa, M., Chouhan, R.S., Vinayaka, A.C., Manonmani, H.K. and Thakur, M.S. 2009. Gold nanoparticles-based dipstick immunoassay for the rapid detection of dichlorodiphenyltrichloroethane: An organochlorine pesticide. *Biosensors and Bioelectronics* 25(1): 224–227.

Liu, B., Tang, D., Zhang, B., Que, X., Yang, H. and Chen, G. 2013. Au (III)-promoted magnetic molecularly imprinted polymer nanospheres for electrochemical determination of streptomycin residues in food. *Biosensors and Bioelectronics* 41: 551–556.

Liu, J. and Lu, Y. 2004. Accelerated color change of gold nanoparticles assembled by DNAzymes for simple and fast colorimetric Pb^{2+} detection. *Journal of the American Chemical Society* 126(39): 12298–12305.

Liu, J., Cao, Z. and Lu, Y. 2009. Functional nucleic acid sensors. *Chemical Reviews* 109(5): 948–1998.

Loaiza, O.A., Lamas-Ardisana, P.J., Añorga, L., Jubete, E., Ruiz, V., Borghei, M., Cabañero, G. and Grande, H.J. 2015. Graphitized carbon nanofiber–Pt nanoparticle hybrids as sensitive tool for preparation of screen printing biosensors. Detection of lactate in wines and ciders. *Bioelectrochemistry* 101: 58–65.

Long, F., Zhu, A., Shi, H., Wang, H. and Liu, J. 2013. Rapid on-site/in-situ detection of heavy metal ions in environmental water using a structure-switching DNA optical biosensor. *Scientific Reports* 3:2308–2318.

Lou, X., Egli, M. and Yang, X. 2016. Determining functional aptamer-protein interaction by biolayer interferometry. *Current Protocols in Nucleic Acid Chemistry* 67(1): 7–25.

Ma, H., Wu, D., Cui, Z., Li, Y., Zhang, Y., Du, B. and Wei, Q. 2013. Graphene-based optical and electrochemical biosensors: A review. *Analytical Letters* 46(1): 1–17.

Ma, X., Jiang, Y., Jia, F., Yu, Y., Chen, J. and Wang, Z. 2014. An aptamer-based electrochemical biosensor for the detection of Salmonella. *Journal of Microbiological Methods* 98: 94–98.

Magalhães, J.M. and Machado, A.A. 1998. Urea potentiometric biosensor based on urease immobilized on chitosan membranes. *Talanta* 47(1): 183–191.

Mascini, M. and Palchetti, I. 2001. Electrochemical biosensor for evaluation of contaminants in food. *Arhiv za higijenu rada i toksikologiju* 52(1): 49–59.

Mauriz, E., Calle, A., Manclús, J.J., Montoya, A., Hildebrandt, A., Barceló, D. and Lechuga, L.M. 2007. Optical immunosensor for fast and sensitive detection of DDT and related compounds in river water samples. *Biosensors and Bioelectronics* 22(7): 1410–1418.

Mazzei, F., Botrè, F. and Favero, G. 2007. Peroxidase based biosensors for the selective determination of D, L-lactic acid and L-malic acid in wines. *Microchemical Journal* 87(1): 81–86.

Mieliauskiene, R., Nistor, M., Laurinavicius, V. and Csöregi, E. 2006. Amperometric determination of acetate with a tri-enzyme based sensor. *Sensors and Actuators B: Chemical* 113(2): 671–676.

Molinero-Abad, B., Alonso-Lomillo, M.A., Domínguez-Renedo, O. and Arcos-Martínez, M.J. 2014. Malate quinone oxidoreductase biosensors based on tetrathiafulvalene and gold nanoparticles modified screen-printed carbon electrodes for malic acid determination in wine. *Sensors and Actuators B: Chemical* 202: 971–975.

Moyo, M., Okonkwo, J.O. and Agyei, N.M. 2014. An amperometric biosensor based on horseradish peroxidase immobilized onto maize tassel-multi-walled carbon nanotubes modified glassy carbon electrode for determination of heavy metal ions in aqueous solution. *Enzyme and Microbial Technology* 56: 28–34.

Murthy, M.R., Mandappa, I.M., Latha, R., Vinayaka, A.C., Thakur, M.S. and Manonmani, H.K. 2010. An immobilized dehydrohalogenase based potentiometric biosensor for the detection of chlorinated pesticides. *Analytical Methods* 2(9): 1355–1359.

Narsaiah, K., Jha, S.N., Bhardwaj, R., Sharma, R. and Kumar, R. 2012. Optical biosensors for food quality and safety assurance-a review. *Journal of Food Science and Technology* 49(4): 383–406.

Nguyen-Boisse, T.T., Saulnier, J., Jaffrezic-Renault, N. and Lagarde, F. 2013. Highly sensitive conductometric biosensors for total lactate, d-and l-lactate determination in dairy products. *Sensors and Actuators B: Chemical* 179: 232–239.

Nikoleli, G.P., Nikolelis, D.P. and Methenitis, C. 2010. Construction of a simple optical sensor based on air stable lipid film with incorporated urease for the rapid detection of urea in milk. *Analytica Chimica Acta* 675(1): 58–63.

Ohk, S.H., Koo, O.K., Sen, T., Yamamoto, C.M. and Bhunia, A.K. 2010. Antibody–aptamer functionalized fibre-optic biosensor for specific detection of Listeria monocytogenes from food. *Journal of Applied Microbiology* 109(3): 808–817.

Ovalle, M., Arroyo, E., Stoytcheva, M., Zlatev, R., Enriquez, L. and Olivas, A. 2015. An amperometric microbial biosensor for the determination of vitamin B 12. *Analytical Methods* 7(19): 8185–8189.

Padmaja, R.J., Akshath, U.S., Abhijith, K.S., Halami, P.M. and Thakur, M.S. 2014. Gold nanoparticle-based immunodetection of Staphylococcus aureus leukotoxin M/F'-PV in subclinical samples of bovine mastitis. *Analytical Methods* 6(14): 5214–5220.

Partanen, T., Heikkilä, P., Hernberg, S., Kauppinen, T., Moneta, G. and Ojajärvi, A. 1991. Renal cell cancer and occupational exposure to chemical agents. *Scandinavian Journal of Work, Environment and Health* 231–239.

Pedreschi, F., Segtnan, V.H. and Knutsen, S.H. 2010. On-line monitoring of fat, dry matter and acrylamide contents in potato chips using near infrared interactance and visual reflectance imaging. *Food Chemistry* 121(2): 616–620.

Pogačnik, L. and Franko, M. 1999. Determination of organophosphate and carbamate pesticides in spiked samples of tap water and fruit juices by a biosensor with photothermal detection. *Biosensors and Bioelectronics* 14(6): 569–578.

Pogačnik, L. and Franko, M. 2003. Detection of organophosphate and carbamate pesticides in vegetable samples by a photothermal biosensor. *Biosensors and Bioelectronics* 18(1): 1–9.

Purbia, R. and Paria, S. 2016. A simple turn on fluorescent sensor for the selective detection of thiamine using coconut water derived luminescent carbon dots. *Biosensors and Bioelectronics* 79: 467–475.

Queirós, R.B., de-Los-Santos-Álvarez, N., Noronha, J.P. and Sales, M.G.F. 2013. A label-free DNA aptamer-based impedance biosensor for the detection of E. coli outer membrane proteins. *Sensors and Actuators B: Chemical* 181: 766–772.

Raghu, P., Reddy, T.M., Reddaiah, K., Swamy, B.K. and Sreedhar, M. 2014. Acetylcholinesterase based biosensor for monitoring of malathion and acephate in food samples: A voltammetric study. *Food Chemistry* 142: 188–196.

Rajamanikandan, R. and Ilanchelian, M. 2016. Simple and visual approach for highly selective biosensing of vitamin B1 based on glutathione coated silver nanoparticles as a colorimetric probe. *Sensors and Actuators B: Chemical* 244: 380–386. doi: 10.1016/j.snb.2016.12.129.

Rama, E.C., Costa-García, A. and Fernández-Abedul, M.T. 2017. Pin-based electrochemical glucose sensor with multiplexing possibilities. *Biosensors and Bioelectronics* 88: 34–40.

Ranjan, R., Rastogi, N.K. and Thakur, M.S. 2012. Development of immobilized biophotonic beads consisting of *Photobacterium leiognathi* for the detection of heavy metals and pesticide. *Journal of Hazardous Materials* 225: 114–123.

Ren, X., Yang, L., Tang, F., Yan, C. and Ren, J. 2010. Enzyme biosensor based on NAD-sensitive quantum dots. *Biosensors and Bioelectronics* 26(1): 271–274.

Rogers, K.R. 2000. Principles of affinity-based biosensors. *Molecular Biotechnology* 14(2): 109–129.

Ronkainen, N.J., Halsall, H.B. and Heineman, W.R. 2010. Electrochemical biosensors. *Chemical Society Reviews* 39(5): 1747–1763.

Sang, S., Wang, Y., Feng, Q., Wei, Y., Ji, J. and Zhang, W. 2016. Progress of new label-free techniques for biosensors: A review. *Critical Reviews in Biotechnology* 36(3): 465–481.

Schuhmann, W. 2002. Amperometric enzyme biosensors based on optimised electron-transfer pathways and non-manual immobilisation procedures. *Reviews in Molecular Biotechnology* 82(4): 425–441.

Scognamiglio, V., Arduini, F., Palleschi, G. and Rea, G. 2014. Biosensing technology for sustainable food safety. *TrAC Trends in Analytical Chemistry* 62: 1–10.

Serna-Cock, L. and Perenguez-Verdugo, J.G. 2011. *Biosensors Applications in Agri-Food Industry* 43–64. London: INTECH Open Access Publisher.

Sharma, R.K., Agrawal, M. and Marshall, F. 2007. Heavy metal contamination of soil and vegetables in suburban areas of Varanasi, India. *Ecotoxicology and Environmental Safety* 66(2): 258–266.

Sharma, T.K., Ramanathan, R., Weerathunge, P., Mohammadtaheri, M., Daima, H.K., Shukla, R. and Bansal, V. 2014. Aptamer-mediated 'turn-off/turn-on' nanozyme activity of gold nanoparticles for kanamycin detection. *Chemical Communications* 50(100): 15856–15859.

Shetty, R.S., Deo, S.K., Liu, Y. and Daunert, S. 2004. Fluorescence-based sensing system for copper using genetically engineered living yeast cells. *Biotechnology and Bioengineering* 88(5): 664–670.

Shim, W.B., Kim, M.J., Mun, H. and Kim, M.G. 2014. An aptamer-based dipstick assay for the rapid and simple detection of aflatoxin B1. *Biosensors and Bioelectronics* 62: 288–294.

Shkotova, L.V., Goriushkina, T.B., Tran-Minh, C., Chovelon, J.M., Soldatkin, A.P. and Dzyadevych, S.V. 2008. Amperometric biosensor for lactate analysis in wine and must during fermentation. *Materials Science and Engineering: C* 28(5): 943–948.

Soldatkin, O.O., Kucherenko, I.S., Pyeshkova, V.M., Kukla, A.L., Jaffrezic-Renault, N., El'Skaya, A.V., Dzyadevych, S.V. and Soldatkin, A.P. 2012. Novel conductometric biosensor based on three-enzyme system for selective determination of heavy metal ions. *Bioelectrochemistry* 83: 25–30.

Song, H., Rajendiran, S., Kim, N., Jeong, S.K., Koo, E., Park, G., Thangadurai, T.D. and Yoon, S. 2012. A tailor designed fluorescent 'turn-on' sensor of formaldehyde based on the BODIPY motif. *Tetrahedron Letters* 53(37): 4913–4916.

Stobiecka, A., Radecka, H. and Radecki, J. 2007. Novel voltammetric biosensor for determining acrylamide in food samples. *Biosensors and Bioelectronics* 22(9): 2165–2170.

Sun, M., Ding, B., Lin, J., Yu, J. and Sun, G. 2011. Three-dimensional sensing membrane functionalized quartz crystal microbalance biosensor for chloramphenicol detection in real time. *Sensors and Actuators B: Chemical* 160(1): 428–434.

Sun, W., Qi, X., Zhang, Y., Yang, H., Gao, H., Chen, Y. and Sun, Z. 2012. Electrochemical DNA biosensor for the detection of Listeria monocytogenes with dendritic nanogold and electrochemical reduced graphene modified carbon ionic liquid electrode. *Electrochimica Acta* 85: 145–151.

Tan, L., Chen, Z., Zhao, Y., Wei, X., Li, Y., Zhang, C., Wei, X., and Hu, X. 2016. Dual channel sensor for detection and discrimination of heavy metal ions based on colorimetric and fluorescence response of the AuNPs-DNA conjugates. *Biosensors and Bioelectronics* 85: 414–421.

Taylor, R.F., Marenchic, I.G. and Cook, E.J. 1988. An acetylcholine receptor-based biosensor for the detection of cholinergic agents. *Analytica Chimica Acta* 213: 131–138.

Tok, J.B.H. and Fischer, N.O. 2008. Single microbead SELEX for efficient ssDNA aptamer generation against botulinum neurotoxin. *Chemical Communications* 16: 1883–1885.

Trivedi, U.B., Lakshminarayana, D., Kothari, I.L., Patel, N.G., Kapse, H.N., Makhija, K.K., Patel, P.B. and Panchal, C.J. 2009. Potentiometric biosensor for urea determination in milk. *Sensors and Actuators B: Chemical* 140(1): 260–266.

Tuerk, C. and Gold, L. 1990. Systematic evolution of ligands by exponential enrichment: RNA ligands to bacteriophage T4 DNA polymerase. *Science* 249(4968): 505–510.

Turner, A.P. 2013. Biosensors: Sense and sensibility. *Chemical Society Reviews* 42(8): 3184–3196.

Valderrama, W.B., Dudley, E.G., Doores, S. and Cutter, C.N. 2016. Commercially available rapid methods for detection of selected food-borne pathogens. *Critical Reviews in Food Science and Nutrition* 56(9): 1519–1531.

Vallee, B.L. and Ulmer, D.D. 1972. Biochemical effects of mercury, cadmium, and lead. *Annual Review of Biochemistry* 41(1): 91–128.

Vaughan, R.D., O'Sullivan, C.K. and Guilbault, G.G. 2001. Development of a quartz crystal microbalance (QCM) immunosensor for the detection of Listeria monocytogenes. *Enzyme and Microbial Technology* 29(10): 635–638.

Vinayaka, A.C. and Thakur, M.S. 2011. Photoabsorption and resonance energy transfer phenomenon in CdTe– protein bioconjugates: An insight into QD– biomolecular interactions. *Bioconjugate Chemistry* 22(5):968–975.

Wang, J. 2001. Glucose biosensors: 40 years of advances and challenges. *Electroanalysis* 13(12): 983.

Wang, J. 2006a. *Analytical Electrochemistry*. Wiley-VCH: John Wiley & Sons.

Wang, J. 2006b. Electrochemical biosensors: Towards point-of-care cancer diagnostics. *Biosensors and Bioelectronics* 21(10): 1887–1892.

Wang, J. 2008. Electrochemical glucose biosensors. *Chemical Reviews* 108(2): 814–825.

Wang, H., Li, Y. and Slavik, M. 2007. Rapid detection of Listeria monocytogenes using quantum dots and nanobeads-based optical biosensor. *Journal of Rapid Methods and Automation in Microbiology* 15(1): 67–76.

Wang, Y., Ye, Z., Si, C. and Ying, Y. 2013a. Monitoring of Escherichia coli O157: H7 in food samples using lectin-based surface plasmon resonance biosensor. *Food Chemistry* 136(3): 1303–1308.

Wang, Z., Etienne, M., Urbanova, V., Kohring, G.W. and Walcarius, A. 2013b. Reagentless d-sorbitol biosensor based on d-sorbitol dehydrogenase immobilized in a sol–gel carbon nanotubes–poly (methylene green) composite. *Analytical and Bioanalytical Chemistry* 405(11): 3899–3906.

Weber, C.C., Link, N., Fux, C., Zisch, A.H., Weber, W. anrd d Fussenegger, M. 2005. Broad-spectrum protein biosensors for class-specific detection of antibiotics. *Biotechnology and Bioengineering* 89(1): 9–17.

Weber, J., Jeedigunta, S. and Kumar, A. 2008. Fabrication and characterization of ZnO nanowire arrays with an investigation into electrochemical sensing capabilities. *Journal of Nanomaterial* 75–80. doi:10.1155/2008/638523.

Wei, H., Li, B., Li, J., Wang, E. and Dong, S. 2007. Simple and sensitive aptamer-based colorimetric sensing of protein using unmodified gold nanoparticle probes. *Chemical Communications* 36: 3735–3737.

Wei, Q., Zhao, Y., Du, B., Wu, D., Li, H. and Yang, M. 2012a. Ultrasensitive detection of kanamycin in animal derived foods by label-free electrochemical immunosensor. *Food Chemistry* 134(3): 1601–1606.

Wei, X., Lin, W., Ma, N., Luo, F., Lin, Z., Guo, L., Qiu, B. and Chen, G. 2012b. Sensitive fluorescence biosensor for folate receptor based on terminal protection of small-molecule-linked DNA. *Chemical Communications* 48(49): 6184–6186.

Wen, G., Li, Z. and Choi, M.M. 2013. Detection of ethanol in food: A new biosensor based on bacteria. *Journal of Food Engineering* 118(1): 56–61.

Wilson, M.S. 2005. Electrochemical immunosensors for the simultaneous detection of two tumor markers. *Analytical Chemistry* 77(5): 1496–1502.

Wu, C.S., Khaing Oo, M.K. and Fan, X. 2010. Highly sensitive multiplexed heavy metal detection using quantum-dot-labeled DNAzymes. *ACS Nano* 4(10): 5897–5904.

Wu, S., Duan, N., Ma, X., Xia, Y., Wang, H. and Wang, Z. 2013. A highly sensitive fluorescence resonance energy transfer aptasensor for staphylococcal enterotoxin B detection based on exonuclease-catalyzed target recycling strategy. *Analytica Chimica Acta* 782: 59–66.

Wu, S., Zhang, H., Shi, Z., Duan, N., Fang, C., Dai, S. and Wang, Z. 2015. Aptamer-based fluorescence biosensor for chloramphenicol determination using upconversion nanoparticles. *Food Control* 50: 597–604.

Wu, X., Chen, X., Gao, F., Ma, W., Xu, L., Kuang, H., Li, A. and Xu, C. 2016. SERS encoded nanoparticle heterodimers for the ultrasensitive detection of folic acid. *Biosensors and Bioelectronics* 75: 55–58.

Xia, C., Wang, N., Lidong, L. and Lin, G. 2008. Synthesis and characterization of waxberry-like microstructures ZnO for biosensors. *Sensors and Actuators B: Chemical* 129(1): 268–273.

Xiao, Y., Patolsky, F., Katz, E., Hainfeld, J.F. and Willner, I. 2003. Plugging into enzymes: Nanowiring of redox enzymes by a gold nanoparticle. *Science* 299(5614): 877–1881.

Xiao, Y., Shlyahovsky, B., Popov, I., Pavlov, V. and Willner, I. 2005. Shape and color of Au nanoparticles follow biocatalytic processes. *Langmuir* 21(13): 5659–5662.

Xie, C., Gao, S., Guo, Q. and Xu, K. 2010. Electrochemical sensor for 2, 4-dichlorophenoxy acetic acid using molecularly imprinted polypyrrole membrane as recognition element. *Microchimica Acta* 169(1–2): 145–152.

Xu, S., Wang, Y., Zhou, D., Kuang, M., Fang, D., Yang, W., Wei, S. and Ma, L. 2016. A novel chemiluminescence sensor for sensitive detection of cholesterol based on the peroxidase-like activity of copper nanoclusters. *Scientific Reports* 6: 39157.

Yakovleva, M., Buzas, O., Matsumura, H., Samejima, M., Igarashi, K., Larsson, P.O., Gorton, L. and Danielsson, B. 2012. A novel combined thermometric and amperometric biosensor for lactose determination based on immobilised cellobiose dehydrogenase. *Biosensors and Bioelectronics* 31(1): 251–256.

Yan, X., Li, H., Hu, T. and Su, X. 2016. A novel fluorimetric sensing platform for highly sensitive detection of organophosphorus pesticides by using egg white-encapsulated gold nanoclusters. *Biosensors and Bioelectronics* 91: 232–237.

Yu, S., Wei, Q., Du, B., Wu, D., Li, H., Yan, L., Ma, H. and Zhang, Y. 2013. Label-free immunosensor for the detection of kanamycin using Ag@Fe_3O_4 nanoparticles and thionine mixed graphene sheet. *Biosensors and Bioelectronics* 48: 224–229.

Zhang, Y., Muench, S.B., Schulze, H., Perz, R., Yang, B., Schmid, R.D. and Bachmann, T.T. 2005. Disposable biosensor test for organophosphate and carbamate insecticides in milk. *Journal of Agricultural and Food Chemistry* 53(13): 5110–5115.

Zhao, Z., Lei, W., Zhang, X., Wang, B. and Jiang, H. 2010. ZnO-based amperometric enzyme biosensors. *Sensors* 10(2): 1216–1231.

16 Characterization Methods for Nanoparticles

R. Gopirajah and C. Anandharamakrishnan

CONTENTS

16.1 INTRODUCTION

Nanoparticles are receiving attention nowadays because of their possible effects on human health and environmental sustainability. A nanoparticle is an object having all three external dimensions in the nanoscale, i.e. 1 nm = 10^{-9} m, although, in practice, particles larger than 100 nm are also termed as a nanoparticle (Cho et al., 2013). The behavior of nanoparticles and the applicability of measurement techniques vary with size and environment. For instance, particles with 500 nm size can either be considered very small or very large depending on the frame of reference (Cho et al., 2013). Nanoparticles created a huge interest due to their tiny size and large surface to volume ratio, and they display novel uniqueness in contrast to the bulk material with large particles (Mittal et al., 2014). As a result of their size, nanoparticles show different physicochemical properties compared to their respective bulk material. For example, changes in optical properties can cause changes in color (e.g. deep red color with golden colloids), thermal behavior, solubility, material strength, and electrical conductivity (Anandharamakrishnan, 2014a, 2015; Anandharamakrishnan and Ishwarya, 2015a,b; Bhushani and Anandharamakrishnan, 2017; Bhushani et al. 2017; Kamat et al., 2002; Hochella, 2002; Burleson et al., 2004).

Nanoparticles can be both natural (e.g. ashes) and human-made entities (e.g. engineered nanoparticles). The two common approaches for nanoparticle synthesis are the bottom-up approach and the top-down approach (Schulze, 2014). In the top-down approach, nanoparticles are formulated using larger ones to direct their assembly, whereas the bottom-up approach is a process that starts building from the molecular level to larger and more complex systems maintaining precise control of molecular structure (Anandharamakrishnan, 2014b; Bhushani and Anandharamakrishnan, 2017; Ezhilarasi et al., 2013; Karthik et al., 2017; Heera and Shanmugam, 2015). For example, the process of converting the bulk material to the fine particle is the top-down approach. In the bottom-up process, the atom is treated to nuclei and finally to nanoparticles. The next step after synthesis is the characterization of nanoparticles. Seeing objects at nanoscale was a dream for scientists that came true in the mid-1980s when a revolutionary instrument known as the scanning tunneling microscope (STM) came into use and, shortly after, the atomic force microscope (AFM) was invented (Scalf and West, 2006). The invention of STM and AFM promoted huge opportunities in the nanoworld. Once scientists were able to see nanoscale objects, they started to analyze them, understand their behavior, and imagine ways of manipulating them for various applications. Imaging nanomaterial is an essential part of nanoscience and nanotechnologies (Fultz et al., 2008).

There are various methods to image nanostructured materials and to characterize their physical and chemical properties (Anandharamakrishnan, 2014c; Tiede et al., 2008). They can be divided in two groups: (1) imaging by microscopy and (2) analysis by spectroscopy. In general, microscopy uses radiation and optics to obtain a magnified image of an object. The imaging resolution of a microscope is limited to the minimum focus of the radiation due to diffraction. The diffraction limit of light microscopy is approximately 1 μm (10^{-6} m), and for high-resolution transmission electron microscopy, the limit is roughly 1 Å (10^{-10} m). Depending on the matrix, microscopy can be used as a direct imaging technique for nanoparticle detection, enabling not only quantitative analysis, but also qualitative information as to size, shape, and aggregation state. The most commonly used microscopic techniques for nanoparticle characterization are optical and electron microscope. An optical microscope uses visible light (i.e. an electromagnetic radiation) and a system of lenses to magnify images of small samples, and therefore it is also called a light microscope. Among several microscopes, optical microscopes are the simplest, and their resolution limit imposes the wavelength of the visible light which is in the range of 400 to 700 nm (Lin et al., 2014). Resolution is usually the ability to differentiate between two closely positioned bright objects and one big object. The optical microscope has a resolving power of around 0.2 microns or 200 nm. However, single objects smaller than this limit of 200 nm are not distinguishable and are seen as "fuzzy objects." The optical microscope resolution restricts itself to the wavelength of visible light, which thus prevents atomic-scale imaging (Smith, 2015). In contrast, a fast-moving electron has a wavelength of less than 1 angstrom (Å), i.e. 1 Å = 10^{-10} m, and provides an enormous improvement in resolution by using an electron beam for imaging. Electron microscopes have been designed to overcome the diffraction limit of optical microscopes, which use electron beams to illuminate the sample. Electron microscopes include the scanning electron microscope (SEM), and the transmission electron microscope (TEM). Conceptually, both electron and optical microscopes use radiation to visualize a sample, whereas photons are used in the optical microscope, and electrons in the electron microscope (Schulze, 2014). Electron microscopes have much greater resolving power than light microscopes and use electromagnetic radiation for higher magnifications of up to 2 million times compared to the best light microscopes limit of 2000 times (Smith, 2015). On the other hand, spectroscopy technique uses absorption, emission, or scattering of electromagnetic radiation by a matter (such as atoms, molecules, atomic or molecular ions, liquids, or solids) to qualitatively or quantitatively study the physical processes (e.g. UV spectroscopy, Fourier-transform infrared spectroscopy). The interaction of radiation with matter can cause redirection of the radiation and transitions between the energy levels of the atoms or molecule.

Nanoparticle formation and characteristics are analyzed by UV-visible spectroscopy, scanning electron microscope (SEM), X-ray diffraction (XRD), Fourier transform-infrared spectroscopy (FTIR), and transmission electron microscope (TEM) (Kavitha et al., 2013; Mittal et al., 2014). Recently, nanoparticles have shown potential application in the medical field including diagnostics and therapeutics (Navalakhe and Nandedkar, 2007; Vadlapudi et al., 2014). In food science, nanotechnology finds great applications in food cultivation, processing, and packaging (Anandharamakrishnan, 2014a; Anu Bhushani et al., 2016). Food manufacturers are now developing nanomaterials to fulfill the consumer demand for tasty food along with food safety and health. Application of nanotechnology in food is called "nanofoods" – which involves the application of different techniques to manufacture nanomaterials and incorporation of those nanomaterials during cultivation, production, processing, or packaging (Sekhon, 2010). The nano foods serve the purposes of maintaining food safety and providing nutrient and flavor enhancement, besides minimizing production and consumer costs. Besides, nanofood provides various benefits such as health-promoting additives, longer shelf lives, and new flavor varieties (Duncan, 2011). For example, nanoencapsulation of bioactive food compounds has great advantages for targeted delivery and efficient absorption of nutrients by the human cells (Ezhilarasi et al., 2013). Food constituents with health benefits known as "bioactive compounds" occur in small quantities with specific foods (Singh, 2016). The bioavailability of food bioactive compounds is a major concern in many areas including drug and nutraceutical delivery, as many bioactive compounds are highly hydrophobic and crystalline in nature, resulting in low water solubility and poor bioavailability (McClements et al., 2015). The control of lipophilic bioactive nanocrystal formation and dissolution can increase the stability and loading capacity, and open new ways to monitor the release of incorporated bioactives and their bioavailability (McClements et al., 2015).

The rapid explosion of nanotechnologies in a wide range of consumer products has also raised some environmental, ethical, safety policy, and regulatory issues (Maynard, 2006). The properties of nanoparticles are often associated with the macroscopic and microscopic regimes, meaning that conventional theories do not necessarily allow to predict their behavior (Rubahn, 2008). This uncertainty creates concern over the environmental and health impact of nanoparticles besides their emerging opportunities in the areas of science and technology. The understanding of the health risk and safety of nanoparticles remains lacking as it demands advanced methodology for detection and characterization of engineered nanoparticles in complex food matrices (Maynard, 2006). Therefore, the use of robust analytical approaches for characterizing nanoparticles and maximize our benefit has become necessary while mitigating their impact (Cho et al., 2013). The characterization methods are expected to provide guidance for ensuring quality control and assessing the safety as well as the toxicity of nanomaterials (Lin et al., 2014). Characteristics such as molecular structure, melting point, chemical composition, solubility, boiling point, flash point, and pH along with physicochemical properties (e.g. size, charge, elemental composition, shape, coating) need to be examined (Lin et al., 2014) for nanomaterials. For the characterization of nanomaterial, several separation techniques such as centrifugation and nanofiltration are used for separating and purifying mixed materials (Sapsford et al., 2011). Analytical ultracentrifugation helps to investigate the conformation, structure, self-aggregation state and stoichiometry of nanomaterials as well as their shape, size distribution, and molecular weight (Inagaki et al., 2013). Electrophoresis methods are generally used to partition and purify biomolecules. For example, gel and capillary electrophoresis are used to provide information on the zeta potential and hydrodynamic size of nanomaterials (Sapsford et al., 2011). The field flow fractionation (FFF) method utilizes an external field such as flow, thermal, electrical, and magnetic fields to reveal the size/size distribution and charge information of the investigated nanomaterials

(Sapsford et al., 2011; Lin et al., 2014). Therefore, this chapter provides an overview of the analytical methods for nanoparticle characterization and their application in the field of food science. A detailed description of each method with its strength and limitations provides a clear picture for selecting a suitable technique for characterization of a potential bioactive nano compound.

16.2 ANALYTICAL METHODS FOR DRIED NANOPARTICLES

Drying is one of the oldest and most cost-effective methods of preserving foods of all varieties. The quality of dried products is strongly dependent on the method of drying and the process conditions (Aprajeeta et al., 2015). During drying, food undergoes various structural changes due to loss of moisture from the inner structure to the surrounding environment. Some studies have investigated the microstructure of dried foods by microscopic techniques (Parada and Aguilera, 2007). The changes in many physical characteristics of food during drying are due to variations in the product microstructure (Wang and Brennan, 1995). Several studies utilize the microscopic technique to visualize the effect of heating and drying methods on food microstructure (Cano-Chauca et al., 2005). A range of microscopic methods are available for the detection and characterization of nanoparticles, and a summary of the information generated by different techniques with their application is given in Table 16.1.

The characteristics of nanomaterials such as size, shape, surface energy, surface composition, and surface charge are crucial and need to be well investigated to better understand the behavior of nanomaterials in *in vivo* human conditions. The characterization of conventional pharmaceuticals and nanomedicines is based on the evaluation of physicochemical properties such as molecular weight, stability, identity, composition, purity, and solubility. Many techniques applying conventional pharmaceuticals can also be used for characterization of nutraceutical foods and nanomedicines (Navalakhe and Nandedkar, 2007). The following section discusses the advanced analytical techniques and instruments used to examine the specific physicochemical properties of dried nanomaterials with their main strengths and limitations.

16.2.1 High-Resolution Transmission Electron Microscopy (HRTEM)

The transmission electron microscope (TEM) has progressed over many years into a highly sophisticated instrument with widespread application across many scientific disciplines (Fultz et al., 2008). TEM provides structural and chemical information over a range of length scale, which makes it an

TABLE 16.1
Overview of Some Nanoparticle Characterization Techniques

Technique	Medium	Principle and Size Range	Advantages/Disadvantages
Laser diffraction	Powders	Ensemble technique Fraunhofer and Mie light scattering. Typical range: 1 μm to 1000 μm	Morphological information limited to aspect ratio. Range excludes particles <1 μm.
Light microscopy	Powders and aerosols	Light scattering Single particle technique Typical Range: >1 μm	Light wavelength limits resolution. Range excludes particles < 1 μm.
SEM	Powders and aerosols	Imaging/particle counting Single particle technique Typical range: 50 nm to 1 cm	Light wavelength limits resolution. Range excludes particles <1 μm.
TEM	Powders and aerosols	Imaging/particle counting Typical range: 5 nm to 500 μm	Sample preparation can be very complex. Must be performed under vacuum. Costly equipment.

Source: Heera and Shanmugam (2015).

indispensable tool for characterizing the properties of nanostructured materials and handling their behavior. High-resolution transmission electron microscopy (HRTEM) is an imaging mode of the TEM that helps in visualizing the crystallographic structure of a sample in atomic scale. The main consideration in HRTEM lies in understanding the phase of the electron wavefront as affected by the objective lens and the specimen. In HRTEM, the specimen itself acts as the object providing a phase shift to the electron wavefront in proportion to its scattering potential. HRTEM requires additional computational simulation to provide quantitative information on the atom arrangements in a material. Also, the HRTEM method demands close attention to the objective lens performance and other microscopic characteristics (Fultz et al., 2008).

16.2.1.1 Working Principle

The basic configuration of TEM is shown in Figure 16.1 and consists of an electron source at the top, which produces a stream of electrons that are accelerated towards the specimen using a positive electron potential. This stream of electrons is then focussed into a thin, monochromatic beam using a condenser system containing metal apertures and magnetic lenses. As the beam strikes the specimen, part of it gets transmitted and focussed again into an image using the objective lens, below the sample. The image is then fed down through a column to the projector lens (consisting of two–three lenses and one–three apertures) for enlarging the image to the set magnification. The imaging surface could be a phosphor screen or the entrance to a CCD camera imaging system. The darker areas of the image denote the denser or thicker sample region Horiuchi, 1994, while the lighter areas represent the thinner or less dense regions of the sample. It is challenging to record the phase of the electron wave, and therefore only the amplitude in the image plane is recorded (Horiuchi, 1994). Also, it is important to understand that the obtained image is not a direct illustration of the samples' crystallographic arrangement.

16.2.1.2 Application

HRTEM with its high resolution forms an invaluable tool to study the nanoscale properties of the crystalline material. Currently, the highest resolution possible is 0.8 Å. At these small scales,

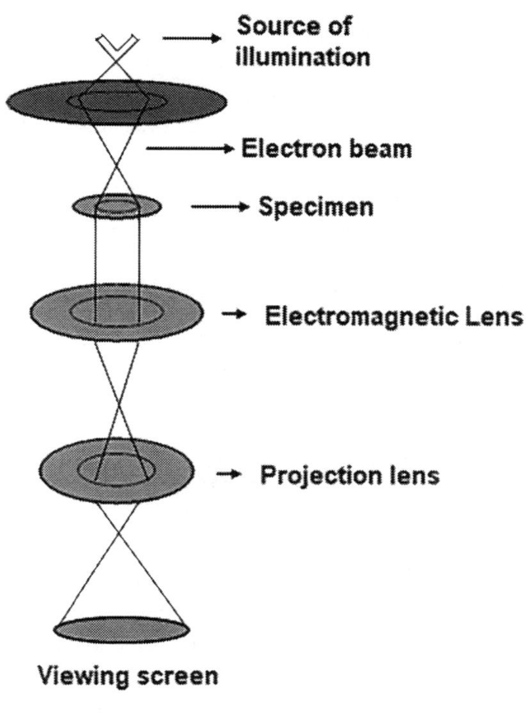

FIGURE 16.1 TEM setup.

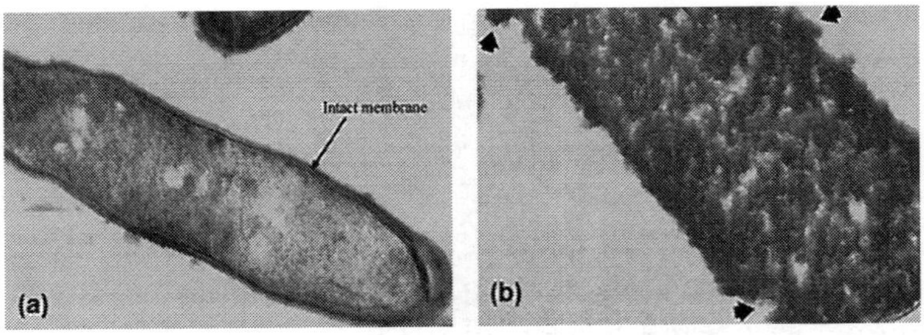

FIGURE 16.2 TEM image of *L. brevis* in orange juice (a) before treatment and (b) after high-intensity pulsed electric field treatment. (Elez-Martinez et al., 2005.)

individual atoms and crystalline defects can be imaged. Elez-Martinez et al. (2005) analyzed the physical changes of *L. brevis* as affected by high-intensity pulsed electric fields in orange juice. The cells exposed to the treatment exhibited thinner or ruptured cell walls and an entirely disorganized cytoplasm as seen using TEM images (Figure 16.2). The effect of ultrasonic assisted curing technology on porcine tissue microstructure was investigated using TEM micrographs by Siró et al. (2009). The TEM has major advantages including an accessible size of less than 1 nm, very high resolution, direct method, and no need of calibration; moreover, it can be directly combined with analytical methods. The TEM has drawbacks such as the need for expensive and complex equipment, high vacuum and care in sample preparation; furthermore, it is time consuming.

16.2.2 Scanning Electron Microscope (SEM)

The scanning electron microscope (SEM) uses electrons instead of light to form an image (Reichelt, 2007). Over a few decades in the early 1950s, scanning electron microscopes have developed new areas of study in the medical and physical science communities (Song et al., 2014). The SEM allowed researchers to study a much wider variety of specimens. It is an electron microscope that uses a high-energy beam of electrons for imaging sample surfaces (Mehnert and Mäder, 2001). The conventional light microscope uses a series of glass lenses to bend light waves to create a magnified image, while the scanning electron microscope creates magnified images by using electrons instead of light waves as seen in the Figure 16.3. The SEM has many advantages over traditional microscopes. These include a large depth of field, which allows more of a specimen to be in focus at one time (Lozano et al., 2012). In addition, the SEM provides much higher resolution in order to magnify closely spaced specimens at much higher levels. It provides precise control over the degree of magnification because it uses electromagnets rather than lenses. All of these benefits, together with remarkable clear images, make the SEM one of the most convenient instruments in research.

16.2.2.1 Working Principle

The schematic arrangement of the SEM is shown in Figure 16.3. A beam of electrons from the top strikes the surface of the specimen and interacts with the atoms in the sample. Signals in the form of secondary electrons, backscattered electrons, and characteristic X-rays are generated. They contain information about the sample's surface topography, composition, etc. (Horst-Günter Rubahn, 2008). The second most common imaging mode for the SEM is the characteristics X-rays. These are used to identify the elemental composition of the sample by a technique known as energy dispersive X-ray (EDX) (Reichelt, 2007). Backscattered electrons (BSE) that come from the sample may also be used to form an image. Such a BSE image along with X-ray characteristic spectra are used in analytical SEM to determine the elemental composition of the sample. The SEM is capable of producing very high-resolution images of a sample surface, revealing details about 1–5 nm in size in its primary detection mode (Reichelt, 2007). The SEM requires that the specimens should be conductive for the electron beam to scan the surface and that the electrons have a path to ground for

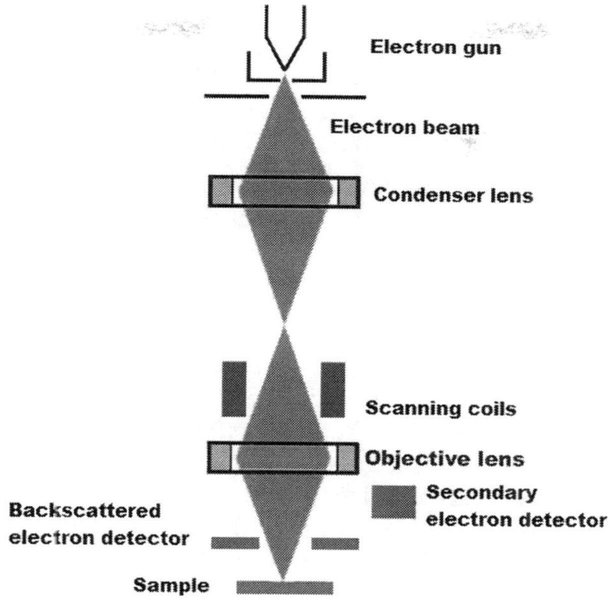

FIGURE 16.3 Scanning electron microscope components.

conventional imaging. Non-conductive solid specimens are generally coated with a layer of conductive material by low vacuum sputter coating or high vacuum evaporation. This is done to prevent the accumulation of static electric charge on the specimen during electron irradiation.

16.2.2.2 Application

To better understand the effects of drying on the morphology of starch granules, electron microscopy was used. Vega-Gálvez et al. (2012) studied the effect of temperature and air velocity on drying kinetics and quality attributes of apple slices during the drying. The hot-air drying process was carried out in a convective dryer and the structural changes of fresh and dried rehydrated apple were observed by cryo-SEM. The effects of spray drying conditions (inlet air temperature and compressed airflow rate) and concentration of drying aids (maltodextrin and gum arabic) on the microstructure of spray dried black mulberry (*Morus nigra*) juice powders were analyzed using SEM (Fazaeli et al., 2012). As shown in Figure 16.4, the microstructure of powders produced with higher maltodextrin concentration shows particles that are larger, amorphous, all piled up with a strong attraction to each other, while when the dextrose equivalent of maltodextrin is decreased by using gum arabic, the particles tend to become more spherical, and more scattered (Fazaeli et al., 2012). Also, Cano-Chauca et al. (2005) studied the effect of different drying aids on the microstructure of mango powder obtained by spray drying. Microstructure analysis obtained by SEM showed that the powder of the mango juices obtained through spray drying using carriers such as maltodextrin, gum arabic, and waxy starch without the addition of cellulose presented surfaces of amorphous particles and lack of crystalline surfaces. When the concentration of cellulose increased, the particles tended to become more spherical and more scattered (Cano-Chauca et al., 2005). Thus, the SEM is effective and helps to understand the microstructure changes of a food product, determining the quality changes during processing.

16.2.3 Atomic Force Microscope (AFM)

The atomic force microscope (AFM) is a kind of scanning probe microscope (SPM). SPMs are designed to measure local properties, such as height, friction, and magnetism, with a probe (Scalf and West, 2006). The AFM acquires the image by using SPM which scans the probe over a small

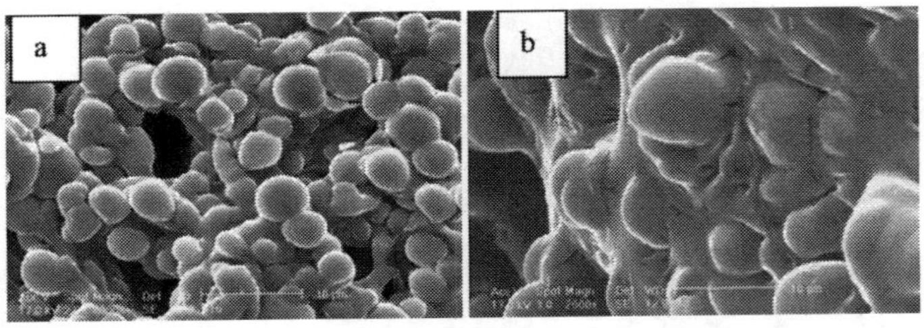

FIGURE 16.4 SEM images of black mulberry powder particles produced with (a) 6% maltodextrin and (b) 8% maltodextrin. (Fazaeli et al., 2012.)

area of the sample, simultaneously measuring its local property. The spring in the AFM is a critical component as it influences the maximum deflection needed for a given force. This requires a spring that is as soft as possible, and, at the same time, a stiff spring with high resonant frequency is necessary to minimize the sensitivity to vibrational noise from the building near 100 Hz. AFMs operate by measuring the force between a probe and the sample. Normally, the probe is a sharp tip, which is a 3–6 μm tall pyramid with a 15–40 nm end radius (Mai, 2015). Although the lateral resolution of AFM is low (approximately 30 nm) due to the convolution, the vertical resolution can be up to 0.1 nm (Mai, 2015; Scalf and West, 2006).

16.2.3.1 Working Principle

AFM image resolution is acquired by measuring the lateral and vertical deflections of the cantilever using the optical lever. As seen in Figure 16.5, the optical lever is operated by reflecting a laser beam off the cantilever (Mai, 2015). The reflected laser beam strikes a position-sensitive photodetector consisting of a four-segment photodetector (Sum et al., 2013). The differences between the segments of the photo-detector of signals specify the laser spot position on the detector thereby indicating the angular deflections of the cantilever. Piezo-tube (PZT) actuators position the AFM tip with high resolution (Sum et al., 2013). Piezoelectric ceramics are a class of materials that contract or expand in the presence of a voltage gradient. Therefore, piezo-ceramics are capable of positioning devices in all the X, Y, Z (three-dimensional) directions with arbitrarily high precision. Thus, the AFM can be used to resolve individual as well as groups of particles. However, the construction of an AFM requires a force sensor to quantify the forces between a small probe and the surface being imaged. The force sensor draws on the relationship between the motion of a cantilever and the applied force as given by Hooke's law in Equation 16.1.

$$F = -K * d \tag{16.1}$$

In this equation, K is a constant and depends on the material and dimensions of the cantilever. D is the motion of the cantilever. The motion of the cantilever can be measured using the "light lever" method, wherein the light is reflected from the back side of the cantilever into a photo-detector (see Figure 16.5). For instance, when the cantilever moves up and down, the light beam moves across the surface of the photo-detector. The motion of the cantilever is then directly proportional to the output of the photo-detector. Motions as small as 1 nm are consistently measured with the light lever method in the AFM (Binnig et al., 1986). Feedback control keeps the motion of an object proportional to another object. Applications of feedback control include automatic control of airplanes and building temperatures. In the AFM, feedback control helps to keep the probe in a "fixed" relationship with the surface while a scan is measured (Bernardes-Filho and Assis, 2005).

16.2.3.2 Application

The AFM offers visualization in three dimensions, whereas the resolution in the vertical, or Z axis is limited by the environmental vibrations. On the other hand, the horizontal, or X-Y axis resolution

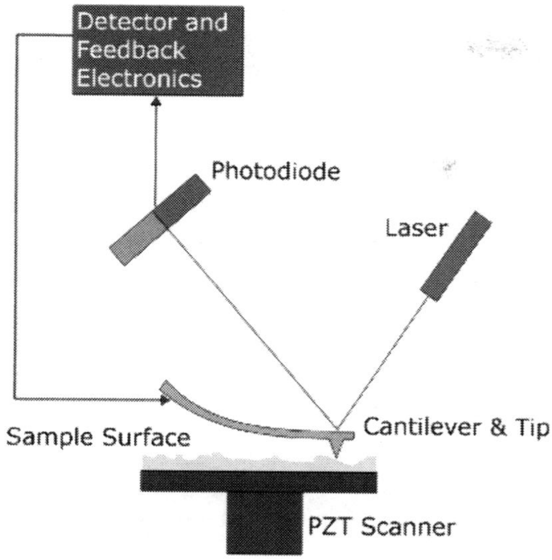

FIGURE 16.5 AFM with an optical lever. (Bernardes-Filho and Assis, 2005.)

is limited by the scanning tip diameter. AFM can be performed in both liquid and gas intermediates. Dry particles can be scanned in both atmospheric air and controlled environments, such as nitrogen gas (Lin et al., 2014). Also, liquid dispersions of particles can be scanned, as long as the dispersant is not corrosive to the probe tip and can be attached to the substrate. The AFM can measure the structural characterization of proteins, polysaccharides, and liposomes (Graveland-Bikker et al., 2006). The molecular structure of R-lactalbumin nanotubes with nanometer resolution in the buffer has been imaged to probe their mechanical properties (Graveland-Bikker et al., 2006). In addition, the AFM has also been utilized to study morphology, size, stability, and dynamic processes of lipid nanocapsules (Palmer et al., 2003) and to characterize natural colloidal matter as a function of pH (Lead et al., 2005) and topography of particles dispersed in a solid matrix (Lozano et al., 2012). All these studies revealed that the AFM is a potential tool for characterizing the nanoparticles of different forms with its unique three-dimensional surface profile.

16.2.4 X-RAY DIFFRACTION (XRD) ANALYSIS

X-ray diffraction is a conventional technique for the determination of crystallographic structure and morphology. This technique is used to establish the metallic nature of particles, which gives information on translational symmetry size and shape of the unit cell from peak positions, and information on electron density inside the unit cell, namely where the atoms are located from peak intensities (Gallego-Urrea et al., 2011). XRD determines the identity of crystalline solids based on their atomic structure. There is an increase or decrease in intensity with the amount of constituent (Hagendorfer et al., 2012). Further crystalline solids consist of regular arrays of atoms, ions, or molecules with interatomic spacing, which results in diffraction of X-rays (Tampo et al., 2005). Bragg's law provides a simplistic model to understand the conditions required for diffraction as shown in Equation 16.2 (Warren, 1941).

$$\lambda = 2dsin\theta \tag{16.2}$$

where "d" is the space between the panes as shown in Figure 16.6. A constructive interference only occurs when the Bragg's law is satisfied. X-ray diffraction peaks are produced by a monochromatic beam of X-rays scattered at specific angle 2θ from a family of planes. The peak intensities are determined by the atomic arrangement within the lattice planes. Consequently, the X-ray diffraction

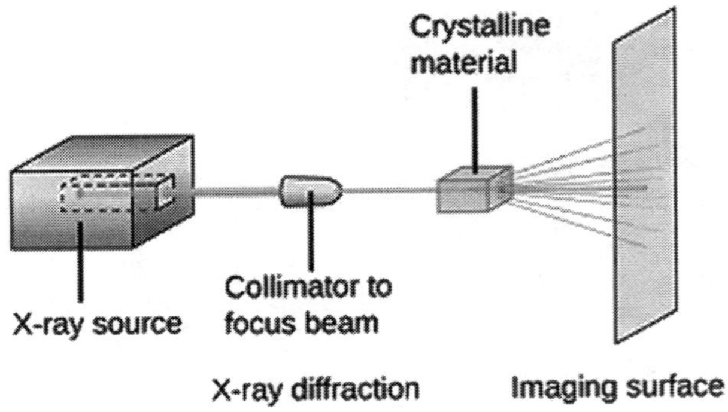

FIGURE 16.6 X-ray diffraction pattern. (OpenStax Content, https://tophat.com/marketplace/beta/oer-openstax-atoms-first/2549/85861.)

pattern is the fingerprint of periodic atomic arrangements in a given material. Standardly available databases on X-ray powder diffraction patterns help in quick phase identification for a large variety of crystalline samples (Chandrasekaran, 1998).

16.2.4.1 Working Principle

Once a sample is irradiated with a parallel beam of monochromatic X-rays, the atomic lattice of the sample acts as a three-dimensional diffraction grating causing the X-ray beam to be diffracted to specific angles (Warren, 1941). As shown in Figure 16.6, a diffractometer can be used to make a diffraction pattern of a crystalline solid (Chandrasekaran, 1998). The essential parts of the diffractometer include X-ray tube (source of X-ray), incident beam optics (conditions the X-ray before hitting the sample), sample holder, and detector (counts the number of X-rays scattered by the sample). During XRD analysis, X-ray beams are reflected off the parallel atomic layers within a molecule over a range of diffraction angles. Because the X-ray beam has a single specific wavelength, constructive or destructive interference can occur. At certain angles, the reflected rays are in phase (constructive interference), and this will give a peak in a diffractogram. From the diffraction pattern, one can identify the molecule or mixture of molecules. More or less like a fingerprint, every molecule has its distinct set of diffraction peaks that can be used to identify it. Identification of the molecules is usually done by comparing the measured diffractogram with a database of known diffraction data.

16.2.4.2 Application

X-ray diffraction (XRD) is a versatile and powerful nondestructive technique for characterizing crystalline materials. XRD gives detailed information about the chemical composition and crystallographic structure of natural and manufactured materials. X-ray crystallography yields unparalleled high-resolution structures of biomolecules and complexes from the solid phase (Dudkiewicz et al., 2011). In fact, with this technique, it is possible to determine the position of the atoms more accurately (factor 100–1000) than with AFM. The melting of starch crystallites was determined as a function of moisture content and temperature (Zobel et al., 1988). The structural complexity of food polysaccharides can be determined by using two-dimensional area detectors, exposure to synchrotron X-ray radiation, computer graphics, and molecular dynamics (Chandrasekaran, 1998). Thus, XRD is nondestructive and efficient in catering for commercial and academic research development. XRD, in association with FTIR and differential scanning calorimetry (DSC), can solve numerous problems encountered in food and pharmaceutical formulations.

16.3 PARTICLE SIZE

The particle size plays a vital role in the measurement of nanoparticle characteristics (Akbari et al., 2011). The quantum effects of nanoparticles with their high surface to volume ratio bring together size-dependent phenomena such as mechanical, chemical, electronic, and magnetic properties. For example, the melting point of nanoparticles is evidently decreased when the size is reached to the nanometer scale (Wang, 2001). The particle size and size distribution of nanoparticles can be determined using numerous commercially available instruments for the analysis of dry powders and powders dispersed in suspension. In general, there are two basic methods of defining particle size. The first method is to inspect the particles and make actual measurements of their dimensions. Microscopic techniques, for example, measure many dimensional parameters from particle images (Jiang et al., 2009). The second method utilizes the relationship between particle behavior and its size. This often implies an assumption of equivalent spherical size developed using a size-dependent property of the particle and relating it to a linear dimension. An example of this method is photon correlation spectroscopy (PCS), where the dynamic fluctuation of the scattered light intensity is the basis for the calculation of the average particle size (Staiger et al., 2002).

16.3.1 Photon Correlation Spectroscopy

Photon correlation spectroscopy (PCS) has become a powerful light-scattering technique for studying the properties of suspensions and solutions of colloids, macromolecules, and polymers, which is absolute, noninvasive and nondestructive (Svanberg and Bergman, 2005). PCS is one of the several different names that have been used historically for this technique. The first name given to it was quasi-elastic light scattering (QELS) because, when photons are scattered by mobile particles, the process is quasi-elastic (Tscharnuter, 2006). QELS measurements yield information on the dynamics of the scatterer, which gave rise to the acronym DLS (dynamic light scattering). PCS or dynamic light scattering (DLS) is a common method for detecting the particle size distribution of nano and submicron liquid dispersions (Svanberg and Bergman, 2005). DLS measures fluctuations in the scattering intensity of a solution as a function of time. Because these fluctuations occur as nanoparticles move through the solution, the timescale over which these changes occur depends upon the rate at which the nanoparticles diffuse. The hydrodynamic diameter (the diameter of a hypothetical nonporous sphere that diffuses at the same rate as the particles being characterized) can be calculated from the time dependence of the scattering intensity measurements (Calzolai et al., 2012). The diameter that is measured in DLS is a value that refers to how a particle diffuses within a fluid, so it is referred to as a hydrodynamic diameter. The diameter that is obtained by this technique is the diameter of a sphere that has the same translational diffusion coefficient as the particle. The translational diffusion coefficient will depend not only on the size of the particle "core" but also on any surface structure, as well as the concentration and type of ions in the medium (Tscharnuter, 2006). The strong particle size dependence of the scattered light will also bias the measured size as a few large particles will have such a large influence that smaller particles will be neglected. The method is not suitable for samples with a multimodal particle size distribution.

16.3.1.1 Working Principle

A typical DLS system may be broken down into several functional components such as a light source, an optical system, a detector system, and a digital correlator as shown in Figure 16.7. The light source is usually a narrow-band, polarized, monochromatic, continuous wave laser, with a wavelength of typically 532 nm, frequency doubled from powerful 1064 nm diode lasers (Mansfield et al., 2015). The optical system consists of a lens focusing the laser beam down into the sample which is enclosed in a temperature-controlled scattering cell surrounded by a refractive index matching liquid. It typically allows experiments to be conducted over a 10°–160° angular range. The detector system has a low dark count and a high gain since most work is done in the single photon counting regime. The digital correlator has become the device of choice to generate the raw data in a DLS experiment, although the early experiments mostly employed "wave analyzers" or "spectrum analyzers."

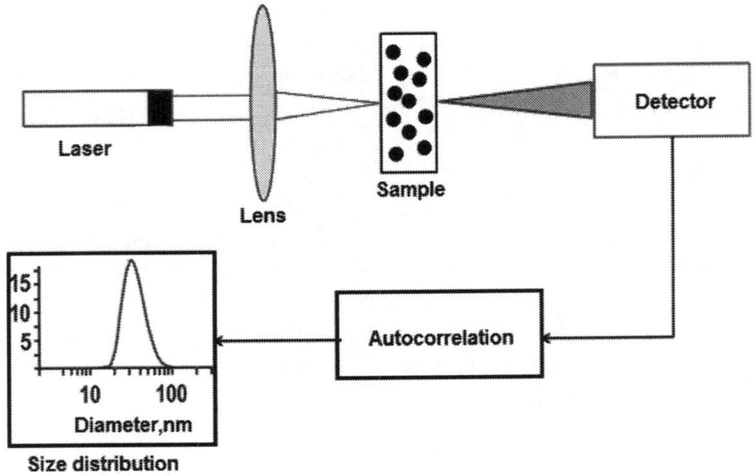

FIGURE 16.7 DLS schematic configuration.

16.3.1.2 Application

The arrangement of the micro and nanocomponents, size distribution, and rheological (mechanical) properties of the food matrix influencing the final stability, texture, and sensory perception of fluid foods can be evaluated using DLS (Mendoza and Lu, 2016). DLS is being extended to study the dynamical process in simple and complex turbid samples based on a diffusing wave spectroscopy (DWS) scheme for determining the mean square displacement of particles in viscoelastic fluid(Mendoza and Lu, 2016). The particle size distribution of cloudy apple juice using photon correlation spectroscopy found particles ranging from about 0.05 to 3 micrometers (Genovese and Lozano, 2000). Also, PCS can be used to determine the microstructural and rheological properties of viscoelastic food materials and their formulations. Overall, DLS has been proven to be effective in determining the behavior of complex fluids including concentrated fruit juices and polymer solutions.

16.3.2 Laser Diffraction

Laser diffraction, otherwise known as low angle laser light scattering (LALLS) is a nondestructive method and can be used for analyzing wet or dry samples. The geometrical dimensions of a nanoparticle in the range of nm to mm size can be measured quickly. In laser diffraction, particle size distributions are calculated by relating a sample's scattering pattern with a suitable optical model (Kowalska and Żbikowska, 2013). Laser diffraction reports the volume of material of a given size since the volume of the measured particle is proportional to the light energy reported by the detector system. This method is in contrast with counting-based techniques, reporting the number of particles of a given size. The differences between the number and volume-based size distributions have been discussed by Choi and Morgenroth (2003). Clearly, the distributions reported by LD techniques will not be the same, especially when polydisperse materials are measured. Thus, although there may be good agreement among various counting-based techniques measurements by LD, these will not always agree with the results of either volume or mass-based techniques. Volume-based distributions will always shift to larger particle sizes in comparison with number distributions. It is difficult to measure polydispersed samples with the LD method.

16.3.2.1 Working Principle

As shown in Figure 16.8a, the laser diffraction method uses a laser beam passing through a sample of particles in suspension (in liquid or air) and collects light intensity data at different scattering

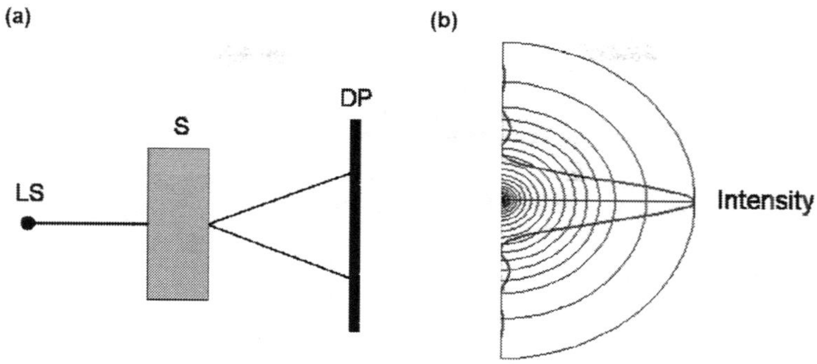

FIGURE 16.8 Schematic configuration of LLS technique showing (a) experimental setup with a laser source (LS), particle suspension (S) and diffraction pattern (DP), and (b) scattered light intensity measured by the light detector. (Kowalska and Żbikowska, 2013.)

angles (32° or more in most instruments) away from the axis of the laser beam (Figure 16.8b). LD measures the particle size distributions by computing the angular variation in the intensity of light scattered as a laser beam passes through a dispersed particulate sample (Pamme et al., 2003). Large particles scatter light at small angles about the laser beam and small particles scatter light at large angles, as illustrated in Figure 16.8. The angular scattering intensity data is then analyzed to calculate the size of the particles responsible for creating the scattering pattern, using the Mie theory of light scattering. The Mie model considers both scattering and diffraction of light around the particle in its medium (Brar and Verma, 2011). Mie light scattering theory calculations describe the interaction of the particle with incident radiation by deriving Maxwell's equation for the incident, scattered, and internal fields (Mie, 1908). The particle size is obtained as a volume equivalent sphere diameter.

16.3.2.2 Application

The particle size of dairy milk fat droplets and other food emulsions can be measured using the LD technique (Kowalska and Żbikowska, 2013). The laser diffraction technique is used to measure the mean diameter and particle size distribution of β-carotene nanodispersions. A volume-weighted average diameter in the range of 60 to 140 nm size β-carotene nanoparticles was measured (Tan and Nakajima, 2005). Also, food properties such as flavor release, mouthfeel, and emulsion stability are evaluated using this LD technique. Overall, the LD technique is effective in measuring physicochemical properties such as mean particle diameter and size distribution in several food nanodispersions.

16.3.3 Field Flow Fractionation (FFF)

Field flow fractionation is a flow-assisted hydrodynamic separation technique that permits physical separation of small quantities (ng–μg range) of macromolecules or particles. The separation is based on the diffusion coefficient in an open flow channel. It is similar to chromatographic techniques but solely based on physical separation without relying on a stationary phase (Hagendorfer et al., 2012). A field is applied perpendicular to the flow of the mixture to cause separation due to differing mobilities of the various components in the field, as shown in Figure 16.9. The field can be gravitational, centrifugal, magnetic, thermal, or a cross flow of fluids (Dulog and Schauer, 1996).

Most of the separation is due to differences in Brownian motion and diffusion after the field has forced the mixture components onto one side (Bechinger et al., 2016). There is a balance between diffusion and the applied force (from the field) on any particular particle in the mixture, which results in differing movement perpendicular to the main flow direction. When the flow is turned on, the particles are exposed to a parabolic velocity profile, and particles at a higher height from

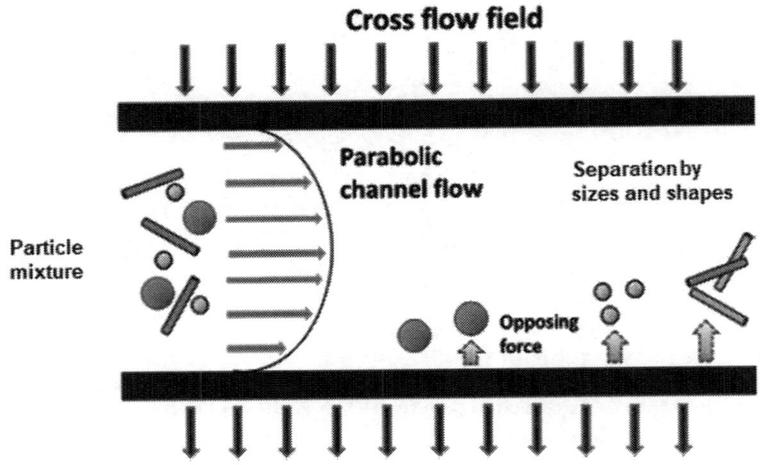

FIGURE 16.9 Crossflow pattern in field flow fractionation. (Joontaek, 2014.)

the channel base (i.e. the smaller particles) will travel faster than those at the bottom, resulting in a separation at the exit of the flow channel. The detection of nanoparticles (NPs) following the separation by FFF can be done by on-line detectors that are usually flow-through HPLC-type detectors, e.g. ultraviolet-visible spectrophotometers. The detectors provide transient signals according to their principles of operation and the properties of the eluting sample that can then be transferred into size distributions. The full range of sizing and separation capability of FFF is achieved by carefully optimizing the operating conditions.

16.3.3.1 Application

Hagendorfer et al. (2012) studied the silver nanoparticle product characteristics using asymmetric flow field flow fractionation and comparing the results with TEM. The method applies a multidetector approach (UV/vis, light scattering, inductively coupled plasma mass spectrometry – ICPMS), and the results show FFF is a powerful and reliable technique for a quantitative-, size-, and mass-specific characterization of polydisperse engineered Ag-NP products. In comparison with TEM, FFF offers faster measurement times and the ability to determine the samples directly in dispersions. FFF has been used to determine the mass of proteins and size of colloidal carriers, including liposomes, injectable emulsions, and particles (Janča, 2006). The limitations of FFF techniques are potential membrane swelling, membrane interactions, the continuous re-equilibration in the channel, the need (in some circumstances) of preconcentration, the additional concentration of the sample during equilibration, and the possibility of aggregation in the channel (Heera and Shanmugam, 2015; Hagendorfer et al., 2012).

16.4 SURFACE CHARGE

The hydrodynamic size and surface charge of nanoparticle dispersions can have a dramatic effect on the way in which an organism responds upon exposure, and these need to be properly characterized. Surface characteristics affect agglomeration in dispersions and the nanoparticle hydrodynamic size distributions can be altered by a small change in particle surface charge. The uptake and translocation of nanoparticles by organisms are also affected by their surface charge (Gatoo et al., 2014). For instance, nanoparticle surface charge has been observed to alter blood-brain barrier integrity and permeability (Lockman et al., 2004). The uptake of anionic nanoparticles was superior to cationic or neutral formulations in the brain (Lockman et al., 2004). The characterization of the hydrodynamic size and surface charge property of a nanoparticle has been proposed in several nanotoxicology review articles (Oberdörster et al., 2005). Murdock et al. (2008) characterized the hydrodynamic size and surface charge of different nanoparticles such as Ag, TiO2, and carbon nanotubes in biologically relevant solutions and found that there is a significant effect on particle surface charge and particle agglomeration.

Zeta potential is the value of the surface charge of a nanoparticle, measured from a double electric layer surrounding a nanoparticle that has some surface charge using the dynamic light scattering technique (Badawy et al., 2010). It is a parameter that represents the charge of a particle. The surface charge creates a stern potential over the nanoparticle in the dispersion medium and a zeta potential layer is formed due to the phase differences between the medium and the surface of the nanoparticle. Increasing ionic strength results in the compression of the electrical double layer (Jiang et al., 2009). The zeta potential decreases with increasing ionic strength. By knowing the zeta potential, we can use Graham's equation (Equation 16.3) to calculate the surface charge density of the particles derived from the Gouy-Chapman theory. Graham's equation assumes an electroneutrality condition and states that the total charge of the double layer must be equal to the negative of the surface charge (Velev et al., 1993; Clogston and Patri, 2011).

$$\sigma = \frac{\in \in_0 \Psi_0}{\lambda_D} \tag{16.3}$$

where:

σ is the electric field caused by the charge on the conductor
$\in$ is the permittivity of the free space
ψ is the potential of the charged surface
λ_D is defined as the Debye length, a charge carrier's net electrostatic effect in solution

The surface charge is controlled by several mechanisms, including surface ionization, ion adsorption, and lattice ion dissolution (Goyne et al., 2002), while the thickness of the electrical double layer is a function of solution ionic strength. It was found that when nanoparticles are dispersed in liquids, their hydrodynamic size is often larger than the primary particle size. Understanding the stability and agglomeration behavior of nanoparticles under different solution conditions will facilitate such investigations. Even if the hydrodynamic sizes of nanoparticles suspended in physiological media are determined to be larger than their primary sizes, it is important to know whether these particles are in an agglomerated (weak bonds between primary particles) or aggregated (hard bonds between primary particles) state, since their corresponding biological fate and effects will be different.

16.4.1 Application

Nanolaminates are used as an edible coating on food which consists of more than one layer, and the materials are in the nanoscale (Ramachandraiah et al., 2014). Layer-by-layer (LBL) deposition is a thin-film fabrication technique that could be used to cover food that has surface charges (Kotov, 2002). The layer formation occurs by depositing alternate layers of oppositely charged materials. An advantage of the LBL technique is that the thickness of the coating can be regulated with accuracy from 1 to 100 nm (Ramachandraiah et al., 2014). Being extremely low in thickness, it is more suitable for coating food than freely standing coatings (Ramachandraiah et al., 2014). Besides serving as a barrier for gas or moisture, these charged layers can also carry antioxidants and antimicrobials. However, it is essential to determine the characteristics of the nanomaterials used in the layers. Proteins, lipids, and polysaccharides are currently being used in such layers. Based on the type of biopolymeric nanoparticles in the coating, the functionality of the film can vary. For example, lipid layers act as barriers to moisture but are not efficient to prevent gases and give poor mechanical strength. Similarly, protein-based layers give good barrier strength against gas but not moisture (Weiss et al., 2010). Therefore, natural edible barriers made of surface-charged nanolaminates can be used to extend the shelf life and nutrition of food products.

16.5 CRYSTAL MORPHOLOGY

Nanocrystals are nanoparticles with a crystalline character. Crystals reveal a large variety of shapes, depending on their chemical composition and structure, and on growth conditions. The shape of the crystals has a direct impact on the separation efficiency and stability of crystalline chemicals, bioavailability, effective delivery of drugs, etc. (Prasanthi et al., 2016). Knowledge of the growth habits and morphological properties of the molecular crystals is of crucial importance in understanding and exploiting many of their physicochemical properties (Compton and Osterloh, 2007). Calculating the crystal growth morphology has diverse applications in the nutraceutical and drug delivery fields (Singh, 2016). Although significant efforts have been made for the last few decades to predict precisely the growth morphology of crystals, it still remains a challenging task till this date. The difficulty in understanding the crystal growth phenomena is due to complex relationships between different processes, which take place on very large time and length scales, particularly in the case of industrial processes. The processes involved during crystal growth are highly complex, involving many-body interactions for an accurate description of the problem. The usefulness of the attempts to compute and predict the growth morphologies of crystals arises for the following reason: crystals grown under a specific set of conditions (temperature, pressure, chemical composition, impurities, pH values, etc.) tend to have a specific morphology at least in the sense of statistics. The symmetry observed in crystals as exhibited by their crystal faces is due to the ordered internal arrangement of atoms in a crystal structure. This arrangement of atoms in crystals is called a lattice. In two dimensions, a plane lattice consists of an orderly array of points. The array is defined by the spacing between points and the directions (or angles) between the points. Thus, the array can be reproduced by specifying the distance and angle to move from point to point. This is referred to as translational symmetry. There are various possibilities to produce nanocrystals in the desired shape and size. Three principles can be used: milling, precipitation methods, and homogenization methods, as well as a combination thereof. The industrially relevant methods are the top-down technologies, i.e. starting from a large-size drug powder to be reduced in size. The bottom-up technologies (i.e. starting from a dissolved molecule, precipitation) are currently – to our knowledge – not used in the production of commercial products. Reasons may include the need for solvent removal, the difficulty in controlling the process, and the fact that many poorly soluble drugs are so not only in aqueous but also in organic media. The main advantages of nanocrystal technology in this product are its user-friendliness and higher bioavailability in comparison to the oral solution.

16.5.1 Application

Gold nanoparticles with calorimetric sensors can be used for detecting food quality and safety for consumption. Colloidal gold (Au) nanoparticles have tunable optoelectronic properties which are researched for their use in high technology applications such as therapeutic, drug delivery, and food safety (Singh, 2016; Kamat et al., 2002). Gold-based nanoparticles are used to detect aflatoxin B1 toxin in milk (Mao et al., 2006). Figure 16.10 shows a gold pentagon nanocrystals image detected using TEM with an average size of 50 nm. The diffraction pattern showed two {200} type reflection as seen in Figure 16.10b (Singh et al., 2017). Thus, characterizing the crystal morphology and its growth using advanced TEM imaging technique helps in several cancer treatments, and in the food and agricultural industries for the detection of contaminants.

16.6 CONCLUSION AND FUTURE PERSPECTIVES

Various analytical techniques and their potential application in food industries were discussed. Clearly, the most appropriate analytical technique depends on the property and nature of the sample to be monitored and on other practical restrictions. Electron microscopy techniques are an

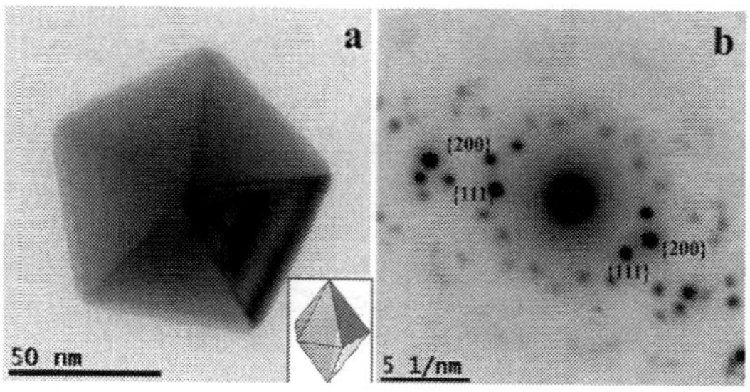

FIGURE 16.10 Nano crystals TEM image: (a) selected area diffraction pattern and (b) dark field image. (Singh et al., 2017.)

essential tool to obtain precise information about the basic structural properties of different kinds of foods. Among the different characterization techniques, the atomic force microscope (AFM) is most suited for three-dimensional visualization along with quantitative and qualitative information on several physical properties such as morphology, size, roughness, and surface texture. Also, statistical information, including surface area, volume, and size distributions can be measured as well. The AFM measures particle size ranging from 1 nanometer to 8 micrometers in a single scan. Moreover, it can characterize nanoparticles in multiple mediums including ambient air, controlled environments, and even liquid dispersions. As the structure of foods can have a profound influence on their nutritional value, the proper conditions that help to retain the quality of the product can be defined by using microscopic techniques. Moreover, this information can help to find a suitable method to confidently detect and characterize engineered nanomaterials (ENMs) released from food.

REFERENCES

Anandharamakrishnan, C. 2014a. *Techniques for Nanoencapsulation of Food Ingredients*. New York: Springer.

Anandharamakrishnan, C. 2014b. "Techniques for formation of nanoemulsions." In *Techniques for Nanoencapsulation of Food Ingredients*, 7–16. New York: Springer.

Anandharamakrishnan, C. 2014c. "Nanoencapsulation of food bioactive compounds." In *Techniques for Nanoencapsulation of Food Ingredients*, 1–6. Springer Briefs in Food, Health, and Nutrition. New York: Springer.

Anandharamakrishnan, C and S. Padma Ishwarya. 2015. *Spray Drying Techniques for Food Ingredient Encapsulation*. Chichester: John Wiley & Sons.

Anandharamakrishnan, C. and S. Padma Ishwarya. 2015a. "Introduction to encapsulation of food ingredients." In *Spray Drying Techniques for Food Ingredient Encapsulation*, 37–64. Chichester: John Wiley & Sons.

Anandharamakrishnan, C. and S. Padma Ishwarya. 2015b. "Spray drying for nanoencapsulation of food components." *Spray Drying Techniques for Food Ingredient Encapsulation*. 180–197. Chichester: John Wiley & Sons.

Anu Bhushani, J., P. Karthik, and C. Anandharamakrishnan. 2016. "Nanoemulsion based delivery system for improved bioaccessibility and caco-2 cell monolayer permeability of green tea catechins." *Food Hydrocolloids,* 56: 372–382.

Aprajeeta, J., R. Gopirajah, and C. Anandharamakrishnan. 2015. "Shrinkage and porosity effects on heat and mass transfer during potato drying." *Journal of Food Engineering,* 144 (January): 119–128.

Akbari, B., M. Pirhadi Tavandashti, and M. Zandrahimi. 2011. "Particle size characterization of nanoparticles – a practical approach." *Iranian Journal of Materials Science and Engineering*, 8 (2): 48–56.

Badawy, Amro M. El, T. P. Luxton, R. G. Silva, K. G. Scheckel, M. T. Suidan, and T. M. Tolaymat. 2010. "Impact of environmental conditions (pH, ionic strength, and electrolyte type) on the surface charge and aggregation of silver nanoparticles suspensions." *Environmental Science and Technology,* 44 (4): 1260–1266.

Bechinger, C., R. Di Leonardo, H. Löwen, C. Reichhardt, G. Volpe, and G. Volpe. 2016. "Active particles in complex and crowded environments." *Reviews of Modern Physics,* 88 (4).

Bernardes-Filho, R. and O. B. G. de Assis. 2005. "Development of an algorithm for tip-related artifacts identification in AFM biological film imaging." *Brazilian Archives of Biology and Technology*, 48 (4): 667–674.

Binnig, G., C. F. Quate, and Ch. Gerber. 1986. "Atomic force microscope." *Physical Review Letters*, 56 (9): 930–933.

Bhushani, A. and C. Anandharamakrishnan. 2017. "Food-grade nanoemulsions for protection and delivery of nutrients." In *Nanoscience in Food and Agriculture*, 4, 99 Encapsulation:139. Swizerland: Springer.

Bhushani, A., U. Harish, and C. Anandharamakrishnan. 2017. "Nanodelivery of nutrients for improved bioavailability." In *Nutrient Delivery*, 369–411. London: Elsevier.

Brar, S. K. and M. Verma. 2011. "Measurement of nanoparticles by light-scattering techniques." *TrAC Trends in Analytical Chemistry, Characterization, Analysis and Risks of Nanomaterials in Environmental and Food Samples*, 30 (1): 4–17.

Burleson, D. J., M. D. Driessen, and R. Lee Penn. 2004. "On the characterization of environmental nanoparticles." *Journal of Environmental Science and Health. Part A, Toxic/Hazardous Substances and Environmental Engineering*, 39 (10): 2707–2753.

Calzolai, L., D. Gilliland, and F. Rossi. 2012. "Measuring Nanoparticles Size Distribution in Food and Consumer Products: A Review." *Food Additives & Contaminants: Part A*, 29 (8):1183–1193.

Cano-Chauca, M., P. C. Stringheta, A. M. Ramos, and J. Cal-Vidal. 2005. "Effect of the carriers on the microstructure of mango powder obtained by spray drying and its functional characterization." *Innovative Food Science and Emerging Technologies*, 6 (4): 420–428.

Chandrasekaran, R. 1998. "X-ray diffraction of food polysaccharides." *Advances in Food and Nutrition Research*, 42: 131–210.

Cho, E. J., H. Holback, K. C. Liu, S. A. Abouelmagd, J. Park, and Y. Yeo. 2013. "Nanoparticle Characterization: State of the Art, Challenges, and Emerging Technologies." *Molecular Pharmaceutics*, 10(6):2093–2110.

Choi, Y. C. and E. Morgenroth. 2003. "Monitoring biofilm detachment under dynamic changes in shear stress using laser-based particle size analysis and mass fractionation." *Water Science and Technology*, 47 (5): 69–76.

Clogston, J. D. and A. K. Patri. 2011. "Zeta potential measurement." *Methods in Molecular Biology* 2011;697: 63–70.

Compton, O. C. and F. E. Osterloh. 2007. "Evolution of size and shape in the colloidal crystallization of gold nanoparticles." *Journal of the American Chemical Society*, 129 (25): 7793–7798.

Dudkiewicz, A., K. Tiede, K. Loeschner, L. H., S. Jensen, E. Jensen, R. Wierzbicki, A. B. A. Boxall, and K. Molhave. 2011. "Characterization of Nanomaterials in Food by Electron Microscopy." *TrAC Trends in Analytical Chemistry, Characterization, Analysis and Risks of Nanomaterials in Environmental and Food Samples*, 30(1):28–43.

Dulog, L. and T. Schauer. 1996. "Field flow fractionation for particle size determination." *Progress in Organic Coatings*, 28 (1): 25–31.

Duncan, Timothy V. 2011. "Applications of nanotechnology in food packaging and food safety: Barrier materials, antimicrobials and sensors." *Journal of Colloid and Interface Science*, 363 (1): 1–24.

Elez-Martinez, P., J. Escola-Hernandez, R. C. Soliva-Fortuny, and O. Martin-Belloso. 2005. "Inactivation of lactobacillus brevis in orange juice by high-intensity pulsed electric fields." *Food Microbiology*. http://agris.fao.org/agris-search/search.do?recordID=US201301000517.

Ezhilarasi, P. N., P. Karthik, N. Chhanwal, and C. Anandharamakrishnan. 2013. "Nanoencapsulation techniques for food bioactive components: A review." *Food and Bioprocess Technology*, 6 (3): 628–647.

Fazaeli, M., Z. Emam-Djomeh, A. K. Ashtari, and M. Omid. 2012. "Effect of spray drying conditions and feed composition on the physical properties of black mulberry juice powder." *Food and Bioproducts Processing*, 90 (4): 667–675.

Fultz, B., J. M. Howe. 2008. "High-resolution TEM imaging." In *Transmission Electron Microscopy and Diffractometry of Materials*, 517–582. Springer, Berlin Heidelberg.

Gallego-Urrea, J. A., J. Tuoriniemi, and M. Hassellöv. 2011. "Applications of particle-tracking analysis to the determination of size distributions and concentrations of nanoparticles in environmental, biological and food samples." *TrAC Trends in Analytical Chemistry*, 30 (3): 473–483.

Gatoo, M. A., S. Naseem, M. Y. Arfat, A. M. Dar, K. Qasim, and S. Zubair. 2014. "Physicochemical properties of nanomaterials: Implication in associated toxic manifestations." *BioMed Research International*, 2014.

Genovese, D. B. and J. E. Lozano. 2000. "Particle size determination of food suspensions: Application to cloudy apple juice." *Journal of Food Process Engineering*, 23 (6): 437–452.

Goyne, K. W., A. R. Zimmerman, B. L. Newalkar, S. Komarneni, S. L. Brantley, and J. Chorover. 2002. "Surface charge of variable porosity Al_2O_3(s) and SiO_2(s) adsorbents." *Journal of Porous Materials,* 9 (4): 243–256.

Graveland-Bikker, J. F., I. A. T. Schaap, C. F. Schmidt, and C. G. de Kruif. 2006. "Structural and mechanical study of a self-assembling protein nanotube." *Nano Letters,* 6 (4): 616–621.

Hagendorfer, H., R. Kaegi, M. Parlinska, B. Sinnet, C. Ludwig, and A. Ulrich. 2012. "Characterization of silver nanoparticle products using asymmetric flow field flow fractionation with a multidetector approach – a comparison to transmission electron microscopy and batch dynamic light scattering." *Analytical Chemistry,* 84 (6): 2678–2685.

Heera, P. and S. Shanmugam. 2015. "Nanoparticle characterization and application: An overview." *International Journal of Current Microbiology and Applied Sciences*, 4: 379–386.

Horiuchi, S. 1994. *Fundamentals of high-resolution transmission electron microscopy.* New York:North Holland.

Mansfield, H., Edward D., K. Sillence, P. Hole, A. C. Williams, and V. V. Khutoryanskiy. 2015. "Pozylation: A new approach to enhance nanoparticle diffusion through mucosal barriers." *Nanoscale,* 7 (32): 13671–13679.

Hochella, M. F. 2002. "Nanoscience and technology: The next revolution in the earth sciences." *Earth and Planetary Science Letters*, 203 (2): 593–605.

Horst-Günter R. 2008. *Basics of Nanotechnology.* 3rd Edition. http://www.wiley.com/WileyCDA/WileyTitle/productCd-3527408002.html.

Inagaki, S., R. Ghirlando, and R. Grisshammer. 2013. "Biophysical characterization of membrane proteins in nanodiscs." *Methods (San Diego, Calif.)* 59 (3): 287–300.

Janča, J. 2006. "Field-flow fractionation in particle size analysis." In *Encyclopedia of Analytical Chemistry.* John Wiley & Sons, Ltd. http://onlinelibrary.wiley.com/doi/10.1002/9780470027318.a1505/abstract.

Jiang, J., G. Oberdörster, and P. Biswas. 2009. "Characterization of size, surface charge, and agglomeration state of nanoparticle dispersions for toxicological studies." *Journal of Nanoparticle Research,* 11 (1): 77–89.

Joontaek, P. 2014. "Shape-based particle separation." *Laboratory Journal – Business Web for Users in Science and Industry.* https://www.laboratory-journal.com.

Kamat, P. V., S. Barazzouk, and S. Hotchandani. 2002. "Electrochemical modulation of fluorophore emission on a nanostructured gold film." *Angewandte Chemie International Edition,* 41 (15): 2764–2767.

Karthik, P., P. N. Ezhilarasi, and C. Anandharamakrishnan. 2017. "Challenges associated in stability of food grade nanoemulsions." *Critical Reviews in Food Science and Nnutrition*, 57 (7):1435–1450.

Kavitha, K. S., S. Baker, D. Rakshith, H. U. Kavitha, H. C. Yashwantha Rao, B. P. Harini, and S. Satish. 2013. "Plants as green source towards synthesis of nanoparticles." *International Research Journal of Biological Sciences,* 2 (6): 66–76.

Kotov, N. A. 2002. "Layer-by-layer assembly of nanoparticles and nanocolloids: intermolecular interactions, structure and materials perspectives." In *Multilayer Thin Films*, edited by Gero Decher and Joseph B. Schlenoff, 207–243. Wiley-VCH Verlag GmbH & Co. KGaA.

Kowalska, M. and A. Żbikowska. 2013. "Application of a laser diffraction method for determination of stability of dispersion systems in food and chemical industry." *Journal of Dispersion Science and Technology,* 34 (10): 1447–1453.

Lead, J. R., D. Muirhead, and C. T. Gibson. 2005. "Characterization of freshwater natural aquatic colloids by Atomic Force Microscopy (AFM)." *Environmental Science and Technology*, 39 (18): 6930–6936.

Lin, P.-C., S. Lin, P. C. Wang, and R. Sridhar. 2014. "Techniques for physicochemical characterization of nanomaterials." *Biotechnology Advances*, 32 (4): 711–726.

Lockman, P. R., J. M. Koziara, R. J. Mumper, and D. D. Allen. 2004. "Nanoparticle surface charges alter blood-brain barrier integrity and permeability." *Journal of Drug Targeting,* 12 (9–10): 635–641.

Lozano, O., J. Mejia, T. Tabarrant, B. Masereel, J.-M. Dogné, O. Toussaint, and S. Lucas. 2012. "Quantification of nanoparticles in aqueous food matrices using particle-induced X-ray emission." *Analytical and Bioanalytical Chemistry,* 403 (10): 2835–2841.

Mai, W. 2015. "Fundamental theory of atomic force microscopy." *Nanosience Research Group. School of Materials Science and Engineering Georgia. Web* 10.

Mao, X., J. Huang, M. F. Leung, Z. Du, L. Ma, Z. Huang, P. Li, and L. Gu. 2006. "Novel core-shell nanoparticles and their application in high-capacity immobilization of enzymes." *Applied Biochemistry and Biotechnology,* 135 (3): 229–239.

Maynard, A. D. 2006. "Nanotechnology: Assessing the risks." *Nano Today*, 1 (2): 22–33.

McClements, D. J., L. Zou, R. Zhang, L. Salvia-Trujillo, T. Kumosani, and H. Xiao. 2015. "Enhancing nutraceutical performance using excipient foods: Designing food structures and compositions to increase bioavailability." *Comprehensive Reviews in Food Science and Food Safety*, 14 (6): 824–847.

Mehnert, W. and K. Mäder. 2001. "Solid lipid nanoparticles." *Advanced Drug Delivery Reviews*, Lipid Assemblies for Drug Delivery, 47 (2): 165–196.

Mendoza, F. and R. Lu. 2016. "Dynamic light scattering for measuring microstructure and rheological properties of food." In *Light Scattering Technology for Food Property, Quality and Safety Assessment*, 331–359. CRC Press. http://www.crcnetbase.com/doi/abs/10.1201/b20220-14.

Mie, Gustav. 1908. "Beiträge Zur Optik Trüber Medien, Speziell Kolloidaler Metallösungen." *Annalen Der Physik*, 330(3):377–445.

Mittal, J., A. Batra, A. Singh, and M. M. Sharma. 2014. "Phytofabrication of nanoparticles through plant as nanofactories." *Advances in Natural Sciences: Nanoscience and Nanotechnology*, 5 (4): 043002.

Murdock, R. C., L. Braydich-Stolle, A. M. Schrand, J. J. Schlager, and S. M. Hussain. 2008. "Characterization of nanomaterial dispersion in solution prior to in vitro exposure using dynamic light scattering technique." *Toxicological Sciences*, 101 (2): 239–253.

Navalakhe, R. and T. Nandedkar. 2007. "*Application of Nanotechnology in Biomedicine*." Indian Journal of Experimental Biology, 45(02):160-165.

Oberdörster, G., A. Maynard, K. Donaldson, V. Castranova, J. Frick, K. Ausman, J. Carter, et al. 2005. "Principles for characterizing the potential human health effects from exposure to nanomaterials: Elements of a screening strategy." *Particle and Fibre Toxicology*, 2 (October): 8.

Palmer, A. F., P. Wingert, and J. Nickels. 2003. "Atomic force microscopy and light scattering of small unilamellar actin-containing liposomes." *Biophysical Journal*, 85 (2): 1233–1247.

Pamme, N., R. Koyama, and A. Manz. 2003. "Counting and sizing of particles and particle agglomerates in a microfluidic device using laser light scattering: application to a particle-enhanced immunoassay." *Lab on a Chip*, 3 (3): 187–192.

Parada, J. and J. M. Aguilera. 2007. "Food microstructure affects the bioavailability of several nutrients." *Journal of Food Science*, 72 (2): R21–R32.

Prasanthi, N. L., M. Sudhir, N. Jyothi, and V. Sri Vajrapriya. 2016. "A review on polymorphism perpetuates pharmaceuticals." *American Journal of Advanced Drug Delivery*, 4 (5).

Ramachandraiah, K., S. G. Han, and K. B. Chin. 2014. "Nanotechnology in meat processing and packaging: Potential applications — a review." *Asian-Australasian Journal of Animal Sciences*, 28 (2): 290–302.

Reichelt, R. 2007. "Scanning electron microscopy." In *Science of Microscopy*, 133–272. Springer, New York, NY.

Rubahn, H.-G. 2008. "*Basics of Nanotechnology*." http://trove.nla.gov.au/version/46355773.

Sapsford, K. E., K. M. Tyner, B. J. Dair, J. R. Deschamps, and I. L. Medintz. 2011. "Analyzing nanomaterial bioconjugates: a review of current and emerging purification and characterization techniques." *Analytical Chemistry*, 83 (12): 4453–4488.

Schulze, K. S. 2014. "The imaging and modelling of the physical processes involved in digestion and absorption." *Acta Physiologica*, n/a-n/a.

Scalf, J., and P. West. 2006. "Part I: Introduction to nanoparticle characterization with AFM." In *Pacific Nanotechnology*. Inc. Santa Clara, CA 95054.

Sekhon, B. S. 2010. "Food nanotechnology – an overview." *Nanotechnology, Science and Applications*, 3 (May): 1–15.

Singh, H. 2016. "Nanotechnology applications in functional foods; opportunities and challenges." *Preventive Nutrition and Food Science*, 21 (1): 1–8.

Singh, M. K., B. Mukherjee, and R. K. Mandal. 2017. "Growth morphology and special diffraction characteristics of multifaceted gold nanoparticles." *Micron*, 94 (March): 46–52.

Siró, I., C. Vén, C. Balla, G. Jónás, I. Zeke, and L. Friedrich. 2009. "Application of an ultrasonic assisted curing technique for improving the diffusion of sodium chloride in porcine meat." *Journal of Food Engineering*, 91 (2): 353–362.

Smith, D. J. 2015. "Chapter 1: Characterization of nanomaterials using transmission electron microscopy." In *Nanocharacterisation, Eds:* Angus I. Kirkland, Sarah J. Haigh, Cambridge: RSC Nanoscience and Nanotechnology, 1–29.

Song, X., R. Li, H. Li, Z. Hu, A. Mustapha, and M. Lin. 2014. "Characterization and quantification of zinc oxide and titanium dioxide nanoparticles in foods." *Food and Bioprocess Technology*, 7 (2): 456–462.

Staiger, M., P. Bowen, J. Ketterer, and J. Bohonek. 2002. "Particle size distribution measurement and assessment of agglomeration of commercial nanosized ceramic particles." *Journal of Dispersion Science and Technology*, 23 (5): 619–360.

Sum, C.-P., A. Shrestha, and A. Kishen. 2013. "Characterizing bacteria adhesion to substrate and early biofilm formation using atomic force microscopy: A review." *Journal of the Indian Institute of Science,* 93 (1): 47–56.

Svanberg, C. and R. Bergman. 2005. "Photon correlation spectroscopy." *Chalmers University of Technology.*

Tampo, H., P. Fons, A. Yamada, K.-K. Kim, H. Shibata, K. Matsubara, S. Niki, H. Yoshikawa, and H. Kanie. 2005. "Determination of crystallographic polarity of ZnO layers." *Applied Physics Letters,* 87 (14): 141904.

Tan, C. P. and M. Nakajima. 2005. "β-carotene nanodispersions: Preparation, characterization and stability evaluation." *Food Chemistry,* 92 (4): 661–671.

Tiede, K., A. B. A. Boxall, S. P. Tear, J. Lewis, H. David, and M. Hassellov. 2008. "Detection and characterization of engineered nanoparticles in food and the environment." *Food Additives and Contaminants. Part A, Chemistry, Analysis, Control, Exposure and Risk Assessment,* 25 (7): 795–821.

Tscharnuter, W. 2006. "Photon correlation spectroscopy in particle sizing." In *Encyclopedia of Analytical Chemistry,* Ed: R.A. Meyers. New Jersey: John Wiley & Sons Ltd.

Vadlapudi, V., D. S. V. G. K. Kaladhar, M. Behara, B. Sujatha, and G. Kishore Naidu. 2014. "Synthesis of green metallic nanoparticles (NPs) and applications." *Oriental Journal of Chemistry,* 29 (4): 1589–1595.

Vega-Gálvez, A., K. Ah-Hen, M. Chacana, J. Vergara, J. Martínez-Monzó, P. García-Segovia, R. Lemus-Mondaca, and K. Di Scala. 2012. "Effect of temperature and air velocity on drying kinetics, antioxidant capacity, total phenolic content, colour, texture and microstructure of apple (Var. Granny Smith) slices." *Food Chemistry,* 132 (1): 51–59.

Velev, O. D., A. D. Nikolov, N. D. Denkov, G. Doxastakis, V. Kiosseoglu, and G. Stalidis. 1993. "Investigation of the mechanisms of stabilization of food emulsions by vegetable proteins." *Food Hydrocolloids,* 7 (1): 55–71.

Wang, Z. L. 2001. "Characterization of nanophase materials." *Particle and Particle Systems Characterization,* 18 (3): 142–165.

Wang, N. and J. G. Brennan. 1995. "A mathematical model of simultaneous heat and moisture transfer during drying of potato." *Journal of Food Engineering,* 24 (1): 47–60.

Warren, B. E. 1941. "X-Ray diffraction methods." *Journal of Applied Physics* 12 (5): 375–384.

Weiss, J., M. Gibis, V. Schuh, and H. Salminen. 2010. "Advances in ingredient and processing systems for meat and meat products." *Meat Science,* 86 (1): 196–213.

Zobel, H. F., S. N. Young, and L. A. Rocca. 1988. "Starch Gelatinization: An X-Ray Diffraction Study." *Cereal Chemistry,* 65(6):443–446.

17 Nanoparticle Synthesis by Plasma Processing

Das Trishitman and C. Anandharamakrishnan

CONTENTS

17.1 INTRODUCTION

Nanoparticles are generally defined as particles of less than 100 nm in size. The size of nanoparticles gives a high surface-to-volume ratio which increases the surface energy compared to the bulk materials. The high surface-to-volume ratio and the quantum effects (due to nanosize) provide nanoparticles with more special mechanical, chemical, optical, and magnetic properties than bulk. These changes in their properties are not only due to the reduction in size but mostly because of the quantum confinement effects. In quantum confinement, confinement defines the motion of the electron to restrict its motion in specific energy levels, and quantum shows the atomic realm of the particles. So, as the size of any particle decreases to nanoscale, it decreases in confining dimensions and makes energy levels discrete and this widens the band gap. In food systems, due to the nanosized particles, the bioavailability of food is increased as the surface volume ratio increases compared to bulk materials (Anandharamakrishnan, 2014; Anandharamakrishnan and Ishwarya, 2015a,b; Bhushani and Anandharamakrishnan, 2017; Bhushani et al., 2017; Ezhilarasi et al., 2013).

There are different processes to synthesize nano-sized particles, but mainly they can be characterized into three phases: solid phase reduction, liquid phase synthesis, or gas phase synthesis. Mechanical solid size reduction can be inexpensive but lacks control over the size, shape, and chemistry. Liquid methods can be expensive because of starting materials, multi-step procedures, and liquid waste which can create undesirable surface chemistry. Also, in liquid synthesis, a new technology/process must often be developed for each new material. Gas phase synthesis cannot give a proper particle size and morphology and is also expensive because of utilization of expensive liquid for feedstock. Both the liquid phase and gas phase synthesis can generate pollutants as wastewater: carbon-di-oxide, carbon monoxides, nitrogen oxides, chlorine, hydrogen chloride, etc. Particle properties (e.g. size, shape, surface chemistry, etc.) are important to their performance in applications. High temperature and anhydrous conditions of plasma can overcome the problems related to size, shape, and surface (Kong and Kawczak, 2008).

The mechanical approach is the only top-down process to produce nanostructured particles in solid phase reduction. It includes mechanical attritions, typically etching through the mask, ball milling, and plastic deformation. It has been used for the generation of magnetic, catalytic, and structural nanoparticles (Koch, 1989). This technology has already been adapted commercially but it is considerably dirty because of a contamination problem during the milling process. The main drawbacks of this technique include low surface area, high polydisperse size distribution, and the partially amorphous state of the as-prepared powders (Kong and Kawczak, 2008).

Liquid/solution processing is a wet chemical synthesis technique for producing nanoparticles by gelation, precipitation, and hydrothermal treatment (Kyprianidou-Leodidou et al., 1994). Better size distribution and stability control can be achieved by manipulating the dopant, or heat treatment. Sonochemical processing, cavitation processing, and microemulsion processing are additional techniques for synthesis of nanoparticles (Kong and Kawczak, 2008). The sonochemical technique can be used to produce a large volume of nanoparticles on an industrial scale. The nanoparticle is generated in hydrodynamic cavitation in sol-gel by purging gas bubbles and creating cavitation. The rapid purging of gas bubbles creating cavitational disturbance and high heating temperature is responsible for nucleation, growth, and quenching of the nanoparticles. The microemulsion technique has been used for the synthesis of metallic, magnetic, semiconductor, and silica nanoparticles by controlling the interfacial surface tension with the addition of co-surfactants. Use of surfactants reduces the contamination problems of mechanical attrition and can be useful for large-scale production of nanoparticles with the use of inexpensive and trouble-free machinery.

Gas phase synthesis and the aerosol method can be separated into the gas-to-particle and the liquid/solid-to-solid routes. The gas-to-particle method is a common method to produce nanoparticle nucleates from a supersaturated vapor through a gas phase process. In the gas phase process, supersaturation can be achieved physically as cooling of hot vapor, or chemically through gas precursors, which will result in the formation of condensable nucleates. Very fine particles can be produced, as the precursor will be in gaseous form, and the final product will have high purity. The degree of agglomeration of the nanoparticle can be controlled by the process parameters (e.g., maximum temperature, initial concentration of the precursor, residence time, and cooling rate). The production rate may be low, multicomponent production of particles can be difficult, and the process can produce a large amount of by-product or pollutant. According to the requirement of supersaturation needed for nanoparticles, gas-to-particle synthesis can be of two types, physical or chemical, although these processes are identical in terms of the aerosol dynamics which occurs once the nucleation is done. The physical vapor process is simple, and no chemical reaction occurs, so the solid precursor and the final products are same, but it needs a higher temperature to produce the vapor, which limits a few materials from processing. In the case of chemical vapor, the production rate is high if a proper precursor with high volatility is available. Composite particles are only processed through the chemical route. Even smaller particles of less than 10 nm in size can be produced with the consistent crystal structure, specific surface structures or properties, and a close particle size distribution using both the techniques. The size variation of particles produced in these processes is about 20%, but to enhance the quantum effects, the size variance should be reduced to 5% (Murray et al., 1993). In the liquid/solid-to-solid technique, particles are formed from droplets of reactant particles via intraparticle reactions. This method can produce single and multi-component particles in homogeneous form. This technique is a continuous process and can be scaled up.

The production yield of nanoparticles can be improved by some new aerosol techniques as combustion flame, plasma, spray pyrolysis, chemical vapour condensation, laser ablation, electrospraying and plasma spray (Swihart, 2003; Gell et al., 2004; Sajti et al., 2010; Tani et al., 2002; Mueller et al., 2003; Karthikeyan et al., 1997; Nakaso et al., 2003; Ye et al., 1997). An overview of nanoparticle synthesis by gas phase is shown in Figure 17.1.

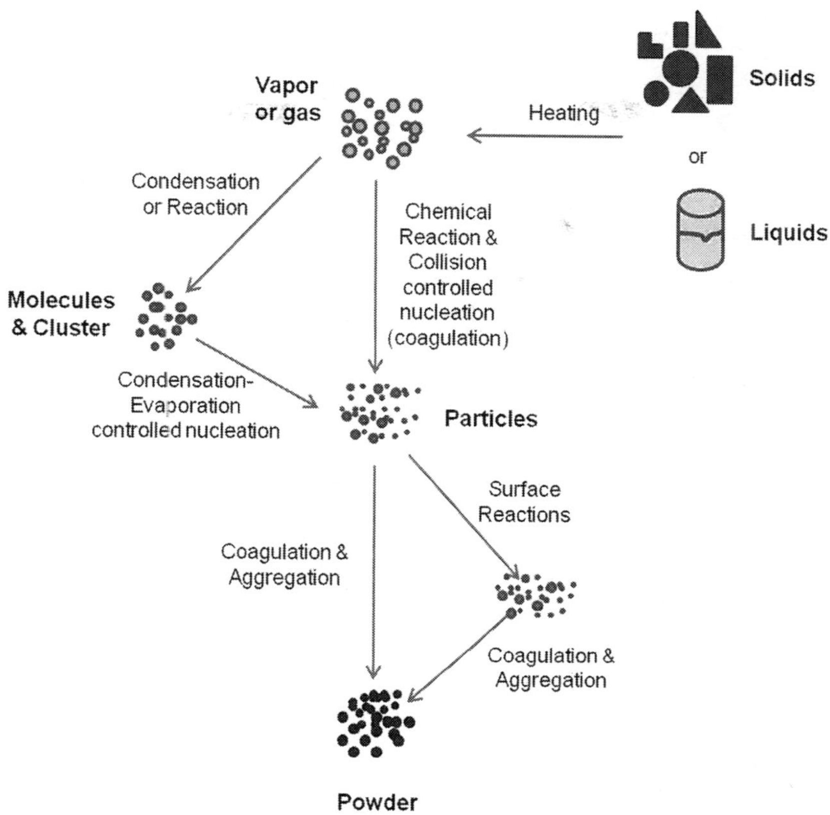

FIGURE 17.1 The general principle behind the gas phase synthesis (redrawn). (Pratsinis, 1998.)

17.2 PLASMA SYNTHESIS

Plasmas are characterized into different species, which are always in non-equilibrium with each other and are of different temperatures; it also varies with different plasmas. Plasma used for processing of materials in nanosize are partially ionized gases, where a few fractions of gas atoms, typically between 10^{-8} and 10^{-3}, are ionized gases. The nature of plasmas is likely to be taken as pervading a fluid mixture of electrons, ions, and neutral molecules, each species having its own transit velocity distribution, and thereafter temperature. The insufficient energy distribution or exchange between the species creates the differences in transitional velocity and temperature. The imbalance in the environment of the plasma is explained in Figure 17.2. In Figure 17.2, a non-equilibrium plasma is created by applying an electrical field to a gas, which can be at low pressure, or in the atmosphere.

These electrical fields accelerate free electrons and provide some of them with sufficient energy to ionize gas atoms or molecules of gases. The reactions or collisions are based on the following equation:

$$A + e_{fast} = A^{+} + 2e_{slow} \tag{17.1}$$

When the electrical field is too high to accelerate a few electrons, it follows Equation 17.1 and a bombardment of accelerated electron starts forming. As the bombardment increases along with the increase of accelerated electrons, it produces more electrons and ions, and in turn, contributes to an exponential growth in electron density. This is a breakdown process at steady state, where these reactions create electrons and ions, and it is balanced by diffusion losses to the reactor walls, by

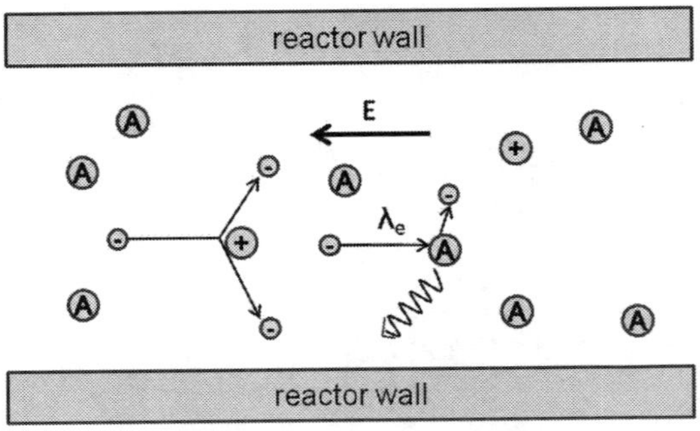

FIGURE 17.2 A schematic of elementary process of plasmas and ranges of some important parameters. Where λ_e is the mean free path of electron, p is the pressure, and E is the electric field strength. The term "O()" in the legend indicates the order of magnitude of the respective properties. (Kortshagen et al., 2016.)

the formation of particles, or by recombination within the plasma volume (Kortshagen et al., 2016). This gives an equal rate of electrons and ions from the plasma. Therefore, the plasma remains electrically neutral with a small exception region, called sheaths, in between the walls in contact with the plasma. This phenomenon is known as plasma "quasi-neutrality,"

$$n_e = n_i \tag{17.2}$$

where:

n_e is the density of the electrons and

n_i is the density of positive ions (Lieberman and Lichtenberg, 2005; Boulos et al., 1994).

Synthesis of nanoparticles by plasma processing is a precursor dependent process, based on the evaporation and condensation principle. The precursor may either be a gas, liquid, or solid and become vaporized or atomized by higher gas temperature in a thermal plasma system. The precursor is injected axially or radially to the plasma processer. Formation of nanoparticles needs a high supersaturation and the plasma process has the ability to create high supersaturation due to a high-temperature gradient. Thus plasma processing can synthesize smaller size particles in homogeneous form. Though the conditions of plasma processing assure the production of homogeneous nanosize particles, some heterogeneous nucleation still can be formed in the wall of the reactor (Gitzhofer, 2009).

There are different ways to induce plasma to form nanoparticles e.g. using microwaves (μW), direct current (DC plasma), or radio frequency (RF plasma). DC plasma jet/DC arc plasma jet, RF induction plasma, and microwave torch plasma systems usually consist of an associated system with electrodes, a power supply (DC or RF or MW), a gas handling apparatus, water cooling, pumping, data acquisition equipment, and various environmental and safety features, such as a gas leak detector, a self-scrubbing pump, and a flare for exhaust gases. The generation of plasma and the formation of nanoparticles depends on major parameters like power supply, the distance between the electrodes, and the type of gas used.

The overall temperature of the gas passing through the microwave plasma is not more than that of the DC plasma or RF plasma (Vollath et al., 1997). Different plasma processes are based on evaporation/condensation principle for nanoparticle synthesis. Inductively coupled plasma (ICP) is one of the new spray pyrolysis modified plasma methods. ICP with high temperature generated under ambient pressure coupled with RF power sources seems to be one of the potential approaches to the plasma synthesis of nanoparticles (Kong and Lau, 2009).

This process has the advantage over other plasma processes as there is no involvement of any electrodes for operation; therefore, the particles are free from contaminants. The allowable use of aggressive precursors and gases in the process and the relatively long residence time of the precursor in the reactor makes uniform nanoparticles, as it is an evaporation and condensation process (Bilodeau and Proulx, 1996; Young and Pfender, 1985). In ICP, the high temperature (up to 10,000°C) and high energy density of plasmas allow the synthesis of metal and ceramic nanoparticles.

Besides all the advantages, ICP synthesis has the disadvantage of low efficiency, due to the evaporation, cooling, and energy conversion process. However, this can be compensated for by synthesis of the more complex precursor, purer and/or typical materials which cannot be produced by other techniques (Fauchais and Baronnet, 2009).

Apart from all the above-mentioned gas phase techniques, a lot of other techniques produce nanoparticles, e.g. laser, sputtering, spark source, exploding wire, expansion cooling, spraying, sol-gel process, liquid plasma generation, mechanical milling, and dynamic shock compression (Ayyub et al., 2001; Márton et al., 2003; Yamada, 2000; Omori, 2000; Wang et al., 2001; Zhang et al., 2002; Nakaso et al., 2003; Trindade et al., 2001; Ashkarran, 2011; Mariotti et al., 2012).

17.3 NANOPARTICLE GENERATION AND GROWTH PHENOMENA

The kinetics of nanoparticle generation or growth using plasma processes comes after the formation of nanoparticle led to contamination problems with the use of many processes. The kinetics of plasma nanoparticles has been proposed by Bouchoule and Boufendi (1993) in a three-phase model system. It consists of (1) nucleation of the nanoparticle, (2) cluster and nanoparticle agglomeration/coagulation, and (3) nanoparticle surface growth.

1. Nucleation of the nanoparticle: It has been carried out in the initial phase by generating clusters of nanoparticles that have been formed by the nucleation mechanism. In the nucleation mechanism, initial nuclei are formed at a dense concentration and are either neutral or slightly negatively charged.
2. Cluster and nanoparticle agglomeration: The particles in dense concentrations are too small and there are too many to be uniformly negatively charged to grow separately to form clusters and agglomerate as single nanoparticles.
3. Nanoparticle surface growth: This stage provides negatively charged nanoparticles with a continuous growth through surface reactions after nanoparticle formation with the remaining precursor in the reactor (Figure 17.3).

This three-phase model system is derived from the chemical vapor deposition reactor system, where nanoparticles can grow continuously and form nanoparticles from 10 to 100 nanometers by electrical fields as a new precursor enters the reactor. This model system is designed for nanoparticle synthesis generated in the gas phase plasma synthesis system.

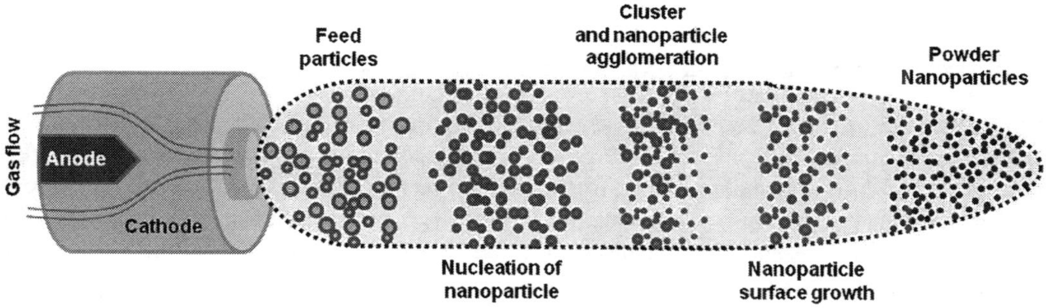

FIGURE 17.3 A schematic representation of nanoparticle generation and growth.

17.4 PLASMA SPRAYED NANOMATERIALS

Plasma sprayed nanomaterials are made in the process of vaporization/plasma formation to condensation. Metallic, inorganic material, and metallic alloy nanoparticles are being produced with different plasma sources (MW, DC, RF). Aluminum oxide, alumina, Fe–TiN, silicon, and silicon carbide were produced by the active plasma reaction method using a DC plasma reactor (Sakka et al., 2002; Rao et al., 1995; Karthikeyan et al., 1997; Ananthapadmanabhan et al., 1996). Crystallized maghemite (γ–Fe_2O_3) particles coated with crystallized zirconia, Nb–N, Nb–N–C, Nb–Ti–N–C, zirconium nitride nanoparticles are formed using microwave plasma, and their physical and chemical properties been changed. Titanium dioxide, SiC, $BaFe_{12}O_{19}$, SiC, Si_3N_4, SiO_2, hydroxyapatite nanoparticles using inductive coupled plasma (Guo et al., 1995; Mizoguchi et al., 1994; Bouyer et al., 2001; Kumar et al., 2001; Akashi, 2009; Pirzada and Yadav, 1998) and ferro- or ferri-magnetic compound (Y–Fe–O) iron nanoparticles using radio frequency plasma were produced (Sugasawa et al., 1998; Girshick et al., 1993).

17.4.1 Silicon Nanoparticles

Silicon (Si) is the most investigated model species used for plasma synthesis of nanoparticles. As a model species, it revealed much information on nanoparticles' physical and chemical identities (Amouroux et al., 2003; Kruis et al., 1993). The melting point of bulk silicon is 1683 K, and it exists as a cubic phase and has a specific density of 2.33 gram per cubic centimeter (Teng et al., 2012). Silicon is mainly being used in the semiconductor industry and photonics. When it is doped with boron (B), an N-type semiconductor is attained, and when it is doped with phosphorus (P), a P-type semiconductor is attained. Even doping of germanium also been done, and silicon germanium nanocrystals can also be made of silicon bulk material (Kortshagen et al., 2016). Si nanoparticles are also used for applications in photonics. Nano-Si is synthesized from the decomposition of Silane (SiH_4), trimethylsilane (TMS) ($(CH_3)_3SiH$), or other Si-containing precursors, which are in gaseous or liquid phase.

17.4.2 Tungsten Carbide Nanoparticles

Tungsten carbide (WC), which was discovered in the late 1890s, has received a lot of attention because of its high refractory property (boiling point of around 6000 K) and very high density (of around 15.6 kg/m^3). WC was more attractive to make microstructure WC, as the hexagonal δ-phase of it gives high hardness and thermal resistivity. These properties make WC useful in wear-resistant coating and reinforcement. Since the early 20th century, it is well-known that the mechanical properties of cemented carbides can be influenced by size reductions (Spriggs, 1995). It has also been found that alloys containing smaller WC particles had a higher hardness than the alloys with larger WC particles. It was also discovered that the carbide's addition to vanadium, chromium, and tantalum can act as a growth inhibitor for tungsten carbide. In the current trend, the WC alloys size is reduced to nanometer range in order to get better physical properties using plasma processing.

17.5 SOLUTION PRECURSOR PLASMA SPRAY (SPPS) SYNTHESIS

Solution precursor plasma spray (SPPS) is a simple, innovative, single-step, rapid technique with a variation of conventional plasma spray process. It is a coating deposited plasma spray method where nano-ceramic or metal materials from solution precursors are used. In the SPPS process, an aqueous or non-aqueous solution containing metallic or ceramic cations is fed to form oxides of interest to deposit as splats. It is considered to be a promising technique for the formation of non-equilibrium phases in multi-component oxide system. It uses molecularly mixed precursors as liquids, enables the synthesis of wide range metal/ceramic oxide powders and coatings, and avoids separate processing methods for the preparation of powders.

In conventional plasma spray 10–100 μm, flowable powders of the coating material are fed into the jet, which further melts them, and are deposited as splats or clogs into the plasma gun. High surface tension among the fine particles leads to inconsistent flow, producing the non-homogeneous and inferior coating. An alternative method is using dry powders as feedstock in solution form, either in form of sols or homogeneously dissolved precursor solutions, in a process known as precursor plasma spray. The former is known as the process of suspension plasma spray (SPS), and the latter is known as solution precursor plasma spray (SPPS). The SPSS is a variation of the conventional plasma spray process and it uses the same plasma spray equipment. The steps in SPPS process consist of droplet breakup, solvent evaporation, particle pyrolysis, melting, and finally deposition. In the SPPS process, the precursor solution comprises a metal precursor in the form of nitrates, sulfates, chlorides, or alkoxides dissolved in an aqueous or non-aqueous solution. Upon injecting the precursor into the plasma flame, the products formed are directed towards the substrate by the carrier gas. Thus SPPS containing cations enables the formation of metal oxides, multi-component composites which are necessary to form the oxide coating in a single step. In the SPPS process, the precursor is a solution that helps in the formation of nanoparticles as coating or powder. SPPS has been used to process many simple as well as complex oxides with a wide range of applications including thermal barrier coating, low thermal conductivity, high thermal stability, phosphor, fuel cells, and magnetic and biomedical coatings. The preparation of alumina, zarconia, and yttria-stabilized zirconia nanoparticles and deposits are being reported. SPPS is a simple versatile process for the preparation of single or mixed oxides nanoparticles, films, and coating with diverse applications and a variety of physical and chemical properties.

In the plasma synthesis process, the liquid precursor is atomized and injected in the high-temperature plasma flame nozzle. At this high temperature, the atomized material undergoes a reaction to form particles that are either collected or deposited as a coating on the top of the substrate or as a powder. Sometimes an electrostatic precipitator has also been used to collect the fine material in powder form. The precursor material is atomized in the high-temperature flame, made into fine droplets, and is accelerated towards the substrate by the carrier gas, leading to various phenomena which can affect the coating process and properties in the substrate. The coating formation in the SPPS process involves multiple sequential steps and is a result of physical and chemical reactions of precursor droplets. The multiple steps in SPPS process consist of: formation of precursor droplets via a liquid atomization; rapidly transferred droplets to a plasma region of high temperature and velocity; due to the interaction in the plasma region, the droplets break into tiny droplets; as a result, of the fine droplets, the solvent evaporates instantly in the hot plasma; and the remaining solute endure thermal decomposition to form related oxide particles. Whether the corresponding oxide particles were either partially or fully melted depends on the residence time of the particle; splat or thin layer on substrate formation; and the continuous process leads to the dense coating on the substrate. The SPPS process has been illustrated in Figure 17.4: it shows the interaction of the precursor, various changes of precursor during the process, and its deposition as a coating in the substrate. Depending upon the mechanism of action during the process; and the interaction between the physical, chemical and thermal reactions results in particles size, crystals, agglomerates, or aggregates. If the atomized droplets stay in the plasma flame for an insufficient time, incomplete evaporation of solvent, incomplete formation of corresponding oxides, and condensation of precursor material will result. As a result, droplets splash on the substrate surface and undergo further reactions depending upon the time, and the available energy forms splats or patchy deposits.

SPPS is a single-step process, and all the reactions are propelled to completion in the atomization process itself, resulting in the formation of nanosize oxide particles. No post-spray calcinations or heat treatment are required in the SPPS process. Even nanomaterials with consistent characteristics require controlled size distribution of atomized liquid droplets and optimum spray parameters. As a result, the diameter of the atomized nozzle, the concentration and flow rate of the liquid feedstock, the pressure and flow rate of atomizing gas, and the droplet size distribution must be monitored for optimum particle size/coating properties. The heat flux of plasma flame is governed by the nature

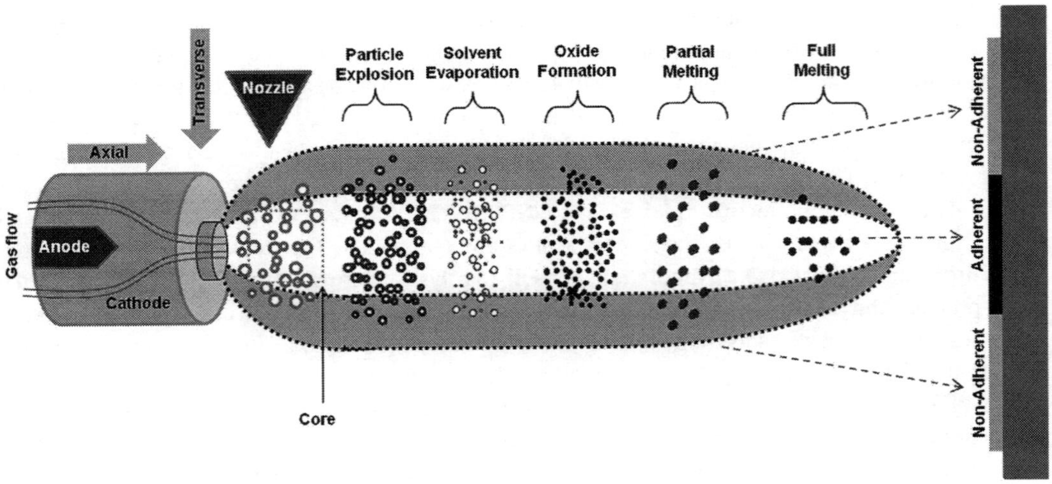

FIGURE 17.4 A schematic representation of solution precursor plasma interaction and associated changes in various parts of the process. Interaction in the core region of the flame leads to adherent coating.

of atomized gases, so the selection of gases for optimum process parameters has to be considered. Generally, argon and nitrogen are used as primary gases while hydrogen is commonly used as a secondary gas. The size, shape, and phase composition of the final nanomaterials depends strongly on the spray feedstock.

The SPPS method for the formation of nanomaterial or coating offers several advantages over conventional plasma spray methods; the evasion of the powder-feedstock preparation step, better control in particle formation; the chemistry of deposit, the ability of equivalent graded deposition and coating with ease; the coating of the evenly distributed nanostructure, porousness, and processing versatility makes the SPPS method superior and more effective. The advantages of the SPPS process also include rapid exploration of composites and fabrication of advanced coating with unique microstructural features. The various materials that have been synthesized by the SPPS process and their potential applications are mentioned in tabulated form in Table 17.1.

17.5.1 Various Applications of SPPS

The superiority of SPPS over conventional methods gives it the potential to deposit a wide range of ceramics for the application of thermal barrier coatings, hard coatings, crystal supports, biomedical implants, and fuel cell components. This makes the SPPS method technologically very attractive and efficient. Through these processes, ceramic–ceramic composite films, coatings, nanostructured ceramic, and metal composites are being developed. Different features of SPPS makes it adaptable for a wide variety of applications, mainly for applications in gas turbine engines and solid fuel cells.

17.5.1.1 Thermal Barrier Coating

Thermal barrier coatings are used in order to shield the component from the high temperatures, which allows the system to run at a higher temperature and thus makes the use of component more efficient. In applications, this technology focuses on turbines, where it reduces the high mechanical stress and environmental factors resulting from heat and extends the component life (Evans et al., 2001a). The thermal barrier coating is directly applied to bond coating with an intermetallic bond coat to protect the load-bearing component of the metal superalloy (Padture et al., 2001). During the thermal cycling, thermally grown oxides grow in between the coating and metal superalloy and protect the load-bearing component from creep or fatigue (Evans et al., 2001b). Although this

TABLE 17.1
Various Materials Synthesized by SPPS and Their Potential Applications

Material System	Coating Thickness (μm)	Particle/ Grain Size (nm)	Porosity Vol. (%)	Potential Applications	References
Al_2O_3	–	10–50	–	Nanopowders	Karthikeyan et al., 1997
ZrO_2		20–80	–	Nanopowders	Karthikeyan et al., 1997
YSZ	–	20–100	–	Nanopowders	Karthikeyan et al., 1997
CeO_2	1	30–50	–	High-temperature oxidation resistance	Viswanathan et al. (n.d.)
ZrO_2-$7Y_2O_3$	250–2000	–	16–30	TBC	Jadhav et al., 2005; Xie et al., 2004c; Xie et al. 2006; Ma et al., 2006; Jadhav et al., 2006; Xie et al., 2004a, 2004b.
Ni-LDC	100–150	–	40–45	Fuel Cell	Ma et al., 2005
LDC	5	–	–	Fuel Cell	Ma et al., 2005
Eu^{3+} Doped TiO_2	–	30–83	–	Optoelectronics	Garcia et al., 2008
Eu^{3+} Doped Y_2O3	20–100	500–5000	–	Luminescent coatings	Devi et al., 2002
Yttrium Iron Garnet	–	90–110	–	Magneto-optic	Devi et al., 2002
Hydroxyapatite	3–5	10–25	–	Biomedical	Garcia et al., 2008 and Loncea et al., 1997
ZrO_2- Al_2O_3	–	10–1000	–	Metastable	Vasiliev et al., 2006b
ZrO_2-Y_2O_3 Al_2O_3	–	4–3000	–	Metastable ceramics	Vasiliev and Padture, 2006a

Source: Brinley et al. (2007).

process is less expensive compared to its application, it gives better thermal resistivity and high strain tolerance.

17.5.1.2 Solid Oxide Fuel Cells

Solid oxide fuel cells (SOFC) are a highly efficient and pollution-free technique in producing electricity through chemical and electrochemical reactions between hydrogen/hydrogen fuels and oxygen. The catalytic activity of an anode is one of the determining factors of performance. The difficulties associated with SOFCs are in the fabrication of cell components, material selection, mechanical and electrochemical reliability, high operating temperatures, and high production cost, which have to be overcome (Ma et al., 2005). The SPPS deposited anode layers in a highly porous microstructure; porosity of 40%–45% with a large surface area and a thickness of 100–150 μm which enhances the interaction of incoming fuel. With the increasing number of the pores in the anode layer, and the active sites at three-phase boundaries (ionic, electronic, and gas phase), catalytic property and gas fuel permeability increases, hence the efficiency of SOFC improves.

17.5.1.3 Metastable Ceramic Composites

Metastable ceramics are formed rapidly by kinetically suppressing equilibrium phases and microstructures. Rapid solidification of materials leads to metastable ceramics, whereas the short time suppresses long-range diffusion; however, both processes result in the formation of powders requiring consolidation, and subsequent sintering forms the component. Rapid pyrolytic decomposition of the liquid precursor is performed by atomizing liquid chemical precursor into the plasma nozzle, and it plays an important role in the process of metastable ceramics. The oxides generated through the solvent evaporation, low-temperature pyrolysis, or decomposition, result in gel or solid crystallization at low temperature. In the process, low homologous temperature involved in pyrolysis and crystallization suppress long-term diffusion, resulting in the formation of transient phase in the case of a single component precursor, or extended solid solution in case of multi-component precursors (Brinley et al., 2007; Jordan et al., 2015).

17.5.1.4 Biomedical Coating

The SPPS process in biomedical coating has been widely used to coat and improve the fixation of implants of bones. Hydroxypatite (HA, $Ca_{10}(PO_4)_6(OH)_2$) has mostly been used in biomedical coating of bone-implants, and it is known as an osteoconductive material which can promote bone growth at the surface. A submicrometer or nanocrystalline HA coating exhibited microstructural features which are formed during the SPPS process and can make melted, or a combination of melted and unmelted, deposits of hollow spheres on the bone-implants (Coyle et al., 2007; Ioncea et al., 1997). Nanocrystalline/submicrometer HA coating improves the resorption of coating in the body without creating an irritant effect which was found with air plasma spray-created larger particles.

17.5.1.5 Luminescent and Magnetic Materials

Various earth materials are doped into oxides or semiconductors to improve the phosphorescence properties, but wet chemical methods are still not sufficient to provide effective doping. As the dopant and the ions differ in ionic sizes as well as chemical properties, this leads to irregular absorption, and variation in precipitation kinetics, and results in accumulation of ions on the surface, rather on the host matrixes. Precursor plasma spray helps in eliminating these problems and increases the doping of the dopant in the crystalline matrixes. Eu^{3+} doped TiO_2 and Y_2O_3 luminescent nanocrystals with efficient nonradiative energy transferred efficiently from the host to the activator Eu^{3+} ions, when synthesized by the SPPS method (Devi et al., 2002; Li et al., 2006). Conventional methods use a higher temperature for making magnetic materials, which results in wider particle size distribution, and affects the magnetic properties as proper maintenance of stoichiometry was difficult. Yttrium iron garnet ($Y_3Fe_5O_{12}$, YIG), a magnetic material used for telecommunication and data storage, can be formed through a single step precursor plasma process (Guo et al., 2005).

17.6 FUTURE TRENDS

Nanoparticle synthesis using plasma is an emerging new class of technique that holds the promise of facilitating breakthrough innovations in almost all energy market segments. Manufacturing of particles, nanomaterials, composites, devices, and the engineered materials using plasma processing are poised to revolutionize many industries and open up many more opportunities. The emergence of plasma processing in synthesis is providing an additional degree of freedom to design and produce nanomaterials that will accelerate application research, and lead to the creation of new innovative technologies and devices.

In the case of SPPS, many possibilities with multiple applications are partially realized and there are many more to explore with existing resources. Thermal barrier coating has already shown a remarkable increase, compared to other precursor-based sprays. As the porosity control is affordable, and the powder formation/agglomeration stage becomes easier and simpler, adoptability and incorporation expected to increase. The strong control of stoichiometry of SPPS opens up many paths for exploration and inclusion of different applications. Since the processing is easier, the desired quality products and applications have increased plasma processing research into nanomaterials, and it is expected to be a staple in the future.

REFERENCES

Akashi, K. 2009. "Progress in thermal plasma deposition of alloys and ceramic fine particles." *Pure and Applied Chemistry* 57(9): 1197–1206.

Amouroux, J., Dresvin, S.V., Morvan, D., Ouvrelle, L., Ivanov, D., Zverev, S.G., Feigenson, O., and Balashov, A. 2003. "Calculation of silicon particles dynamics, heat and mass transfers in thermal plasmas. effect of particles vaporization." *High Temperature Material Processes: An International Quarterly of High-Technology Plasma Processes* 7(1): 93–105.

Anandharamakrishnan, C. 2014. "Liquid-based nanoencapsulation techniques." In *Techniques for Nanoencapsulation of Food Ingredients*, 29–41. New York: Springer.

Anandharamakrishnan, C. and Padma Ishwarya, S. 2015a. "Introduction to encapsulation of food ingredients." In *Spray Drying Techniques for Food Ingredient Encapsulation*, 37–64. Chichester: Wiley-Blackwell.

Anandharamakrishnan, C. and Padma Ishwarya, S. 2015b. "Spray drying for nanoencapsulation of food components." In *Spray Drying Techniques for Food Ingredient Encapsulation*, 180–197. Chichester, West Sussex: Wiley-Blackwell.

Ananthapadmanabhan, P. V., Sreekumar, K. P., Venkatramani, N., Sinha, P. K., and Taylor, P.R. 1996. "Characterization of plasma-synthesized alumina." *Journal of Alloys and Compounds* 244(1): 70–74.

Ashkarran, A. A. 2011. "Metal and metal oxide nanostructures prepared by electrical arc discharge method in liquids." *Journal of Cluster Science* 22(2): 233.

Ayyub, P., Chandra, R., Taneja, P., Sharma, A. K., and Pinto, R. 2001. "Synthesis of nanocrystalline material by sputtering and laser ablation at low temperatures." *Applied Physics A* 73(1): 67–73.

Bhushani, A. and Anandharamakrishnan, C. 2017. "Food-grade nanoemulsions for protection and delivery of nutrients." In *Nanoscience in Food and Agriculture*, 4, eds. S. Ranjan, N. Dasgupta, and E. Lichtfouse, 99–139. New York: Springer.

Bhushani, A., Harish, U., and Chinnaswamy, A. 2017. "Nanodelivery of nutrients for improved bioavailability." In *Nutrient Delivery*, ed. A. M. Grumezescu, 369–411. Cambridge: Elsevier.

Bilodeau, J.-F. and Proulx, P. 1996. "A mathematical model for ultrafine iron powder growth in a thermal plasma." *Aerosol Science and Technology* 24(3): 175–189.

Bouchoule, A. and Boufendi, L. 1993. "Particulate formation and dusty plasma behaviour in argon-silane RF discharge." *Plasma Sources Science and Technology* 2(3): 204.

Boulos, M. I., Fauchais, P., and Pfender, E. 1994. "The plasma state." In *Thermal Plasmas*, 1–47. Heidelberg: Springer.

Bouyer, E., Schiller, G., Müller, M., and Henne, R. H. 2001. "Thermal plasma chemical vapor deposition of Si-based ceramic coatings from liquid precursors." *Plasma Chemistry and Plasma Processing* 21(4): 523–546.

Brinley, E., Babu, K. S., and Seal, S. 2007. "The solution precursor plasma spray processing of nanomaterials." *JOM* 59(7): 54–59.

Coyle, T. W., Garcia, E., Zhang, Z., and Gan, L. 2007. "Plasma spray deposition of hydroxyapatite coatings from sol precursors." *Materials Science Forum* 2007.

Devi, P. S., Margolis J., Parise, J. B., Grey, C. P., Sampath, S., Herman, H., and Gafney, H.D. 2002. "Single-step deposition of Eu-doped Y2O3 phosphor coatings through a precursor plasma spraying technique." *Journal of Materials Research* 17(11): 2771–2774.

Evans, A. G., He, M. Y., and Hutchinson, J. W. 2001a. "Mechanics-based scaling laws for the durability of thermal barrier coatings." *Progress in Materials Science* 46(3): 249–271.

Evans, A.G., Mumm, D. R., Hutchinson, J. W., Meier, G. H., and Pettit, F. S. 2001b. "Mechanisms controlling the durability of thermal barrier coatings." *Progress in Materials Science* 46(5): 505–553.

Ezhilarasi, P. N., Karthik, P., Chhanwal, N., and Anandharamakrishnan, C. 2013. "Nanoencapsulation techniques for food bioactive components: A review." *Food and Bioprocess Technology* 6(3): 628–647.

Fauchais, P. and Baronnet, J.-M.. 2009. "State of the art of plasma chemical synthesis of homogenous and heterogenous products." *Pure and Applied Chemistry* 52(7): 1669–1705.

Garcia, E., Zhang, Z. B., Coyle, T. W., Gan, L., and Pilliar, R. 2008. "Hydroxyapatite coatings produced by plasma spraying of organic based solution precursor." In *Advances in Bioceramics and* Biocomposites II, Ceramic Engineering *and Science Proceedings*, eds. M. Mizuno, A. Wereszczak, E. Lara-Curzio, 103–110. Hoboken: Wiley-Blackwell.

Gell, M., Xie, L., Ma, X., Jordan, E. H., and Padture, N. P. 2004. "Highly durable thermal barrier coatings made by the solution precursor plasma spray process." *Surface and Coatings Technology, Proceedings of the 30th International Conference on Metallurgical Coatings and Thin Films*, 177–178 (January): 97–102.

Girshick, S. L., Chiu, C.-P., Muno, R., Wu, C. Y., Yang, L., Singh, S. K., and McMurry, P.H. 1993. "Thermal plasma synthesis of ultrafine iron particles." *Journal of Aerosol Science, Aerosols in* Materials Processing 24(3): 367–382.

Gitzhofer, F. 2009. "Induction plasma synthesis of ultrafine SiC." *Pure and Applied Chemistry* 68(5): 1113–1120.

Guo, J.Y., Gitzhofer, F., and Boulos, M. I. 1995. "Induction plasma synthesis of ultrafine SiC powders from silicon and CH4." *Journal of Materials Science* 30(22): 5589–5599.

Guo, X. Z., Ravi, B. G., Devi, P. S., Hanson J. C., Margolies, J., Gambino, R. J., Parise, J. B., and Sampath, S. 2005. "Synthesis of Yttrium Iron Garnet (YIG) by citrate–nitrate gel combustion and precursor plasma spray processes." *Journal of Magnetism and Magnetic Materials* 295(2): 145–154.

Ioncea, A., Novac, C., Radu, D., Florea, C., Dumitrescu, O., and Ionescu, D. 1997. "Bioactive coatings based on hydroxyapatite." *Key Engineering Materials* 132–136 (April): 1532–1535.

Jadhav, A., Padture, N. P., Wu, F., Jordan, E. H., and Gell, M. 2005. "Thick ceramic thermal barrier coatings with high durability deposited using solution-precursor plasma spray." *Materials Science and Engineering: A* 405(1): 313–320.

Jadhav, A. D., Padture, N. P., Jordan, E. H., Gell, M., Miranzo, P., and Fuller, E. R. 2006. "Low-thermal-conductivity plasma-sprayed thermal barrier coatings with engineered microstructures." *Acta Materialia* 54(12): 3343–3349.

Jordan, E. H., Jiang, C., and Gell, M. 2015. "The Solution Precursor Plasma Spray (SPPS) process: A review with energy considerations." *Journal of Thermal Spray Technology* 24(7): 1153–1165.

Karthikeyan, J., Berndt, C. C., Tikkanen, J., Reddy, S., and Herman, H. 1997. "Plasma spray synthesis of nanomaterial powders and deposits." *Materials Science and Engineering: A* 238(2): 275–286.

Koch, C. C. 1989. "Materials synthesis by mechanical alloying." *Annual Review of Materials Science* 19(1): 121–143.

Kong, P. and Kawczak, A. 2008. "Plasma synthesis of nanoparticles for nanocomposite energy applications." In *8th World Congress Nanocomposites,* San Diego, CA, September 15–18.

Kong, P. C. and Lau, Y. C. 1990. "Plasma synthesis of ceramic powders." *Pure and Applied Chemistry* 62(9): 1809–1816.

Kortshagen, U. R., Sankaran, R. M., Pereira, R. N., Girshick, S. L., Wu, J. J., and Aydil, E. S. 2016. "Nonthermal plasma synthesis of nanocrystals: fundamental principles, materials, and applications." *Chemical Reviews* 116(18): 11061–11127.

Kruis, F. E., Kusters, K. A., Pratsinis, S. E., and Scarlett, B. 1993. "A simple model for the evolution of the characteristics of aggregate particles undergoing coagulation and sintering." *Aerosol Science and Technology* 19(4): 514–526.

Kumar, R., Cheang, P., and Khor, K. A. 2001. "RF plasma processing of ultra-fine hydroxyaptite powders." *Journal of Materials Processing Technology, 5th Asia Pacific Conference on Materials Processing* 113(1–3): 456–462.

Kyprianidou-Leodidou, T., Caseri, W., and Suter, U. W. 1994. "Size variation of PbS particles in high-refractive-index nanocomposites." *The Journal of Physical Chemistry* 98(36): 8992–8997.

Li, J.-G., Wang, X., Watanabe, K., and Ishigaki, T. 2006. "Phase structure and luminescence properties of Eu3+-doped TiO2 nanocrystals synthesized by Ar/O2 radio frequency thermal plasma oxidation of liquid precursor mists." *The Journal of Physical Chemistry B* 110(3): 1121–1127.

Lieberman, M. A. and Lichtenberg, A. J. 2005. "Basic plasma equations and equilibrium." In *Principles of Plasma Discharges and Materials Processing*, 23–42. Hoboken: John Wiley & Sons, Inc.

Ma, X., Dai, J., Zhang, H., Roth, J., Xiao, T. D., and Reisner, D. E. 2005. "Solid oxide fuel cell development by using novel plasma spray techniques." *Journal of Fuel Cell Science and Technology* 2(3): 190–196.

Ma, X., Wu, F., Roth, J., Gell, M., and Jordan, E. H. 2006. "Low thermal conductivity thermal barrier coating deposited by the solution plasma spray process." *Surface and Coatings Technology, Proceedings of the 33rd International Conference on Metallurgical Coatings and Thin Films* 201(7): 4447–4452.

Mariotti, D., Patel, J., Švrček, V., and Maguire, P. 2012. "Plasma–liquid interactions at atmospheric pressure for nanomaterials synthesis and surface engineering." *Plasma Processes and Polymers* 9(11–12): 1074–1085.

Márton, Z. S., Landström, L., Boman, M., and Heszler, P. 2003. "A comparative study of size distribution of nanoparticles generated by laser ablation of graphite and tungsten." *Materials Science and Engineering: C Current Trends in Nanotechnologies: From Materials to Systems. Proceedings of Symposium Q, E-MRS Spring Meeting* 2002. June 18–21 2002, 23(1): 225–228.

Mizoguchi, Y., Kagawa, M., Suzuki, M., Syono, Y., and Hirai, T. 1994. "Synthesis of ultrafine particles and thin films of BaFe12O19 by the spray-ICP technique." *Nanostructured Materials*, Vapor Phase Synthesis of Nanostructured Materials 4(5): 591–596.

Mueller, R., Mädler, L., and Pratsinis, S.E. 2003. "Nanoparticle synthesis at high production rates by flame spray pyrolysis." *Chemical Engineering Science* 58(10): 1969–1976.

Murray, C. B., Norris, D. J., and Bawendi, M. G. 1993. "Synthesis and characterization of nearly monodisperse CdE (E = sulfur, selenium, tellurium) semiconductor nanocrystallites." *Journal of the American Chemical Society* 115(19): 8706–8715.

Nakaso, K., Han, B., Ahn, K. H., Choi, M., and Okuyama, K. 2003. "Synthesis of non-agglomerated nanoparticles by an Electrospray Assisted Chemical Vapor Deposition (ES-CVD) method." *Journal of Aerosol Science* 34(7): 869–881.

Omori, M. 2000. "Sintering, consolidation, reaction and crystal growth by the Spark Plasma System (SPS)." *Materials Science and Engineering: A* 287(2): 183–188.

Padture, N. P., Schlichting, K. W., Bhatia, T., Ozturk, A., Cetegen, B., Jordan, E. H., Gell, M., Jiang, S., Xiao, T. D., and Strutt, P. R. 2001. "Towards durable thermal barrier coatings with novel microstructures deposited by solution-precursor plasma spray." *Acta Materialia* 49(12): 2251–2257.

Pirzada, S. and Yadav, T. 1998. Method of producing nanoscale powders by quenching of vapors. US5788738 A, patent filed September 3, 1996, and issued August 4, 1998.

Pratsinis, S. E. 1998. "Flame aerosol synthesis of ceramic powders." *Progress in Energy and Combustion Science* 24(3): 197–219.

Rao, N., Girshick, S., Heberlein, J., McMurry, P., Jones, S., Hansen, D., and Micheel, B. 1995. "Nanoparticle formation using a plasma expansion process." *Plasma Chemistry and Plasma Processing* 15(4): 581–606.

Sajti, C. L., Sattari, R., Chichkov, B. N., and Barcikowski, S. 2010. "Gram scale synthesis of pure ceramic nanoparticles by laser ablation in liquid." *The Journal of Physical Chemistry C* 114(6): 2421–2427.

Sakka, Y., Okuyama, H., Uchikoshi, T., and Ohno, S. 2002. "Synthesis and characterization of Fe and composite Fe–TiN nanoparticles by Dc Arc-Plasma." *Journal of Alloys and Compounds* 346(1): 285–291.

Spriggs, G. E. 1995. "A history of fine grained hardmetal." *International Journal of Refractory Metals and Hard Materials*, Special Issue on Fine Grained Hardmetals, 13(5): 241–255.

Sugasawa, M., Kikukawa, N., Ishikawa, N., Kayano, N., and Kimura, T. 1998. "Synthesis of Y–Fe–O ultrafine particles using inductively coupled plasma." *Journal of Aerosol Science* 29(5–6): 675–686.

Swihart, M. T. 2003. "Vapor-phase synthesis of nanoparticles." *Current Opinion in Colloid & Interface Science* 8(1): 127–133.

Tani, T., Mädler, L., and Pratsinis, S. E. 2002. "Homogeneous ZnO nanoparticles by flame spray pyrolysis." *Journal of Nanoparticle Research* 4(4): 337–343.

Teng, R., Zhou, Q., Dai, X., Wu, Z., Xu, W., Xiao, Q., Wu, X., and Guo, X. 2012. "Optimization of heat shield for single silicon crystal growth by using numerical simulation." *Rare Metals* 31 (5): 489–493.

Trindade, T., O'Brien, P., and Pickett, N. L. 2001. "Nanocrystalline semiconductors: Synthesis, properties, and perspectives." *Chemistry of Materials* 13(11): 3843–3858.

Vasiliev, A. L. and Padture, N. P. 2006a. "Coatings of metastable ceramics deposited by solution-precursor plasma spray: II. ternary ZrO_2–Y_2O_3–Al_2O_3 system." *Acta Materialia* 54(18): 4921–4928.

Vasiliev, A. L., Padture, N. P., and Ma, X. 2006b. "Coatings of metastable ceramics deposited by solution-precursor plasma spray: I. binary ZrO_2–$A12O_3$ system." *Acta Materialia* 54(18): 4913–4920.

Viswanathan, V., Filmalter, R., Patil, S., Deshpande, S., and Seal, S. n.d. "High-temperature oxidation behavior of solution precursor plasma sprayed nanoceria coating on martensitic steels." *Journal of the American Ceramic Society* 90(3): 870–877.

Vollath, D., Vinga Szabó, D., and Hauβelt, J. 1997. "Synthesis and properties of ceramic nanoparticles and nanocomposites." *Journal of the European Ceramic Society* 17(11): 1317–1324.

Wang, Q., Yang, H., Shi, J., and Zou, G. 2001. "Preparation and characterization of nanocrystalline powders of Cu–Zn alloy by wire electrical explosion method." *Materials Science and Engineering: A* 307(1): 190–194.

Xie, L., Jordan, E. H., Padture, N. P., and Gell, M. 2004a. "Phase and microstructural stability of solution precursor plasma sprayed thermal barrier coatings." *Materials Science and Engineering: A* 381(1): 189–195.

Xie, L., Ma, X., Jordan, E. H., Padture, N. P., Xiao, D.T., and Gell, M. 2004b. "Deposition mechanisms of thermal barrier coatings in the solution precursor plasma spray process." *Surface and Coatings Technology, Proceedings of the 30th International Conference on Metallurgical Coatings and Thin Films,* 177–178 (January): 103–107.

Xie, L., Ma, X., Jordan, E. H., Padture, N. P., Xiao, D. T., and Gell, M. 2004c. "Deposition of thermal barrier coatings using the solution precursor plasma spray process." *Journal of Materials Science* 39(5): 1639–1646.

Xie, L., Chen, D., Jordan, E. H., Ozturk, A., Wu, F., Ma, X., Cetegen, B. M., and Gell, M. 2006. "Formation of vertical cracks in solution-precursor plasma-sprayed thermal barrier coatings." *Surface and Coatings Technology* 201(3): 1058–1064.

Yamada, K. 2000. "Synthesis of tungsten carbide by dynamic shock compression of a tungsten–acetylene black powder mixture." *Journal of Alloys and Compounds* 305(1): 253–258.

Ye, T., Guiwen, Z., Weiping, Z., and Shangda, X. 1997. "Combustion synthesis and photoluminescence of nanocrystalline Y_2O_3: Eu phosphors." *Materials Research Bulletin* 32(5): 501–506.

Young, R. M. and Pfender, E. 1985. "Generation and behavior of fine particles in thermal plasmas—a review." *Plasma Chemistry and Plasma Processing* 5(1): 1–37.

Zhang, L., Ranade, M. B., and Gentry, J. W. 2002. "Synthesis of nanophase silver particles using an aerosol reactor." *Journal of Aerosol Science* 33(11): 1559–1575.

18 Multilayer Encapsulation Techniques

Sayantani Dutta, J.A. Moses, and C. Anandharamakrishnan

CONTENTS

18.1 INTRODUCTION

Encapsulation is a useful technique for long-term preservation and site-specific targeted delivery of bioactives. The encapsulate consists of two parts, the inner "core" and the outer "coat." The core is also known as the active component, fill, internal phase, or payload phase, while the coat is synonymously called the coating, membrane, shell, carrier material, wall material, external phase, or matrix. The technique of encapsulation entails surrounding the core material (the active constituents) with a coat. The coat renders the core long-term protection against environmental hazards. This is the procedure for "single layer encapsulation" (SLE), which comprises entrapment of the core in an individual coat or a composite of two or more coats, acting as one (through one-time coating) (Figure 18.1a).

The SLE technique is well-developed and accepted in pharmaceutical, nutraceutical, food, and cosmetic industries, and therefore is the most-used technique. In food products, it is used for the aroma compound vanillin (Hundre et al., 2015), *Lactobacillus plantarum* (Rajam et al., 2012; Dolly et al., 2011; Rajam et al., 2015; Rajam and Anandharamakrishnan, 2015), hydroxycitric acids (Ezhilarasi et al., 2016; Parthasarathi et al., 2013; Ezhilarasi et al., 2014; Ezhilarasi et al., 2013), DHA (Karthik and Anandharamakrishnan, 2013, 2016a,b), green tea polyphenols (Pasrija et al., 2015; Bhushani et al., 2016, 2017), and vitamin E (Parthasarathi and Anandharamakrishnan, 2016; Parthasarathi et al., 2016). SLE is basically a reservoir type of encapsulation technique, with the coat (capsule) encasing the core material. Accordingly, the technique is also called single-core,

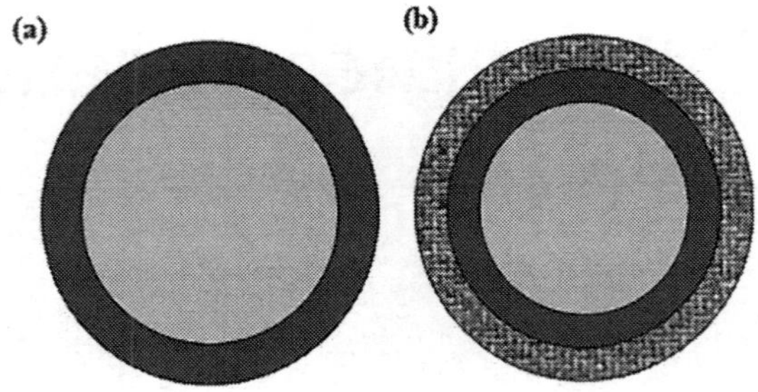

FIGURE 18.1 Single coat encapsulate (a) and multi coat encapsulate (b). (Anandharamakrishnan, 2015.)

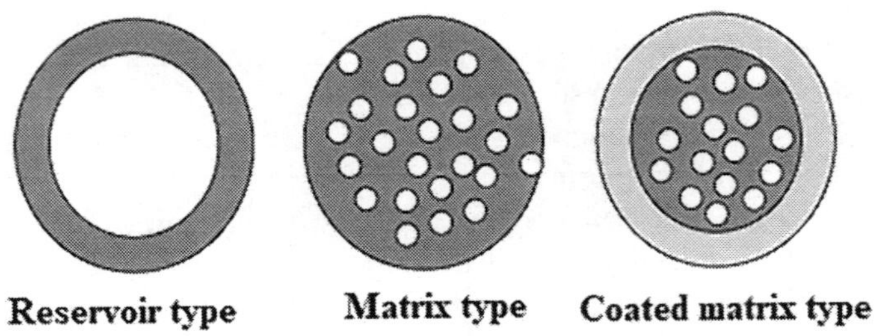

FIGURE 18.2 Reservoir type (left), matrix type (middle), and coated matrix type (right) encapsulates. The active component is indicated in white, and the carrier material in gray. The active component in the matrix type of encapsulates might be in the form of tiny droplets or be dispersed at the molecular level throughout the particle. (Zuidam and Shimoni, 2010.)

mono-core, or core-shell type. Besides single-core, multi/poly-core composition of the core are also possible in SLE. Another type is the matrix type encapsulation, wherein the core material is more homogenously distributed within the shell, with parts of the core being also present at the surface of the coat, in contrast with the former type (Figure 18.2). Predominantly, encapsulates in either form are spherical in shape, but they also can be cylindrical, oval, or in irregular shapes (Zuidam and Shimoni, 2010).

"Multilayer encapsulation" is an augmented approach of the conventional encapsulation process (Figure 18.1b). This technique was first reported by researchers Lim and Sun in the 1980s for the encapsulation of pancreatic islets, employing alginate/polylysine coating (Lim and Sun, 1980). This method is alternatively known as "multilayer emulsion." At the outset, the said method was primarily aimed at meeting the medical requirements of clinical studies of encapsulated drugs. This was felt to be necessary since unilayer encapsulation of drugs at that time met with difficulties in matching the medical requisites, such as absolute biocompatibility or precisely modified permeability, along with a provision for a reasonably high mechanical stability of the resultant complex (Schneider et al., 2001). In medical applications, it has been reported that for an encapsulate, besides delivering efficacy at the targeted site, it is equally important to avoid detrimental effects to the patient's humoral and cellular immunity (de Vos et al., 2006). Understandably, multilayer encapsulation offers greater advantages than unilayer encapsulation and is thus an emerging area of current research.

This chapter aims to collate important findings in multilayer encapsulation and present an explanatory overview of the same, highlighting the various facets, techniques, and utilities of the said encapsulation.

18.2 FABRICATION OF MULTILAYER CAPSULE

There are many materials for the development of multilayers in capsules, such as soy oil coated with β-lactoglobulin and pectin for encapsulation of VOCs (Benjamin et al., 2012); employing spray drying for the design of dry emulsion of pea protein isolate, using an additional layer of pectin (for enhanced stability of the encapsulate) (Gharsallaoui et al., 2012); encapsulation of β-lactoglobulin in carrageenan and gelatin, in the form of primary, secondary, and tertiary emulsion layers, respectively (Gu et al., 2005); preservation of fish oil and citrem dissolved in an acetate buffer and subjected to encapsulation in multilayered chitosan and sodium alginate (Gudipati et al., 2010); entrapment of corn oil in a primary layer of an acetate buffer and a secondary layer of chitosan (Mun et al., 2005); and the employment of sodium dodecyl sulfate and chitosan in aggregate, multilayered for encapsulation of limonene (Djordjevic et al., 2007). Clearly, the materials and their applications of multilayered encapsulation are diverse and offer flexible combinations (Table 18.1). We shall be discussing the versatility and application of multilayer encapsulates on a case-by-case basis in this chapter.

18.2.1 Layer-by-Layer Encapsulation

One of the most widely used encapsulation technique for multilayer encapsulation is "layer-by-layer encapsulation" (LBL). It allows employing natural and/or synthetic polyelectrolytes to achieve ultrathin films, as a coat, with desirable functional properties (Fukui et al., 2010). In the LBL technique, a primary layer of polyelectrolyte (polysaccharide), composed of ionic emulsifiers such as proteins (encasing the core, as an emulsion), gets layered upon by an oppositely charged polyelectrolyte. The combination provides strong electrostatic and repulsive forces that provide stability to the emulsion from factors such as pH, temperature and salt content. The difficulty is to accurately maintain the ratio of layered materials (Guzey et al., 2004; Moreau et al., 2003). In consideration, the emulsion encapsulation has been particularly developed as a favorable technique typically for lipophilic compounds such as drugs, volatile organic compounds (VOCs), and similar items (Bhujbal et al., 2014).

The LBL technique was first reported in the 1960s, when adsorption of charged colloidal particles on the oppositely charged solid surface was attempted for loading cationic molecules, such as surfactants and proteins, on glass (Iler, 1966). However, the utility of the technique in food engineering was brought out in the research of Decher and colleagues, which involved electrostatic multilayer adsorption of charged polymers on solid surfaces. The surface charge reversal of the assembling technique was involved, without requiring special equipment or procedures (Decher and Hong, 1991; Decher et al., 1992).

LBL also employs emulsifiers and aids, for the production of small lipid droplets (upon homogenization, for liposome formation), to avoid particle aggregation and post homogenization. The protocol requires preparation of a basic primary emulsion with small droplets (by homogenization of the mother lipid solution) followed by addition of an oppositely charged polyelectrolyte solution. The electrolyte solution forms a secondary emulsion, with droplets now coated with the bilayer interface. In fact, this procedure could be emulated for multiple layers as well. The process basically exploits the strong electrostatic attraction between the surface and the oppositely charged polyelectrolyte molecules in solution, ensuring that there is no free polyelectrolyte present in the aqueous solution. The method of preparing multilayered colloidal suspension has been shown in Figure 18.3.

This phenomenon of bilayerization is made effective owing to a "charge reversal" mechanism, since the charges at the surface of droplets are outnumbered by those in the polyelectrolyte solution, thereby nullifying any incidence of neutralization. Thus owing to electrostatic repulsion between the adsorbed and the free polyelectrolytes, further deposition is arrested. Even if deposition occurs, it happens atop the existing layer, as a monolayer of the polyelectrolyte. For example, in the complex S–P1–P2, S is the substrate, and P1 and P2 are two oppositely charged polyelectrolytes; and this can be replicated multiple times. The polyelectrolyte present at the outermost layer contributes to the net charge on the overall interface (Decher, 1997). The process of preparation of multilayered encapsulates on a planar surface by LBL technique has been illustrated in Figure 18.4. The encapsulates reportedly have higher stability under environmental stresses such as variable pH, ionic fluctuations and thermal exposure (Guzey and McClements, 2006).

TABLE 18.1
Different Techniques for Multilayer Encapsulation

Method of Encapsulation	Compound Encapsulated	Type of Wall Material	Mean Size of Encapsulate	References
Layer-by-layer assembly	Insulin	Chitosan, heparin	~1 μm	Song et al. (2014)
	Oil in water emulsion	Ovalbumin and highmethoxyl pectin	—	Humblet-Hua et al. (2011)
	Sunflower oil	Poly(allylamine hydrochloride), poly(sodium 4-styrenesulfonate)	Thickness of shell 14 nm to 26 nm	Sadovoy et al. (2011)
	Indomethacin	Multilayer film of sodium alginate and chitosan	—	Wang et al. (2007)
	Tuna oil	Chitosan and lecithin	260–440 nm	Klinkesorn et al. (2005)
	Horseradish peroxidase	Dextran sulfate and protamine	8–10 μm	Balabushevich et al. (2003)
	Glucose oxidase	Polyallylamine/polystyrenesulfonate	5–50 μm	Trau and Renneberg (2003)
	Chymotrypsin	Sodium poly(styrene sulfonate)/poly(allylamine hydrochloride)	100–300 nm	Balabushevitch et al. (2001)
	Pyrene and fluorescein diacetate	Poly(allylamine hydrochloride), poly(sodium 4-styrenesulfonate)	25–30 nm	Caruso et al. (2000)
	Polysterenesulfonate, melamine formaldehyde	Tb^{3+}/polysterenesulfonate, 4-pyrene sulfate/polyallylamine	10–11 μm	Radtchenko et al. (2000)
	Cytochrome C	Alginate microcapsules coated with poly(N-vinylamine) and poly(acrylic acid) or water soluble anionic cellulose derivatives	—	Rilling et al. (1997)
Phase coacervation	Linseed oil	Chitosan and sodium dodecyl sulfate	1.99–3.51 μm	Chatterjee et al. (2014)
Complex coacervation	Aspartame	Gelatin and gum Arabic	84–102 μm	Rocha-Selmi et al. (2013)
	Biphenyl	Gelatin and acacia	>250 μm	Ijichi et al. (1997)
Spray drying	Green coffee oil	Lecithin and chitosan	1.35–3.70 μm	Carvalho et al. (2014)
	Fish oil	Lecithin and chitosan	13.7 μm	Serfert et al. (2011)
	ω-3 fatty acids	Corn oil, chitosan, EDTA	25 μM	Shaw et al. (2007)
	Tuna oil	Lecithin and chitosan	5–30 μm	Klinkesorn et al. (2006)
Gelation	Ampicillin	Chitosan-alginate with and without tripolyphosphate	450–885 μm	Anal and Stevens (2005)
	Bovine serum albumin	Chitosan, alginate	563–920 μm	Anal et al. (2003)

(Continued)

TABLE 18.1 (CONTINUED)
Different Techniques for Multilayer Encapsulation

Method of Encapsulation	Compound Encapsulated	Type of Wall Material	Mean Size of Encapsulate	References
Electrospraying	Green fluorescent polystyrene microspheres, albumin-fluorescein isothiocyanate conjugate, tetramethylrhodamine isothiocyanate-dextran, yeast cells	Alginate/chitosan microcapsules	130–80 μm	Fukui et al. (2010)
Electrospinning	α-tocopherol	Whey protein isolate, zein, soy protein isolate; coated on wheat gluten film	Coating: ~1.15 ± 0.05 mg/cm	Fabra et al. (2016)
Liposomal encapsulation	Anthocyanin	Chitosan and pectin	200 nm	Gibis et al. (2014)
Fluidized bed drying	*Lactobacillus acidophilus*	sodium alginate-calcium chloride matrix; egg albumin and stearic acid composite	450 μm	Pitigraisorn et al. (2017)
Polyelectrolyte multilayer	Glucose oxidase	Calcium alginate	20–30 μm	Brown et al. (2005)

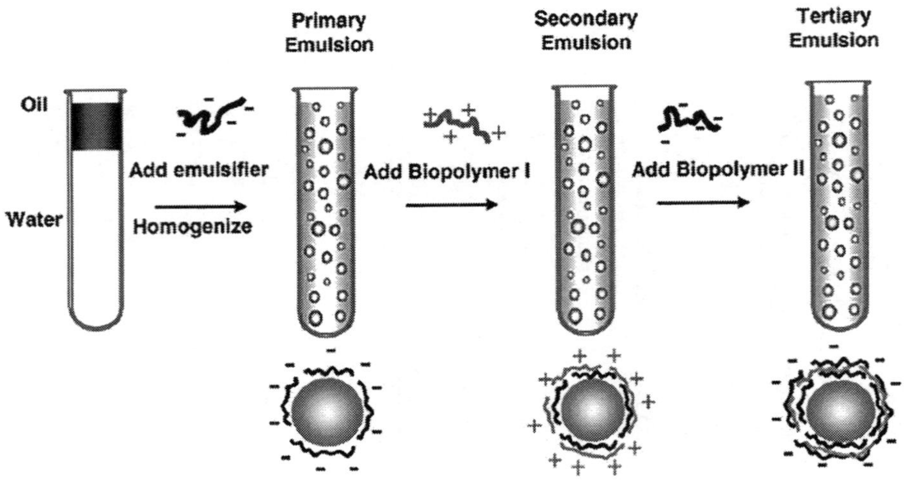

FIGURE 18.3 Preparation of oil in water emulsion by layer-by-layer method. (Ogawa et al., 2004.)

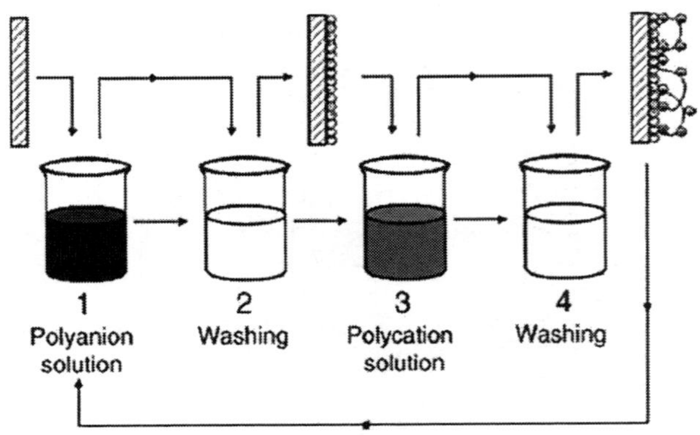

FIGURE 18.4 Layer by layer deposition of oppositely charged polyelectrolytes on a planar surface. (Guzey and McClements, 2006.)

Improved oxidative stability of limonene in M-O/W emulsions has been found as a result of LBL encapsulation (Djordjevic et al., 2007). Research has also revealed that in LBL emulsions, owing to a complex interface of the bilayers, there is effective regulation of the release of active constituents from the complex. Further, the release can also be suitably regulated by modifications in the ingredients of the emulsion, as well as by manipulating the physical properties (e.g. permeability, thickness, charge). Besides, employing specific environmental factors, such as a change in the pH, disengage the secondary layer, owing to weaker electrostatic attraction (Gu et al., 2005).

Despite the advantages, the main cause of concern for the encapsulates prepared by LBL technique is averting droplet adhesion, which affects the stability of the microcapsules. This could be achieved by employing filtration, saturation, and centrifugation methods (Voigt et al., 1999). It should also be ensured that the quantity of the polyelectrolyte concentration is enough to monocoat all the droplets, but not high to cause depletion flocculation. Caution is also warranted so that coating occurs quickly, particularly in scenarios wherein inter droplet collision occurs. Also, the repulsion among the coated particles should be sufficient to avoid droplet coalescence. In addition, it is worth considering that the size and concentration of both polyelectrolytes and emulsion droplets, and other factors such as pH, ionic strength, and dielectric constant; are suitably fixed for stable encapsulates. In fact, for incorporation of multiple layers, the additional unit operation of homogenization or ultrasound treatment is advisable, after the addition of each layer, to further eliminate risks of aggregation.

Stable microemulsions have been prepared by several methods (Voigt et al., 1999), some of which have been summarized below:

1. *Saturation method*: This method requires the addition of sufficient polyelectrolytes to cover the surface of all the microparticles present in the mother aqueous solution, i.e. to saturate the particles with polyelectrolyte, hence the name. This technique necessitates accurate estimation of polyelectrolyte quantity required for achieving saturation. The methods for ascertaining this are diverse and include processes such as estimation of ζ potential, thus avoiding chances of bridging flocculation, as well as depletion flocculation. The latter is particularly risky and originates from too high a concentration of the polyelectrolytes. Often there develops an osmotic force overcoming the much-desired repulsion among the coated particles, causing inter-particle aggregation.
2. *Centrifugation method*: This method uses centrifugal force to separate the unabsorbed polyelectrolytes from the resultant solution. These unabsorbed particles are those that are in excess of that required to coat the particles in the colloidal mother solution. In a bi or multilayered encapsulation, this method could be repeated many times, ensuring no free polyelectrolyte of a particular type left unabsorbed, before introducing the next type.
3. *Filtration method*: This is another technique for separation of unbound free polyelectrolytes in the resultant solution. It involves filtering the excess polyelectrolytes. The process is pressure driven and involves the addition of a buffer (avoiding proximity among polyelectrolyte adsorbed particles). Post-filtration, the unbound polyelectrolytes are flushed out of the semipermeable membrane of the filter. Besides being a more benign technique compared to centrifugation, this method offers the flexibility of having the densities of the aqueous solution and the polyelectrolyte to be similar or, at least close (Voigt et al., 1999).

18.2.1.1 Encapsulation and Release of Multilayer Food Bioactives in LBL Encapsulation

LBL encapsulation is widely popular for its thin film, especially because that film forms firm and insoluble complexes around the core, such as anionic emulsifier. There are several factors which contribute to the stability of the emulsion and its layers. The key factors include elements such as pH, the ionic strength of the solution, solvent characteristics, emulsifier type and characteristics, and polyelectrolyte type and characteristics. The technique offers the flexibility to choose between synthetic and natural polyelectrolytes, producing a new category of ultrathin films. The attributes of such films viz. a controlled release of bioactives, self-rupturing ability, compatibility with biological matrices, are achieved through variations in composite(s) of polyelectrolytes (Bhujbal et al., 2014).

The technique also offers ease of release of bioactives, simply by modification in pH, and ionic strengths into the media intended. Augmentation of this technique has been attempted to design encapsulates using a "sacrifice template." However, the literature shows limitations in the same emerging from low encapsulation efficiency, principally owing to difficulties in the management of electrostatic interaction between aqueous and organic phases when using polyelectrolytes (Orr et al., 2006; Elosegui-Artola et al., 2014; Garcia and Garcia, 2014). Another endeavor with curcumin-in-β-cyclodextrin-in-nano magnetoliposomes has demonstrated considerable encapsulation efficiency (71%) and high antioxidant activity, compared with conventional liposomes and iron oxide nanoparticles (Aadinath et al., 2016). Similar success has also been achieved for liposome encapsulated cyclodextrin inclusion complex for hydrophobic drugs, contributing to the encapsulates' higher drug loading, encapsulation efficiency, and efficient release of the drug (Chen et al., 2007).

The LBL encapsulates find application in various foods, owing to unique functionalities which include the ability to avoid post-layering inter-particle aggregation, aversion to lipolysis, and controlled release of the core or the active compound. Multilayered protection of the active compound from environmental stresses including fluctuations in ionic equilibria, pH, mechanical, thermal, and electrochemical shocks, is another advantage over single layer emulsions. In one of the studies, researchers administered *Lactobacillus acidophilus*, a probiotic, as a nutritional supplement to orange juice. Owing to risks of bacterial death during pasteurization, the microorganism was encapsulated by the emulsification technique in multilayers of calcium alginate. Authors observed that in contrast to viability loss in free bacteria at 5 log cycles post-pasteurization, the loss of bacteria in multilayer encapsulate was only 2.5 log cycles (Dowlatabadi et al., 2016). This endeavor was therefore successful in preserving bacterial viability, even post-pasteurization.

18.2.2 Multilayered Encapsulation by Complex Coacervation

Another technique for multilayered encapsulation that has been explored by many authors is coacervation. The process is accomplished in three steps viz. formation of three immiscible chemical phases, deposition of coating layers on the core material, and strengthening of the coat under continuous agitation. The technique has been classified into simple and complex coacervation. The first method necessitates the addition of a hydrophobic substance to a colloidal solution, making the resultant solution biphasic. The second method involves modification to the acid–basic nature of the solution. Depending upon the choice of acidity or alkalinity, the microparticles are produced, but only beyond a critical pH of that system. Complex coacervation offers the flexibility of involving multiple colloids as multilayers, when needed (Kiyoyama et al., 2003; Sinha and Trehan, 2003). The complex coacervates are insoluble in water, thermo-resistant, and allow a controlled release of the core (Dong et al., 2011).

In one of the studies, the encapsulation of aspartame was conducted by coacervation. Aspartame is a methyl ester of the aspartic acid/phenylalanine dipeptide, is about 200 times sweeter than sugar, and has reportedly been protected from environmental rigor by double emulsion, along with subsequent complex coacervation, for enhanced stability, sweetness, and sustained release (Rocha-Selmi et al., 2013). The said endeavor was conducted on sunflower oil acting as a primary emulsion (to dissolve aspartame), using gelatin and gum arabic as the wall materials. This formulation was found to be stable even at 80°C, and provided slow and sustained release of the core, as desired (Figure 18.5). The authors opined that the said encapsulation could allow application of encapsulated aspartame for high-temperature applications such as baking and frying, which was otherwise difficult. Also, researchers have shown that the casein-protein hydrolysate could be designed with double emulsion and coacervation technique (Mendanha et al., 2009).

In another study, a chitosan-based multilayered capsule was designed for the loading of indomethacin microcrystals by coacervation (Ye et al., 2005a). In a similar study, a multilayer of chitosan and sodium dodecyl sulfate (SDS) was investigated for entrapment of linseed oil by combined emulsification and phase coacervation techniques, leveraging the ionic interaction between chitosan and SDS (Chatterjee et al., 2014). To begin with, these authors prepared an emulsion (o/w) and employed SDS as an anionic emulsifier. The method involved rinsing the microparticles with deionized water to reduce the viscosity of the microcapsules, as well as mitigating chances of repulsion among the positively charged amine group of chitosan that were located at the surface of the encapsulates. The rinsing reportedly improved particle dispersion in the system after 11 cycles of addition of chitosan and SDS. The microparticles were thereafter subjected to further treatment by an alkali, to ensure solidification of the outer shells and enhancement of tensile strength of the capsules. Post-addition, they were applied on woven polyester by padding (Chatterjee et al., 2014).

Another study on the same technique focused on encapsulation of biphenyl with multilayered gelatin and acacia by coacervation (Ijichi et al., 1997). The work showed good results on the dependency of the stability of multilayers on chitosan and SDS (Figure 18.6). The authors have obtained encapsulates of size >250 μm by multilayered complex coacervation. They have reported the radial growth of coacervation on the oil to be stepwise, in the form of a tentacle (Figure 18.7). This observation affirmed the importance of the affinity of the coacervation material for the core, i.e. the oil in this case, for radial expansion of the complex (Ijichi et al., 1997).

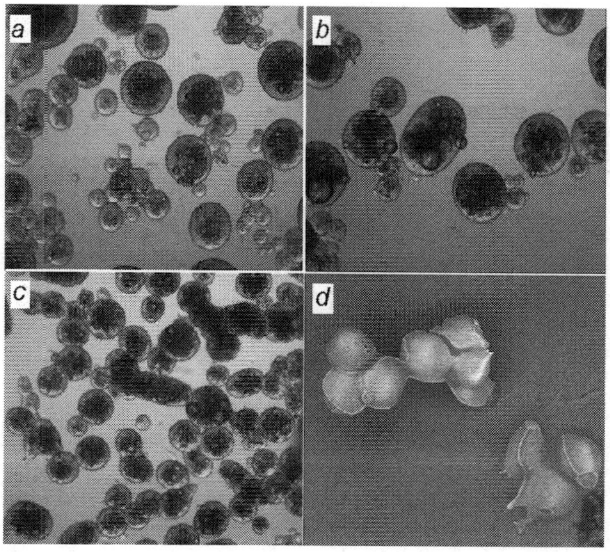

FIGURE 18.5 Microscopic images of aspartame microcapsules: (a) Formulation A (2.5% gelatin and gum arabic and 50% core material); (b) Formulation B (2.5% gelatin and gum arabic and 75% core material); (c) Formulation C (2.5% gelatin and gum arabic and 100% core material); (d) SEM image with 500× magnification. (Rocha-Selmi et al., 2013.)

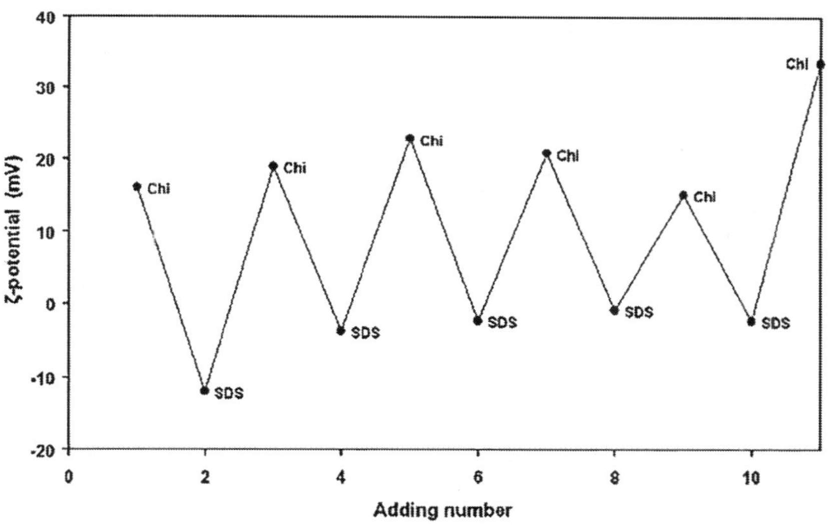

FIGURE 18.6 Effect of multilayer of chitosan and SDS on the stability of the encapsulate of linseed oil by coacervation technique, affirmed by zeta potential. (Chatterjee et al., 2014.)

18.2.2.1 Limitations of the Coacervation Technique

Despite being a likely candidate for multilayered encapsulation, coacervation techniques unfortunately suffer from several limitations. Substances such as glutaraldehyde in the wall are often used for encapsulation, which is undesirable for certain applications. Besides, the process itself is cumbersome, complex, and cost-inefficient, compared with other similar techniques. In fact, certain polymers that are often employed in this technique are unsafe for the design of consumer products. To overcome these limitations, the incorporation of enzymes such as transglutaminase in the coat

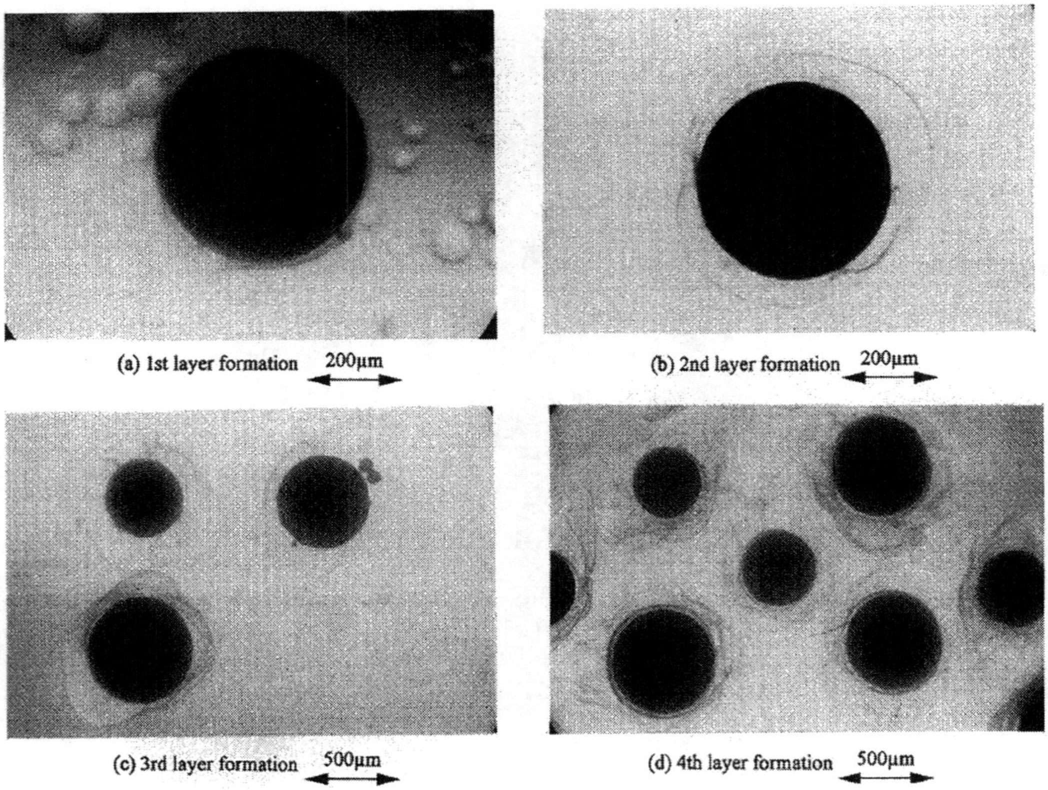

FIGURE 18.7 Stepwise membrane formation in multilayered complex coacervate. (Ijichi et al., 1997.)

has been recommended, to cross-link proteins in the shell (Soper and Thomas, 2000). According to these authors, the enzyme when added to the encapsulation solution at 10°C, neutral pH reportedly requires a little over 16 hours for the formation of a hardened coat on the core being entrapped.

18.2.3 Spray Drying Technique for Multilayered Encapsulates

Spray drying, an extensively practiced method of microencapsulation, has also been successfully employed for the development of multilayered encapsulates, as shown by a number of studies. The basic approach for emulsion preparation remains the same. Encapsulation by the said technique has been reported for corn oil-in-water emulsion (5%, w/w) stabilized by Tween 80, using maltodextrin as wall material (Klinkesorn et al., 2004).

In one of the studies, tuna oil-in-water emulsion has been spray dried with an electrostatic layer-by-layer coating of lecithin and chitosan, to provide oxidative stability to the oil (Klinkesorn et al., 2006). SEM micrographs illustrate the morphology of the powder particles thus formed (Figures 18.8 a and b). The spray-dried powder has a diameter of 5–30 μm, <3% moisture, encapsulation efficiency >85%, and rapid dispensability (<1 min). The said authors have also proven the utility of corn syrup as a coat for encapsulation of tuna oil during spray drying. This microencapsulated tuna oil was found to have more oxidative stability than bulk tuna oil when stored at 37°C.

When employing a two-layered interfacial membrane comprised of a composite of chitosan and lecithin for tuna oil, the resultant emulsion was found to be both cationic and thick. This aggregate reportedly restricts lipolysis post-spray drying, compared with the non-encapsulated oil (Klinkesorn et al., 2005). Another experiment for encapsulating ω-3 fatty acids using corn oil (5–20%), lecithin, and chitosan, with a small quantity of EDTA (25 μM), enhanced oxidative stability of the oil

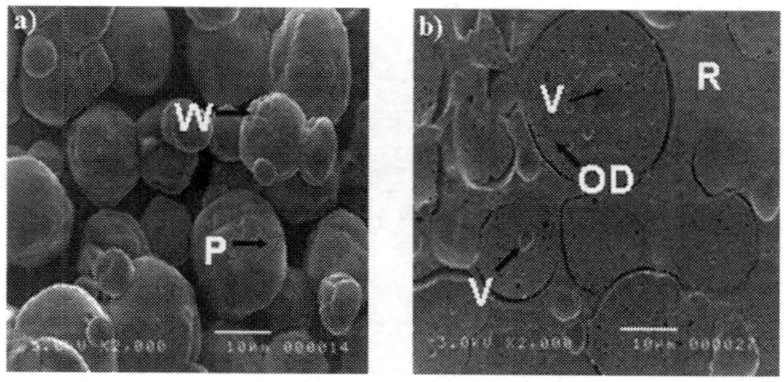

FIGURE 18.8 SEM photomicrographs of spray dried tuna oil at 165°C showing (a) outer structure and (b) inner structure, W: wrinkle, P: pore, V: void, R: resin, OD: oil droplet or air cell. (Klinkesorn et al., 2006.)

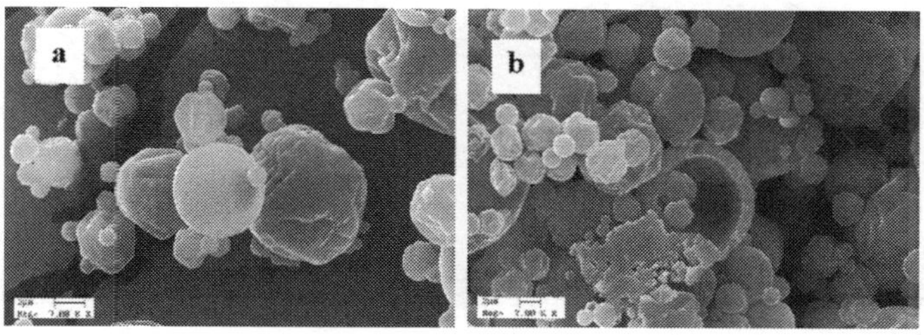

FIGURE 18.9 SEM images of spray dried coffee oil showing (a) outer and (b) inner morphology, when using corn syrup and Hi-Cap 100 E composite (1:1) for microencapsulation. (Carvalho et al., 2014.)

(Shaw et al., 2007). Also, spray dried squid oil in an LBL multi-coat of gelatin, maltodextrin, and caseinate has been found to be successful in suitably protecting the oil. In this study, Avicel was also added as an additive to further enhance the stability. Synergism between α-tocopherol in the oil and coats provides durability for the encapsulates (Lin et al., 1995).

A successful attempt has also been made at encapsulation of emulsified green coffee oil by spray drying (Carvalho et al., 2014). These authors stabilized the green coffee oil emulsion by an LBL technique using lecithin and chitosan as a coat. The optimized emulsion also contained corn syrup and modified Hi-Cap 100 (50:50) for stability. The said stabilized emulsion had particles of a diameter between 1.35–3.70 μm and ζ potential from −2.25 to +40.40 mV. The spray dried powder had moisture between 0.40%–2.49% (wet basis) and achieved 86%–97% encapsulation efficiency. The best results in terms of encapsulation efficiency (97.5%) and protection from oxidative rancidity were provided by the composite coating of corn syrup and Hi-Cap 100 E composite (1:1) (Figure 18.9) (Carvalho et al., 2014).

Researchers have also spray dried fish oil emulsion in a multilayered coat composed of lecithin and chitosan (Serfert et al., 2011). Interestingly, these authors reported lower encapsulation efficiency when using a multilayered coat, compared with a single coat. They opined that coat was responsible for the lower oxidative stability of the encapsulate.

18.2.4 Liposomal Encapsulation

Liposomal encapsulation is yet another technique widely employed for the encapsulation of bioactives. The most stable liposomes are known to be prepared with lecithin, cholesterol, and negatively charged phospholipids. Multilayered liposomal encapsulation has been investigated by many researchers to obtain improved encapsulation efficiency and storability of the active ingredients. In one of the studies, anthocyanin from hibiscus was subjected to multilayered encapsulation consisting of layers of chitosan and pectin as an interface (Gibis et al., 2014). The work brought out the

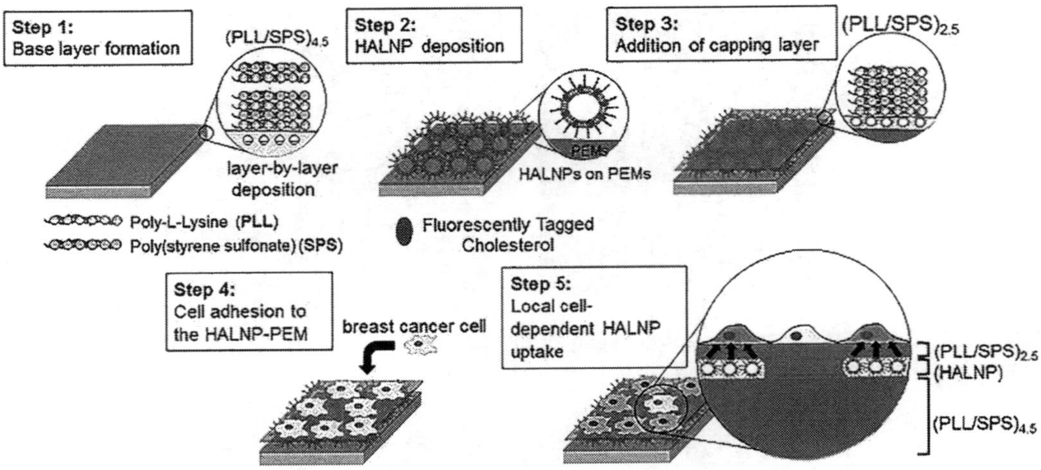

FIGURE 18.10 Flow sheet illustration of ionic driven embedment of hyaluronic acid coated lipid nanoparticles (HALNPs) in the PEM Platform (HALNP-PEM). (Hayward et al., 2015.)

suitability of the said wall materials in achieving encapsulation with an increased particle diameter of 65 nm after the addition of the first layer of chitosan, which further increased to 200 nm after the incorporation of the second layer of pectin. Both of these particle sizes were higher than the size of the control liposome particle. The encapsulates (size <46 nm) had an efficiency between 63%–72%, ζ potential from −2.25 to +40.40 mV, and were stable for about 30 days (under room temperature). In turn, the hibiscus extract, owing to the inherent antioxidants, also conferred improved stability to the soya lecithin liposome.

In another investigation, multilayered liposomes have been prepared by dehydration/rehydration cycles of the lyophilized material, ascorbic acid; using egg phosphatidylcholine, an edible phospholipid (Kirby et al., 1991). The encapsulation yielded significantly improved stability of ascorbic acid and paved the path for incorporation of the same in edible emulsions without decomposition during food processing.

Estimation of the efficiency of one component multilayered liposome (by an LBL technique) with potential biomedical applications, using synthetic polyvinyl alcohol (PVA) has also been recently conducted (Zasada et al., 2015). The researchers have shown considerable adsorption capacity of the film produced by functional modifications therein through cationic and anionic derivatives. The liposomes formed were free from any cytotoxicity and safe for applications such as the delivery of pharmaceuticals and other biologicals. Another study on the therapeutic application of multilayered liposomal entrapment reported use of hyaluronic acid covalently functionalized liposome (HALNPs) embedded into polyelectrolyte multilayer (PEM) through ionic stabilization (Hayward et al., 2015) (Figure 18.10). The study claimed to be the first of its kind, allowing delivery of both hydrophobic and hydrophilic drugs and presented a report on the encapsulation of doxorubicin, a chemotherapeutic drug, as an example.

Researchers have worked on the therapeutic potency of iRGD (a tumor-penetrating peptide) cross-linked multilayer nanoliposomes for improved drug delivery (Liu et al., 2013). They employed cross-linked multi-lamellar liposomal vesicle (cMLV) with iRGD. The intervention displayed enhanced antitumor activity in breast cancer cells, including efficacy on multidrug-resistant strains, owing to higher penetrability of the conjugate.

More recently, the possibility of developing albumin-encapsulated multilayered liposomes has been revealed in the research conducted on encapsulation of bovine serum albumin (BSA) (Okamoto et al., 2018). The efficiency of eight layered BSA-liposome in loading hydrophobic drugs into the inner layers has been the novelty of the study, which is claimed to be a pioneering work. Authors found that water solubility of a drug-enhanced, when bound to albumin at the inner aqueous core of the liposome, rendered significantly high drug loading capacity.

18.2.5 Fluidized Bed Drying

Encapsulation by fluidized bed drying is one of the approaches for encapsulation of bioactives. A study on *Lactobacillus acidophilus* (a heat-sensitive probiotic) showed improved stability upon moist-heat exposure when the bacteria were subjected to a combination treatment of fluidized bed drying and electrospinning (Pitigraisorn et al., 2017). The experiment involved first coating the said core with sodium alginate-calcium chloride matrix, and then topping it with egg albumin and stearic acid composite coat, as a further shell coating. In fact, without the shell coating, the microstructure could have collapsed during fluidized bed drying, as experienced by the authors in the control set. During drying of this layer, cassava pearls were used as a drying aid, and cassava starch was additionally applied as a layer onto encapsulates. The granules from the pearls during drying adhered to the surface of the particles.

The encapsulate had an average particle size of 450 μm with an encapsulation efficiency of more than 90%, cell payload 12.3 log CFU/g_{dry} of microorganism (for coat of egg albumin: stearic acid: 1: 0.25). SEM images of the particles displaying their morphology are shown in Figure 18.11. Authors observed that after the moist heat treatment (70°C, 100% RH, 30 min) and drying operation, a higher number of bacterial colonies survived when the amount of stearic acid was increased in the composite. Interestingly, when exposed to moist heat, the multilayered encapsulate showed a loss of only 0.6 log CFU/g of the bacteria. This result was achieved for the first time for the bacterial strain, thereby providing a means to produce thermo-tolerant bacterial encapsulate. The authors

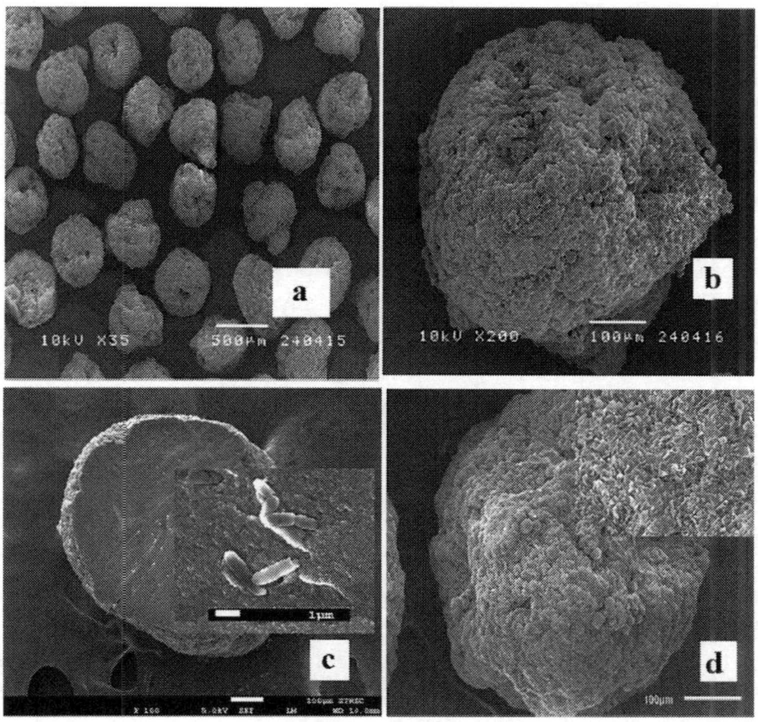

FIGURE 18.11 SEM photographs of encapsulated *L. acidophilus* showing egg albumin and stearic acid coat (1: 0.25) at (a) 35×, and (b) 200× magnification; (c) cross sectional view of a microparticle at 100× magnification (image in inset was at 100× magnification); and (d) encapsulate after being exposed to moist heat treatment at 200× magnification. (Pitigraisorn et al., 2017.)

have opined that such an endeavor could be useful for fortifying probiotics in thermo-processed foods and feeds as well.

18.2.6 Atomic Layer Deposition (ALD)

Atomic layer deposition (ALD) is a relatively new technique that offers the formation of ultrathin films, one atomic layer at a time. The ingredients are added one by one, with intermittent pump/purge cycles. As a result, it produces a self-saturating surface reaction, limited to a single layer on the surface exposed. The technique produces a film in each layer, and repetitive layers produce multilayered films. The impurity content during layer deposition is also strikingly low (Andrade et al., 2012).

18.2.7 Ionotropic Gelation

Ionotropic gelation technique uniquely allows production of encapsulation as beads, with soft wall materials, which can be subsequently dried for hardening (Lim et al., 1997). Encapsulation by ionotropic gelation has been carried out for entrapment of ampicillin in chitosan-alginate multilayered beads (Anal and Stevens, 2005). Reportedly, the multilayered beads were composed of alginate encasement around ampicillin, reinforced with chitosan and cross-linked with tripolyphosphate (TPP). It has been reported that chitosan aids maintenance of bead structure, and impermeability so that the drug does not leak out, besides helping to achieve the necessary swell in the bead wall to compensate for the difference in osmotic pressure when injected. Compared with the chitosan-alginate and chitosan-alginate-TPP beads included in this study, the release of the antibiotic from the said multilayered encapsulate in simulated intestinal fluid (SIF) (at pH 7.5) was more sustained, and did not require any pre-incubation with simulated gastric fluid (SGF) (at pH 1.2). The study also highlighted the utility of the said encapsulation technique for sustained release of oral, low molecular weight, highly hydrophobic compounds. The capsule released only 30% of the drug in 24 hours. In fact, the authors opined that such a capsule can sustain the low pH of gastric fluids and prevent the substantial loss of the drug (bioactive) otherwise experienced.

In a similar study, chitosan reinforced alginate beads have been used to encapsulate bovine serum albumin (BSA) (Anal et al., 2003). The inclusion of chitosan was found to provide higher uptake of BSA in the capsule during entrapment. The release of BSA was analyzed in both SGF and SIF. Interestingly, even incorporation of pepsin and pancreatin into SGF did not alter the leakage or any alteration in the beads formed. The beads released only 40%–50% of BSA into SIF in more than 12 hours, without prior incubation in SGF; which was a unique achievement, as reported. Even after 24 hours, the classical beads got eroded in SIF, while those with chitosan were remarkably intact. This data provided a strong justification to consider such encapsulates for dispensing proteins, protein drugs, and related therapeutics. The encapsulates can also be employed for site-specific delivery and protein delivery systems.

18.2.8 Electrospraying

The electrospray technique is another method engaged in nanoencapsulation, wherein the feed ejects as a mist of droplets from the equipment employed when being transformed into a gas phase at a high speed and then falls onto a collector vessel (Figure 18.12). The technique produces uniformly dispersed microparticles when the core is mixed with anionic and cationic polyelectrolytes. The process also provides high encapsulation efficiency of the active compounds. The said technique was first adopted by Bugarski et al. for the preparation of size-controlled microparticles (Bugarski et al., 1994). The polyelectrolytes could be materials such as poly(ethyleneimine) (PEI), poly(acrylic acid) (PAA), chitosan, alginate for encapsulation of active materials such as bioactives, enzymes, proteins, and live cells (Ye et al., 2005b; Lee et al., 1997; Nedović et al., 2001; Manojlovic et al., 2006;

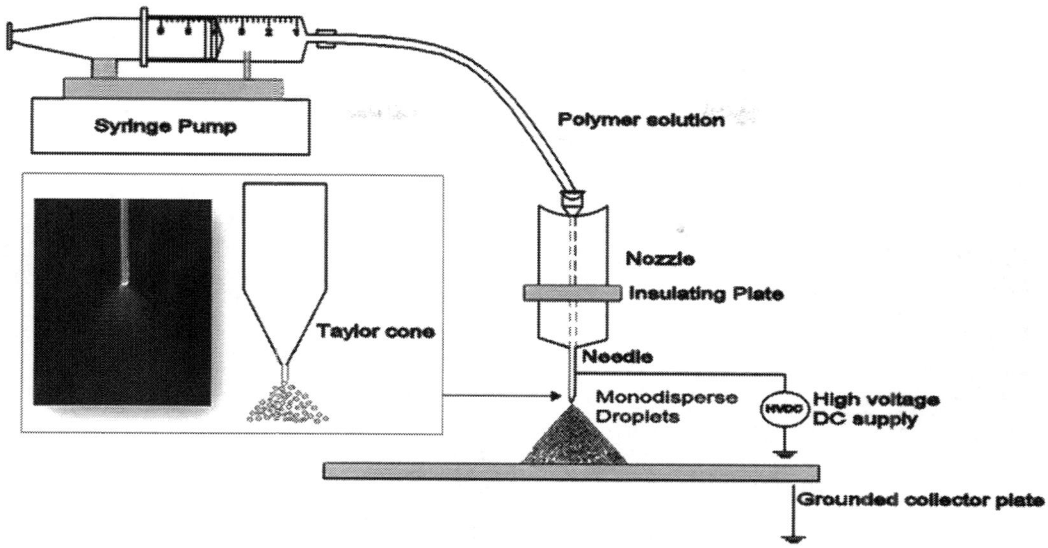

FIGURE 18.12 Schematic representation of electrospraying setup. Inset shows the digital image and representation of electrospraying process. (Bhushani and Anandharamakrishnan, 2014.)

Xie and Wang, 2007; Pommersheim et al., 1994). In fact, the polyelectrolytes at times themselves have therapeutic values that have been highlighted in various reports (Huang et al., 2008; Volodkin et al., 2008; Wood et al., 2005).

Few scientists have reported that polyelectrolytes such as chitosan and alginate could be used for encapsulation by electrospray technique (Fukui et al., 2010; Ye et al., 2005b). Fukui and team have successfully microencapsulated proteins, dextran and polymeric microspheres by this technique with a laudable efficiency (>99%). The produced microparticles are of about 100 μm. The authors have also carried out the encapsulation of yeast cells through this technique of encapsulation (Fukui et al., 2010). Other materials for encapsulation of bioactives by electrospraying include whey protein isolate, whey protein concentrate, soy protein isolate, egg albumen, collagen, gelatin, zein, and casein (protein based), as well as cellulose and its derivatives, and dextran (polysaccharide based) (Bhushani and Anandharamakrishnan, 2014).

Research has also indicated that uniformly sized droplets of feed are produced when it is emitted from the syringe and it does not experience the common phenomenon of droplet coalescence during the transit (atomizer to drying unit). This observation has been attributed to the inter-particle repulsion among the charged droplets (Fukui et al., 2010). The repulsion among droplets is voltage-dependent, along with feed drop size and the needle to collector distance. The dried particle size reduces as the distance reduces (and voltage increases). Moreover, the particle size also depends on applied electrical potential, flow rate of the feed, molecular weight, and the microstructure of the polymer and ambient conditions (Bhushani and Anandharamakrishnan, 2014).

18.2.9 Electrospinning

Electrospinning is a widely used technology for the production of dried fibers (encapsulated) in the submicron range using natural and synthetic polymers. Electrospun fibers have versatile applications, including nanocatalysis, tissue engineering scaffolds, protective clothing, filtration, biomedical, pharmaceutical, optical electronics, healthcare, biotechnology, defense and security, and environmental engineering (Bhardwaj and Kundu, 2010).

The electrospinning or "electrostatic spinning" technique was first conducted by Rayleigh in 1897 (Bhardwaj and Kundu, 2010). In this method, the polymer solution passes through a capillary

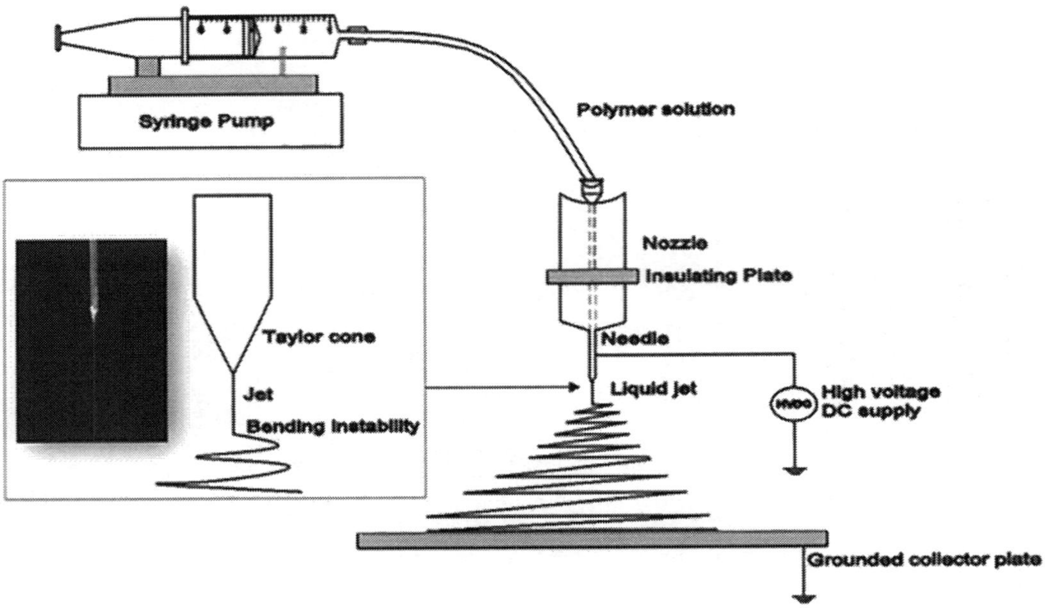

FIGURE 18.13 Schematic representation of electrospinning setup. Inset shows the digital image and representation of electrospinning process. (Bhushani and Anandharamakrishnan, 2014.)

maintained at a high voltage. The presence of two electrostatic forces (repulsion and Coulombic force) at the tip of the capillary renders distortion in the droplet of the feed solution, resulting in the formation of a conical shape (a Taylor cone) of the emanating droplets. Due to the counteraction of the electrostatic force and surface tension, the charged polymer solution moves in the form of jet towards the collector, and as a result of the uneven distribution of the charge, the motion of the jet gets whipped or bent (Bhushani and Anandharamakrishnan, 2014) (Figure 18.13).

Using sequential electrospinning, multilayer scaffolds were formulated in the work of Pham and co-workers where (ε-caprolactone) microfibers and poly (ε-caprolactone) nanofibers were used to utilize both the advantages of micro and nanofibers on rat marrow stromal cell attachment, spreading, and migration (Pham et al., 2006). A similar method was also used to prepare a prototype for artificial grafts (Kidoaki et al., 2005). These authors have studied the formation of trilayered electrospun mesh consisting of type I collagen, styrenated gelatin, and segmented polyurethane, and a bilayered tubular construct using SPU microfiber mesh as the outer layer and type I collagen nanofiber mesh as the inner layer (Figure 18.14a–d).

Development of multilayered electrospun fibers for food systems is less known. Fabra and team have studied an approach of the formulation of bilayer packaging where α-tocopherol was encapsulated by electrospinning using different coating materials such as whey protein isolate, zein, and soy protein isolate onto wheat gluten film (Fabra et al., 2016). This bilayer packaging material exhibited improved water vapor barrier efficiency and delayed the release of α-tocopherol, which justifies the use of this material for food preservation as well as encapsulation for bioactives. A biodegradable packaging material with antimicrobial properties has also been developed using electrospinning, by coating polyhydroxybutyrate-co-valerate (PHBV) film with cinnamaldehyde-loaded zein fibers (Cerqueira et al., 2016). Similarly to the previous study, authors observed the controlled release of active components.

18.2.10 Nanotechnology and Multilayered Encapsulation

Nanotechnology is a notable contributor in the science of multilayered encapsulation and has been also extensively studied for the same. The technique employs two immiscible liquids, with one

FIGURE 18.14 Fabricated bilayered tubular construct of SPU/collagen. (a) Scanning electron micrograph of the tube; (b) magnified image of region 3 in (a); (c) magnified images of the outer layer of region 4 in (a); (d) magnified images of the inner layer of region 5 in (a). (Kidoaki et al., 2005.)

being the dispersed phase in another, having particles between 50–1000 nm. This procedure offers high encapsulation efficiency, high penetrability at the target site, and increased stability to the active constituent (owing to nanosized droplets contributing to high kinetic stability) and could be employed for formation of multilayers (Sanguansri and Augustin, 2006; Solans et al., 2005). The said technique for emulsion-based nanoencapsulation is particularly suitable for active compounds such as phytosterols, carotenoids, dietary fats, entrapped in oil-in-water based emulsions and other supplements. On the other hand, the water-in-oil approach could be used for polyphenols. In fact, both these nanoemulsions offer the flexibility of being dispensed as a liquid or a spray/freeze-dried product (Zuidam and Shimoni, 2010; Sonneville-Aubrun et al., 2004; Jafari et al., 2008).

Nanoemulsions being energy intensive at the formulation stage are produced by high-energy emulsification processes, involving high-speed or high-pressure homogenization, high shear stirring, ultrasonication, and microfluidization. A unique feat of multilayered emulsion nanoencapsulation is the design of "hollow nanocapsules," produced by adsorption of polymers on a colloidal base (template), and then removal of the colloid using physical/chemical means. The "core-less" hollow encapsulate's characteristics could be modified by chemical means. These flexible attributes of this encapsulate include variable permeability and wall thickness, to name a few (Caruso et al., 2002). Hollow capsules have been considered useful when complete removal of the core is required for a specific function, such as for intravenous or subcutaneous administration of medications, for releasing the drug(s) into the blood (Schüler and Caruso, 2001).

In other interesting research, investigators have designed a glucose biosensor with microspheres coated with polyelectrolyte multilayers (PEMs) (Brown et al., 2005). Here, glucose oxidase was encapsulated within calcium alginate microspheres with encasement of an oxygen-quenched ruthenium compound in the same microstructure. This resulted in the formation of multilayered nanofilms at the surface of the microspheres. As a result, the entrapped enzyme was suitably stabilized and rendered a controlled release feature. These would be implantable biosensors, nicknamed "smart tattoos."

Researchers have also designed hydrogel-based nanofiber enabled encapsulation devices (NEEDs) for multifarious applications. Primary designs envisaged the development of such devices from highly porous nanofibrous membranes of preformed tubular or planar devices and saturating the same with hydrogel precursor solutions, followed by cross-linking. The device reportedly displayed biocompatibility (derived from the hydrogel) and mechanical robustness (contributed by nanofibers) (An et al., 2015). The authors have pointed out that NEEDs comprises multiple compartments which could be fabricated into a single compartment, acting as multilayers, and become suitable for carrying out co-encapsulation, co-culture, and co-delivery of bioactives. They have demonstrated that the device is efficient in Type 1 diabetic mouse model when working on primary rat islets.

The suitability of multi-layered encapsulation for employing water-dispersible superparamagnetic iron oxide nanoparticles for cancer therapy has also been explored. This method has been found suitable for applications of magnetic resonance imaging (MRI) and hyperthermia (Yallapu et al., 2011). The nanoparticles consisted of iron salts, coated with β-cyclodextrin and pluronic polymer (F127). These nanoparticles showed improved hypothermia effects and magnetic resonance imaging (MRI) characteristics. Thereafter the nanoparticles were used for entrapment of curcumin as the active core, for anti-cancer activity. The curcumin loaded nanoparticles (F127250-CUR) revealed anti-cancer activity on MDA-MB-231 (breast), A2780CP (ovarian), and PC-3 (prostate) cancer cells, with hemocompatibility, despite having only 40% release of the core material. The effects were validated by transmission electron microscope (TEM) and Western blot data. Reportedly, the nanoencapsulate also aided initiation of apoptosis in cancerous cells.

18.3 CHARACTERIZATION OF ENCAPSULATES

The success in designing a feasible and stable multilayered encapsulate largely depends on a multitude of factors. These factors include the charge on the droplets (pre-encapsulation), encapsulate coat type and thickness, flocculation during encapsulation, post-preparation droplet aggregation, creaming, and Ostwald ripening, to name a few. In addition, factors such as the interaction of the encapsulate with the surroundings' attributes, such as those of pH, solvent, ionic strength, and temperature, also need to be taken into account.

The structural characteristics of the multilayered buildup could be examined by different techniques, such as UV-Vis spectrophotometer, FTIR, X-ray refractrometry are suitable for the determination of the layer thickness of encapsulates (Caruso et al., 1998a,b). On the other hand, the multilayer buildup on a colloidal solution has been testified by methods of scanning electron microscope (SEM), TEM, atomic force microscope (AFM), and Fourier transform infrared (FTIR) spectroscopy. Similarly, the quantity of polyelectrolyte adsorbed could be estimated by FTIR and zeta (ζ) potential. Likewise, the electrical potential at the surface of the encapsulate could be estimated by dynamic light scattering (DLS), whereas particle growth on the bilayers could be examined by single particle light scattering (SPLS).

AFM is particularly suitable in producing 3D topographical information about the surface, thickness of the layer(s), aggregation, homogeneity, visualization of the bilayers(s), and virtual characterization of salt concentrations during or post-encapsulation. Being a non-destructive technique, AFM is appropriate for studying liquid samples, such as colloidal suspension of multilayered encapsulates (Park et al., 2001; Mermut et al., 2003). The permeability of multilayers could be assessed by fluorescence spectroscopy (Antipov et al., 2001; Ibarz et al., 2002). Creaming in emulsions could be accounted for by visual or instrumental assessment of cream separation upon standing when stored (in test tubes). Correspondingly, rheological studies provide insight into flocculation in the samples, wherein higher apparent viscosity and pseudo-plastic behavior represent higher flocs during encapsulation (McClements, 2015).

18.4 CONCLUSION AND FUTURE PROSPECTS

Multilayered encapsulation is indeed a preferable approach to monolayer encapsulation for enhanced stability and targeted delivery of bioactives. There are numerous methods to carry out this encapsulation, and the versatility of methods illustrates the wide possibilities with myriad ingredients and coat materials. This discussion provides a comprehensive overview of various advantages and applications of multilayered encapsulation, overcoming the limitations of monolayer encapsulation. Considering the widespread applicability, and the possibilities of coat materials and preparation methods, it is fair to believe that multilayered encapsulation will be a subject of interest in current as well as future researches, and will gain more importance over single layer encapsulation in time to come.

REFERENCES

Aadinath, W., Bhushani, A., and Anandharamakrishnan, C. 2016. Synergistic radical scavenging potency of curcumin-in-β-cyclodextrin-in-nanomagnetoliposomes. *Materials Science and Engineering: C* 64: 293–302.

An, D., Ji, Y., Chiu, A., Lu, .Y-C., Song, W., Zhai, L., Qi, L., Luo, D., and Ma, M. 2015. Developing robust, hydrogel-based, nanofiber-enabled encapsulation devices (NEEDs) for cell therapies. *Biomaterials* 37: 40–48.

Anal, A.K, Bhopatkar, D., Tokura, S., Tamura, H., and Stevens, W.F. 2003. Chitosan-alginate multilayer beads for gastric passage and controlled intestinal release of protein. *Drug Development and Industrial Pharmacy* 29 (6): 713–724.

Anal, A.K. and Stevens, W.F. 2005. Chitosan–alginate multilayer beads for controlled release of ampicillin. *International Journal of Pharmaceutics* 290 (1–2): 45–54.

Anandharamakrishnan, C. and Padma Ishwarya, S. 2015. Synergistic spray drying techniques for encapsulation. In *Spray Drying Techniques for Food Ingredient Encapsulation*, ed. Anandharamakrishnan, C. and Padma Ishwarya, S., 252–274. Chichester, United Kingdom: John Wiley & Sons.

Andrade, R.D., Skurtys, O., and Osorio, F.A. 2012. Atomizing spray systems for application of edible coatings. *Comprehensive Reviews in Food Science and Food Safety* 11 (3): 323–337.

Antipov, A.A, Sukhorukov, G.B., Donath, E., and Möhwald, H. 2001. Sustained release properties of polyelectrolyte multilayer capsules. *The Journal of Physical Chemistry B* 105 (12): 2281–2284.

Balabushevitch, N.G., Sukhorukov, G.B., Moroz, N.A., Volodkin, D.V., Larionova, N.I., Donath, E., and Mohwald, H. 2001. Encapsulation of proteins by layer-by-layer adsorption of polyelectrolytes onto protein aggregates: Factors regulating the protein release. *Biotechnology and Bioengineering* 76: 207–213.

Benjamin, O., Silcock, P., Leus, L., and Everett, D.W. 2012. Multilayer emulsions as delivery systems for controlled release of volatile compounds using pH and salt triggers. *Food Hydrocolloids* 27(1): 109–118.

Bhardwaj, N. and Kundu, S.C. 2010. Electrospinning: A fascinating fiber fabrication technique. *Biotechnology Advances* 28(3): 325–347.

Bhujbal, S.V., Haan, B., Niclou, S.P., and De Vos, P. 2014. A novel multilayer immunoisolating encapsulation system overcoming protrusion of cells. *Scientific Reports* 4 (6856): 1–8.

Bhushani, J.A. and Anandharamakrishnan, C.. 2014. Electrospinning and electrospraying techniques: Potential food based applications. *Trends in Food Science and Technology* 38(1): 21–33.

Bhushani, Anu, J., Karthik, P., and Anandharamakrishnan, C. 2016. Nanoemulsion based delivery system for improved bioaccessibility and Caco-2 cell monolayer permeability of green tea catechins. *Food Hydrocolloids* 56: 372–382.

Bhushani, J.A., Kurrey, N.K., and Anandharamakrishnan, C. 2017. Nanoencapsulation of green tea catechins by electrospraying technique and its effect on controlled release and in-vitro permeability. *Journal of Food Engineering* 199: 82–92.

Brown, J.Q., Srivastava, R., and McShane, M.J. 2005. Encapsulation of glucose oxidase and an oxygen-quenched fluorophore in polyelectrolyte-coated calcium alginate microspheres as optical glucose sensor systems. *Biosensors and Bioelectronics* 21(1): 212–216.

Bugarski, B., Li, Q., Goosen, M.F.A., Poncelet, D., Neufeld, R.J., and Vunjak, G. 1994. Electrostatic droplet generation: Mechanism of polymer droplet formation. *AIChE Journal* 40(6): 1026–1031.

Caruso, F., Furlong, D.N., Ariga, K., Ichinose, I., and Kunitake, T. 1998a. Characterization of polyelectrolyte– protein multilayer films by atomic force microscopy, scanning electron microscopy, and Fourier transform infrared reflection– absorption spectroscopy. *Langmuir* 14 (16): 4559–4565.

Caruso, F., Caruso, R.A., Donath, E., Möhwald, H., and Sukhorukov, G. 2002. Fabrication of multilayer-coated particles and hollow shells via electrostatic self-assembly of nanocomposite multilayers on decomposable colloidal templates. Google Patents, WO/1999/047253, 23.09.1999, US.

Caruso, F., Kurth, D.G., Volkmer, D., Koop, M.J., and Müller, A. 1998b. Ultrathin molybdenum polyoxometalate–polyelectrolyte multilayer films. *Langmuir* 14 (13): 3462–3465.

Caruso, F., Yang, W., Trau, D., and Renneberg, R. 2000. Microencapsulation of uncharged low molecular weight organic materials by polyelectrolyte multilayer self-assembly. *Langmuir* 16: 8932–8936.

Carvalho, A.G.S, Silva, V.M., and Hubinger, M.D. 2014. Microencapsulation by spray drying of emulsified green coffee oil with two-layered membranes. *Food Research International* 61: 236–245.

Cerqueira, M.A., Fabra, M.J., Castro-Mayorga, J.L., Bourbon, A.I., Pastrana, L.M., Vicente, A.A., and Lagaron, J.M. 2016. Use of electrospinning to develop antimicrobial biodegradable multilayer systems: Encapsulation of cinnamaldehyde and their physicochemical characterization. *Food and Bioprocess Technology* 9(11): 1874–1884.

Chatterjee, S., Salaün, F., and Campagne, C. 2014. Development of multilayer microcapsules by a phase coacervation method based on ionic interactions for textile applications. *Pharmaceutics* 6(2): 281–297.

Chen, H., Gao, J., Wang, F., and Liang, W. 2007. Preparation, characterization and pharmacokinetics of liposomes-encapsulated cyclodextrins inclusion complexes for hydrophobic drugs. *Drug Delivery* 14(4): 201–208.

de Vos, P., Faas, M.M., Strand, B., and Calafiore, R. 2006. Alginate-based microcapsules for immunoisolation of pancreatic islets. *Biomaterials* 27(32): 5603–5617.

Decher, G. 1997. Fuzzy nanoassemblies: Toward layered polymeric multicomposites. *Science* 277(5330): 1232–1237.

Decher, G.H.J.D., Hong, J.D. and Schmitt, J. 1992. Buildup of ultrathin multilayer films by a self-assembly process: III. Consecutively alternating adsorption of anionic and cationic polyelectrolytes on charged surfaces. *Thin Solid Films* 210: 831–835.

Decher, G. and Hong, J.D. 1991. Buildup of ultrathin multilayer films by a self-assembly process: II. Consecutive adsorption of anionic and cationic bipolar amphiphiles and polyelectrolytes on charged surfaces. *Berichte der Bunsengesellschaft für physikalische Chemie* 95(11): 1430–1434.

Djordjevic, D., Cercaci, L., Alamed, J., McClements, D.J., and Decker, E.A. 2007. Chemical and physical stability of citral and limonene in sodium dodecyl sulfate–chitosan and gum arabic-stabilized oil-in-water emulsions. *Journal of Agricultural and Food Chemistry* 55(9): 3585–3591.

Dolly, P., Anishaparvin, A., Joseph. G.S., and Anandharamakrishnan, C. 2011. Microencapsulation of Lactobacillus plantarum (mtcc 5422) by spray-freeze-drying method and evaluation of survival in simulated gastrointestinal conditions. *Journal of Microencapsulation* 28(6): 568–574.

Dong, Z., MA, Y., Hayat, K., Jia, C., Xia, S., and Zhang, X. 2011. Morphology and release profile of microcapsules encapsulating peppermint oil by complex coacervation. *Journal of Food Engineering* 104(3): 455–460.

Dowlatabadi, M., Mokhtarian, M., Mortazavi, A., and Rad, A.H.E. 2016. Investigation of multilayer encapsulation by method of external gelation on the survival of probiotic bacteria undergoing orange juice pasteurization. *Iranian Journal of Nutrition Sciences and Food Technology* 11(3): 93–102.

Elosegui-Artola, A, Bazellieres, E., Allen, M.D., Andreu, I., R Oria, Sunyer, R., Gomm, J.J., Marshall, J.F., Jones, J.L., Trepat, X., and Roca-Cusachs, P. 2014. Rigidity sensing and adaptation through regulation of integrin types. *Nature Materials* 13(6): 631–637.

Ezhilarasi, P.N., Indrani, D., Jena, B.S., and Anandharamakrishnan, C. 2013. Freeze drying technique for microencapsulation of Garcinia fruit extract and its effect on bread quality. *Journal of Food Engineering* 117(4): 513–520.

Ezhilarasi, P.N., Indrani, D., Jena, B.S., and Anandharamakrishnan, C. 2014. Microencapsulation of Garcinia fruit extract by spray drying and its effect on bread quality. *Journal of the Science of Food and Agriculture* 94(6): 1116–1123.

Ezhilarasi, PN, Muthukumar, S.P., and Anandharamakrishnan, C. 2016. Solid lipid nanoparticle enhances bioavailability of hydroxycitric acid compared to a microparticle delivery system. *RSC Advances* 6(59): 53784–53793.

Fabra, M.J., López-Rubio. A., and Lagaron, J.M. 2016. Use of the electrohydrodynamic process to develop active/bioactive bilayer films for food packaging applications. *Food Hydrocolloids* 55: 11–18.

Fukui, Y., Maruyama, T., Iwamatsu, Y., Fujii, A., Tanaka, T., Ohmukai, Y., and Matsuyama, H. 2010. Preparation of monodispersed polyelectrolyte microcapsules with high encapsulation efficiency by an electrospray technique. *Colloids and Surfaces A: Physicochemical and Engineering Aspects* 370(1): 28–34.

Garcia, J.R. and Garcia, A.J. 2014. Cellular mechanotransduction: Sensing rigidity. *Nature Materials* 13(6): 539–540. doi:10.1038/nmat3996.

Gharsallaoui, A., Roudaut, G., Beney, L., Chambin, O., Voilley, A., and Saurel, R.. 2012. Properties of spray-dried food flavours microencapsulated with two-layered membranes: Roles of interfacial interactions and water. *Food Chemistry* 132(4): 1713–1720.

Gibis, M., Zeeb, B., and Weiss, J. 2014. Formation, characterization, and stability of encapsulated hibiscus extract in multilayered liposomes. *Food Hydrocolloids* 38: 28–39.

Gu, Y.S., Decker, E.D., and McClements, J. 2005. Production and characterization of oil-in-water emulsions containing droplets stabilized by multilayer membranes consisting of β-lactoglobulin, ι-carrageenan and gelatin. *Langmuir* 21(13): 5752–5760.

Gudipati, V., Sandra, S., McClements, D.J., and Decker, E.A. 2010. Oxidative stability and in vitro digestibility of fish oil-in-water emulsions containing multilayered membranes. *Journal of Agricultural and Food Chemistry* 58(13): 8093–8099.

Guzey, D., Kim, H.J., and McClements, D.J. 2004. Factors influencing the production of o/w emulsions stabilized by β-lactoglobulin–pectin membranes. *Food Hydrocolloids* 18(6): 967–975.

Guzey, D., and McClements, D.J. 2006. Formation, stability and properties of multilayer emulsions for application in the food industry. *Advances in Colloid and Interface Science* 128: 227–248.

Hayward, S.L., Francis, D.M., Sis, M.J., and Kidambi, S. 2015. Ionic driven embedment of hyaluronic acid coated liposomes in polyelectrolyte multilayer films for local therapeutic delivery. *Scientific Reports* 5: 14683.

Huang, H., Pierstorff, E., Osawa, E., and Ho, D. 2008. Protein-mediated assembly of nanodiamond hydrogels into a biocompatible and biofunctional multilayer nanofilm. *ACS Nano* 2(2): 203–212.

Humblet-Hua, K.N.P., Scheltens, G., Linden, E. van der, and Sagis, L.M.C. 2011. Encapsulation systems based on ovalbumin fibrils and high methoxyl pectin. *Food Hydrocolloids* 25: 569-576.

Hundre, S.Y., Karthik, P., and Anandharamakrishnan, P. 2015. Effect of whey protein isolate and β-cyclodextrin wall systems on stability of microencapsulated vanillin by spray–freeze drying method. *Food Chemistry* 174: 16–24.

Ibarz, G., Dähne, L., Donath, E., and Möhwald, H. 2002. Controlled permeability of polyelectrolyte capsules via defined annealing. *Chemistry of Materials* 14(10): 4059–4062.

Ijichi, K., Yoshizawa, H., Uemura, Y., Hatate, Y., and Kawano, Y. 1997. Multi-layered gelatin/acacia microcapsules by complex coacervation method. *Journal of Chemical Engineering of Japan* 30(5): 793–798. doi:10.1252/jcej.30.793.

Iler, R.K. 1966. Multilayers of colloidal particles. *Journal of Colloid and Interface Science* 21(6): 569–594.

Jafari, S.M., Assadpoor, E., Bhandari, B., and He, Y. 2008. Nano-particle encapsulation of fish oil by spray drying. *Food Research International* 41(2): 172–183.

Karthik, P. and Anandharamakrishnan, C. 2013. Microencapsulation of docosahexaenoic acid by spray-freeze-drying method and comparison of its stability with spray-drying and freeze-drying methods. *Food and Bioprocess Technology* 6(10): 2780–2790.

Karthik, P. and Anandharamakrishnan, C. 2016a. Enhancing omega-3 fatty acids nanoemulsion stability and in-vitro digestibility through emulsifiers. *Journal of Food Engineering* 187: 92–105.

Karthik, P. and Anandharamakrishnan, C. 2016b. Fabrication of a nutrient delivery system of docosahexaenoic acid nanoemulsions via high energy techniques. *RSC Advances* 6(5): 3501–3513.

Kidoaki, S., Kwon, I.K., and Matsuda, T. 2005. Mesoscopic spatial designs of nano-and microfiber meshes for tissue-engineering matrix and scaffold based on newly devised multilayering and mixing electrospinning techniques. *Biomaterials* 26(1): 37–46.

Kirby, C.J., Whittle, C.J., Rigby, N., Coxon, D.T, and Law, B.A. 1991. Stabilization of ascorbic acid by microencapsulation in liposomes. *International Journal of Food Science and Technology* 26(5): 437–449.

Kiyoyama, S., Shiomori, K., Kawano, Y., and Hatate, Y. 2003. Preparation of microcapsules and control of their morphology. *Journal of Microencapsulation* 20(4): 497–508.

Klinkesorn, U., Sophanodora, P., Chinachoti, P., and McClements, D.J. 2004. Stability and rheology of corn oil-in-water emulsions containing maltodextrin. *Food Research International* 37(9): 851–859.

Klinkesorn, U., Sophanodora, P., Chinachoti, P., McClements, D.J., and Decker, E.A. 2005. Increasing the oxidative stability of liquid and dried tuna oil-in-water emulsions with electrostatic layer-by-layer deposition technology. *Journal of Agricultural and Food Chemistry* 53(11): 4561–4566.

Klinkesorn, U., Sophanodora, P., Pavinee P., Decker, E.A., and McClements, D.J. 2006. Characterization of spray-dried tuna oil emulsified in two-layered interfacial membranes prepared using electrostatic layer-by-layer deposition. *Food Research International* 39(4): 449–457.

Lee, K.Y., Park, W.H., and Ha, W. S. 1997. Polyelectrolyte complexes of sodium alginate with chitosan or its derivatives for microcapsules. *Journal of Applied Polymer Science* 63(4): 425–432.

Lim, F. and Sun, A.M. 1980. Microencapsulated islets as bioartificial endocrine pancreas. *Science*, 210 (4472):908–910.

Lim, L.Y., Wan, L.S.C., and Thai, P.Y. 1997. Chitosan microspheres prepared by emulsification and ionotropic gelation. *Drug Development and Industrial Pharmacy* 23(10): 981–985.

Lin, C.-C., Lin, S.-Y., and Hwang, L. S. 1995. Microencapsulation of squid oil with hydrophilic macromolecules for oxidative and thermal stabilization. *Journal of Food Science* 60(1): 36–39.

Liu, Y., Ji, M., Wong, M.K., and Wang, P. 2013. Enhanced therapeutic efficacy of iRGD-conjugated crosslinked multilayer liposomes for drug delivery. *BioMed Research International* 2013 (Article ID 378380): 1–11.

Manojlovic, V., Djonlagic, J., Obradovic, B., Nedovic, V., and Bugarski, B. 2006. Investigations of cell immobilization in alginate: Rheological and electrostatic extrusion studies. *Journal of Chemical Technology and Biotechnology* 81(4): 505–510.

McClements, D.J. 2015. *Food Emulsions: Principles, Practices, and Techniques*. Boca Raton, FL: CRC Press.

Mendanha, D.V., Molina Ortiz, S.E., Favaro-Trindade, C.S., Mauri, A., Monterrey-Quintero, E.S., and Thomazini, M. 2009. Microencapsulation of casein hydrolysate by complex coacervation with SPI/pectin. *Food Research International* 42(8): 1099–1104.

Mermut, O., Lefebvre, J., Gray, D.G., and Barrett, C.J. 2003. Structural and mechanical properties of polyelectrolyte multilayer films studied by AFM. *Macromolecules* 36 (23): 8819–8824.

Moreau, L., Kim, H.-J., Decker, E.A., and McClements, D.J. 2003. Production and characterization of oil-in-water emulsions containing droplets stabilized by β-lactoglobulin–pectin membranes. *Journal of Agricultural and Food Chemistry* 51(22): 6612–6617.

Mun, S., Decker, E.A., and McClements, D.J. 2005. Influence of droplet characteristics on the formation of oil-in-water emulsions stabilized by surfactant–chitosan layers. *Langmuir* 21(14): 6228–6234.

Nedović, V.A., Obradović, B., Leskošek-Čukalović, I., Trifunović, O., Pešić, R., and Bugarski, B. 2001. Electrostatic generation of alginate microbeads loaded with brewing yeast. *Process Biochemistry*, 37(1): 17–22.

Ogawa, S., Decker, E.A., and McClements, D.J. 2003. Production and characterization of O/W emulsions containing cationic droplets stabilized by lecithin-chitosan membranes. *Journal of Agricultural and Food Chemistry* 51: 2806–2812.

Okamoto, Y., Taguchi, K., Yamasaki, K., Sakuragi, M., Kuroda, S., and Otagiri, M. 2018. Albumin-encapsulated liposomes: A novel drug delivery carrier with hydrophobic drugs encapsulated in the inner aqueous core. *Journal of Pharmaceutical Sciences* 107(1): 436–445.

Orr, A.W., Helmke, B.P., Blackman, B.R., and Schwartz, M.A. 2006. Mechanisms of mechanotransduction. *Developmental Cell* 10(1): 11–20.

Park, M.-K., Xia. C., Advincula, R.C., Schütz, P., and Caruso, F. 2001. Cross-linked, luminescent spherical colloidal and hollow-shell particles. *Langmuir* 17(24):7670–7674.

Parthasarathi, S. and Anandharamakrishnan, C. 2016. Enhancement of oral bioavailability of vitamin E by spray-freeze drying of whey protein microcapsules. *Food and Bioproducts Processing* 100: 469–476.

Parthasarathi, S., Ezhilarasi, P.N., Jena, B.S., and Anandharamakrishnan, C. 2013. A comparative study on conventional and microwave-assisted extraction for microencapsulation of Garcinia fruit extract. *Food and Bioproducts Processing* 91(2): 103–110.

Parthasarathi, S., Muthukumar, S.P., and Anandharamakrishnan, C. 2016. The influence of droplet size on the stability, in vivo digestion, and oral bioavailability of vitamin E emulsions. *Food and Function* 7(5): 2294–2302.

Pasrija, D., Ezhilarasi, P.N., Indrani, D., and Anandharamakrishnan, C. 2015. Microencapsulation of green tea polyphenols and its effect on incorporated bread quality. *LWT-Food Science and Technology* 64(1): 289–296.

Pham, Q.P., Sharma, U., and Mikos, A.G. 2006. Electrospun poly (ε-caprolactone) microfiber and multilayer nanofiber/microfiber scaffolds: Characterization of scaffolds and measurement of cellular infiltration. *Biomacromolecules* 7 (10): 2796–2805.

Pitigraisorn, P., Srichaisupakit, K., Wongpadungkiat, N., and Wongsasulak. S. 2017. Encapsulation of Lactobacillus acidophilus in moist-heat-resistant multilayered microcapsules. *Journal of Food Engineering* 192: 11–18.

Pommersheim, R., Schrezenmeir, J., and Vogt, W. 1994. Immobilization of enzymes by multilayer microcapsules. *Macromolecular Chemistry and Physics* 195(5): 1557–1567.

Radtchenko, I.L., Sukhorukov, G.B., Leporatti, S., Khomutov, G.B., Donath, E., and Möhwald, H. 2000. Assembly of alternated multivalent ion/polyelectrolyte layers on colloidal particles. stability of the multilayers and encapsulation of macromolecules into polyelectrolyte capsules. *Journal of Colloid and Interface Science* 230: 272–280.

Rajam, R. and Anandharamakrishnan, C. 2015. Microencapsulation of Lactobacillus plantarum (MTCC 5422) with fructooligosaccharide as wall material by spray drying. *LWT-Food Science and Technology* 60(2): 773–780.

Rajam, R., Karthik, P., Parthasarathi, S., Joseph, G.S., and Anandharamakrishnan, C. 2012. Effect of whey protein–alginate wall systems on survival of microencapsulated Lactobacillus plantarum in simulated gastrointestinal conditions. *Journal of Functional Foods* 4(4): 891–898.

Rajam, R., Bharat Kumar, S., Prabhasankar, P., and Anandharamakrishnan, C. 2015. Microencapsulation of Lactobacillus plantarum MTCC 5422 in fructooligosaccharide and whey protein wall systems and its impact on noodle quality. *Journal of Food Science and Technology* 52(7): 4029–4041.

Rilling, R., Walter, T., Pommersheim, R., and Vogt, W. 1997. Encapsulation of cytochrome C by multilayer microcapsules. A model for improved enzyme immobilization. *Journal of Membrane Science* 129: 283–287.

Rocha-Selmi, G.A., Bozza, F.T., Thomazini, M., Bolini, H.M.A., and Fávaro-Trindade, C.S. 2013. Microencapsulation of aspartame by double emulsion followed by complex coacervation to provide protection and prolong sweetness. *Food Chemistry* 139(1–4): 72–78.

Sadovoy, A.V., Kiryukhin, M.V., Sukhorukovab, G.B., and Antipina, M.N. 2011. Kinetic stability of water-dispersed oil droplets encapsulated in a polyelectrolyte multilayer shell. *Physical Chemistry Chemical Physics* 13, 4005–4012.

Sanguansri, P. and Augustin, M.A. 2006. Nanoscale materials development–a food industry perspective. *Trends in Food Science and Technology* 17(10): 547–556.

Schneider, S., Feilen, P.J., Slotty, V., Kampfner, D., Preuss, S., Berger, S., Beyer, J., and Pommersheim, R. 2001. Multilayer capsules: A promising microencapsulation system for transplantation of pancreatic islets. *Biomaterials* 22(14): 1961–1970.

Schüler, C. and Caruso, F. 2001. Decomposable hollow biopolymer-based capsules. *Biomacromolecules* 2(3): 921–926.

Serfert, Y., Schröder, J., Mescher, A., Laackmann, J., Drusch, S., and Schwarz, K. 2011. Characterization of spray-dried layer-by-layer emulsions. *International Congress on Engineering and Food*. May 22–26, 2011, Athens.

Shaw, L.A., McClements, D.J., and Decker, E.A. 2007. Spray-dried multilayered emulsions as a delivery method for ω-3 fatty acids into food systems. *Journal of Agricultural and Food Chemistry* 55(8): 3112–3119.

Sinha, V.R. and Trehan, A. 2003. Biodegradable microspheres for protein delivery. *Journal of Controlled Release* 90(3): 261–280.

Solans, C., Izquierdo, P., Nolla, J., Azemar, N., and Garcia-Celma, M.J. 2005. Nano-emulsions. *Current Opinion in Colloid and Interface Science* 10(3–4): 102–110.

Song, L., Zhi, Z., and Pickup, J.C. 2014. Nanolayer encapsulation of insulin-chitosan complexes improves efficiency of oral insulin delivery. *International Journal of Nanomedicine* 2014(9): 2127–2136.

Sonneville-Aubrun, O., Simonnet, J.-T., and L'alloret, F. 2004. Nanoemulsions: A new vehicle for skincare products. *Advances in Colloid and Interface Science* 108: 145–149.

Soper, J.C. and Thomas, M.T. 2000. Enzymatically protein encapsulating oil particles by complex coacervation. Google Patents, 6,039,901, March 21, 2000, US.

Voigt, A., Lichtenfeld, H., Sukhorukov, G.B., Zastrow, H., Donath, E., Bäumler, H., and Möhwald, H. 1999. Membrane filtration for microencapsulation and microcapsules fabrication by layer-by-layer polyelectrolyte adsorption. *Industrial and Engineering Chemistry Research* 38(10): 4037–4043.

Volodkin, D., Arntz, Y., Schaaf, P., Moehwald, H., Voegel, J.C., and Ball, V. 2008. Composite multilayered biocompatible polyelectrolyte films with intact liposomes: Stability and temperature triggered dye release. *Soft Matter* 4(1): 122–130.

Wang, C., Ye, S., Dai, L., Liu, X., and Tong, Z. 2007. Enhanced resistance of polyelectrolyte multilayer microcapsules to pepsin erosion and release properties of encapsulated indomethacin. *Biomacromolecules* 8: 1739–1744.

Wood, K.C., Boedicker, J.Q., Lynn, D.M., and Hammond, P.T. 2005. Tunable drug release from hydrolytically degradable layer-by-layer thin films. *Langmuir* 21(4): 1603–1609.

Xie, J. and Wang, C.-H. 2007. Electrospray in the dripping mode for cell microencapsulation. *Journal of Colloid and Interface Science* 312(2): 247–255.

Yallapu, M.M., Shadi F Othman, S.F., Curtis, E.T., Gupta, B.K., Jaggi, M., and Chauhan, S.C. 2011. Multifunctional magnetic nanoparticles for magnetic resonance imaging and cancer therapy. *Biomaterials* 32(7): 1890–1905.

Ye, S., Wang, C., Liu, X., and Tong, Z. 2005a. Deposition temperature effect on release rate of indomethacin microcrystals from microcapsules of layer-by-layer assembled chitosan and alginate multilayer films. *Journal of Controlled Release* 106(3): 319–328.

Ye, S., Wang. C., Liu, X., and Tong, Z. 2005b. Multilayer nanocapsules of polysaccharide chitosan and alginate through layer-by-layer assembly directly on PS nanoparticles for release. *Journal of Biomaterials Science, Polymer Edition* 16(7): 909–923.

Zasada, K., Łukasiewicz-Atanasov, M., Kłysik, K., Lewandowska-Łańcucka, J., Gzyl-Malcher, B., Puciul-Malinowska, A., Karewicz, A., and Nowakowska, M. 2015. One-component ultrathin multilayer films based on poly (vinyl alcohol) as stabilizing coating for phenytoin-loaded liposomes. *Colloids and Surfaces B: Biointerfaces* 135: 133–142.

Zuidam, N.J. and Shimoni, E. 2010. Overview of microencapsulates for use in food products or processes and methods to make them. In *Encapsulation Technologies for Active Food Ingredients and Food Processing,* eds. Zuidam, N.J. and Nedovic, V., 3–29. New York: Springer.

Index